Calculus
for the Managerial, Life, and Social Sciences
FIFTH EDITION

Calculus
for the Managerial, Life, and Social Sciences
FIFTH EDITION

S. T. Tan
Stonehill College

Brooks/Cole
Thomson Learning™

Australia • Canada • Mexico • Singapore • Spain • United Kingdom • United States

Sponsoring Editor: *Curt Hinrichs*
Marketing Team: *Karin Sandberg, Beth Kroenke, Laura Hubrich*
Editorial Assistant: *JoAnne von Zastrow, Emily Davidson*
Production Coordinator: *Marjorie Z. Sanders*
Production Service: *Cecile Joyner, The Cooper Company*
Interior Design: *Delgado Design, Inc.*
Cover Design: *Lisa Henry*
Cover Illustration: *Judith Harkness*
Photo Editor: *Terry Powell*
Manuscript Editor: *Betty Duncan*
Print Buyer: *Vena Dyer*
Typesetting: *Bi-Comp/The PRD Group*
Cover Printing: *Phoenix Color Corp.*
Printing and Binding: *World Color Corp., Versailles*

Photo Credits: page xv (top and bottom), Terry Powell/The
Photographer's Window; **p. xvi** (top and bottom), Nathan
Benn/Corbis; **p. xvii** (top), Jonathan Blair/Corbis; (bottom),
Paul A. Souders/Corbis; **p. 3,** PhotoEdit; **p. 59,** Terry
Powell/The Photographer's Window; **p. 183,** Bob Western;
p. 266, NASA; **p. 277,** PhotoDisc; **p. 373,** Bob Western;
p. 439, PhotoDisc; **p. 541,** Roy Corral/Corbis; **p. 601,** Image
Bank.

Printed in the United States of America

10 9 8 7 6 5 4 3 2 1

Library of Congress Cataloging-in-Publication Data

Tan, Soo Tang.
    Calculus for the managerial, life, and social sciences/
S. T. Tan—5th ed.
       p. cm.
    Includes index.
    ISBN 0-534-36586-8
    1. Calculus. I. Title.

QA303.T142 2000   99-047846
515—dc21

To Pat, Bill, and Michael

CONTENTS

* Sections marked with an asterisk are not prerequisites for later material.

PREFACE

Calculus for the Managerial, Life, and Social Sciences, Fifth Edition, is suitable for use in a one-semester or two-quarter introductory calculus course for students in the managerial life, and social sciences. As with the previous editions, the objective of *Calculus for the Managerial, Life, and Social Sciences, Fifth Edition,* is twofold: (1) to write a textbook that is readable by students and (2) to make the book a useful teaching tool for instructors. We hope that with the present edition we have come one step closer to realizing our goal. The fifth edition of this text incorporates many suggestions by users of the earlier editions.

FEATURES

The following list includes some of the many important features of the book:

■ **Coverage of Topics**   This book contains more than enough material for the usual applied calculus course. Optional sections have been marked with an asterisk in the table of contents, thereby allowing the instructor to be flexible in choosing the topics most suitable for his or her course.

■ **Approach**   The problem-solving approach is stressed throughout the book. Numerous examples and solved problems are used to amplify each new concept or result in order to facilitate students' comprehension of the material. Figures are used extensively to help students visualize concepts and ideas.

■ **Level of Presentation**   Our approach is intuitive, and we state the results informally. However, we have taken special care to ensure that this approach does not compromise the mathematical content and accuracy. Proofs of certain results are given, but they may be omitted if desired.

■ **Applications**   The text is application oriented. Many interesting, relevant, and up-to-date applications are drawn from the fields of business, economics, social and behavioral sciences, life sciences, physical sciences, and other fields of general interest. Some of these applications have their source in newspapers, weekly periodicals, and other magazines. Applications are found in the illustrative examples in the main body of the text as well as in the exercise sets. In fact, one goal of the text is to include at least one real-life application in each section (whenever feasible).

■ **Sources**   We have included sources for those applications that are based on real-life data.

■ **Exercises**   Each section of the text is accompanied by an extensive set of exercises containing an ample set of problems of a routine, computational nature that will help students master new techniques. The routine problems are followed by an extensive set of application-oriented problems that test students' mastery of the topics.

■ **Self-Check Exercises** Every section has self-check exercises, with solutions, to help students monitor their own progress.

■ **Portfolios** These interviews are designed to convey to the student the real-world experiences of professionals who have a background in mathematics and use it in their professions.

■ **Group Discussion Questions** These are optional questions, appearing throughout the main body of the text, that can be discussed in class or assigned as homework. These questions generally require more thought and effort than the usual exercises. Complete solutions to these exercises are given in the *Complete Solutions Manual.*

TECHNOLOGY

Exploring with Technology Questions

These optional questions appear throughout the main body of the text and serve to enhance the student's understanding of the concepts and theory presented. Complete solutions to these exercises are given in the *Complete Solutions Manual.*

Using Technology Subsections

These pages contain optional material and are placed at the end of the sections for which their use is appropriate. The subsections are written in the traditional example–exercise format, with answers given at the back of the book. They may be used in the classroom if desired or as material for self-study by the student.

As many up-to-date and relevant applications have been introduced in these subsections, they provide students with an opportunity to interpret results in a real-life setting.

Student Resources on the Web

Students and instructors will now have access to these additional materials at the Brooks/Cole World Wide Web site:
http://www.brookscole.com/math/authors/tans

■ Review material and practice chapter quizzes and tests

■ Group projects and extended problems for each chapter

■ Instructions, including keystrokes, for the procedures referenced in the text for specific calculators (TI-82, TI-83, TI-85, TI-86, and other popular models)

■ Modified *Using Technology* sections for CAS systems, including the command statements for Mathematica, Maple, and other popular systems

NEW IN THE FIFTH EDITION

■ Exercises that emphasize the understanding of concepts and theory have been added to most sections. These new exercises are usually near the end of the exercise sets and take the form of true or false questions. More questions asking for the interpretation of graphs have also been added.

■ More real-life applications, with sources, have been added; for example the narrowing gender gap, net-connected computers in Europe, the growth of the Indian gaming industry, accumulation years for baby boomers, the growth of online shopping, online banking usage, the document management business, portable phone usage, growth of international e-mail, the growth of the prison population, marijuana arrests, blood alcohol level after drinking, senior citizen's health care costs, and the spread of HIV. Examples of other new applications are a coast guard patrol search mission, the yield of an apple orchard, blowing soap bubbles, absorption of drugs, loan amortization, and the frequency of road repairs.

■ The section on mathematical modeling has been expanded and strengthened. It now includes a new subsection on constructing mathematical models. A *Using Technology* section, *Constructing Mathematical Models from Raw Data,* has also been added.

■ More Group Discussion questions and Exploring with Technology questions have been added.

■ More Using Technology sections and exercises have been added.

■ Coverage of the equation of a circle has been added to Chapter 1. A section on double integrals and their applications has been added to Chapter 8.

■ Instructions for the TI-83 and TI-86 calculators have been added to the Web site.

SUPPLEMENTS

■ *Student's Solutions Manual,* available to both students and instructors, includes the solutions to odd-numbered exercises. ISBN 0-534-37001-2

■ *Complete Solutions Manual* includes solutions to all exercises. ISBN 0-534-37002-0

■ *Printed Test Bank with Chapter Tests,* free to adopters of the book, contains sample tests for each chapter. ISBN 0-534-37009-8

■ *Total Testing Tools (T3)* is a revolutionary, Internet-ready testing suite allowing instructors to customize exams and track student progress in a modern and accessible format. Questions can be selected by concept, problem type, or level of difficulty from an extensive base of over 700 different concepts. Algorithmic questions also allow instructors to regenerate the values of variables and calculations, adding to an extraordinarily diverse set of exam options.

ACKNOWLEDGMENTS

I wish to express my personal appreciation to each of the following reviewers, whose many suggestions have helped make a much improved book.

Daniel Anderson
University of Iowa

Jim Bruening
Southeast Missouri State University

Connie Carruthers
Scottsdale Community College

William J. Carsrud
Gonzaga University

Charles Clever
South Dakota State University

William Coppage
Wright State University

Margaret Crider
Tomball College

Lyle Dixon
Kansas State University

Bruce Edwards
University of Florida at Gainesville

Charles S. Frady
Georgia State University

Howard Frisinger
Colorado State University

Larry Gerstein
University of California at Santa Barbara

Kedrick Hartfield
Mercer University

Karen J. Hay
Phoenix College

Jeff Knisley
East Tennessee State University

Michael Lambe
Grossmont College

Lowell Leake
University of Cincinnati

Maurice Monahan
South Dakota State University

Richard Nadel
Florida International University

Karla V. Neal
Louisiana State University

Lloyd Olson
North Dakota State University

Richard Porter
Northeastern University

Priscilla Putnam-Haindl
Jersey City State College

Georgia B. Pyrros
University of Delaware

Nathan P. Ritchey
Youngstown State University

Thomas N. Roe
South Dakota State University

Ivan Schukei
University of South Carolina—Beaufort

Nora Schukei
University of South Carolina—Beaufort

Barbara J. Shabell
California State Polytechnic University, Pomona

Gary D. Shaffer
Allegany Community College

Shai Simonson
Stonehill College

W. Allen Smith
Georgia State University

P. L. Sperry
University of South Carolina

Joseph F. Stokes
Western Kentucky University

I wish to thank Andy Bulman-Fleming for his assistance in preparing the Test Banks for this edition. My thanks also go to the editorial, production, and marketing staffs of Brooks/Cole: Curt Hinrichs, Emily Davidson, Marjorie Sanders, Vernon Boes, Karin Sandberg, Beth Kroenke, and Samantha Cabaluna for their thoughtful contributions and patient assistance and cooperation during the development and production of this book. Finally, I wish to thank Cecile Joyner of The Cooper Company and Betty Duncan, for doing an excellent job ensuring the accuracy and readability of this sixth edition, and Delgado Design for the design of the interior of the book.

S. T. Tan

A PPLICATIONS

Walking vs. Running The oxygen consumption (in ml/lb/min) for a person walking at x mph is approximated by the function

$$f(x) = \frac{5}{3}x^2 + \frac{5}{3}x + 10 \qquad (0 \le x \le 9)$$

whereas the oxygen consumption for a runner at x mph is approximated by the function

$$g(x) = 11x + 10 \qquad (4 \le x \le 9).$$

a. Sketch the graphs of f and g.
b. At what speed is the oxygen consumption the same for a walker as it is for a runner? What is the level of oxygen consumption at that speed?
c. What happens to the oxygen consumption of the walker and the runner at speeds beyond that found in (b)?

Source: Exercise Physiology by William McArdley, Frank Katch, and Victor Katch

Online Shopping Retail revenue per year from Internet shopping is approximated by the function

$$f(t) = 0.075t^3 + 0.025t^2 + 2.45t + 2.4 \quad (0 \le t \le 4)$$

where $f(t)$ is measured in billions of dollars and t is measured in years with $t = 0$ corresponding to the beginning of 1997.

a. Find an expression giving the rate of change of the retail revenue per year from Internet shopping at any time t.
b. How fast was the retail revenue per year from Internet shopping changing at the beginning of the year 2000?
c. What was the retail revenue per year from Internet shopping at the beginning of the year 2000?

Source: Forrester Research

Spread of HIV in Children The estimated number of children newly "infected" with HIV through mother-to-child contact worldwide is given by

$$f(t) = -0.2083t^3 + 3.0357t^2 + 44.0476t + 200.2857$$
$$(0 \le t \le 12)$$

where $f(t)$ is measured in thousands and t is measured in years with $t = 0$ corresponding to the beginning of 1990.

a. Use a graphing utility to plot the graph of f in the viewing rectangle $[0, 12] \times [0, 800]$.

b. How fast was the estimated number of children newly infected with HIV through mother-to-child contact worldwide increasing at the beginning of the year 2000?

Source: United Nations

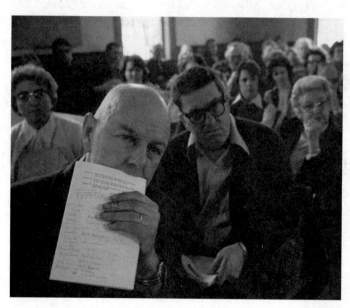

Solvency of the Social Security System Unless payroll taxes are increased significantly and/or benefits are scaled back drastically, it is a matter of time before the current Social Security system goes broke. Based on data from the Board of Trustees of the Social Security Administration, the assets of the system—the Social Security "trust fund"—may be approximated by

$$f(t) = -0.0129t^4 + 0.3087t^3 + 2.1760t^2$$
$$+ 62.8466t + 506.2955 \quad (0 \le t \le 35)$$

where $f(t)$ is measured in millions of dollars and t is measured in year with $t = 0$ corresponding to 1995.

a. Use a graphing calculator to sketch the graph of f.

b. Based on this model, when can the Social Security system be expected to go broke?

Source: Social Security Administration

Marijuana Arrests The number of arrests for marijuana sales and possession in New York City grew at the rate of approximately

$$f(t) = 0.0125^4 - 0.01389t^3 + 0.55417t^2 + 0.53294t + 4.95238 \quad (0 \le t \le 5)$$

thousand per year where t is measured in years with $t = 0$ corresponding to the beginning of 1992. Find the approximate number of marijuana arrests in the city from the beginning of 1992 to the end of 1997.

Source: State Division of Criminal Justice Services

Blood Alcohol Level The percentage of alcohol in a person's bloodstream t hours after drinking 8 fluid ounces of whiskey is given by

$$A(t) = 0.23te^{-0.4t} \quad (0 \le t \le 12)$$

a. What is the percentage of alcohol in a person's bloodstream after 1/2 hour? After 8 hours?
b. How fast is the percentage of alcohol in a person's bloodstream changing after 1/2 hour? After 8 hours?

Source: Encyclopedia Britannica

Calculus
for the Managerial, Life, and Social Sciences
Fifth Edition

PRELIMINARIES

1

The first two sections of this chapter contain a brief review of algebra. We then introduce the Cartesian coordinate system, which allows us to represent points in the plane in terms of ordered pairs of real numbers. This in turn enables us to compute the distance between two points algebraically. This chapter also covers straight lines. The slope of a straight line plays an important role in the study of calculus.

What sales figure can be predicted for next year? In Example 10, page 46, you will see how the manager of a local sporting goods store used sales figures from the previous years to predict the sales level for next year.

1.1 Precalculus Review I

Sections 1.1 and 1.2 review some of the basic concepts and techniques of algebra that are essential in the study of calculus. The material in this review will help you work through the examples and exercises in this book. You can read through this material now and do the exercises in areas where you feel a little "rusty," or you can review the material on an as-needed basis as you study the text. We begin our review with a discussion of real numbers.

THE REAL NUMBER LINE

The real number system is made up of the set of real numbers together with the usual operations of addition, subtraction, multiplication, and division.

Real numbers may be represented geometrically by points on a line. Such a line is called the **real number,** or **coordinate, line** and can be constructed as follows. Arbitrarily select a point on a straight line to represent the number zero. This point is called the **origin.** If the line is horizontal, then a point at a convenient distance to the right of the origin is chosen to represent the number 1. This determines the scale for the number line. Each positive real number lies at an appropriate distance to the right of the origin, and each negative real number lies at an appropriate distance to the left of the origin (Figure 1.1).

FIGURE 1.1
The real number line

A *one-to-one correspondence* is set up between the set of all real numbers and the set of points on the number line; that is, exactly one point on the line is associated with each real number. Conversely, exactly one real number is associated with each point on the line. The real number that is associated with a point on the real number line is called the **coordinate** of that point.

INTERVALS

Throughout this book, we will often restrict our attention to certain subsets of the set of real numbers. For example, if x denotes the number of cars rolling off a plant assembly line each day, then x must be nonnegative—that is, $x \geq 0$. Further, suppose management decides that the daily production must not exceed 200 cars. Then, x must satisfy the inequality $0 \leq x \leq 200$.

More generally, we will be interested in the following subsets of real numbers: open intervals, closed intervals, and half-open intervals. The set of all real numbers that lie *strictly* between two fixed numbers a and b is called an **open interval** (a, b). It consists of all real numbers x that satisfy the inequalities $a < x < b$, and it is called "open" because neither of its end points is included in the interval. A **closed interval** contains *both* of its end points. Thus, the set of all real numbers x that satisfy the inequalities $a \le x \le b$ is the closed interval $[a, b]$. Notice that square brackets are used to indicate that the end points are included in this interval. **Half-open intervals** contain only *one* of their end points. Thus, the interval $[a, b)$ is the set of all real numbers x that satisfy $a \le x < b$, whereas the interval $(a, b]$ is described by the inequalities $a < x \le b$. Examples of these **finite intervals** are illustrated in Table 1.1.

Table 1.1 Finite Intervals

Interval	Graph	Example
Open (a, b)		$(-2, 1)$
Closed $[a, b]$		$[-1, 2]$
Half-open $(a, b]$		$(\frac{1}{2}, 3]$
Half-open $[a, b)$		$[-\frac{1}{2}, 3)$

In addition to finite intervals, we will encounter **infinite intervals.** Examples of infinite intervals are the half lines (a, ∞), $[a, \infty)$, $(-\infty, a)$, and $(-\infty, a]$ defined by the set of all real numbers that satisfy $x > a$, $x \ge a$, $x < a$, and $x \le a$, respectively. The symbol ∞, called *infinity,* is not a real number. It is used here only for notational purposes in conjunction with the definition of infinite intervals. The notation $(-\infty, \infty)$ is used for the set of all real numbers x since, by definition, the inequalities $-\infty < x < \infty$ hold for any real number x. Infinite intervals are illustrated in Table 1.2.

Table 1.2 Infinite Intervals

Interval	Graph	Example
(a, ∞)		$(2, \infty)$
$[a, \infty)$		$[-1, \infty)$
$(-\infty, a)$		$(-\infty, 1)$
$(-\infty, a]$		$(-\infty, -\frac{1}{2}]$

PROPERTIES OF INEQUALITIES

In practical applications, intervals are often found by solving one or more inequalities involving a variable. In such situations, the following properties may be used to advantage.

Properties of Inequalities

If a, b, and c are any real numbers, then

		Example
Property 1	If $a < b$ and $b < c$, then $a < c$.	$2 < 3$ and $3 < 8$, so $2 < 8$
Property 2	If $a < b$, then $a + c < b + c$.	$-5 < -3$, so $-5 + 2 < -3 + 2$; that is, $-3 < -1$
Property 3	If $a < b$ and $c > 0$, then $ac < bc$.	$-5 < -3$, and since $2 > 0$, we have $(-5)(2) < (-3)(2)$; that is, $-10 < -6$
Property 4	If $a < b$ and $c < 0$, then $ac > bc$.	$-2 < 4$, and since $-3 < 0$, we have $(-2)(-3) > (4)(-3)$; that is, $6 > -12$

Similar properties hold if each inequality sign, $<$, between a and b is replaced by $\geq$, $>$, or $\leq$.

A real number is a *solution of an inequality* involving a variable if a true statement is obtained when the variable is replaced by that number. The set of all real numbers satisfying the inequality is called the *solution set*.

EXAMPLE 1   Find the set of real numbers that satisfy $-1 \leq 2x - 5 < 7$.

SOLUTION ✔   Add 5 to each member of the given double inequality, obtaining

$$4 \leq 2x < 12$$

Next, multiply each member of the resulting double inequality by 1/2, yielding

$$2 \leq x < 6$$

Thus, the solution is the set of all values of x lying in the interval $[2, 6)$.

■ ■ ■ ■

EXAMPLE 2   The management of Corbyco, a giant conglomerate, has estimated that x thousand dollars is needed to purchase

$$100{,}000(-1 + \sqrt{1 + 0.001x})$$

shares of common stock of the Starr Communications Company. Determine how much money Corbyco needs in order to purchase at least 100,000 shares of Starr's stock.

SOLUTION ✔   The amount of cash Corbyco needs to purchase at least 100,000 shares is found by solving the inequality

$$100{,}000(-1 + \sqrt{1 + 0.001x}) \geq 100{,}000$$

Proceeding, we find

$$-1 + \sqrt{1 + 0.001x} \geq 1$$
$$\sqrt{1 + 0.001x} \geq 2$$
$$1 + 0.001x \geq 4 \qquad \text{(Square both sides.)}$$
$$0.001x \geq 3$$
$$x \geq 3000$$

so Corbyco needs at least \$3,000,000.   ■ ■ ■ ■

ABSOLUTE VALUE

Absolute Value

> The **absolute value** of a number a is denoted by $|a|$ and is defined by
>
> $$|a| = \begin{cases} a & \text{if } a \geq 0 \\ -a & \text{if } a < 0 \end{cases}$$

Since $-a$ is a positive number when a is negative, it follows that the absolute value of a number is always nonnegative. For example, $|5| = 5$ and $|-5| = -(-5) = 5$. Geometrically, $|a|$ is the distance between the origin and the point on the number line that represents the number a (Figure 1.2).

FIGURE 1.2
The absolute value of a number

(a)   (b)

Absolute Value Properties

If a and b are any real numbers, then

		Example
Property 5	$\lvert -a \rvert = \lvert a \rvert$	$\lvert -3 \rvert = -(-3) = 3 = \lvert 3 \rvert$
Property 6	$\lvert ab \rvert = \lvert a \rvert\, \lvert b \rvert$	$\lvert (2)(-3) \rvert = \lvert -6 \rvert = 6$
		$= \lvert 2 \rvert\, \lvert -3 \rvert$
Property 7	$\left\lvert \dfrac{a}{b} \right\rvert = \dfrac{\lvert a \rvert}{\lvert b \rvert} \quad (b \neq 0)$	$\left\lvert \dfrac{(-3)}{(-4)} \right\rvert = \left\lvert \dfrac{3}{4} \right\rvert = \dfrac{3}{4} = \dfrac{\lvert -3 \rvert}{\lvert -4 \rvert}$
Property 8	$\lvert a + b \rvert \leq \lvert a \rvert + \lvert b \rvert$	$\lvert 8 + (-5) \rvert = \lvert 3 \rvert = 3$
		$\leq \lvert 8 \rvert + \lvert -5 \rvert$
		$= 13$

Property 8 is called the **triangle inequality.**

EXAMPLE 3

Evaluate each of the following expressions:

a. $\lvert \pi - 5 \rvert + 3$ **b.** $\lvert \sqrt{3} - 2 \rvert + \lvert 2 - \sqrt{3} \rvert$

SOLUTION ✔

a. Since $\pi - 5 < 0$, we see that $\lvert \pi - 5 \rvert = -(\pi - 5)$. Therefore,

$$\lvert \pi - 5 \rvert + 3 = -(\pi - 5) + 3 = 8 - \pi$$

b. Since $\sqrt{3} - 2 < 0$, we see that $\lvert \sqrt{3} - 2 \rvert = -(\sqrt{3} - 2)$. Next, observe that $2 - \sqrt{3} > 0$, so $\lvert 2 - \sqrt{3} \rvert = 2 - \sqrt{3}$. Therefore,

$$\lvert \sqrt{3} - 2 \rvert + \lvert 2 - \sqrt{3} \rvert = -(\sqrt{3} - 2) + (2 - \sqrt{3})$$
$$= 4 - 2\sqrt{3} = 2(2 - \sqrt{3}) \qquad ■■■■$$

EXPONENTS AND RADICALS

Recall that if b is any real number and n is a positive integer, then the expression b^n (read "b to the power n") is defined as the number

$$b^n = \underbrace{b \cdot b \cdot b \cdots \cdot b}_{n \text{ factors}}$$

The number b is called the **base,** and the superscript n is called the **power** of the exponential expression b^n. For example,

$$2^5 = 2 \cdot 2 \cdot 2 \cdot 2 \cdot 2 = 32 \qquad \text{and} \qquad \left(\frac{2}{3}\right)^3 = \left(\frac{2}{3}\right)\left(\frac{2}{3}\right)\left(\frac{2}{3}\right) = \frac{8}{27}$$

If $b \neq 0$, we define

$$b^0 = 1$$

For example, $2^0 = 1$ and $(-\pi)^0 = 1$, but the expression 0^0 is undefined.

Next, recall that if n is a positive integer, then the expression $b^{1/n}$ is defined to be the number that, when raised to the nth power, is equal to b. Thus,

$$(b^{1/n})^n = b$$

Such a number, if it exists, is called the **nth root of b,** also written $\sqrt[n]{b}$.

Observe that the nth root of a negative number is not defined when n is even. For example, the square root of -2 is not defined because there is no real number b such that $b^2 = -2$. Also, given a number b, more than one number might satisfy our definition of the nth root. For example, both 3 and -3 squared equal 9, and each is a square root of 9. So, to avoid ambiguity, we define $b^{1/n}$ to be the positive nth root of b whenever it exists. Thus, $\sqrt{9} = 9^{1/2} = 3$.

Next, recall that if p/q (p, q, positive integers with $q \neq 0$) is a rational number in lowest terms, then the expression $b^{p/q}$ is defined as the number $(b^{1/q})^p$ or, equivalently, $\sqrt[q]{b^p}$, whenever it exists. Expressions involving negative rational exponents are taken care of by the definition

$$b^{-p/q} = \frac{1}{b^{p/q}}$$

Table 1.3

Definition of a^n ($a > 0$)	Example	Definition of a^n ($a > 0$)	Example
Integer exponent: If n is a positive integer, then $a^n = a \cdot a \cdot a \cdots \cdot a$ (n factors of a)	$2^5 = 2 \cdot 2 \cdot 2 \cdot 2 \cdot 2$ (5 factors) $= 32$	**Fractional exponent:** **a.** If n is a positive integer, then $a^{1/n}$　or　$\sqrt[n]{a}$ denotes the **nth root** of a.	$16^{1/2} = \sqrt{16}$ $= 4$
Zero exponent: If n is equal to zero, then $a^0 = 1$ (0^0 is not defined.)	$7^0 = 1$	**b.** If m and n are positive integers, then $a^{m/n} = \sqrt[n]{a^m} = (\sqrt[n]{a})^m$	$8^{2/3} = (\sqrt[3]{8})^2$ $= 4$
Negative exponent: If n is a positive integer, then $a^{-n} = \dfrac{1}{a^n}$　　$(a \neq 0)$	$6^{-2} = \dfrac{1}{6^2}$ $= \dfrac{1}{36}$	**c.** If m and n are positive integers, then $a^{-m/n} = \dfrac{1}{a^{m/n}}$　　$(a \neq 0)$	$9^{-3/2} = \dfrac{1}{9^{3/2}}$ $= \dfrac{1}{27}$

Examples are

$$2^{3/2} = (2^{1/2})^3 \approx (1.4142)^3 \approx 2.8283$$

and

$$4^{-5/2} = \frac{1}{4^{5/2}} = \frac{1}{(4^{1/2})^5} = \frac{1}{2^5} = \frac{1}{32}$$

The rules defining the exponential expression a^n, where $a > 0$ for all rational values of n, are given in Table 1.3.

The first three definitions in Table 1.3 are also valid for negative values of a, whereas the fourth definition is valid only for negative values of a when n is odd. Thus,

$$(-8)^{1/3} = \sqrt[3]{-8} = -2 \qquad (n \text{ is odd.})$$
$$(-8)^{1/2} \text{ has no real value} \qquad (n \text{ is even.})$$

Finally, note that it can be shown that a^n has meaning for *all* real numbers n. For example, using a pocket calculator with a "y^x" key, we see that $2^{\sqrt{2}} \approx 2.665144$.

The five laws of exponents are listed in Table 1.4.

Table 1.4 The Laws of Exponents	
Law	Example
1. $a^m \cdot a^n = a^{m+n}$	$x^2 \cdot x^3 = x^{2+3} = x^5$
2. $\dfrac{a^m}{a^n} = a^{m-n} \qquad (a \neq 0)$	$\dfrac{x^7}{x^4} = x^{7-4} = x^3$
3. $(a^m)^n = a^{m \cdot n}$	$(x^4)^3 = x^{4 \cdot 3} = x^{12}$
4. $(ab)^n = a^n \cdot b^n$	$(2x)^4 = 2^4 \cdot x^4 = 16x^4$
5. $\left(\dfrac{a}{b}\right)^n = \dfrac{a^n}{b^n} \qquad (b \neq 0)$	$\left(\dfrac{x}{2}\right)^3 = \dfrac{x^3}{2^3} = \dfrac{x^3}{8}$

These laws are valid for any real numbers a, b, m, and n whenever the quantities are defined.

Remember, $(x^2)^3 \neq x^5$. The correct equation is $(x^2)^3 = x^{2 \cdot 3} = x^6$.

The next several examples illustrate the use of the laws of exponents.

EXAMPLE 4   Simplify the expressions:

   a. $(3x^2)(4x^3)$ **b.** $\dfrac{16^{5/4}}{16^{1/2}}$ **c.** $(6^{2/3})^3$ **d.** $(x^3 y^{-2})^{-2}$ **e.** $\left(\dfrac{y^{3/2}}{x^{1/4}}\right)^{-2}$

SOLUTION ✔   **a.** $(3x^2)(4x^3) = 12x^{2+3} = 12x^5$ (Law 1)

b. $\dfrac{16^{5/4}}{16^{1/2}} = 16^{5/4-1/2} = 16^{3/4} = (\sqrt[4]{16})^3 = 2^3 = 8$ (Law 2)

c. $(6^{2/3})^3 = 6^{2/3 \cdot 3} = 6^{6/3} = 6^2 = 36$ (Law 3)

d. $(x^3 y^{-2})^{-2} = (x^3)^{-2}(y^{-2})^{-2} = x^{(3)(-2)} y^{(-2)(-2)} = x^{-6} y^4 = \dfrac{y^4}{x^6}$ (Law 4)

e. $\left(\dfrac{y^{3/2}}{x^{1/4}}\right)^{-2} = \dfrac{y^{(3/2)(-2)}}{x^{(1/4)(-2)}} = \dfrac{y^{-3}}{x^{-1/2}} = \dfrac{x^{1/2}}{y^3}$ (Law 5)   ■■■■

We can also use the laws of exponents to simplify expressions involving radicals, as illustrated in the next example.

EXAMPLE 5   Simplify the expressions. (Assume x, y, and n are positive.)

a. $\sqrt[4]{16x^4 y^8}$ **b.** $\sqrt{12m^3 n} \cdot \sqrt{3m^5 n}$ **c.** $\dfrac{\sqrt[3]{-27x^6}}{\sqrt[3]{8y^3}}$

SOLUTION ✔   **a.** $\sqrt[4]{16x^4 y^8} = (16x^4 y^8)^{1/4} = 16^{1/4} \cdot x^{4/4} y^{8/4} = 2xy^2$

b. $\sqrt{12m^3 n} \cdot \sqrt{3m^5 n} = \sqrt{36m^8 n^2} = (36m^8 n^2)^{1/2} = 36^{1/2} \cdot m^{8/2} n^{2/2} = 6m^4 n$

c. $\dfrac{\sqrt[3]{-27x^6}}{\sqrt[3]{8y^3}} = \dfrac{(-27x^6)^{1/3}}{(8y^3)^{1/3}} = \dfrac{-27^{1/3} x^{6/3}}{8^{1/3} y^{3/3}} = -\dfrac{3x^2}{2y}$ ■■■■

When a radical appears in the numerator or denominator of an algebraic expression, we often try to simplify the expression by eliminating the radical from the numerator or denominator. This process, called **rationalization,** is illustrated in the next two examples.

EXAMPLE 6   Rationalize the denominator of the expression $\dfrac{3x}{2\sqrt{x}}$.

SOLUTION ✔ $\dfrac{3x}{2\sqrt{x}} = \dfrac{3x}{2\sqrt{x}} \cdot \dfrac{\sqrt{x}}{\sqrt{x}} = \dfrac{3x\sqrt{x}}{2\sqrt{x^2}} = \dfrac{3x\sqrt{x}}{2x} = \dfrac{3}{2}\sqrt{x}$ ■■■■

EXAMPLE 7   Rationalize the numerator of the expression $\dfrac{3\sqrt{x}}{2x}$.

SOLUTION ✔ $\dfrac{3\sqrt{x}}{2x} = \dfrac{3\sqrt{x}}{2x} \cdot \dfrac{\sqrt{x}}{\sqrt{x}} = \dfrac{3\sqrt{x^2}}{2x\sqrt{x}} = \dfrac{3x}{2x\sqrt{x}} = \dfrac{3}{2\sqrt{x}}$ ■■■■

1.1 Exercises

In Exercises 1–4, determine whether the statement is true or false.

1. $-3 < -20$ **2.** $-5 \le -5$ **3.** $\dfrac{2}{3} > \dfrac{5}{6}$

4. $-\dfrac{5}{6} < -\dfrac{11}{12}$

In Exercises 5–10, show the given interval on a number line.

5. $(3, 6)$ **6.** $(-2, 5]$ **7.** $[-1, 4)$

8. $\left[-\tfrac{6}{5}, -\tfrac{1}{2}\right]$ **9.** $(0, \infty)$ **10.** $(-\infty, 5]$

In Exercises 11–20, find the values of x that satisfy the inequality (inequalities).

11. $2x + 4 < 8$ **12.** $-6 > 4 + 5x$

13. $-4x \ge 20$ **14.** $-12 \le -3x$

15. $-6 < x - 2 < 4$ **16.** $0 \le x + 1 \le 4$

17. $x + 1 > 4$ or $x + 2 < -1$

18. $x + 1 > 2$ or $x - 1 < -2$

19. $x + 3 > 1$ and $x - 2 < 1$

20. $x - 4 \le 1$ and $x + 3 > 2$

In Exercises 21–30, evaluate the expression.

21. $|-6 + 2|$ **22.** $4 + |-4|$

23. $\dfrac{|-12 + 4|}{|16 - 12|}$ **24.** $\left|\dfrac{0.2 - 1.4}{1.6 - 2.4}\right|$

25. $\sqrt{3}|-2| + 3|-\sqrt{3}|$ **26.** $|-1| + \sqrt{2}|-2|$

27. $|\pi - 1| + 2$ **28.** $|\pi - 6| - 3$

29. $|\sqrt{2} - 1| + |3 - \sqrt{2}|$

30. $|2\sqrt{3} - 3| - |\sqrt{3} - 4|$

In Exercises 31–36, suppose a and b are real numbers other than zero and that $a > b$. State whether the inequality is true or false.

31. $b - a > 0$ **32.** $\dfrac{a}{b} > 1$

33. $a^2 > b^2$ **34.** $\dfrac{1}{a} > \dfrac{1}{b}$

35. $a^3 > b^3$ **36.** $-a < -b$

In Exercises 37–42, determine whether the statement is true for all real numbers a and b.

37. $|-a| = a$ **38.** $|b^2| = b^2$

39. $|a - 4| = |4 - a|$ **40.** $|a + 1| = |a| + 1$

41. $|a + b| = |a| + |b|$ **42.** $|a - b| = |a| - |b|$

In Exercises 43–58, evaluate the expression.

43. $27^{2/3}$ **44.** $8^{-4/3}$

45. $\left(\dfrac{1}{\sqrt{3}}\right)^0$ **46.** $(7^{1/2})^4$

47. $\left[\left(\dfrac{1}{8}\right)^{1/3}\right]^{-2}$ **48.** $\left[\left(-\dfrac{1}{3}\right)^2\right]^{-3}$

49. $\left(\dfrac{7^{-5} \cdot 7^2}{7^{-2}}\right)^{-1}$ **50.** $\left(\dfrac{9}{16}\right)^{-1/2}$

51. $(125^{2/3})^{-1/2}$ **52.** $\sqrt[3]{2^6}$

53. $\dfrac{\sqrt{32}}{\sqrt{8}}$ **54.** $\sqrt[3]{\dfrac{-8}{27}}$

55. $\dfrac{16^{5/8}16^{1/2}}{16^{7/8}}$ **56.** $\left(\dfrac{9^{-3} \cdot 9^5}{9^{-2}}\right)^{-1/2}$

57. $16^{1/4} \cdot (8)^{-1/3}$ **58.** $\dfrac{6^{2.5} \cdot 6^{-1.9}}{6^{-1.4}}$

In Exercises 59–68, determine whether the statement is true or false. Give a reason for your choice.

59. $x^4 + 2x^4 = 3x^4$ **60.** $3^2 \cdot 2^2 = 6^2$

61. $x^3 \cdot 2x^2 = 2x^6$ **62.** $3^3 + 3 = 3^4$

63. $\dfrac{2^{4x}}{1^{3x}} = 2^{4x-3x}$ **64.** $(2^2 \cdot 3^2)^2 = 6^4$

65. $\dfrac{1}{4^{-3}} = \dfrac{1}{64}$ **66.** $\dfrac{4^{3/2}}{2^4} = \dfrac{1}{2}$

67. $(1.2^{1/2})^{-1/2} = 1$ **68.** $5^{2/3} \cdot (25)^{2/3} = 25$

In Exercises 69–74, rewrite the expression using positive exponents only.

69. $(xy)^{-2}$

70. $3s^{1/3} \cdot s^{-7/3}$

71. $\dfrac{x^{-1/3}}{x^{1/2}}$

72. $\sqrt{x^{-1}} \cdot \sqrt{9x^{-3}}$

73. $12^0 (s + t)^{-3}$

74. $(x - y)(x^{-1} + y^{-1})$

In Exercises 75–90, simplify the expression. (Assume x, y, r, s, and t are positive.)

75. $\dfrac{x^{7/3}}{x^{-2}}$

76. $(49x^{-2})^{-1/2}$

77. $(x^2 y^{-3})(x^{-5} y^3)$

78. $\dfrac{5x^6 y^3}{2x^2 y^7}$

79. $\dfrac{x^{3/4}}{x^{-1/4}}$

80. $\left(\dfrac{x^3 y^2}{z^2}\right)^2$

81. $\left(\dfrac{x^3}{-27y^{-6}}\right)^{-2/3}$

82. $\left(\dfrac{e^x}{e^{x-2}}\right)^{-1/2}$

83. $\left(\dfrac{x^{-3}}{y^{-2}}\right)^2 \left(\dfrac{y}{x}\right)^4$

84. $\dfrac{(r^n)^4}{r^{5-2n}}$

85. $\sqrt[3]{x^{-2}} \cdot \sqrt{4x^5}$

86. $\sqrt{81x^6 y^{-4}}$

87. $-\sqrt[4]{16x^4 y^8}$

88. $\sqrt[3]{x^{3a+b}}$

89. $\sqrt[6]{64x^8 y^3}$

90. $\sqrt[3]{27r^6} \cdot \sqrt{s^2 t^4}$

In Exercises 91–94, use the fact that $2^{1/2} \approx 1.414$ and $3^{1/2} \approx 1.732$ to evaluate the expression without using a calculator.

91. $2^{3/2}$ **92.** $8^{1/2}$ **93.** $9^{3/4}$ **94.** $6^{1/2}$

In Exercises 95–98, use the fact that $10^{1/2} \approx 3.162$ and $10^{1/3} \approx 2.154$ to evaluate the expression without using a calculator.

95. $10^{3/2}$ **96.** $1000^{3/2}$ **97.** $10^{2.5}$

98. $(0.0001)^{-1/3}$

In Exercises 99–104, rationalize the denominator of the expression.

99. $\dfrac{3}{2\sqrt{x}}$

100. $\dfrac{3}{\sqrt{xy}}$

101. $\dfrac{2y}{\sqrt{3y}}$

102. $\dfrac{5x^2}{\sqrt{3x}}$

103. $\dfrac{1}{\sqrt[3]{x}}$

104. $\sqrt{\dfrac{2x}{y}}$

In Exercises 105–110, rationalize the numerator of the expression.

105. $\dfrac{2\sqrt{x}}{3}$

106. $\dfrac{\sqrt[3]{x}}{24}$

107. $\sqrt{\dfrac{2y}{x}}$

108. $\sqrt[3]{\dfrac{2x}{3y}}$

109. $\dfrac{\sqrt[3]{x^2 z}}{y}$

110. $\dfrac{\sqrt[3]{x^2 y}}{2x}$

111. DRIVING RANGE OF A CAR An advertisement for a certain car states that the EPA fuel economy is 20 mpg city and 27 mpg highway and that the car's fuel-tank capacity is 18.1 gal. Assuming ideal driving conditions, determine the driving range for the car from the foregoing data.

112. Find the minimum cost C (in dollars), given that

$$5(C - 25) \geq 1.75 + 2.5C$$

113. Find the maximum profit P (in dollars) given that

$$6(P - 2500) \leq 4(P + 2400)$$

114. CELSIUS AND FAHRENHEIT TEMPERATURES The relationship between Celsius (°C) and Fahrenheit (°F) temperatures is given by the formula

$$C = \frac{5}{9}(F - 32)$$

a. If the temperature range for Montreal during the month of January is $-15° < °C < -5°$, find the range in degrees Fahrenheit in Montreal for the same period.
b. If the temperature range for New York City during the month of June is $63° < °F < 80°$, find the range in degrees Celsius in New York City for the same period.

115. MEETING SALES TARGETS A salesman's monthly commission is 15% on all sales over $12,000. If his goal is to make a commission of at least $1500 per month, what minimum monthly sales figures must he attain?

116. MARKUP ON A CAR The markup on a used car was at least 30% of its current wholesale price. If the car was sold for $5600, what was the maximum wholesale price?

117. QUALITY CONTROL The PAR Manufacturing Company manufactures steel rods. Suppose the rods ordered by a customer are manufactured to a specification of 0.5 in. and are acceptable only if they are within the *tolerance limits* of 0.49 in. and 0.51 in. Letting x denote the diameter of a rod, write an inequality using absolute value to express a criterion involving x that must be satisfied in order for a rod to be acceptable.

118. QUALITY CONTROL The diameter x (in inches) of a batch of ball bearings manufactured by PAR Manufacturing satisfies the inequality

$$|x - 0.1| \leq 0.01$$

What is the smallest diameter a ball bearing in the batch can have? The largest diameter?

119. **MEETING PROFIT GOALS**   A manufacturer of a certain commodity has estimated that her profit in thousands of dollars is given by the expression

$$-6x^2 + 30x - 10$$

where x (in thousands) is the number of units produced. What production range will enable the manufacturer to realize a profit of at least $14,000 on the commodity?

120. **DISTRIBUTION OF INCOMES**   The distribution of income in a certain city can be described by the exponential model $y = (2.8 \cdot 10^{11})(x)^{-1.5}$, where y is the number of families with an income of x or more dollars.
a. How many families in this city have an income of $20,000 or more?

b. How many families have an income of $40,000 or more?

c. How many families have an income of $100,000 or more?

In Exercises 121–124, determine whether the statement is true or false. If it is true, explain why it is true. If it is false, give an example to show why it is false.

121. If $a < b$, then $a - c > b - c$.

122. $|a - b| = |b - a|$

123. $|a - b| \le |b| + |a|$

124. $\sqrt{a^2 - b^2} = |a| - |b|$

1.2 Precalculus Review II

OPERATIONS WITH ALGEBRAIC EXPRESSIONS

In calculus we often work with algebraic expressions such as

$$2x^{4/3} - x^{1/3} + 1, \qquad 2x^2 - x - \frac{2}{\sqrt{x}}, \qquad \frac{3xy + 2}{x + 1}, \qquad 2x^3 + 2x + 1$$

An algebraic expression of the form ax^n, where the coefficient a is a real number and n is a nonnegative integer, is called a **monomial,** meaning it consists of one term. For example, $7x^2$ is a monomial. A **polynomial** is a monomial or the sum of two or more monomials. For example,

$$x^2 + 4x + 4, \qquad x^3 + 5, \qquad x^4 + 3x^2 + 3, \qquad x^2y + xy + y$$

are all polynomials.

Constant terms and terms containing the same variable factor are called **like,** or **similar, terms.** Like terms may be combined by adding or subtracting their numerical coefficients. For example,

$$3x + 7x = 10x \qquad \text{and} \qquad \frac{1}{2}xy + 3xy = \frac{7}{2}xy$$

The distributive property of the real number system,

$$ab + ac = a \cdot (b + c)$$

is used to justify this procedure.

To add or subtract two or more algebraic expressions, first remove the parentheses and then combine like terms. The resulting expression is written in order of decreasing degree from left to right.

 * EXAMPLE 1

a. $(2x^4 + 3x^3 + 4x + 6) - (3x^4 + 9x^3 + 3x^2)$
$$= 2x^4 + 3x^3 + 4x + 6 - 3x^4 - 9x^3 - 3x^2 \qquad \text{(Remove parentheses.)}$$
$$= 2x^4 - 3x^4 + 3x^3 - 9x^3 - 3x^2 + 4x + 6$$
$$= -x^4 - 6x^3 - 3x^2 + 4x + 6 \qquad \text{(Combine like terms.)}$$

b. $2t^3 - \{t^2 - [t - (2t - 1)] + 4\}$
$$= 2t^3 - \{t^2 - [t - 2t + 1] + 4\}$$
$$= 2t^3 - \{t^2 - [-t + 1] + 4\} \qquad \text{(Remove parentheses and combine like terms within brackets.)}$$
$$= 2t^3 - \{t^2 + t - 1 + 4\} \qquad \text{(Remove brackets.)}$$
$$= 2t^3 - \{t^2 + t + 3\} \qquad \text{(Combine like terms within the braces.)}$$
$$= 2t^3 - t^2 - t - 3 \qquad \text{(Remove braces.)} \qquad ■■■■$$

An algebraic expression is said to be **simplified** if none of its terms are similar. Observe that when the algebraic expression in Example 1b was simplified, the innermost grouping symbols were removed first; that is, the parentheses ( ) were removed first, the brackets [ ] second, and the braces { } third.

When algebraic expressions are multiplied, each term of one algebraic expression is multiplied by each term of the other. The resulting algebraic expression is then simplified.

 EXAMPLE 2

Perform the indicated operations:

a. $(x^2 + 1)(3x^2 + 10x + 3)$ b. $(e^t + e^{-t})e^t - e^t(e^t - e^{-t})$

SOLUTION ✔

a. $(x^2 + 1)(3x^2 + 10x + 3) = x^2(3x^2 + 10x + 3) + 1(3x^2 + 10x + 3)$
$$= 3x^4 + 10x^3 + 3x^2 + 3x^2 + 10x + 3$$
$$= 3x^4 + 10x^3 + 6x^2 + 10x + 3$$

b. $(e^t + e^{-t})e^t - e^t(e^t - e^{-t}) = e^{2t} + e^0 - e^{2t} + e^0$
$$= e^{2t} - e^{2t} + e^0 + e^0$$
$$= 1 + 1 \qquad \text{(Recall that } e^0 = 1.)$$
$$= 2 \qquad ■■■■$$

Certain product formulas that are frequently used in algebraic computations are given in Table 1.5.

---

* The symbol ⓡ indicates that these examples were selected from the calculus portion of the text in order to help you review the algebraic computations you will *actually* be using in calculus.

Table 1.5

Formula	Example
$(a + b)^2 = a^2 + 2ab + b^2$	$(2x + 3y)^2 = (2x)^2 + 2(2x)(3y) + (3y)^2$
	$= 4x^2 + 12xy + 9y^2$
$(a - b)^2 = a^2 - 2ab + b^2$	$(4x - 2y)^2 = (4x)^2 - 2(4x)(2y) + (2y)^2$
	$= 16x^2 - 16xy + 4y^2$
$(a + b)(a - b) = a^2 - b^2$	$(2x + y)(2x - y) = (2x)^2 - (y)^2$
	$= 4x^2 - y^2$

FACTORING

Factoring is the process of expressing an algebraic expression as a product of other algebraic expressions. For example, by applying the distributive property, we may write

$$3x^2 - x = x(3x - 1)$$

The first step in factoring an algebraic expression is to check to see whether it contains any common terms. If it does, the greatest common term is then factored out. For example, the common factor of the algebraic expression $2a^2x + 4ax + 6a$ is $2a$, because

$$2a^2x + 4ax + 6a = 2a \cdot ax + 2a \cdot 2x + 2a \cdot 3 = 2a(ax + 2x + 3)$$

EXAMPLE 3   Factor out the greatest common factor in each of the following expressions:

a. $-0.3t^2 + 3t$ **b.** $2x^{3/2} - 3x^{1/2}$ **c.** $2ye^{xy^2} + 2xy^3e^{xy^2}$

d. $4x(x + 1)^{1/2} - 2x^2\left(\dfrac{1}{2}\right)(x + 1)^{-1/2}$

SOLUTION ✔
a. $-0.3t^2 + 3t = -0.3t(t - 10)$
b. $2x^{3/2} - 3x^{1/2} = x^{1/2}(2x - 3)$
c. $2ye^{xy^2} + 2xy^3e^{xy^2} = 2ye^{xy^2}(1 + xy^2)$

d. $4x(x + 1)^{1/2} - 2x^2\left(\dfrac{1}{2}\right)(x + 1)^{-1/2} = 4x(x + 1)^{1/2} - x^2(x + 1)^{-1/2}$

$$= x(x + 1)^{-1/2}[4(x + 1)^{1/2}(x + 1)^{1/2} - x]$$
$$= x(x + 1)^{-1/2}[4(x + 1) - x]$$
$$= x(x + 1)^{-1/2}(4x + 4 - x) = x(x + 1)^{-1/2}(3x + 4)$$

Here we select $(x + 1)^{-1/2}$ as the common factor because it is "contained" in each algebraic term. In particular, observe that

$$(x + 1)^{-1/2}(x + 1)^{1/2}(x + 1)^{1/2} = (x + 1)^{1/2}$$ ■■■■

Sometimes an algebraic expression may be factored by regrouping and rearranging its terms and then factoring out a common term. This technique is illustrated in Example 4.

EXAMPLE 4

Factor:

a. $2ax + 2ay + bx + by$ **b.** $3x\sqrt{y} - 4 - 2\sqrt{y} + 6x$

SOLUTION ✔

a. First, factor the common term $2a$ from the first two terms and the common term b from the last two terms. Thus,

$$2ax + 2ay + bx + by = 2a(x + y) + b(x + y)$$

Since $(x + y)$ is common to both terms of the polynomial, we may factor it out. Hence,

$$2a(x + y) + b(x + y) = (x + y)(2a + b)$$

b. $3x\sqrt{y} - 4 - 2\sqrt{y} + 6x = 3x\sqrt{y} - 2\sqrt{y} + 6x - 4$
$$= \sqrt{y}(3x - 2) + 2(3x - 2)$$
$$= (3x - 2)(\sqrt{y} + 2)$$ ■ ■ ■ ■

The first step in factoring a polynomial is to find the common factors. The next step is to express the polynomial as the product of a constant and/or one or more prime polynomials.

Certain product formulas that are useful in factoring binomials and trinomials are listed in Table 1.6.

Table 1.6

Formula	Example
Difference of two squares	
$x^2 - y^2 = (x + y)(x - y)$	$x^2 - 36 = (x + 6)(x - 6)$
	$8x^2 - 2y^2 = 2(4x^2 - y^2)$
	$\qquad\qquad = 2(2x + y)(2x - y)$
	$9 - a^6 = (3 + a^3)(3 - a^3)$
Perfect-square trinomial	
$x^2 + 2xy + y^2 = (x + y)^2$	$x^2 + 8x + 16 = (x + 4)^2$
$x^2 - 2xy + y^2 = (x - y)^2$	$4x^2 - 4xy + y^2 = (2x - y)^2$
Sum of two cubes	
$x^3 + y^3 = (x + y)(x^2 - xy + y^2)$	$z^3 + 27 = z^3 + (3)^3$
	$\qquad\qquad = (z + 3)(z^2 - 3z + 9)$
Difference of two cubes	
$x^3 - y^3 = (x - y)(x^2 + xy + y^2)$	$8x^3 - y^6 = (2x)^3 - (y^2)^3$
	$\qquad\qquad = (2x - y^2)(4x^2 + 2xy^2 + y^4)$

The factors of the second-degree polynomial with integral coefficients

$$px^2 + qx + r$$

are $(ax + b)(cx + d)$, where $ac = p$, $ad + bc = q$, and $bd = r$. Since only a limited number of choices are possible, we use a trial-and-error method to factor polynomials having this form.

For example, to factor $x^2 - 2x - 3$, we first observe that the only possible first-degree terms are

$$(x \qquad)(x \qquad) \qquad \text{(Since the coefficient of } x^2 \text{ is 1)}$$

Next, we observe that the product of the constant terms is (-3). This gives us the following possible factors:

$$(x - 1)(x + 3)$$
$$(x + 1)(x - 3)$$

Looking once again at the polynomial $x^2 - 2x - 3$, we see that the coefficient of x is -2. Checking to see which set of factors yields -2 for the coefficient of x, we find that

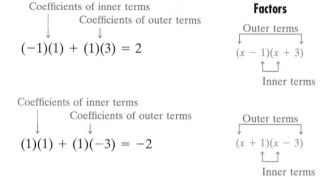

and we conclude that the correct factorization is

$$x^2 - 2x - 3 = (x + 1)(x - 3)$$

With practice, you will soon find that you can perform many of these steps mentally and the need to write out each step will be eliminated.

EXAMPLE 5

Factor:

a. $3x^2 + 4x - 4$ **b.** $3x^2 - 6x - 24$

SOLUTION ✔

a. Using trial and error, we find that the correct factorization is

$$3x^2 + 4x - 4 = (3x - 2)(x + 2)$$

b. Since each term has the common factor 3, we have

$$3x^2 - 6x - 24 = 3(x^2 - 2x - 8)$$

Using the trial-and-error method of factorization, we find that

$$x^2 - 2x - 8 = (x - 4)(x + 2)$$

Thus, we have

$$3x^2 - 6x - 24 = 3(x - 4)(x + 2)$$ ■ ■ ■ ■

ROOTS OF POLYNOMIAL EQUATIONS

A polynomial equation of degree n in the variable x is an equation of the form

$$a_n x^n + a_{n-1} x^{n-1} + \cdots + a_0 = 0$$

where n is a nonnegative integer and $a_0, a_1, \ldots, a_n$ are real numbers with $a_n \neq 0$. For example, the equation

$$-2x^5 + 8x^3 - 6x^2 + 3x + 1 = 0$$

is a polynomial equation of degree 5 in x.

The **roots of a polynomial equation** are precisely the values of x that satisfy the given equation.* One way of finding the roots of a polynomial equation is to first factor the polynomial and then solve the resulting equation. For example, the polynomial equation

$$x^3 - 3x^2 + 2x = 0$$

may be rewritten in the form

$$x(x^2 - 3x + 2) = 0 \qquad \text{or} \qquad x(x - 1)(x - 2) = 0$$

Since the product of two real numbers can be equal to zero if and only if one (or both) of the factors is equal to zero, we have

$$x = 0, \qquad x - 1 = 0, \qquad \text{or} \qquad x - 2 = 0$$

from which we see that the desired roots are $x = 0, 1,$ and 2.

THE QUADRATIC FORMULA

In general, the problem of finding the roots of a polynomial equation is a difficult one. But the roots of a quadratic equation (a polynomial equation of degree 2) are easily found either by factoring or by using the following quadratic formula.

Quadratic Formula

> The solutions of the equation $ax^2 + bx + c = 0$ $(a \neq 0)$ are given by
>
> $$x = \frac{-b \pm \sqrt{b^2 - 4ac}}{2a}$$

---

* In this book, we are interested only in the *real* roots of an equation.

EXAMPLE 6

Solve each of the following quadratic equations:

a. $2x^2 + 5x - 12 = 0$ **b.** $x^2 = -3x + 8$

SOLUTION ✔

a. The equation is in standard form, with $a = 2$, $b = 5$, and $c = -12$. Using the quadratic formula, we find

$$x = \frac{-b \pm \sqrt{b^2 - 4ac}}{2a} = \frac{-5 \pm \sqrt{5^2 - 4(2)(-12)}}{2(2)}$$

$$= \frac{-5 \pm \sqrt{121}}{4} = \frac{-5 \pm 11}{4}$$

$$= -4 \text{ or } \frac{3}{2}$$

This equation can also be solved by factoring. Thus,

$$2x^2 + 5x - 12 = (2x - 3)(x + 4) = 0$$

from which we see that the desired roots are $x = 3/2$ or $x = -4$, as obtained earlier.

b. We first rewrite the given equation in the standard form $x^2 + 3x - 8 = 0$, from which we see that $a = 1$, $b = 3$, and $c = -8$. Using the quadratic formula, we find

$$x = \frac{-b \pm \sqrt{b^2 - 4ac}}{2a} = \frac{-3 \pm \sqrt{3^2 - 4(1)(-8)}}{2(1)}$$

$$= \frac{-3 \pm \sqrt{41}}{2}$$

That is, the solutions are

$$\frac{-3 + \sqrt{41}}{2} \approx 1.7 \quad \text{and} \quad \frac{-3 - \sqrt{41}}{2} \approx -4.7$$

In this case, the quadratic formula proves quite handy!   ■ ■ ■ ■

RATIONAL EXPRESSIONS

Quotients of polynomials are called **rational expressions.** Examples of rational expressions are

$$\frac{6x - 1}{2x + 3}, \qquad \frac{3x^2y^3 - 2xy}{4x}, \qquad \frac{2}{5ab}$$

Since rational expressions are quotients in which the variables represent real numbers, the properties of the real numbers apply to rational expressions as well, and operations with rational fractions are performed in the same manner as operations with arithmetic fractions. For example, using the properties of the real number system, we may write

$$\frac{ac}{bc} = \frac{a}{b} \cdot \frac{c}{c} = \frac{a}{b} \cdot 1 = \frac{a}{b}$$

where a, b, and c are any real numbers and b and c are not zero.

Similarly, using the same properties of real numbers, we may write

$$\frac{(x+2)(x-3)}{(x-2)(x-3)} = \frac{x+2}{x-2} \qquad (x \neq 2, 3)$$

after "canceling" the common factors.

 An example of incorrect cancellation is

$$\frac{\cancel{3}+4x}{\cancel{3}} \neq 1 + 4x$$

Instead, we need to write

$$\frac{3+4x}{3} = \frac{3}{3} + \frac{4x}{3} = 1 + \frac{4x}{3}$$

A rational expression is simplified, or in lowest terms, when the numerator and denominator have no common factors other than 1 and −1 and the expression contains no negative exponents.

 EXAMPLE 7   Simplify the following expressions:

a. $\dfrac{x^2 + 2x - 3}{x^2 + 4x + 3}$ **b.** $\dfrac{[(t^2 + 4)(2t - 4) - (t^2 - 4t + 4)(2t)]}{(t^2 + 4)^2}$

SOLUTION ✔
a. $\dfrac{x^2 + 2x - 3}{x^2 + 4x + 3} = \dfrac{(x+3)(x-1)}{(x+3)(x+1)} = \dfrac{x-1}{x+1}$

b. $\dfrac{[(t^2 + 4)(2t - 4) - (t^2 - 4t + 4)(2t)]}{(t^2 + 4)^2}$

$$= \frac{2t^3 - 4t^2 + 8t - 16 - 2t^3 + 8t^2 - 8t}{(t^2 + 4)^2} \quad \text{(Carry out the indicated multiplication.)}$$

$$= \frac{4t^2 - 16}{(t^2 + 4)^2} \quad \text{(Combine like terms.)}$$

$$= \frac{4(t^2 - 4)}{(t^2 + 4)^2} \quad \text{(Factor.)} \qquad ■■■■$$

The operations of multiplication and division are performed with algebraic fractions in the same manner as with arithmetic fractions (Table 1.7).

Table 1.7

Operation	Example
If P, Q, R, and S are polynomials, then	
Multiplication	
$\dfrac{P}{Q} \cdot \dfrac{R}{S} = \dfrac{PR}{QS} \qquad (Q, S \neq 0)$	$\dfrac{2x}{y} \cdot \dfrac{(x+1)}{(y-1)} = \dfrac{2x(x+1)}{y(y-1)} = \dfrac{2x^2 + 2x}{y^2 - y}$
Division	
$\dfrac{P}{Q} \div \dfrac{R}{S} = \dfrac{P}{Q} \cdot \dfrac{S}{R} = \dfrac{PS}{QR} \qquad (Q, R, S \neq 0)$	$\dfrac{x^2 + 3}{y} \div \dfrac{y^2 + 1}{x} = \dfrac{x^2 + 3}{y} \cdot \dfrac{x}{y^2 + 1} = \dfrac{x^3 + 3x}{y^3 + y}$

When rational expressions are multiplied and divided, the resulting expressions should be simplified.

EXAMPLE 8 Perform the indicated operations and simplify:

$$\frac{2x - 8}{x + 2} \cdot \frac{x^2 + 4x + 4}{x^2 - 16}$$

SOLUTION ✔

$$\frac{2x - 8}{x + 2} \cdot \frac{x^2 + 4x + 4}{x^2 - 16}$$

$$= \frac{2(x - 4)}{x + 2} \cdot \frac{(x + 2)^2}{(x + 4)(x - 4)}$$

$$= \frac{2(x - 4)(x + 2)(x + 2)}{(x + 2)(x + 4)(x - 4)} \qquad \text{[Cancel the common factors } (x + 2)(x - 4).\text{]}$$

$$= \frac{2(x + 2)}{x + 4}$$

■ ■ ■ ■

For rational expressions, the operations of addition and subtraction are performed by finding a common denominator of the fractions and then adding or subtracting the fractions. Table 1.8 shows the rules for fractions with equal denominators.

Table 1.8

Operation	Example
If P, Q, and R are polynomials, then	
Addition	
$\dfrac{P}{R} + \dfrac{Q}{R} = \dfrac{P + Q}{R} \qquad (R \neq 0)$	$\dfrac{2x}{x + 2} + \dfrac{6x}{x + 2} = \dfrac{2x + 6x}{x + 2} = \dfrac{8x}{x + 2}$
Subtraction	
$\dfrac{P}{R} - \dfrac{Q}{R} = \dfrac{P - Q}{R} \qquad (R \neq 0)$	$\dfrac{3y}{y - x} - \dfrac{y}{y - x} = \dfrac{3y - y}{y - x} = \dfrac{2y}{y - x}$

To add or subtract fractions that have different denominators, first find a common denominator, preferably the least common denominator (LCD). Then carry out the indicated operations following the procedure described in Table 1.8.

To find the LCD of two or more rational expressions:

1. *Find the prime factors of each denominator.*
2. *Form the product of the different prime factors that occur in the denominators. Each prime factor in this product should be raised to the highest power of that factor appearing in the denominators.*

$$\frac{x}{2 + y} \neq \frac{x}{2} + \frac{x}{y}$$

EXAMPLE 9   Simplify:

$$\textbf{a.}\ \frac{2x}{x^2+1}+\frac{6(3x^2)}{x^3+2} \qquad \textbf{b.}\ \frac{1}{x+h}-\frac{1}{x}$$

SOLUTION ✔

$$\textbf{a.}\ \frac{2x}{x^2+1}+\frac{6(3x^2)}{x^3+2}$$

$$=\frac{2x(x^3+2)+6(3x^2)(x^2+1)}{(x^2+1)(x^3+2)} \qquad [\text{LCD}=(x^2+1)(x^3+2)]$$

$$=\frac{2x^4+4x+18x^4+18x^2}{(x^2+1)(x^3+2)}$$

$$=\frac{20x^4+18x^2+4x}{(x^2+1)(x^3+2)}$$

$$=\frac{2x(10x^3+9x+2)}{(x^2+1)(x^3+2)}$$

$$\textbf{b.}\ \frac{1}{x+h}-\frac{1}{x}=\frac{1}{x+h}\cdot\frac{x}{x}-\frac{1}{x}\cdot\frac{x+h}{x+h} \qquad [\text{LCD}=(x)(x+h)]$$

$$=\frac{x}{x(x+h)}-\frac{x+h}{x(x+h)}$$

$$=\frac{x-x-h}{x(x+h)}$$

$$=\frac{-h}{x(x+h)}$$

■ ■ ■ ■

OTHER ALGEBRAIC FRACTIONS

The techniques used to simplify rational expressions may also be used to simplify algebraic fractions in which the numerator and denominator are not polynomials, as illustrated in Example 10.

EXAMPLE 10   Simplify:

$$\textbf{a.}\ \frac{1+\dfrac{1}{x+1}}{x-\dfrac{4}{x}} \qquad \textbf{b.}\ \frac{x^{-1}+y^{-1}}{x^{-2}-y^{-2}}$$

SOLUTION ✔

a.
$$\frac{1 + \dfrac{1}{x+1}}{x - \dfrac{4}{x}} = \frac{1 \cdot \dfrac{x+1}{x+1} + \dfrac{1}{x+1}}{x \cdot \dfrac{x}{x} - \dfrac{4}{x}} = \frac{\dfrac{x+1+1}{x+1}}{\dfrac{x^2-4}{x}}$$

$$= \frac{x+2}{x+1} \cdot \frac{x}{x^2-4} = \frac{x+2}{x+1} \cdot \frac{x}{(x+2)(x-2)}$$

$$= \frac{x}{(x+1)(x-2)}$$

b.
$$\frac{x^{-1} + y^{-1}}{x^{-2} - y^{-2}} = \frac{\dfrac{1}{x} + \dfrac{1}{y}}{\dfrac{1}{x^2} - \dfrac{1}{y^2}} = \frac{\dfrac{y+x}{xy}}{\dfrac{y^2-x^2}{x^2y^2}} \qquad \left(x^{-n} = \frac{1}{x^n} \right)$$

$$= \frac{y+x}{xy} \cdot \frac{x^2y^2}{y^2-x^2} = \frac{y+x}{xy} \cdot \frac{(xy)^2}{(y+x)(y-x)}$$

$$= \frac{xy}{y-x}$$

■■■■

EXAMPLE 11

Perform the given operations and simplify:

a. $\dfrac{x^2(2x^2+1)^{1/2}}{x-1} \cdot \dfrac{4x^3 - 6x^2 + x - 2}{x(x-1)(2x^2+1)}$ **b.** $\dfrac{12x^2}{\sqrt{2x^2+3}} + 6\sqrt{2x^2+3}$

SOLUTION ✔

a. $\dfrac{x^2(2x^2+1)^{1/2}}{x-1} \cdot \dfrac{4x^3 - 6x^2 + x - 2}{x(x-1)(2x^2+1)} = \dfrac{x(4x^3 - 6x^2 + x - 2)}{(x-1)^2(2x^2+1)^{1-1/2}}$

$$= \frac{x(4x^3 - 6x^2 + x - 2)}{(x-1)^2(2x^2+1)^{1/2}}$$

b. $\dfrac{12x^2}{\sqrt{2x^2+3}} + 6\sqrt{2x^2+3} = \dfrac{12x^2}{(2x^2+3)^{1/2}} + 6(2x^2+3)^{1/2}$

$$= \frac{12x^2 + 6(2x^2+3)^{1/2}(2x^2+3)^{1/2}}{(2x^2+3)^{1/2}}$$

$$= \frac{12x^2 + 6(2x^2+3)}{(2x^2+3)^{1/2}}$$

$$= \frac{24x^2 + 18}{(2x^2+3)^{1/2}} = \frac{6(4x^2+3)}{\sqrt{2x^2+3}}$$

■■■■

RATIONALIZING ALGEBRAIC FRACTIONS

When the denominator of an algebraic fraction contains sums or differences involving radicals, we may **rationalize the denominator**—that is, transform the fraction into an equivalent one with a denominator that does not contain radicals. In doing so, we make use of the fact that

$$(\sqrt{a} + \sqrt{b})(\sqrt{a} - \sqrt{b}) = (\sqrt{a})^2 - (\sqrt{b})^2$$
$$= a - b$$

This procedure is illustrated in Example 12.

EXAMPLE 12   Rationalize the denominator: $\dfrac{1}{1 + \sqrt{x}}$.

SOLUTION ✔   Upon multiplying the numerator and the denominator by $(1 - \sqrt{x})$, we obtain

$$\frac{1}{1 + \sqrt{x}} = \frac{1}{1 + \sqrt{x}} \cdot \frac{1 - \sqrt{x}}{1 - \sqrt{x}}$$
$$= \frac{1 - \sqrt{x}}{1 - (\sqrt{x})^2}$$
$$= \frac{1 - \sqrt{x}}{1 - x}$$

In other situations, it may be necessary to rationalize the numerator of an algebraic expression. In calculus, for example, one encounters the following problem.

EXAMPLE 13   Rationalize the numerator: $\dfrac{\sqrt{1 + h} - 1}{h}$.

SOLUTION ✔

$$\frac{\sqrt{1 + h} - 1}{h} = \frac{\sqrt{1 + h} - 1}{h} \cdot \frac{\sqrt{1 + h} + 1}{\sqrt{1 + h} + 1}$$
$$= \frac{(\sqrt{1 + h})^2 - (1)^2}{h(\sqrt{1 + h} + 1)}$$
$$= \frac{1 + h - 1}{h(\sqrt{1 + h} + 1)} \qquad [(\sqrt{1+h})^2 = \sqrt{1+h} \cdot \sqrt{1+h}$$
$$\qquad\qquad\qquad\qquad = 1 + h]$$
$$= \frac{h}{h(\sqrt{1 + h} + 1)}$$
$$= \frac{1}{\sqrt{1 + h} + 1}$$

1.2 Exercises

In Exercises 1–22, perform the indicated operations and simplify each expression.

1. $(7x^2 - 2x + 5) + (2x^2 + 5x - 4)$

2. $(3x^2 + 5xy + 2y) + (4 - 3xy - 2x^2)$

3. $(5y^2 - 2y + 1) - (y^2 - 3y - 7)$

4. $3(2a - b) - 4(b - 2a)$

5. $x - \{2x - [-x - (1 - x)]\}$

6. $3x^2 - \{x^2 + 1 - x[x - (2x - 1)]\} + 2$

7. $\left(\dfrac{1}{3} - 1 + e\right) - \left(-\dfrac{1}{3} - 1 + e^{-1}\right)$

8. $-\dfrac{3}{4}y - \dfrac{1}{4}x + 100 + \dfrac{1}{2}x + \dfrac{1}{4}y - 120$

9. $3\sqrt{8} + 8 - 2\sqrt{y} + \dfrac{1}{2}\sqrt{x} - \dfrac{3}{4}\sqrt{y}$

10. $\dfrac{8}{9}x^2 + \dfrac{2}{3}x + \dfrac{16}{3}x^2 - \dfrac{16}{3}x - 2x + 2$

11. $(x + 8)(x - 2)$ **12.** $(5x + 2)(3x - 4)$

13. $(a + 5)^2$ **14.** $(3a - 4b)^2$

15. $(x + 2y)^2$ **16.** $(6 - 3x)^2$

17. $(2x + y)(2x - y)$ **18.** $(3x + 2)(2 - 3x)$

19. $(x^2 - 1)(2x) - x^2(2x)$

20. $(x^{1/2} + 1)\left(\dfrac{1}{2}x^{-1/2}\right) - (x^{1/2} - 1)\left(\dfrac{1}{2}x^{-1/2}\right)$

21. $2(t + \sqrt{t})^2 - 2t^2$ **22.** $2x^2 + (-x + 1)^2$

In Exercises 23–30, factor out the greatest common factor from each expression.

23. $4x^5 - 12x^4 - 6x^3$

24. $4x^2y^2z - 2x^5y^2 + 6x^3y^2z^2$

25. $7a^4 - 42a^2b^2 + 49a^3b$

26. $3x^{2/3} - 2x^{1/3}$ **27.** $e^{-x} - xe^{-x}$

28. $2ye^{xy^2} + 2xy^3e^{xy^2}$ **29.** $2x^{-5/2} - \dfrac{3}{2}x^{-3/2}$

30. $\dfrac{1}{2}\left(\dfrac{2}{3}u^{3/2} - 2u^{1/2}\right)$

In Exercises 31–44, factor each expression.

31. $6ac + 3bc - 4ad - 2bd$

32. $3x^3 - x^2 + 3x - 1$

33. $4a^2 - b^2$ **34.** $12x^2 - 3y^2$

35. $10 - 14x - 12x^2$ **36.** $x^2 - 2x - 15$

37. $3x^2 - 6x - 24$ **38.** $3x^2 - 4x - 4$

39. $12x^2 - 2x - 30$ **40.** $(x + y)^2 - 1$

41. $9x^2 - 16y^2$ **42.** $8a^2 - 2ab - 6b^2$

43. $x^6 + 125$ **44.** $x^3 - 27$

In Exercises 45–52, perform the indicated operations and simplify each expression.

45. $(x^2 + y^2)x - xy(2y)$ **46.** $2kr(R - r) - kr^2$

47. $2(x - 1)(2x + 2)^3[4(x - 1) + (2x + 2)]$

48. $5x^2(3x^2 + 1)^4(6x) + (3x^2 + 1)^5(2x)$

49. $4(x - 1)^2(2x + 2)^3(2) + (2x + 2)^4(2)(x - 1)$

50. $(x^2 + 1)(4x^3 - 3x^2 + 2x) - (x^4 - x^3 + x^2)(2x)$

51. $(x^2 + 2)^2[5(x^2 + 2)^2 - 3](2x)$

52. $(x^2 - 4)(x^2 + 4)(2x + 8) - (x^2 + 8x - 4)(4x^3)$

In Exercises 53–58, find the real roots of each equation by factoring.

53. $x^2 + x - 12 = 0$ **54.** $3x^2 - x - 4 = 0$

55. $4t^2 + 2t - 2 = 0$ **56.** $-6x^2 + x + 12 = 0$

57. $\dfrac{1}{4}x^2 - x + 1 = 0$ **58.** $\dfrac{1}{2}a^2 + a - 12 = 0$

In Exercises 59–64, solve the equation by using the quadratic formula.

59. $4x^2 + 5x - 6 = 0$ **60.** $3x^2 - 4x + 1 = 0$

61. $8x^2 - 8x - 3 = 0$ **62.** $x^2 - 6x + 6 = 0$

63. $2x^2 + 4x - 3 = 0$ **64.** $2x^2 + 7x - 15 = 0$

In Exercises 65–70, simplify the expression.

65. $\dfrac{x^2 + x - 2}{x^2 - 4}$

66. $\dfrac{2a^2 - 3ab - 9b^2}{2ab^2 + 3b^3}$

67. $\dfrac{12t^2 + 12t + 3}{4t^2 - 1}$

68. $\dfrac{x^3 + 2x^2 - 3x}{-2x^2 - x + 3}$

69. $\dfrac{(4x - 1)(3) - (3x + 1)(4)}{(4x - 1)^2}$

70. $\dfrac{(1 + x^2)^2(2) - 2x(2)(1 + x^2)(2x)}{(1 + x^2)^4}$

In Exercises 71–88, perform the indicated operations and simplify each expression.

71. $\dfrac{2a^2 - 2b^2}{b - a} \cdot \dfrac{4a + 4b}{a^2 + 2ab + b^2}$

72. $\dfrac{x^2 - 6x + 9}{x^2 - x - 6} \cdot \dfrac{3x + 6}{2x^2 - 7x + 3}$

73. $\dfrac{3x^2 + 2x - 1}{2x + 6} \div \dfrac{x^2 - 1}{x^2 + 2x - 3}$

74. $\dfrac{3x^2 - 4xy - 4y^2}{x^2 y} \div \dfrac{(2y - x)^2}{x^3 y}$

75. $\dfrac{58}{3(3t + 2)} + \dfrac{1}{3}$

76. $\dfrac{a + 1}{3a} + \dfrac{b - 2}{5b}$

77. $\dfrac{2x}{2x - 1} - \dfrac{3x}{2x + 5}$

78. $\dfrac{-xe^x}{x + 1} + e^x$

79. $\dfrac{4}{x^2 - 9} - \dfrac{5}{x^2 - 6x + 9}$

80. $\dfrac{x}{1 - x} + \dfrac{2x + 3}{x^2 - 1}$

81. $\dfrac{1 + \dfrac{1}{x}}{1 - \dfrac{1}{x}}$

82. $\dfrac{\dfrac{1}{x} + \dfrac{1}{y}}{1 - \dfrac{1}{xy}}$

83. $\dfrac{4x^2}{2\sqrt{2x^2 + 7}} + \sqrt{2x^2 + 7}$

84. $6(2x + 1)^2 \sqrt{x^2 + x} + \dfrac{(2x + 1)^4}{2\sqrt{x^2 + x}}$

85. $\dfrac{2x(x + 1)^{-1/2} - (x + 1)^{1/2}}{x^2}$

86. $\dfrac{(x^2 + 1)^{1/2} - 2x^2(x^2 + 1)^{-1/2}}{1 - x^2}$

87. $\dfrac{(2x + 1)^{1/2} - (x + 2)(2x + 1)^{-1/2}}{2x + 1}$

88. $\dfrac{2(2x - 3)^{1/3} - (x - 1)(2x - 3)^{-2/3}}{(2x - 3)^{2/3}}$

In Exercises 89–94, rationalize the denominator of each expression.

89. $\dfrac{1}{\sqrt{3} - 1}$

90. $\dfrac{1}{\sqrt{x} + 5}$

91. $\dfrac{1}{\sqrt{x} - \sqrt{y}}$

92. $\dfrac{a}{1 - \sqrt{a}}$

93. $\dfrac{\sqrt{a} + \sqrt{b}}{\sqrt{a} - \sqrt{b}}$

94. $\dfrac{2\sqrt{a} + \sqrt{b}}{2\sqrt{a} - \sqrt{b}}$

In Exercises 95–100, rationalize the numerator of each expression.

95. $\dfrac{\sqrt{x}}{3}$

96. $\dfrac{\sqrt[3]{y}}{x}$

97. $\dfrac{1 - \sqrt{3}}{3}$

98. $\dfrac{\sqrt{x} - 1}{x}$

99. $\dfrac{1 + \sqrt{x + 2}}{\sqrt{x + 2}}$

100. $\dfrac{\sqrt{x + 3} - \sqrt{x}}{3}$

In Exercises 101–104, determine whether the statement is true or false. If it is true, explain why it is true. If it is false, give an example to show why it is false.

101. If $b^2 - 4ac > 0$, then $ax^2 + bx + c = 0$, $a \neq 0$, has two real roots.

102. If $b^2 - 4ac < 0$, then $ax^2 + bx + c = 0$, $a \neq 0$, has no real roots.

103. $\dfrac{a}{b + c} = \left(\dfrac{a}{b} + \dfrac{a}{c}\right)$

104. $\sqrt{(a + b)(b - a)} = \sqrt{b^2 - a^2}$ for all real numbers a and b.

1.3 The Cartesian Coordinate System

FIGURE **1.3**
The Cartesian coordinate system

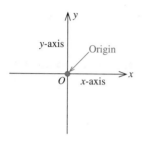

FIGURE **1.4**
An ordered pair (x, y)

THE CARTESIAN COORDINATE SYSTEM

In Section 1.1 we saw how a one-to-one correspondence between the set of real numbers and the points on a straight line leads to a coordinate system on a line (a one-dimensional space).

A similar representation for points in a plane (a two-dimensional space) is realized through the **Cartesian coordinate system,** which is constructed as follows: Take two perpendicular lines, one of which is normally chosen to be horizontal. These lines intersect at a point O, called the **origin** (Figure 1.3). The horizontal line is called the **x-axis,** and the vertical line is called the **y-axis.** A number scale is set up along the x-axis, with the positive numbers lying to the right of the origin and the negative numbers lying to the left of it. Similarly, a number scale is set up along the y-axis, with the positive numbers lying above the origin and the negative numbers lying below it.

The number scales on the two axes need not be the same. Indeed, in many applications different quantities are represented by x and y. For example, x may represent the number of typewriters sold and y the total revenue resulting from the sales. In such cases it is often desirable to choose different number scales to represent the different quantities. Note, however, that the zeros of both number scales coincide at the origin of the two-dimensional coordinate system.

A point in the plane can now be represented uniquely in this coordinate system by an **ordered pair** of numbers—that is, a pair (x, y), where x is the first number and y the second. To see this, let P be any point in the plane (Figure 1.4). Draw perpendiculars from P to the x-axis and y-axis, respectively. Then the number x is precisely the number that corresponds to the point on the x-axis at which the perpendicular through P hits the x-axis. Similarly, y is the number that corresponds to the point on the y-axis at which the perpendicular through P crosses the y-axis.

Conversely, given an ordered pair (x, y) with x as the first number and y the second, a point P in the plane is uniquely determined as follows: Locate the point on the x-axis represented by the number x and draw a line through that point parallel to the y-axis. Next, locate the point on the y-axis represented by the number y and draw a line through that point parallel to the x-axis. The point of intersection of these two lines is the point P (see Figure 1.4).

In the ordered pair (x, y), x is called the **abscissa,** or **x-coordinate,** y is called the **ordinate,** or **y-coordinate,** and x and y together are referred to as the **coordinates** of the point P.

Letting (a, b) denote the point P with x-coordinate a and y-coordinate b, the points $A = (2, 3)$, $B = (-2, 3)$, $C = (-2, -3)$, $D = (2, -3)$, $E = (3, 2)$, $F = (4, 0)$, and $G = (0, -5)$ are plotted in Figure 1.5. The fact that, in general, $(x, y) \neq (y, x)$ is clearly illustrated by points A and E.

The axes divide the plane into four quadrants. Quadrant I consists of the points (x, y) that satisfy $x > 0$ and $y > 0$; Quadrant II, the points (x, y), where

FIGURE 1.5
Several points in the Cartesian plane

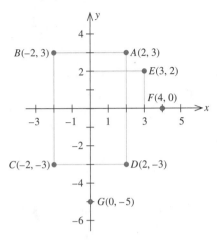

FIGURE 1.6
The four quadrants in the Cartesian plane

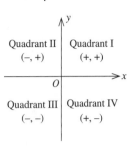

FIGURE 1.7
The distance d between the points (x_1, y_1) and (x_2, y_2)

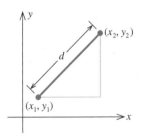

$x < 0$ and $y > 0$; Quadrant III, the points (x, y), where $x < 0$ and $y < 0$; and Quadrant IV, the points (x, y), where $x > 0$ and $y < 0$ (Figure 1.6).

THE DISTANCE FORMULA

One immediate benefit that arises from using the Cartesian coordinate system is that the distance between any two points in the plane may be expressed solely in terms of their coordinates. Suppose, for example, (x_1, y_1) and (x_2, y_2) are any two points in the plane (Figure 1.7). Then the distance between these two points can be computed using the following formula.

Distance Formula

The distance d between two points $P_1(x_1, y_1)$ and $P_2(x_2, y_2)$ in the plane is given by
$$d = \sqrt{(x_2 - x_1)^2 + (y_2 - y_1)^2} \tag{1}$$

For a proof of this result, see Exercise 46, page 36.
    In what follows, we give several applications of the distance formula.

EXAMPLE 1   Find the distance between the points $(-4, 3)$ and $(2, 6)$.

SOLUTION ✔   Let $P_1(-4, 3)$ and $P_2(2, 6)$ be points in the plane. Then, we have

$$x_1 = -4, \qquad y_1 = 3, \qquad x_2 = 2, \qquad y_2 = 6$$

Using Formula (1), we have

$$d = \sqrt{[2 - (-4)]^2 + (6 - 3)^2}$$
$$= \sqrt{6^2 + 3^2}$$
$$= \sqrt{45} = 3\sqrt{5}$$

■ ■ ■ ■

> **Group Discussion**
> Refer to Example 1. Suppose we label the point $(2, 6)$ as P_1 and the point $(-4, 3)$ as P_2. (1) Show that the distance d between the two points is the same as that obtained earlier. (2) Prove that, in general, the distance d in formula (1) is independent of the way we label the two points.

EXAMPLE 2   Let $P(x, y)$ denote a point lying on the circle with radius r and center $C(h, k)$ (Figure 1.8). Find a relationship between x and y.

FIGURE 1.8
A circle with radius r and center $C(h, k)$

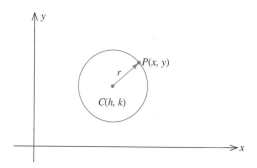

SOLUTION ✔   By the definition of a circle, the distance between $C(h, k)$ and $P(x, y)$ is r. Using Formula (1), we have

$$\sqrt{(x - h)^2 + (y - k)^2} = r$$

which, upon squaring both sides, gives the equation

$$(x - h)^2 + (y - k)^2 = r^2$$

that must be satisfied by the variables x and y.

■ ■ ■ ■

A summary of the result obtained in Example 2 follows.

Equation of a Circle

> An equation of the circle with center $C(h, k)$ and radius r is given by
>
> $$(x - h)^2 + (y - k)^2 = r^2 \qquad \textbf{(2)}$$

EXAMPLE 3   Find an equation of the circle with:

a. Radius 2 and center $(-1, 3)$.
b. Radius 3 and center located at the origin.

SOLUTION ✔   **a.** We use Formula (2) with $r = 2$, $h = -1$, and $k = 3$, obtaining

$$[x - (-1)]^2 + (y - 3)^2 = 2^2 \quad \text{or} \quad (x + 1)^2 + (y - 3)^2 = 4$$

(Figure 1.9a).
b. Using Formula (2) with $r = 3$ and $h = k = 0$, we obtain

$$x^2 + y^2 = 3^2 \quad \text{or} \quad x^2 + y^2 = 9$$

(Figure 1.9b).

■ ■ ■ ■

FIGURE 1.9

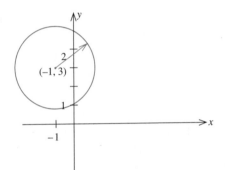

(a) The circle with radius 2 and center $(-1, 3)$ **(b)** The circle with radius 3 and center $(0, 0)$

Group Discussion

1. Use the distance formula to help you describe the set of points in the xy-plane satisfying each of the following inequalities.
   a. $(x - h)^2 + (y - k)^2 \leq r^2$ **b.** $(x - h)^2 + (y - k)^2 < r^2$
   c. $(x - h)^2 + (y - k)^2 \geq r^2$ **d.** $(x - h)^2 + (y - k)^2 > r^2$
2. Consider the equation $x^2 + y^2 = 4$.
   a. Show that $y = \pm\sqrt{4 - x^2}$.
   b. Describe the set of points (x, y) in the xy-plane satisfying the following equations:
   (i) $y = \sqrt{4 - x^2}$
   (ii) $y = -\sqrt{4 - x^2}$

APPLICATION

EXAMPLE 4   In the following diagram (Figure 1.10), S represents the position of a power relay station located on a straight coastal highway, and M shows the location of a marine biology experimental station on an island. A cable is to be laid

FIGURE 1.10
Cable connecting relay station S to experimental station M

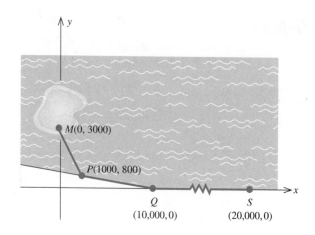

connecting the relay station with the experimental station. If the cost of running the cable on land is $2 per running foot and the cost of running the cable underwater is $6 per running foot, find the total cost for laying the cable.

SOLUTION ✔ The length of cable required on land is given by the distance from P to Q plus the distance from Q to S. The distance is

$$\sqrt{(10,000 - 1000)^2 + (0 - 800)^2} + \sqrt{(20,000 - 10,000)^2 + (0 - 0)^2}$$
$$= \sqrt{9000^2 + 800^2} + 10,000$$
$$= \sqrt{81,640,000} + 10,000$$
$$\approx 19,035.49$$

or approximately 19,035.49 feet. Next, we see that the length of cable required underwater is given by the distance from M to P. This distance is

$$\sqrt{(0 - 1000)^2 + (3000 - 800)^2} = \sqrt{1000^2 + 2200^2}$$
$$= \sqrt{5,840,000}$$
$$\approx 2416.61$$

or approximately 2416.61 feet. Therefore, the total cost for laying the cable is

$$2(19,035.49) + 6(2416.61) = 52,570.64$$

or approximately $52,571.   ▪ ▪ ▪ ▪

Group Discussion

In the Cartesian coordinate system, the two axes are perpendicular to each other. Consider a coordinate system in which the x- and y-axes are noncollinear (that is, the axes do not lie along a straight line) and are not perpendicular to each other (see the accompanying figure).

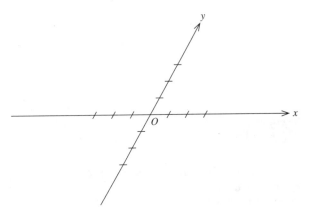

1. Describe how a point is represented in this coordinate system by an ordered pair (x, y) of real numbers. Conversely, show how an ordered pair (x, y) of real numbers uniquely determines a point in the plane.

2. Suppose you want to find a formula for the distance between two points $P_1(x_1, y_1)$ and $P_2(x_2, y_2)$ in the plane. What is the advantage that the Cartesian coordinate system has over the coordinate system under consideration? Comment on your answer.

SELF-CHECK EXERCISES 1.3

1. a. Plot the points $A(4, -2)$, $B(2, 3)$, and $C(-3, 1)$.

   b. Find the distance between the points A and B; between B and C; between A and C.

   c. Use the Pythagorean theorem to show that the triangle with vertices A, B, and C is a right triangle.

2. The figure at the top of page 34 shows the location of cities A, B, and C. Suppose a pilot wishes to fly from city A to city C but must make a mandatory stopover in city B. If the single-engine light plane has a range of 650 miles, can she make the trip without refueling in city B?

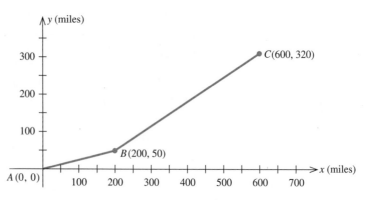

Solutions to Self-Check Exercises 1.3 can be found on page 36.

1.3 Exercises

In Exercises 1–6, refer to the following figure and determine the coordinates of each point and the quadrant in which it is located.

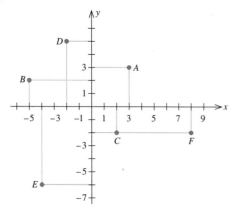

1. *A*   **2.** *B*   **3.** *C*
4. *D*   **5.** *E*   **6.** *F*

In Exercises 7–12, refer to the following figure.

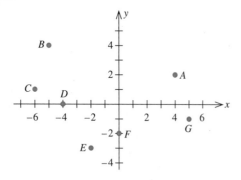

7. Which point has coordinates $(4, 2)$?

8. What are the coordinates of point *B*?

9. Which points have negative *y*-coordinates?

10. Which point has a negative *x*-coordinate and a negative *y*-coordinate?

11. Which point has an *x*-coordinate that is equal to zero?

12. Which point has a *y*-coordinate that is equal to zero?

In Exercises 13–20, sketch a set of coordinate axes and plot each point.

13. $(-2, 5)$ **14.** $(1, 3)$

15. $(3, -1)$ **16.** $(3, -4)$

17. $(8, -7/2)$ **18.** $(-5/2, 3/2)$

19. $(4.5, -4.5)$ **20.** $(1.2, -3.4)$

In Exercises 21–24, find the distance between the given points.

21. $(1, 3)$ and $(4, 7)$ **22.** $(1, 0)$ and $(4, 4)$

23. $(-1, 3)$ and $(4, 9)$ **24.** $(-2, 1)$ and $(10, 6)$

25. Find the coordinates of the points that are 10 units away from the origin and have a *y*-coordinate equal to -6.

26. Find the coordinates of the points that are 5 units away from the origin and have an *x*-coordinate equal to 3.

27. Show that the points $(3, 4)$, $(-3, 7)$, $(-6, 1)$, and $(0, -2)$ form the vertices of a square.

28. Show that the triangle with vertices $(-5, 2)$, $(-2, 5)$, and $(5, -2)$ is a right triangle.

In Exercises 29–34, find an equation of the circle that satisfies the given conditions.

29. Radius 5 and center $(2, -3)$

30. Radius 3 and center $(-2, -4)$

31. Radius 5 and center at the origin

32. Center at the origin and passes through $(2, 3)$

33. Center $(2, -3)$ and passes through $(5, 2)$

34. Center $(-a, a)$ and radius $2a$

35. DISTANCE TRAVELED A grand tour of four cities begins at city A and makes successive stops at cities B, C, and D before returning to city A. If the cities are located as shown in the following figure, find the total distance covered on the tour.

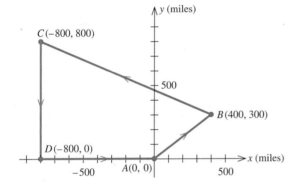

36. DELIVERY CHARGES A furniture store offers free setup and delivery services to all points within a 25-mi radius of its warehouse distribution center. If you live 20 mi east and 14 mi south of the warehouse, will you incur a delivery charge? Justify your answer.

37. TRAVEL TIME Towns A, B, C, and D are located as shown in the following figure. Two highways link town A to

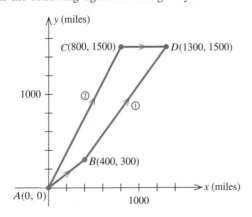

town D. Route 1 runs from town A to town D via town B, and route 2 runs from town A to town D via town C. If a salesman wishes to drive from town A to town D and traffic conditions are such that he could expect to average the same speed on either route, which highway should he take in order to arrive in the shortest time?

38. MINIMIZING SHIPPING COSTS Refer to the figure for Exercise 37. Suppose a fleet of 100 automobiles are to be shipped from an assembly plant in town A to town D. They may be shipped either by freight train along Route 1 at a cost of 11 cents/mile per automobile or by truck along Route 2 at a cost of $10\frac{1}{2}$ cents/mile per automobile. Which means of transportation minimizes the shipping cost? What is the net savings?

39. CONSUMER DECISIONS Ivan wishes to determine which antenna he should purchase for his home. The TV store has supplied him with the following information:

Range in Miles			
VHF	UHF	Model	Price
30	20	A	$40
45	35	B	$50
60	40	C	$60
75	55	D	$70

Ivan wishes to receive channel 17 (VHF), which is located 25 mi east and 35 mi north of his home, and channel 38 (UHF), which is located 20 mi south and 32 mi west of his home. Which model will allow him to receive both channels at the least cost? (Assume that the terrain between Ivan's home and both broadcasting stations is flat.)

40. CALCULATING THE COST OF LAYING CABLE In the following diagram, S represents the position of a power relay station located on a coastal highway, and M shows the location of a marine biology experimental station on an

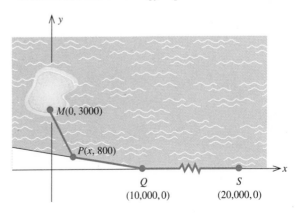

island. A cable is to be laid connecting the relay station with the experimental station. If the cost of running the cable on land is \$2/running foot and the cost of running cable under water is \$6/running foot, find an expression in terms of x that gives the total cost for laying the cable. Use this expression to find the total cost when $x = 900$; when $x = 1000$.

41. Two ships leave port at the same time. Ship A sails north at a speed of 20 mph while ship B sails east at a speed of 30 mph.
a. Find an expression in terms of the time t (in hours) giving the distance between the two ships.
b. Using the expression obtained in part (a), find the distance between the two ships 2 hr after leaving port.

In Exercises 42–45, determine whether the statement is true or false. If it is true, explain why it is true. If it is false, give an example to show why it is false.

42. The point $(-a, b)$ is symmetric to the point (a, b) with respect to the y-axis.

43. The point $(-a, -b)$ is symmetric to the point (a, b) with respect to the origin.

44. If the distance between the points $P_1(a, b)$ and $P_2(c, d)$ is D, then the distance between the points $P_1(a, b)$ and $P_3(kc, kd)$, $(k \neq 0)$, is given by $|k|D$.

45. The circle with equation $kx^2 + ky^2 = a^2$ lies inside the circle with equation $x^2 + y^2 = a^2$, provided $k > 1$.

46. Let (x_1, y_1) and (x_2, y_2) be two points lying in the xy-plane. Show that the distance between the two points is given by

$$d = \sqrt{(x_2 - x_1)^2 + (y_2 - y_1)^2}$$

Hint: Refer to the accompanying figure and use the Pythagorean theorem.

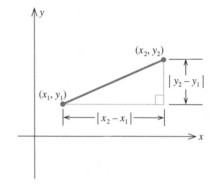

SOLUTIONS TO SELF-CHECK EXERCISES 1.3

1. a. The points are plotted in the following figure:

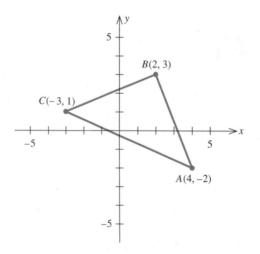

b. The distance between A and B is

$$d(A, B) = \sqrt{(2 - 4)^2 + [3 - (-2)]^2}$$
$$= \sqrt{(-2)^2 + 5^2} = \sqrt{4 + 25} = \sqrt{29}$$

The distance between B and C is

$$d(B, C) = \sqrt{(-3 - 2)^2 + (1 - 3)^2}$$
$$= \sqrt{(-5)^2 + (-2)^2} = \sqrt{25 + 4} = \sqrt{29}$$

The distance between A and C is

$$d(A, C) = \sqrt{(-3 - 4)^2 + [1 - (-2)]^2}$$
$$= \sqrt{(-7)^2 + 3^2} = \sqrt{49 + 9} = \sqrt{58}$$

c. We will show that

$$[d(A, C)]^2 = [d(A, B)]^2 + [d(B, C)]^2$$

From part (b), we see that $[d(A, B)]^2 = 29$, $[d(B, C)]^2 = 29$, and $[d(A, C)]^2 = 58$, and the desired result follows.

2. The distance between city A and city B is

$$d(A, B) = \sqrt{200^2 + 50^2} \approx 206$$

or 206 mi. The distance between city B and city C is

$$d(B, C) = \sqrt{[600 - 200]^2 + [320 - 50]^2}$$
$$= \sqrt{400^2 + 270^2} \approx 483$$

or 483 mi. Therefore, the total distance the pilot would have to cover is 689 mi, so she must refuel in city B.

1.4 Straight Lines

FIGURE 1.11

Linear depreciation of an asset

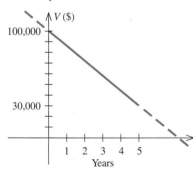

In computing income tax, business firms are allowed by law to depreciate certain assets such as buildings, machines, furniture, automobiles, and so on, over a period of time. Linear depreciation, or the straight-line method, is often used for this purpose. The graph of the straight line shown in Figure 1.11 describes the book value V of a computer that has an initial value of $100,000 and that is being depreciated linearly over 5 years with a scrap value of $30,000. Note that only the solid portion of the straight line is of interest here.

The book value of the computer at the end of year t, where t lies between 0 and 5, can be read directly from the graph. But there is one shortcoming in this approach: The result depends on how accurately you draw and read the graph. A better and more accurate method is based on finding an *algebraic* representation of the depreciation line.

Slope of a Line

To see how a straight line in the xy-plane may be described algebraically, we need to first recall certain properties of straight lines. Let L denote the unique straight line that passes through the two distinct points (x_1, y_1) and (x_2, y_2). If $x_1 = x_2$, then L is a vertical line, and the slope is undefined (Figure 1.12).

FIGURE 1.12
m is undefined

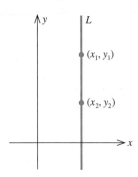

If $x_1 \ne x_2$, we define the slope of L as follows:

Slope of a Nonvertical Line

If (x_1, y_1) and (x_2, y_2) are any two distinct points on a nonvertical line L, then the slope m of L is given by

$$m = \frac{\Delta y}{\Delta x} = \frac{y_2 - y_1}{x_2 - x_1} \qquad (3)$$

See Figure 1.13.

FIGURE 1.13

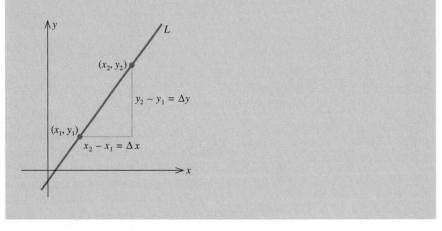

Observe that the slope of a straight line is a constant whenever it is defined. The number $\Delta y = y_2 - y_1$ (Δy is read "delta y") is a measure of the vertical change in y, and $\Delta x = x_2 - x_1$ is a measure of the horizontal change in x, as shown in Figure 1.13. From this figure we can see that the slope m of a straight line L is a measure of the *rate of change of y with respect to x*.

Figure 1.14a shows a straight line L_1 with slope 2. Observe that L_1 has the property that a unit increase in x results in a 2-unit increase in y. To see

FIGURE 1.14

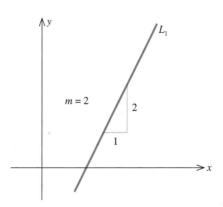

(a) The line rises ($m > 0$).

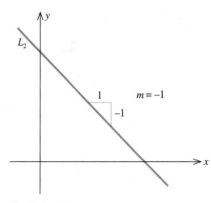

(b) The line falls ($m < 0$).

FIGURE 1.15
A family of straight lines

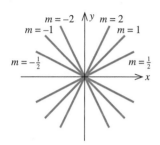

this, let $\Delta x = 1$ in Formula (3) so that $m = \Delta y$. Since $m = 2$, we conclude that $\Delta y = 2$. Similarly, Figure 1.14b shows a line L_2 with slope -1. Observe that a straight line with positive slope slants upward from left to right (y increases as x increases), whereas a line with negative slope slants downward from left to right (y decreases as x increases). Finally, Figure 1.15 shows a family of straight lines passing through the origin with indicated slopes.

EXAMPLE 1   Sketch the straight line that passes through the point $(-2, 5)$ and has slope $-4/3$.

SOLUTION ✔   First, plot the point $(-2, 5)$ (Figure 1.16). Next, recall that a slope of $-4/3$ indicates that an increase of 1 unit in the x-direction produces a *decrease* of $4/3$ units in the y-direction, or equivalently, a 3-unit increase in the x-direction produces a $3(4/3)$, or 4-unit, decrease in the y-direction. Using this information, we plot the point $(1, 1)$ and draw the line through the two points.

FIGURE 1.16
L has slope $-4/3$ and passes through $(-2, 5)$.

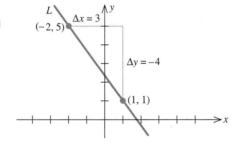

Group Discussion

   Show that the slope of a nonvertical line is independent of the two distinct points $P_1(x_1, y_1)$ and $P_2(x_2, y_2)$ used to compute it.

Hint: Suppose we pick two other distinct points, $P_3(x_3, y_3)$ and $P_4(x_4, y_4)$ lying on L. Draw a picture and use similar triangles to demonstrate that using P_3 and P_4 gives the same value as that obtained using P_1 and P_2.

EXAMPLE 2   Find the slope m of the line that passes through the points $(-1, 1)$ and $(5, 3)$.

SOLUTION ✔   Choose (x_1, y_1) to be the point $(-1, 1)$ and (x_2, y_2) to be the point $(5, 3)$. Then, with $x_1 = -1$, $y_1 = 1$, $x_2 = 5$, and $y_2 = 3$, we find

$$m = \frac{y_2 - y_1}{x_2 - x_1} = \frac{3 - 1}{5 - (-1)} = \frac{1}{3} \qquad \text{[Using (3)]}$$

(Figure 1.17). Try to verify that the result obtained would have been the same had we chosen the point $(-1, 1)$ to be (x_2, y_2) and the point $(5, 3)$ to be (x_1, y_1).

FIGURE 1.17
L passes through (5, 3) and (−1, 1).

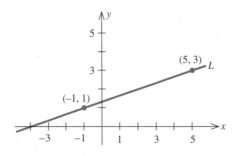

EXAMPLE 3   Find the slope of the line that passes through the points $(-2, 5)$ and $(3, 5)$.

SOLUTION ✔   The slope of the required line is given by

$$m = \frac{5 - 5}{3 - (-2)} = \frac{0}{5} = 0$$

(Figure 1.18).

FIGURE 1.18
The slope of the horizontal line L is 0.

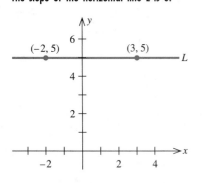

REMARK   In general, the slope of a horizontal line is zero.   ■ ■ ■

We can use the slope of a straight line to determine whether a line is parallel to another line.

Parallel Lines

Two distinct lines are **parallel** if and only if their slopes are equal or their slopes are undefined.

EXAMPLE 4   Let L_1 be a line that passes through the points $(-2, 9)$ and $(1, 3)$ and let L_2 be the line that passes through the points $(-4, 10)$ and $(3, -4)$. Determine whether L_1 and L_2 are parallel.

SOLUTION ✔ The slope m_1 of L_1 is given by

$$m_1 = \frac{3-9}{1-(-2)} = -2$$

The slope m_2 of L_2 is given by

$$m_2 = \frac{-4-10}{3-(-4)} = -2$$

Since $m_1 = m_2$, the lines L_1 and L_2 are in fact parallel (Figure 1.19). ■ ■ ■ ■

FIGURE 1.19

L_1 and L_2 have the same slope and hence are parallel.

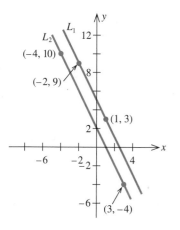

EQUATIONS OF LINES

We will now show that every straight line lying in the xy-plane may be represented by an equation involving the variables x and y. One immediate benefit of this is that problems involving straight lines may be solved algebraically.

Let L be a straight line parallel to the y-axis (perpendicular to the x-axis) (Figure 1.20). Then, L crosses the x-axis at some point $(a, 0)$ with the x-coordinate given by $x = a$, where a is some real number. Any other point on L has the form $(a, \bar{y})$, where $\bar{y}$ is an appropriate number. Therefore, the vertical line L is described by the sole condition

$$x = a$$

and this is, accordingly, the equation of L. For example, the equation $x = -2$ represents a vertical line 2 units to the left of the y-axis, and the equation $x = 3$ represents a vertical line 3 units to the right of the y-axis (Figure 1.21).

Next, suppose L is a nonvertical line so that it has a well-defined slope m. Suppose (x_1, y_1) is a fixed point lying on L and (x, y) is a variable point on L distinct from (x_1, y_1) (Figure 1.22).

FIGURE 1.20

The vertical line $x = a$

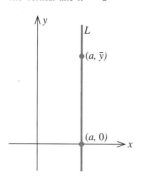

FIGURE 1.21

The vertical lines $x = -2$ and $x = 3$

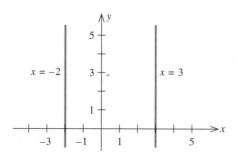

FIGURE 1.22

L passes through (x_1, y_1) and has slope m.

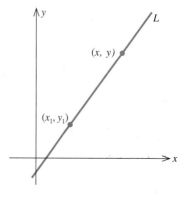

Using Formula (3) with the point $(x_2, y_2) = (x, y)$, we find that the slope of L is given by

$$m = \frac{y - y_1}{x - x_1}$$

Upon multiplying both sides of the equation by $x - x_1$, we obtain Formula (4).

Point-Slope Form

An equation of the line that has slope m and passes through the point (x_1, y_1) is given by

$$y - y_1 = m(x - x_1) \tag{4}$$

Equation (4) is called the **point-slope form of the equation of a line** since it utilizes a given point (x_1, y_1) on a line and the slope m of the line.

EXAMPLE 5   Find an equation of the line that passes through the point $(1, 3)$ and has slope 2.

SOLUTION ✔   Using the point-slope form of the equation of a line with the point $(1, 3)$ and $m = 2$, we obtain

$$y - 3 = 2(x - 1) \qquad [(y - y_1) = m(x - x_1)]$$

which, when simplified, becomes

$$2x - y + 1 = 0$$

FIGURE 1.23
L passes through $(1, 3)$ and has slope 2.

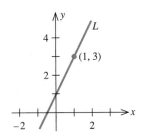

(Figure 1.23). ■■■■

EXAMPLE 6   Find an equation of the line that passes through the points $(-3, 2)$ and $(4, -1)$.

SOLUTION ✔   The slope of the line is given by

$$m = \frac{-1 - 2}{4 - (-3)} = -\frac{3}{7}$$

Using the point-slope form of the equation of a line with the point $(4, -1)$ and the slope $m = -3/7$, we have

$$y + 1 = -\frac{3}{7}(x - 4) \qquad [(y - y_1) = m(x - x_1)]$$

$$7y + 7 = -3x + 12$$

$$3x + 7y - 5 = 0$$

(Figure 1.24).

FIGURE 1.24
L passes through $(-3, 2)$ and $(4, -1)$.

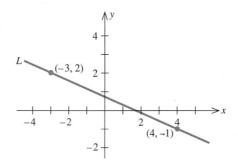

Group Discussion
Consider the slope-intercept form of a straight line $y = mx + b$. Describe the family of straight lines obtained by keeping:

1. The value of m fixed and allowing the value of b to vary.
2. The value of b fixed and allowing the value of m to vary.

We can use the slope of a straight line to determine whether a line is perpendicular to another line.

Perpendicular Lines

If L_1 and L_2 are two distinct nonvertical lines that have slopes m_1 and m_2, respectively, then L_1 is **perpendicular** to L_2 (written $L_1 \perp L_2$) if and only if

$$m_1 = -\frac{1}{m_2}$$

If the line L_1 is vertical (so that its slope is undefined), then L_1 is perpendicular to another line, L_2, if and only if L_2 is horizontal (so that its slope is zero). For a proof of these results, see Exercise 83, page 52.

EXAMPLE **7**   Find an equation of the line that passes through the point (3, 1) and is perpendicular to the line of Example 5.

SOLUTION ✔   Since the slope of the line in Example 5 is 2, the slope of the required line is given by $m = -1/2$, the negative reciprocal of 2. Using the point-slope form of the equation of a line, we obtain

$$y - 1 = -\frac{1}{2}(x - 3) \qquad [(y - y_1) = m(x - x_1)]$$

$$2y - 2 = -x + 3$$

$$x + 2y - 5 = 0$$

(See Figure 1.25).

FIGURE 1.25
L_2 is perpendicular to L_1 and passes through (3, 1).

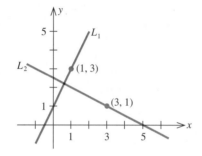

FIGURE 1.26
The line L has x-intercept a and y-intercept b.

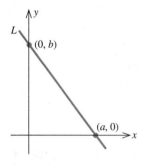

A straight line L that is neither horizontal nor vertical cuts the x-axis and the y-axis at, say, points $(a, 0)$ and $(0, b)$, respectively (Figure 1.26). The numbers a and b are called the **x-intercept** and **y-intercept**, respectively, of L.

Exploring with Technology

1. Use a graphing utility to plot the straight lines L_1 and L_2 with equations $2x + y - 5 = 0$ and $41x + 20y - 11 = 0$ on the same set of axes using the standard viewing rectangle.
   a. Can you tell if the lines L_1 and L_2 are parallel to each other?
   b. Verify your observations by computing the slopes of L_1 and L_2 algebraically.
2. Use a graphing utility to plot the straight lines L_1 and L_2 with equations $x + 2y - 5 = 0$ and $5x - y + 5 = 0$ on the same set of axes using the standard viewing rectangle.
   a. Can you tell if the lines L_1 and L_2 are perpendicular to each other?
   b. Verify your observation by computing the slopes of L_1 and L_2 algebraically.

Now, let L be a line with slope m and y-intercept b. Using Formula (4), the point-slope form of the equation of a line, with the point $(0, b)$ and slope m, we have

$$y - b = m(x - 0)$$
$$y = mx + b$$

Slope-Intercept Form

> The equation of the line that has slope m and intersects the y-axis at the point $(0, b)$ is given by
>
> $$y = mx + b \qquad\qquad \textbf{(5)}$$

EXAMPLE 8   Find an equation of the line that has slope 3 and y-intercept -4.

SOLUTION ✔   Using Equation (5) with $m = 3$ and $b = -4$, we obtain the required equation

$$y = 3x - 4 \qquad\qquad ■ ■ ■ ■$$

EXAMPLE 9   Determine the slope and y-intercept of the line whose equation is $3x - 4y = 8$.

SOLUTION ✔   Rewrite the given equation in the slope-intercept form and obtain

$$y = \frac{3}{4}x - 2$$

Comparing this result with Equation (5), we find $m = 3/4$ and $b = -2$, and we conclude that the slope and y-intercept of the given line are $3/4$ and -2, respectively. ■ ■ ■ ■

Exploring with Technology

1. Use a graphing utility to plot the straight lines with equations $y = -2x + 3$, $y = -x + 3$, $y = x + 3$, and $y = 2.5x + 3$ on the same set of axes using the standard viewing rectangle. What effect does changing the coefficient m of x in the equation $y = mx + b$ have on its graph?
2. Use a graphing utility to plot the straight lines with equations $y = 2x - 2$, $y = 2x - 1$, $y = 2x$, $y = 2x + 1$, and $y = 2x + 4$ on the same set of axes using the standard viewing rectangle. What effect does changing the constant b in the equation $y = mx + b$ have on its graph?
3. Describe in words the effect of changing both m and b in the equation $y = mx + b$.

APPLICATIONS

EXAMPLE 10

The sales manager of a local sporting goods store plotted sales versus time for the last 5 years and found the points to lie approximately along a straight line (Figure 1.27). By using the points corresponding to the first and fifth years, find an equation of the trend line. What sales figure can be predicted for the sixth year?

SOLUTION ✔

Using Formula (3) with the points (1, 20) and (5, 60), we find that the slope of the required line is given by

$$m = \frac{60 - 20}{5 - 1} = 10$$

FIGURE 1.27
Sales of a sporting goods store

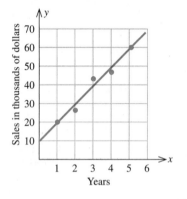

Next, using the point-slope form of the equation of a line with the point (1, 20) and $m = 10$, we obtain

$$y - 20 = 10(x - 1) \qquad [(y - y_1) = m(x - x_1)]$$
$$y = 10x + 10$$

as the required equation.

The sales figure for the sixth year is obtained by letting $x = 6$ in the last equation, giving

$$y = 70$$

or $70,000.   ■■■■

EXAMPLE 11

Suppose an art object purchased for $50,000 is expected to appreciate in value at a constant rate of $5000 per year for the next 5 years. Use Formula (5) to write an equation predicting the value of the art object in the next several years. What will its value be 3 years from the date of purchase?

SOLUTION ✔

Let x denote the time (in years) that has elapsed since the date the object was purchased and let y denote the object's value (in dollars). Then, $y = 50,000$ when $x = 0$. Furthermore, the slope of the required equation is given by $m = 5000$, since each unit increase in x (1 year) implies an increase of 5000 units (dollars) in y. Using (5) with $m = 5000$ and $b = 50,000$, we obtain

$$y = 5000x + 50,000 \qquad (y = mx + b)$$

Three years from the date of purchase, the value of the object will be given by

$$y = 5000(3) + 50,000$$

or $65,000.   ■■■■

Group Discussion
Refer to Example 11. Can the equation predicting the value of the art object be used to predict long-term growth?

GENERAL EQUATION OF A LINE

We have considered several forms of the equation of a straight line in the plane. These different forms of the equation are equivalent to each other. In fact, each is a special case of the following equation.

General Form of a Linear Equation

The equation

$$Ax + By + C = 0 \tag{6}$$

where A, B, and C are constants and A and B are not both zero, is called the general form of a linear equation in the variables x and y.

We will now state (without proof) an important result concerning the algebraic representation of straight lines in the plane.

THEOREM 1

An equation of a straight line is a linear equation; conversely, every linear equation represents a straight line.

This result justifies the use of the adjective *linear* describing Equation (6).

EXAMPLE 12   Sketch the straight line represented by the equation

$$3x - 4y - 12 = 0$$

SOLUTION ✔   Since every straight line is uniquely determined by two distinct points, we need find only two such points through which the line passes in order to sketch it. For convenience let us compute the x- and y-intercepts. Setting $y = 0$, we find $x = 4$; thus, the x-intercept is 4. Setting $x = 0$ gives $y = -3$, and the y-intercept is -3. A sketch of the line appears in Figure 1.28.

FIGURE 1.28
To sketch $3x - 4y - 12 = 0$, first find the x-intercept, 4, and the y-intercept, -3.

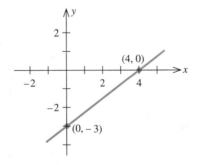

Following is a summary of the common forms of the equations of straight lines discussed in this section.

Equations of Straight Lines

Vertical line:	$x = a$
Horizontal line:	$y = b$
Point-slope form:	$y - y_1 = m(x - x_1)$
Slope-intercept form:	$y = mx + b$
General form:	$Ax + By + C = 0$

SELF-CHECK EXERCISES 1.4

1. Determine the number a so that the line passing through the points $(a, 2)$ and $(3, 6)$ is parallel to a line with slope 4.

2. Find an equation of the line that passes through the point $(3, -1)$ and is perpendicular to a line with slope $-1/2$.

3. Does the point $(3, -3)$ lie on the line with equation $2x - 3y - 12 = 0$? Sketch the graph of the line.

4. The percentage of people over age 65 who have high school diplomas is summarized in the following table:

Year, x	1960	1965	1970	1975	1980	1985	1990
Percentage with Diplomas, y	20	25	30	36	42	47	52

Source: The World Almanac

   a. Plot the percentage of people over age 65 who have high school diplomas (y) versus the year (x).
   b. Draw the straight line L through the points $(1960, 20)$ and $(1990, 52)$.
   c. Find an equation of the line L.
   d. If the trend continues, estimate the percentage of people over age 65 who will have high school diplomas by the year 2000.

Solutions to Self-Check Exercises 1.4 can be found on page 53.

1.4 Exercises

In Exercises 1–6, match the statement with one of the graphs (a)–(f).

1. The slope of the line is zero.

2. The slope of the line is undefined.

3. The slope of the line is positive, and its y-intercept is positive.

4. The slope of the line is positive, and its y-intercept is negative.

5. The slope of the line is negative, and its x-intercept is negative.

6. The slope of the line is negative, and its x-intercept is positive.

a.

b.

f.

c.

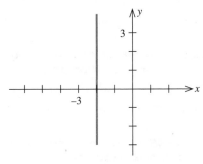

In Exercises 7–10, find the slope of the line shown in each figure.

7.

d.

8.

e.

9.

10.

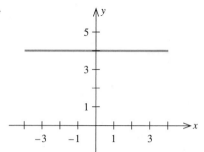

In Exercises 11–16, find the slope of the line that passes through each pair of points.

11. $(4, 3)$ and $(5, 8)$ **12.** $(4, 5)$ and $(3, 8)$

13. $(-2, 3)$ and $(4, 8)$ **14.** $(-2, -2)$ and $(4, -4)$

15. (a, b) and (c, d)

16. $(-a + 1, b - 1)$ and $(a + 1, -b)$

17. Given the equation $y = 4x - 3$, answer the following questions:
a. If x increases by 1 unit, what is the corresponding change in y?
b. If x decreases by 2 units, what is the corresponding change in y?

18. Given the equation $2x + 3y = 4$, answer the following questions:
a. Is the slope of the line described by this equation positive or negative?
b. As x increases in value, does y increase or decrease?
c. If x decreases by 2 units, what is the corresponding change in y?

In Exercises 19 and 20, determine whether the line through each pair of points is parallel.

19. $A(1, -2)$, $B(-3, -10)$ and $C(1, 5)$, $D(-1, 1)$

20. $A(2, 3)$, $B(2, -2)$ and $C(-2, 4)$, $D(-2, 5)$

In Exercises 21 and 22, determine whether the line through each pair of points is perpendicular.

21. $A(-2, 5)$, $B(4, 2)$ and $C(-1, -2)$, $D(3, 6)$

22. $A(2, 0)$, $B(1, -2)$ and $C(4, 2)$, $D(-8, 4)$

23. If the line passing through the points $(1, a)$ and $(4, -2)$ is parallel to the line passing through the points $(2, 8)$ and $(-7, a + 4)$, what is the value of a?

24. If the line passing through the points $(a, 1)$ and $(5, 8)$ is parallel to the line passing through the points $(4, 9)$ and $(a + 2, 1)$, what is the value of a?

25. Find an equation of the horizontal line that passes through $(-4, -3)$.

26. Find an equation of the vertical line that passes through $(0, 5)$.

In Exercises 27–30, find an equation of the line that passes through the point and has the indicated slope m.

27. $(3, -4)$; $m = 2$ **28.** $(2, 4)$; $m = -1$

29. $(-3, 2)$; $m = 0$ **30.** $(1, 2)$; $m = -1/2$

In Exercises 31–34, find an equation of the line that passes through the given points.

31. $(2, 4)$ and $(3, 7)$ **32.** $(2, 1)$ and $(2, 5)$

33. $(1, 2)$ and $(-3, -2)$ **34.** $(-1, -2)$ and $(3, -4)$

In Exercises 35–38, find an equation of the line that has slope m and y-intercept b.

35. $m = 3$; $b = 4$ **36.** $m = -2$; $b = -1$

37. $m = 0$; $b = 5$ **38.** $m = -1/2$; $b = 3/4$

In Exercises 39–44, write the equation in the slope-intercept form and then find the slope and y-intercept of the corresponding line.

39. $x - 2y = 0$ **40.** $y - 2 = 0$

41. $2x - 3y - 9 = 0$ **42.** $3x - 4y + 8 = 0$

43. $2x + 4y = 14$ **44.** $5x + 8y - 24 = 0$

45. Find an equation of the line that passes through the point $(-2, 2)$ and is parallel to the line $2x - 4y - 8 = 0$.

46. Find an equation of the line that passes through the point $(2, 4)$ and is perpendicular to the line $3x + 4y - 22 = 0$.

In Exercises 47–52, find an equation of the line that satisfies the given condition.

47. The line parallel to the x-axis and 6 units below it

48. The line passing through the origin and parallel to the line joining the points $(2, 4)$ and $(4, 7)$

49. The line passing through the point (a, b) with slope equal to zero

50. The line passing through $(-3, 4)$ and parallel to the x-axis

51. The line passing through $(-5, -4)$ and parallel to the line joining $(-3, 2)$ and $(6, 8)$

52. The line passing through (a, b) with undefined slope

53. Given that the point $P(-3, 5)$ lies on the line $kx + 3y + 9 = 0$, find k.

54. Given that the point $P(2, -3)$ lies on the line $-2x + ky + 10 = 0$, find k.

In Exercises 55–60, sketch the straight line defined by the given linear equation by finding the x- and y-intercepts.
Hint: See Example 12, page 47.

55. $3x - 2y + 6 = 0$ **56.** $2x - 5y + 10 = 0$

57. $x + 2y - 4 = 0$ **58.** $2x + 3y - 15 = 0$

59. $y + 5 = 0$ **60.** $-2x - 8y + 24 = 0$

61. Show that an equation of a line through the points $(a, 0)$ and $(0, b)$ with $a \neq 0$ and $b \neq 0$ can be written in the form

$$\frac{x}{a} + \frac{y}{b} = 1$$

(Recall that the numbers a and b are the x- and y-intercepts, respectively, of the line. This form of an equation of a line is called the **intercept form.**)

In Exercises 62–65, use the results of Exercise 61 to find an equation of a line with the given x- and y-intercepts.

62. x-intercept 3; y-intercept 4

63. x-intercept -2; y-intercept -4

64. x-intercept $-\frac{1}{2}$; y-intercept $\frac{3}{4}$

65. x-intercept 4; y-intercept $-\frac{1}{2}$

In Exercises 66 and 67, determine whether the given points lie on a straight line.

66. $A(-1, 7)$, $B(2, -2)$, and $C(5, -9)$

67. $A(-2, 1)$, $B(1, 7)$, and $C(4, 13)$

68. SOCIAL SECURITY CONTRIBUTIONS For wages less than the maximum taxable wage base, Social Security contributions by employees are 7.65% of the employee's wages.

a. Find an equation that expresses the relationship between the wages earned (x) and the Social Security taxes paid (y) by an employee who earns less than the maximum taxable wage base.
b. For each additional dollar that an employee earns, by how much is his or her Social Security contribution increased? (Assume that the employee's wages are less than the maximum taxable wage base.)
c. What Social Security contributions will an employee who earns \$35,000 (which is less than the maximum taxable wage base) be required to make?

69. COLLEGE ADMISSIONS Using data compiled by the Admissions Office at Faber University, college admissions officers estimate that 55% of the students who are offered admission to the freshman class at the university will actually enroll.
a. Find an equation that expresses the relationship between the number of students who actually enroll (y) and the number of students who are offered admission to the university (x).
b. If the desired freshman class size for the upcoming academic year is 1100 students, how many students should be admitted?

70. WEIGHT OF WHALES The equation $W = 3.51L - 192$, expressing the relationship between the length L (in feet) and the expected weight W (in British tons) of adult blue whales, was adopted in the late 1960s by the International Whaling Commission.
a. What is the expected weight of an 80-ft blue whale?
b. Sketch the straight line that represents the equation.

71. THE NARROWING GENDER GAP Since the founding of the Equal Employment Opportunity Commission and the passage of equal-pay laws, the gulf between men's and women's earnings has continued to close gradually. At the beginning of 1990 ($t = 0$), women's wages were 68% of men's wages. However, women's wages are projected to be 80% of men's wages by the beginning of the year 2000 ($t = 10$). If this gap between women's and men's wages continues to narrow *linearly*, what percentage of men's wages can we expect women's wages to be at the beginning of 2002?
Source: Journal of Economic Perspectives

72. IDEAL HEIGHTS AND WEIGHTS FOR WOMEN The Venus Health Club for Women provides its members with the following table, which gives the average desirable weight (in pounds) for women of a certain height (in inches):

Height, x	60	63	66	69	72
Weight, y	108	118	129	140	152

a. Plot the weight (y) versus the height (x).
b. Draw a straight line L through the points corresponding to heights of 5 ft and 6 ft.
c. Derive an equation of the line L.
d. Using the equation of part (c), estimate the average desirable weight for a woman who is 5 ft 5 in. tall.

73. COST OF A COMMODITY A manufacturer obtained the following data relating the cost y (in dollars) to the number of units (x) of a commodity produced:

No. of Units Produced, x	0	20	40	60	80	100
Cost, y	200	208	222	230	242	250

a. Plot the cost (y) versus the quantity produced (x).
b. Draw the straight line through the points $(0, 200)$ and $(100, 250)$.
c. Derive an equation of the straight line of part (b).
d. Taking this equation to be an approximation of the relationship between the cost and the level of production, estimate the cost of producing 54 units of the commodity.

74. DIGITAL TV SERVICES The percentage of homes with digital TV services, which stood at 5% at the beginning of 1999 ($t = 0$) is projected to grow linearly so that at the beginning of 2003 ($t = 4$) the percentage of such homes is projected to be 25%.
a. Derive an equation of the line passing through the points $A(0, 5)$ and $B(4, 25)$.
b. Plot the line with the equation found in part (a).
c. Using the equation found in part (a), find the percentage of homes with digital TV services at the beginning of 2001.
Source: Paul Kagan Associates

75. SALES GROWTH Metro Department Store's annual sales (in millions of dollars) during the past 5 yr were:

Annual Sales, x	5.8	6.2	7.2	8.4	9.0
Year, y	1	2	3	4	5

a. Plot the annual sales (y) versus the year (x).
b. Draw a straight line L through the points corresponding to the first and fifth years.
c. Derive an equation of the line L.
d. Using the equation found in part (c), estimate Metro's annual sales 4 yr from now ($x = 9$).

In Exercises 76–80, determine whether the statement is true or false. If it is true, explain why it is true. If it is false, give an example to show why it is false.

76. Suppose the slope of a line L is $-1/2$ and P is a given point on L. If Q is the point on L lying 4 units to the left of P, then Q is situated 2 units above P.

77. The line with equation $Ax + By + C = 0$, $(B \neq 0)$, and the line with equation $ax + by + c = 0$, $(b \neq 0)$, are parallel if $Ab - aB = 0$.

78. If the slope of the line L_1 is positive, then the slope of a line L_2 perpendicular to L_1 may be positive or negative.

79. The lines with equations $ax + by + c_1 = 0$ and $bx - ay + c_2 = 0$, where $a \neq 0$ and $b \neq 0$, are perpendicular to each other.

80. If L is the line with equation $Ax + By + C = 0$, where $A \neq 0$, then L crosses the x-axis at the point $\left(-\dfrac{C}{A}, 0\right)$.

81. Is there a difference between the statements "The slope of a straight line is zero" and "The slope of a straight line does not exist (is not defined)"? Explain your answer.

82. Show that two distinct lines with equations $a_1x + b_1y + c_1 = 0$ and $a_2x + b_2y + c_2 = 0$, respectively, are parallel if and only if $a_1b_2 - b_1a_2 = 0$.

Hint: Write each equation in the slope-intercept form and compare.

83. Prove that if a line L_1 with slope m_1 is perpendicular to a line L_2 with slope m_2, then $m_1m_2 = -1$.

Hint: Refer to the following figure. Show that $m_1 = b$ and $m_2 = c$. Next, apply the Pythagorean theorem to triangles OAC, OCB, and OBA to show that $1 = -bc$.

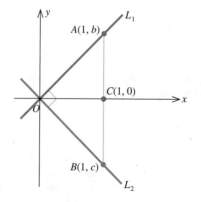

SOLUTIONS TO SELF-CHECK EXERCISES 1.4

1. The slope of the line that passes through the points $(a, 2)$ and $(3, 6)$ is

$$m = \frac{6 - 2}{3 - a}$$

$$= \frac{4}{3 - a}$$

Since this line is parallel to a line with slope 4, m must be equal to 4; that is,

$$\frac{4}{3 - a} = 4$$

or, upon multiplying both sides of the equation by $3 - a$,

$$4 = 4(3 - a)$$
$$4 = 12 - 4a$$
$$4a = 8$$
$$a = 2$$

2. Since the required line L is perpendicular to a line with slope $-1/2$, the slope of L is

$$-\frac{1}{-1/2} = 2$$

Next, using the point-slope form of the equation of a line, we have

$$y - (-1) = 2(x - 3)$$
$$y + 1 = 2x - 6$$
$$y = 2x - 7$$

3. Substituting $x = 3$ and $y = -3$ into the left-hand side of the given equation, we find

$$2(3) - 3(-3) - 12 = 3$$

which is not equal to zero (the right-hand side). Therefore, $(3, -3)$ does not lie on the line with equation $2x - 3y - 12 = 0$.

Setting $x = 0$, we find $y = -4$, the y-intercept. Next, setting $y = 0$ gives $x = 6$, the x-intercept. We now draw the line passing through the points $(0, -4)$ and $(6, 0)$ as shown.

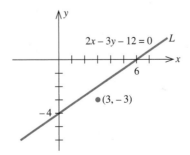

4. a. and **b.** See the accompanying figure.

c. The slope of L is

$$m = \frac{52 - 20}{1990 - 1960} = \frac{32}{30} = \frac{16}{15}$$

Using the point-slope form of the equation of a line with the point (1960, 20), we find

$$y - 20 = \frac{16}{15}(x - 1960) = \frac{16}{15}x - \frac{(16)(1960)}{15}$$

$$y = \frac{16}{15}x - \frac{6272}{3} + 20$$

$$= \frac{16}{15}x - \frac{6212}{3}$$

d. To estimate the percentage of people over age 65 who will have high school diplomas by the year 2000, let $x = 2000$ in the equation obtained in part (c). Thus, the required estimate is

$$y = \frac{16}{15}(2000) - \frac{6212}{3} \approx 62.67$$

or approximately 63%.

Group projects for this chapter can be found at the Brooks/Cole Web site:
http://www.brookscole.com/math/authors/tans/

CHAPTER 1 Summary of Principal Formulas and Terms

Formulas

1.	Quadratic formula	$x = \dfrac{-b \pm \sqrt{b^2 - 4ac}}{2a}$
2.	Distance between two points	$d = \sqrt{(x_2 - x_1)^2 + (y_2 - y_1)^2}$
3.	Slope of a line	$m = \dfrac{y_2 - y_1}{x_2 - x_1}$
4.	Equation of a vertical line	$x = a$
5.	Equation of a horizontal line	$y = b$
6.	Point-slope form of the equation of a line	$y - y_1 = m(x - x_1)$
7.	Slope-intercept form of the equation of a line	$y = mx + b$
8.	General equation of a line	$Ax + By + C = 0$

Terms

real number (coordinate) line	triangle inequality
open interval	polynomial
closed interval	roots of a polynomial equation
half-open interval	Cartesian coordinate system
finite interval	ordered pair
infinite interval	parallel lines
absolute value	perpendicular lines

CHAPTER 1 REVIEW EXERCISES

In Exercises 1–4, find the values of x that satisfy the inequality (inequalities).

1. $-x + 3 \le 2x + 9$ 2. $-2 \le 3x + 1 \le 7$

3. $x - 3 > 2$ or $x + 3 < -1$

4. $2x^2 > 50$

In Exercises 5–8, evaluate the expression.

5. $|-5 + 7| + |-2|$ 6. $\left| \dfrac{5 - 12}{-4 - 3} \right|$

7. $|2\pi - 6| - \pi$

8. $|\sqrt{3} - 4| + |4 - 2\sqrt{3}|$

In Exercises 9–14, evaluate the expression.

9. $\left(\dfrac{9}{4} \right)^{3/2}$ 10. $\dfrac{5^6}{5^4}$

11. $(3 \cdot 4)^{-2}$ 12. $(-8)^{5/3}$

13. $\dfrac{(3 \cdot 2^{-3})(4 \cdot 3^5)}{2 \cdot 9^3}$ 14. $\dfrac{3\sqrt[3]{54}}{\sqrt[3]{18}}$

In Exercises 15–19, simplify the expression.

15. $\dfrac{4(x^2 + y)^3}{x^2 + y}$ 16. $\dfrac{a^6 b^{-5}}{(a^3 b^{-2})^{-3}}$

17. $\dfrac{\sqrt[4]{16x^5 yz}}{\sqrt[4]{81xyz^5}}$ (x, y, and z positive)

18. $(2x^3)(-3x^{-2})\left(\dfrac{1}{6} x^{-1/2} \right)$

19. $\left(\dfrac{3xy^2}{4x^3 y} \right)^{-2} \left(\dfrac{3xy^3}{2x^2} \right)^3$

In Exercises 20–23, factor the expression.

20. $-2\pi^2 r^3 + 100\pi r^2$ 21. $2v^3 w + 2vw^3 + 2u^2 vw$

22. $16 - x^2$ 23. $12t^3 - 6t^2 - 18t$

In Exercises 24–27, solve the equation by factoring.

24. $8x^2 + 2x - 3 = 0$ 25. $-6x^2 - 10x + 4 = 0$

26. $-x^3 - 2x^2 + 3x = 0$ 27. $2x^4 + x^2 = 1$

In Exercises 28 and 29, use the quadratic formula to solve the quadratic equation.

28. $x^2 - 2x - 5 = 0$ 29. $2x^2 + 8x + 7 = 0$

In Exercises 30–33, perform the indicated operations and simplify the expression.

30. $\dfrac{(t + 6)(60) - (60t + 180)}{(t + 6)^2}$

31. $\dfrac{6x}{2(3x^2 + 2)} + \dfrac{1}{4(x + 2)}$

32. $\dfrac{2}{3}\left(\dfrac{4x}{2x^2 - 1} \right) + 3\left(\dfrac{3}{3x - 1} \right)$

33. $\dfrac{-2x}{\sqrt{x + 1}} + 4\sqrt{x + 1}$

34. Rationalize the numerator:

$$\dfrac{\sqrt{x} - 1}{x - 1}$$

35. Rationalize the denominator:

$$\dfrac{\sqrt{x} - 1}{2\sqrt{x}}$$

In Exercises 36 and 37, find the distance between the two points.

36. $(-2, -3)$ and $(1, -7)$

37. $\left(\dfrac{1}{2}, \sqrt{3} \right)$ and $\left(-\dfrac{1}{2}, 2\sqrt{3} \right)$

In Exercises 38–43, find an equation of the line L that passes through the point (−2, 4) and satisfies the condition.

38. L is a vertical line.

39. L is a horizontal line.

40. L passes through the point $\left(3, \dfrac{7}{2} \right)$.

41. The x-intercept of L is 3.

42. L is parallel to the line $5x - 2y = 6$.

43. L is perpendicular to the line $4x + 3y = 6$.

44. Find an equation of the straight line that passes through the point $(2, 3)$ and is parallel to the line with equation $3x + 4y - 8 = 0$.

45. Find an equation of the straight line that passes through the point $(-1, 3)$ and is parallel to the line passing through the points $(-3, 4)$ and $(2, 1)$.

46. Find an equation of the line that passes through the point $(-2, -4)$ and is perpendicular to the line with equation $2x - 3y - 24 = 0$.

47. Sketch the graph of the equation $3x - 4y = 24$.

48. Sketch the graph of the line that passes through the point $(3, 2)$ and has slope $-2/3$.

49. Find the minimum cost C (in dollars) given that

$$2(1.5C + 80) \leq 2(2.5C - 20)$$

50. Find the maximum revenue R (in dollars) given that

$$12(2R - 320) \leq 4(3R + 240)$$

 Additional study hints and sample chapter tests can be found at the Brooks/Cole Web site: http://www.brookscole.com/math/authors/tans/

FUNCTIONS, LIMITS, AND THE DERIVATIVE

2

In this chapter we define a *function*, a special relationship between two variables. The concept of a function enables us to describe many relationships that exist in applications. We also begin the study of differential calculus. Historically, differential calculus was developed in response to the problem of finding the tangent line to an arbitrary curve. But it quickly became apparent that solving this problem provided mathematicians with a method for solving many practical problems involving the rate of change of one quantity with respect to another. The basic tool used in differential calculus is the *derivative* of a function. The concept of the derivative is based, in turn, on a more fundamental notion—that of the *limit* of a function.

How does the change in the demand for a certain make of tires affect the unit price of the tires? The management of the Titan Tire Company has determined the demand function that relates the unit price of its Super Titan tires to the quantity demanded. In Example 7, page 165, you will see how this function can be used to compute the rate of change of the unit price of the Super Titan tires with respect to the quantity demanded.

2.1 Functions and Their Graphs

FUNCTIONS

A manufacturer would like to know how his company's profit is related to its production level; a biologist would like to know how the size of the population of a certain culture of bacteria will change over time; a psychologist would like to know the relationship between the learning time of an individual and the length of a vocabulary list; and a chemist would like to know how the initial speed of a chemical reaction is related to the amount of substrate used. In each instance we are concerned with the same question: How does one quantity depend upon another? The relationship between two quantities is conveniently described in mathematics by using the concept of a function.

Function

> A **function** is a rule that assigns to each element in a set A one and only one element in a set B.

The set A is called the **domain** of the function. It is customary to denote a function by a letter of the alphabet, such as the letter f. If x is an element in the domain of a function f, then the element in B that f associates with x is written $f(x)$ (read "f of x") and is called the value of f at x. The set comprising all the values assumed by $y = f(x)$ as x takes on all possible values in its domain is called the **range** of the function f.

We can think of a function f as a machine. The domain is the set of inputs (raw material) for the machine, the rule describes how the input is to be processed, and the value(s) of the function are the outputs of the machine (Figure 2.1).

We can also think of a function f as a mapping in which an element x in the domain of f is mapped onto a unique element $f(x)$ in B (Figure 2.2).

FIGURE 2.1
A function machine

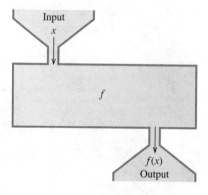

FIGURE 2.2
The function f viewed as a mapping

REMARKS

1. It is important to understand that the output $f(x)$ associated with an input x is unique. To appreciate the importance of this uniqueness property, consider a rule that associates with each item x in a department store its selling price y. Then, each x must correspond to *one and only one y*. Notice, however, that different x's may be associated with the same y. In the context of the present example, this says that different items may have the same price.
2. Although the sets A and B that appear in the definition of a function may be quite arbitrary, in this book they will denote sets of real numbers.

■ ■ ■

An example of a function may be taken from the familiar relationship between the area of a circle and its radius. Letting x and y denote the radius and area of a circle, respectively, we have, from elementary geometry,

$$y = \pi x^2 \tag{1}$$

Equation (1) defines y as a function of x since for each admissible value of x (that is, for each nonnegative number representing the radius of a certain circle) there corresponds precisely one number $y = \pi x^2$ that gives the area of the circle. The rule defining this "area function" may be written as

$$f(x) = \pi x^2 \tag{2}$$

To compute the area of a circle of radius 5 inches, we simply replace x in Equation (2) with the number 5. Thus, the area of the circle is

$$f(5) = \pi 5^2 = 25\pi$$

or 25π square inches

In general, to evaluate a function at a specific value of x, we replace x with that value, as illustrated in Examples 1 and 2.

EXAMPLE 1   Let the function f be defined by the rule $f(x) = 2x^2 - x + 1$. Compute:

a. $f(1)$ **b.** $f(-2)$ **c.** $f(a)$ **d.** $f(a + h)$

SOLUTION ✔
a. $f(1) = 2(1)^2 - (1) + 1 = 2 - 1 + 1 = 2$
b. $f(-2) = 2(-2)^2 - (-2) + 1 = 8 + 2 + 1 = 11$
c. $f(a) = 2(a)^2 - (a) + 1 = 2a^2 - a + 1$
d. $f(a + h) = 2(a + h)^2 - (a + h) + 1 = 2a^2 + 4ah + 2h^2 - a - h + 1$

■ ■ ■ ■

EXAMPLE 2   The Thermo-Master Company manufactures an indoor–outdoor thermometer at its Mexican subsidiary. Management estimates that the profit (in dollars) realizable by Thermo-Master in the manufacture and sale of x thermometers per week is

$$P(x) = -0.001x^2 + 8x - 5000$$

Find Thermo-Master's weekly profit if its level of production is (a) 1000 thermometers per week and (b) 2000 thermometers per week.

SOLUTION ✔ **a.** The weekly profit realizable by Thermo-Master when the level of production is 1000 units per week is found by evaluating the profit function P at $x = 1000$. Thus,

$$P(1000) = -0.001(1000)^2 + 8(1000) - 5000 = 2000$$

or $2000.

b. When the level of production is 2000 units per week, the weekly profit is given by

$$P(2000) = -0.001(2000)^2 + 8(2000) - 5000 = 7000$$

or $7000.   ■ ■ ■ ■

DETERMINING THE DOMAIN OF A FUNCTION

Suppose we are given the function $y = f(x)$.* Then, the variable x is called the **independent variable.** The variable y, whose value depends on x, is called the **dependent variable.**

In determining the domain of a function, we need to find what restrictions, if any, are to be placed on the independent variable x. In many practical applications the domain of a function is dictated by the nature of the problem, as illustrated in Example 3.

EXAMPLE 3   An open box is to be made from a rectangular piece of cardboard 16 inches long and 10 inches wide by cutting away identical squares (x by x inches) from each corner and folding up the resulting flaps (Figure 2.3). Find an expression that gives the volume V of the box as a function of x. What is the domain of the function?

FIGURE 2.3

(a) The box is constructed by cutting x- by x-inch squares from each corner.

(b) The dimensions of the resulting box are $(10 - 2x)$ by $(16 - 2x)$ by x inches.

---

* It is customary to refer to a function f as $f(x)$ or by the equation $y = f(x)$ defining it.

SOLUTION ✔ The dimensions of the box are $(16 - 2x)$ inches long, $(10 - 2x)$ inches wide, and x inches high, so its volume (in cubic inches) is given by

$$V = f(x) = (16 - 2x)(10 - 2x)x \quad \text{(Length · width · height)}$$
$$= (160 - 52x + 4x^2)x$$
$$= 4x^3 - 52x^2 + 160x$$

Since the length of each side of the box must be greater than or equal to zero, we see that

$$16 - 2x \geq 0, \quad 10 - 2x \geq 0, \quad x \geq 0$$

simultaneously; that is,

$$x \leq 8, \quad x \leq 5, \quad x \geq 0$$

All three inequalities are satisfied simultaneously provided that $0 \leq x \leq 5$. Thus, the domain of the function f is the interval $[0, 5]$. ■ ■ ■ ■

In general, if a function is defined by a rule relating x to $f(x)$ without specific mention of its domain, it is understood that the domain will consist of all values of x for which $f(x)$ is a real number. In this connection, you should keep in mind that (1) division by zero is not permitted and (2) the square root of a negative number is not defined.

EXAMPLE 4 Find the domain of each of the functions defined by the following equations:

a. $f(x) = \sqrt{x - 1}$ **b.** $f(x) = \dfrac{1}{x^2 - 4}$ **c.** $f(x) = x^2 + 3$

SOLUTION ✔
a. Since the square root of a negative number is undefined, it is necessary that $x - 1 \geq 0$. The inequality is satisfied by the set of real numbers $x \geq 1$. Thus, the domain of f is the interval $[1, \infty)$.
b. The only restriction on x is that $x^2 - 4$ be different from zero since division by zero is not allowed. But $(x^2 - 4) = (x + 2)(x - 2) = 0$ if $x = -2$ or $x = 2$. Thus, the domain of f in this case consists of the intervals $(-\infty, -2)$, $(-2, 2)$, and $(2, \infty)$.
c. Here, any real number satisfies the equation, so the domain of f is the set of all real numbers. ■ ■ ■ ■

GRAPHS OF FUNCTIONS

If f is a function with domain A, then corresponding to each real number x in A there is precisely one real number $f(x)$. We can also express this fact by using **ordered pairs** of real numbers. Write each number x in A as the first member of an ordered pair and each number $f(x)$ corresponding to x as the second member of the ordered pair. This gives exactly one ordered pair $(x, f(x))$ for each x in A. This observation leads to an alternative definition of a function f:

> A function f with domain A is the set of all ordered pairs $(x, f(x))$ where x belongs to A.

Observe that the condition that there be one and only one number $f(x)$ corresponding to each number x in A translates into the requirement that *no two ordered pairs have the same first number.*

Since ordered pairs of real numbers correspond to points in the plane, we have found a way to exhibit a function graphically.

Graph of a Function of One Variable

> The **graph of a function** f is the set of all points (x, y) in the xy-plane such that x is in the domain of f and $y = f(x)$.

Figure 2.4 shows the graph of a function f. Observe that the y-coordinate of the point (x, y) on the graph of f gives the height of that point (the distance above the x-axis), if $f(x)$ is positive. If $f(x)$ is negative, then $-f(x)$ gives the depth of the point (x, y) (the distance below the x-axis). Also, observe that the domain of f is a set of real numbers lying on the x-axis, whereas the range of f lies on the y-axis.

FIGURE 2.4
The graph of f

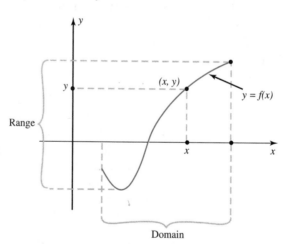

EXAMPLE 5

The graph of a function f is shown in Figure 2.5.

a. What is the value of $f(3)$? The value of $f(5)$?
b. What is the height or depth of the point $(3, f(3))$ from the x-axis? The point $(5, f(5))$ from the x-axis?
c. What is the domain of f? The range of f?

SOLUTION ✔

a. From the graph of f, we see that $y = -2$ when $x = 3$ and conclude that $f(3) = -2$. Similarly, we see that $f(5) = 3$.
b. Since the point $(3, -2)$ lies below the x-axis, we see that the depth of the point $(3, f(3))$ is $-f(3) = -(-2) = 2$ units below the x-axis. The point $(5, f(5))$ lies above the x-axis and is located at a height of $f(5)$, or 3 units above the x-axis.

FIGURE 2.5

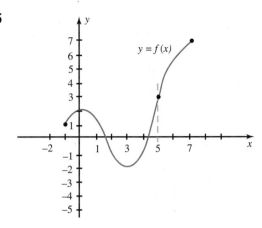

c. Observe that x may take on all values between $x = -1$ and $x = 7$, inclusive, and so the domain of f is $[-1, 7]$. Next, observe that as x takes on all values in the domain of f, $f(x)$ takes on all values between -2 and 7, inclusive. (You can easily see this by running your index finger along the x-axis from $x = -1$ to $x = 7$ and observing the corresponding values assumed by the y-coordinate of each point of the graph of f.) Therefore, the range of f is $[-2, 7]$.   ■ ■ ■ ■

   Much information about the graph of a function can be gained by plotting a few points on its graph. Later on we will develop more systematic and sophisticated techniques for graphing functions.

EXAMPLE 6   Sketch the graph of the function defined by the equation $y = x^2 + 1$. What is the range of f?

SOLUTION ✔   The domain of the function is the set of all real numbers. By assigning several values to the variable x and computing the corresponding values for y, we obtain the following solutions to the equation $y = x^2 + 1$:

FIGURE 2.6
The graph of $y = x^2 + 1$ is a parabola.

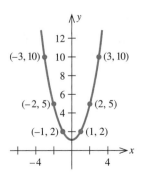

x	-3	-2	-1	0	1	2	3
y	10	5	2	1	2	5	10

By plotting these points and then connecting them with a smooth curve, we obtain the graph of $y = f(x)$, which is a parabola (Figure 2.6). To determine the range of f, we observe that $x^2 \geq 0$, if x is any real number and so $x^2 + 1 \geq 1$ for all real numbers x. We conclude that the range of f is $[1, \infty)$. The graph of f confirms this result visually.   ■ ■ ■ ■

   Some functions are defined in a piecewise fashion, as Examples 7 and 8 show.

Group Discussion
Let $f(x) = x^2$.

1. Plot the graphs of $F(x) = x^2 + c$ on the same set of axes for $c = -2, -1, -\frac{1}{2}, 0, \frac{1}{2}, 1, 2$.
2. Plot the graphs of $G(x) = (x + c)^2$ on the same set of axes for $c = -2, -1, -\frac{1}{2}, 0, \frac{1}{2}, 1, 2$.
3. Plot the graphs of $H(x) = cx^2$ on the same set of axes for $c = -2, -1, -\frac{1}{2}, -\frac{1}{4}, 0, \frac{1}{4}, \frac{1}{2}, 1, 2$.
4. Study the family of graphs in parts 1–3 and describe the relationship between the graph of a function f and the graphs of the functions defined by (i) $y = f(x) + c$, (ii) $y = f(x + c)$, and (iii) $y = cf(x)$, where c is a constant.

EXAMPLE 7   The Madison Finance Company plans to open two branch offices 2 years from now in two separate locations: an industrial complex and a newly developed commercial center in the city. As a result of these expansion plans, Madison's total deposits during the next 5 years are expected to grow in accordance with the rule

$$f(x) = \begin{cases} \sqrt{2x} + 20 & \text{if } 0 \le x \le 2 \\ \dfrac{1}{2}x^2 + 20 & \text{if } 2 < x \le 5 \end{cases}$$

where $y = f(x)$ gives the total amount of money (in millions of dollars) on deposit with Madison in year x ($x = 0$ corresponds to the present). Sketch the graph of the function f.

SOLUTION ✔   The function f is defined in a piecewise fashion on the interval $[0, 5]$. In the subdomain $[0, 2]$, the rule for f is given by $f(x) = \sqrt{2x} + 20$. The values of $f(x)$ corresponding to $x = 0, 1$, and 2 may be tabulated as follows:

x	0	1	2
f(x)	20	21.4	22

Next, in the subdomain $(2, 5]$, the rule for f is given by $f(x) = \frac{1}{2}x^2 + 20$. The values of $f(x)$ corresponding to $x = 3, 4$, and 5 are shown in the following table:

x	3	4	5
f(x)	24.5	28	32.5

Using the values of $f(x)$ in this table, we sketch the graph of the function f as shown in Figure 2.7.

FIGURE 2.7
We obtain the graph of the function
$y = f(x)$ by graphing
$y = \sqrt{2x} + 20$ over [0, 2]
and $y = \frac{1}{2}x^2 + 20$ over (2, 5].

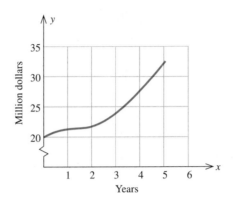

■ ■ ■ ■

EXAMPLE **8**   Sketch the graph of the function f defined by

$$f(x) = \begin{cases} -x & \text{if } x < 0 \\ \sqrt{x} & \text{if } x \geq 0 \end{cases}$$

SOLUTION ✔   The function f is defined in a piecewise fashion on the set of all real numbers. In the subdomain $(-\infty, 0)$, the rule for f is given by $f(x) = -x$. The equation $y = -x$ is a linear equation in the slope-intercept form (with slope -1 and intercept 0). Therefore, the graph of f corresponding to the subdomain $(-\infty, 0)$ is the half line shown in Figure 2.8. Next, in the subdomain $[0, \infty)$,

FIGURE 2.8
The graph of $y = f(x)$ is obtained
by graphing
$y = -x$ over $(-\infty, 0)$
and $y = \sqrt{x}$ over $[0, \infty)$.

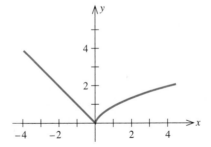

the rule for f is given by $f(x) = \sqrt{x}$. The values of $f(x)$ corresponding to $x = 0, 1, 2, 3, 4, 9,$ and 16 are shown in the following table:

x	0	1	2	3	4	9	16
f(x)	0	1	$\sqrt{2}$	$\sqrt{3}$	2	3	4

Using these values, we sketch the graph of the function f as shown in Figure 2.8.

■ ■ ■ ■

THE VERTICAL-LINE TEST

Although it is true that every function f of a variable x has a graph in the xy-plane, it is important to realize that not every curve in the xy-plane is the graph of a function. For example, consider the curve depicted in Figure 2.9. This is the graph of the equation $y^2 = x$. In general, the **graph of an equation** is the set of all ordered pairs (x, y) that satisfy the given equation. Observe that the points $(9, -3)$ and $(9, 3)$ both lie on the curve. This implies that the number $x = 9$ is associated with *two* numbers: $y = -3$ and $y = 3$. But this clearly violates the uniqueness property of a function. Thus, we conclude that the curve under consideration cannot be the graph of a function.

FIGURE 2.9
Since a vertical line passes through the curve at more than one point, we deduce that it is not the graph of a function.

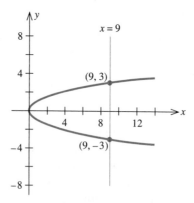

This example suggests the following test for determining when a curve is the graph of a function.

Vertical-Line Test

> A curve in the xy-plane is the graph of a function $y = f(x)$ if and only if each vertical line intersects it in at most one point.

EXAMPLE 9

Determine which of the curves shown in Figure 2.10 are the graphs of functions of x.

SOLUTION ✔

The curves depicted in Figure 2.10a, c, and d are graphs of functions because each curve satisfies the requirement that each vertical line intersects the curve in at most one point. Note that the vertical line shown in Figure 2.10c does not intersect the graph because the point on the x-axis through which this line passes does not lie in the domain of the function. The curve depicted in Figure 2.10b is *not* the graph of a function because the vertical line shown there intersects the graph at three points.

FIGURE 2.10
The vertical-line test can be used to determine which of these curves are graphs of functions.

(a)

(b)

(c)

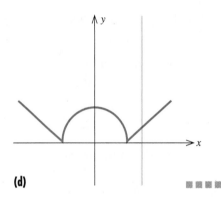

(d)

SELF-CHECK EXERCISES 2.1

1. Let f be the function defined by

$$f(x) = \frac{\sqrt{x+1}}{x}$$

   a. Find the domain of f.
   b. Compute $f(3)$.
   c. Compute $f(a + h)$.

2. Statistics obtained by Amoco Corporation show that more and more motorists are pumping their own gas. The following function gives self-serve sales as a percentage of all U.S. gas sales:

$$f(t) = \begin{cases} 6t + 17 & \text{if } 0 \le t \le 6 \\ 15.98(t-6)^{1/4} + 53 & \text{if } 6 < t \le 20 \end{cases}$$

Here t is measured in years, with $t = 0$ corresponding to the beginning of 1974.
   a. Sketch the graph of the function f.
   b. What percentage of all gas sales at the beginning of 1978 were self-serve? At the beginning of 1994?
   Source: Amoco Corporation

3. Let $f(x) = \sqrt{2x+1} + 2$. Determine whether the point $(4, 6)$ lies on the graph of f.

Solutions to Self-Check Exercises 2.1 can be found on page 74.

2.1 Exercises

1. Let f be the function defined by $f(x) = 5x + 6$. Find $f(3), f(-3), f(a), f(-a)$, and $f(a + 3)$.

2. Let f be the function defined by $f(x) = 4x - 3$. Find $f(4), f(\frac{1}{4}), f(0), f(a)$, and $f(a + 1)$.

3. Let g be the function defined by $g(x) = 3x^2 - 6x - 3$. Find $g(0), g(-1), g(a), g(-a)$, and $g(x + 1)$.

4. Let h be the function defined by $h(x) = x^3 - x^2 + x + 1$. Find $h(-5), h(0), h(a)$, and $h(-a)$.

5. Let s be the function defined by $s(t) = \dfrac{2t}{t^2 - 1}$. Find $s(4), s(0), s(a), s(2 + a)$, and $s(t + 1)$.

6. Let g be the function defined by $g(u) = (3u - 2)^{3/2}$. Find $g(1), g(6), g(\frac{11}{3})$, and $g(u + 1)$.

7. Let f be the function defined by $f(t) = \dfrac{2t^2}{\sqrt{t - 1}}$. Find $f(2), f(a), f(x + 1)$, and $f(x - 1)$.

8. Let f be the function defined by $f(x) = 2 + 2\sqrt{5 - x}$. Find $f(-4), f(1), f(\frac{11}{4})$, and $f(x + 5)$.

9. Let f be the function defined by
$$f(x) = \begin{cases} x^2 + 1 & \text{if } x \le 0 \\ \sqrt{x} & \text{if } x > 0 \end{cases}$$
Find $f(-2), f(0)$, and $f(1)$.

10. Let g be the function defined by
$$g(x) = \begin{cases} -\dfrac{1}{2}x + 1 & \text{if } x < 2 \\ \sqrt{x - 2} & \text{if } x \ge 2 \end{cases}$$
Find $g(-2), g(0), g(2)$, and $g(4)$.

11. Let f be the function defined by
$$f(x) = \begin{cases} -\dfrac{1}{2}x^2 + 3 & \text{if } x < 1 \\ 2x^2 + 1 & \text{if } x \ge 1 \end{cases}$$
Find $f(-1), f(0), f(1)$, and $f(2)$.

12. Let f be the function defined by
$$f(x) = \begin{cases} 2 + \sqrt{1 - x} & \text{if } x \le 1 \\ \dfrac{1}{1 - x} & \text{if } x > 1 \end{cases}$$
Find $f(0), f(1)$, and $f(2)$.

13. Refer to the graph of the function f in the following figure.

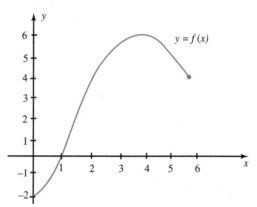

a. Find the value of $f(0)$.
b. Find the value of x for which (i) $f(x) = 3$ and (ii) $f(x) = 0$.
c. Find the domain of f.
d. Find the range of f.

14. Refer to the graph of the function f in the following figure.

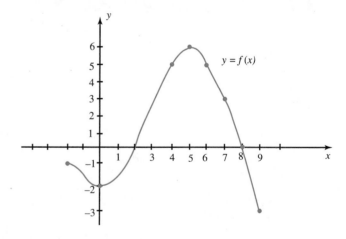

a. Find the value of $f(7)$.
b. Find the values of x corresponding to the point on the graph of f located at a height of 5 units from the x-axis.
c. Find the points on the x-axis at which the graph of f crosses it. What are the values of $f(x)$ at those points?
d. Find the domain and range of f.

In Exercises 15–18, determine whether the point lies on the graph of the function.

15. $(2, \sqrt{3})$; $g(x) = \sqrt{x^2 - 1}$

16. $(3, 3)$; $f(x) = \dfrac{x + 1}{\sqrt{x^2 + 7}} + 2$

17. $(-2, -3)$; $f(t) = \dfrac{|t - 1|}{t + 1}$

18. $\left(-3, -\dfrac{1}{13}\right)$; $h(t) = \dfrac{|t + 1|}{t^3 + 1}$

In Exercises 19–32, find the domain of the function.

19. $f(x) = x^2 + 3$

20. $f(x) = 7 - x^2$

21. $f(x) = \dfrac{3x + 1}{x^2}$

22. $g(x) = \dfrac{2x + 1}{x - 1}$

23. $f(x) = \sqrt{x^2 + 1}$

24. $f(x) = \sqrt{x - 5}$

25. $f(x) = \sqrt{5 - x}$

26. $g(x) = \sqrt{2x^2 + 3}$

27. $f(x) = \dfrac{x}{x^2 - 1}$

28. $f(x) = \dfrac{1}{x^2 + x - 2}$

29. $f(x) = (x + 3)^{3/2}$

30. $g(x) = 2(x - 1)^{5/2}$

31. $f(x) = \dfrac{\sqrt{1 - x}}{x^2 - 4}$

32. $f(x) = \dfrac{\sqrt{x - 1}}{(x + 2)(x - 3)}$

33. Let f be a function defined by the rule $f(x) = x^2 - x - 6$.
   a. Find the domain of f.
   b. Compute $f(x)$ for $x = -3, -2, -1, 0, \frac{1}{2}, 1, 2, 3$.
   c. Use the results obtained in parts (a) and (b) to sketch the graph of f.

34. Let f be a function defined by the rule $f(x) = 2x^2 + x - 3$.
   a. Find the domain of f.
   b. Compute $f(x)$ for $x = -3, -2, -1, -\frac{1}{2}, 0, 1, 2, 3$.
   c. Use the results obtained in parts (a) and (b) to sketch the graph of f.

In Exercises 35–46, sketch the graph of the function with the given rule. Find the domain and range of the function.

35. $f(x) = 2x^2 + 1$

36. $f(x) = 9 - x^2$

37. $f(x) = 2 + \sqrt{x}$

38. $g(x) = 4 - \sqrt{x}$

39. $f(x) = \sqrt{1 - x}$

40. $f(x) = \sqrt{x - 1}$

41. $f(x) = |x| - 1$

42. $f(x) = |x| + 1$

43. $f(x) = \begin{cases} x & \text{if } x < 0 \\ 2x + 1 & \text{if } x \geq 0 \end{cases}$

44. $f(x) = \begin{cases} 4 - x & \text{if } x < 2 \\ 2x - 2 & \text{if } x \geq 2 \end{cases}$

45. $f(x) = \begin{cases} -x + 1 & \text{if } x \leq 1 \\ x^2 - 1 & \text{if } x > 1 \end{cases}$

46. $f(x) = \begin{cases} -x - 1 & \text{if } x < -1 \\ 0 & \text{if } -1 \leq x \leq 1 \\ x + 1 & \text{if } x > 1 \end{cases}$

In Exercises 47–54, use the vertical-line test to determine whether the graph represents y as a function of x.

47.

48.

49.

50.

51.

52.

53.   **54.**

55. The circumference of a circle is given by $C(r) = 2\pi r$, where r is the radius of the circle. What is the circumference of a circle with a 5-in. radius?

56. The volume of a sphere of radius r is given by $V(r) = \frac{4}{3}\pi r^3$. Compute $V(2.1)$ and $V(2)$. What does the quantity $V(2.1) - V(2)$ measure?

57. GROWTH OF A CANCEROUS TUMOR The volume of a spherical cancer tumor is given by the function

$$V(r) = \tfrac{4}{3}\pi r^3$$

where r is the radius of the tumor in centimeters. By what factor is the volume of the tumor increased if its radius is doubled?

58. GROWTH OF A CANCEROUS TUMOR The surface area of a spherical cancer tumor is given by the function

$$S(r) = 4\pi r^2$$

where r is the radius of the tumor in centimeters. After extensive chemotherapy treatment, the surface area of the tumor is reduced by 75%. What is the radius of the tumor after treatment?

59. SALES OF PRERECORDED MUSIC The following graphs show the sales y of prerecorded music (in billions of dollars) by format as a function of time t (in years) with $t = 0$ corresponding to 1985.

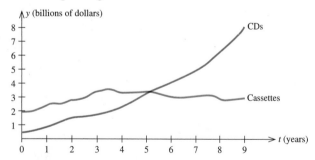

a. In what years were the sales of prerecorded cassettes greater than those of prerecorded CDs?
b. In what years were the sales of prerecorded CDs greater than those of prerecorded cassettes?
c. In what year were the sales of prerecorded cassettes

the same as those of prerecorded CDs? Estimate the level of sales in each format at that time.
Source: Recording Industry Association of America

60. THE GENDER GAP The following graph shows the ratio of women's earnings to men's from 1960 through 1990.

a. Write the rule for the function f giving the ratio of women's earnings to men's in year t, with $t = 0$ corresponding to 1960.
Hint: The function f is defined piecewise and is linear over each of three subintervals.

b. In what decade(s) was the gender gap expanding? Shrinking?
c. Refer to part (b). How fast was the gender gap (the ratio/year) expanding or shrinking in each of these decades?
Source: U.S. Bureau of Labor Statistics

61. CLOSING THE GENDER GAP IN EDUCATION The following graph shows the ratio of bachelor's degrees earned by women to men from 1960 through 1990.

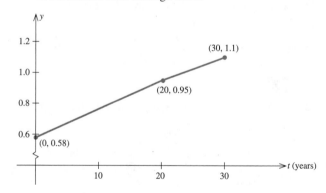

a. Write the rule for the function f giving the ratio of bachelor's degrees earned by women to men in year t, with $t = 0$ corresponding to 1960.
Hint: The function f is defined piecewise and is linear over each of two subintervals.

b. How fast was the ratio changing in the period from 1960 through 1980? The decade from 1980 to 1990?
c. In what year (approximately) was the number of

bachelor's degrees earned by women equal for the first time to that earned by men?
Source: Department of Education

62. **CONSUMPTION FUNCTION** The consumption function in a certain economy is given by the equation

$$C(y) = 0.75y + 6$$

where $C(y)$ is the personal consumption expenditure, y is the disposable personal income, and both $C(y)$ and y are measured in billions of dollars. Find $C(0)$, $C(50)$, and $C(100)$.

63. **SALES TAXES** In a certain state the sales tax T on the amount of taxable goods is 6% of the value of the goods purchased (x), where both T and x are measured in dollars.
   a. Express T as a function of x.
   b. Find $T(200)$ and $T(5.65)$.

64. **SURFACE AREA OF A SINGLE-CELLED ORGANISM** The surface area S of a single-celled organism may be found by multiplying 4π times the square of the radius r of the cell. Express S as a function of r.

65. **FRIEND'S RULE** Friend's rule, a method for calculating pediatric drug dosages, is based on a child's age. If a denotes the adult dosage (in milligrams) and if t is the age of the child (in years), then the child's dosage is given by

$$D(t) = \frac{2}{25} ta$$

If the adult dose of a substance is 500 mg, how much should a 4-yr-old child receive?

66. **COLAs** Social Security recipients receive an automatic cost-of-living adjustment (COLA) once each year. Their monthly benefit is increased by the amount that consumer prices increased during the preceding year. Suppose that consumer prices increased by 5.3% during the preceding year.
   a. Express the adjusted monthly benefit of a Social Security recipient as a function of his or her current monthly benefit.
   b. If Harrington's monthly Social Security benefit is now $620, what will be his adjusted monthly benefit?

67. **COST OF RENTING A TRUCK** The Ace Truck Leasing Company leases a certain size truck at $30/day and $.15/mi, whereas the Acme Truck Leasing Company leases the same size truck at $25/day and $.20/mi.
   a. Find the daily cost of leasing from each company as a function of the number of miles driven.
   b. Sketch the graphs of the two functions on the same set of axes.
   c. Which company should a customer rent a truck from for 1 day if she plans to drive at most 70 mi and wishes to minimize her cost?

68. **LINEAR DEPRECIATION** A new machine was purchased by the National Textile Company for $120,000. For income tax purposes, the machine is depreciated linearly over 10 yr; that is, the book value of the machine decreases at a constant rate, so that at the end of 10 yr the book value is zero.
   a. Express the book value of the machine (V) as a function of the age, in years, of the machine (n).
   b. Sketch the graph of the function in part (a).
   c. Find the book value of the machine at the end of the sixth year.
   d. Find the rate at which the machine is being depreciated each year.

69. **LINEAR DEPRECIATION** Refer to Exercise 68. An office building worth $1 million when completed in 1984 was depreciated linearly over 50 years. What was the book value of the building in 1999? What will the book value be in 2003? In 2007? (Assume that the book value of the building will be zero at the end of the 50th year.)

70. **BOYLE'S LAW** As a consequence of Boyle's law, the pressure P of a fixed sample of gas held at a constant temperature is related to the volume V of the gas by the rule

$$P = f(V) = \frac{k}{V}$$

where k is a constant. What is the domain of the function f? Sketch the graph of the function f.

71. **POISEUILLE'S LAW** According to a law discovered by the nineteenth-century physician Poiseuille, the velocity (in centimeters/second) of blood r centimeters from the central axis of an artery is given by

$$v(r) = k(R^2 - r^2)$$

where k is a constant and R is the radius of the artery. Suppose that for a certain artery, $k = 1000$ and $R = 0.2$ so that $v(r) = 1000(0.04 - r^2)$.
   a. What is the domain of the function $v(r)$?
   b. Compute $v(0)$, $v(0.1)$, and $v(0.2)$ and interpret your results.

72. **POPULATION GROWTH** A study prepared for a certain Sunbelt town's Chamber of Commerce projected that the population of the town in the next 3 yr will grow according to the rule

$$P(x) = 50,000 + 30x^{3/2} + 20x$$

where $P(x)$ denotes the population x mo from now. By how much will the population increase during the next 9 mo? During the next 16 mo?

73. **WORKER EFFICIENCY** An efficiency study conducted for the Elektra Electronics Company showed that the number of "Space Commander" walkie-talkies assembled by the

average worker t hr after starting work at 8:00 A.M. is given by

$$N(t) = -t^3 + 6t^2 + 15t \qquad (0 \le t \le 4)$$

How many walkie-talkies can an average worker be expected to assemble between 8:00 and 9:00 A.M.? Between 9:00 and 10:00 A.M.?

74. LEARNING CURVES The Emory Secretarial School finds from experience that the average student taking advanced typing will progress according to the rule

$$N(t) = \frac{60t + 180}{t + 6}$$

where $N(t)$ measures the number of words/minute the student can type after t wk in the course. How fast can the average student be expected to type after 2 wk in the course? After 4 wk in the course?

75. POLITICS Political scientists have discovered the following empirical rule, known as the "cube rule," which gives the relationship between the proportion of seats in the House of Representatives won by Democratic candidates $s(x)$ and the proportion of popular votes x received by the Democratic presidential candidate:

$$s(x) = \frac{x^3}{x^3 + (1 - x)^3} \qquad (0 \le x \le 1)$$

Compute $s(0.6)$ and interpret your result.

76. HOME SHOPPING INDUSTRY According to industry sources, revenue from the home shopping industry for the years since its inception may be approximated by the function

$$R(t) = \begin{cases} -0.03t^3 + 0.25t^2 - 0.12t & \text{if } 0 \le t \le 3 \\ 0.57t - 0.63 & \text{if } 3 < t \le 11 \end{cases}$$

where $R(t)$ measures the revenue in billions of dollars and t is measured in years, with $t = 0$ corresponding to the beginning of 1984. What was the revenue at the beginning of 1985? At the beginning of 1993?
Source: Paul Kagan Associates

77. POSTAL REGULATIONS The postage for first-class mail is 33 cents for the first ounce or fraction thereof and 23 cents for each additional ounce or fraction thereof. Any parcel

not exceeding 12 ounces may be sent by first-class mail. Letting x denote the weight of a parcel in ounces and $f(x)$ the postage in cents, complete the following description of the "postage function" f:

$$f(x) = \begin{cases} 33 & \text{if } 0 < x \le 1 \\ 56 & \text{if } 1 < x \le 2 \\ \ldots \\ ? & \text{if } 11 < x \le 12 \end{cases}$$

a. What is the domain of f?
b. Sketch the graph of f.

78. HARBOR CLEANUP The amount of solids discharged from the MWRA (Massachusetts Water Resources Authority) sewage treatment plant on Deer Island (near Boston Harbor) is given by the function

$$f(t) = \begin{cases} 130 & \text{if } 0 \le t \le 1 \\ -30t + 160 & \text{if } 1 < t \le 2 \\ 100 & \text{if } 2 < t \le 4 \\ -5t^2 + 25t + 80 & \text{if } 4 < t \le 6 \\ 1.25t^2 - 26.25t + 162.5 & \text{if } 6 < t \le 10 \end{cases}$$

where $f(t)$ is measured in tons/day and t is measured in years, with $t = 0$ corresponding to 1989.
Source: Metropolitan District Commission
a. What amount of solids were discharged per day in 1989? In 1992? In 1996?
b. Sketch the graph of f.

In Exercises 79–82, determine whether the statement is true or false. If it is true, explain why it is true. If it is false, give an example to show why it is false.

79. If $a = b$, then $f(a) = f(b)$.

80. If $f(a) = f(b)$, then $a = b$.

81. If f is a function, then $f(a + b) = f(a) + f(b)$.

82. A vertical line must intersect the graph of $y = f(x)$ at exactly one point.

SOLUTIONS TO SELF-CHECK EXERCISES 2.1

1. a. The expression under the radical sign must be nonnegative, so $x + 1 \ge 0$ or $x \ge -1$. Also, $x \ne 0$ because division by zero is not permitted. Therefore, the domain of f is $[-1, 0) \cup (0, \infty)$.

b. $f(3) = \dfrac{\sqrt{3+1}}{3} = \dfrac{\sqrt{4}}{3} = \dfrac{2}{3}$

c. $f(a+h) = \dfrac{\sqrt{(a+h)+1}}{a+h} = \dfrac{\sqrt{a+h+1}}{a+h}$

2. a. For t in the subdomain $[0, 6]$, the rule for f is given by $f(t) = 6t + 17$. The equation $y = 6t + 17$ is a linear equation, so that portion of the graph of f is the line segment joining the points $(0, 17)$ and $(6, 53)$. Next, in the subdomain $(6, 20]$, the rule for f is given by $f(t) = 15.98(t - 6)^{1/4} + 53$. Using a calculator, we construct the following table of values of $f(t)$ for selected values of t.

t	6	8	10	12	14	16	18	20
$f(t)$	53	72	75.6	78	79.9	81.4	82.7	83.9

We have included $t = 6$ in the table, although it does not lie in the subdomain of the function under consideration, in order to help us obtain a better sketch of that portion of the graph of f in the subdomain $(6, 20]$. The graph of f is as follows:

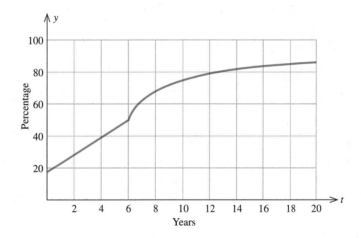

b. The percentage of all self-serve gas sales at the beginning of 1978 is found by evaluating f at $t = 4$. Since this point lies in the interval $[0, 6]$, we use the rule $f(t) = 6t + 17$ and find

$$f(4) = 6(4) + 17$$

giving 41% as the required figure. The percentage of all self-serve gas sales at the beginning of 1994 is given by

$$f(20) = 15.98(20 - 6)^{1/4} + 53$$

or approximately 83.9%.

3. A point (x, y) lies on the graph of the function f if and only if the coordinates satisfy the equation $y = f(x)$. Now,

$$f(4) = \sqrt{2(4) + 1} + 2 = \sqrt{9} + 2 = 5 \neq 6$$

and we conclude that the given point does not lie on the graph of f.

GRAPHING A FUNCTION

Most of the graphs of functions in this book can be plotted with the help of a graphing utility. Furthermore, a graphing utility can be used to analyze the nature of a function. However, the amount and accuracy of the information obtained using a graphing utility depend on the experience and sophistication of the user. As you progress through this book, you will see that the more knowledge of calculus you gain, the more effective the graphing utility will prove as a tool in problem solving.

FINDING A SUITABLE VIEWING RECTANGLE

The first step in plotting the graph of a function with a graphing utility is to select a suitable viewing rectangle. We usually do this by experimenting. For example, you might first plot the graph using the *standard viewing rectangle* $[-10, 10]$ by $[-10, 10]$. If necessary, you then might adjust the viewing rectangle by enlarging it or reducing it to obtain a sufficiently complete view of the graph or at least the portion of the graph that is of interest.

EXAMPLE 1

Plot the graph of $f(x) = 2x^2 - 4x - 5$ in the standard viewing rectangle.

SOLUTION ✔

The graph of f, shown in Figure T1, is a parabola. From our previous work (Example 6, Section 2.1), we know that the figure does give a good view of the graph.

FIGURE T1
The graph of $y = 2x^2 - 4x - 5$ in the viewing rectangle $[-10, 10]$ $\times [-10, 10]$

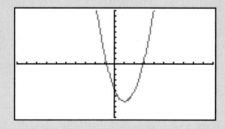

■ ■ ■ ■

EXAMPLE 2

Let $f(x) = x^3(x - 3)^4$.

a. Plot the graph of f in the standard viewing rectangle.
b. Plot the graph of f in the rectangle $[-1, 5] \times [-40, 40]$.

SOLUTION ✔

a. The graph of f in the standard viewing rectangle is shown in Figure T2a. Since the graph does not appear to be complete, we need to adjust the viewing rectangle.

An incomplete sketch of $f(x) = x^3(x - 3)^4$ is shown in (a), and a complete sketch is shown in (b).

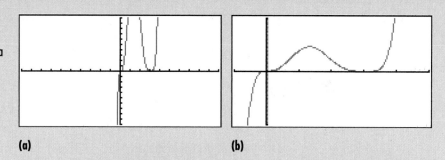

(a)   (b)

b. The graph of f in the rectangle $[-1, 5] \times [-40, 40]$, shown in Figure T2b, is an improvement over the previous graph. (Later we will be able to show that the figure does in fact give a rather complete view of the graph of f.)

▪ ▪ ▪ ▪

EVALUATING A FUNCTION

A graphing utility can be used to find the value of a function with minimal effort, as the next example shows.

EXAMPLE **3**

Let $f(x) = x^3 - 4x^2 + 4x + 2$.

a. Plot the graph of f in the standard viewing rectangle.
b. Find $f(3)$ and verify your result by direct computation.
c. Find $f(4.215)$.

SOLUTION ✔

a. The graph of f is shown in Figure T3.

FIGURE T3
The graph of
$f(x) = x^3 - 4x^2 + 4x + 2$
in the standard viewing rectangle

b. Using the evaluation function of the graphing utility and the value 3 for x, we find $y = 5$. This result is verified by computing

$$f(3) = 3^3 - 4(3^2) + 4(3) + 2 = 27 - 36 + 12 + 2 = 5$$

c. Using the evaluation function of the graphing utility and the value 4.215 for x, we find $y = 22.679738375$. Thus, $f(4.215) = 22.679738375$. The efficacy of the graphing utility is clearly demonstrated here! ■ ■ ■ ■

EXAMPLE **4**

The anticipated rise in the number of Alzheimer's patients in the United States is given by

$$f(t) = -0.0277t^4 + 0.3346t^3 - 1.1261t^2 + 1.7575t + 3.7745 \qquad (0 \le t \le 6)$$

where $f(t)$ is measured in millions and t is measured in decades, with $t = 0$ corresponding to the beginning of 1990.

a. Use a graphing utility to plot the graph of f in the viewing rectangle $[0, 7] \times [0, 12]$.
b. What is the anticipated number of Alzheimer's patients in the United States at the beginning of the year 2000 $(t = 1)$? At the beginning of 2030 $(t = 4)$?

Source: Alzheimer's Association

SOLUTION ✔   **a.** The graph of f in the viewing rectangle $[0, 7] \times [0, 12]$ is shown in Figure T4.

FIGURE T4
The graph of f in the viewing rectangle $[0, 7] \times [0, 12]$

b. Using the evaluation function of the graphing utility and the value 1 for x, we see that the anticipated number of Alzheimer's patients at the beginning of the year 2000 is given by

$$f(1) = 4.7128$$

or approximately 4.7 million. The anticipated number of Alzheimer's patients at the beginning of 2030 is given by

$$f(4) = 7.1101$$

or approximately 7.1 million. ■ ■ ■ ■

In Exercises 1–8, plot the graph of the function f in the standard viewing window.

1. $f(x) = 2x^2 - 16x + 29$ **2.** $f(x) = -x^2 - 10x - 20$

3. $f(x) = x^3 - 2x^2 + x - 2$

4. $f(x) = -2.01x^3 + 1.21x^2 - 0.78x + 1$

5. $f(x) = 0.2x^4 - 2.1x^2 + 1$

6. $f(x) = -0.4x^4 + 1.2x - 1.2$

7. $f(x) = 2x\sqrt{x^2 + 1}$ **8.** $f(x) = \dfrac{\sqrt{x} + 1}{\sqrt{x} - 1}$

In Exercises 9–20, plot the graph of the function f in (a) the standard viewing window and (b) the indicated window.

9. $f(x) = 2x^2 - 32x + 125;\ [5, 15] \times [-5, 10]$

10. $f(x) = x^2 + 20x + 95;\ [-20, 10] \times [-15, -5]$

11. $f(x) = x^3 - 20x^2 + 8x - 10;\ [-20, 20] \times [-1200, 100]$

12. $f(x) = -2x^3 + 10x^2 - 15x - 5;\ [-10, 10] \times [-100, 100]$

13. $f(x) = x^4 - 2x^2 + 8;\ [-2, 2] \times [6, 10]$

14. $f(x) = x^4 - 2x^3;\ [-1, 3] \times [-2, 2]$

15. $f(x) = x + \dfrac{1}{x};\ [-1, 3] \times [-5, 5]$

16. $f(x) = \dfrac{4}{x^2 - 8};\ [-5, 5] \times [-5, 5]$

17. $f(x) = 2 - \dfrac{1}{x^2 + 1};\ [-3, 3] \times [0, 3]$

18. $f(x) = x - 2\sqrt{x};\ [0, 20] \times [-2, 10]$

19. $f(x) = x\sqrt{4 - x^2};\ [-3, 3] \times [-2, 2]$

20. $f(x) = \dfrac{\sqrt{x} - 1}{x};\ [0, 50] \times [-0.25, 0.25]$

In Exercises 21–30, plot the graph of the function f in an appropriate viewing window. (*Note:* The answer is not unique.)

21. $f(x) = x^2 - 4x + 16$ **22.** $f(x) = -x^2 + 2x - 11$

23. $f(x) = 2x^3 - 10x^2 + 5x - 10$

24. $f(x) = -x^3 + 5x^2 - 14x + 20$

25. $f(x) = 2x^4 - 3x^3 + 5x^2 - 20x + 40$

26. $f(x) = -2x^4 + 5x^2 - 4$

27. $f(x) = \dfrac{x^3}{x^3 + 1}$ **28.** $f(x) = \dfrac{2x^4 - 3x}{x^2 - 1}$

29. $f(x) = 0.2\sqrt{x} - 0.3x^3$ **30.** $f(x) = \sqrt{x}(2x - 1)^3$

In Exercises 31–34, use the evaluation function of your graphing utility to find the value of f at the given value of x and verify your result by direct computation.

31. $f(x) = -3x^3 + 5x^2 - 2x + 8;\ x = -1$

32. $f(x) = 2x^4 - 3x^3 + 2x^2 + x - 5;\ x = 2$

33. $f(x) = \dfrac{x^4 - 3x^2}{x - 2};\ x = 1$ **34.** $f(x) = \dfrac{\sqrt{x^2 - 1}}{3x + 4};\ x = 2$

In Exercises 35–42, use the evaluation function of your graphing utility to find the value of f at the indicated value of x. Express your answer accurate to four decimal places.

35. $f(x) = 3x^3 - 2x^2 + x - 4;\ x = 2.145$

36. $f(x) = 2x^3 + 5x^2 + 3x + 1;\ x = -0.27$

37. $f(x) = 5x^4 - 2x^2 + 8x - 3;\ x = 1.28$

38. $f(x) = 4x^4 - 3x^3 + 1;\ x = -2.42$

39. $f(x) = \dfrac{2x^3 - 3x + 1}{3x - 2};\ x = 2.41$

40. $f(x) = \dfrac{2x + 5}{3x^2 - 4x + 1};\ x = -1.72$

41. $f(x) = \sqrt{2x^2 + 1} + \sqrt{3x^2 - 1};\ x = 0.62$

42. $f(x) = 2x(3x^3 + 5)^{1/3};\ x = -6.24$

43. MANUFACTURING CAPACITY Data obtained from the Federal Reserve show that the annual increase in manufacturing capacity between 1988 and 1994 is given by

$$f(t) = 0.0388889t^3 - 0.283333t^2 + 0.477778t$$
$$+ 2.04286 \qquad (0 \le t \le 6)$$

where $f(t)$ is a percentage and t is measured in years, with $t = 0$ corresponding to the beginning of 1988.

a. Use a graphing utility to plot the graph of f in the viewing rectangle $[0, 8] \times [0, 4]$.

b. What was the annual increase in manufacturing capacity at the beginning of 1990 ($t = 2$)? At the beginning of 1992 ($t = 4$)?

Source: Federal Reserve

44. DECLINE OF UNION MEMBERSHIP The total union membership as a percentage of the private workforce is given by

$$f(t) = 0.00017t^4 - 0.00921t^3 + 0.15437t^2$$
$$- 1.360723t + 16.8028 \qquad (0 \le t \le 10)$$

where t is measured in years, with $t = 0$ corresponding to the beginning of 1983.

a. Plot the graph of f in the viewing rectangle $[0, 11] \times [8, 20]$.

b. What was the total union membership as a percentage of the private force at the beginning of 1986? At the beginning of 1993?

Source: American Federation of Labor and Congress of Industrial Organizations

45. KEEPING WITH THE TRAFFIC FLOW By driving at a speed to match the prevailing traffic speed, you decrease the chances of an accident. According to a University of Virginia School of Engineering and Applied Science study, the number of accidents per 100 million vehicle miles, y, is related to the deviation from the mean speed, x, in mph by the equation

$$y = 1.05x^3 - 21.95x^2 + 155.9x - 327.3 \qquad (6 \le x \le 11)$$

a. Plot the graph of y in the viewing rectangle $[6, 11] \times [20, 150]$.

b. What is the number of accidents per 100 million vehicle miles if the deviation from the mean speed is 6 mph, 8 mph, and 11 mph?

Source: University of Virginia School of Engineering and Applied Science

2.2 THE ALGEBRA OF FUNCTIONS

The Sum, Difference, Product, and Quotient of Functions

Let $S(t)$ and $R(t)$ denote, respectively, the federal government's spending and revenue at any time t, measured in billions of dollars. The graphs of these functions for the period between 1981 and 1991 are shown in Figure 2.11.

FIGURE 2.11
$S(t) - R(t)$ gives the federal budget deficit at any time t.

Source: Office of Management and Budget

The budget deficit at any time t is given by $S(t) - R(t)$ billion dollars. This observation suggests that we can define a function D whose value at any time t is given by $D(t) = S(t) - R(t)$. The function D, the *difference* of the two functions S and R, is written $D = S - R$, and may be called the "deficit function" since it gives the budget deficit at any time t. It has the same domain as the functions S and R.

Most functions are built up from other, generally simpler, functions. For example, we may view the function $f(x) = 2x + 4$ as the sum of the two functions $g(x) = 2x$ and $h(x) = 4$. The function $g(x) = 2x$ may in turn be viewed as the product of the functions $p(x) = 2$ and $q(x) = x$.

In general, given the functions f and g, we define the sum $f + g$, the difference $f - g$, the product fg, and the quotient f/g of f and g as follows.

The Sum, Difference, Product, and Quotient of Functions

Let f and g be functions with domains A and B, respectively. Then the **sum** $f + g$, **difference** $f - g$, and **product** fg of f and g are functions with domain $A \cap B$* and rule given by

$$(f + g)(x) = f(x) + g(x) \quad \text{(Sum)}$$
$$(f - g)(x) = f(x) - g(x) \quad \text{(Difference)}$$
$$(fg)(x) = f(x)g(x) \quad \text{(Product)}$$

The **quotient** f/g of f and g has domain $A \cap B$ excluding all points x such that $g(x) = 0$ and rule given by

$$\left(\frac{f}{g}\right)(x) = \frac{f(x)}{g(x)} \quad \text{(Quotient)}$$

────────────────────────────────────

*$A \cap B$ is read "A intersected with B" and denotes the set of all points common to both A and B.

EXAMPLE 1   Let $f(x) = \sqrt{x + 1}$ and $g(x) = 2x + 1$. Find the sum s, the difference d, the product p, and the quotient q of the functions f and g.

SOLUTION ✔   Since the domain of f is $A = [-1, \infty)$ and the domain of g is $B = (-\infty, \infty)$, we see that the domain of s, d, and p is $A \cap B = [-1, \infty)$. The rules follow.

$$s(x) = (f + g)(x) = f(x) + g(x) = \sqrt{x + 1} + 2x + 1$$
$$d(x) = (f - g)(x) = f(x) - g(x) = \sqrt{x + 1} - (2x + 1) = \sqrt{x + 1} - 2x - 1$$
$$p(x) = (fg)(x) = f(x)g(x) = \sqrt{x + 1}(2x + 1) = (2x + 1)\sqrt{x + 1}$$

The quotient function q has rule

$$q(x) = \left(\frac{f}{g}\right)(x) = \frac{f(x)}{g(x)} = \frac{\sqrt{x + 1}}{2x + 1}$$

Its domain is $[-1, \infty)$ together with the restriction $x \neq -\frac{1}{2}$. We denote this by $[-1, -\frac{1}{2}) \cup (-\frac{1}{2}, \infty)$.   ■ ■ ■ ■

APPLICATIONS

The mathematical formulation of a problem arising from a practical situation often leads to an expression that involves the combination of functions. Consider, for example, the costs incurred in operating a business. Costs that remain more or less constant regardless of the firm's level of activity are called **fixed costs.** Examples of fixed costs are rental fees and executive salaries. On the other hand, costs that vary with production or sales are called **variable costs.** Examples of variable costs are wages and costs of raw materials. The **total cost** of operating a business is thus given by the *sum* of the variable costs and the fixed costs, as illustrated in the next example.

EXAMPLE **2**

Suppose Puritron, a manufacturer of water filters, has a monthly fixed cost of $10,000 and a variable cost of

$$-0.0001x^2 + 10x \qquad (0 \le x \le 40{,}000)$$

dollars, where x denotes the number of filters manufactured per month. Find a function C that gives the total cost incurred by Puritron in the manufacture of x filters.

SOLUTION ✔

Puritron's monthly fixed cost is always $10,000, regardless of the level of production, and it is described by the constant function $F(x) = 10{,}000$. Next, the variable cost is described by the function $V(x) = -0.0001x^2 + 10x$. Since the total cost incurred by Puritron at any level of production is the sum of the variable cost and the fixed cost, we see that the required total cost function is given by

$$\begin{aligned} C(x) &= V(x) + F(x) \\ &= -0.0001x^2 + 10x + 10{,}000 \qquad (0 \le x \le 40{,}000) \end{aligned}$$ ■ ■ ■ ■

Next, the **total profit** realized by a firm in operating a business is the *difference* between the total revenue realized and the total cost incurred; that is, $P(x) = R(x) - C(x)$.

EXAMPLE **3**

Refer to Example 2. Suppose that the total revenue realized by Puritron from the sale of x water filters is given by the total revenue function

$$R(x) = -0.0005x^2 + 20x \qquad (0 \le x \le 40{,}000)$$

a. Find the total profit function—that is, the function that describes the total profit Puritron realizes in manufacturing and selling x water filters per month.
b. What is the profit when the level of production is 10,000 filters per month?

SOLUTION ✔

a. The total profit realized by Puritron in manufacturing and selling x water filters per month is the difference between the total revenue realized and the total cost incurred. Thus, the required total profit function is given by

$$\begin{aligned} P(x) &= R(x) - C(x) \\ &= (-0.0005x^2 + 20x) - (-0.0001x^2 + 10x + 10{,}000) \\ &= -0.0004x^2 + 10x - 10{,}000 \end{aligned}$$

b. The profit realized by Puritron when the level of production is 10,000 filters per month is

$$P(10{,}000) = -0.0004(10{,}000)^2 + 10(10{,}000) - 10{,}000 = 50{,}000$$

or $50,000 per month. ■ ■ ■ ■

COMPOSITION OF FUNCTIONS

Another way to build up a function from other functions is through a process known as the *composition of functions*. Consider, for example, the function

h, whose rule is given by $h(x) = \sqrt{x^2 - 1}$. Let f and g be functions defined by the rules $f(x) = x^2 - 1$ and $g(x) = \sqrt{x}$. Evaluating the function g at the point $f(x)$ [remember that for each real number x in the domain of f, $f(x)$ is simply a real number], we find that

$$g(f(x)) = \sqrt{f(x)} = \sqrt{x^2 - 1}$$

which is just the rule defining the function h!

In general, the composition of a function g with a function f is defined as follows.

The Composition of Two Functions

> Let f and g be functions. Then the composition of g and f is the function $g \circ f$ defined by
>
> $$(g \circ f)(x) = g(f(x))$$
>
> The domain of $g \circ f$ is the set of all x in the domain of f such that $f(x)$ lies in the domain of g.

The function $g \circ f$ (read "g circle f") is also called a **composite function.** The interpretation of the function $h = g \circ f$ as a machine is illustrated in Figure 2.12, and its interpretation as a mapping is shown in Figure 2.13.

FIGURE 2.12
The composite function $h = g \circ f$ is found by evaluating f at x and then g at $f(x)$, in that order.

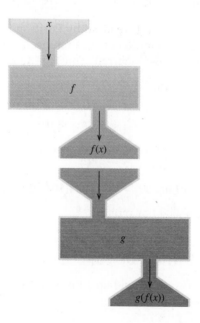

FIGURE 2.13
The function $h = g \circ f$ viewed as a mapping

EXAMPLE 4   Let $f(x) = x^2 - 1$ and $g(x) = \sqrt{x} + 1$. Compute:

a. The rule for the composite function $g \circ f$.
b. The rule for the composite function $f \circ g$.

SOLUTION ✔ **a.** To find the rule for the composite function $g \circ f$, evaluate the function g at $f(x)$. Therefore,

$$(g \circ f)(x) = g(f(x)) = \sqrt{f(x)} + 1 = \sqrt{x^2 - 1} + 1$$

b. To find the rule for the composite function $f \circ g$, evaluate the function f at $g(x)$. Thus,

$$(f \circ g)(x) = f(g(x)) = (g(x))^2 - 1 = (\sqrt{x} + 1)^2 - 1$$
$$= x + 2\sqrt{x} + 1 - 1 = x + 2\sqrt{x}$$ ■ ■ ■ ■

 Example 4 reminds us that in general $g \circ f$ is different from $f \circ g$, so care must be taken regarding the order when computing a composite function.

 Group Discussion

Let $f(x) = \sqrt{x} + 1$ for $x \geq 0$ and let $g(x) = (x - 1)^2$ for $x \geq 1$.

1. Show that $(g \circ f)(x)$ and $(f \circ g)(x) = x$. (*Remark:* The function g is said to be the *inverse* of f and vice versa.)
2. Plot the graphs of f and g together with the straight line $y = x$. Describe the relationship between the graphs of f and g.

 EXAMPLE 5   An environmental impact study conducted for Oxnard's City Environmental Management Department indicates that, under existing environmental protection laws, the level of carbon monoxide present in the air due to pollution from automobile exhaust will be $0.01x^{2/3}$ parts per million (ppm) when the number of motor vehicles is x thousand. A separate study conducted by a state government agency estimates that t years from now the number of motor vehicles in Oxnard will be $0.2t^2 + 4t + 64$ thousand.

a. Find an expression for the concentration of carbon monoxide in the air due to automobile exhaust t years from now.
b. What will be the level of concentration 5 years from now?

SOLUTION ✔ **a.** The level of carbon monoxide present in the air due to pollution from automobile exhaust is described by the function $g(x) = 0.01x^{2/3}$, where x is the number (in thousands) of motor vehicles. But the number of motor vehicles x (in thousands) t years from now may be estimated by the rule $f(t) = 0.2t^2 + 4t + 64$. Therefore, the concentration of carbon monoxide due to automobile exhaust t years from now is given by

$$C(t) = (g \circ f)(t) = g(f(t)) = 0.01(0.2t^2 + 4t + 64)^{2/3}$$

parts per million.

MICHAEL MARCHLIK

TITLE: Project Manager
INSTITUTION: Ebasco Services Incorporated

Calculus wasn't Mike Marchlik's favorite college subject. In fact, it was not until he started his first job that the "lights went on" and he realized how using calculus allowed him to solve real problems in his everyday work.

Marchlik emphasizes that he doesn't do "number crunching" himself. "I don't work out integrals, but in my work I use computer models that do that." The important issue is not computation but how the answers relate to client problems.

Marchlik's clients typically process highly toxic or explosive materials. Ebasco evaluates a client site, such as a chemical plant, to determine how safety systems might fail and what the probable consequences might be.

To avoid a major disaster like the one that occurred in Bhopal, India, in 1984, Marchlik and his team might be asked to determine how quickly a poisonous chemical would spread if a leak occurred. Or they might help avert a disaster like the one that rocked the Houston area in 1989. In that incident, hydrocarbon vapor exploded at a Phillips 66 chemical plant, shaking office buildings in Houston, 12 miles away. Several people were killed, and hundreds were injured. Property damage totaled $1.39 billion.

In assessing risks for a fuel-storage depot, Ebasco considers variables such as weather conditions, including probable wind speed, the flow properties of a gas, and possible ignition sources. With today's more powerful computers, the models can involve a system of very complex equations to project likely scenarios.

Mathematical models vary, however. One model might forecast how much gas will flow out of a hole and how quickly it will disperse. Another model might project where the gas will go, depending on local factors such as temperature and wind speed. Choosing the right model is essential. A model based on flat terrain when the client's storage depot is set among hills is going to produce the wrong answer.

Marchlik and his team run several models together to come up with their projections. Each model uses "equations that have to be integrated to come up with solutions." The bottom line? Marchlik stresses that "calculus is at the very heart" of Ebasco's risk-assessment work.

Ebasco is a diversified engineering and construction company. In addition to designing and building electric generating facilities, the company provides clients with studies. Their studies include analyses of potential hazards, recommendations for safe work practices, and evaluations of responses to emergencies in chemical plants, fuel-storage depots, nuclear facilities, and hazardous waste sites. As a project manager, Marchlik deals directly with clients to help them understand and implement Ebasco's recommendations to ensure a safer environment.

b. The level of concentration 5 years from now will be
$$C(5) = 0.01[0.2(5)^2 + 4(5) + 64]^{2/3}$$
$$= (0.01)89^{2/3} \approx 0.20$$

parts per million.

■■■■

SELF-CHECK EXERCISES **2.2**

1. Let f and g be functions defined by the rules
$$f(x) = \sqrt{x} + 1 \quad \text{and} \quad g(x) = \frac{x}{1+x}$$
respectively. Find the rules for
a. the sum s, the difference d, the product p, and the quotient q of f and g.
b. the composite functions $f \circ g$ and $g \circ f$.

2. Health-care spending per person by the private sector includes payments by individuals, corporations, and their insurance companies and is approximated by the function
$$f(t) = 2.5t^2 + 31.3t + 406 \qquad (0 \le t \le 20)$$
where $f(t)$ is measured in dollars and t is measured in years, with $t = 0$ corresponding to the beginning of 1975. The corresponding government spending—including expenditures for Medicaid, Medicare, and other federal, state, and local government public health care—is
$$g(t) = 1.4t^2 + 29.6t + 251 \qquad (0 \le t \le 20)$$
where t has the same meaning as before.
a. Find a function that gives the difference between private and government health-care spending per person at any time t.
b. What was the difference between private and government expenditures per person at the beginning of 1985? At the beginning of 1993?
Source: Health Care Financing Administration

Solutions to Self-Check Exercises 2.2 can be found on page 89.

2.2 Exercises

In Exercises 1–8, let $f(x) = x^3 + 5$, $g(x) = x^2 - 2$, and $h(x) = 2x + 4$. Find the rule for each function.

1. $f + g$ **2.** $f - g$ **3.** fg **4.** gf

5. $\dfrac{f}{g}$ **6.** $\dfrac{f-g}{h}$ **7.** $\dfrac{fg}{h}$ **8.** fgh

In Exercises 9–18, let $f(x) = x - 1$, $g(x) = \sqrt{x+1}$, and $h(x) = 2x^3 - 1$. Find the rule for each function.

9. $f + g$ **10.** $g - f$ **11.** fg **12.** gf

13. $\dfrac{g}{h}$ **14.** $\dfrac{h}{g}$ **15.** $\dfrac{fg}{h}$ **16.** $\dfrac{fh}{g}$

17. $\dfrac{f-h}{g}$ **18.** $\dfrac{gh}{g-f}$

In Exercises 19–24, find the functions $f + g$, $f - g$, fg, and f/g.

19. $f(x) = x^2 + 5$; $g(x) = \sqrt{x} - 2$

20. $f(x) = \sqrt{x-1}$; $g(x) = x^3 + 1$

21. $f(x) = \sqrt{x+3}$; $g(x) = \dfrac{1}{x-1}$

22. $f(x) = \dfrac{1}{x^2+1}$; $g(x) = \dfrac{1}{x^2-1}$

23. $f(x) = \dfrac{x+1}{x-1}$; $g(x) = \dfrac{x+2}{x-2}$

24. $f(x) = x^2 + 1$; $g(x) = \sqrt{x+1}$

In Exercises 25–30, find the rules for the composite functions $f \circ g$ and $g \circ f$.

25. $f(x) = x^2 + x + 1$; $g(x) = x^2$

26. $f(x) = 3x^2 + 2x + 1$; $g(x) = x + 3$

27. $f(x) = \sqrt{x} + 1$; $g(x) = x^2 - 1$

28. $f(x) = 2\sqrt{x} + 3$; $g(x) = x^2 + 1$

29. $f(x) = \dfrac{x}{x^2 + 1}$; $g(x) = \dfrac{1}{x}$

30. $f(x) = \sqrt{x + 1}$; $g(x) = \dfrac{1}{x - 1}$

In Exercises 31–34, evaluate $h(2)$, where $h = g \circ f$.

31. $f(x) = x^2 + x + 1$; $g(x) = x^2$

32. $f(x) = \sqrt[3]{x^2 - 1}$; $g(x) = 3x^3 + 1$

33. $f(x) = \dfrac{1}{2x + 1}$; $g(x) = \sqrt{x}$

34. $f(x) = \dfrac{1}{x - 1}$; $g(x) = x^2 + 1$

In Exercises 35–42, find functions f and g such that $h = g \circ f$. (*Note:* The answer is not unique.)

35. $h(x) = (2x^3 + x^2 + 1)^5$ **36.** $h(x) = (3x^2 - 4)^{-3}$

37. $h(x) = \sqrt{x^2 - 1}$ **38.** $h(x) = (2x - 3)^{3/2}$

39. $h(x) = \dfrac{1}{x^2 - 1}$ **40.** $h(x) = \dfrac{1}{\sqrt{x^2 - 4}}$

41. $h(x) = \dfrac{1}{(3x^2 + 2)^{3/2}}$

42. $h(x) = \dfrac{1}{\sqrt{2x + 1}} + \sqrt{2x + 1}$

In Exercises 43–46, find $f(a + h) - f(a)$ for each function. Simplify your answer.

43. $f(x) = 3x + 4$ **44.** $f(x) = -\dfrac{1}{2}x + 3$

45. $f(x) = 4 - x^2$ **46.** $f(x) = x^2 - 2x + 1$

47. If $f(x) = x^2 + 1$, find and simplify

$$\frac{f(a + h) - f(a)}{h} \qquad (h \neq 0)$$

48. If $f(x) = 1/x$, find and simplify

$$\frac{f(a + h) - f(a)}{h} \qquad (h \neq 0)$$

49. Manufacturing Costs   TMI, Inc., a manufacturer of blank audiocassette tapes, has a monthly fixed cost of $12,100 and a variable cost of $.60/tape. Find a function C that gives the total cost incurred by TMI in the manufacture of x tapes/month.

50. Cost of Producing Electronic Organizers   Apollo, Inc., manufactures its electronic organizers at a variable cost of

$$V(x) = 0.000003x^3 - 0.03x^2 + 200x$$

dollars, where x denotes the number of units manufactured per month. The monthly fixed cost attributable to the division that produces these electronic organizers is $100,000. Find a function C that gives the total cost incurred by the manufacture of x electronic organizers. What is the total cost incurred in producing 2000 units/month?

51. Cost of Producing Electronic Organizers   Refer to Exercise 50. Suppose the total revenue realized by Apollo from the sale of x electronic organizers is given by the total revenue function

$$R(x) = 0.1x^2 + 500x \qquad (0 \leq x \leq 5000)$$

where x *is* measured in dollars.
a. Find the total profit function.
b. What is the profit when 1500 units are produced and sold each month?

52. Overcrowding of Prisons   The 1980s saw a trend toward old-fashioned punitive deterrence as opposed to the more liberal penal policies and community-based corrections popular in the 1960s and early 1970s. As a result, prisons became more crowded, and the gap between the number of people in prison and the prison capacity widened. Based on figures from the U.S. Department of Justice, the number of prisoners (in thousands) in federal and state prisons is approximated by the function

$$N(t) = 3.5t^2 + 26.7t + 436.2 \qquad (0 \leq t \leq 10)$$

where t is measured in years and $t = 0$ corresponds to 1983. The number of inmates for which prisons were designed is given by

$$C(t) = 24.3t + 365 \qquad (0 \leq t \leq 10)$$

where $C(t)$ is measured in thousands and t has the same meaning as before.
a. Find an expression that shows the gap between the number of prisoners and the number of inmates for which the prisons were designed at any time t.
b. Find the gap at the beginning of 1983 and at the beginning of 1986.
Source: U.S. Department of Justice

53. EFFECT OF MORTGAGE RATES ON HOUSING STARTS A study prepared for the National Association of Realtors estimated that the number of housing starts per year over the next 5 yr will be

$$N(r) = \frac{7}{1 + 0.02r^2}$$

million units, where r (percent) is the mortgage rate. Suppose the mortgage rate over the next t mo is

$$r(t) = \frac{10t + 150}{t + 10} \qquad (0 \le t \le 24)$$

percent/year.

a. Find an expression for the number of housing starts per year as a function of t, t months from now.

b. Using the result from part (a), determine the number of housing starts at present, 12 mo from now, and 18 mo from now.

54. HOTEL OCCUPANCY RATE The occupancy rate of the all-suite Wonderland Hotel, located near an amusement park, is given by the function

$$r(t) = \frac{10}{81}t^3 - \frac{10}{3}t^2 + \frac{200}{9}t + 55 \qquad (0 \le t \le 11)$$

where t is measured in months and $t = 0$ corresponds to the beginning of January. Management has estimated that the monthly revenue (in thousands of dollars) is approximated by the function

$$R(r) = -\frac{3}{5000}r^3 + \frac{9}{50}r^2 \qquad (0 \le r \le 100)$$

where r is the occupancy rate.

a. What is the hotel's occupancy rate at the beginning of January? At the beginning of June?

b. What is the hotel's monthly revenue at the beginning of January? At the beginning of June?

Hint: Compute $R(r(0))$ and $R(r(5))$.

55. HOUSING STARTS AND CONSTRUCTION JOBS The president of a major housing construction firm reports that the number of construction jobs (in millions) created is given by

$$N(x) = 1.42x$$

where x denotes the number of housing starts. Suppose the number of housing starts in the next t mo is expected to be

$$x(t) = \frac{7(t + 10)^2}{(t + 10)^2 + 2(t + 15)^2}$$

million units/year. Find an expression for the number of jobs created per month in the next t mo. How many jobs will have been created 6 months and 12 mo from now?

In Exercises 56–59, determine whether the statement is true or false. If it is true, explain why it is true. If it is false, give an example to show why it is false.

56. If f and g are functions with domain D, then $f + g = g + f$.

57. If $g \circ f$ is defined at $x = a$, then $f \circ g$ must also be defined at $x = a$.

58. If f and g are functions, then $f \circ g = g \circ f$.

59. If f is a function, then $f \circ f = f^2$.

SOLUTIONS TO SELF-CHECK EXERCISES 2.2

1. a. $\quad s(x) = f(x) + g(x) = \sqrt{x} + 1 + \dfrac{x}{1 + x}$

$$d(x) = f(x) - g(x) = \sqrt{x} + 1 - \frac{x}{1 + x}$$

$$p(x) = f(x)g(x) = (\sqrt{x} + 1) \cdot \frac{x}{1 + x} = \frac{x(\sqrt{x} + 1)}{1 + x}$$

$$q(x) = \frac{f(x)}{g(x)} = \frac{\sqrt{x} + 1}{\dfrac{x}{1 + x}} = \frac{(\sqrt{x} + 1)(1 + x)}{x}$$

b. $(f \circ g)(x) = f(g(x)) = \sqrt{\dfrac{x}{1+x}} + 1$

$(g \circ f)(x) = g(f(x)) = \dfrac{\sqrt{x} + 1}{1 + (\sqrt{x} + 1)} = \dfrac{\sqrt{x} + 1}{\sqrt{x} + 2}$

2. a. The difference between private and government health-care spending per person at any time t is given by the function d with the rule

$$d(t) = f(t) - g(t) = (2.5t^2 + 31.3t + 406)$$
$$- (1.4t^2 + 29.6t + 251)$$
$$= 1.1t^2 + 1.7t + 155$$

b. The difference between private and government expenditures per person at the beginning of 1985 is given by

$$d(10) = 1.1(10)^2 + 1.7(10) + 155$$

or \$282/person.

The difference between private and government expenditures per person at the beginning of 1993 is given by

$$d(18) = 1.1(18)^2 + 1.7(18) + 155$$

or \$542/person.

2.3 Functions and Mathematical Models

MATHEMATICAL MODELS

Before we can apply mathematics to solving real-world problems, we must be able to formulate those problems in the language of mathematics. This process is referred to as **mathematical modeling.** A *mathematical model* may describe precisely the problem under consideration, or, more likely than not, it may provide only an acceptable approximation of the problem. For example, the accumulated amount A at the end of t years when a sum of P dollars is deposited in a fixed bank account, earning interest at the rate of r percent per year compounded m times a year, is given *exactly* by the formula (model)

$$A = P\left(1 + \frac{r}{m}\right)^{mt}$$

However, the size of a cancer tumor may be *approximated* by the volume of a sphere,

$$V = \frac{4}{3}\pi r^3$$

where r is the radius of the tumor in centimeters.

The many techniques used in constructing mathematical models of practical problems range from theoretical consideration of the problem on the one extreme to an interpretation of data associated with the problem on the other. The model for the accumulated amount of a fixed bank account mentioned earlier may be derived theoretically (see Chapter 5). Later we will see how linear equations (models) can be constructed from a given set of data points.

In calculus we are concerned primarily with how one (dependent) variable depends on one or more (independent) variables. Consequently, most of our mathematical models will involve functions of one or more variables.* Once a function has been constructed to describe a specific real-world problem, a host of questions pertaining to the problem may be answered by analyzing the function (mathematical model). For example, if we have a function that gives the population of a certain culture of bacteria at any time t, then we can determine how fast the population is increasing or decreasing at any time t, and so on. Conversely, if we have a model that gives the rate of change of the cost of producing a certain item as a function of the level of production and if we know the fixed cost incurred in producing this item, then we can find the total cost incurred in producing a certain number of those items.

Before going on, let us look at two mathematical models. The first one is used to estimate spending by business on computer security, and the second is used to project the growth of the number of people enrolled in health maintenance organizations (HMOs). These models are derived from data using the least-squares technique. In the Using Technology section on page 105, you can see how mathematical models are constructed from raw data.

EXAMPLE 1

The estimated spending (in billions of dollars) by businesses on computer security equipment and services from 1987 to 1993 is given in the following table. The figures include spending for protection against computer criminals who steal, erase, or alter data, along with protection against fires, electrical failures, and natural disasters.

Year	1987	1988	1989	1990	1991	1992	1993
Spending	0.49	0.59	0.66	0.73	0.81	0.93	1.02

A mathematical model approximating the amount of spending over the period in question is given by

$$S(t) = 0.0864t + 0.4879$$

where t is measured in years, with $t = 0$ corresponding to 1987.

a. Sketch the graph of the function S and the given data on the same set of axes.
b. Assuming that this trend continued, what was the spending by business on computer security equipment and services in 1995 ($t = 8$).
c. What is the rate of increase of the annual expenditure over the period in question?

Source: Frost & Sullivan, Inc.

---

* Functions of more than one variable will be studied later.

FIGURE 2.14
Estimated spending by businesses on computer security equipment and services

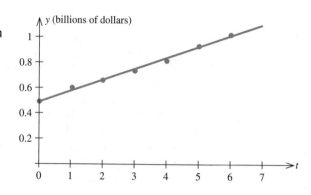

SOLUTION ✔

a. The graph of S is shown in Figure 2.14.
b. The estimated spending in 1995 is

$$S(8) = 0.0864(8) + 0.4879$$
$$\approx 1.1791$$

or approximately $1.18 billion.

c. The function S is linear, and so we conclude that the annual increase in the expenditure is given by the slope of the straight line represented by S, which is approximately $0.09 billion per year.   ■ ■ ■ ■

 EXAMPLE 2

The number of people (in millions) enrolled in HMOs from 1988 to 1998 is given in the following table.

Year	1988	1990	1992	1994	1996	1998
No. of People	32.7	36.5	41.4	51.1	66.5	78.0

A mathematical model approximating the number of people, $N(t)$, enrolled in HMOs during this period is

$$N(t) = -0.0258t^3 + 0.7465t^2 - 0.3491t + 33.1444 \qquad (0 \le t \le 10)$$

where t is measured in years and $t = 0$ corresponds to 1988.

a. Sketch the graph of the function N to see how the model compares with the actual data.
b. Assume that this trend continues and use the model to predict how many people will be enrolled in HMOs at the beginning of 2002.

SOLUTION ✔

a. The graph of the function N is shown in Figure 2.15.
b. The number of people that will be enrolled in HMOs at the beginning of 2002 is given by

$$N(14) = -0.0258(14)^3 + 0.7465(14)^2 - 0.3481(14) + 33.1444$$
$$= 103.7898$$

or approximately 103.8 million people.

FIGURE 2.15
The graph of $y = N(t)$ approximates the number of people enrolled in HMOs from 1988 to 1998.

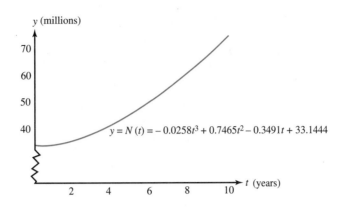

POLYNOMIAL FUNCTIONS

We begin by recalling a special class of functions, polynomial functions.

Polynomial Function

> A **polynomial function** of degree n is a function of the form
>
> $$f(x) = a_0x^n + a_1x^{n-1} + \cdots + a_{n-1}x + a_n \qquad (a_0 \neq 0)$$
>
> where $a_0, a_1, \ldots, a_n$ are constants and n is a nonnegative integer.

For example, the functions

$$f(x) = 4x^5 - 3x^4 + x^2 - x + 8$$
$$g(x) = 0.001x^3 - 2x^2 + 20x + 400$$

are polynomial functions of degrees 5 and 3, respectively. Observe that a polynomial function is defined everywhere so that it has domain $(-\infty, \infty)$.

A polynomial function of degree 1 $(n = 1)$

$$f(x) = a_0x + a_1 \qquad (a_0 \neq 0)$$

is the equation of a straight line in the slope-intercept form with slope $m = a_0$ and y-intercept $b = a_1$ (see Section 1.4). For this reason, a polynomial function of degree 1 is called a **linear function.** For example, the linear function $f(x) = 2x + 3$ may be written as a linear equation in x and y—namely, $y = 2x + 3$ or $2x - y + 3 = 0$. Conversely, the linear equation $2x - 3y + 4 = 0$ can be solved for y in terms of x to yield the linear function $y = f(x) = \frac{2}{3}x + \frac{4}{3}$.

A polynomial function of degree 2 is referred to as a **quadratic function.** A polynomial function of degree 3 is called a **cubic function,** and so on. The mathematical model in Example 2 involves a cubic function.

RATIONAL AND POWER FUNCTIONS

Another important class of functions is rational functions. A **rational function** is simply the quotient of two polynomials. Examples of rational functions are

$$F(x) = \frac{3x^3 + x^2 - x + 1}{x - 2}$$

$$G(x) = \frac{x^2 + 1}{x^2 - 1}$$

In general, a rational function has the form

$$R(x) = \frac{f(x)}{g(x)}$$

where $f(x)$ and $g(x)$ are polynomial functions. Since division by zero is not allowed, we conclude that the domain of a rational function is the set of all real numbers except the zeros of g—that is, the roots of the equation $g(x) = 0$. Thus, the domain of the function F is the set of all numbers except $x = 2$, whereas the domain of the function G is the set of all numbers except those that satisfy $x^2 - 1 = 0$ or $x = \pm 1$.

Functions of the form

$$f(x) = x^r$$

where r is any real number, are called **power functions.** We encountered examples of power functions earlier in our work. For example, the functions

$$f(x) = \sqrt{x} = x^{1/2} \quad \text{and} \quad g(x) = \frac{1}{x^2} = x^{-2}$$

are power functions.

Many of the functions we will encounter later will involve combinations of the functions introduced here. For example, the following functions may be viewed as suitable combinations of such functions:

$$f(x) = \sqrt{\frac{1 - x^2}{1 + x^2}}$$

$$g(x) = \sqrt{x^2 - 3x + 4}$$

$$h(x) = (1 + 2x)^{1/2} + \frac{1}{(x^2 + 2)^{3/2}}$$

As with polynomials of degree 3 or greater, analyzing the properties of these functions is facilitated by using the tools of calculus, to be developed later.

EXAMPLE 3

A study of driving costs based on 1992 model compact (six-cylinder) cars found that the average cost (car payments, gas, insurance, upkeep, and depreciation), measured in cents per mile, is approximated by the function

$$C(x) = \frac{2095}{x^{2.2}} + 20.08$$

where x (in thousands) denotes the number of miles the car is driven in 1 year. Using this model, estimate the average cost of driving a compact car 10,000 miles a year and 20,000 miles a year.

Source: Runzheimer International Study

SOLUTION ✔

The average cost of driving a compact car 10,000 miles a year is given by

$$C(10) = \frac{2095}{10^{2.2}} + 20.08$$
$$\approx 33.3$$

or approximately 33 cents per mile. The average cost of driving 20,000 miles a year is given by

$$C(20) = \frac{2095}{20^{2.2}} + 20.08$$
$$\approx 23.0$$

or approximately 23 cents per mile. ■ ■ ■ ■

SOME ECONOMIC MODELS

In the remainder of this section, we look at some economic models.

In a free market economy, consumer demand for a particular commodity depends on the commodity's unit price. A **demand equation** expresses the relationship between the unit price and the quantity demanded. The graph of the demand equation is called a **demand curve.** In general, the quantity demanded of a commodity decreases as the commodity's unit price increases, and vice versa. Accordingly, a demand function defined by $p = f(x)$, where p measures the unit price and x measures the number of units of the commodity in question, is generally characterized as a decreasing function of x; that is, $p = f(x)$ decreases as x increases. Since both x and p assume only nonnegative values, the demand curve is that part of the graph of $f(x)$ that lies in the first quadrant (Figure 2.16).

In a competitive market a relationship also exists between the unit price of a commodity and the commodity's availability in the market. In general, an increase in the commodity's unit price induces the producer to increase the supply of the commodity. Conversely, a decrease in the unit price generally leads to a drop in the supply. The equation that expresses the relation between

FIGURE 2.16
A demand curve

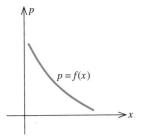

FIGURE 2.17
A supply curve

FIGURE 2.18
Market equilibrium corresponds to (x_0, p_0), the point at which the supply and demand curves intersect.

the unit price and the quantity supplied is called a **supply equation,** and its graph is called a **supply curve.** A supply function defined by $p = f(x)$ is generally characterized as an increasing function of x; that is, $p = f(x)$ increases as x increases. Since both x and p assume only nonnegative values, the supply curve is that part of the graph of $f(x)$ that lies in the first quadrant (Figure 2.17).

Under pure competition, the price of a commodity will eventually settle at a level dictated by the following condition: that the supply of the commodity be equal to the demand for it. If the price is too high, the consumer will not buy, and if the price is too low, the supplier will not produce. **Market equilibrium** prevails when the quantity produced is equal to the quantity demanded. The quantity produced at market equilibrium is called the **equilibrium quantity,** and the corresponding price is called the **equilibrium price.**

Market equilibrium corresponds to the point at which the demand curve and the supply curve intersect. In Figure 2.18 x_0 represents the equilibrium quantity and p_0 the equilibrium price. The point (x_0, p_0) lies on the supply curve and therefore satisfies the supply equation. At the same time it also lies on the demand curve and therefore satisfies the demand equation. Thus, to find the point (x_0, p_0), and hence the equilibrium quantity and price, we solve the demand and supply equations simultaneously for x and p. For meaningful solutions, x and p must both be positive.

EXAMPLE 4   The demand function for a certain brand of videocassette is given by

$$p = d(x) = -0.01x^2 - 0.2x + 8$$

and the corresponding supply function is given by

$$p = s(x) = 0.01x^2 + 0.1x + 3$$

where p is expressed in dollars and x is measured in units of a thousand. Find the equilibrium quantity and price.

SOLUTION ✔   We solve the following system of equations:

$$p = -0.01x^2 - 0.2x + 8$$
$$p = \ \ \ 0.01x^2 + 0.1x + 3$$

FIGURE 2.19
The supply curve and the demand curve intersect at the point (10, 5).

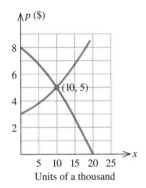

p ($)

(10, 5)

x

5 10 15 20 25
Units of a thousand

Substituting the first equation into the second yields

$$-0.01x^2 - 0.2x + 8 = 0.01x^2 + 0.1x + 3$$

which is equivalent to

$$0.02x^2 + 0.3x - 5 = 0$$
$$2x^2 + 30x - 500 = 0$$
$$x^2 + 15x - 250 = 0$$
$$(x + 25)(x - 10) = 0$$

Thus, $x = -25$ or $x = 10$. Since x must be nonnegative, the root $x = -25$ is rejected. Therefore, the equilibrium quantity is 10,000 videocassettes. The equilibrium price is given by

$$p = 0.01(10)^2 + 0.1(10) + 3 = 5$$

or $5 per videocassette (Figure 2.19).   ■ ■ ■ ■

Exploring with Technology

1. **a.** Use a graphing utility to plot the straight lines L_1 and L_2 with equations $y = 2x - 1$ and $y = 2.1x + 3$, respectively, on the same set of axes using the standard viewing rectangle. Do the lines appear to intersect?
   b. Plot the straight lines L_1 and L_2 using the viewing rectangle $[-100, 100] \times [-100, 100]$. Do the lines appear to intersect? Can you find the point of intersection using TRACE and ZOOM? Using the "intersection" function of your graphing utility?
   c. Find the point of intersection of L_1 and L_2 algebraically.
   d. Comment on the effectiveness of the methods of solutions in parts (b) and (c).
2. **a.** Use a graphing utility to plot the straight lines L_1 and L_2 with equations $y = 3x - 2$ and $y = -2x + 3$, respectively, on the same set of axes using the standard viewing rectangle. Then use TRACE and ZOOM to find the point of intersection of L_1 and L_2. Repeat using the "intersection" function of your graphing utility.
   b. Find the point of intersection of L_1 and L_2 algebraically.
   c. Comment on the effectiveness of the methods.

CONSTRUCTING MATHEMATICAL MODELS

We close this section by showing how some mathematical models can be constructed using elementary geometric and algebraic arguments.

EXAMPLE 5   The owner of the Rancho Los Feliz has 3000 yards of fencing material with which to enclose a rectangular piece of grazing land along the straight portion of a river. Fencing is not required along the river. Letting x denote the width of the rectangle, find a function f in the variable x giving the area of the grazing land if she uses all of the fencing material (Figure 2.20).

FIGURE 2.20
The rectangular grazing land has width x and length y.

SOLUTION ✔ The area of the rectangular grazing land is $A = xy$. Next, observe that the amount of fencing is $2x + y$ and this must be equal to 3000 since all the fencing material is used; that is,

$$2x + y = 3000$$

From the equation we see that $y = 3000 - 2x$. Substituting this value of y into the expression for A gives

$$A = xy = x(3000 - 2x) = 3000x - 2x^2$$

Finally, observe that both x and y must be nonnegative since they represent the width and length of a rectangle, respectively. Thus, $x \geq 0$ and $y \geq 0$. But the latter is equivalent to $3000 - 2x \geq 0$, or $x \leq 1500$. So the required function is $f(x) = 3000x - 2x^2$ with domain $0 \leq x \leq 1500$.   ■ ■ ■ ■

REMARK Observe that if we view the function $f(x) = 3000x - 2x^2$ strictly as a mathematical entity, then its domain is the set of all real numbers. But physical consideration dictates that its domain should be restricted to the interval [0, 1500].   ■ ■ ■

EXAMPLE 6 If exactly 200 people sign up for a charter flight, the Leisure World Travel Agency charges $300 per person. However, if more than 200 people sign up for the flight (assume this is the case), then each fare is reduced by $1 for each additional person. Letting x denote the number of passengers above 200, find a function giving the revenue realized by the company.

SOLUTION ✔ If there are x passengers above 200, then the number of passengers signing up for the flight is $200 + x$. Furthermore, the fare will be $\$(300 - x)$ per passenger. Therefore, the revenue will be

$$R = (200 + x)(300 - x) \qquad \text{(Number of passengers times the fare}$$
$$= -x^2 + 100x + 60{,}000 \qquad \text{per passenger)}$$

Clearly, x must be nonnegative, and $300 - x \geq 0$, or $x \leq 300$. So the required function is $f(x) = -x^2 + 100x + 60{,}000$ with domain [0, 300].   ■ ■ ■ ■

SELF-CHECK EXERCISES 2.3

1. Thomas Young has suggested the following rule for calculating the dosage of medicine for children from ages 1 to 12 years. If a denotes the adult dosage (in milligrams) and t is the age of the child (in years), then the child's dosage is given by

$$D(t) = \frac{at}{t + 12}$$

If the adult dose of a substance is 500 mg, how much should a 4-yr-old child receive?

2. The demand function for Mrs. Baker's cookies is given by

$$d(x) = -\frac{2}{15}x + 4$$

where $d(x)$ is the wholesale price in dollars per pound and x is the quantity demanded each week, measured in thousands of pounds. The supply function for the cookies is given by

$$s(x) = \frac{1}{75}x^2 + \frac{1}{10}x + \frac{3}{2}$$

where $s(x)$ is the wholesale price in dollars per pound and x is the quantity, in thousands of pounds, that will be made available in the market per week by the supplier.

a. Sketch the graphs of the functions d and s.
b. Find the equilibrium quantity and price.

Solutions to Self-Check Exercises 2.3 can be found on page 108.

2.3 Exercises

In Exercises 1–8, determine whether the equation defines y as a linear function of x. If so, write it in the form y = mx + b.

1. $2x + 3y = 6$

2. $-2x + 4y = 7$

3. $x = 2y - 4$

4. $2x = 3y + 8$

5. $2x - 4y + 9 = 0$

6. $3x - 6y + 7 = 0$

7. $2x^2 - 8y + 4 = 0$

8. $3\sqrt{x} + 4y = 0$

In Exercises 9–14, determine whether the given function is a polynomial function, a rational function, or some other function. State the degree of each polynomial function.

9. $f(x) = 3x^6 - 2x^2 + 1$

10. $f(x) = \frac{x^2 - 9}{x - 3}$

11. $G(x) = 2(x^2 - 3)^3$

12. $H(x) = 2x^{-3} + 5x^{-2} + 6$

13. $f(t) = 2t^2 + 3\sqrt{t}$

14. $f(r) = \frac{6r}{(r^3 - 8)}$

15. Find the constants m and b in the linear function $f(x) = mx + b$ so that $f(0) = 2$ and $f(3) = -1$.

16. Find the constants m and b in the linear function $f(x) = mx + b$ so that $f(2) = 4$ and the straight line represented by f has slope -1.

17. A manufacturer has a monthly fixed cost of $40,000 and a production cost of $8 for each unit produced. The product sells for $12/unit.
a. What is the cost function?
b. What is the revenue function?
c. What is the profit function?
d. Compute the profit (loss) corresponding to production levels of 8000 and 12,000 units.

18. A manufacturer has a monthly fixed cost of $100,000 and a production cost of $14 for each unit produced. The product sells for $20/unit.
    a. What is the cost function?
    b. What is the revenue function?
    c. What is the profit function?
    d. Compute the profit (loss) corresponding to production levels of 12,000 and 20,000 units.

19. **DISPOSABLE INCOME** Economists define the *disposable annual income* for an individual by the equation $D = (1 - r)T$, where T is the individual's total income and r is the net rate at which he or she is taxed. What is the disposable income for an individual whose income is $40,000 and whose net tax rate is 28%?

20. **DRUG DOSAGES** A method sometimes used by pediatricians to calculate the dosage of medicine for children is based on the child's surface area. If a denotes the adult dosage (in milligrams) and S is the surface area of the child (in square meters), then the child's dosage is given by

$$D(S) = \frac{Sa}{1.7}$$

If the adult dose of a substance is 500 mg, how much should a child whose surface area is 0.4 m² receive?

21. **COWLING'S RULE** Cowling's rule is a method for calculating pediatric drug dosages. If a denotes the adult dosage (in milligrams) and t is the age of the child (in years), then the child's dosage is given by

$$D(t) = \left(\frac{t + 1}{24}\right)a$$

If the adult dose of a substance is 500 mg, how much should a 4-yr-old child receive?

22. **WORKER EFFICIENCY** An efficiency study showed that the average worker at Delphi Electronics assembled cordless telephones at the rate of

$$f(t) = -\frac{3}{2}t^2 + 6t + 10 \qquad (0 \le t \le 4)$$

phones/hour, t hr after starting work during the morning shift. At what rate does the average worker assemble telephones 2 hr after starting work?

23. **REVENUE FUNCTIONS** The revenue (in dollars) realized by Apollo, Inc., from the sale of its inkjet printers is given by

$$R(x) = -0.1x^2 + 500x$$

where x denotes the number of units manufactured per month. What is Apollo's revenue when 1000 units are produced?

24. **EFFECT OF ADVERTISING ON SALES** The quarterly profit of Cunningham Realty depends on the amount of money x spent on advertising per quarter according to the rule

$$P(x) = -\frac{1}{8}x^2 + 7x + 30 \qquad (0 \le x \le 50)$$

where $P(x)$ and x are measured in thousands of dollars. What is Cunningham's profit when its quarterly advertising budget is $28,000?

25. **E-MAIL USAGE** The number of international e-mailings per day (in millions) is approximated by the function

$$f(t) = 38.57t^2 - 24.29t + 79.14 \qquad (0 \le t \le 4)$$

where t is measured in years with $t = 0$ corresponding to the beginning of 1998.
    a. Sketch the graph of f.
    b. How many international e-mailings per day were there at the beginning of the year 2000?
    Source: Pioneer Consulting

26. **DOCUMENT MANAGEMENT** The size (measured in millions of dollars) of the document-management business is described by the function

$$f(t) = 0.22t^2 + 1.4t + 3.77 \qquad (0 \le t \le 6)$$

where t is measured in years with $t = 0$ corresponding to the beginning of 1996.
    a. Sketch the graph of f.
    b. What was the size of the document-management business at the beginning of the year 2000?
    Source: Sun Trust Equitable Securities

27. **REACTION OF A FROG TO A DRUG** Experiments conducted by A. J. Clark suggest that the response $R(x)$ of a frog's heart muscle to the injection of x units of acetylcholine (as a percentage of the maximum possible effect of the drug) may be approximated by the rational function

$$R(x) = \frac{100x}{b + x} \qquad (x \ge 0)$$

where b is a positive constant that depends on the particular frog.
    a. If a concentration of 40 units of acetylcholine produces a response of 50% for a certain frog, find the "response function" for this frog.
    b. Using the model found in part (a), find the response of the frog's heart muscle when 60 units of acetylcholine are administered.

28. **FORECASTING SALES** The annual sales of the Crimson Drug Store are expected to be given by

$$S(t) = 2.3 + 0.4t$$

million dollars t yr from now, whereas the annual sales of the Cambridge Drug Store are expected to be given by

$$S(t) = 1.2 + 0.6t$$

million dollars t yr from now. When will the annual sales of the Cambridge Drug Store first surpass the annual sales of the Crimson Drug Store?

29. **CRICKET CHIRPING AND TEMPERATURE** Entomologists have discovered that a linear relationship exists between the number of chirps of crickets of a certain species and the air temperature. When the temperature is 70°F, the crickets chirp at the rate of 120 times/minute, and when the temperature is 80°F, they chirp at the rate of 160 times/minute.
a. Find an equation giving the relationship between the air temperature T and the number of chirps per minute, N, of the crickets.
b. Find N as a function of T and use this formula to determine the rate at which the crickets chirp when the temperature is 102°F.

30. **LINEAR DEPRECIATION** In computing income tax, business firms are allowed by law to depreciate certain assets such as buildings, machines, furniture, automobiles, and so on, over a period of time. The linear depreciation, or straight-line method, is often used for this purpose. Suppose an asset has an initial value of $\$C$ and is to be depreciated linearly over n years with a scrap value of $\$S$. Show that the book value of the asset at any time t ($0 \le t \le n$) is given by the linear function

$$V(t) = C - \frac{(C - S)}{n} t$$

Hint: Find an equation of the straight line that passes through the points $(0, C)$ and (n, S). Then rewrite the equation in the slope-intercept form.

31. **LINEAR DEPRECIATION** Using the linear depreciation model of Exercise 30, find the book value of a printing machine at the end of the second year if its initial value is $100,000 and it is depreciated linearly over 5 years with a scrap value of $30,000.

32. **PRICE OF IVORY** According to the World Wildlife Fund, a group in the forefront of the fight against illegal ivory trade, the price of ivory (in dollars per kilo) compiled from a variety of legal and black market sources is approximated by the function

$$f(t) = \begin{cases} 8.37t + 7.44 & \text{if } 0 \le t \le 8 \\ 2.84t + 51.68 & \text{if } 8 < t \le 30 \end{cases}$$

where t is measured in years and $t = 0$ corresponds to the beginning of 1970.

a. Sketch the graph of the function f.
b. What was the price of ivory at the beginning of 1970? At the beginning of 1990?

33. **SALES OF DIGITAL TVs** The number of homes with digital TVs is expected to grow according to the function

$$f(t) = 0.1714t^2 + 0.6657t + 0.7143 \qquad (0 \le t \le 6)$$

where t is measured in years with $t = 0$ corresponding to the beginning of the year 2000 and $f(t)$ is measured in millions of homes.
a. How many homes had digital TVs at the beginning of the year 2000?
b. How many homes will have digital TVs at the beginning of 2005?
Source: Consumer Electronics Manufacturers Association

34. **SENIOR CITIZENS' HEALTH CARE** According to a study conducted for the Senate Select Committee on Aging, the out-of-pocket cost to senior citizens for health care, $f(t)$ (as a percentage of income), in year t where $t = 0$ corresponds to 1977, is given by

$$y = \begin{cases} \frac{2}{7}t + 12 & \text{if } 0 \le t \le 7 \\ t + 7 & \text{if } 7 < t \le 10 \\ \frac{3}{5}t + 11 & \text{if } 10 < t < 20 \end{cases}$$

a. Sketch the graph of f.
b. What was the out-of-pocket cost to senior citizens for health care in 1982? In 1992?
Source: Senate Select Committee on Aging, AARP

35. **PRICE OF AUTOMOBILE PARTS** For years, automobile manufacturers had a monopoly on the replacement-parts market, particularly for sheet metal parts such as fenders, doors, and hoods, the parts most often damaged in a crash. Beginning in the late 1970s, however, competition appeared on the scene. In a report conducted by an insurance company to study the effects of the competition, the price of an OEM (original equipment manufacturer) fender for a particular 1983 model car was found to be

$$f(t) = \frac{110}{\frac{1}{2}t + 1} \qquad (0 \le t \le 2)$$

where $f(t)$ is measured in dollars and t is in years. Over the same period of time, the price of a non-OEM fender for the car was found to be

$$g(t) = 26\left(\frac{1}{4}t^2 - 1\right)^2 + 52 \qquad (0 \le t \le 2)$$

where $g(t)$ is also measured in dollars. Find a function $h(t)$ that gives the difference in price between an OEM fender and a non-OEM fender. Compute $h(0), h(1)$, and $h(2)$. What does the result of your computation seem to say about the price gap between OEM and non-OEM fenders over the 2 yr?

For the demand equations in Exercises 36–39, where x represents the quantity demanded in units of a thousand and p is the unit price in dollars, (a) sketch the demand curve and (b) determine the quantity demanded when the unit price is set at $p.

36. $p = -x^2 + 36; p = 11$ **37.** $p = -x^2 + 16; p = 7$

38. $p = \sqrt{9 - x^2}; p = 2$ **39.** $p = \sqrt{18 - x^2}; p = 3$

For the supply equations in Exercises 40–43, where x is the quantity supplied in units of a thousand and p is the unit price in dollars, (a) sketch the supply curve and (b) determine the price at which the supplier will make 2000 units of the commodity available in the market.

40. $p = 2x^2 + 18$ **41.** $p = x^2 + 16x + 40$

42. $p = x^3 + x + 10$ **43.** $p = x^3 + 2x + 3$

44. DEMAND FOR CLOCK RADIOS   In the accompanying figure, L_1 is the demand curve for the model A clock radios manufactured by Ace Radio, Inc., and L_2 is the demand curve for their model B clock radios. Which line has the greater slope? Interpret your results.

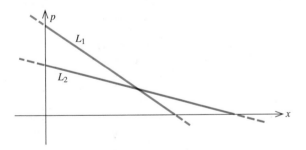

45. SUPPLY OF CLOCK RADIOS   In the accompanying figure, L_1 is the supply curve for the model A clock radios manufactured by Ace Radio, Inc., and L_2 is the supply curve for their model B clock radios. Which line has the greater slope? Interpret your results.

46. DEMAND FOR SMOKE ALARMS   The demand function for the Sentinel smoke alarm is given by

$$p = \frac{30}{0.02x^2 + 1} \qquad (0 \le x \le 10)$$

where x (measured in units of a thousand) is the quantity demanded per week and p is the unit price in dollars. Sketch the graph of the demand function. What is the unit price that corresponds to a quantity demanded of 10,000 units?

47. DEMAND FOR COMMODITIES   Assume that the demand function for a certain commodity has the form

$$p = \sqrt{-ax^2 + b} \qquad (a \ge 0, b \ge 0)$$

where x is the quantity demanded, measured in units of a thousand, and p is the unit price in dollars. Suppose the quantity demanded is 6000 ($x = 6$) when the unit price is $8 and 8000 ($x = 8$) when the unit price is $6. Determine the demand equation. What is the quantity demanded when the unit price is set at $7.50?

48. SUPPLY FUNCTIONS   The supply function for the Luminar desk lamp is given by

$$p = 0.1x^2 + 0.5x + 15$$

where x is the quantity supplied (in thousands) and p is the unit price in dollars. Sketch the graph of the supply function. What unit price will induce the supplier to make 5000 lamps available in the marketplace?

49. SUPPLY FUNCTIONS   Suppliers of transistor radios will market 10,000 units when the unit price is $20 and 62,500 units when the unit price is $35. Determine the supply function if it is known to have the form

$$p = a\sqrt{x} + b \qquad (a \ge 0, b \ge 0)$$

where x is the quantity supplied and p is the unit price in dollars. Sketch the graph of the supply function. What unit price will induce the supplier to make 40,000 transistor radios available in the marketplace?

50. Suppose the demand and supply equations for a certain commodity are given by $p = ax + b$ and $p = cx + d$, respectively, where $a < 0$, $c > 0$, and $b > d > 0$ (see the accompanying figure).

a. Find the equilibrium quantity and equilibrium price in terms of a, b, c, and d.

b. Use part (a) to determine what happens to the market equilibrium if c is increased while a, b, and d remain fixed. Interpret your answer in economic terms.

c. Use part (a) to determine what happens to the market equilibrium if b is decreased while a, c, and d remain fixed. Interpret your answer in economic terms.

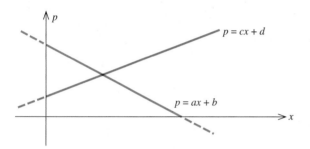

For each pair of supply and demand equations in Exercises 51–54, where x represents the quantity demanded in units of a thousand and p the unit price in dollars, find the equilibrium quantity and the equilibrium price.

51. $p = -2x^2 + 80$ and $p = 15x + 30$

52. $p = -x^2 - 2x + 100$ and $p = 8x + 25$

53. $11p + 3x - 66 = 0$ and $2p^2 + p - x = 10$

54. $p = 60 - 2x^2$ and $p = x^2 + 9x + 30$

55. MARKET EQUILIBRIUM The weekly demand and supply functions for Sportsman 5×7 tents are given by

$$p = -0.1x^2 - x + 40$$
$$p = 0.1x^2 + 2x + 20$$

respectively, where p is measured in dollars and x is measured in units of a hundred. Find the equilibrium quantity and price.

56. MARKET EQUILIBRIUM The management of the Titan Tire Company has determined that the weekly demand and supply functions for their Super Titan tires are given by

$$p = 144 - x^2$$
$$p = 48 + \frac{1}{2}x^2$$

respectively, where p is measured in dollars and x is measured in units of a thousand. Find the equilibrium quantity and price.

57. WALKING VERSUS RUNNING The oxygen consumption (in milliliter/pound/minute) for a person walking at x mph is approximated by the function

$$f(x) = \frac{5}{3}x^2 + \frac{5}{3}x + 10 \qquad (0 \le x \le 9)$$

where the oxygen consumption for a runner at x mph is approximated by the function

$$g(x) = 11x + 10 \qquad (4 \le x \le 9)$$

a. Sketch the graphs of f and g.

b. At what speed is the oxygen consumption the same for a walker as it is for a runner? What is the level of oxygen consumption at that speed?

c. What happens to the oxygen consumption of the walker and the runner at speeds beyond that found in part (b)?

Source: Exercise Physiology, by William McArdley, Frank Katch, and Victor Katch

58. ENCLOSING AN AREA Patricia wishes to have a rectangular-shaped garden in her backyard. She has 80 ft of fencing material with which to enclose her garden. Letting x denote the width of the garden, find a function f in the variable x giving the area of the garden. What is its domain?

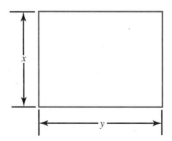

59. ENCLOSING AN AREA Patricia's neighbor, Juanita, also wishes to have a rectangular-shaped garden in her backyard. But Juanita wants her garden to have an area of 250 ft². Letting x denote the width of the garden, find a function f in the variable x giving the length of the fencing material required to construct the garden. What is the domain of the function?

Hint: Refer to the figure for Exercise 58. The amount of fencing material required is equal to the perimeter of the rectangle, which is twice the width plus twice the length of the rectangle.

(continued on p. 108)

FINDING THE POINTS OF INTERSECTION OF TWO GRAPHS AND MODELING

A graphing utility can be used to find the point(s) of intersection of the graphs of two functions.

EXAMPLE 1

Find the points of intersection of the graphs of

$$f(x) = 0.3x^2 - 1.4x - 3 \quad \text{and} \quad g(x) = -0.4x^2 + 0.8x + 6.4$$

SOLUTION ✔

The graphs of both f and g in the standard viewing rectangle are shown in Figure T1. Using **TRACE** and **ZOOM** or the function for finding the points of intersection of two graphs on your graphing utility, we find the point(s) of intersection, accurate to four decimal places, to be $(-2.4158, 2.1329)$ and $(5.5587, -1.5125)$.

FIGURE T1
The graphs of f and g in the standard viewing window

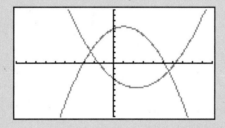

EXAMPLE 2

Consider the demand and supply functions

$$p = d(x) = -0.01x^2 - 0.2x + 8 \quad \text{and} \quad p = s(x) = 0.01x^2 + 0.1x + 3$$

of Example 4 in Section 2.3.

a. Plot the graphs of d and s in the viewing rectangle $[0, 15] \times [0, 10]$.
b. Verify that the equilibrium point is $(10, 5)$, as obtained in Example 4.

SOLUTION ✔

a. The graphs of d and s are shown in Figure T2.

FIGURE T2
The graphs of d and s in the window $[0, 15] \times [0, 10]$

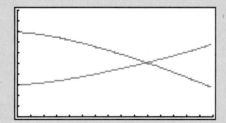

b. Using **TRACE** and **ZOOM** or the function for finding the point of intersection of two graphs, we see that $x = 10$ and $y = 5$, so the equilibrium point is $(10, 5)$, as obtained before.

CONSTRUCTING MATHEMATICAL MODELS FROM RAW DATA

A graphing utility can sometimes be used to construct mathematical models from sets of data. For example, if the points corresponding to the given data are scattered about a straight line, then one uses **LINR** (linear regression) from the **STAT CALC** (statistical calculation) menu of the graphing utility to obtain a function (model) that approximates the data at hand. If the points seem to be scattered along a parabola (the graph of a quadratic function), then one uses **P2REG** (second-order polynomial regression), and so on.

Details for using the items in the STAT CALC menu can be found at the Web site: http://www.brookscole.com/math/authors/tans/

EXAMPLE 3

Indian Gaming Industry

The following data gives the estimated gross revenues (in billions of dollars) from the Indian gaming industries from 1990 ($t = 0$) to 1997 ($t = 7$).

Year	0	1	2	3	4	5	6	7
Revenue	0.5	0.7	1.6	2.6	3.4	4.8	5.6	6.8

a. Use a graphing utility to find a polynomial function f of degree 4 that models the data.
b. Plot the graph of the function f, using the viewing rectangle $[0, 8] \times [0, 10]$.
c. Use the function evaluation capability of the graphing utility to compute $f(0), f(1), \ldots, f(7)$ and compare these values with the original data.

Source: Christiansen/Cummings Associates

SOLUTION ✔

a. Choosing **P4REG** (fourth-order polynomial regression) from the **STAT CALC** (statistical calculations) menu of a graphing utility, we find

$$f(t) = 0.00379t^4 - 0.06616t^3 + 0.41667t^2 - 0.07291t + 0.48333$$

b. The graph of f is shown in Figure T3.

FIGURE T3
The graph of f in the viewing rectangle $[0, 8] \times [0, 10]$

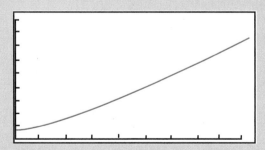

c. The required values, which compare favorably with the given data, follow:

t	0	1	2	3	4	5	6	7
$f(t)$	0.5	0.8	1.5	2.5	3.6	4.6	5.7	6.8

Exercises

In Exercises 1–6, find the points of intersection of the graphs of the functions. Express your answer accurate to four decimal places.

1. $f(x) = 1.2x + 3.8$; $g(x) = -0.4x^2 + 1.2x + 7.5$

2. $f(x) = 0.2x^2 - 1.3x - 3$; $g(x) = -1.3x + 2.8$

3. $f(x) = 0.3x^2 - 1.7x - 3.2$; $g(x) = -0.4x^2 + 0.9x + 6.7$

4. $f(x) = -0.3x^2 + 0.6x + 3.2$; $g(x) = 0.2x^2 - 1.2x - 4.8$

5. $f(x) = 0.3x^3 - 1.8x^2 + 2.1x - 2$; $g(x) = 2.1x - 4.2$

6. $f(x) = -0.2x^3 + 1.2x^2 - 1.2x + 2$; $g(x) = -0.2x^2 + 0.8x + 2.1$

7. The monthly demand and supply functions for a certain brand of wall clock are given by

$$p = -0.2x^2 - 1.2x + 50$$
$$p = 0.1x^2 + 3.2x + 25$$

respectively, where p is measured in dollars and x is measured in units of a hundred.
a. Plot the graphs of both functions in an appropriate viewing rectangle.
b. Find the equilibrium quantity and price.

8. The quantity demanded x (in units of a hundred) of Mikado miniature cameras per week is related to the unit price p (in dollars) by

$$p = -0.2x^2 + 80$$

The quantity x (in units of a hundred) that the supplier is willing to make available in the market is related to the unit price p (in dollars) by

$$p = 0.1x^2 + x + 40$$

a. Plot the graphs of both functions in an appropriate viewing rectangle.
b. Find the equilibrium quantity and price.

In Exercises 9–14, use the STAT CALC menu of a graphing utility to construct a mathematical model associated with given data.

9. SALES OF DIGITAL SIGNAL PROCESSORS The projected sales (in billions of dollars) of digital signal processors (DSPs) follow:

Year	1997	1998	1999	2000	2001	2002
Sales	3.1	4	5	6.2	8	10

a. Use P2REG to find a second-degree polynomial regression model for the data. Let $t = 0$ correspond to 1997.
b. Plot the graph of the function f found in part (a), using the viewing rectangle $[0, 5] \times [0, 12]$.
c. Compute the values of $f(t)$ for $t = 0, 1, 2, 3, 4$, and 5. How does your model compare with the given data?
Source: A. G. Edwards & Sons, Inc.

10. PRISON POPULATION The following data gives the past, present, and projected U.S. prison population (in millions) from 1980 through 2005.

Year	1980	1985	1990	1995	2000	2005
Population	0.52	0.77	1.18	1.64	2.23	3.20

a. Letting $t = 0$ correspond to the beginning of 1980 and supposing t is measured in 5-yr intervals, use P2REG to find a second-degree polynomial regression model based on the given data.
b. Plot the graph of the function f found in part (a), using the viewing rectangle $[0, 5] \times [0, 3.5]$.
c. Compute $f(0), f(1), f(2), f(3), f(4)$, and $f(5)$. Compare these values with the given data.

11. DIGITAL TV SHIPMENTS The estimated number of digital TV shipments between the year 2000 and 2006 (in millions of units) is given in the following table:

Year	2000	2001	2002	2003	2004	2005	2006
Units Shipped	0.63	1.43	2.57	4.1	6	8.1	10

a. Use **P3REG** to find a third-degree polynomial regression model for the data. Let $t = 0$ correspond to the year 2000.
b. Plot the graph of the function f found in part (a), using the viewing rectangle $[0, 6] \times [0, 11]$.
c. Compute the values of $f(t)$ for $t = 0, 1, 2, 3, 4, 5$, and 6.

Source: Consumer Electronics Manufacturers Association

12. **ON-LINE SHOPPING** The following data gives the revenue per year (in billions of dollars) from Internet shopping.

Year	1997	1998	1999	2000	2001
Revenue	2.4	5	8	12	17.4

a. Use **P3REG** to find a third-degree polynomial regression model for the data. Let $t = 0$ correspond to 1997.
b. Plot the graph of the function f found in part (a), using the viewing rectangle $[0, 4] \times [0, 20]$.
c. Compare the values of f at $t = 0, 1, 2, 3$, and 4 with the given data.

Source: Forrester Research, Inc.

13. **CABLE AD REVENUE** The past and projected revenues (in billions of dollars) from cable advertisement for the years 1995 through the year 2000 follow:

Year	1995	1996	1997	1998	1999	2000
Ad Revenue	5.1	6.6	8.1	9.4	11.1	13.7

a. Use **P3REG** to find a third-degree polynomial regression model for the data. Let $t = 0$ correspond to 1995.

b. Plot the graph of the function f found in part (a), using the viewing rectangle $[0, 6] \times [0, 14]$.
c. Compare the values of f at $t = 1, 2, 3, 4$, and 5 with the given data.

Source: National Cable Television Association

14. **ON-LINE SPENDING** The following data gives the world-wide spending and projected spending (in billions of dollars) on the Web from 1997 through 2002.

Year	1997	1998	1999	2000	2001	2002
Spending	5.0	10.5	20.5	37.5	60	95

a. Choose **P4REG** to find a fourth-degree polynomial regression model for the data. Let $t = 0$ correspond to 1997.
b. Plot the graph of the function f found in part (a), using the viewing rectangle $[0, 5] \times [0, 100]$.

Source: International Data Corporation

15. **MARIJUANA ARRESTS** The number of arrests (in thousands) for marijuana sales and possession in New York City from 1992 through 1997 is given below.

Year	1992	1993	1994	1995	1996	1997
No. of Arrests	5.0	5.8	8.8	11.7	18.5	27.5

a. Use **P4REG** to find a fourth-degree polynomial regression model for the data. Let $t = 0$ correspond to 1992.
b. Plot the graph of the function f found in part (a), using the viewing rectangle $[0, 5] \times [0, 30]$.
c. Compare the values of f at $t = 0, 1, 2, 3, 4$, and 5 with the given data.

Source: New York State Division of Criminal Justice Services

60. PACKAGING   By cutting away identical squares from each corner of a rectangular piece of cardboard and folding up the resulting flaps, an open box may be made. If the cardboard is 15 in. long and 8 in. wide and the square cutaways have dimensions of x in. by x in., find a function giving the volume of the resulting box.

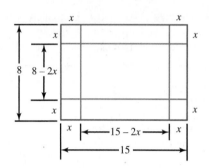

61. CONSTRUCTION COSTS   A rectangular box is to have a square base and a volume of 20 ft³. The material for the base costs 30 cents/ft², the material for the sides costs 10 cents/ft², and the material for the top costs 20 cents/ft². Letting x denote the length of one side of the base, find a function in the variable x giving the cost of constructing the box.

62. AREA OF A NORMAN WINDOW   A Norman window has the shape of a rectangle surmounted by a semicircle (see the accompanying figure). Suppose a Norman window is to have a perimeter of 28 ft; find a function in the variable x giving the area of the window.

63. YIELD OF AN APPLE ORCHARD   An apple orchard has an average yield of 36 bushels of apples/tree if tree density is 22 trees/acre. For each unit increase in tree density, the yield decreases by 2 bushels. Letting x denote the number of trees beyond 22/acre, find a function in x that gives the yield of apples.

In Exercises 64–67, determine whether the statement is true or false. If it is true, explain why it is true. If it is false, give an example to show why it is false.

64. A polynomial function is a sum of constant multiples of power functions.

65. A polynomial function is a rational function, but the converse is false.

66. If $r > 0$, then the power function $f(x) = x^r$ is defined for all nonnegative values of x.

67. The function $f(x) = 2^x$ is a power function.

SOLUTIONS TO SELF-CHECK EXERCISES 2.3

1. Since the adult dose of the substance is 500 mg, $a = 500$; thus, the rule in this case is

$$D(t) = \frac{500t}{t + 12}$$

A 4-yr-old should receive

$$D(4) = \frac{500(4)}{4 + 12}$$

or 125 mg of the substance.

2. a. The graphs of the functions d and s are shown in the following figure:

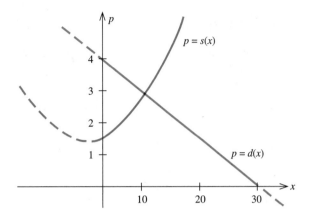

b. Solve the following system of equations:

$$p = -\frac{2}{15}x + 4$$

$$p = \frac{1}{75}x^2 + \frac{1}{10}x + \frac{3}{2}$$

Substituting the first equation into the second yields

$$\frac{1}{75}x^2 + \frac{1}{10}x + \frac{3}{2} = -\frac{2}{15}x + 4$$

$$\frac{1}{75}x^2 + \left(\frac{1}{10} + \frac{2}{15}\right)x - \frac{5}{2} = 0$$

$$\frac{1}{75}x^2 + \frac{7}{30}x - \frac{5}{2} = 0$$

Multiplying both sides of the last equation by 150, we have

$$2x^2 + 35x - 375 = 0$$
$$(2x - 15)(x + 25) = 0$$

Thus, $x = -25$ or $x = 15/2 = 7.5$. Since x must be nonnegative, we take $x = 7.5$, and the equilibrium quantity is 7500 pounds. The equilibrium price is given by

$$p = -\frac{2}{15}\left(\frac{15}{2}\right) + 4$$

or \$3/pound.

2.4 Limits

INTRODUCTION TO CALCULUS

Historically, the development of calculus by Isaac Newton (1642–1727) and Gottfried Wilhelm Leibniz (1646–1716) resulted from the investigation of the following problems:

1. Finding the tangent line to a curve at a given point on the curve (Figure 2.21a)
2. Finding the area of a planar region bounded by an arbitrary curve (Figure 2.21b)

FIGURE 2.21

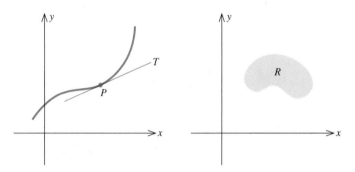

(a) What is the slope of the tangent line *T* at point *P*?

(b) What is the area of the region *R*?

The tangent-line problem might appear to be unrelated to any practical applications of mathematics, but as you will see later, the problem of finding the *rate of change* of one quantity with respect to another is mathematically equivalent to the geometric problem of finding the slope of the *tangent line* to a curve at a given point on the curve. It is precisely the discovery of the relationship between these two problems that spurred the development of calculus in the seventeenth century and made it such an indispensable tool for solving practical problems. The following are a few examples of such problems:

■ Finding the velocity of an object
■ Finding the rate of change of a bacteria population with respect to time
■ Finding the rate of change of a company's profit with respect to time
■ Finding the rate of change of a travel agency's revenue with respect to the agency's expenditure for advertising

The study of the tangent-line problem led to the creation of *differential calculus,* which relies on the concept of the *derivative* of a function. The study of the area problem led to the creation of *integral calculus,* which relies on the concept of the *antiderivative,* or *integral,* of a function. (The derivative of a function and the integral of function are intimately related, as you will see in Section 6.4.) Both the derivative of a function and the integral of a function are defined in terms of a more fundamental concept—the limit—our next topic.

A REAL-LIFE EXAMPLE

From data obtained in a test run conducted on a prototype of a maglev (magnetic levitation train), which moves along a straight monorail track, engineers have determined that the position of the maglev (in feet) from the origin at time t is given by

$$s = f(t) = 4t^2 \qquad (0 \le t \le 30) \tag{3}$$

where f is called the **position function** of the maglev. The position of the maglev at time $t = 0, 1, 2, 3, \ldots, 10$, measured from its initial position, is

$$f(0) = 0, \quad f(1) = 4, \quad f(2) = 16, \quad f(3) = 36, \ldots, \quad f(10) = 400$$

feet (Figure 2.22).

FIGURE 2.22
A maglev moving along an elevated monorail track

Suppose we want to find the velocity of the maglev at $t = 2$. This is just the velocity of the maglev as shown on its speedometer at that precise instant of time. Offhand, calculating this quantity using only Equation (3) appears to be an impossible task; but consider what quantities we *can* compute using this relationship. Obviously, we can compute the position of the maglev at any time t as we did earlier for some selected values of t. Using these values, we can then compute the *average velocity* of the maglev over an interval of time. For example, the average velocity of the train over the time interval [2, 4] is given by

$$\frac{\text{Distance covered}}{\text{Time elapsed}} = \frac{f(4) - f(2)}{4 - 2}$$
$$= \frac{4(4^2) - 4(2^2)}{2}$$
$$= \frac{64 - 16}{2} = 24$$

or 24 feet/second.

Although this is not quite the velocity of the maglev at $t = 2$, it does provide us with an approximation of its velocity at that time.

Can we do better? Intuitively, the smaller the time interval we pick (with $t = 2$ as the left end point), the better the average velocity over that time interval will approximate the actual velocity of the maglev at $t = 2$.*

Now, let's describe this process in general terms. Let $t > 2$. Then, the average velocity of the maglev over the time interval $[2, t]$ is given by

$$\frac{f(t) - f(2)}{t - 2} = \frac{4t^2 - 4(2^2)}{t - 2} = \frac{4(t^2 - 4)}{t - 2} \qquad \textbf{(4)}$$

By choosing the values of t closer and closer to 2, we obtain a sequence of numbers that gives the average velocities of the maglev over smaller and smaller time intervals. As we observed earlier, this sequence of numbers should approach the *instantaneous velocity* of the train at $t = 2$.

Let's try some sample calculations. Using Equation (4) and taking the sequence $t = 2.5, 2.1, 2.01, 2.001$, and 2.0001, which approaches 2, we find

The average velocity over $[2, 2.5]$ is $\dfrac{4(2.5^2 - 4)}{2.5 - 2} = 18$, or 18 feet/second

The average velocity over $[2, 2.1]$ is $\dfrac{4(2.1^2 - 4)}{2.1 - 2} = 16.4$, or 16.4 feet/second

and so forth. These results are summarized in Table 2.1.

Table 2.1

	t approaches 2 from the right.				
t	2.5	2.1	2.01	2.001	2.0001
Average Velocity over [2, t]	18	16.4	16.04	16.004	16.0004

Average velocity approaches 16 from the right.

From Table 2.1, we see that the average velocity of the maglev seems to approach the number 16 as it is computed over smaller and smaller time intervals. These computations suggest that the instantaneous velocity of the train at $t = 2$ is 16 feet/second.

REMARK Notice that we cannot obtain the instantaneous velocity for the maglev at $t = 2$ by substituting $t = 2$ into Equation (2) because this value of t is not in the domain of the average velocity function.   ■ ■ ■

INTUITIVE DEFINITION OF A LIMIT

Consider the function g defined by

$$g(t) = \frac{4(t^2 - 4)}{t - 2}$$

---

* Actually, any interval containing $t = 2$ will do.

which gives the average velocity of the maglev [see Equation (4)]. Suppose we are required to determine the value that $g(t)$ approaches as t approaches the (fixed) number 2. If we take the sequence of values of t approaching 2 from the right-hand side, as we did earlier, we see that $g(t)$ approaches the number 16. Similarly, if we take a sequence of values of t approaching 2 from the left, such as $t = 1.5$, 1.9, 1.999, and 1.9999, we obtain the results shown in Table 2.2.

Table 2.2

			t approaches 2 from the left.		
t	1.5	1.9	1.99	1.999	1.9999
$g(t)$	14	15.6	15.96	15.996	15.9996

$g(t)$ approaches 16 from the left.

Observe that $g(t)$ approaches the number 16 as t approaches 2—this time from the left-hand side. In other words, as t approaches 2 from *either* side of 2, $g(t)$ approaches 16. In this situation, we say that the limit of $g(t)$ as t approaches 2 is 16, written

$$\lim_{t \to 2} g(t) = \lim_{t \to 2} \frac{4(t^2 - 4)}{t - 2} = 16$$

The graph of the function g, shown in Figure 2.23, confirms this observation.

FIGURE 2.23

As t approaches $t = 2$ from either direction, $g(t)$ approaches $y = 16$.

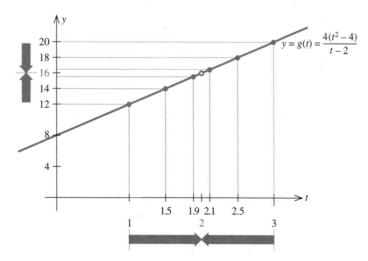

Observe that the point $t = 2$ is not in the domain of the function g [for this reason, the point (2, 16) is missing from the graph of g]. This, however, is inconsequential because the value, if any, of $g(t)$ at $t = 2$ plays no role in computing the limit.

This example leads to the following informal definition.

Limit of a Function

> The function f has the **limit** L as x approaches a, written
>
> $$\lim_{x \to a} f(x) = L$$
>
> if the value $f(x)$ can be made as close to the number L as we please by taking x sufficiently close to (but not equal to) a.

Exploring with Technology

1. Use a graphing utility to plot the graph of

$$g(x) = \frac{4(x^2 - 4)}{x - 2}$$

   in the viewing rectangle $[0, 3] \times [0, 20]$.
2. Use ZOOM and TRACE to describe what happens to the values of $f(x)$ as x approaches 2, first from the right and then from the left.
3. What happens to the y-value when x takes on the value 2? Explain.
4. Reconcile your results with those of the preceding example.

EVALUATING THE LIMIT OF A FUNCTION

Let us now consider some examples involving the computation of limits.

EXAMPLE 1   Let $f(x) = x^3$ and evaluate $\lim_{x \to 2} f(x)$.

SOLUTION ✔   The graph of f is shown in Figure 2.24. You can see that $f(x)$ can be made as close to the number 8 as we please by taking x sufficiently close to 2. Therefore,

$$\lim_{x \to 2} x^3 = 8$$

FIGURE 2.24
$f(x)$ is close to 8 whenever x is close to 2.

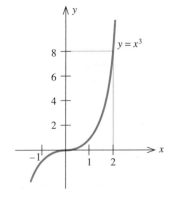

EXAMPLE 2

Let

$$g(x) = \begin{cases} x+2 & \text{if } x \neq 1 \\ 1 & \text{if } x = 1 \end{cases}$$

Evaluate $\lim\limits_{x \to 1} g(x)$.

SOLUTION ✔

The domain of g is the set of all real numbers. From the graph of g shown in Figure 2.25, we see that $g(x)$ can be made as close to 3 as we please by taking x sufficiently close to 1. Therefore,

$$\lim\limits_{x \to 1} g(x) = 3$$

FIGURE 2.25
$\lim\limits_{x \to 1} g(x) = 3$

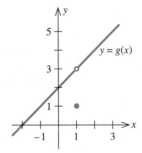

Observe that $g(1) = 1$, which is not equal to the limit of the function g as x approaches 1. [Once again, the value of $g(x)$ at $x = 1$ has no bearing on the existence or value of the limit of g as x approaches 1.] ■ ■ ■ ■

EXAMPLE 3

Evaluate the limit of the following functions as x approaches the indicated point.

a. $f(x) = \begin{cases} -1 & \text{if } x < 0 \\ 1 & \text{if } x \geq 0 \end{cases}; x = 0$ **b.** $g(x) = \dfrac{1}{x^2}; x = 0$

SOLUTION ✔

The graphs of the functions f and g are shown in Figure 2.26.

FIGURE 2.26

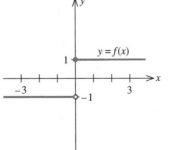

(a) $\lim\limits_{x \to 0} f(x)$ does not exist.   **(b)** $\lim\limits_{x \to 0} g(x)$ does not exist.

a. Referring to Figure 2.26a, we see that no matter how close x is to $x = 0$, $f(x)$ takes on the values 1 or -1, depending on whether x is positive or negative. Thus, there is no *single* real number L that $f(x)$ approaches as *x* approaches zero. We conclude that the limit of $f(x)$ does *not* exist as *x* approaches zero.

b. Referring to Figure 2.26b, we see that as *x* approaches $x = 0$ (from either side), $g(x)$ increases without bound and thus does not approach any specific real number. We conclude, accordingly, that the limit of $g(x)$ does *not* exist as *x* approaches zero.   ■■■■

Group Discussion

Consider the graph of the function *h* whose graph is depicted in the following figure:

It has the property that as *x* approaches 0 from either the right or the left, the curve oscillates more and more frequently between the lines $y = -1$ and $y = 1$.

1. Explain why $\lim_{x \to 0} h(x)$ does not exist.

2. Compare this function with those in Example 3. More specifically, discuss the different ways the functions fail to have a limit at $x = 0$.

Until now, we have relied on knowing the actual values of a function or the graph of a function near $x = a$ to help us evaluate the limit of the function $f(x)$ as *x* approaches *a*. The following properties of limits, which we list without proof, enable us to evaluate limits of functions algebraically.

THEOREM 1

Properties of Limits
Suppose

$$\lim_{x \to a} f(x) = L \qquad \text{and} \qquad \lim_{x \to a} g(x) = M$$

Then,

1. $\lim_{x \to a}[f(x)]^r = [\lim_{x \to a} f(x)]^r = L^r$ (*r*, a real number)

2. $\lim_{x \to a} cf(x) = c \lim_{x \to a} f(x) = cL$ (*c*, a real number)

3. $\lim_{x \to a}[f(x) \pm g(x)] = \lim_{x \to a} f(x) \pm \lim_{x \to a} g(x) = L \pm M$

4. $\lim_{x \to a}[f(x)g(x)] = [\lim_{x \to a} f(x)][\lim_{x \to a} g(x)] = LM$

5. $\lim_{x \to a} \dfrac{f(x)}{g(x)} = \dfrac{\lim_{x \to a} f(x)}{\lim_{x \to a} g(x)} = \dfrac{L}{M}$ (Provided $M \neq 0$)

EXAMPLE 4   Use Theorem 1 to evaluate the following limits.

a. $\lim\limits_{x \to 2} x^3$ **b.** $\lim\limits_{x \to 4} 5x^{3/2}$ **c.** $\lim\limits_{x \to 1} (5x^4 - 2)$

d. $\lim\limits_{x \to 3} 2x^3 \sqrt{x^2 + 7}$ **e.** $\lim\limits_{x \to 2} \dfrac{2x^2 + 1}{x + 1}$

SOLUTION ✔   **a.** $\lim\limits_{x \to 2} x^3 = [\lim\limits_{x \to 2} x]^3$ (Property 1)

$\qquad\qquad = 2^3 = 8 \qquad (\lim\limits_{x \to 2} x = 2)$

b. $\lim\limits_{x \to 4} 5x^{3/2} = 5[\lim\limits_{x \to 4} x^{3/2}] \qquad$ (Property 2)

$\qquad\qquad = 5(4)^{3/2} = 40$

c. $\lim\limits_{x \to 1}(5x^4 - 2) = \lim\limits_{x \to 1} 5x^4 - \lim\limits_{x \to 1} 2 \qquad$ (Property 3)

To evaluate $\lim\limits_{x \to 1} 2$, observe that the constant function $g(x) = 2$ has value 2 for all values of x. Therefore, $g(x)$ must approach the limit 2 as x approaches $x = 1$ (or any other point for that matter!). Therefore,

$$\lim\limits_{x \to 1}(5x^4 - 2) = 5(1)^4 - 2 = 3$$

d. $\lim\limits_{x \to 3} 2x^3 \sqrt{x^2 + 7} = 2 \lim\limits_{x \to 3} x^3 \sqrt{x^2 + 7} \qquad$ (Property 2)

$\qquad\qquad = 2 \lim\limits_{x \to 3} x^3 \lim\limits_{x \to 3} \sqrt{x^2 + 7} \qquad$ (Property 4)

$\qquad\qquad = 2(3)^3 \sqrt{3^2 + 7} \qquad$ (Property 1)

$\qquad\qquad = 2(27) \sqrt{16} = 216$

e. $\lim\limits_{x \to 2} \dfrac{2x^2 + 1}{x + 1} = \dfrac{\lim\limits_{x \to 2}(2x^2 + 1)}{\lim\limits_{x \to 2}(x + 1)} \qquad$ (Property 5)

$\qquad\qquad = \dfrac{2(2)^2 + 1}{2 + 1} = \dfrac{9}{3} = 3$ ■ ■ ■ ■

INDETERMINATE FORMS

Let's emphasize once again that Property 5 of limits is valid only when the limit of the function that appears in the denominator is not equal to zero at the point in question.

   If the numerator has a limit different from zero and the denominator has a limit equal to zero, then the limit of the quotient does not exist at the point in question. This is the case with the function $g(x) = 1/x^2$ in Example 3b. Here, as x approaches zero, the numerator approaches 1 but the denominator approaches zero, so the quotient becomes arbitrarily large. Thus, as observed earlier, the limit does not exist.

Next, consider

$$\lim_{x \to 2} \frac{4(x^2 - 4)}{x - 2}$$

which we evaluated earlier by looking at the values of the function for x near $x = 2$. If we attempt to evaluate this expression by applying Property 5 of limits, we see that both the numerator and denominator of the function

$$\frac{4(x^2 - 4)}{x - 2}$$

approach zero as x approaches 2; that is, we obtain an expression of the form 0/0. In this event, we say that the limit of the quotient $f(x)/g(x)$ as x approaches 2 has the **indeterminate form 0/0.**

We will need to evaluate limits of this type when we discuss the derivative of a function, a fundamental concept in the study of calculus. As the name suggests, the meaningless expression 0/0 does not provide us with a solution to our problem. One strategy that can be used to solve this type of problem follows.

Strategy for Evaluating Indeterminate Forms

> **1.** Replace the given function with an appropriate one that takes on the same values as the original function everywhere except at $x = a$.
> **2.** Evaluate the limit of this function as x approaches a.

Examples 5 and 6 illustrate this strategy.

EXAMPLE 5   Evaluate:

$$\lim_{x \to 2} \frac{4(x^2 - 4)}{x - 2}$$

SOLUTION ✔   Since both the numerator and the denominator of this expression approach zero as x approaches 2, we have the indeterminate form 0/0. We rewrite

$$\frac{4(x^2 - 4)}{x - 2} = \frac{4(x - 2)(x + 2)}{(x - 2)}$$

which, upon canceling the common factors, is equivalent to $4(x + 2)$. Next, we replace $4(x^2 - 4)/(x - 2)$ with $4(x + 2)$ and take the limit as x approaches 2, obtaining

$$\lim_{x \to 2} \frac{4(x^2 - 4)}{x - 2} = \lim_{x \to 2} 4(x + 2) = 16$$

The graphs of the functions

$$f(x) = \frac{4(x^2 - 4)}{x - 2} \qquad \text{and} \qquad g(x) = 4(x + 2)$$

are shown in Figure 2.27a–b. Observe that the graphs are identical except when $x = 2$. The function g is defined for all values of x and, in particular, its value at $x = 2$ is $g(2) = 4(2 + 2) = 16$. Thus, the point $(2, 16)$ is on the graph of g. However, the function f is not defined at $x = 2$. Since $f(x) = g(x)$ for all values of x except $x = 2$, it follows that the graph of f must look exactly like the graph of g, with the exception that the point $(2, 16)$ is missing from the graph of f. This illustrates graphically why we can evaluate the limit of f by evaluating the limit of the "equivalent" function g.

FIGURE 2.27
The graphs of $f(x)$ and $g(x)$ are identical except at the point (2, 16).

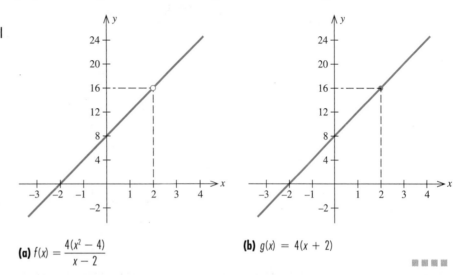

(a) $f(x) = \dfrac{4(x^2 - 4)}{x - 2}$

(b) $g(x) = 4(x + 2)$

REMARK Notice that the limit in Example 5 is the same limit that we evaluated earlier when we discussed the instantaneous velocity of a maglev at a specified time.

Exploring with Technology

1. Use a graphing utility to plot the graph of

$$f(x) = \frac{4(x^2 - 4)}{x - 2}$$

in the viewing rectangle $[0, 3] \times [0, 20]$. Then use **ZOOM** and **TRACE** to find

$$\lim_{x \to 2} \frac{4(x^2 - 4)}{x - 2}$$

2. Use a graphing utility to plot the graph of $g(x) = 4(x + 2)$ in the viewing rectangle $[0, 3] \times [0, 20]$. Then use **ZOOM** and **TRACE** to find $\lim_{x \to 2} 4(x + 2)$. What happens to the y-value when x takes on the value 2? Explain.

3. Can you distinguish between the graphs of f and g?

4. Reconcile your results with those of Example 5.

EXAMPLE 6   Evaluate:

$$\lim_{h \to 0} \frac{\sqrt{1+h} - 1}{h}$$

SOLUTION ✔   Letting h approach zero, we obtain the indeterminate form 0/0. Next, we rationalize the numerator of the quotient (see page 25) by multiplying both the numerator and the denominator by the expression $(\sqrt{1+h} + 1)$, obtaining

$$\frac{\sqrt{1+h} - 1}{h} = \frac{(\sqrt{1+h} - 1)(\sqrt{1+h} + 1)}{h(\sqrt{1+h} + 1)}$$

$$= \frac{1 + h - 1}{h(\sqrt{1+h} + 1)} \qquad [(\sqrt{a} - \sqrt{b})(\sqrt{a} + \sqrt{b}) = a - b]$$

$$= \frac{h}{h(\sqrt{1+h} + 1)}$$

$$= \frac{1}{\sqrt{1+h} + 1}$$

Therefore,

$$\lim_{h \to 0} \frac{\sqrt{1+h} - 1}{h} = \lim_{h \to 0} \frac{1}{\sqrt{1+h} + 1}$$

$$= \frac{1}{\sqrt{1} + 1} = \frac{1}{2}$$ ■ ■ ■ ■

Exploring with Technology

1. Use a graphing utility to plot the graph of

$$g(x) = \frac{\sqrt{1+x} - 1}{x}$$

in the viewing rectangle $[-1, 2] \times [0, 1]$. Then use **ZOOM** and **TRACE** to find

$$\lim_{x \to 0} \frac{\sqrt{1+x} - 1}{x}$$

by observing the values of $g(x)$ as x approaches 0 from the left and from the right.
2. Use a graphing utility to plot the graph of

$$f(x) = \frac{1}{\sqrt{1+x} + 1}$$

in the viewing rectangle $[-1, 2] \times [0, 1]$. Then use **ZOOM** and **TRACE** to find

$$\lim_{x \to 0} \frac{1}{\sqrt{1+x} + 1}$$

What happens to the y-value when x takes on the value 0? Explain.
3. Can you distinguish between the graphs of f and g?
4. Reconcile your results with those of Example 6.

LIMITS AT INFINITY

Up to now we have studied the limit of a function as x approaches a (finite) number a. There are occasions, however, when we want to know whether $f(x)$ approaches a unique number as x increases without bound. Consider, for example, the function P, giving the number of fruit flies (*Drosophila*) in a container under controlled laboratory conditions, as a function of a time t. The graph of P is shown in Figure 2.28. You can see from the graph of P that, as t increases without bound (gets larger and larger), $P(t)$ approaches the number 400. This number, called the *carrying capacity* of the environment, is determined by the amount of living space and food available, as well as other environmental factors.

FIGURE 2.28

The graph of $P(t)$ gives the population of fruit flies in a laboratory experiment.

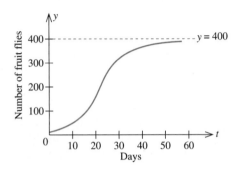

As another example, suppose we are given the function

$$f(x) = \frac{2x^2}{1 + x^2}$$

and we want to determine what happens to $f(x)$ as x gets larger and larger. Picking the sequence of numbers 1, 2, 5, 10, 100, and 1000 and computing the corresponding values of $f(x)$, we obtain the following table of values:

x	1	2	5	10	100	1000
$f(x)$	1	1.6	1.92	1.98	1.9998	1.999998

From the table, we see that as x gets larger and larger, $f(x)$ gets closer and closer to 2. The graph of the function f shown in Figure 2.29 confirms this observation. We call the line $y = 2$ a **horizontal asymptote.***

FIGURE 2.29

The graph of $y = \dfrac{2x^2}{1 + x^2}$ has a horizontal asymptote at $y = 2$.

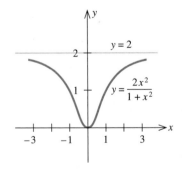

---

* We will discuss asymptotes in greater detail in Section 4.3.

In this situation we say that the limit of the function

$$f(x) = \frac{2x^2}{1 + x^2}$$

as x increases without bound is 2, written

$$\lim_{x \to \infty} \frac{2x^2}{1 + x^2} = 2$$

In the general case, the following definition is applicable:

Limit of a Function at Infinity

> The function f has the limit L as x increases without bound (or, as x approaches infinity), written
>
> $$\lim_{x \to \infty} f(x) = L$$
>
> if $f(x)$ can be made arbitrarily close to L by taking x large enough.
>   Similarly, the function f has the limit M as x decreases without bound (or as x approaches negative infinity), written
>
> $$\lim_{x \to -\infty} f(x) = M$$
>
> if $f(x)$ can be made arbitrarily close to M by taking x to be negative and sufficiently large in absolute value.

EXAMPLE 7   Let f and g be the functions

$$f(x) = \begin{cases} -1 & \text{if } x < 0 \\ 1 & \text{if } x \geq 0 \end{cases} \qquad \text{and} \qquad g(x) = \frac{1}{x^2}$$

Evaluate:

a. $\lim_{x \to \infty} f(x)$ and $\lim_{x \to -\infty} f(x)$ **b.** $\lim_{x \to \infty} g(x)$ and $\lim_{x \to -\infty} g(x)$

SOLUTION ✔   The graphs of $f(x)$ and $g(x)$ are shown in Figure 2.30. Referring to the graphs of the respective functions, we see that

a. $\lim_{x \to \infty} f(x) = 1$ and $\lim_{x \to -\infty} f(x) = -1$

b. $\lim_{x \to \infty} \frac{1}{x^2} = 0$ and $\lim_{x \to -\infty} \frac{1}{x^2} = 0$

FIGURE **2.30**

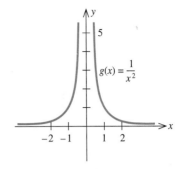

(a) $\lim\limits_{x \to \infty} f(x) = 1$ and $\lim\limits_{x \to -\infty} f(x) = -1$ **(b)** $\lim\limits_{x \to \infty} g(x) = 0$ and $\lim\limits_{x \to -\infty} g(x) = 0$

All the properties of limits listed in Theorem 1 are valid when a is replaced by ∞ or $-\infty$. In addition, we have the following property for the limit at infinity.

THEOREM 2

For all $n > 0$,

$$\lim_{x \to \infty} \frac{1}{x^n} = 0 \quad \text{and} \quad \lim_{x \to -\infty} \frac{1}{x^n} = 0$$

provided that $\dfrac{1}{x^n}$ is defined.

Exploring with Technology

1. Use a graphing utility to plot the graphs of

$$y_1 = \frac{1}{x^{0.5}}, \quad y_2 = \frac{1}{x}, \quad \text{and} \quad y_3 = \frac{1}{x^{1.5}}$$

in the viewing rectangle $[0, 200] \times [0, 0.5]$. What can you say about $\lim\limits_{x \to \infty} \dfrac{1}{x^n}$ if $n = 0.5$, $n = 1$, and $n = 1.5$? Are these results predicted by Theorem 2?

2. Use a graphing utility to plot the graphs of

$$y_1 = \frac{1}{x} \quad \text{and} \quad y_2 = \frac{1}{x^{5/3}}$$

in the viewing rectangle $[-50, 0] \times [-0.5, 0]$. What can you say about $\lim\limits_{x \to -\infty} \dfrac{1}{x^n}$ if $n = 1$ and $n = \dfrac{5}{3}$?

Are these results predicted by Theorem 2?

Hint: To graph y_2, write it in the form $y2 = 1/(x^{(1/3)})^5$.

In evaluating the limit at infinity of a rational function, the following technique is often used: *Divide the numerator and denominator of the expression by x^n, where n is the highest power present in the denominator of the expression.*

EXAMPLE 8   Evaluate:

$$\lim_{x \to \infty} \frac{x^2 - x + 3}{2x^3 + 1}$$

SOLUTION ✔   Since the limits of both the numerator and the denominator do not exist as x approaches infinity, the property pertaining to the limit of a quotient (Property 5) is not applicable. Let us divide the numerator and denominator of the rational expression by x^3, obtaining

$$\lim_{x \to \infty} \frac{x^2 - x + 3}{2x^3 + 1} = \lim_{x \to \infty} \frac{\dfrac{1}{x} - \dfrac{1}{x^2} + \dfrac{3}{x^3}}{2 + \dfrac{1}{x^3}}$$

$$= \frac{0 - 0 + 0}{2 + 0} = \frac{0}{2} \qquad \text{(Using Theorem 2)}$$

$$= 0$$

■ ■ ■ ■

EXAMPLE 9   Let

$$f(x) = \frac{3x^2 + 8x - 4}{2x^2 + 4x - 5}$$

Compute $\lim_{x \to \infty} f(x)$ if it exists.

SOLUTION ✔   Again, we see that Property 5 is not applicable. Dividing the numerator and the denominator by x^2, we obtain

$$\lim_{x \to \infty} \frac{3x^2 + 8x - 4}{2x^2 + 4x - 5} = \lim_{x \to \infty} \frac{3 + \dfrac{8}{x} - \dfrac{4}{x^2}}{2 + \dfrac{4}{x} - \dfrac{5}{x^2}}$$

$$= \frac{\lim\limits_{x \to \infty} 3 + 8 \lim\limits_{x \to \infty} \dfrac{1}{x} - 4 \lim\limits_{x \to \infty} \dfrac{1}{x^2}}{\lim\limits_{x \to \infty} 2 + 4 \lim\limits_{x \to \infty} \dfrac{1}{x} - 5 \lim\limits_{x \to \infty} \dfrac{1}{x^2}}$$

$$= \frac{3 + 0 - 0}{2 + 0 - 0}$$

$$= \frac{3}{2} \qquad \text{(Using Theorem 2)}$$

■ ■ ■ ■

EXAMPLE 10   Let $f(x) = \dfrac{2x^3 - 3x^2 + 1}{x^2 + 2x + 4}$ and evaluate:

a. $\lim\limits_{x \to \infty} f(x)$ **b.** $\lim\limits_{x \to -\infty} f(x)$

SOLUTION ✔   **a.** Dividing the numerator and the denominator of the rational expression by x^2, we obtain

$$\lim_{x \to \infty} \frac{2x^3 - 3x^2 + 1}{x^2 + 2x + 4} = \lim_{x \to \infty} \frac{2x - 3 + \dfrac{1}{x^2}}{1 + \dfrac{2}{x} + \dfrac{4}{x^2}}$$

Since the numerator becomes arbitrarily large whereas the denominator approaches 1 as x approaches infinity, we see that the quotient $f(x)$ gets larger and larger as x approaches infinity. In other words, the limit does not exist. We indicate this by writing

$$\lim_{x \to \infty} \frac{2x^3 - 3x^2 + 1}{x^2 + 2x + 4} = \infty$$

b. Once again, dividing both the numerator and the denominator by x^2, we obtain

$$\lim_{x \to -\infty} \frac{2x^3 - 3x^2 + 1}{x^2 + 2x + 4} = \lim_{x \to -\infty} \frac{2x - 3 + \dfrac{1}{x^2}}{1 + \dfrac{2}{x} + \dfrac{4}{x^2}}$$

In this case the numerator becomes arbitrarily large in magnitude but negative in sign, whereas the denominator approaches 1 as x approaches negative infinity. Therefore, the quotient $f(x)$ decreases without bound, and the limit does not exist. We indicate this by writing

$$\lim_{x \to -\infty} \frac{2x^3 - 3x^2 + 1}{x^2 + 2x + 4} = -\infty$$ ■■■■

     Example 11 gives an application of the concept of the limit of a function at infinity.

EXAMPLE 11   The Custom Office Company makes a line of executive desks. It is estimated that the total cost of making x Senior Executive Model desks is $C(x) = 100x + 200{,}000$ dollars per year, so that the average cost of making x desks is given by

$$\overline{C}(x) = \frac{C(x)}{x} = \frac{100x + 200{,}000}{x} = 100 + \frac{200{,}000}{x}$$

dollars per desk. Evaluate $\lim\limits_{x \to \infty} \overline{C}(x)$ and interpret your results.

SOLUTION ✓

$$\lim_{x \to \infty} \overline{C}(x) = \lim_{x \to \infty} \left(100 + \frac{200{,}000}{x} \right)$$

$$= \lim_{x \to \infty} 100 + \lim_{x \to \infty} \frac{200{,}000}{x} = 100$$

A sketch of the graph of the function $\overline{C}(x)$ appears in Figure 2.31. The result we obtained is fully expected if we consider its economic implications. Note that as the level of production increases, the fixed cost per desk produced, represented by the term $(200{,}000/x)$, drops steadily. The average cost should approach a constant unit cost of production—$100 in this case.

FIGURE 2.31
As the level of production increases, the average cost approaches $100 per desk.

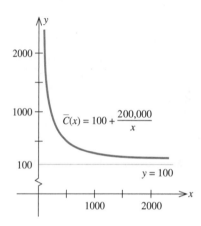

$\overline{C}(x) = 100 + \dfrac{200{,}000}{x}$

$y = 100$

■ ■ ■ ■

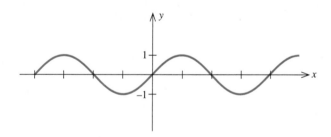

Group Discussion

Consider the graph of the function f depicted in the following figure:

It has the property that the curve oscillates between $y = -1$ and $y = 1$ indefinitely in either direction.

1. Explain why $\lim_{x \to -\infty} f(x)$ and $\lim_{x \to \infty} f(x)$ do not exist.

2. Compare this function with those of Example 10. More specifically, discuss the different ways each function fails to have a limit at infinity or minus infinity.

1. Find the indicated limit if it exists.

a. $\lim\limits_{x\to 3}\dfrac{\sqrt{x^2+7}+\sqrt{3x-5}}{x+2}$

b. $\lim\limits_{x\to -1}\dfrac{x^2-x-2}{2x^2-x-3}$

2. The average cost per disc (in dollars) incurred by the Herald Record Company in pressing x compact audio discs is given by the average cost function

$$\overline{C}(x)=1.8+\dfrac{3000}{x}$$

Evaluate $\lim\limits_{x\to\infty}\overline{C}(x)$ and interpret your result.

Solutions to Self-Check Exercises 2.4 can be found on page 135.

2.4 Exercises

In Exercises 1–8, use the graph of the given function f to determine $\lim\limits_{x\to a} f(x)$ at the indicated value of a, if it exists.

1. $a=-2$

2. $a=1$

3. $a=3$

7. $a=-2$

8. $a=0$

In Exercises 9–16, complete the table by computing $f(x)$ at the given values of x. Use these results to estimate the indicated limit (if it exists).

9. $f(x)=x^2+1;\ \lim\limits_{x\to 2} f(x)$

x	1.9	1.99	1.999	2.001	2.01	2.1
$f(x)$						

4. $a=1$

5. $a=-2$

6. 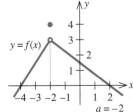 $a=-2$

10. $f(x)=2x^2-1;\ \lim\limits_{x\to 1} f(x)$

x	0.9	0.99	0.999	1.001	1.01	1.1
$f(x)$						

11. $f(x) = \dfrac{|x|}{x}; \lim\limits_{x \to 0} f(x)$

x	-0.1	-0.01	-0.001	0.001	0.01	0.1
$f(x)$						

12. $f(x) = \dfrac{|x - 1|}{x - 1}; \lim\limits_{x \to 1} f(x)$

x	0.9	0.99	0.999	1.001	1.01	1.1
$f(x)$						

13. $f(x) = \dfrac{1}{(x - 1)^2}; \lim\limits_{x \to 1} f(x)$

x	0.9	0.99	0.999	1.001	1.01	1.1
$f(x)$						

14. $f(x) = \dfrac{1}{x - 2}; \lim\limits_{x \to 2} f(x)$

x	1.9	1.99	1.999	2.001	2.01	2.1
$f(x)$						

15. $f(x) = \dfrac{x^2 + x - 2}{x - 1}; \lim\limits_{x \to 1} f(x)$

x	0.9	0.99	0.999	1.001	1.01	1.1
$f(x)$						

16. $f(x) = \dfrac{x - 1}{x - 1}; \lim\limits_{x \to 1} f(x)$

x	0.9	0.99	0.999	1.001	1.01	1.1
$f(x)$						

In Exercises 17–22, sketch the graph of the function f and evaluate $\lim\limits_{x \to a} f(x)$, if it exists, for the given values of a.

17. $f(x) = \begin{cases} x - 1 & \text{if } x \le 0 \\ -1 & \text{if } x > 0 \end{cases}$ $(a = 0)$

18. $f(x) = \begin{cases} x - 1 & \text{if } x \le 3 \\ -2x + 8 & \text{if } x > 3 \end{cases}$ $(a = 3)$

19. $f(x) = \begin{cases} x & \text{if } x < 1 \\ 0 & \text{if } x = 1 \\ -x + 2 & \text{if } x > 1 \end{cases}$ $(a = 1)$

20. $f(x) = \begin{cases} -2x + 4 & \text{if } x < 1 \\ 4 & \text{if } x = 1 \\ x^2 + 1 & \text{if } x > 1 \end{cases}$ $(a = 1)$

21. $f(x) = \begin{cases} |x| & \text{if } x \ne 0 \\ 1 & \text{if } x = 0 \end{cases}$ $(a = 0)$

22. $f(x) = \begin{cases} |x - 1| & \text{if } x \ne 1 \\ 0 & \text{if } x = 1 \end{cases}$ $(a = 1)$

In Exercises 23–40, find the indicated limit.

23. $\lim\limits_{x \to 2} 3$

24. $\lim\limits_{x \to -2} -3$

25. $\lim\limits_{x \to 3} x$

26. $\lim\limits_{x \to -2} -3x$

27. $\lim\limits_{x \to 1}(1 - 2x^2)$

28. $\lim\limits_{t \to 3}(4t^2 - 2t + 1)$

29. $\lim\limits_{x \to 1}(2x^3 - 3x^2 + x + 2)$

30. $\lim\limits_{x \to 0}(4x^5 - 20x^2 + 2x + 1)$

31. $\lim\limits_{s \to 0}(2s^2 - 1)(2s + 4)$

32. $\lim\limits_{x \to 2}(x^2 + 1)(x^2 - 4)$

33. $\lim\limits_{x \to 2} \dfrac{2x + 1}{x + 2}$

34. $\lim\limits_{x \to 1} \dfrac{x^3 + 1}{2x^3 + 2}$

35. $\lim\limits_{x \to 2} \sqrt{x + 2}$

36. $\lim\limits_{x \to -2} \sqrt[3]{5x + 2}$

37. $\lim\limits_{x \to -3} \sqrt{2x^4 + x^2}$

38. $\lim\limits_{x \to 2} \sqrt{\dfrac{2x^3 + 4}{x^2 + 1}}$

39. $\lim\limits_{x \to -1} \dfrac{\sqrt{x^2 + 8}}{2x + 4}$

40. $\lim\limits_{x \to 3} \dfrac{x\sqrt{x^2 + 7}}{2x - \sqrt{2x + 3}}$

In Exercises 41–48, find the indicated limit given that $\lim\limits_{x \to a} f(x) = 3$ and $\lim\limits_{x \to a} g(x) = 4$.

41. $\lim\limits_{x \to a} [f(x) - g(x)]$

42. $\lim\limits_{x \to a} 2f(x)$

43. $\lim\limits_{x \to a} [2f(x) - 3g(x)]$

44. $\lim\limits_{x \to a} [f(x)g(x)]$

45. $\lim_{x \to a} \sqrt{g(x)}$

46. $\lim_{x \to a} \sqrt[3]{5f(x) + 3g(x)}$

47. $\lim_{x \to a} \dfrac{2f(x) - g(x)}{f(x)g(x)}$

48. $\lim_{x \to a} \dfrac{g(x) - f(x)}{f(x) + \sqrt{g(x)}}$

In Exercises 49–62, find the indicated limit, if it exists.

49. $\lim_{x \to 1} \dfrac{x^2 - 1}{x - 1}$

50. $\lim_{x \to -2} \dfrac{x^2 - 4}{x + 2}$

51. $\lim_{x \to 0} \dfrac{x^2 - x}{x}$

52. $\lim_{x \to 0} \dfrac{2x^2 - 3x}{x}$

53. $\lim_{x \to -5} \dfrac{x^2 - 25}{x + 5}$

54. $\lim_{b \to -3} \dfrac{b + 1}{b + 3}$

55. $\lim_{x \to 1} \dfrac{x}{x - 1}$

56. $\lim_{x \to 2} \dfrac{x + 2}{x - 2}$

57. $\lim_{x \to -2} \dfrac{x^2 - x - 6}{x^2 + x - 2}$

58. $\lim_{z \to 2} \dfrac{z^3 - 8}{z - 2}$

59. $\lim_{x \to 1} \dfrac{\sqrt{x} - 1}{x - 1}$

Hint: Multiply by $\dfrac{\sqrt{x} + 1}{\sqrt{x} + 1}$.

60. $\lim_{x \to 4} \dfrac{x - 4}{\sqrt{x} - 2}$

Hint: See Exercise 59.

61. $\lim_{x \to 1} \dfrac{x - 1}{x^3 + x^2 - 2x}$

62. $\lim_{x \to -2} \dfrac{4 - x^2}{2x^2 + x^3}$

In Exercises 63–68, use the graph of the function f to determine $\lim_{x \to \infty} f(x)$ and $\lim_{x \to -\infty} f(x)$, if they exist.

63.

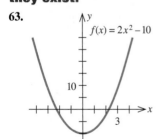

$f(x) = 2x^2 - 10$

64.

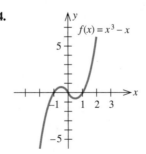

$f(x) = x^3 - x$

65.

$f(x) = \dfrac{1}{x^2}$

66.

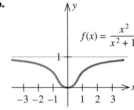

$f(x) = \dfrac{x^2}{x^2 + 1}$

67.

$f(x) = 2 - |x|$

68.

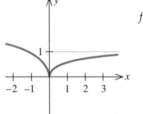

$f(x) = \begin{cases} \sqrt{-x} & \text{if } x \leq 0 \\ \dfrac{x}{x + 1} & \text{if } x > 0 \end{cases}$

In Exercises 69–72, complete the table by computing $f(x)$ at the given values of x. Use the results to guess at the indicated limits, if they exist.

69. $f(x) = \dfrac{1}{x^2 + 1}$; $\lim_{x \to \infty} f(x)$ and $\lim_{x \to -\infty} f(x)$

x	1	10	100	1000
$f(x)$				

x	-1	-10	-100	-1000
$f(x)$				

70. $f(x) = \dfrac{2x}{x + 1}$; $\lim_{x \to \infty} f(x)$ and $\lim_{x \to -\infty} f(x)$

x	1	10	100	1000
$f(x)$				

x	-5	-10	-100	-1000
$f(x)$				

(continued on p. 133)

FINDING THE LIMIT OF A FUNCTION

A graphing utility can be used to help us find the limit of a function, if it exists, as illustrated in the following examples.

EXAMPLE 1   Let $f(x) = \dfrac{x^3 - 1}{x - 1}$.

a. Plot the graph of f in the viewing rectangle $[-2, 2] \times [0, 4]$.

b. Use **ZOOM** to find $\lim\limits_{x \to 1} \dfrac{x^3 - 1}{x - 1}$.

c. Verify your result by evaluating the limit algebraically.

SOLUTION ✔   **a.** The graph of f in the viewing rectangle $[-2, 2] \times [0, 4]$ is shown in Figure T1.

FIGURE T1

The graph of $f(x) = \dfrac{x^3 - 1}{x - 1}$ in the viewing rectangle $[-2, 2] \times [0, 4]$

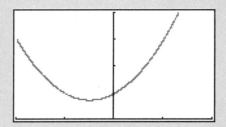

b. Using **ZOOM-IN** repeatedly, we see that the y-value approaches 3 as the x-value approaches 1. We conclude, accordingly, that

$$\lim_{x \to 1} \frac{x^3 - 1}{x - 1} = 3$$

c. We compute

$$\lim_{x \to 1} \frac{x^3 - 1}{x - 1} = \lim_{x \to 1} \frac{(x - 1)(x^2 + x + 1)}{x - 1}$$
$$= \lim_{x \to 1} (x^2 + x + 1) = 3$$

■ ■ ■ ■

REMARK If you attempt to find the limit in Example 1 by using the evaluation function of your graphing utility to find the value of $f(x)$ when $x = 1$, you will see that the graphing utility does not display the y-value. This happens because the point $x = 1$ is not in the domain of f.

EXAMPLE **2**	Use **ZOOM** to find $\lim\limits_{x \to 0} (1 + x)^{1/x}$.

SOLUTION ✔ We first plot the graph of $f(x) = (1 + x)^{1/x}$ in a suitable viewing rectangle. Figure T2 shows a plot of f in the rectangle $[-1, 1] \times [0, 4]$. Using **ZOOM-IN** repeatedly, we see that $\lim\limits_{x \to 0} (1 + x)^{1/x} \approx 2.71828$.

FIGURE T2
The graph of $f(x) = (1 + x)^{1/x}$ in the viewing rectangle $[-1, 1] \times [0, 4]$

The limit of $f(x) = (1 + x)^{1/x}$ as x approaches zero, denoted by the letter e, plays a very important role in the study of mathematics and its applications (see Section 5.6). Thus,

$$\lim_{x \to 0} (1 + x)^{1/x} = e$$

where, as we have just seen, $e \approx 2.71828$. ▪▪▪▪

EXAMPLE 3 When organic waste is dumped into a pond, the oxidation process that takes place reduces the pond's oxygen content. However, given time, nature will restore the oxygen content to its natural level. Suppose that the oxygen content t days after the organic waste has been dumped into the pond is given by

$$f(t) = 100 \left(\frac{t^2 + 10t + 100}{t^2 + 20t + 100} \right)$$

percent of its normal level.

a. Plot the graph of f in the viewing rectangle $[0, 200] \times [70, 100]$.
b. What can you say about $f(t)$ when t is very large?
c. Verify your observation in part (b) by evaluating $\lim\limits_{t \to \infty} f(t)$.

SOLUTION ✔ **a.** The graph of f is shown in Figure T3.
b. From the graph of f it appears that $f(t)$ approaches 100 steadily as t gets

131

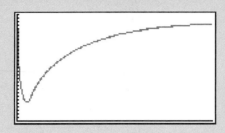

larger and larger. This observation tells us that eventually the oxygen content of the pond will be restored to its natural level.

c. To verify the observation made in part (b), we compute

$$\lim_{t \to \infty} f(t) = \lim_{t \to \infty} 100 \left(\frac{t^2 + 10t + 100}{t^2 + 20t + 100} \right)$$

$$= 100 \lim_{t \to \infty} \left(\frac{1 + \dfrac{10}{t} + \dfrac{100}{t^2}}{1 + \dfrac{20}{t} + \dfrac{100}{t^2}} \right) = 100$$

■■■■

Exercises

In Exercises 1–10, use a graphing utility to find the indicated limit by first plotting the graph of the function in a suitable viewing rectangle and then using the zoom-in feature of the calculator.

1. $\displaystyle\lim_{x \to 1} \frac{2x^3 - 2x^2 + 3x - 3}{x - 1}$

2. $\displaystyle\lim_{x \to -2} \frac{2x^3 + 3x^2 - x + 2}{x + 2}$

3. $\displaystyle\lim_{x \to -1} \frac{x^3 + 1}{x + 1}$

4. $\displaystyle\lim_{x \to -1} \frac{x^4 - 1}{x - 1}$

5. $\displaystyle\lim_{x \to 1} \frac{x^3 - x^2 - x + 1}{x^3 - 3x + 2}$

6. $\displaystyle\lim_{x \to 2} \frac{x^3 + 2x^2 - 16}{2x^3 - x^2 + 2x - 16}$

7. $\displaystyle\lim_{x \to 0} \frac{\sqrt{x + 1} - 1}{x}$

8. $\displaystyle\lim_{x \to 0} \frac{(x + 4)^{3/2} - 8}{x}$

9. $\displaystyle\lim_{x \to 0} (1 + 2x)^{1/x}$

10. $\displaystyle\lim_{x \to 0} \frac{2^x - 1}{x}$

11. Use a graphing utility to show that $\displaystyle\lim_{x \to 3} \frac{2}{x - 3}$ does not exist.

12. Use a graphing utility to show that $\displaystyle\lim_{x \to 2} \frac{x^3 - 2x + 1}{x - 2}$ does not exist.

13. CITY PLANNING A major developer is building a 5000-acre complex of homes, offices, stores, schools, and churches in the rural community of Marlboro. As a result of this development, the planners have estimated that Marlboro's population (in thousands) t yr from now will be given by

$$P(t) = \frac{25t^2 + 125t + 200}{t^2 + 5t + 40}$$

a. Plot the graph of P in the viewing rectangle $[0, 50] \times [0, 30]$.

b. What will be the population of Marlboro in the long run?
Hint: Find $\displaystyle\lim_{t \to \infty} P(t)$.

71. $f(x) = 3x^3 - x^2 + 10;$ $\lim\limits_{x \to \infty} f(x)$ and $\lim\limits_{x \to -\infty} f(x)$

x	1	5	10	100	1000
$f(x)$					

x	-1	-5	-10	-100	-1000
$f(x)$					

72. $f(x) = \dfrac{|x|}{x};$ $\lim\limits_{x \to \infty} f(x)$ and $\lim\limits_{x \to -\infty} f(x)$

x	1	10	100	-1	-10	-100
$f(x)$						

In Exercises 73–80, find the indicated limits, if they exist.

73. $\lim\limits_{x \to \infty} \dfrac{3x+2}{x-5}$

74. $\lim\limits_{x \to -\infty} \dfrac{4x^2-1}{x+2}$

75. $\lim\limits_{x \to -\infty} \dfrac{3x^3+x^2+1}{x^3+1}$

76. $\lim\limits_{x \to \infty} \dfrac{2x^2+3x+1}{x^4-x^2}$

77. $\lim\limits_{x \to -\infty} \dfrac{x^4+1}{x^3-1}$

78. $\lim\limits_{x \to \infty} \dfrac{4x^4-3x^2+1}{2x^4+x^3+x^2+x+1}$

79. $\lim\limits_{x \to \infty} \dfrac{x^5-x^3+x-1}{x^6+2x^2+1}$

80. $\lim\limits_{x \to \infty} \dfrac{2x^2-1}{x^3+x^2+1}$

81. TOXIC WASTE A city's main well was recently found to be contaminated with trichloroethylene, a cancer-causing chemical, as a result of an abandoned chemical dump leaching chemicals into the water. A proposal submitted to city council members indicates that the cost, measured in millions of dollars, of removing x percent of the toxic pollutant is given by

$$C(x) = \frac{0.5x}{100-x} \qquad (0 < x < 100)$$

a. Find the cost of removing 50%, 60%, 70%, 80%, 90%, and 95% of the pollutant.

b. Evaluate

$$\lim_{x \to 100} \frac{0.5x}{100-x}$$

and interpret your result.

82. A DOOMSDAY SITUATION The population of a certain breed of rabbits introduced into an isolated island is given by

$$P(t) = \frac{72}{9-t} \qquad (0 < t < 9)$$

where t is measured in months.
a. Find the number of rabbits present in the island initially.
b. Show that the population of rabbits is increasing without bound.
c. Sketch the graph of the function P.
(*Comment:* This phenomenon is referred to as a *doomsday situation.*)

83. AVERAGE COST The average cost per disc in dollars incurred by the Herald Record Company in pressing x video discs is given by the average cost function

$$\overline{C}(x) = 2.2 + \frac{2500}{x}$$

Evaluate $\lim\limits_{x \to \infty} \overline{C}(x)$ and interpret your result.

84. CONCENTRATION OF A DRUG IN THE BLOODSTREAM The concentration of a certain drug in a patient's bloodstream t hr after injection is given by

$$C(t) = \frac{0.2t}{t^2+1}$$

mg/cm^3. Evaluate $\lim\limits_{t \to \infty} C(t)$ and interpret your result.

85. BOX OFFICE RECEIPTS The total worldwide box office receipts for a long-running blockbuster movie are approximated by the function

$$T(x) = \frac{120x^2}{x^2+4}$$

where $T(x)$ is measured in millions of dollars and x is the number of months since the movie's release.
a. What are the total box office receipts after the first month? The second month? The third month?
b. What will the movie gross in the long run?

86. POPULATION GROWTH   A major corporation is building a 4325-acre complex of homes, offices, stores, schools, and churches in the rural community of Glen Cove. As a result of this development, the planners have estimated that Glen Cove's population (in thousands) t yr from now will be given by

$$P(t) = \frac{25t^2 + 125t + 200}{t^2 + 5t + 40}$$

a. What is the current population of Glen Cove?
b. What will the population be in the long run?

87. DRIVING COSTS   A study of driving costs of 1992 model subcompact (four-cylinder) cars found that the average cost (car payments, gas, insurance, upkeep, and depreciation), measured in cents per mile, is approximated by the function

$$C(x) = \frac{2010}{x^{2.2}} + 17.80$$

where x denotes the number of miles (in thousands) the car is driven in a year.
a. What is the average cost of driving a subcompact car 5000 mi/yr? 10,000 mi/yr? 15,000 mi/yr? 20,000 mi/yr? 25,000 mi/yr?
b. Use part (a) to help sketch the graph of the function C.
c. What happens to the average cost as the number of miles driven increases without bound?

In Exercises 89–94, determine whether the statement is true or false. If it is true, explain why it is true. If it is false, give an example to show why it is false.

89. If $\lim_{x \to a} f(x)$ exists, then f is defined at $x = a$.

90. If $\lim_{x \to 0} f(x) = 4$ and $\lim_{x \to 0} g(x) = 0$, then $\lim_{x \to 0} f(x)g(x) = 0$.

91. If $\lim_{x \to 2} f(x) = 3$ and $\lim_{x \to 2} g(x) = 0$, then $\lim_{x \to 2} [f(x)]/[g(x)]$ does not exist.

92. If $\lim_{x \to 3} f(x) = 0$ and $\lim_{x \to 3} g(x) = 0$, then $\lim_{x \to 3} [f(x)]/[g(x)]$ does not exist.

93. $\lim_{x \to 2} \left(\frac{x}{x + 1} + \frac{3}{x - 1} \right) = \lim_{x \to 2} \frac{x}{x + 1} + \lim_{x \to 2} \frac{3}{x - 1}$

94. $\lim_{x \to 1} \left(\frac{2x}{x - 1} - \frac{2}{x - 1} \right) = \lim_{x \to 1} \frac{2x}{x - 1} - \lim_{x \to 1} \frac{2}{x - 1}$

95. SPEED OF A CHEMICAL REACTION   Certain proteins, known as enzymes, serve as catalysts for chemical reactions in living things. In 1913 Leonor Michaelis and L. M. Menten discovered the following formula giving the initial speed V (in moles/liter/second) at which the reaction begins in terms of the amount of substrate x (the substance being acted upon, measured in moles/liter):

$$V = \frac{ax}{x + b}$$

where a and b are positive constants. Evaluate

$$\lim_{x \to \infty} \frac{ax}{x + b}$$

and interpret your result.

96. Show by means of an example that $\lim_{x \to a} [f(x) + g(x)]$ may exist even though neither $\lim_{x \to a} f(x)$ nor $\lim_{x \to a} g(x)$ exists. Does this example contradict Theorem 1?

97. Show by means of an example that $\lim_{x \to a} [f(x)g(x)]$ may exist even though neither $\lim_{x \to a} f(x)$ nor $\lim_{x \to a} g(x)$ exists. Does this example contradict Theorem 1?

SOLUTIONS TO SELF-CHECK EXERCISES 2.4

1. a. $\lim\limits_{x\to 3}\dfrac{\sqrt{x^2+7}+\sqrt{3x-5}}{x+2} = \dfrac{\sqrt{9+7}+\sqrt{3(3)-5}}{3+2}$

$$= \dfrac{\sqrt{16}+\sqrt{4}}{5}$$

$$= \dfrac{6}{5}$$

b. Letting x approach -1 leads to the indeterminate form $0/0$. Thus, we proceed as follows:

$$\lim_{x\to -1}\dfrac{x^2-x-2}{2x^2-x-3} = \lim_{x\to -1}\dfrac{(x+1)(x-2)}{(x+1)(2x-3)}$$

$$= \lim_{x\to -1}\dfrac{x-2}{2x-3} \qquad \text{(Canceling the common factors)}$$

$$= \dfrac{-1-2}{2(-1)-3}$$

$$= \dfrac{3}{5}$$

2. $\lim\limits_{x\to\infty}\overline{C}(x) = \lim\limits_{x\to\infty}\left(1.8+\dfrac{3000}{x}\right)$

$$= \lim_{x\to\infty}1.8 + \lim_{x\to\infty}\dfrac{3000}{x}$$

$$= 1.8$$

Our computation reveals that, as the production of audio discs increases "without bound," the average cost drops and approaches a unit cost of \$1.80/disc.

2.5 One-Sided Limits and Continuity

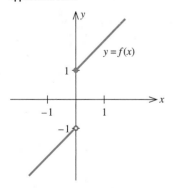

ONE-SIDED LIMITS

Consider the function f defined by

$$f(x) = \begin{cases} x-1 & \text{if } x < 0 \\ x+1 & \text{if } x \geq 0 \end{cases}$$

From the graph of f shown in Figure 2.32, we see that the function f does not have a limit as x approaches zero because no matter how close x is to zero, $f(x)$ takes on values that are close to 1 if x is positive and values that are close to -1 if x is negative. Therefore, $f(x)$ cannot be close to a single number L—no matter how close x is to zero. Now, if we restrict x to be greater than zero (to the right of zero), then we see that $f(x)$ can be made as close to the number 1 as we please by taking x sufficiently close to zero. In this situation we say that the right-hand limit of f as x approaches zero (from the right) is

1, written

$$\lim_{x \to 0^+} f(x) = 1$$

Similarly, we see that $f(x)$ can be made as close to the number -1 as we please by taking x sufficiently close to, but to the left of, zero. In this situation we say that the left-hand limit of f as x approaches zero (from the left) is -1, written

$$\lim_{x \to 0^-} f(x) = -1$$

These limits are called **one-sided limits.** More generally, we have the following definitions.

One-Sided Limits

> The function f has the **right-hand limit** L as x approaches a from the right, written
>
> $$\lim_{x \to a^+} f(x) = L$$
>
> if the values $f(x)$ can be made as close to L as we please by taking x sufficiently close to (but not equal to) a and to the right of a.
>   Similarly, the function f has the **left-hand limit** M as x approaches a from the left, written
>
> $$\lim_{x \to a^-} f(x) = M$$
>
> if the values $f(x)$ can be made as close to M as we please by taking x sufficiently close to (but not equal to) a and to the left of a.

The connection between one-sided limits and the two-sided limit defined earlier is given by the following theorem.

THEOREM 3

Let f be a function that is defined for all values of x close to $x = a$ with the possible exception of a itself. Then,

$$\lim_{x \to a} f(x) = L \quad \text{if and only if} \quad \lim_{x \to a^+} f(x) = \lim_{x \to a^-} f(x) = L$$

Thus, the two-sided limit exists if and only if the one-sided limits exist and are equal.

EXAMPLE 1   Let

$$f(x) = \begin{cases} \sqrt{x} & \text{if } x > 0 \\ -x & \text{if } x \le 0 \end{cases} \quad \text{and} \quad g(x) = \begin{cases} -1 & \text{if } x < 0 \\ 1 & \text{if } x \ge 0 \end{cases}$$

a. Show that $\lim_{x \to 0} f(x)$ exists by studying the one-sided limits of f as x approaches $x = 0$.

b. Show that $\lim_{x \to 0} g(x)$ does not exist.

SOLUTION ✔   **a.** For $x > 0$, we find

$$\lim_{x \to 0^+} f(x) = \lim_{x \to 0^+} \sqrt{x} = 0$$

and for $x \leq 0$

$$\lim_{x \to 0^-} f(x) = \lim_{x \to 0^-} (-x) = 0$$

Thus,

$$\lim_{x \to 0} f(x) = 0$$

(Figure 2.33a).

b. We have

$$\lim_{x \to 0^-} g(x) = -1 \qquad \text{and} \qquad \lim_{x \to 0^+} g(x) = 1$$

and since these one-sided limits are not equal, we conclude that $\lim_{x \to 0} g(x)$ does not exist (Figure 2.33b).

FIGURE 2.33

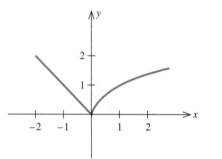

(a) $\lim_{x \to 0} f(x)$ exists.

(b) $\lim_{x \to 0} g(x)$ does not exist.   ■ ■ ■ ■

CONTINUOUS FUNCTIONS

Continuous functions will play an important role throughout most of our study of calculus. Loosely speaking, a function is continuous at a point if the graph of the function at that point is devoid of holes, gaps, jumps, or breaks. Consider, for example, the graph of the function f depicted in Figure 2.34.

Let's take a closer look at the behavior of f at or near each of the points $x = a$, $x = b$, $x = c$, and $x = d$. First, note that f is not defined at $x = a$; that is, the point $x = a$ is not in the domain of f, thereby resulting in a "hole" in the graph of f. Next, observe that the value of f at b, $f(b)$, is not equal to the

FIGURE 2.34
The graph of this function is not continuous
at $x = a$, $x = b$, $x = c$, and $x = d$.

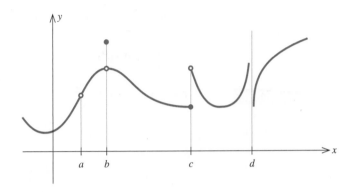

limit of $f(x)$ as x approaches b, resulting in a "jump" in the graph of f at that point. The function f does not have a limit at $x = c$ since the left-hand and right-hand limits of $f(x)$ are not equal, also resulting in a jump in the graph of f at that point. Finally, the limit of f does not exist at $x = d$, resulting in a break in the graph of f. The function f is *discontinuous* at each of these points. It is *continuous* at all other points.

Continuity at a Point

A function f is **continuous at the point** $x = a$ if the following conditions are satisfied:

1. $f(a)$ is defined.   **2.** $\lim_{x \to a} f(x)$ exists.   **3.** $\lim_{x \to a} f(x) = f(a)$

Thus, a function f is continuous at the point $x = a$ if the limit of f at the point $x = a$ exists and has the value $f(a)$. Geometrically, f is continuous at the point $x = a$ if proximity of x to a implies the proximity of $f(x)$ to $f(a)$.

If f is not continuous at $x = a$, then f is said to be **discontinuous** at $x = a$. Also, f is **continuous on an interval** if f is continuous at every point in the interval.

Figure 2.35 depicts the graph of a continuous function on the interval (a, b). Notice that the graph of the function over the stated interval can be sketched without lifting one's pencil from the paper.

FIGURE 2.35
The graph of f is continuous on the interval
(a, b).

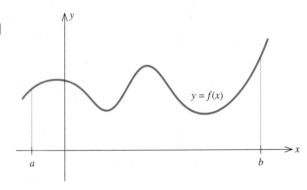

EXAMPLE **2**

Find the values of x for which each of the following functions is continuous.

a. $f(x) = x + 2$ **b.** $g(x) = \dfrac{x^2 - 4}{x - 2}$ **c.** $h(x) = \begin{cases} x + 2 & \text{if } x \neq 2 \\ 1 & \text{if } x = 2 \end{cases}$

d. $F(x) = \begin{cases} -1 & \text{if } x < 0 \\ 1 & \text{if } x \geq 0 \end{cases}$ **e.** $G(x) = \begin{cases} \dfrac{1}{x} & \text{if } x > 0 \\ -1 & \text{if } x \leq 0 \end{cases}$

The graph of each function is shown in Figure 2.36.

FIGURE **2.36**

(a)   (b)   (c)

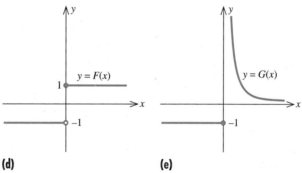

(d)   (e)

SOLUTION ✔

a. The function f is continuous everywhere because the three conditions for continuity are satisfied for all values of x.

b. The function g is discontinuous at the point $x = 2$ because g is not defined at that point. It is continuous everywhere else.

c. The function h is discontinuous at $x = 2$ because the third condition for continuity is violated; the limit of $h(x)$ as x approaches 2 exists and has the value 4, but this limit is not equal to $h(2) = 1$. It is continuous for all other values of x.

d. The function F is continuous everywhere except at the point $x = 0$, where the limit of $F(x)$ fails to exist as x approaches zero (see Example 3a, Section 2.4).

e. Since the limit of $G(x)$ does not exist as x approaches zero, we conclude that G fails to be continuous at $x = 0$. The function G is continuous at all other points.

■ ■ ■ ■

PROPERTIES OF CONTINUOUS FUNCTIONS

The following properties of continuous functions follow directly from the definition of continuity and the corresponding properties of limits. They are stated without proof.

Properties of Continuous Functions

1. The constant function $f(x) = c$ is continuous everywhere.
2. The identity function $f(x) = x$ is continuous everywhere.

If f and g are continuous at x = a, then

3. $[f(x)]^n$, where n is a real number, is continuous at $x = a$ whenever it is defined at that point.
4. $f \pm g$ is continuous at $x = a$.
5. fg is continuous at $x = a$.
6. f/g is continuous at $x = a$ provided $g(a) \neq 0$.

Using these properties of continuous functions, we can prove the following results. (A proof is sketched in Exercise 102, page 154.)

Continuity of Polynomial and Rational Functions

1. A polynomial function $y = P(x)$ is continuous at every point x.
2. A rational function $R(x) = p(x)/q(x)$ is continuous at every point x where $q(x) \neq 0$.

EXAMPLE 3   Find the values of x for which each of the following functions is continuous.

a. $f(x) = 3x^3 + 2x^2 - x + 10$ **b.** $g(x) = \dfrac{8x^{10} - 4x + 1}{x^2 + 1}$

c. $h(x) = \dfrac{4x^3 - 3x^2 + 1}{x^2 - 3x + 2}$

SOLUTION ✔   **a.** The function f is a polynomial function of degree 3, so $f(x)$ is continuous for all values of x.

b. The function g is a rational function. Observe that the denominator of g—namely, $x^2 + 1$—is never equal to zero. Therefore, we conclude that g is continuous for all values of x.

c. The function h is a rational function. In this case, however, the denominator of h is equal to zero at $x = 1$ and $x = 2$, which can be seen by factoring it. Thus,

$$x^2 - 3x + 2 = (x - 2)(x - 1)$$

We therefore conclude that h is continuous everywhere except at $x = 1$ and $x = 2$, where it is discontinuous.   ■ ■ ■ ■

APPLICATIONS

Up to this point, most of the applications we have discussed involved functions that are continuous everywhere. In Example 4 we consider an application from the field of educational psychology that involves a discontinuous function.

EXAMPLE 4

FIGURE 2.37
A learning curve that is discontinuous at $t = t_1$

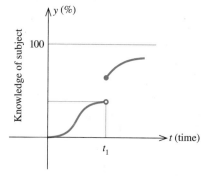

Figure 2.37 depicts the learning curve associated with a certain individual. Beginning with no knowledge of the subject being taught, the individual makes steady progress toward understanding it over the time interval $0 \leq t < t_1$. In this instance, the individual's progress slows as we approach time t_1 because he fails to grasp a particularly difficult concept. All of a sudden, a breakthrough occurs at time t_1, propelling his knowledge of the subject to a higher level. The curve is discontinuous at t_1.

■ ■ ■ ■

INTERMEDIATE VALUE THEOREM

Let's look again at our model of the motion of the maglev on a straight stretch of track. We know that the train cannot vanish at any instant of time and it cannot skip portions of the track and reappear someplace else. To put it another way, the train cannot occupy the positions s_1 and s_2 without at least, at some time, occupying an intermediate position (Figure 2.38).

FIGURE 2.38
The position of the maglev

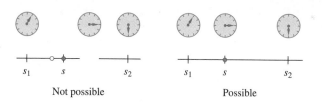

To state this fact mathematically, recall that the position of the maglev as a function of time is described by

$$f(t) = 4t^2 \qquad (0 \leq t \leq 10)$$

Suppose the position of the maglev is s_1 at some time t_1 and its position is s_2 at some time t_2 (Figure 2.39).

FIGURE 2.39
If $s_1 \leq s_3 \leq s_2$, then there must be at least one t_3 ($t_1 \leq t_3 \leq t_2$) such that $f(t_3) = s_3$.

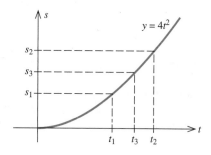

Then, if s_3 is any number between s_1 and s_2 giving an intermediate position of the maglev, there must be at least one t_3 between t_1 and t_2 giving the time at which the train is at s_3—that is, $f(t_3) = s_3$.

This discussion carries the gist of the intermediate value theorem. The proof of this theorem can be found in most advanced calculus texts.

THEOREM 4

The Intermediate Value Theorem

If f is a continuous function on a closed interval $[a, b]$ and M is any number between $f(a)$ and $f(b)$, then there is at least one number c in $[a, b]$ such that $f(c) = M$ (Figure 2.40).

FIGURE 2.40

(a) $f(c) = M$ (b) $f(c_1) = f(c_2) = f(c_3) = M$

To illustrate the intermediate value theorem, let's look at the example involving the motion of the maglev again (see Figure 2.22, page 111). Notice that the initial position of the train is $f(0) = 0$ and the position at the end of its test run is $f(10) = 400$. Furthermore, the function f is continuous on $[0, 10]$. So, the intermediate value theorem guarantees that if we arbitrarily pick a number between 0 and 400—say, 100—giving the position of the maglev, there must be a $\bar{t}$ (read "t bar") between 0 and 10 at which time the train is at the position $s = 100$. To find the value of $\bar{t}$, we solve the equation $f(\bar{t}) = s$, or

$$4\bar{t}^2 = 100$$

giving $\bar{t} = 5$ (t must lie between 0 and 10).

 It is important to remember when we use Theorem 4 that the function f must be continuous. The conclusion of the intermediate value theorem may not hold if f is not continuous (see Exercise 103).

The next theorem is an immediate consequence of the intermediate value theorem. It not only tells us when a zero of a function f [root of the equation $f(x) = 0$] exists but also provides the basis for a method of approximating it.

THEOREM 5

Existence of Zeros of a Continuous Function
If f is a continuous function on a closed interval $[a, b]$, and if $f(a)$ and $f(b)$ have opposite signs, then there is at least one solution of the equation $f(x) = 0$ in the interval (a, b) (Figure 2.41).

FIGURE 2.41
If $f(a)$ and $f(b)$ have opposite signs, there must be at least one number c $(a < c < b)$ such that $f(c) = 0$.

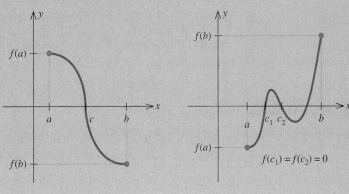

Geometrically, this property states that if the graph of a continuous function goes from above the x-axis to below the x-axis, or vice versa, it must *cross* the x-axis. This is not necessarily true if the function is discontinuous (Figure 2.42).

FIGURE 2.42
$f(a) < 0$ and $f(b) > 0$, but the graph of f does not cross the x-axis between a and b because f is discontinuous.

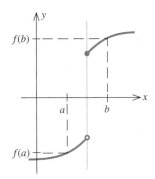

EXAMPLE 5

Let $f(x) = x^3 + x + 1$.

a. Show that f is continuous for all values of x.
b. Compute $f(-1)$ and $f(1)$ and use the results to deduce that there must be at least one point $x = c$, where c lies in the interval $(-1, 1)$ and $f(c) = 0$.

SOLUTION ✔

a. The function f is a polynomial function of degree 3 and is therefore continuous everywhere.
b. $f(-1) = (-1)^3 + (-1) + 1 = -1$
  $f(1) = 1^3 + 1 + 1 = 3$

Since $f(-1)$ and $f(1)$ have opposite signs, Theorem 5 tells us that there must be at least one point $x = c$ with $-1 < c < 1$ such that $f(c) = 0$.   ■ ■ ■ ■

The next example shows how the intermediate value theorem can be used to help us find the zero of a function.

EXAMPLE 6   Let $f(x) = x^3 + x - 1$. Since f is a polynomial function, it is continuous everywhere. Observe that $f(0) = -1$ and $f(1) = 1$ so that Theorem 5 guarantees the existence of at least one root of the equation $f(x) = 0$ in $(0, 1)$.*

We can locate the root more precisely by using Theorem 5 once again as follows: Evaluate $f(x)$ at the midpoint of $[0, 1]$, obtaining

$$f(0.5) = -0.375$$

Because $f(0.5) < 0$ and $f(1) > 0$, Theorem 5 now tells us that the root must lie in $(0.5, 1)$.

Repeat the process: Evaluate $f(x)$ at the midpoint of $[0.5, 1]$, which is

$$\frac{0.5 + 1}{2} = 0.75$$

Thus,

$$f(0.75) = 0.1719$$

Because $f(0.5) < 0$ and $f(0.75) > 0$, Theorem 5 tells us that the root is in $(0.5, 0.75)$. This process can be continued. Table 2.3 summarizes the results of our computations through nine steps.

From Table 2.3 we see that the root is approximately 0.68, accurate to two decimal places. By continuing the process through a sufficient number of steps, we can obtain as accurate an approximation to the root as we please.   ■ ■ ■ ■

Table 2.3

Step	Root of $f(x) = 0$ Lies In
1	(0, 1)
2	(0.5, 1)
3	(0.5, 0.75)
4	(0.625, 0.75)
5	(0.625, 0.6875)
6	(0.65625, 0.6875)
7	(0.671875, 0.6875)
8	(0.6796875, 0.6875)
9	(0.6796875, 0.6835937)

REMARK The process of finding the root of $f(x) = 0$ used in Example 6 is called the **method of bisection.** It is crude but effective.   ■ ■ ■

SELF-CHECK EXERCISES 2.5

1. Evaluate $\lim\limits_{x \to -1^-} f(x)$ and $\lim\limits_{x \to -1^+} f(x)$, where

$$f(x) = \begin{cases} 1 & \text{if } x < -1 \\ 1 + \sqrt{x+1} & \text{if } x \geq -1 \end{cases}$$

Does $\lim\limits_{x \to -1} f(x)$ exist?

* It can be shown that f has precisely one zero in (0, 1) (see Exercise 96, Section 4.1.).

2. Determine the values of x for which the given function is discontinuous. At each point of discontinuity, indicate which condition(s) for continuity are violated. Sketch the graph of the function.

a. $f(x) = \begin{cases} -x^2 + 1 & \text{if } x \le 1 \\ x - 1 & \text{if } x > 1 \end{cases}$

b. $g(x) = \begin{cases} -x + 1 & \text{if } x < -1 \\ 2 & \text{if } -1 < x \le 1 \\ -x + 3 & \text{if } x > 1 \end{cases}$

Solutions to Self-Check Exercises 2.5 can be found on page 154.

2.5 Exercises

In Exercises 1–8, use the graph of the function f to find $\lim\limits_{x \to a^-} f(x)$, $\lim\limits_{x \to a^+} f(x)$, and $\lim\limits_{x \to a} f(x)$ at the indicated value of a, if the limit exists.

1.

2.

3.

4.

5.

6.

7.

8.

In Exercises 9–14, refer to the graph of the function f and determine whether each statement is true or false.

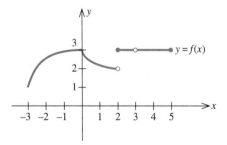

9. $\lim\limits_{x \to -3^+} f(x) = 1$

10. $\lim\limits_{x \to 0} f(x) = f(0)$

11. $\lim\limits_{x \to 2^-} f(x) = 2$

12. $\lim\limits_{x \to 2^+} f(x) = 3$

13. $\lim\limits_{x \to 3} f(x)$ does not exist.

14. $\lim\limits_{x \to 5^-} f(x) = 3$

In Exercises 15–20, refer to the graph of the function *f* and determine whether each statement is true or false.

15. $\lim\limits_{x\to -3^+} f(x) = 2$

16. $\lim\limits_{x\to 0} f(x) = 2$

17. $\lim\limits_{x\to 2} f(x) = 1$

18. $\lim\limits_{x\to 4^-} f(x) = 3$

19. $\lim\limits_{x\to 4^+} f(x)$ does not exist.

20. $\lim\limits_{x\to 4} f(x) = 2$

In Exercises 21–42, find the indicated one-sided limit, if it exists.

21. $\lim\limits_{x\to 1^+}(2x + 4)$

22. $\lim\limits_{x\to 1^-}(3x - 4)$

23. $\lim\limits_{x\to 2^-}\dfrac{x - 3}{x + 2}$

24. $\lim\limits_{x\to 1^+}\dfrac{x + 2}{x + 1}$

25. $\lim\limits_{x\to 0^+}\dfrac{1}{x}$

26. $\lim\limits_{x\to 0^-}\dfrac{1}{x}$

27. $\lim\limits_{x\to 0^+}\dfrac{x - 1}{x^2 + 1}$

28. $\lim\limits_{x\to 2^+}\dfrac{x + 1}{x^2 - 2x + 3}$

29. $\lim\limits_{x\to 0^+}\sqrt{x}$

30. $\lim\limits_{x\to 2^+} 2\sqrt{x - 2}$

31. $\lim\limits_{x\to -2^+}(2x + \sqrt{2 + x})$

32. $\lim\limits_{x\to -5^+} x(1 + \sqrt{5 + x})$

33. $\lim\limits_{x\to 1^-}\dfrac{1 + x}{1 - x}$

34. $\lim\limits_{x\to 1^+}\dfrac{1 + x}{1 - x}$

35. $\lim\limits_{x\to 2^-}\dfrac{x^2 - 4}{x - 2}$

36. $\lim\limits_{x\to -3^+}\dfrac{\sqrt{x + 3}}{x^2 + 1}$

37. $\lim\limits_{x\to 3^+}\dfrac{x^2 - 9}{x + 3}$

38. $\lim\limits_{x\to -2^-}\dfrac{\sqrt[3]{x + 10}}{2x^2 + 1}$

39. $\lim\limits_{x\to 0^+} f(x)$ and $\lim\limits_{x\to 0^-} f(x)$, where

$$f(x) = \begin{cases} 2x & \text{if } x < 0 \\ x^2 & \text{if } x \ge 0 \end{cases}$$

40. $\lim\limits_{x\to 0^+} f(x)$ and $\lim\limits_{x\to 0^-} f(x)$, where

$$f(x) = \begin{cases} -x + 1 & \text{if } x \le 0 \\ 2x + 3 & \text{if } x > 0 \end{cases}$$

41. $\lim\limits_{x\to 1^+} f(x)$ and $\lim\limits_{x\to 1^-} f(x)$, where

$$f(x) = \begin{cases} \sqrt{x + 3} & \text{if } x \ge 1 \\ 2 + \sqrt{x} & \text{if } x < 1 \end{cases}$$

42. $\lim\limits_{x\to 1^+} f(x)$ and $\lim\limits_{x\to 1^-} f(x)$, where

$$f(x) = \begin{cases} x + 2\sqrt{x - 1} & \text{if } x \ge 1 \\ 1 - \sqrt{1 - x} & \text{if } x < 1 \end{cases}$$

In Exercises 43–50, determine the values of *x*, if any, at which each function is discontinuous. At each point of discontinuity, state the condition(s) for continuity that are violated.

43. $f(x) = \begin{cases} 2x - 4 & \text{if } x \le 0 \\ 1 & \text{if } x > 0 \end{cases}$

44. $f(x) = \begin{cases} x^2 + 1 & \text{if } x \ne 0 \\ 0 & \text{if } x = 0 \end{cases}$

45. 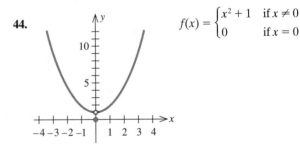 $f(x) = \begin{cases} x + 5 & \text{if } x \le 0 \\ -x^2 + 5 & \text{if } x > 0 \end{cases}$

46. 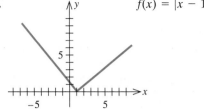 $f(x) = |x - 1|$

47. $f(x) = \begin{cases} x + 5 & \text{if } x < 0 \\ 2 & \text{if } x = 0 \\ -x^2 + 5 & \text{if } x > 0 \end{cases}$

48. $f(x) = \begin{cases} \dfrac{x^2 - 1}{x + 1} & \text{if } x \neq -1 \\ 1 & \text{if } x = -1 \end{cases}$

49. $f(x) = \begin{cases} -|x| + 1 & \text{if } x \neq 0 \\ 0 & \text{if } x = 0 \end{cases}$

50. $f(x) = \begin{cases} \dfrac{1}{x^2} & \text{if } x \neq 0 \\ 1 & \text{if } x = 0 \end{cases}$

In Exercises 51–66, find the values of *x* for which each function is continuous.

51. $f(x) = 2x^2 + x - 1$

52. $f(x) = x^3 - 2x^2 + x - 1$

53. $f(x) = \dfrac{2}{x^2 + 1}$ **54.** $f(x) = \dfrac{x}{2x^2 + 1}$

55. $f(x) = \dfrac{2}{2x - 1}$ **56.** $f(x) = \dfrac{x + 1}{x - 1}$

57. $f(x) = \dfrac{2x + 1}{x^2 + x - 2}$ **58.** $f(x) = \dfrac{x - 1}{x^2 + 2x - 3}$

59. $f(x) = \begin{cases} x & \text{if } x \leq 1 \\ 2x - 1 & \text{if } x > 1 \end{cases}$

60. $f(x) = \begin{cases} -x + 1 & \text{if } x \leq -1 \\ x + 1 & \text{if } x > -1 \end{cases}$

61. $f(x) = \begin{cases} -2x + 1 & \text{if } x < 0 \\ x^2 + 1 & \text{if } x \geq 0 \end{cases}$

62. $f(x) = \begin{cases} x + 1 & \text{if } x \leq 1 \\ -x^2 + 1 & \text{if } x > 1 \end{cases}$

63. $f(x) = \begin{cases} \dfrac{x^2 - 1}{x - 1} & \text{if } x \neq 1 \\ 2 & \text{if } x = 1 \end{cases}$

64. $f(x) = \begin{cases} \dfrac{x^2 - 4}{x + 2} & \text{if } x \neq -2 \\ 1 & \text{if } x = -2 \end{cases}$

65. $f(x) = |x + 1|$ **66.** $f(x) = \dfrac{|x - 1|}{x - 1}$

In Exercises 67–70, determine all values of *x* at which the function is discontinuous.

67. $f(x) = \dfrac{2x}{x^2 - 1}$ **68.** $f(x) = \dfrac{1}{(x - 1)(x - 2)}$

69. $f(x) = \dfrac{x^2 - 2x}{x^2 - 3x + 2}$ **70.** $f(x) = \dfrac{x^2 - 3x + 2}{x^2 - 2x}$

71. THE POSTAGE FUNCTION The graph of the "postage function"

$$f(x) = \begin{cases} 33 & \text{if } 0 < x \leq 1 \\ 56 & \text{if } 1 < x \leq 2 \\ \vdots \\ 286 & \text{if } 11 < x \leq 12 \end{cases}$$

where x denotes the weight of a parcel in ounces and $f(x)$ the postage in cents, is shown in the figure on page 148. Determine the values of x for which f is discontinuous.

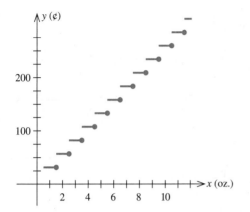

72. INVENTORY CONTROL As part of an optimal inventory policy, the manager of an office supply company orders 500 reams of photocopy paper every 20 days. The accompanying graph shows the *actual* inventory level of paper in an office supply store during the first 60 business days of 1996. Determine the values of t for which the "inventory function" is discontinuous and give an interpretation of the graph.

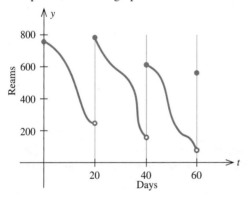

73. LEARNING CURVES The following graph describes the progress Michael made in solving a problem correctly during a mathematics quiz. Here, y denotes the percentage of work completed, and x is measured in minutes. Give an interpretation of the graph.

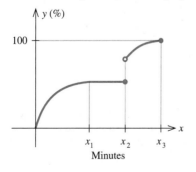

74. AILING FINANCIAL INSTITUTIONS The Franklin Savings and Loan Company acquired two ailing financial institutions in 1992. One of them was acquired at time $t = T_1$, and the other was acquired at time $t = T_2$ ($t = 0$ corresponds to the beginning of 1992). The following graph shows the total amount of money on deposit with Franklin. Explain the significance of the discontinuities of the function at T_1 and T_2.

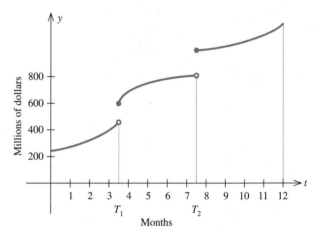

75. ENERGY CONSUMPTION The following graph shows the amount of home heating oil remaining in a 200-gallon tank over a 120-day period ($t = 0$ corresponds to October 1). Explain why the function is discontinuous at $t = 40$, $t = 70$, $t = 95$, and $t = 110$.

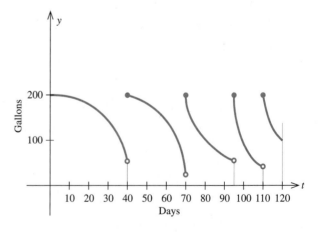

76. PRIME INTEREST RATE The function P, whose graph follows, gives the prime rate (the interest rate banks charge their best corporate customers) as a function of time for the first 32 wk in 1989. Determine the values of t for which P is discontinuous and interpret your results.

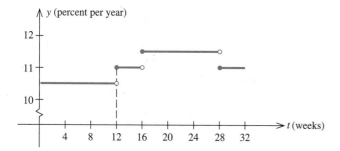

77. **ADMINISTRATION OF AN INTRAVENOUS SOLUTION** A dextrose solution is being administered to a patient intravenously. The 1-liter (L) bottle holding the solution is removed and replaced by another as soon as the contents drop to approximately 5% of the initial (1-L) amount. The rate of discharge is constant, and it takes 6 hr to discharge 95% of the contents of a full bottle. Draw a graph showing the amount of dextrose solution in a bottle in the IV system over a 24-hr period, assuming that we started with a full bottle.

78. **COMMISSIONS** The base salary of a salesman working on commission is $12,000. For each $50,000 of sales beyond $100,000, he is paid a $1000 commission. Sketch a graph showing his earnings as a function of the level of his sales x. Determine the values of x for which the function f is discontinuous.

79. **PARKING FEES** The fee charged per car in a downtown parking lot is $1 for the first half hour and $.50 for each additional half hour or part thereof, subject to a maximum of $5. Derive a function f relating the parking fee to the length of time a car is left in the lot. Sketch the graph of f and determine the values of x for which the function f is discontinuous.

80. **COMMODITY PRICES** The function that gives the cost of a certain commodity is defined by

$$C(x) = \begin{cases} 5x & \text{if } 0 < x < 10 \\ 4x & \text{if } 10 \le x < 30 \\ 3.5x & \text{if } 30 \le x < 60 \\ 3.25x & \text{if } x \ge 60 \end{cases}$$

where x is the number of pounds of a certain commodity sold and $C(x)$ is measured in dollars. Sketch the graph of the function C and determine the values of x for which the function C is discontinuous.

81. **ENERGY EXPENDED BY A FISH** Suppose that a fish swimming a distance of L ft at a speed of v ft/sec relative to the water and against a current flowing at the rate of u ft/sec ($u < v$) expends a total energy given by

$$E(v) = \frac{aLv^3}{v - u}$$

where E is measured in foot-pounds (ft-lb) and a is a constant.

   a. Evaluate $\lim_{v \to u^+} E(v)$ and interpret your result.

   b. Evaluate $\lim_{v \to \infty} E(v)$ and interpret your result.

82. Let

$$f(x) = \begin{cases} x + 2 & \text{if } x \le 1 \\ kx^2 & \text{if } x > 1 \end{cases}$$

Find the value of k that will make f continuous on $(-\infty, \infty)$.

83. Let

$$f(x) = \begin{cases} \dfrac{x^2 - 4}{x + 2} & \text{if } x \ne -2 \\ k & \text{if } x = -2 \end{cases}$$

For what value of k will f be continuous on $(-\infty, \infty)$?

84. **a.** Suppose f is continuous at a and g is discontinuous at a. Is the sum $f + g$ discontinuous at a? Explain.

   b. Suppose f and g are both discontinuous at a. Is the sum $f + g$ necessarily discontinuous at a? Explain.

85. **a.** Suppose f is continuous at a and g is discontinuous at a. Is the product fg necessarily discontinuous at a? Explain.

   b. Suppose f and g are both discontinuous at a. Is the product fg necessarily discontinuous at a? Explain.

In Exercises 86–89, (a) show that the function f is continuous for all values of x in the interval $[a, b]$ and (b) prove that f must have at least one zero in the interval (a, b) by showing that $f(a)$ and $f(b)$ have opposite signs.

86. $f(x) = x^2 - 6x + 8$; $a = 1$, $b = 3$

87. $f(x) = x^3 - 2x^2 + 3x + 2$; $a = -1$, $b = 1$

88. $f(x) = 2x^3 - 3x^2 - 36x + 14$; $a = 0$, $b = 1$

89. $f(x) = 2x^{5/3} - 5x^{4/3}$; $a = 14$, $b = 16$

(continued on p. 154)

Using Technology

FINDING THE POINTS OF DISCONTINUITY OF A FUNCTION

You can very often recognize the points of discontinuity of a function f by examining its graph. For example, Figure T1 shows the graph of $f(x) = x/(x^2 - 1)$ obtained using a graphing utility. It is evident that f is discontinuous at $x = -1$ and $x = 1$. This observation is also borne out by the fact that both these points are not in the domain of f.

FIGURE T1

The graph of $f(x) = \dfrac{x}{x^2 - 1}$ in the viewing rectangle $[-4, 4] \times [-10, 10]$

Consider the function

$$g(x) = \frac{2x^3 + x^2 - 7x - 6}{x^2 - x - 2}$$

Using a graphing utility we obtain the graph of g shown in Figure T2. An examination of this graph does not reveal any points of discontinuity. However, if we factor both the numerator and the denominator of the rational expression, we see that

$$g(x) = \frac{(x + 1)(x - 2)(2x + 3)}{(x + 1)(x - 2)}$$
$$= 2x + 3$$

provided $x \neq -1$ and $x \neq 2$, so that its graph in fact looks like that shown in Figure T3.

FIGURE T2

The graph of $g(x) = \dfrac{2x^3 + x^2 - 7x - 6}{x^2 - x - 2}$ in the standard viewing rectangle

150

This example shows the limitation of the graphing utility and reminds us of the importance of studying functions analytically!

GRAPHING FUNCTIONS DEFINED PIECEWISE

The following example illustrates how to plot the graphs of functions defined in a piecewise manner on a graphing utility.

EXAMPLE 1   Plot the graph of

$$f(x) = \begin{cases} x + 1 & \text{if } x \leq 1 \\ \dfrac{2}{x} & \text{if } x > 1 \end{cases}$$

SOLUTION ✔   We enter the function

$$y1 = (x + 1)(x \leq 1) + (2/x)(x > 1)$$

The graph of the function in the viewing rectangle $[-5, 5] \times [-2, 4]$ is shown in Figure T4.

FIGURE T4
The graph of f in the viewing rectangle $[-5, 5] \times [-2, 4]$

EXAMPLE 2 The percentage of U.S. households, $P(t)$, watching television during weekdays between the hours of 4 P.M. and 4 A.M. is given by

$$P(t) = \begin{cases} 0.01354t^4 - 0.49375t^3 + 2.58333t^2 + 3.8t + 31.60704 & \text{if } 0 \le t \le 8 \\ 1.35t^2 - 33.05t + 208 & \text{if } 8 < t \le 12 \end{cases}$$

where t is measured in hours, with $t = 0$ corresponding to 4 P.M. Plot the graph of P in the viewing rectangle $[0, 12] \times [0, 80]$.
Source: A. C. Nielsen Co.

SOLUTION ✔ We enter the function

$$y2 = (0.01354x^4 - 0.49375x^3 + 2.58333x^2 + 3.8x + 31.60704)(x \ge 0)(x \le 8)$$
$$+ (1.35x^2 - 33.05x + 208)(x > 8)(x \le 12)$$

The graph of P is shown in Figure T5.

FIGURE T5
The graph of P in the viewing rectangle $[0, 12] \times [0, 80]$

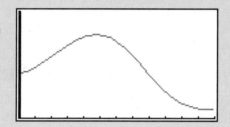

Exercises

In Exercises 1–10, use a graphing utility to plot the graph of *f* and to spot the points of discontinuity of *f*. Then use analytical means to verify your observation and find all points of discontinuity.

1. $f(x) = \dfrac{2}{x^2 - x}$

2. $f(x) = \dfrac{2x + 1}{x^2 + x - 2}$

3. $f(x) = \dfrac{\sqrt{x}}{x^2 - x - 2}$

4. $f(x) = \dfrac{3}{\sqrt{x}(x + 1)}$

5. $f(x) = \dfrac{6x^3 + x^2 - 2x}{2x^2 - x}$

6. $f(x) = \dfrac{2x^3 - x^2 - 13x - 6}{2x^2 - 5x - 3}$

7. $f(x) = \dfrac{2x^4 - 3x^3 - 2x^2}{2x^2 - 3x - 2}$

8. $f(x) = \dfrac{6x^4 - x^3 + 5x^2 - 1}{6x^2 - x - 1}$

9. $f(x) = \dfrac{x^3 + x^2 - 2x}{x^4 + 2x^3 - x - 2}$

Hint: $x^4 + 2x^3 - x - 2 = (x^3 - 1)(x + 2)$

10. $f(x) = \dfrac{x^3 - x}{x^{4/3} - x + x^{1/3} - 1}$

Hint: $x^{4/3} - x + x^{1/3} - 1 = (x^{1/3} - 1)(x + 1)$
Can you explain why part of the graph is missing?

In Exercises 11–14, use a graphing utility to plot the graph of *f* in the indicated viewing rectangle.

11. $f(x) = \begin{cases} -1 & \text{if } x \le 1 \\ x+1 & \text{if } x > 1; [-5,5] \times [-2,8] \end{cases}$

12. $f(x) = \begin{cases} \dfrac{1}{3}x^2 - 2x & \text{if } x \le 3 \\ -x+6 & \text{if } x > 3; [0,7] \times [-5,5] \end{cases}$

13. $f(x) = \begin{cases} 2 & \text{if } x \le 0 \\ \sqrt{4-x^2} & \text{if } x > 0; [-2,2] \times [-4,4] \end{cases}$

14. $f(x) = \begin{cases} -x^2 + x + 2 & \text{if } x \le 1 \\ 2x^3 - x^2 - 4 & \text{if } x > 1; [-4,4] \times [-5,5] \end{cases}$

15. **FLIGHT PATH OF A PLANE** The function

$$f(x) = \begin{cases} 0 & \text{if } 0 \le x < 1 \\ -0.00411523x^3 + 0.0679012x^2 & \\ \quad - 0.123457x + 0.0596708 & \text{if } 1 \le x < 10 \\ 1.5 & \text{if } 10 \le x \le 100 \end{cases}$$

where both x and $f(x)$ are measured in units of 1000 ft, describes the flight path of a plane taking off from the origin and climbing to an altitude of 15,000 ft. Plot the graph of f to visualize the trajectory of the plane.

16. **HOME SHOPPING INDUSTRY** According to industry sources, revenue from the home shopping industry for the years since its inception may be approximated by the function

$$R(t) = \begin{cases} -0.03t^3 + 0.25t^2 - 0.12t & \text{if } 0 \le t \le 3 \\ 0.57t - 0.63 & \text{if } 3 < t \le 11 \end{cases}$$

where $R(t)$ measures the revenue in billions of dollars and t is measured in years, with $t = 0$ corresponding to the beginning of 1984. Plot the graph of R.

Source: Paul Kagan Associates

In Exercises 90–91, use the intermediate value theorem to find the value of *c* such that *f*(*c*) = *M*.

90. $f(x) = x^2 - x + 1$ on $[-1, 4]$; $M = 7$

91. $f(x) = x^2 - 4x + 6$ on $[0, 3]$; $M = 2$

92. Use the method of bisection (see Example 6) to find the root of the equation $x^3 - x + 1 = 0$ accurate to two decimal places.

93. Use the method of bisection to find the root of the equation $x^5 + 2x - 7 = 0$ accurate to two decimal places.

94. Joan is looking straight out a window of an apartment building at a height of 32 ft from the ground. A boy throws a tennis ball straight up by the side of the building where the window is located. Suppose the height of the ball (measured in feet) from the ground at time *t* is $h(t) = 4 + 64t - 16t^2$.
   a. Show that $h(0) = 4$ and $h(2) = 68$.
   b. Use the intermediate value theorem to conclude that the ball must cross Joan's line of sight at least once.
   c. At what time(s) does the ball cross Joan's line of sight? Interpret your results.

In Exercises 95–99, determine whether the statement is true or false. If it is true, explain why it is true. If it is false, give an example to show why it is false.

95. Suppose the function *f* is defined on the interval $[a, b]$. If $f(a)$ and $f(b)$ have the same sign, then *f* has no zero in $[a, b]$.

96. If $\lim_{x \to a} f(x) = L$, then $\lim_{x \to a^+} f(x) - \lim_{x \to a^-} f(x) \neq 0$.

97. If $\lim_{x \to a^-} f(x) = L$ and $\lim_{x \to a^+} f(x) = L$, then $f(a) = L$.

98. If $\lim_{x \to a} f(x) = L$ and $g(a) = M$, then $\lim_{x \to a} f(x)g(x) = LM$.

99. If *f* is continuous on $[-2, 3]$, $f(-2) = 3$ and $f(3) = 1$, then there exists at least one number *c* in $[-2, 3]$ such that $f(c) = 2$.

100. Let $f(x) = x - \sqrt{1 - x^2}$.
   a. Show that *f* is continuous for all values of *x* in the interval $[-1, 1]$.
   b. Show that *f* has at least one zero in $[-1, 1]$.
   c. Find the zeros of *f* in $[-1, 1]$ by solving the equation $f(x) = 0$.

101. Let $f(x) = \dfrac{x^2}{x^2 + 1}$.
   a. Show that *f* is continuous for all values of *x*.
   b. Show that $f(x)$ is nonnegative for all values of *x*.
   c. Show that *f* has a zero at $x = 0$. Does this contradict Theorem 5?

102. a. Prove that a polynomial function $y = P(x)$ is continuous at every point *x*. Follow these steps:
   (1) Use Properties 2 and 3 of continuous functions to establish that the function $g(x) = x^n$, where *n* is a positive integer, is continuous everywhere.
   (2) Use Properties 1 and 5 to show that $f(x) = cx^n$, where *c* is a constant and *n* is a positive integer, is continuous everywhere.
   (3) Use Property 4 to complete the proof of the result.
   b. Prove that a rational function $R(x) = p(x)/q(x)$ is continuous at every point *x*, where $q(x) \neq 0$.

   Hint: Use the result of part (a) and Property 6.

103. Show that the conclusion of the intermediate value theorem does not hold if *f* is discontinuous on $[a, b]$.

SOLUTIONS TO SELF-CHECK EXERCISES 2.5

1. For $x < -1, f(x) = 1$, and so

$$\lim_{x \to -1^-} f(x) = \lim_{x \to -1^-} 1 = 1$$

For $x \geq -1, f(x) = 1 + \sqrt{x + 1}$, and so

$$\lim_{x \to -1^+} f(x) = \lim_{x \to -1^+} (1 + \sqrt{x + 1}) = 1$$

Since the left-hand and right-hand limits of f exist as x approaches $x = -1$ and both are equal to 1, we conclude that

$$\lim_{x \to -1} f(x) = 1$$

2. a. The graph of f is as follows:

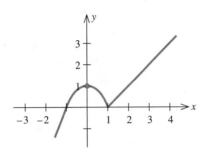

We see that f is continuous everywhere.
b. The graph of g is as follows:

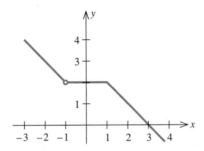

Since g is not defined at $x = -1$, it is discontinuous there. It is continuous everywhere else.

2.6 The Derivative

FIGURE 2.43
Graph showing the position s of a maglev at time t

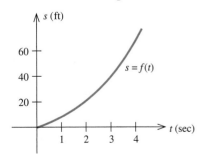

AN INTUITIVE EXAMPLE

We mentioned in Section 2.4 that the problem of finding the *rate of change* of one quantity with respect to another is mathematically equivalent to the problem of finding the *slope of the tangent line* to a curve at a given point on the curve. Before going on to establish this relationship, let's show its plausibility by looking at it from an intuitive point of view.

Consider the motion of the maglev discussed in Section 2.4. Recall that the position of the maglev at any time t is given by

$$s = f(t) = 4t^2 \qquad (0 \le t \le 30)$$

where s is measured in feet and t in seconds. The graph of the function f is sketched in Figure 2.43.

Observe that the graph of f rises slowly at first but more rapidly as t increases, reflecting the fact that the speed of the maglev is increasing with time. This observation suggests a relationship between the speed of the maglev at any time t and the *steepness* of the curve at the point corresponding to this value of t. Thus, it would appear that we can solve the problem of finding the speed of the maglev at any time if we can find a way to measure the steepness of the curve at any point on the curve.

To discover a yardstick that will measure the steepness of a curve, consider the graph of a function f such as the one shown in Figure 2.44a. Think of the curve as representing a stretch of roller coaster track (Figure 2.44b). When the car is at the point P on the curve, a passenger sitting erect in the car and looking straight ahead will have a line of sight that is parallel to the line T, the tangent to the curve at P.

As Figure 2.44a suggests, the steepness of the curve—that is, the rate at which y is increasing or decreasing with respect to x—is given by the slope of the tangent line to the graph of f at the point $P(x, f(x))$. But for now we will show how this relationship can be used to estimate the rate of change of a function from its graph.

FIGURE 2.44

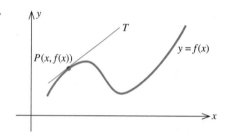

(a) T is the tangent line to the curve at P.

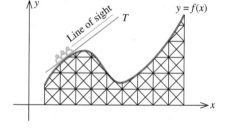

(b) T is parallel to the line of sight.

EXAMPLE 1   The graph of the function $y = N(t)$, shown in Figure 2.45, gives the number of Social Security beneficiaries from the beginning of 1990 ($t = 0$) through the year 2045 ($t = 55$).

FIGURE 2.45
The number of Social Security beneficiaries from 1990 through 2045. We can use the slope of the tangent line at the indicated points to estimate the rate at which the number of Social Security beneficiaries will be changing.

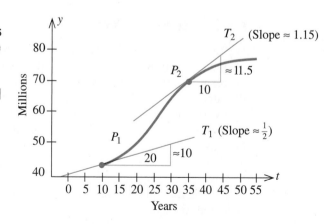

Use the graph of $y = N(t)$ to estimate the rate at which the number of Social Security beneficiaries was growing at the beginning of the year 2000 ($t = 10$). How fast will the number be growing at the beginning of 2025 ($t = 35$)? [Assume that the rate of change of the function N at any value of t is given by the slope of the tangent line at the point $P(t, N(t))$.]
Source: Social Security Administration

SOLUTION ✔

From the figure, we see that the slope of the tangent line T_1 to the graph of $y = N(t)$ at $P_1(10, 44.7)$ is approximately 0.5. This tells us that the quantity y is increasing at the rate of 1/2 unit per unit increase in t, when $t = 10$. In other words, at the beginning of the year 2000, the number of Social Security beneficiaries was increasing at the rate of approximately 0.5 million, or 500,000, per year.

The slope of the tangent line T_2 at $P_2(35, 71.9)$ is approximately 1.15. This tells us that at the beginning of 2025 the number of Social Security beneficiaries will be growing at the rate of approximately 1.15 million, or 1,150,000, per year.

■ ■ ■ ■

SLOPE OF A TANGENT LINE

In Example 1 we answered the questions raised by drawing the graph of the function N and estimating the position of the tangent lines. Ideally, however, we would like to solve a problem analytically whenever possible. To do this we need a precise definition of the slope of a tangent line to a curve.

To define the tangent line to a curve C at a point P on the curve, fix P and let Q be any point on C distinct from P (Figure 2.46). The straight line passing through P and Q is called a **secant line.**

FIGURE 2.46

As Q approaches P along the curve C, the secant lines approach the tangent line T.

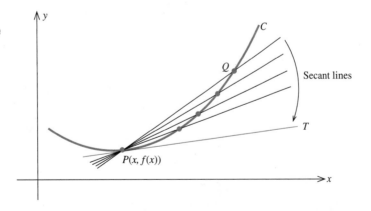

Now, as the point Q is allowed to move toward P along the curve, the secant line through P and Q rotates about the fixed point P and approaches a fixed line through P. This fixed line, which is the limiting position of the secant lines through P and Q as Q approaches P, is the **tangent line to the graph of f** at the point P.

FIGURE 2.47

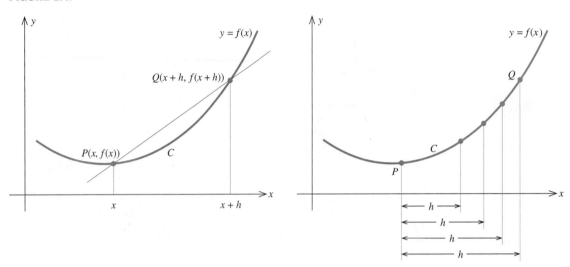

(a) The points $P(x, f(x))$ and $Q(x + h, f(x + h))$

(b) As h approaches zero, Q approaches P.

We can describe the process more precisely as follows. Suppose the curve C is the graph of a function f defined by $y = f(x)$. Then the point P is described by $P(x, f(x))$ and the point Q by $Q(x + h, f(x + h))$, where h is some appropriate nonzero number (Figure 2.47a). Observe that we can make Q approach P along the curve C by letting h approach zero (Figure 2.47b).

Next, using the formula for the slope of a line, we can write the slope of the secant line passing through $P(x, f(x))$ and $Q(x + h, f(x + h))$ as

$$\frac{f(x + h) - f(x)}{(x + h) - x} = \frac{f(x + h) - f(x)}{h} \tag{5}$$

As observed earlier, Q approaches P, and therefore the secant line through P and Q approaches the tangent line T as h approaches zero. Consequently, we might expect that the slope of the secant line would approach the slope of the tangent line T as h approaches zero. This leads to the following definition.

Slope of a Tangent Line

The slope of the tangent line to the graph f at the point $P(x, f(x))$ is given by

$$\lim_{h \to 0} \frac{f(x + h) - f(x)}{h} \tag{6}$$

if it exists.

RATES OF CHANGE

We now show that the problem of finding the slope of the tangent line to the graph of a function f at the point $P(x, f(x))$ is mathematically equivalent to the problem of finding the rate of change of f at x. To see this, suppose we are given a function f that describes the relationship between the two quantities x and y:

$$y = f(x)$$

The number $f(x + h) - f(x)$ measures the change in y that corresponds to a change h in x (Figure 2.48).

FIGURE 2.48

$f(x + h) - f(x)$ is the change in y that corresponds to a change h in x.

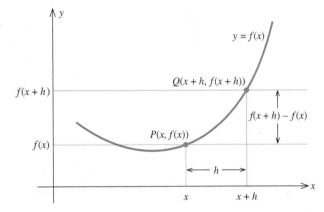

Then, the **difference quotient**

$$\frac{f(x + h) - f(x)}{h} \tag{7}$$

measures the **average rate of change of y with respect to x** over the interval $[x, x + h]$. For example, if y measures the position of a car at time x, then quotient (7) gives the average velocity of the car over the time interval $[x, x + h]$.

Observe that the difference quotient (7) is the same as (5). We conclude that the difference quotient (7) also measures the slope of the secant line that passes through the two points $P(x, f(x))$ and $Q(x + h, f(x + h))$ lying on the graph of $y = f(x)$. Next, by taking the limit of the difference quotient (7) as h goes to zero—that is, by evaluating

$$\lim_{h \to 0} \frac{f(x + h) - f(x)}{h} \tag{8}$$

we obtain the **rate of change of f at x**. For example, if y measures the position of a car at time x, then the limit (8) gives the velocity of the car at time x. For emphasis, the rate of change of a function f at x is often called the **instantaneous rate of change of f at x.** This distinguishes it from the average rate of change of f, which is computed over an *interval* $[x, x + h]$ rather than at a *point x*.

Observe that the limit (8) is the same as (6). Therefore, the limit of the difference quotient also measures the slope of the tangent line to the graph of $y = f(x)$ at the point $(x, f(x))$.

The following summarizes this discussion.

Average and Instantaneous Rates of Change

The **average rate of change** of f over the interval $[x, x + h]$ or **slope of the secant line** to the graph of f through the points $(x, f(x))$ and $(x + h, f(x + h))$ is

$$\frac{f(x + h) - f(x)}{h} \qquad (9)$$

The **instantaneous rate of change** of f at x or **slope of the tangent line** to the graph of f at $(x, f(x))$ is

$$\lim_{h \to 0} \frac{f(x + h) - f(x)}{h} \qquad (10)$$

> **Group Discussion**
> Explain the difference between the average rate of change of a function and the instantaneous rate of change of a function.

THE DERIVATIVE

The limit (6), or (10), which measures both the slope of the tangent line to the graph of $y = f(x)$ at the point $P(x, f(x))$ and the (instantaneous) rate of change of f at x is given a special name: the **derivative of f at x.**

Derivative of a Function

The derivative of a function f with respect to x is the function f' (read "f prime"), defined by

$$f'(x) = \lim_{h \to 0} \frac{f(x + h) - f(x)}{h} \qquad (11)$$

The domain of f' is the set of all x where the limit exists.

Thus, the derivative of a function f is a function f' that gives the slope of the tangent line to the graph of f at *any* point $(x, f(x))$ and also the rate of change of f at x (Figure 2.49).

FIGURE 2.49
The slope of the tangent line at $P(x, f(x))$ is $f'(x)$; f changes at the rate of $f'(x)$ units per unit change in x at x.

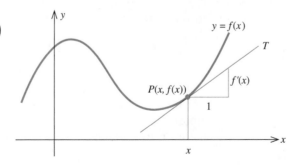

Other notations for the derivative of f include:

$$D_x f(x) \qquad \text{(Read "d sub x of f of x")}$$

$$\frac{dy}{dx} \qquad \text{(Read "d y d x")}$$

$$y' \qquad \text{(Read "y prime")}$$

The last two are used when the rule for f is written in the form $y = f(x)$.

The calculation of the derivative of f is facilitated using the following four-step process.

Four-Step Process for Finding $f'(x)$

1. Compute $f(x + h)$.
2. Form the difference $f(x + h) - f(x)$.
3. Form the quotient $\dfrac{f(x + h) - f(x)}{h}$.
4. Compute $f'(x) = \lim\limits_{h \to 0} \dfrac{f(x + h) - f(x)}{h}$.

EXAMPLE 2   Find the slope of the tangent line to the graph of $f(x) = 3x + 5$ at any point $(x, f(x))$.

SOLUTION ✔   The slope of the tangent line at any point on the graph of f is given by the derivative of f at x. To find the derivative, we use the four-step process:

Step 1 $f(x + h) = 3(x + h) + 5 = 3x + 3h + 5$
Step 2 $f(x + h) - f(x) = (3x + 3h + 5) - (3x + 5) = 3h$
Step 3 $\dfrac{f(x + h) - f(x)}{h} = \dfrac{3h}{h} = 3$
Step 4 $f'(x) = \lim\limits_{h \to 0} \dfrac{f(x + h) - f(x)}{h} = \lim\limits_{h \to 0} 3 = 3$

We expect this result since the tangent line to any point on a straight line must coincide with the line itself and therefore must have the same slope as the line. In this case the graph of f is a straight line with slope 3.   ■ ■ ■ ■

EXAMPLE 3   Let $f(x) = x^2$.

a. Compute $f'(x)$.
b. Compute $f'(2)$ and interpret your result.

SOLUTION ✔

a. To find $f'(x)$, we use the four-step process:

Step 1 $f(x + h) = (x + h)^2 = x^2 + 2xh + h^2$

Step 2 $f(x + h) - f(x) = x^2 + 2xh + h^2 - x^2 = 2xh + h^2 = h(2x + h)$

Step 3 $\dfrac{f(x + h) - f(x)}{h} = \dfrac{h(2x + h)}{h} = 2x + h$

Step 4 $f'(x) = \lim\limits_{h \to 0} \dfrac{f(x + h) - f(x)}{h} = \lim\limits_{h \to 0} (2x + h) = 2x$

b. $f'(2) = 2(2) = 4$. This result tells us that the slope of the tangent line to the graph of f at the point $(2, 4)$ is 4. It also tells us that the function f is changing at the rate of 4 units per unit change in x at $x = 2$. The graph of f and the tangent line at $(2, 4)$ are shown in Figure 2.50.   ■ ■ ■ ■

FIGURE 2.50
The tangent line to the graph of $f(x) = x^2$ at $(2, 4)$

Exploring with Technology

1. Consider the function $f(x) = x^2$ of Example 3. Suppose we want to compute $f'(2)$ using Equation (11). Thus,

$$f'(2) = \lim_{h \to 0} \frac{f(2 + h) - f(2)}{h} = \lim_{h \to 0} \frac{(2 + h)^2 - 2^2}{h}$$

Use a graphing utility to plot the graph of

$$g(x) = \frac{(2 + x)^2 - 4}{x}$$

in the viewing rectangle $[-3, 3] \times [-2, 6]$.

2. Use ZOOM and TRACE to find $\lim\limits_{x \to 0} g(x)$.

3. Explain why the limit found in part 2 is $f'(2)$.

EXAMPLE **4**

Let $f(x) = x^2 - 4x$.

a. Compute $f'(x)$.

b. Find the point on the graph of f where the tangent line to the curve is horizontal.

c. Sketch the graph of f and the tangent line to the curve at the point found in part (b).

d. What is the rate of change of f at this point?

SOLUTION ✔

a. To find $f'(x)$, we use the four-step process:

Step 1 $f(x + h) = (x + h)^2 - 4(x + h) = x^2 + 2xh + h^2 - 4x - 4h$

Step 2 $f(x + h) - f(x) = x^2 + 2xh + h^2 - 4x - 4h - (x^2 - 4x)$

$$= 2xh + h^2 - 4h = h(2x + h - 4)$$

FIGURE 2.51
The tangent line to the graph of $y = x^2 - 4x$ at $(2, -4)$ is $y = -4$.

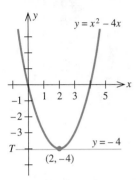

Step 3 $\dfrac{f(x + h) - f(x)}{h} = \dfrac{h(2x + h - 4)}{h} = 2x + h - 4$

Step 4 $f'(x) = \lim\limits_{h \to 0} \dfrac{f(x + h) - f(x)}{h} = \lim\limits_{h \to 0} (2x + h - 4) = 2x - 4$

b. At a point on the graph of f where the tangent line to the curve is horizontal and hence has slope zero, the derivative f' of f is zero. Accordingly, to find such point(s) we set $f'(x) = 0$, which gives $2x - 4 = 0$, or $x = 2$. The corresponding value of y is given by $y = f(2) = -4$, and the required point is $(2, -4)$.

c. The graph of f and the tangent line are shown in Figure 2.51.

d. The rate of change of f at $x = 2$ is zero.   ■ ■ ■ ■

EXAMPLE 5   Let $f(x) = 1/x$.

a. Compute $f'(x)$.

b. Find the slope of the tangent line T to the graph of f at the point where $x = 1$.

c. Find an equation of the tangent line T in part (b).

SOLUTION ✔   **a.** To find $f'(x)$, we use the four-step process:

Step 1 $f(x + h) = \dfrac{1}{x + h}$

Group Discussion
Can the tangent line to the graph of a function intersect the graph at more than one point? Explain your answer using illustrations.

Step 2 $f(x + h) - f(x) = \dfrac{1}{x + h} - \dfrac{1}{x} = \dfrac{x - (x + h)}{x(x + h)} = -\dfrac{h}{x(x + h)}$

Step 3 $\dfrac{f(x + h) - f(x)}{h} = -\dfrac{h}{x(x + h)} \cdot \dfrac{1}{h} = -\dfrac{1}{x(x + h)}$

Step 4 $f'(x) = \lim\limits_{h \to 0} \dfrac{f(x + h) - f(x)}{h} = \lim\limits_{h \to 0} -\dfrac{1}{x(x + h)} = -\dfrac{1}{x^2}$

FIGURE 2.52
The tangent line to the graph of $f(x) = 1/x$ at $(1, 1)$

b. The slope of the tangent line T to the graph of f where $x = 1$ is given by $f'(1) = -1$.

c. When $x = 1$, $y = f(1) = 1$ and T is tangent to the graph of f at the point $(1, 1)$. From part (b), we know that the slope of T is -1. Thus, an equation of T is

$$y - 1 = -1(x - 1)$$
$$y = -x + 2$$

(Figure 2.52).   ■ ■ ■ ■

Exploring with Technology

1. Use the results of Example 5 to draw the graph of $f(x) = 1/x$ and its tangent line at the point $(1, 1)$ by plotting the graphs of $y_1 = 1/x$ and $y_2 = -x + 2$ in the viewing rectangle $[-4, 4] \times [-4, 4]$.
2. Some graphing utilities draw the tangent line to the graph of a function at a given point automatically—you need only specify the function and give the x-coordinate of the point of tangency. If your graphing utility has this feature, verify the result of part 1 without finding an equation of the tangent line.

Group Discussion

Consider the following alternative approach to the definition of the derivative of a function: Let h be a positive number and suppose $P(x - h, f(x - h))$ and $Q(x + h, f(x + h))$ are two points on the graph of f.

1. Give a geometric and a physical interpretation of the quotient

$$\frac{f(x + h) - f(x - h)}{2h}$$

Make a sketch to illustrate your answer.

2. Give a geometric and a physical interpretation of the limit

$$\lim_{h \to 0} \frac{f(x + h) - f(x - h)}{2h}$$

Make a sketch to illustrate your answer.

3. Explain why it makes sense to define

$$f'(x) = \lim_{h \to 0} \frac{f(x + h) - f(x - h)}{2h}$$

4. Using the definition given in part (c), formulate a four-step process for finding $f'(x)$ similar to that given on page 161 and use it to find the derivative of $f(x) = x^2$. Compare your answer with that obtained in Example 3 on page 162.

APPLICATIONS

EXAMPLE 6   Suppose the distance (in feet) covered by a car moving along a straight road t seconds after starting from rest is given by the function $f(t) = 2t^2$ ($0 \leq t \leq 30$).

a. Calculate the average velocity of the car over the time intervals [22, 23], [22, 22.1], and [22, 22.01].
b. Calculate the (instantaneous) velocity of the car when $t = 22$.
c. Compare the results obtained in part (a) with that obtained in part (b).

SOLUTION ✔

a. We first compute the average velocity (average rate of change of f) over the interval $[t, t + h]$ using Formula (9). We find

$$\frac{f(t + h) - f(t)}{h} = \frac{2(t + h)^2 - 2t^2}{h}$$

$$= \frac{2t^2 + 4th + 2h^2 - 2t^2}{h}$$

$$= 4t + 2h$$

Next, using $t = 22$ and $h = 1$, we find that the average velocity of the car over the time interval $[22, 23]$ is

$$4(22) + 2(1) = 90$$

or 90 feet per second. Similarly, using $t = 22$, $h = 0.1$, and $h = 0.01$, we find that its average velocities over the time intervals $[22, 22.1]$ and $[22, 22.01]$ are 88.2 and 88.02 feet per second, respectively.

b. Using the limit (10), we see that the instantaneous velocity of the car at any time t is given by

$$\lim_{h \to 0} \frac{f(t + h) - f(t)}{h} = \lim_{h \to 0}(4t + 2h) \qquad \text{[Using the results from part (a)]}$$

$$= 4t$$

In particular, the velocity of the car 22 seconds from rest ($t = 22$) is given by

$$v = 4(22)$$

or 88 feet per second.

c. The computations in part (a) show that, as the time intervals over which the average velocity of the car are computed become smaller and smaller, the average velocities over these intervals do approach 88 feet per second, the instantaneous velocity of the car at $t = 22$.   ■ ■ ■ ■

EXAMPLE 7

The management of the Titan Tire Company has determined that the weekly demand function for their Super Titan tires is given by

$$p = f(x) = 144 - x^2$$

where p is measured in dollars and x is measured in units of a thousand (Figure 2.53).

FIGURE 2.53
The graph of the demand function $p = 144 - x^2$

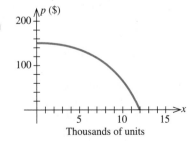

Thousands of units

a. Find the average rate of change in the unit price of a tire if the quantity demanded is between 5000 and 6000 tires, between 5000 and 5100 tires, and between 5000 and 5010 tires.

b. What is the instantaneous rate of change of the unit price when the quantity demanded is 5000 units?

SOLUTION ✔

a. The average rate of change of the unit price of a tire if the quantity demanded is between x and $x + h$ is

$$\frac{f(x + h) - f(x)}{h} = \frac{[144 - (x + h)^2] - (144 - x^2)}{h}$$

$$= \frac{144 - x^2 - 2x(h) - h^2 - 144 + x^2}{h}$$

$$= -2x - h$$

To find the average rate of change of the unit price of a tire when the quantity demanded is between 5000 and 6000 tires (that is, over the interval $[5, 6]$), we take $x = 5$ and $h = 1$, obtaining

$$-2(5) - 1 = -11$$

or $-\$11$ per 1000 tires. (Remember, x is measured in units of a thousand.) Similarly, taking $h = 0.1$ and $h = 0.01$ with $x = 5$, we find that the average rates of change of the unit price when the quantities demanded are between 5000 and 5100 and between 5000 and 5010 are $-\$10.10$ and $-\$10.01$ per 1000 tires, respectively.

b. The instantaneous rate of change of the unit price of a tire when the quantity demanded is x units is given by

$$\lim_{h \to 0} \frac{f(x + h) - f(x)}{h} = \lim_{h \to 0} (-2x - h) \qquad \text{[Using the results from part (a)]}$$

$$= -2x$$

In particular, the instantaneous rate of change of the unit price per tire when the quantity demanded is 5000 is given by $-2(5)$, or $-\$10$ per 1000 tires.   ■ ■ ■ ■

The derivative of a function provides us with a tool for measuring the rate of change of one quantity with respect to another. Table 2.4 lists several other applications involving this limit.

Table 2.4

x stands for	y stands for	$\dfrac{f(a + h) - f(a)}{h}$ measures the	$\displaystyle\lim_{h \to 0} \dfrac{f(a + h) - f(a)}{h}$ measures the
Time	**Concentration of a drug** in the bloodstream at time x	Average rate of change in the concentration of the drug over the time interval $[a, a + h]$	Instantaneous rate of change in the concentration of the drug in the bloodstream at time $x = a$
Number of items sold	**Revenue** at a sales level of x units	Average rate of change in the revenue when the sales level is between $x = a$ and $x = a + h$	Instantaneous rate of change in the revenue when the sales level is a units

Table 2.4   (continued)

x stands for	y stands for	$\dfrac{f(a + h) - f(a)}{h}$ measures the	$\displaystyle\lim_{h \to 0} \dfrac{f(a + h) - f(a)}{h}$ measures the
Time	**Volume of sales** at time x	Average rate of change in the volume of sales over the time interval $[a, a + h]$	Instantaneous rate of change in the volume of sales at time $x = a$
Time	**Population** of *Drosophila* (fruit flies) at time x	Average rate of growth of the fruit fly population over the time interval $[a, a + h]$	Instantaneous rate of change of the fruit fly population at time $x = a$
Temperature in a chemical reaction	**Amount of product formed in the chemical reaction** when the temperature is x degrees	Average rate of formation of chemical product over the temperature range $[a, a + h]$	Instantaneous rate of formation of chemical product when the temperature is a degrees

DIFFERENTIABILITY AND CONTINUITY

In practical applications, one encounters functions that fail to be **differentiable**—that is, do not have a derivative at certain values in the domain of the function f. It can be shown that a continuous function f fails to be differentiable at a point $x = a$ when the graph of f makes an abrupt change of direction at that point. We call such a point a "corner." A function also fails to be differentiable at a point where the tangent line is vertical since the slope of a vertical line is undefined. These cases are illustrated in Figure 2.54.

FIGURE 2.54

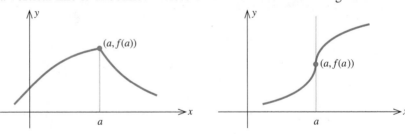

(a) The graph makes an abrupt change of direction at $x = a$.

(b) The slope at $x = a$ is undefined.

The next example illustrates a function that is not differentiable at a point.

EXAMPLE 8

Mary works at the B&O department store, where, on a weekday, she is paid $6 per hour for the first 8 hours and $9 per hour for overtime. The function

$$f(x) = \begin{cases} 6x & \text{if } 0 \le x \le 8 \\ 9x - 24 & \text{if } 8 < x \end{cases}$$

gives Mary's earnings on a weekday in which she worked x hours. Sketch the graph of the function f and explain why it is not differentiable at $x = 8$.

SOLUTION ✔   The graph of f is shown in Figure 2.55. Observe that the graph of f has a corner at $x = 8$ and consequently is not differentiable at $x = 8$.

FIGURE 2.55
The function f is not differentiable at
(8, 48).

We close this section by mentioning the connection between the continuity and the differentiability of a function at a given value $x = a$ in the domain of f. By reexamining the function of Example 8, it becomes clear that f is continuous everywhere and, in particular, when $x = 8$. This shows that in general the continuity of a function at a point $x = a$ does not necessarily imply the differentiability of the function at that point. The converse, however, is true: If a function f is differentiable at a point $x = a$, then it is continuous there.

Exploring with Technology

1. Use a graphing utility to plot the graph of $f(x) = x^{1/3}$ in the viewing rectangle $[-2, 2] \times [-2, 2]$.
2. Use the graphing utility to draw the tangent line to the graph of f at the point (0, 0). Can you explain why the process breaks down?

Differentiability and Continuity

If a function is differentiable at $x = a$, then it is continuous at $x = a$.

For a proof of this result, see Exercise 59, page 174.

EXAMPLE 9   Figure 2.56 depicts a portion of the graph of a function. Explain why the function fails to be differentiable at each of the points $x = a, b, c, d, e, f$, and g.

FIGURE 2.56
The graph of this function is not differentiable at the points a–g.

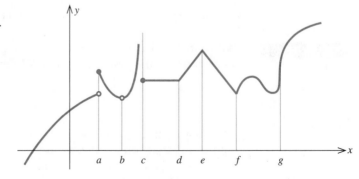

SOLUTION ✔ The function fails to be differentiable at the points $x = a$, b, and c because it is discontinuous at each of these points. The derivative of the function does not exist at $x = d$, e, and f because it has a kink at each of these points. Finally, the function is not differentiable at $x = g$ because the tangent line is vertical at that point.   ■ ■ ■ ■

> **Group Discussion**
>
> Suppose a function f is differentiable at $x = a$. Can there be two tangent lines to the graphs of f at the point $(a, f(a))$? Explain your answer.

SELF-CHECK EXERCISES 2.6

1. Let $f(x) = -x^2 - 2x + 3$.
   a. Find the derivative f' of f, using the definition of the derivative.
   b. Find the slope of the tangent line to the graph of f at the point $(0, 3)$.
   c. Find the rate of change of f when $x = 0$.
   d. Find an equation of the tangent line to the graph of f at the point $(0, 3)$.
   e. Sketch the graph of f and the tangent line to the curve at the point $(0, 3)$.

2. The losses (in millions of dollars) due to bad loans extended chiefly in agriculture, real estate, shipping, and energy by the Franklin Bank are estimated to be

$$A = f(t) = -t^2 + 10t + 30 \qquad (0 \le t \le 10)$$

   where t is the time in years ($t = 0$ corresponds to the beginning of 1994). **How fast were the losses mounting at the beginning of 1997? At the beginning of 1999? How fast will the losses be mounting at the beginning of 2001? Interpret your results.**

Solutions to Self-Check Exercises 2.6 can be found on page 174.

2.6 Exercises

1. **AVERAGE WEIGHT OF AN INFANT** The following graph shows the weight measurements of the average infant from the time of birth ($t = 0$) through age 2 ($t = 24$). By computing the slopes of the respective tangent lines, estimate the rate of change of the average infant's weight when $t = 3$ and when $t = 18$. What is the average rate of change in the average infant's weight over the first year of life?

2. **FORESTRY** The following graph shows the volume of wood produced in a single-species forest. Here $f(t)$ is

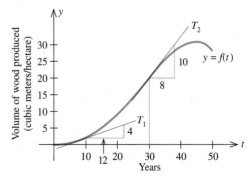

measured in cubic meters per hectare and t is measured in years. By computing the slopes of the respective tangent lines, estimate the rate at which the wood grown is changing at the beginning of year 10 and at the beginning of year 30.

Source: The Random House Encyclopedia

3. **TV-VIEWING PATTERNS** The following graph shows the percentage of U.S. households watching television during a 24-hr period on a weekday ($t = 0$ corresponds to 6 A.M.). By computing the slopes of the respective tangent lines, estimate the rate of change of the percentage of households watching television at 4 P.M. and 11 P.M.

Source: A. C. Nielsen Company

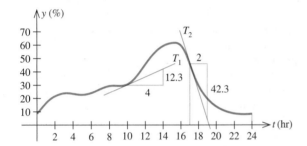

4. **CROP YIELD** Productivity and yield of cultivated crops are often reduced by insect pests. The following graph shows the relationship between the yield of a certain crop, $f(x)$, as a function of the density of aphids x. (Aphids are small insects that suck plant juices.) Here, $f(x)$ is measured in kilograms/4000 square meters, and x is measured in hundreds of aphids per bean stem. By computing the slopes of the respective tangent lines, estimate the rate of change of the crop yield with respect to the density of aphids when that density is 200 aphids/bean stem and when it is 800 aphids/bean stem.

Source: The Random House Encyclopedia

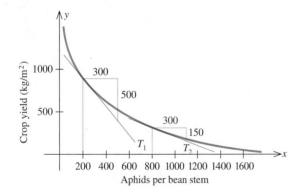

5. The position of car A and car B, starting out side by side and traveling along a straight road, is given by $s = f(t)$ and $s = g(t)$, respectively, where s is measured in feet and t is measured in seconds (see the accompanying figure).

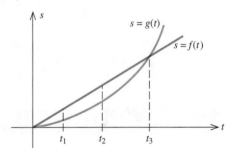

a. Which car is traveling faster at t_1?
b. What can you say about the speed of the cars at t_2?
Hint: Compare tangent lines.
c. Which car is traveling faster at t_3?
d. What can you say about the positions of the cars at t_3?

6. The velocity of car A and car B, starting out side by side and traveling along a straight road, is given by $v = f(t)$ and $v = g(t)$, respectively, where v is measured in feet/second and t is measured in seconds (see the accompanying figure).

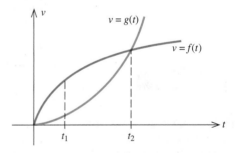

a. What can you say about the velocity and acceleration of the two cars at t_1? (Acceleration is the rate of change of velocity.)
b. What can you say about the velocity and acceleration of the two cars at t_2?

7. **EFFECT OF A BACTERICIDE ON BACTERIA** In the following figure, $f(t)$ gives the population P_1 of a certain bacteria culture at time t after a portion of bactericide A was introduced into the population at $t = 0$. The graph of g gives the population P_2 of a similar bacteria culture at time t after a portion of bactericide B was introduced into the population at $t = 0$.

a. Which population is decreasing faster at t_1?
b. Which population is decreasing faster at t_2?

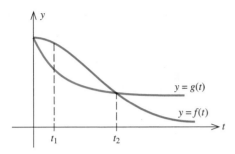

c. Which bactericide is more effective in reducing the population of bacteria in the short run? In the long run?

8. MARKET SHARE The following figure shows the devastating effect the opening of a new discount department store had on an established department store in a small town. The revenue of the discount store at time t (in months) is given by $f(t)$ million dollars, whereas the revenue of the established department store at time t is given by $g(t)$ million dollars. Answer the following questions by giving the value of t at which the specified event took place.

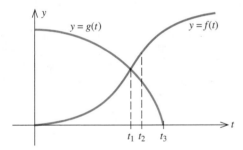

a. The revenue of the established department store is decreasing at the slowest rate.
b. The revenue of the established department store is decreasing at the fastest rate.
c. The revenue of the discount store first overtakes that of the established store.
d. The revenue of the discount store is increasing at the fastest rate.

In Exercises 9–16, use the four-step process to find the slope of the tangent line to the graph of the given function at any point.

9. $f(x) = 13$ **10.** $f(x) = -6$

11. $f(x) = 2x + 7$ **12.** $f(x) = 8 - 4x$

13. $f(x) = 3x^2$ **14.** $f(x) = -\frac{1}{2}x^2$

15. $f(x) = -x^2 + 3x$ **16.** $f(x) = 2x^2 + 5x$

In Exercises 17–22, find the slope of the tangent line to the graph of each function at the given point and determine an equation of the tangent line.

17. $f(x) = 2x + 7$ at $(2, 11)$

18. $f(x) = -3x + 4$ at $(-1, 7)$

19. $f(x) = 3x^2$ at $(1, 3)$

20. $f(x) = 3x - x^2$ at $(-2, -10)$

21. $f(x) = -\frac{1}{x}$ at $\left(3, -\frac{1}{3}\right)$

22. $f(x) = \frac{3}{2x}$ at $\left(1, \frac{3}{2}\right)$

23. Let $f(x) = 2x^2 + 1$.
a. Find the derivative f' of f.
b. Find an equation of the tangent line to the curve at the point $(1, 3)$.
c. Sketch the graph of f.

24. Let $f(x) = x^2 + 6x$.
a. Find the derivative f' of f.
b. Find the point on the graph of f where the tangent line to the curve is horizontal.
Hint: Find the value of x for which $f'(x) = 0$.
c. Sketch the graph of f and the tangent line to the curve at the point found in part (b).

25. Let $f(x) = x^2 - 2x + 1$.
a. Find the derivative f' of f.
b. Find the point on the graph of f where the tangent line to the curve is horizontal.
c. Sketch the graph of f and the tangent line to the curve at the point found in part (b).
d. What is the rate of change of f at this point?

26. Let $f(x) = \frac{1}{x-1}$.
a. Find the derivative f' of f.
b. Find an equation of the tangent line to the curve at the point $(-1, -\frac{1}{2})$.
c. Sketch the graph of f.

27. Let $y = f(x) = x^2 + x$.
a. Find the average rate of change of y with respect to x in the interval from $x = 2$ to $x = 3$, from $x = 2$ to $x = 2.5$, and from $x = 2$ to $x = 2.1$.
b. Find the (instantaneous) rate of change of y at $x = 2$.
c. Compare the results obtained in part (a) with that of part (b).

28. Let $y = f(x) = x^2 - 4x$.
   a. Find the average rate of change of y with respect to x in the interval from $x = 3$ to $x = 4$, from $x = 3$ to $x = 3.5$, and from $x = 3$ to $x = 3.1$.
   b. Find the (instantaneous) rate of change of y at $x = 3$.
   c. Compare the results obtained in part (a) with that of part (b).

29. VELOCITY OF A CAR Suppose the distance s (in feet) covered by a car moving along a straight road t sec after starting from rest is given by the function $f(t) = 2t^2 + 48t$.
   a. Calculate the average velocity of the car over the time intervals [20, 21], [20, 20.1], and [20, 20.01].
   b. Calculate the (instantaneous) velocity of the car when $t = 20$.
   c. Compare the results of part (a) with that of part (b).

30. VELOCITY OF A BALL THROWN INTO THE AIR A ball is thrown straight up with an initial velocity of 128 ft/sec, so that its height (in feet) after t sec is given by $s(t) = 128t - 16t^2$.
   a. What is the average velocity of the ball over the time intervals [2, 3], [2, 2.5], and [2, 2.1]?
   b. What is the instantaneous velocity at time $t = 2$?
   c. What is the instantaneous velocity at time $t = 5$? Is the ball rising or falling at this time?
   d. When will the ball hit the ground?

31. During the construction of a high-rise building, a worker accidentally dropped his portable electric screwdriver from a height of 400 ft. After t sec, the screwdriver had fallen a distance of $s = 16t^2$ ft.
   a. How long did it take the screwdriver to reach the ground?
   b. What was the average velocity of the screwdriver between the time it was dropped and the time it hit the ground?
   c. What was the velocity of the screwdriver at the time it hit the ground?

32. A hot air balloon rises vertically from the ground so that its height after t sec is $h = \frac{1}{2}t^2 + \frac{1}{2}t$ ft ($0 \leq t \leq 60$).
   a. What is the height of the balloon at the end of 40 sec?
   b. What is the average velocity of the balloon between $t = 0$ and $t = 40$?
   c. What is the velocity of the balloon at the end of 40 sec?

33. At a temperature of 20°C, the volume V (in liters) of 1.33 g of O_2 is related to its pressure p (in atmospheres) by the formula $V = 1/p$.

   a. What is the average rate of change of V with respect to p as p increases from $p = 2$ to $p = 3$?
   b. What is the rate of change of V with respect to p when $p = 2$?

34. COST OF PRODUCING SURFBOARDS The total cost $C(x)$ (in dollars) incurred by the Aloha Company in manufacturing x surfboards a day is given by

$$C(x) = -10x^2 + 300x + 130 \qquad (0 \leq x \leq 15)$$

   a. Find $C'(x)$.
   b. What is the rate of change of the total cost when the level of production is ten surfboards a day?
   c. What is the average cost Aloha incurs in manufacturing ten surfboards a day?

35. EFFECT OF ADVERTISING ON PROFIT The quarterly profit (in thousands of dollars) of Cunningham Realty is given by

$$P(x) = -\frac{1}{3}x^2 + 7x + 30 \qquad (0 \leq x \leq 50)$$

where x (in thousands of dollars) is the amount of money Cunningham spends on advertising per quarter.
   a. Find $P'(x)$.
   b. What is the rate of change of Cunningham's quarterly profit if the amount it spends on advertising is $10,000/quarter ($x = 10$) and $30,000/quarter ($x = 30$)?

36. DEMAND FOR TENTS The demand function for the Sportsman 5×7 tents is given by

$$p = f(x) = -0.1x^2 - x + 40$$

where p is measured in dollars and x is measured in units of a thousand.
   a. Find the average rate of change in the unit price of a tent if the quantity demanded is between 5000 and 5050 tents; between 5000 and 5010 tents.
   b. What is the rate of change of the unit price if the quantity demanded is 5000?

37. A COUNTRY'S GDP The gross domestic product (GDP) of a certain country is projected to be

$$N(t) = t^2 + 2t + 50 \qquad (0 \leq t \leq 5)$$

billion dollars t yr from now. What will be the rate of change of the country's GDP 2 yr and 4 yr from now?

38. GROWTH OF BACTERIA Under a set of controlled laboratory conditions, the size of the population of a certain bacteria culture at time t (in minutes) is described by the function

$$P = f(t) = 3t^2 + 2t + 1$$

Find the rate of population growth at $t = 10$ min.

In Exercises 39–43, let *x* and *f(x)* represent the given quantities. Fix *x = a* and let *h* be a small positive number. Give an interpretation of the quantities

$$\frac{f(a+h) - f(a)}{h} \quad \textbf{and} \quad \lim_{h \to 0} \frac{f(a+h) - f(a)}{h}$$

39. *x* denotes time, and $f(x)$ denotes the population of seals at time *x*.

40. *x* denotes time, and $f(x)$ denotes the prime interest rate at time *x*.

41. *x* denotes time, and $f(x)$ denotes a country's industrial production.

42. *x* denotes the level of production of a certain commodity, and $f(x)$ denotes the total cost incurred in producing *x* units of the commodity.

43. *x* denotes altitude, and $f(x)$ denotes atmospheric pressure.

In each of Exercises 44–49, the graph of a function is shown. For each function, state whether or not (a) *f(x)* has a limit at *x = a*, (b) *f(x)* is continuous at *x = a*, and (c) *f(x)* is differentiable at *x = a*. Justify your answers.

44.

45.

46.

47.

48.
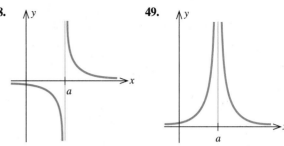

49.

50. The distance *s* (in feet) covered by a motorcycle traveling in a straight line and starting from rest in *t* sec is given by the function

$$s(t) = -0.1t^3 + 2t^2 + 24t$$

Calculate the motorcycle's average velocity over the time interval $[2, 2 + h]$ for $h = 1, 0.1, 0.01, 0.001, 0.0001$, and 0.00001 and use your results to guess at the motorcycle's instantaneous velocity at $t = 2$.

51. The daily total cost $C(x)$ incurred by Trappee and Sons, Inc., for producing *x* cases of Texa-Pep hot sauce is given by

$$C(x) = 0.000002x^3 + 5x + 400$$

Calculate

$$\frac{C(100 + h) - C(100)}{h}$$

for $h = 1, 0.1, 0.01, 0.001$, and 0.0001 and use your results to estimate the rate of change of the total cost function when the level of production is 100 cases/day.

In Exercises 52 and 53, determine whether the statement is true or false. If it is true, explain why it is true. If it is false, give an example to show why it is false.

52. If *f* is continuous at $x = a$, then *f* is differentiable at $x = a$.

53. If *f* is continuous at $x = a$ and *g* is differentiable at $x = a$, then $\lim_{x \to a} f(x)g(x) = f(a)g(a)$.

54. Sketch the graph of the function $f(x) = |x + 1|$ and show that the function does not have a derivative at $x = -1$.

55. Sketch the graph of the function $f(x) = 1/(x - 1)$ and show that the function does not have a derivative at $x = 1$.

56. Let

$$f(x) = \begin{cases} x^2 & \text{if } x \le 1 \\ ax + b & \text{if } x > 1 \end{cases}$$

Find the values of a and b so that f is continuous and has a derivative at $x = 1$. Sketch the graph of f.

57. Sketch the graph of the function $f(x) = x^{2/3}$. Is the function continuous at $x = 0$? Does $f'(0)$ exist? Why, or why not?

58. Prove that the derivative of the function $f(x) = |x|$ for $x \ne 0$ is given by

$$f'(x) = \begin{cases} 1 & \text{if } x > 0 \\ -1 & \text{if } x < 0 \end{cases}$$

Hint: Recall the definition of the absolute value of a number.

59. Show that if a function f is differentiable at a point $x = a$, then f must be continuous at that point.
Hint: Write

$$f(x) - f(a) = \left[\frac{f(x) - f(a)}{x - a} \right](x - a)$$

Use the product rule for limits and the definition of the derivative to show that

$$\lim_{x \to a}[f(x) - f(a)] = 0$$

SOLUTIONS TO SELF-CHECK EXERCISES 2.6

1. a. $f'(x) = \lim\limits_{h \to 0} \dfrac{f(x + h) - f(x)}{h}$

$$= \lim_{h \to 0} \frac{[-(x + h)^2 - 2(x + h) + 3] - (-x^2 - 2x + 3)}{h}$$

$$= \lim_{h \to 0} \frac{-x^2 - 2xh - h^2 - 2x - 2h + 3 + x^2 + 2x - 3}{h}$$

$$= \lim_{h \to 0} \frac{h(-2x - h - 2)}{h}$$

$$= \lim_{h \to 0} (-2x - h - 2) = -2x - 2$$

b. From the result of part (a), we see that the slope of the tangent line to the graph of f at any point $(x, f(x))$ is given by

$$f'(x) = -2x - 2$$

In particular, the slope of the tangent line to the graph of f at $(0, 3)$ is

$$f'(0) = -2$$

c. The rate of change of f when $x = 0$ is given by $f'(0) = -2$, or -2 units/unit change in x.

d. Using the result from part (b), we see that an equation of the required tangent line is

$$y - 3 = -2(x - 0)$$
$$y = -2x + 3$$

e.

$$y = -x^2 - 2x + 3$$

$$y = -2x + 3$$

2. The rate of change of the losses at any time t is given by

$$f'(t) = \lim_{h \to 0} \frac{f(t+h) - f(t)}{h}$$

$$= \lim_{h \to 0} \frac{[-(t+h)^2 + 10(t+h) + 30] - (-t^2 + 10t + 30)}{h}$$

$$= \lim_{h \to 0} \frac{-t^2 - 2th - h^2 + 10t + 10h + 30 + t^2 - 10t - 30}{h}$$

$$= \lim_{h \to 0} \frac{h(-2t - h + 10)}{h}$$

$$= \lim_{h \to 0} (-2t - h + 10)$$

$$= -2t + 10$$

Therefore, the rate of change of the losses suffered by the bank at the beginning of 1997 ($t = 3$) was

$$f'(3) = -2(3) + 10 = 4$$

That is, the losses were increasing at the rate of $4 million/year. At the beginning of 1999 ($t = 5$),

$$f'(5) = -2(5) + 10 = 0$$

and we see that the growth in losses due to bad loans was zero at this point. At the beginning of 2001 ($t = 7$),

$$f'(7) = -2(7) + 10 = -4$$

and we conclude that the losses will be decreasing at the rate of $4 million/year.

Group projects for this chapter can be found at the Brooks/Cole Web site:
http://www.brookscole.com/math/authors/tans/

Using Technology

GRAPHING A FUNCTION AND ITS TANGENT LINES

We can use a graphing utility to plot the graph of a function f and the tangent line at any point on the graph.

EXAMPLE 1   Let $f(x) = x^2 - 4x$.

a. Find an equation of the tangent line to the graph of f at the point $(3, -3)$.
b. Plot both the graph of f and the tangent line found in part (a) on the same set of axes.

SOLUTION ✔   **a.** The slope of the tangent line at any point on the graph of f is given by $f'(x)$. But from Example 4 (page 162) we find $f'(x) = 2x - 4$. Using this result, we see that the slope of the required tangent line is

$$f'(3) = 2(3) - 4 = 2$$

Finally, using the point-slope form of the equation of a line, we find that an equation of the tangent line is

$$y - (-3) = 2(x - 3)$$
$$y + 3 = 2x - 6$$
$$y = 2x - 9.$$

b. The graph of f in the standard viewing rectangle and the tangent line of interest are shown in Figure T1.

FIGURE T1
The graph of $f(x) = x^2 - 4x$ and the tangent line $y = 2x - 9$ in the standard viewing rectangle

REMARK Some graphing utilities will draw both the graph of a function f and the tangent line to the graph of f at a specified point when the function and the specified value of x are entered.

FINDING THE DERIVATIVE OF A FUNCTION AT A GIVEN POINT

The numerical derivative operation of a graphing utility can be used to give an approximate value of the derivative of a function for a given value of x.

EXAMPLE 2

Let $f(x) = \sqrt{x}$.

a. Use the numerical derivative operation of a graphing utility to find the derivative of f at $(4, 2)$.
b. Find an equation of the tangent line to the graph of f at $(4, 2)$.
c. Plot the graph of f and the tangent line on the same set of axes.

SOLUTION ✔

a. Using the numerical derivative operation of a graphing utility, we find that

$$f'(4) = \frac{1}{4}$$

b. An equation of the required tangent line is

$$y - 2 = \frac{1}{4}(x - 4)$$

$$y = \frac{1}{4}x + 1$$

c. The graph of f and the tangent line in the viewing rectangle $[0, 15] \times [0, 4]$ is shown in Figure T2.

FIGURE T2
The graph of $f(x) = \sqrt{x}$ and the tangent line $y = \frac{1}{4}x + 1$ in the viewing rectangle $[0, 15] \times [0, 4]$

In Exercises 1–10, (a) find an equation of the tangent line to the graph of *f* at the indicated point and (b) use a graphing utility to plot the graph of *f* and the tangent line on the same set of axes. Use a suitable viewing rectangle.

1. $f(x) = 4x - 3;\ (2, 5)$

2. $f(x) = -2x + 5;\ (1, 3)$

3. $f(x) = 2x^2 + x;\ (-2, 6)$

4. $f(x) = -x^2 + 2x;\ (1, 1)$

5. $f(x) = 2x^2 + x - 3;\ (2, 7)$

6. $f(x) = -3x^2 + 2x - 1;\ (1, -2)$

7. $f(x) = x + \dfrac{1}{x};\ (1, 2)$

8. $f(x) = x - \dfrac{1}{x};\ (1, 0)$

9. $f(x) = \sqrt{x};\ (4, 2)$

10. $f(x) = \dfrac{1}{\sqrt{x}};\ \left(4, \dfrac{1}{2}\right)$

In Exercises 11–20, (a) use the numerical derivative operation of a graphing utility to find the derivative of *f* for the given value of *x* (to two desired places of accuracy), (b) find an equation of the tangent line to the graph of *f* at the indicated point, and (c) plot the graph of *f* and the tangent line on the same set of axes. Use a suitable viewing rectangle.

11. $f(x) = x^3 + x + 1;\ x = 1;\ (1, 3)$

12. $f(x) = -2x^3 + 3x^2 + 2;\ x = -1;\ (-1, 7)$

13. $f(x) = x^4 - 3x^2 + 1;\ x = 2;\ (2, 5)$

14. $f(x) = -x^4 + 3x + 1;\ x = 1;\ (1, 3)$

15. $f(x) = x - \sqrt{x};\ x = 4;\ (4, 2)$

16. $f(x) = x^{3/2} - x;\ x = 4;\ (4, 4)$

17. $f(x) = \dfrac{1}{x+1};\ x = 1;\ \left(1, \dfrac{1}{2}\right)$

18. $f(x) = \dfrac{x}{x+1};\ x = 3;\ \left(3, \dfrac{3}{4}\right)$

19. $f(x) = x\sqrt{x^2 + 1};\ x = 2;\ (2, 2\sqrt{5})$

20. $f(x) = \dfrac{x}{\sqrt{x^2 + 1}};\ x = 1;\ \left(1, \dfrac{\sqrt{2}}{2}\right)$

CHAPTER 2 Summary of Principal Formulas and Terms

Formulas

1. Average rate of change of f over $[x, x + h]$
   or
   Slope of the secant line to the graph of f through $(x, f(x))$ and $(x + h, f(x + h))$
   or
   Difference quotient

$$\frac{f(x + h) - f(x)}{h}$$

2. Instantaneous rate of change of f at $(x, f(x))$
   or
   Slope of tangent line to the graph of f at $(x, f(x))$ at x
   or
   Derivative of f

$$\lim_{h \to 0} \frac{f(x + h) - f(x)}{h}$$

Terms

function	power function
domain	demand function
range	supply function
independent variable	market equilibrium
dependent variable	equilibrium quantity
ordered pairs	equilibrium price
graph of a function	limit of a function
graph of an equation	indeterminate form
vertical-line test	limit of a function at infinity
composite function	right-hand limit of a function
polynomial function	left-hand limit of a function
linear function	continuity of a function at a point
quadratic function	secant line
cubic function	tangent line to the graph of f
rational function	differentiable function

CHAPTER 2 REVIEW EXERCISES

1. Find the domain of each function:

   a. $f(x) = \sqrt{9 - x}$ **b.** $f(x) = \dfrac{x + 3}{2x^2 - x - 3}$

2. Let $f(x) = 3x^2 + 5x - 2$. Find:
   a. $f(-2)$ **b.** $f(a + 2)$
   c. $f(2a)$ **d.** $f(a + h)$

3. Let $y^2 = 2x + 1$.
   a. Sketch the graph of this equation.
   b. Is y a function of x? Why?
   c. Is x a function of y? Why?

4. Sketch the graph of the function defined by

$$f(x) = \begin{cases} x + 1 & \text{if } x < 1 \\ -x^2 + 4x - 1 & \text{if } x \geq 1 \end{cases}$$

5. Let $f(x) = 1/x$ and $g(x) = 2x + 3$. Find:
   a. $f(x)g(x)$ b. $f(x)/g(x)$
   c. $f(g(x))$ d. $g(f(x))$

In Exercises 6–19, find the indicated limits, if they exist.

6. $\lim\limits_{x \to 0} (5x - 3)$ 7. $\lim\limits_{x \to 1}(x^2 + 1)$

8. $\lim\limits_{x \to -1}(3x^2 + 4)(2x - 1)$

9. $\lim\limits_{x \to 3} \dfrac{x - 3}{x + 4}$ 10. $\lim\limits_{x \to 2} \dfrac{x + 3}{x^2 - 9}$

11. $\lim\limits_{x \to -2} \dfrac{x^2 - 2x - 3}{x^2 + 5x + 6}$ 12. $\lim\limits_{x \to 3} \sqrt{2x^3 - 5}$

13. $\lim\limits_{x \to 3} \dfrac{4x - 3}{\sqrt{x + 1}}$ 14. $\lim\limits_{x \to 1^+} \dfrac{x - 1}{x(x - 1)}$

15. $\lim\limits_{x \to 1^-} \dfrac{\sqrt{x} - 1}{x - 1}$ 16. $\lim\limits_{x \to \infty} \dfrac{x^2}{x^2 - 1}$

17. $\lim\limits_{x \to -\infty} \dfrac{x + 1}{x}$ 18. $\lim\limits_{x \to \infty} \dfrac{3x^2 + 2x + 4}{2x^2 - 3x + 1}$

19. $\lim\limits_{x \to -\infty} \dfrac{x^2}{x + 1}$

20. Sketch the graph of the function

$$f(x) = \begin{cases} 2x - 3 & \text{if } x \leq 2 \\ -x + 3 & \text{if } x > 2 \end{cases}$$

and evaluate $\lim\limits_{x \to a^+} f(x)$, $\lim\limits_{x \to a^-} f(x)$, and $\lim\limits_{x \to a} f(x)$ at the point $a = 2$, if the limits exist.

21. Sketch the graph of the function

$$f(x) = \begin{cases} 4 - x & \text{if } x \leq 2 \\ x + 2 & \text{if } x > 2 \end{cases}$$

and evaluate $\lim\limits_{x \to a^+} f(x)$, $\lim\limits_{x \to a^-} f(x)$, and $\lim\limits_{x \to a} f(x)$ at the point $a = 2$, if the limits exist.

In Exercises 22–25, determine all values of x for which each function is discontinuous.

22. $g(x) = \begin{cases} x + 3 & \text{if } x \neq 2 \\ 0 & \text{if } x = 2 \end{cases}$

23. $f(x) = \dfrac{3x + 4}{4x^2 - 2x - 2}$

24. $f(x) = \begin{cases} \dfrac{1}{(x + 1)^2} & \text{if } x \neq -1 \\ 2 & \text{if } x = -1 \end{cases}$

25. $f(x) = \dfrac{|2x|}{x}$

26. Let $y = x^2 + 2$.
   a. Find the average rate of change of y with respect to x in the intervals $[1, 2]$, $[1, 1.5]$, and $[1, 1.1]$.
   b. Find the (instantaneous) rate of change of y at $x = 1$.

27. Use the definition of the derivative to find the slope of the tangent line to the graph of the function $f(x) = 3x + 5$ at any point $P(x, f(x))$ on the graph.

28. Use the definition of the derivative to find the slope of the tangent line to the graph of the function $f(x) = -1/x$ at any point $P(x, f(x))$ on the graph.

29. Use the definition of the derivative to find the slope of the tangent line to the graph of the function $f(x) = \frac{3}{2}x + 5$ at the point $(-2, 2)$ and determine an equation of the tangent line.

30. Use the definition of the derivative to find the slope of the tangent line to the graph of the function $f(x) = -x^2$ at the point $(2, -4)$ and determine an equation of the tangent line.

31. The graph of the function f is shown in the accompanying figure.
   a. Is f continuous at $x = a$? Why?
   b. Is f differentiable at $x = a$? Justify your answers.

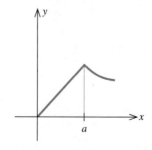

32. Sales of a certain clock radio are approximated by the relationship $S(x) = 6000x + 30{,}000$ $(0 \le x \le 5)$, where $S(x)$ denotes the number of clock radios sold in year x ($x = 0$ corresponds to the year 1996). Find the number of clock radios expected to be sold in the year 2000.

33. A company's total sales (in millions of dollars) are approximately linear as a function of time (in years). Sales in 1994 were $2.4 million, whereas sales in 1999 amounted to $7.4 million.
a. Find an equation that gives the company's sales as a function of time.
b. What were the sales in 1997?

34. A company has a fixed cost of $30,000 and a production cost of $6 for each unit it manufactures. A unit sells for $10.
a. What is the cost function?
b. What is the revenue function?
c. What is the profit function?
d. Compute the profit (loss) corresponding to production levels of 6000, 8000, and 12,000 units, respectively.

35. Find the point of intersection of the two straight lines having the equations $y = \frac{3}{4}x + 6$ and $3x - 2y + 3 = 0$.

36. The cost and revenue functions for a certain firm are given by $C(x) = 12x + 20{,}000$ and $R(x) = 20x$, respectively. Find the company's break-even point.

37. Given the demand equation $3x + p - 40 = 0$ and the supply equation $2x - p + 10 = 0$, where p is the unit price in dollars and x represents the quantity in units of a thousand, determine the equilibrium quantity and the equilibrium price.

38. Clark's rule is a method for calculating pediatric drug dosages based on a child's weight. If a denotes the adult dosage (in milligrams) and if w is the weight of the child (in pounds), then the child's dosage is given by

$$D(w) = \frac{aw}{150}$$

If the adult dose of a substance is 500 mg, how much should a child who weighs 35 lb receive?

39. The monthly revenue R (in hundreds of dollars) realized in the sale of Royal electric shavers is related to the unit price p (in dollars) by the equation

$$R(p) = -\frac{1}{2}p^2 + 30p$$

Find the revenue when an electric shaver is priced at $30.

40. The membership of the newly opened Venus Health Club is approximated by the function

$$N(x) = 200(4 + x)^{1/2} \qquad (1 \le x \le 24)$$

where $N(x)$ denotes the number of members x months after the club's grand opening. Find $N(0)$ and $N(12)$ and interpret your results.

41. Psychologist L. L. Thurstone discovered the following model for the relationship between the learning time T and the length of a list n:

$$T = f(n) = An\sqrt{n - b}$$

where A and b are constants that depend on the person and the task. Suppose that, for a certain person and a certain task, $A = 4$ and $b = 4$. Compute $f(4), f(5), \ldots,$ $f(12)$ and use this information to sketch the graph of the function f. Interpret your results.

42. The monthly demand and supply functions for the Luminar desk lamp are given by

$$p = d(x) = -1.1x^2 + 1.5x + 40$$
$$p = s(x) = 0.1x^2 + 0.5x + 15$$

respectively, where p is measured in dollars and x in units of a thousand. Find the equilibrium quantity and price.

43. The Photo-Mart transfers movie films to videocassettes. The fees charged for this service are shown in the following table. Find a function C relating the cost $C(x)$ to the number of feet x of film transferred. Sketch the graph of the function C and discuss its continuity.

Length of Film in Feet, x	Price ($) for Conversion
$1 \le x \le 100$	5.00
$100 < x \le 200$	9.00
$200 < x \le 300$	12.50
$300 < x \le 400$	15.00
$x > 400$	$7 + 0.02x$

44. The average cost (in dollars) of producing x units of a certain commodity is given by

$$\overline{C}(x) = 20 + \frac{400}{x}$$

Evaluate $\lim_{x \to \infty} \overline{C}(x)$ and interpret your results.

 Additional study hints and sample chapter tests can be found at the Brooks/Cole Web site: http://www.brookscole.com/math/authors/tans/

DIFFERENTIATION

3

This chapter gives several rules that will greatly simplify the task of finding the derivative of a function, thus enabling us to study how fast one quantity is changing with respect to another in many real-world situations. For example, we will be able to find how fast the population of an endangered species of whales grows after certain conservation measures have been implemented, how fast an economy's consumer price index (CPI) is changing at any time, and how fast the learning time changes with respect to the length of a list. We also see how these rules of differentiation facilitate the study of marginal analysis, the study of the rate of change of economic quantities. Finally, we introduce the notion of the differential of a function. Using differentials is a relatively easy way of approximating the change in one quantity due to a small change in a related quantity.

How is a pond's oxygen content affected by organic waste? In Example 7, page 202, you will see how to find the rate at which oxygen is being restored to the pond after organic waste has been dumped into it.

3.1 Basic Rules of Differentiation

FOUR BASIC RULES

The method used in Chapter 2 for computing the derivative of a function is based on a faithful interpretation of the definition of the derivative as the limit of a quotient. Thus, to find the rule for the derivative f' of a function f, we first computed the difference quotient

$$\frac{f(x + h) - f(x)}{h}$$

and then evaluated its limit as h approached zero. As you have probably observed, this method is tedious even for relatively simple functions.

The main purpose of this chapter is to derive certain rules that will simplify the process of finding the derivative of a function. Throughout this book we will use the notation

$$\frac{d}{dx}[f(x)]$$

[read "d, d x of f of x"] to mean "the derivative of f with respect to x at x." In stating the rules of differentiation, we assume that the functions f and g are differentiable.

Rule 1: Derivative of a Constant

$$\frac{d}{dx}(c) = 0 \qquad (c, \text{ a constant})$$

The derivative of a constant function is equal to zero.

We can see this from a geometric viewpoint by recalling that the graph of a constant function is a straight line parallel to the x-axis (Figure 3.1). Since the tangent line to a straight line at any point on the line coincides with the straight line itself, its slope [as given by the derivative of $f(x) = c$] must be zero. We can also use the definition of the derivative to prove this result by computing

$$f'(x) = \lim_{h \to 0} \frac{f(x + h) - f(x)}{h}$$

$$= \lim_{h \to 0} \frac{c - c}{h}$$

$$= \lim_{h \to 0} 0 = 0$$

FIGURE 3.1

The slope of the tangent line to the graph of $f(x) = c$, where c is a constant, is zero.

EXAMPLE 1

a. If $f(x) = 28$, then

$$f'(x) = \frac{d}{dx}(28) = 0$$

b. If $f(x) = -2$, then

$$f'(x) = \frac{d}{dx}(-2) = 0$$

■ ■ ■ ■

Rule 2: The Power Rule

If n is any real number, then $\dfrac{d}{dx}(x^n) = nx^{n-1}$.

Let's verify the power rule for the special case $n = 2$. If $f(x) = x^2$, then

$$f'(x) = \frac{d}{dx}(x^2) = \lim_{h \to 0} \frac{f(x + h) - f(x)}{h}$$

$$= \lim_{h \to 0} \frac{(x + h)^2 - x^2}{h}$$

$$= \lim_{h \to 0} \frac{x^2 + 2xh + h^2 - x^2}{h}$$

$$= \lim_{h \to 0} \frac{2xh + h^2}{h} = \lim_{h \to 0} (2x + h) = 2x$$

as we set out to show.

    The proof of the power rule for the general case is not easy to prove and will be omitted. However, you will be asked to prove the rule for the special case $n = 3$ in Exercise 72, page 196.

EXAMPLE 2

a. If $f(x) = x$, then

$$f'(x) = \frac{d}{dx}(x) = 1 \cdot x^{1-1} = x^0 = 1$$

b. If $f(x) = x^8$, then

$$f'(x) = \frac{d}{dx}(x^8) = 8x^7$$

c. If $f(x) = x^{5/2}$, then

$$f'(x) = \frac{d}{dx}(x^{5/2}) = \frac{5}{2}x^{3/2}$$

■ ■ ■ ■

To differentiate a function whose rule involves a radical, we first rewrite the rule using fractional powers. The resulting expression can then be differentiated using the power rule.

EXAMPLE 3

Find the derivative of the following functions:

a. $f(x) = \sqrt{x}$ **b.** $g(x) = \dfrac{1}{\sqrt[3]{x}}$

SOLUTION ✔

a. Rewriting $\sqrt{x}$ in the form $x^{1/2}$, we obtain

$$f'(x) = \frac{d}{dx}(x^{1/2})$$

$$= \frac{1}{2}x^{-1/2} = \frac{1}{2x^{1/2}} = \frac{1}{2\sqrt{x}}$$

b. Rewriting $\dfrac{1}{\sqrt[3]{x}}$ in the form $x^{-1/3}$, we obtain

$$g'(x) = \frac{d}{dx}(x^{-1/3})$$

$$= -\frac{1}{3}x^{-4/3} = -\frac{1}{3x^{4/3}}$$ ■ ■ ■ ■

Rule 3: Derivative of a Constant Multiple of a Function

$$\frac{d}{dx}[cf(x)] = c\frac{d}{dx}[f(x)] \qquad (c, \text{a constant})$$

The derivative of a constant times a differentiable function is equal to the constant times the derivative of the function.

This result follows from the following computations.
    If $g(x) = cf(x)$, then

$$g'(x) = \lim_{h \to 0}\frac{g(x + h) - g(x)}{h} = \lim_{h \to 0}\frac{cf(x + h) - cf(x)}{h}$$

$$= c\lim_{h \to 0}\frac{f(x + h) - f(x)}{h}$$

$$= cf'(x)$$

EXAMPLE 4   **a.** If $f(x) = 5x^3$, then

$$f'(x) = \frac{d}{dx}(5x^3) = 5\frac{d}{dx}(x^3)$$
$$= 5(3x^2) = 15x^2$$

b. If $f(x) = \dfrac{3}{\sqrt{x}}$, then

$$f'(x) = \frac{d}{dx}(3x^{-1/2})$$
$$= 3\left(-\frac{1}{2}x^{-3/2}\right) = -\frac{3}{2x^{3/2}}$$ ■ ■ ■ ■

Rule 4: The Sum Rule

$$\frac{d}{dx}[f(x) \pm g(x)] = \frac{d}{dx}[f(x)] \pm \frac{d}{dx}[g(x)]$$

The derivative of the sum (difference) of two differentiable functions is equal to the sum (difference) of their derivatives.

This result may be extended to the sum and difference of any finite number of differentiable functions. Let's verify the rule for a sum of two functions.
     If $s(x) = f(x) + g(x)$, then

$$s'(x) = \lim_{h \to 0} \frac{s(x + h) - s(x)}{h}$$
$$= \lim_{h \to 0} \frac{[f(x + h) + g(x + h)] - [f(x) + g(x)]}{h}$$
$$= \lim_{h \to 0} \frac{[f(x + h) - f(x)] + [g(x + h) - g(x)]}{h}$$
$$= \lim_{h \to 0} \frac{f(x + h) - f(x)}{h} + \lim_{h \to 0} \frac{g(x + h) - g(x)}{h}$$
$$= f'(x) + g'(x)$$

EXAMPLE 5   Find the derivatives of the following functions:

a. $f(x) = 4x^5 + 3x^4 - 8x^2 + x + 3$ **b.** $g(t) = \dfrac{t^2}{5} + \dfrac{5}{t^3}$

SOLUTION ✔   **a.** $f'(x) = \dfrac{d}{dx}(4x^5 + 3x^4 - 8x^2 + x + 3)$

$$= \dfrac{d}{dx}(4x^5) + \dfrac{d}{dx}(3x^4) - \dfrac{d}{dx}(8x^2) + \dfrac{d}{dx}(x) + \dfrac{d}{dx}(3)$$

$$= 20x^4 + 12x^3 - 16x + 1$$

b. Here, the independent variable is t instead of x, so we differentiate with respect to t. Thus,

$$g'(t) = \dfrac{d}{dt}\left(\dfrac{1}{5}t^2 + 5t^{-3}\right) \qquad \left(\text{Rewriting } \dfrac{1}{t^3} \text{ as } t^{-3}\right)$$

$$= \dfrac{2}{5}t - 15t^{-4}$$

$$= \dfrac{2t^5 - 75}{5t^4} \qquad \left(\text{Rewriting } t^{-4} \text{ as } \dfrac{1}{t^4} \text{ and simplifying}\right) \quad ■■■■$$

EXAMPLE 6   Find the slope and an equation of the tangent line to the graph of $f(x) = 2x + 1/\sqrt{x}$ at the point $(1, 3)$.

SOLUTION ✔   The slope of the tangent line at any point on the graph of f is given by

$$f'(x) = \dfrac{d}{dx}\left(2x + \dfrac{1}{\sqrt{x}}\right)$$

$$= \dfrac{d}{dx}(2x + x^{-1/2}) \qquad \left(\text{Rewriting } \dfrac{1}{\sqrt{x}} = \dfrac{1}{x^{1/2}} = x^{-1/2}\right)$$

$$= 2 - \dfrac{1}{2}x^{-3/2} \qquad \text{(Using the sum rule)}$$

$$= 2 - \dfrac{1}{2x^{3/2}}$$

In particular, the slope of the tangent line to the graph of f at $(1, 3)$ (where $x = 1$) is

$$f'(1) = 2 - \dfrac{1}{2(1^{3/2})} = 2 - \dfrac{1}{2} = \dfrac{3}{2}$$

Using the point-slope form of the equation of a line with slope $3/2$ and the point $(1, 3)$, we see that an equation of the tangent line is

$$y - 3 = \dfrac{3}{2}(x - 1) \qquad [(y - y_1) = m(x - x_1)]$$

or, upon simplification,

$$y = \dfrac{3}{2}x + \dfrac{3}{2}$$

$■■■■$

APPLICATIONS

EXAMPLE **7**

A group of marine biologists at the Neptune Institute of Oceanography recommended that a series of conservation measures be carried out over the next decade to save a certain species of whale from extinction. After implementing the conservation measures, the population of this species is expected to be

$$N(t) = 3t^3 + 2t^2 - 10t + 600 \qquad (0 \le t \le 10)$$

where $N(t)$ denotes the population at the end of year t. Find the rate of growth of the whale population when $t = 2$ and $t = 6$. How large will the whale population be 8 years after implementing the conservation measures?

SOLUTION ✔

The rate of growth of the whale population at any time t is given by

$$N'(t) = 9t^2 + 4t - 10$$

In particular, when $t = 2$ and $t = 6$, we have

$$N'(2) = 9(2)^2 + 4(2) - 10$$
$$= 34$$
$$N'(6) = 9(6)^2 + 4(6) - 10$$
$$= 338$$

FIGURE 3.2
The whale population after year t is given by $N(t)$.

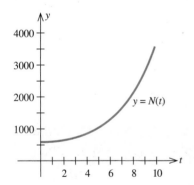

so the whale population's rate of growth will be 34 whales per year after 2 years and 338 per year after 6 years.

The whale population at the end of the eighth year will be

$$N(8) = 3(8)^3 + 2(8)^2 - 10(8) + 600$$
$$= 2184 \text{ whales}$$

The graph of the function N appears in Figure 3.2. Note the rapid growth of the population in the later years, as the conservation measures begin to pay off, compared with the growth in the early years.   ■ ■ ■ ■

EXAMPLE **8**

The altitude of a rocket (in feet) t seconds into flight is given by

$$s = f(t) = -t^3 + 96t^2 + 195t + 5 \qquad (t \ge 0)$$

a. Find an expression v for the rocket's velocity at any time t.
b. Compute the rocket's velocity when $t = 0, 30, 50, 65,$ and 70. Interpret your results.
c. Using the results from the solution to part (b) and the observation that at the highest point in its trajectory the rocket's velocity is zero, find the maximum altitude attained by the rocket.

SOLUTION ✔   **a.** The rocket's velocity at any time t is given by

$$v = f'(t) = -3t^2 + 192t + 195$$

b. The rocket's velocity when $t = 0, 30, 50, 65,$ and 70 is given by

$$f'(0) = -3(0)^2 + 192(0) + 195 = 195$$
$$f'(30) = -3(30)^2 + 192(30) + 195 = 3255$$
$$f'(50) = -3(50)^2 + 192(50) + 195 = 2295$$
$$f'(65) = -3(65)^2 + 192(65) + 195 = 0$$
$$f'(70) = -3(70)^2 + 192(70) + 195 = -1065$$

or $195, 3255, 2295, 0,$ and -1065 feet per second (ft/sec).

Thus, the rocket has an initial velocity of 195 ft/sec at $t = 0$ and accelerates to a velocity of 3255 ft/sec at $t = 30$. Fifty seconds into the flight, the rocket's velocity is 2295 ft/sec, which is less than the velocity at $t = 30$. This means that the rocket begins to decelerate after an initial period of acceleration. (Later on we will learn how to determine the rocket's maximum velocity.)

The deceleration continues: The velocity is 0 ft/sec at $t = 65$ and -1065 ft/sec when $t = 70$. This number tells us that 70 seconds into flight the rocket is heading back to Earth with a speed of 1065 ft/sec.

c. The results of part (b) show that the rocket's velocity is zero when $t = 65$. At this instant, the rocket's maximum altitude is

$$s = f(65) = -(65)^3 + 96(65)^2 + 195(65) + 5$$
$$= 143,655 \text{ feet}$$

A sketch of the graph of f appears in Figure 3.3.   ■ ■ ■ ■

FIGURE 3.3
The rocket's altitude t seconds into flight is given by $f(t)$.

Exploring with Technology

Refer to Example 8.

1. Use a graphing utility to plot the graph of the velocity function
$$v = f'(t) = -3t^2 + 192t + 195$$
using the viewing rectangle $[0, 120] \times [-5000, 5000]$. Then, using ZOOM and TRACE or the root-finding capability of your graphing utility, verify that $f'(65) = 0$.

2. Plot the graph of the position function of the rocket
$$s = f(t) = -t^3 + 96t^2 + 195t + 5$$
using the viewing rectangle $[0, 120] \times [0, 150,000]$. Then, using ZOOM and TRACE repeatedly, verify that the maximum altitude of the rocket is 143,655 feet.

3. Use ZOOM and TRACE or the root-finding capability of your graphing utility to find when the rocket returns to Earth.

SELF-CHECK EXERCISES 3.1

1. Find the derivative of each of the following functions using the rules of differentiation.
   a. $f(x) = 1.5x^2 + 2x^{1.5}$
   b. $g(x) = 2\sqrt{x} + \dfrac{3}{\sqrt{x}}$

2. Let $f(x) = 2x^3 - 3x^2 + 2x - 1$.
   a. Compute $f'(x)$.
   b. What is the slope of the tangent line to the graph of f when $x = 2$?
   c. What is the rate of change of the function f at $x = 2$?

3. A certain country's gross domestic product (GDP) (in millions of dollars) is described by the function

$$G(t) = -2t^3 + 45t^2 + 20t + 6000 \qquad (0 \le t \le 11)$$

where $t = 0$ corresponds to the beginning of 1990.
   a. At what rate was the GDP changing at the beginning of 1995? At the beginning of 1997? At the beginning of 2000?
   b. What was the average rate of growth of the GDP over the period 1995–2000?

Solutions to Self-Check Exercises 3.1 can be found on page 197.

3.1 Exercises

In Exercises 1–34, find the derivative of the function f by using the rules of differentiation.

1. $f(x) = -3$

2. $f(x) = 365$

3. $f(x) = x^5$

4. $f(x) = x^7$

5. $f(x) = x^{2.1}$

6. $f(x) = x^{0.8}$

7. $f(x) = 3x^2$

8. $f(x) = -2x^3$

9. $f(r) = \pi r^2$

10. $f(r) = \frac{4}{3}\pi r^3$

11. $f(x) = 9x^{1/3}$

12. $f(x) = \frac{5}{4}x^{4/5}$

13. $f(x) = 3\sqrt{x}$

14. $f(u) = \dfrac{2}{\sqrt{u}}$

15. $f(x) = 7x^{-12}$

16. $f(x) = 0.3x^{-1.2}$

17. $f(x) = 5x^2 - 3x + 7$

18. $f(x) = x^3 - 3x^2 + 1$

19. $f(x) = -x^3 + 2x^2 - 6$

20. $f(x) = x^4 - 2x^2 + 5$

21. $f(x) = 0.03x^2 - 0.4x + 10$

22. $f(x) = 0.002x^3 - 0.05x^2 + 0.1x - 20$

23. $f(x) = \dfrac{x^3 - 4x^2 + 3}{x}$

24. $f(x) = \dfrac{x^3 + 2x^2 + x - 1}{x}$

25. $f(x) = 4x^4 - 3x^{5/2} + 2$

26. $f(x) = 5x^{4/3} - \dfrac{2}{3}x^{3/2} + x^2 - 3x + 1$

27. $f(x) = 3x^{-1} + 4x^{-2}$

28. $f(x) = -\dfrac{1}{3}(x^{-3} - x^6)$

29. $f(t) = \dfrac{4}{t^4} - \dfrac{3}{t^3} + \dfrac{2}{t}$

30. $f(x) = \dfrac{5}{x^3} - \dfrac{2}{x^2} - \dfrac{1}{x} + 200$

31. $f(x) = 2x - 5\sqrt{x}$

32. $f(t) = 2t^2 + \sqrt{t^3}$

33. $f(x) = \dfrac{2}{x^2} - \dfrac{3}{x^{1/3}}$

34. $f(x) = \dfrac{3}{x^3} + \dfrac{4}{\sqrt{x}} + 1$

35. Let $f(x) = 2x^3 - 4x$. Find:
   a. $f'(-2)$ b. $f'(0)$ c. $f'(2)$

36. Let $f(x) = 4x^{5/4} + 2x^{3/2} + x$. Find:
   a. $f'(0)$ b. $f'(16)$

(continued on p. 194)

Using Technology

FINDING THE RATE OF CHANGE OF A FUNCTION

We can use the numerical derivative operation of a graphing utility to obtain the value of the derivative at a given value of x. Since the derivative of a function $f(x)$ measures the rate of change of the function with respect to x, the numerical derivative operation can be used to answer questions pertaining to the rate of change of one quantity y with respect to another quantity x, where $y = f(x)$, for a specific value of x.

EXAMPLE 1 Let $y = 3t^3 + 2\sqrt{t}$.

a. Use the numerical derivative operation of a graphing utility to find how fast y is changing with respect to t when $t = 1$.
b. Verify the result of part (a), using the rules of differentiation of this section.

SOLUTION ✔ **a.** Write $f(t) = 3t^3 + 2\sqrt{t}$. Using the numerical derivative operation of a graphing utility, we find that the rate of change of y with respect to t when $t = 1$ is given by $f'(1) = 10$.

b. Here, $f(t) = 3t^3 + 2t^{1/2}$ and

$$f'(t) = 9t^2 + 2\left(\frac{1}{2}t^{-1/2}\right) = 9t^2 + \frac{1}{\sqrt{t}}$$

Using this result, we see that when $t = 1$, y is changing at the rate of

$$f'(1) = 9(1^2) + \frac{1}{\sqrt{1}} = 10$$

units per unit change in t, as obtained earlier. ■■■■

EXAMPLE 2 According to the U.S. Department of Energy and the Shell Development Company, a typical car's fuel economy depends on the speed it is driven and is approximated by the function

$$f(x) = 0.00000310315x^4 - 0.000455174x^3$$
$$+ 0.00287869x^2 + 1.25986x \qquad (0 \le x \le 75)$$

where x is measured in mph and $f(x)$ is measured in miles per gallon (mpg).

a. Use a graphing utility to graph the function f on the interval $[0, 75]$.
b. Find the rate of change of f when $x = 20$ and $x = 50$.
c. Interpret your results.
Source: U.S. Department of Energy and the Shell Development Company

FIGURE T1

SOLUTION ✔ **a.** The result is shown in Figure T1.
b. Using the numerical derivative operation of a graphing utility, we see that $f'(20) = 0.9280996$. The rate of change of f when $x = 50$ is given by $f'(50) = -0.314501$.
c. The results of part (b) tell us that when a typical car is being driven at 20 mph, its fuel economy increases at the rate of approximately 0.9 mpg per 1 mph increase in its speed. At a speed of 50 mph, its fuel economy decreases at the rate of approximately 0.3 mpg per 1 mph increase in its speed. ■■■■

In Exercises 1–6, use the numerical derivative operation of a graphing utility to find the rate of change of f(x) at the given value of x. Give your answer accurate to four decimal places.

1. $f(x) = 4x^5 - 3x^3 + 2x^2 + 1; x = 0.5$

2. $f(x) = -x^5 + 4x^2 + 3; x = 0.4$

3. $f(x) = x - 2\sqrt{x}; x = 3$

4. $f(x) = \dfrac{\sqrt{x} - 1}{x}; x = 2$

5. $f(x) = x^{1/2} - x^{1/3}; x = 1.2$

6. $f(x) = 2x^{5/4} + x; x = 2$

7. **CARBON MONOXIDE IN THE ATMOSPHERE** The projected average global atmosphere concentration of carbon monoxide is approximated by the function

$$f(t) = 0.881443t^4 - 1.45533t^3 + 0.695876t^2$$
$$+ 2.87801t + 293 \qquad (0 \le t \le 4)$$

where t is measured in 40-yr intervals, with $t = 0$ corresponding to the beginning of 1860 and $f(t)$ is measured in parts per million by volume.
a. Use a graphing utility to plot the graph of f in the viewing rectangle $[0, 4] \times [280, 400]$.
b. Use a graphing utility to estimate how fast the projected average global atmospheric concentration of carbon monoxide was changing at the beginning of the year 1900 ($t = 1$) and at the beginning of 2000 ($t = 3.5$).
Source: "Beyond the Limits," Meadows et al.

8. **GROWTH OF HMOs** Based on data compiled by the Group Health Association of America, the number of people receiving their care in an HMO (Health Maintenance Organization) from the beginning of 1984 through 1994 is approximated by the function

$$f(t) = 0.0514t^3 - 0.853t^2 + 6.8147t$$
$$+ 15.6524 \qquad (0 \le t \le 11)$$

where $f(t)$ gives the number of people in millions and t is measured in years, with $t = 0$ corresponding to the beginning of 1984.
a. Use a graphing utility to plot the graph of f in the viewing window $[0, 12] \times [0, 80]$.
b. How fast was the number of people receiving their care in an HMO growing at the beginning of 1992?
Source: Group Health Association of America

9. **HOME SALES** According to the Greater Boston Real Estate Board—Multiple Listing Service, the average number of days a single-family home remains for sale from listing to accepted offer is approximated by the function

$$f(t) = 0.0171911t^4 - 0.662121t^3 + 6.18083t^2$$
$$- 8.97086t + 53.3357 \qquad (0 \le t \le 10)$$

where t is measured in years, with $t = 0$ corresponding to the beginning of 1984.
a. Use a graphing utility to plot the graph of f in the viewing rectangle $[0, 12] \times [0, 120]$.
b. How fast was the average number of days a single-family home remained for sale from listing to accepted offer changing at the beginning of 1984 ($t = 0$)? At the beginning of 1988 ($t = 4$)?
Source: Greater Boston Real Estate Board—Multiple Listing Service

10. **SPREAD OF HIV** The estimated number of children newly infected with HIV through mother-to-child contact worldwide is given by

$$f(t) = -0.2083t^3 + 3.0357t^2 + 44.0476t$$
$$+ 200.2857 \qquad (0 \le t \le 12)$$

where $f(t)$ is measured in thousands and t is measured in years with $t = 0$ corresponding to the beginning of 1990.
a. Use a graphing utility to plot the graph of f in the viewing rectangle $[0, 12] \times [0, 800]$.
b. How fast was the estimated number of children newly infected with HIV through mother-to-child contact worldwide increasing at the beginning of the year 2000?
Source: United Nations

11. **MANUFACTURING CAPACITY** Data obtained from the Federal Reserve shows that the annual change in manufacturing capacity between 1988 and 1994 is given by

$$f(t) = 0.0388889t^3 - 0.283333t^2$$
$$+ 0.477778t + 2.04286 \qquad (0 \le t \le 6)$$

where $f(t)$ is a percentage and t is measured in years, with $t = 0$ corresponding to the beginning of 1988.
a. Use a graphing utility to plot the graph of f in the viewing rectangle $[0, 8] \times [0, 4]$.
b. How fast was $f(t)$ changing at the beginning of 1990 ($t = 2$)? At the beginning of 1992 ($t = 4$)?
Source: Federal Reserve

In Exercises 37–40, find the given limit by eval-uating the derivative of a suitable function at an appropriate point.

Hint: Look at the definition of the derivative.

37. $\lim\limits_{h \to 0} \dfrac{(1 + h)^3 - 1}{h}$

38. $\lim\limits_{x \to 1} \dfrac{x^5 - 1}{x - 1}$

Hint: Let $h = x - 1$.

39. $\lim\limits_{h \to 0} \dfrac{3(2 + h)^2 - (2 + h) - 10}{h}$

40. $\lim\limits_{t \to 0} \dfrac{1 - (1 + t)^2}{t(1 + t)^2}$

In Exercises 41–44, find the slope and an equa-tion of the tangent line to the graph of the func-tion *f* at the specified point.

41. $f(x) = 2x^2 - 3x + 4;\ (2, 6)$

42. $f(x) = -\dfrac{5}{3} x^2 + 2x + 2;\ \left(-1, -\dfrac{5}{3}\right)$

43. $f(x) = x^4 - 3x^3 + 2x^2 - x + 1;\ (1, 0)$

44. $f(x) = \sqrt{x} + \dfrac{1}{\sqrt{x}};\ \left(4, \dfrac{5}{2}\right)$

45. Let $f(x) = x^3$.
a. Find the point on the graph of *f* where the tangent line is horizontal.
b. Sketch the graph of *f* and draw the horizontal tan-gent line.

46. Let $f(x) = x^3 - 4x^2$. Find the point(s) on the graph of *f* where the tangent line is horizontal.

47. Let $f(x) = x^3 + 1$.
a. Find the point(s) on the graph of *f* where the slope of the tangent line is equal to 12.
b. Find the equation(s) of the tangent line(s) of part (a).
c. Sketch the graph of *f* showing the tangent line(s).

48. Let $f(x) = \frac{2}{3}x^3 + x^2 - 12x + 6$. Find the values of *x* for which:
a. $f'(x) = -12$ **b.** $f'(x) = 0$
c. $f'(x) = 12$

49. Let $f(x) = \frac{1}{4}x^4 - \frac{1}{3}x^3 - x^2$. Find the point(s) on the graph of *f* where the slope of the tangent line is equal to:
a. $-2x$ **b.** 0   **c.** $10x$

50. A straight line perpendicular to and passing through the point of tangency of the tangent line is called the *normal*

to the curve. Find an equation of the tangent line and the normal to the curve $y = x^3 - 3x + 1$ at the point $(2, 3)$.

51. GROWTH OF A CANCEROUS TUMOR   The volume of a spherical cancer tumor is given by the function

$$V(r) = \frac{4}{3}\pi r^3$$

where *r* is the radius of the tumor in centimeters. Find the rate of change in the volume of the tumor when:
a. $r = \dfrac{2}{3}$ cm   **b.** $r = \dfrac{5}{4}$ cm

52. VELOCITY OF BLOOD IN AN ARTERY   The velocity (in centime-ters per second) of blood *r* centimeters from the central axis of an artery is given by

$$v(r) = k(R^2 - r^2)$$

where *k* is a constant and *R* is the radius of the artery (see the accompanying figure). Suppose that $k = 1000$ and $R = 0.2$ cm. Find $v(0.1)$ and $v'(0.1)$ and interpret your results.

Blood vessel

53. EFFECT OF STOPPING ON AVERAGE SPEED   According to data from a study by General Motors, the average speed of your trip *A* (in mph) is related to the number of stops per mile you make on the trip *x* by the equation

$$A = \frac{26.5}{x^{0.45}}$$

Compute dA/dx for $x = 0.25$ and $x = 2$ and interpret your results.
Source: General Motors

54. WORKER EFFICIENCY   An efficiency study conducted for the Elektra Electronics Company showed that the num-ber of "Space Commander" walkie-talkies assembled by the average worker *t* hr after starting work at 8 A.M. is given by

$$N(t) = -t^3 + 6t^2 + 15t$$

a. Find the rate at which the average worker will be assembling walkie-talkies *t* hr after starting work.
b. At what rate will the average worker be assembling walkie-talkies at 10 A.M.? At 11 A.M.?
c. How many walkie-talkies will the average worker as-semble between 10 A.M. and 11 A.M.?

55. CONSUMER PRICE INDEX An economy's consumer price index (CPI) is described by the function

$$I(t) = -0.2t^3 + 3t^2 + 100 \qquad (0 \le t \le 10)$$

where $t = 0$ corresponds to 1989.
a. At what rate was the CPI changing in 1994? In 1996? In 1999?
b. What was the average rate of increase in the CPI over the period from 1994 to 1999?

56. EFFECT OF ADVERTISING ON SALES The relationship between the amount of money x that the Cannon Precision Instruments Corporation spends on advertising and the company's total sales $S(x)$ is given by the function

$$S(x) = -0.002x^3 + 0.6x^2 + x + 500 \qquad (0 \le x \le 200)$$

where x is measured in thousands of dollars. Find the rate of change of the sales with respect to the amount of money spent on advertising. Are Cannon's total sales increasing at a faster rate when the amount of money spent on advertising is (a) $100,000 or (b) $150,000?

57. POPULATION GROWTH A study prepared for a Sunbelt town's chamber of commerce projected that the town's population in the next 3 yr will grow according to the rule

$$P(t) = 50,000 + 30t^{3/2} + 20t$$

where $P(t)$ denotes the population t months from now. How fast will the population be increasing 9 mo and 16 mo from now?

58. CURBING POPULATION GROWTH Five years ago, the government of a Pacific island state launched an extensive propaganda campaign toward curbing the country's population growth. According to the Census Department, the population (measured in thousands of people) for the following 4 yr was

$$P(t) = -\frac{1}{3}t^3 + 64t + 3000$$

where t is measured in years and $t = 0$ at the start of the campaign. Find the rate of change of the population at the end of years 1, 2, 3, and 4. Was the plan working?

59. CONSERVATION OF SPECIES A certain species of turtle faces extinction because dealers collect truckloads of turtle eggs to be sold as aphrodisiacs. After severe conservation measures are implemented, it is hoped that the turtle population will grow according to the rule

$$N(t) = 2t^3 + 3t^2 - 4t + 1000 \qquad (0 \le t \le 10)$$

where $N(t)$ denotes the population at the end of year t.

Find the rate of growth of the turtle population when $t = 2$ and $t = 8$. What will be the population 10 yr after the conservation measures are implemented?

60. FLIGHT OF A ROCKET The altitude (in feet) of a rocket t sec into flight is given by

$$s = f(t) = -2t^3 + 114t^2 + 480t + 1 \qquad (t \ge 0)$$

a. Find an expression v for the rocket's velocity at any time t.
b. Compute the rocket's velocity when $t = 0, 20, 40,$ and 60. Interpret your results.
c. Using the results from the solution to part (b), find the maximum altitude attained by the rocket.
Hint: At its highest point, the velocity of the rocket is zero.

61. STOPPING DISTANCE OF A RACING CAR During a test by the editors of an auto magazine, the stopping distance s (in feet) of the MacPherson X-2 racing car conformed to the rule

$$s = f(t) = 120t - 15t^2 \qquad (t \ge 0)$$

where t was the time (in seconds) after the brakes were applied.
a. Find an expression for the car's velocity v at any time t.
b. What was the car's velocity when the brakes were first applied?
c. What was the car's stopping distance for that particular test?
Hint: The stopping time is found by setting $v = 0$.

62. DEMAND FUNCTIONS The demand function for the Luminar desk lamp is given by

$$p = f(x) = -0.1x^2 - 0.4x + 35$$

where x is the quantity demanded (measured in thousands) and p is the unit price in dollars.
a. Find $f'(x)$.
b. What is the rate of change of the unit price when the quantity demanded is 10,000 units ($x = 10$)? What is the unit price at that level of demand?

63. INCREASE IN TEMPORARY WORKERS According to the Labor Department, the number of temporary workers (in millions) is estimated to be

$$N(t) = 0.025t^2 + 0.255t + 1.505 \qquad (0 \le t \le 5)$$

where t is measured in years, with $t = 0$ corresponding to 1991.
a. How many temporary workers were there at the beginning of 1994?
b. How fast was the number of temporary workers growing at the beginning of 1994?
Source: Labor Department

64. SALES OF DIGITAL SIGNAL PROCESSORS   The sales of digital signal processors (DSPs) in billions of dollars is projected to be

$$S(t) = 0.14t^2 + 0.68t + 3.1 \qquad (0 \le t \le 6)$$

where t is measured in years, with $t = 0$ corresponding to the beginning of 1997.
a. What were the sales of DSPs at the beginning of 1997? What will be the sales at the beginning of 2002?
b. How fast was the level of sales increasing at the beginning of 1997? How fast will the level of sales be increasing at the beginning of 2002?
Source: World Semiconductor Trade Statistics

65. SUPPLY FUNCTIONS   The supply function for a certain make of transistor radio is given by

$$p = f(x) = 0.0001x^{5/4} + 10$$

where x is the quantity supplied and p is the unit price in dollars.
a. Find $f'(x)$.
b. What is the rate of change of the unit price if the quantity supplied is 10,000 transistor radios?

66. PORTABLE PHONES   The percentage of the U.S. population with portable phones is projected to be

$$P(t) = 24.4t^{0.34} \qquad (1 \le t \le 10)$$

where t is measured in years, with $t = 1$ corresponding to the beginning of 1998.
a. What percentage of the U.S. population is expected to have portable phones by the beginning of 2006?
b. How fast is the percentage of the U.S. population with portable phones expected to be changing at the beginning of 2006?
Source: BancAmerica Robertson Stephens

67. AVERAGE SPEED OF A VEHICLE ON A HIGHWAY   The average speed of a vehicle on a stretch of Route 134 between 6 A.M. and 10 A.M. on a typical weekday is approximated by the function

$$f(t) = 20t - 40\sqrt{t} + 50 \qquad (0 \le t \le 4)$$

where $f(t)$ is measured in miles per hour and t is measured in hours, $t = 0$ corresponding to 6 A.M.
a. Compute $f'(t)$.
b. Compute $f(0)$, $f(1)$, and $f(2)$ and interpret your results.
c. Compute $f'(\frac{1}{2})$, $f'(1)$, and $f'(2)$ and interpret your results.

68. HEALTH-CARE SPENDING   Despite efforts at cost containment, the cost of the Medicare program is increasing at a high rate. Two major reasons for this increase are an aging population and the constant development and extensive use by physicians of new technologies. Based on data from the Health Care Financing Administration and the U.S. Census Bureau, health-care spending through the year 2000 may be approximated by the function

$$S(t) = 0.02836t^3 - 0.05167t^2 + 9.60881t + 41.9 \qquad (0 \le t \le 35)$$

where $S(t)$ is the spending in billions of dollars and t is measured in years, with $t = 0$ corresponding to the beginning of 1965.
a. Find an expression for the rate of change of health-care spending at any time t.
b. How fast was health-care spending changing at the beginning of 1980? How fast was health-care spending changing at the beginning of 2000?
c. What was the amount of health-care spending at the beginning of 1980? What was the amount of health-care spending at the beginning of 2000?
Source: Health Care Financing Administration

69. ON-LINE SHOPPING   Retail revenue per year from Internet shopping is approximated by the function

$$f(t) = 0.075t^3 + 0.025t^2 + 2.45t + 2.4 \qquad (0 \le t \le 4)$$

where $f(t)$ is measured in billions of dollars and t is measured in years with $t = 0$ corresponding to the beginning of 1997.
a. Find an expression giving the rate of change of the retail revenue per year from Internet shopping at any time t.
b. How fast was the retail revenue per year from Internet shopping changing at the beginning of the year 2000?
c. What was the retail revenue per year from Internet shopping at the beginning of the year 2000?
Source: Forrester Research, Inc.

In Exercises 70 and 71, determine whether the statement is true or false. If it is true, explain why it is true. If it is false, give an example to show why it is false.

70. If f and g are differentiable, then

$$\frac{d}{dx}[2f(x) - 5g(x)] = 2f'(x) - 5g'(x)$$

71. If $f(x) = \pi^x$, then $f'(x) = x\pi^{x-1}$
72. Prove the power rule (Rule 2) for the special case $n = 3$.

Hint: Compute $\lim\limits_{h \to 0} \left[\dfrac{(x+h)^3 - x^3}{h} \right]$.

SOLUTIONS TO SELF-CHECK EXERCISES 3.1

1. a. $f'(x) = \dfrac{d}{dx}(1.5x^2) + \dfrac{d}{dx}(2x^{1.5})$

$= (1.5)(2x) + (2)(1.5x^{0.5})$

$= 3x + 3\sqrt{x} = 3(x + \sqrt{x})$

b. $g'(x) = \dfrac{d}{dx}(2x^{1/2}) + \dfrac{d}{dx}(3x^{-1/2})$

$= (2)\left(\dfrac{1}{2}x^{-1/2}\right) + (3)\left(-\dfrac{1}{2}x^{-3/2}\right)$

$= x^{-1/2} - \dfrac{3}{2}x^{-3/2}$

$= \dfrac{1}{2}x^{-3/2}(2x - 3) = \dfrac{2x - 3}{2x^{3/2}}$

2. a. $f'(x) = \dfrac{d}{dx}(2x^3) - \dfrac{d}{dx}(3x^2) + \dfrac{d}{dx}(2x) - \dfrac{d}{dx}(1)$

$= (2)(3x^2) - (3)(2x) + 2$

$= 6x^2 - 6x + 2$

b. The slope of the tangent line to the graph of f when $x = 2$ is given by

$$f'(2) = 6(2)^2 - 6(2) + 2 = 14$$

c. The rate of change of f at $x = 2$ is given by $f'(2)$. Using the results of part (b), we see that $f'(2)$ is 14 units/unit change in x.

3. a. The rate at which the GDP was changing at any time t $(0 < t < 11)$ is given by

$$G'(t) = -6t^2 + 90t + 20$$

In particular, the rates of change of the GDP at the beginning of the years 1995 $(t = 5)$, 1997 $(t = 7)$, and 2000 $(t = 10)$ are given by

$$G'(5) = 320, \quad G'(7) = 356, \quad \text{and} \quad G'(10) = 320$$

respectively—that is, by \$320 million/year, \$356 million/year, and \$320 million/year, respectively.

b. The average rate of growth of the GDP over the period from the beginning of 1995 $(t = 5)$ to the beginning of 2000 $(t = 10)$ is given by

$$\frac{G(10) - G(5)}{10 - 5} = \frac{[-2(10)^3 + 45(10)^2 + 20(10) + 6000]}{5}$$

$$- \frac{[-2(5)^3 + 45(5)^2 + 20(5) + 6000]}{5}$$

$$= \frac{8700 - 6975}{5}$$

or \$345 million/year.

3.2 The Product and Quotient Rules

In this section we study two more rules of differentiation: the **product rule** and the **quotient rule.**

THE PRODUCT RULE

The derivative of the product of two differentiable functions is given by the following rule:

Rule 5: The Product Rule

$$\frac{d}{dx}[f(x)g(x)] = f(x)g'(x) + g(x)f'(x)$$

The derivative of the product of two functions is the first function times the derivative of the second plus the second function times the derivative of the first.

The product rule may be extended to the case involving the product of any finite number of functions (see Exercise 63, page 207). We prove the product rule at the end of this section.

 The derivative of the product of two functions is *not* given by the product of the derivatives of the functions; that is, in general

$$\frac{d}{dx}[f(x)g(x)] \neq f'(x)g'(x)$$

EXAMPLE 1   Find the derivative of the function

$$f(x) = (2x^2 - 1)(x^3 + 3)$$

SOLUTION ✔   By the product rule,

$$\begin{aligned}
f'(x) &= (2x^2 - 1)\frac{d}{dx}(x^3 + 3) + (x^3 + 3)\frac{d}{dx}(2x^2 - 1) \\
&= (2x^2 - 1)(3x^2) + (x^3 + 3)(4x) \\
&= 6x^4 - 3x^2 + 4x^4 + 12x \\
&= 10x^4 - 3x^2 + 12x \\
&= x(10x^3 - 3x + 12)
\end{aligned}$$
■ ■ ■ ■

EXAMPLE 2   Differentiate (that is, find the derivative of) the function

$$f(x) = x^3(\sqrt{x} + 1)$$

SOLUTION ✔ First, we express the function in exponential form, obtaining

$$f(x) = x^3(x^{1/2} + 1)$$

By the product rule,

$$f'(x) = x^3 \frac{d}{dx}(x^{1/2} + 1) + (x^{1/2} + 1)\frac{d}{dx}x^3$$

$$= x^3\left(\frac{1}{2}x^{-1/2}\right) + (x^{1/2} + 1)(3x^2)$$

$$= \frac{1}{2}x^{5/2} + 3x^{5/2} + 3x^2$$

$$= \frac{7}{2}x^{5/2} + 3x^2$$

■ ■ ■ ■

REMARK We can also solve the problem by first expanding the product before differentiating f. Examples for which this is not possible will be considered in Section 3.3, where the true value of the product rule will be appreciated.

■ ■ ■

THE QUOTIENT RULE

The derivative of the quotient of two differentiable functions is given by the following rule:

Rule 6: The Quotient Rule

$$\frac{d}{dx}\left[\frac{f(x)}{g(x)}\right] = \frac{g(x)f'(x) - f(x)g'(x)}{[g(x)]^2} \qquad (g(x) \neq 0)$$

As an aid to remembering this expression, observe that it has the following form:

$$\frac{d}{dx}\left[\frac{f(x)}{g(x)}\right]$$

$$= \frac{(\text{Denominator})\left(\begin{array}{c}\text{Derivative of}\\\text{numerator}\end{array}\right) - (\text{Numerator})\left(\begin{array}{c}\text{Derivative of}\\\text{denominator}\end{array}\right)}{(\text{Square of denominator})}$$

For a proof of the quotient rule, see Exercise 64, page 207.

 The derivative of a quotient is *not* equal to the quotient of the derivatives; that is,

$$\frac{d}{dx}\left[\frac{f(x)}{g(x)}\right] \neq \frac{f'(x)}{g'(x)}$$

For example, if $f(x) = x^3$ and $g(x) = x^2$, then

$$\frac{d}{dx}\left[\frac{f(x)}{g(x)}\right] = \frac{d}{dx}\left(\frac{x^3}{x^2}\right) = \frac{d}{dx}(x) = 1$$

which is *not* equal to

$$\frac{f'(x)}{g'(x)} = \frac{\dfrac{d}{dx}(x^3)}{\dfrac{d}{dx}(x^2)} = \frac{3x^2}{2x} = \frac{3}{2}x$$

EXAMPLE 3   Find $f'(x)$ if $f(x) = \dfrac{x}{2x - 4}$.

SOLUTION ✔   Using the quotient rule, we obtain

$$f'(x) = \frac{(2x - 4)\dfrac{d}{dx}(x) - x\dfrac{d}{dx}(2x - 4)}{(2x - 4)^2}$$

$$= \frac{(2x - 4)(1) - x(2)}{(2x - 4)^2}$$

$$= \frac{2x - 4 - 2x}{(2x - 4)^2} = -\frac{4}{(2x - 4)^2}$$ ■■■■

EXAMPLE 4   Find $f'(x)$ if $f(x) = \dfrac{x^2 + 1}{x^2 - 1}$.

SOLUTION ✔   By the quotient rule,

$$f'(x) = \frac{(x^2 - 1)\dfrac{d}{dx}(x^2 + 1) - (x^2 + 1)\dfrac{d}{dx}(x^2 - 1)}{(x^2 - 1)^2}$$

$$= \frac{(x^2 - 1)(2x) - (x^2 + 1)(2x)}{(x^2 - 1)^2}$$

$$= \frac{2x^3 - 2x - 2x^3 - 2x}{(x^2 - 1)^2}$$

$$= -\frac{4x}{(x^2 - 1)^2}$$ ■■■■

EXAMPLE 5   Find $h'(x)$ if

$$h(x) = \frac{\sqrt{x}}{x^2 + 1}$$

SOLUTION ✔ Rewrite $h(x)$ in the form $h(x) = \dfrac{x^{1/2}}{x^2 + 1}$. By the quotient rule, we find

$$h'(x) = \frac{(x^2 + 1)\dfrac{d}{dx}(x^{1/2}) - x^{1/2}\dfrac{d}{dx}(x^2 + 1)}{(x^2 + 1)^2}$$

$$= \frac{(x^2 + 1)\left(\dfrac{1}{2}x^{-1/2}\right) - x^{1/2}(2x)}{(x^2 + 1)^2}$$

$$= \frac{\dfrac{1}{2}x^{-1/2}(x^2 + 1 - 4x^2)}{(x^2 + 1)^2} \qquad \text{(Factoring out } \tfrac{1}{2}x^{-1/2}$$
$$\text{from the numerator)}$$

$$= \frac{1 - 3x^2}{2\sqrt{x}(x^2 + 1)^2} \qquad\qquad\qquad\qquad ■■■■$$

APPLICATIONS

EXAMPLE 6   The sales (in millions of dollars) of a laser disc recording of a hit movie t years from the date of release is given by

$$S(t) = \frac{5t}{t^2 + 1}$$

a. Find the rate at which the sales are changing at time t.

b. How fast are the sales changing at the time the laser discs are released ($t = 0$)? Two years from the date of release?

SOLUTION ✔ **a.** The rate at which the sales are changing at time t is given by $S'(t)$. Using the quotient rule, we obtain

$$S'(t) = \frac{d}{dt}\left[\frac{5t}{t^2 + 1}\right] = 5\frac{d}{dt}\left[\frac{t}{t^2 + 1}\right]$$

$$= 5\left[\frac{(t^2 + 1)(1) - t(2t)}{(t^2 + 1)^2}\right]$$

$$= 5\left[\frac{t^2 + 1 - 2t^2}{(t^2 + 1)^2}\right] = \frac{5(1 - t^2)}{(t^2 + 1)^2}$$

b. The rate at which the sales are changing at the time the laser discs are released is given by

$$S'(0) = \frac{5(1 - 0)}{(0 + 1)^2} = 5$$

That is, they are increasing at the rate of $5 million per year.

Two years from the date of release, the sales are changing at the rate of

$$S'(2) = \frac{5(1-4)}{(4+1)^2} = -\frac{3}{5} = -0.6$$

That is, they are decreasing at the rate of \$600,000 per year.
The graph of the function S is shown in Figure 3.4.

FIGURE 3.4
After a spectacular rise, the sales begin to taper off.

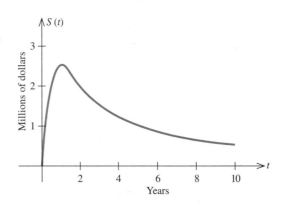

Exploring with Technology

Refer to Example 6.

1. Use a graphing utility to plot the graph of the function S using the viewing rectangle $[0, 10] \times [0, 3]$.
2. Use **TRACE** and **ZOOM** to determine the coordinates of the highest point on the graph of S in the interval $[0, 10]$. Interpret your results.

Group Discussion
Suppose the revenue of a company is given by $R(x) = xp(x)$, where x is the number of units of the product sold at a unit price of $p(x)$ dollars.
1. Compute $R'(x)$ and explain, in words, the relationship between $R'(x)$ and $p(x)$ and/or its derivative.
2. What can you say about $R'(x)$ if $p(x)$ is constant? Is this expected?

EXAMPLE 7

When organic waste is dumped into a pond, the oxidation process that takes place reduces the pond's oxygen content. However, given time, nature will restore the oxygen content to its natural level. Suppose the oxygen content t days after organic waste has been dumped into the pond is given by

$$f(t) = 100 \left[\frac{t^2 + 10t + 100}{t^2 + 20t + 100} \right] \qquad (0 < t < \infty)$$

percent of its normal level.

a. Derive a general expression that gives the rate of change of the pond's oxygen level at any time t.

b. How fast is the pond's oxygen content changing 1 day, 10 days, and 20 days after the organic waste has been dumped?

SOLUTION ✔   **a.** The rate of change of the pond's oxygen level at any time t is given by the derivative of the function f. Thus, the required expression is

$$f'(t) = 100 \frac{d}{dt} \left[\frac{t^2 + 10t + 100}{t^2 + 20t + 100} \right]$$

$$= 100 \left[\frac{(t^2 + 20t + 100) \dfrac{d}{dt}(t^2 + 10t + 100) - (t^2 + 10t + 100) \dfrac{d}{dt}(t^2 + 20t + 100)}{(t^2 + 20t + 100)^2} \right]$$

$$= 100 \left[\frac{(t^2 + 20t + 100)(2t + 10) - (t^2 + 10t + 100)(2t + 20)}{(t^2 + 20t + 100)^2} \right]$$

$$= 100 \left[\frac{2t^3 + 10t^2 + 40t^2 + 200t + 200t + 1000 - 2t^3 - 20t^2 - 20t^2 - 200t - 200t - 2000}{(t^2 + 20t + 100)^2} \right]$$

$$= 100 \left[\frac{10t^2 - 1000}{(t^2 + 20t + 100)^2} \right]$$

b. The rate at which the pond's oxygen content is changing 1 day after the organic waste has been dumped is given by

$$f'(1) = 100 \left[\frac{10 - 1000}{(1 + 20 + 100)^2} \right] = -6.76$$

That is, it is dropping at the rate of 6.8% per day. After 10 days the rate is

$$f'(10) = 100 \left[\frac{10(10)^2 - 1000}{(100 + 200 + 100)^2} \right] = 0$$

That is, it is neither increasing nor decreasing. After 20 days the rate is

$$f'(20) = 100 \left[\frac{10(20)^2 - 1000}{(400 + 400 + 100)^2} \right] = 0.37$$

That is, the oxygen content is increasing at the rate of 0.37% per day, and the restoration process has indeed begun.   ■ ■ ■ ■

Group Discussion

Consider a particle moving along a straight line. Newton's second law of motion states that the external force F acting on the particle is equal to the rate of change of its momentum. Thus,

$$F = \frac{d}{dt}(mv)$$

where m, the mass of the particle, and v, its velocity, are both functions of time t.

1. Use the product rule to show that

$$F = m\frac{dv}{dt} + v\frac{dm}{dt}$$

and explain the expression on the right-hand side in words.

2. Use the results of part 1 to show that if the mass of a particle is constant, then $F = ma$, where a is the acceleration of the particle.

VERIFICATION OF THE PRODUCT RULE

We will now verify the product rule. If $p(x) = f(x)g(x)$, then

$$p'(x) = \lim_{h \to 0} \frac{p(x+h) - p(x)}{h}$$

$$= \lim_{h \to 0} \frac{f(x+h)g(x+h) - f(x)g(x)}{h}$$

By adding $-f(x+h)g(x) + f(x+h)g(x)$ (which is zero!) to the numerator and factoring, we have

$$p'(x) = \lim_{h \to 0} \frac{f(x+h)[g(x+h) - g(x)] + g(x)[f(x+h) - f(x)]}{h}$$

$$= \lim_{h \to 0} \left\{ f(x+h)\left[\frac{g(x+h) - g(x)}{h}\right] + g(x)\left[\frac{f(x+h) - f(x)}{h}\right] \right\}$$

$$= \lim_{h \to 0} f(x+h)\left[\frac{g(x+h) - g(x)}{h}\right]$$

$$+ \lim_{h \to 0} g(x)\left[\frac{f(x+h) - f(x)}{h}\right] \qquad \text{(By Property 3 of limits)}$$

$$= \lim_{h \to 0} f(x+h) \cdot \lim_{h \to 0} \frac{g(x+h) - g(x)}{h}$$

$$+ \lim_{h \to 0} g(x) \cdot \lim_{h \to 0} \frac{f(x+h) - f(x)}{h} \qquad \text{(By Property 4 of limits)}$$

$$= f(x)g'(x) + g(x)f'(x)$$

Observe that in the second from the last link in the chain of equalities, we have used the fact that $\lim_{h \to 0} f(x+h) = f(x)$ because f is continuous at x.

SELF-CHECK EXERCISES 3.2

1. Find the derivative of $f(x) = \dfrac{2x+1}{x^2-1}$.

2. What is the slope of the tangent line to the graph of

$$f(x) = (x^2+1)(2x^3 - 3x^2 + 1)$$

at the point $(2, 25)$? How fast is the function f changing when $x = 2$?

 3. The total sales of the Security Products Corporation in its first 2 yr of operation are given by

$$S = f(t) = \frac{0.3t^3}{1 + 0.4t^2} \qquad (0 \le t \le 2)$$

where S is measured in millions of dollars and $t = 0$ corresponds to the date Security Products began operations. How fast were the sales increasing at the beginning of the company's second year of operation?

Solutions to Self-Check Exercises 3.2 can be found on page 210.

3.2 Exercises

In Exercises 1–30, find the derivative of the given function.

1. $f(x) = 2x(x^2 + 1)$ **2.** $f(x) = 3x^2(x - 1)$

3. $f(t) = (t - 1)(2t + 1)$

4. $f(x) = (2x + 3)(3x - 4)$

5. $f(x) = (3x + 1)(x^2 - 2)$

6. $f(x) = (x + 1)(2x^2 - 3x + 1)$

7. $f(x) = (x^3 - 1)(x + 1)$

8. $f(x) = (x^3 - 12x)(3x^2 + 2x)$

9. $f(w) = (w^3 - w^2 + w - 1)(w^2 + 2)$

10. $f(x) = \dfrac{1}{5}x^5 + (x^2 + 1)(x^2 - x - 1) + 28$

11. $f(x) = (5x^2 + 1)(2\sqrt{x} - 1)$

12. $f(t) = (1 + \sqrt{t})(2t^2 - 3)$

13. $f(x) = (x^2 - 5x + 2)\left(x - \dfrac{2}{x}\right)$

14. $f(x) = (x^3 + 2x + 1)\left(2 + \dfrac{1}{x^2}\right)$

15. $f(x) = \dfrac{1}{x-2}$ **16.** $g(x) = \dfrac{3}{2x+4}$

17. $f(x) = \dfrac{x-1}{2x+1}$ **18.** $f(t) = \dfrac{1-2t}{1+3t}$

19. $f(x) = \dfrac{1}{x^2+1}$ **20.** $f(u) = \dfrac{u}{u^2+1}$

21. $f(s) = \dfrac{s^2-4}{s+1}$ **22.** $f(x) = \dfrac{x^3-2}{x^2+1}$

23. $f(x) = \dfrac{\sqrt{x}}{x^2+1}$ **24.** $f(x) = \dfrac{x^2+1}{\sqrt{x}}$

25. $f(x) = \dfrac{x^2+2}{x^2+x+1}$ **26.** $f(x) = \dfrac{x+1}{2x^2+2x+3}$

27. $f(x) = \dfrac{(x+1)(x^2+1)}{x-2}$

28. $f(x) = (3x^2 - 1)\left(x^2 - \dfrac{1}{x}\right)$

29. $f(x) = \dfrac{x}{x^2-4} - \dfrac{x-1}{x^2+4}$

30. $f(x) = \dfrac{x+\sqrt{3x}}{3x-1}$

In Exercises 31–34, suppose f and g are functions that are differentiable at $x = 1$ and that $f(1) = 2$, $f'(1) = -1$, $g(1) = -2$, $g'(1) = 3$. Find the value of $h'(1)$.

31. $h(x) = f(x)g(x)$ **32.** $h(x) = (x^2 + 1)g(x)$

33. $h(x) = \dfrac{xf(x)}{x+g(x)}$ **34.** $h(x) = \dfrac{f(x)g(x)}{f(x)-g(x)}$

In Exercises 35–38, find the derivative of each of the given functions and evaluate $f'(x)$ at the given value of x.

35. $f(x) = (2x - 1)(x^2 + 3); \ x = 1$

36. $f(x) = \dfrac{2x + 1}{2x - 1}; \ x = 2$

37. $f(x) = \dfrac{x}{x^4 - 2x^2 - 1}; \ x = -1$

38. $f(x) = (\sqrt{x} + 2x)(x^{3/2} - x); \ x = 4$

In Exercises 39–42, find the slope and an equation of the tangent line to the graph of the function f at the specified point.

39. $f(x) = (x^3 + 1)(x^2 - 2); \ (2, 18)$

40. $f(x) = \dfrac{x^2}{x + 1}; \ \left(2, \dfrac{4}{3}\right)$

41. $f(x) = \dfrac{x + 1}{x^2 + 1}; \ (1, 1)$

42. $f(x) = \dfrac{1 + 2x^{1/2}}{1 + x^{3/2}}; \ \left(4, \dfrac{5}{9}\right)$

43. Find an equation of the tangent line to the graph of the function $f(x) = (x^3 + 1)(3x^2 - 4x + 2)$ at the point $(1, 2)$.

44. Find an equation of the tangent line to the graph of the function $f(x) = \dfrac{3x}{x^2 - 2}$ at the point $(2, 3)$.

45. Let $f(x) = (x^2 + 1)(2 - x)$. Find the point(s) on the graph of f where the tangent line is horizontal.

46. Let $f(x) = \dfrac{x}{x^2 + 1}$. Find the point(s) on the graph of f where the tangent line is horizontal.

47. Find the point(s) on the graph of the function $f(x) = (x^2 + 6)(x - 5)$ where the slope of the tangent line is equal to -2.

48. Find the point(s) on the graph of the function $f(x) = \dfrac{x + 1}{x - 1}$ where the slope of the tangent line is equal to $-1/2$.

49. A straight line perpendicular to and passing through the point of tangency of the tangent line is called the *normal* to the curve. Find the equation of the tangent line and the normal to the curve $y = \dfrac{1}{1 + x^2}$ at the point $(1, \frac{1}{2})$.

50. CONCENTRATION OF A DRUG IN THE BLOODSTREAM The concentration of a certain drug in a patient's bloodstream t hr after injection is given by

$$C(t) = \frac{0.2t}{t^2 + 1}$$

a. Find the rate at which the concentration of the drug is changing with respect to time.
b. How fast is the concentration changing $\frac{1}{2}$ hr, 1 hr, and 2 hr after the injection?

51. COST OF REMOVING TOXIC WASTE A city's main well was recently found to be contaminated with trichloroethylene, a cancer-causing chemical, as a result of an abandoned chemical dump leaching chemicals into the water. A proposal submitted to the city's council members indicates that the cost, measured in millions of dollars, of removing x percent of the toxic pollutant is given by

$$C(x) = \frac{0.5x}{100 - x}$$

Find $C'(80)$, $C'(90)$, $C'(95)$, and $C'(99)$ and interpret your results.

52. DRUG DOSAGES Thomas Young has suggested the following rule for calculating the dosage of medicine for children 1 to 12 yr old. If a denotes the adult dosage (in milligrams) and if t is the child's age (in years), then the child's dosage is given by

$$D(t) = \frac{at}{t + 12}$$

Suppose that the adult dosage of a substance is 500 mg. Find an expression that gives the rate of change of a child's dosage with respect to the child's age. What is the rate of change of a child's dosage with respect to his or her age for a 6-yr-old child? For a 10-yr-old child?

53. EFFECT OF BACTERICIDE The number of bacteria $N(t)$ in a certain culture t min after an experimental bactericide is introduced obeys the rule

$$N(t) = \frac{10{,}000}{1 + t^2} + 2000$$

Find the rate of change of the number of bacteria in the culture 1 minute after and 2 minutes after the bactericide is introduced. What is the population of the bacteria in the culture 1 min and 2 min after the bactericide is introduced?

54. DEMAND FUNCTIONS The demand function for the Sicard wristwatch is given by

$$d(x) = \frac{50}{0.01x^2 + 1} \qquad (0 \le x \le 20)$$

where x (measured in units of a thousand) is the quantity demanded per week and $d(x)$ is the unit price in dollars.

a. Find $d'(x)$.
b. Find $d'(5)$, $d'(10)$, and $d'(15)$ and interpret your results.

55. **LEARNING CURVES** From experience, the Emory Secretarial School knows that the average student taking Advanced Typing will progress according to the rule

$$N(t) = \frac{60t + 180}{t + 6} \qquad (t \geq 0)$$

where $N(t)$ measures the number of words per minute the student can type after t wk in the course.
a. Find an expression for $N'(t)$.
b. Compute $N'(t)$ for $t = 1$, 3, 4, and 7 and interpret your results.
c. Sketch the graph of the function N. Does it confirm the results obtained in part (b)?
d. What will be the average student's typing speed at the end of the 12-wk course?

56. **BOX OFFICE RECEIPTS** The total worldwide box office receipts for a long-running movie are approximated by the function

$$T(x) = \frac{120x^2}{x^2 + 4}$$

where $T(x)$ is measured in millions of dollars and x is the number of years since the movie's release. How fast are the total receipts changing 1 yr, 3 yr, and 5 yr after its release?

57. **FORMALDEHYDE LEVELS** A study on formaldehyde levels in 900 homes indicates that emissions of various chemicals can decrease over time. The formaldehyde level (parts per million) in an average home in the study is given by

$$f(t) = \frac{0.055t + 0.26}{t + 2} \qquad (0 \leq t \leq 12)$$

where t is the age of the house in years. How fast is the formaldehyde level of the average house dropping when it is new? At the beginning of its fourth year?
Source: Bonneville Power Administration

58. **POPULATION GROWTH** A major corporation is building a 4325-acre complex of homes, offices, stores, schools, and churches in the rural community of Glen Cove. As a result of this development, the planners have estimated that Glen Cove's population (in thousands) t yr from now will be given by

$$P(t) = \frac{25t^2 + 125t + 200}{t^2 + 5t + 40}$$

a. Find the rate at which Glen Cove's population is changing with respect to time.

b. What will be the population after 10 yr? At what rate will the population be increasing when $t = 10$?

In Exercises 59–62, determine whether the statement is true or false. If it is true, explain why it is true. If it is false, give an example to show why it is false.

59. If f and g are differentiable, then

$$\frac{d}{dx}[f(x)g(x)] = f'(x)g'(x).$$

60. If f is differentiable, then

$$\frac{d}{dx}[xf(x)] = f(x) + xf'(x).$$

61. If f is differentiable, then

$$\frac{d}{dx}\left[\frac{f(x)}{x^2}\right] = \frac{f'(x)}{2x}.$$

62. If f, g, and h are differentiable, then

$$\frac{d}{dx}\left[\frac{f(x)g(x)}{h(x)}\right]$$
$$= \frac{f'(x)g(x)h(x) + f(x)g'(x)h(x) - f(x)g(x)h'(x)}{[h(x)]^2}.$$

63. Extend the product rule for differentiation to the following case involving the product of three differentiable functions: Let $h(x) = u(x)v(x)w(x)$ and show that $h'(x) = u(x)v(x)w'(x) + u(x)v'(x)w(x) + u'(x)v(x)w(x)$.
Hint: Let $f(x) = u(x)v(x)$, $g(x) = w(x)$, and $h(x) = f(x)g(x)$, and apply the product rule to the function h.

64. Prove the quotient rule for differentiation (Rule 6).
Hint: Verify the following steps:

a. $\dfrac{k(x + h) - k(x)}{h} = \dfrac{f(x + h)g(x) - f(x)g(x + h)}{hg(x + h)g(x)}$

b. By adding $[-f(x)g(x) + f(x)g(x)]$ to the numerator and simplifying,

$$\frac{k(x + h) - k(x)}{h} = \frac{1}{g(x + h)g(x)}$$
$$\times \left\{ \left[\frac{f(x + h) - f(x)}{h}\right] \cdot g(x) \right.$$
$$\left. - \left[\frac{g(x + h) - g(x)}{h}\right] \cdot f(x) \right\}$$

c. $k'(x) = \lim\limits_{h \to 0} \dfrac{k(x + h) - k(x)}{h}$

$$= \frac{g(x)f'(x) - f(x)g'(x)}{[g(x)]^2}$$

THE PRODUCT AND QUOTIENT RULES

EXAMPLE 1

Let $f(x) = (2\sqrt{x} + 0.5x)\left(0.3x^3 + 2x - \dfrac{0.3}{x}\right)$. Find $f'(0.2)$.

SOLUTION ✔

Using the numerical derivative operation of a graphing utility, we find

$$f'(0.2) = 6.47974948127$$

■ ■ ■ ■

EXAMPLE 2

Importance of Time in Treating Heart Attacks

According to the American Heart Association, the treatment benefit for heart attacks depends on the time to treatment and is described by the function

$$f(t) = \frac{-16.94t + 203.28}{t + 2.0328} \qquad (0 \le t \le 12)$$

where t is measured in hours and $f(t)$ is a percentage.

a. Use a graphing utility to graph the function f using the viewing rectangle $[0, 13] \times [0, 120]$.
b. Use a graphing utility to find the derivative of f when $t = 0$ and $t = 2$.
c. Interpret the results obtained in part (b).
Source: American Heart Association

SOLUTION ✔

a. The graph of f is shown in Figure T1.

FIGURE T1

b. Using the numerical derivative operation of a graphing utility, we find

$$f'(0) \approx -57.5266$$

$$f'(2) \approx -14.6165$$

c. The results of part (b) show that the treatment benefit drops off at the rate of 58% per hour at the time when the heart attack first occurs and falls off at the rate of 15% per hour when the time to treatment is 2 hours. Thus, it is extremely urgent that a patient suffering a heart attack receive medical attention as soon as possible. ■■■■

Exercises

In Exercises 1–6, use the numerical derivative operation of a graphing utility to find the rate of change of $f(x)$ at the given value of x. Give your answer accurate to four decimal places.

1. $f(x) = (2x^2 + 1)(x^3 + 3x + 4); x = -0.5$

2. $f(x) = (\sqrt{x} + 1)(2x^2 + x - 3); x = 1.5$

3. $f(x) = \dfrac{\sqrt{x} - 1}{\sqrt{x} + 1}; x = 3$

4. $f(x) = \dfrac{\sqrt{x}(x^2 + 4)}{x^3 + 1}; x = 4$

5. $f(x) = \dfrac{\sqrt{x}(1 + x^{-1})}{x + 1}; x = 1$

6. $f(x) = \dfrac{x^2(2 + \sqrt{x})}{1 + \sqrt{x}}; x = 1$

7. NEW CONSTRUCTION JOBS The president of a major housing construction company claims that the number of construc-

tion jobs created in the next t months is given by

$$f(t) = 1.42 \left(\frac{7t^2 + 140t + 700}{3t^2 + 80t + 550} \right)$$

where $f(t)$ is measured in millions of jobs per year. At what rate will construction jobs be created 1 yr from now, assuming her projection is correct?

8. POPULATION GROWTH A major corporation is building a 4325-acre complex of homes, offices, stores, schools, and churches in the rural community of Glen Cove. As a result of this development, the planners have estimated that Glen Cove's population (in thousands) t yr from now will be given by

$$P(t) = \frac{25t^2 + 125t + 200}{t^2 + 5t + 40}$$

a. What will be the population 10 yr from now?
b. At what rate will the population be increasing 10 yr from now?

SOLUTIONS TO SELF-CHECK EXERCISES 3.2

1. We use the quotient rule to obtain

$$
\begin{aligned}
f'(x) &= \frac{(x^2-1)\frac{d}{dx}(2x+1)-(2x+1)\frac{d}{dx}(x^2-1)}{(x^2-1)^2}\\
&= \frac{(x^2-1)(2)-(2x+1)(2x)}{(x^2-1)^2}\\
&= \frac{2x^2-2-4x^2-2x}{(x^2-1)^2}\\
&= \frac{-2x^2-2x-2}{(x^2-1)^2}\\
&= -\frac{2(x^2+x+1)}{(x^2-1)^2}
\end{aligned}
$$

2. The slope of the tangent line to the graph of f at any point is given by

$$
\begin{aligned}
f'(x) &= (x^2+1)\frac{d}{dx}(2x^3-3x^2+1)\\
&\quad + (2x^3-3x^2+1)\frac{d}{dx}(x^2+1)\\
&= (x^2+1)(6x^2-6x)+(2x^3-3x^2+1)(2x)
\end{aligned}
$$

In particular, the slope of the tangent line to the graph of f when $x=2$ is

$$
\begin{aligned}
f'(2) &= (2^2+1)[6(2^2)-6(2)]\\
&\quad + [2(2^3)-3(2^2)+1][2(2)]\\
&= 60+20=80
\end{aligned}
$$

Note that it is not necessary to simplify the expression for $f'(x)$ since we are required only to evaluate the expression at $x=2$. We also conclude, from this result, that the function f is changing at the rate of 80 units/unit change in x when $x=2$.

3. The rate at which the company's total sales are changing at any time t is given by

$$
\begin{aligned}
S'(t) &= \frac{(1+0.4t^2)\frac{d}{dt}(0.3t^3)-(0.3t^3)\frac{d}{dt}(1+0.4t^2)}{(1+0.4t^2)^2}\\
&= \frac{(1+0.4t^2)(0.9t^2)-(0.3t^3)(0.8t)}{(1+0.4t^2)^2}
\end{aligned}
$$

Therefore, at the beginning of the second year of operation, Security Products' sales were increasing at the rate of

$$
S'(1) = \frac{(1+0.4)(0.9)-(0.3)(0.8)}{(1+0.4)^2} = 0.520408
$$

or \$520,408/year.

3.3 The Chain Rule

This section introduces another rule of differentiation called the **chain rule.** When used in conjunction with the rules of differentiation developed in the last two sections, the chain rule enables us to greatly enlarge the class of functions we are able to differentiate.

THE CHAIN RULE

Consider the function $h(x) = (x^2 + x + 1)^2$. If we were to compute $h'(x)$ using only the rules of differentiation from the previous sections, then our approach might be to expand $h(x)$. Thus,

$$h(x) = (x^2 + x + 1)^2 = (x^2 + x + 1)(x^2 + x + 1)$$
$$= x^4 + 2x^3 + 3x^2 + 2x + 1$$

from which we find

$$h'(x) = 4x^3 + 6x^2 + 6x + 2$$

But what about the function $H(x) = (x^2 + x + 1)^{100}$? The same technique may be used to find the derivative of the function H, but the amount of work involved in this case would be prodigious! Consider, also, the function $G(x) = \sqrt{x^2 + 1}$. For each of the two functions H and G, the rules of differentiation of the previous sections cannot be applied directly to compute the derivatives H' and G'.

Observe that both H and G are **composite functions;** that is, each is composed of, or built up from, simpler functions. For example, the function H is composed of the two simpler functions $f(x) = x^2 + x + 1$ and $g(x) = x^{100}$ as follows:

$$H(x) = g[f(x)] = [f(x)]^{100}$$
$$= (x^2 + x + 1)^{100}$$

In a similar manner, we see that the function G is composed of the two simpler functions $f(x) = x^2 + 1$ and $g(x) = \sqrt{x}$. Thus,

$$G(x) = g[f(x)] = \sqrt{f(x)}$$
$$= \sqrt{x^2 + 1}$$

As a first step toward finding the derivative h' of a composite function $h = g \circ f$ defined by $h(x) = g[f(x)]$, we write

$$u = f(x) \qquad \text{and} \qquad y = g[f(x)] = g(u)$$

The dependency of h on g and f is illustrated in Figure 3.5. Since u is a function of x, we may compute the derivative of u with respect to x, if f is a differentiable function, obtaining $du/dx = f'(x)$. Next, if g is a differentiable function of u, we may compute the derivative of g with respect to u, obtaining $dy/du = g'(u)$. Now, since the function h is composed of the function g and the function

FIGURE 3.5
The composite function $h(x) = g[\,f(x)]$

f, we might suspect that the rule $h'(x)$ for the derivative h' of h will be given by an expression that involves the rules for the derivatives of f and g. But how do we combine these derivatives to yield h'?

This question can be answered by interpreting the derivative of each function as giving the rate of change of that function. For example, suppose $u = f(x)$ changes three times as fast as x—that is,

$$f'(x) = \frac{du}{dx} = 3$$

And suppose $y = g(u)$ changes twice as fast as u—that is,

$$g'(u) = \frac{dy}{du} = 2$$

Then, we would expect $y = h(x)$ to change six times as fast as x—that is,

$$h'(x) = g'(u)f'(x) = (2)(3) = 6$$

or equivalently,

$$\frac{dy}{dx} = \frac{dy}{du} \cdot \frac{du}{dx} = (2)(3) = 6$$

This observation suggests the following result, which we state without proof.

Rule 7: The Chain Rule

If $h(x) = g[\,f(x)]$, then

$$h'(x) = \frac{d}{dx} g(f(x)) = g'(f(x))f'(x) \qquad (1)$$

Equivalently, if we write $y = h(x) = g(u)$, where $u = f(x)$, then

$$\frac{dy}{dx} = \frac{dy}{du} \cdot \frac{du}{dx} \qquad (2)$$

REMARKS

1. If we label the composite function h in the following manner

Inside function
↓
$$h(x) = g[\,f(x)]$$
↑
Outside function

then $h'(x)$ is just the *derivative* of the "outside function" *evaluated at* the "inside function" times the *derivative* of the "inside function."

2. Equation (2) can be remembered by observing that if we "cancel" the du's, then

$$\frac{dy}{dx} = \frac{dy}{du} \cdot \frac{du}{dx} = \frac{dy}{dx}$$

■ ■ ■

THE CHAIN RULE FOR POWERS OF FUNCTIONS

Many composite functions have the special form $h(x) = g(f(x))$, where g is defined by the rule $g(x) = x^n$ (n, a real number)—that is,

$$h(x) = [f(x)]^n$$

In other words, the function h is given by the power of a function f. The functions

$$h(x) = (x^2 + x + 1)^2, \qquad H = (x^2 + x + 1)^{100}, \qquad G = \sqrt{x^2 + 1}$$

discussed earlier are examples of this type of composite function. By using the following corollary of the chain rule, the general power rule, we can find the derivative of this type of function much more easily than by using the chain rule directly.

The General Power Rule

If the function f is differentiable and $h(x) = [f(x)]^n$ (n, a real number), then

$$h'(x) = \frac{d}{dx}[f(x)]^n = n[f(x)]^{n-1}f'(x) \tag{3}$$

To see this, we observe that $h(x) = g(f(x))$, where $g(x) = x^n$, so that, by virtue of the chain rule, we have

$$h'(x) = g'(f(x))f'(x)$$
$$= n[f(x)]^{n-1}f'(x)$$

since $g'(x) = nx^{n-1}$.

EXAMPLE 1 Find $H'(x)$, if

$$H(x) = (x^2 + x + 1)^{100}$$

SOLUTION ✔ Using the general power rule, we obtain

$$H'(x) = 100(x^2 + x + 1)^{99}\frac{d}{dx}(x^2 + x + 1)$$

$$= 100(x^2 + x + 1)^{99}(2x + 1)$$

Observe the bonus that comes from using the chain rule: The answer is completely factored.

■ ■ ■ ■

EXAMPLE 2   Differentiate the function $G(x) = \sqrt{x^2 + 1}$.

SOLUTION ✔   We rewrite the function $G(x)$ as

$$G(x) = (x^2 + 1)^{1/2}$$

and apply the general power rule, obtaining

$$G'(x) = \frac{1}{2}(x^2 + 1)^{-1/2}\frac{d}{dx}(x^2 + 1)$$

$$= \frac{1}{2}(x^2 + 1)^{-1/2} \cdot 2x = \frac{x}{\sqrt{x^2 + 1}}$$ ◼◼◼◼

EXAMPLE 3   Differentiate the function $f(x) = x^2(2x + 3)^5$.

SOLUTION ✔   Applying the product rule followed by the general power rule, we obtain

$$f'(x) = x^2\frac{d}{dx}(2x + 3)^5 + (2x + 3)^5\frac{d}{dx}(x^2)$$

$$= (x^2)5(2x + 3)^4 \cdot \frac{d}{dx}(2x + 3) + (2x + 3)^5(2x)$$

$$= 5x^2(2x + 3)^4(2) + 2x(2x + 3)^5$$

$$= 2x(2x + 3)^4(5x + 2x + 3) = 2x(7x + 3)(2x + 3)^4$$ ◼◼◼◼

EXAMPLE 4   Find $f'(x)$ if $f(x) = \dfrac{1}{(4x^2 - 7)^2}$.

SOLUTION ✔   Rewriting $f(x)$ and then applying the general power rule, we obtain

$$f'(x) = \frac{d}{dx}\left[\frac{1}{(4x^2 - 7)^2}\right] = \frac{d}{dx}(4x^2 - 7)^{-2}$$

$$= -2(4x^2 - 7)^{-3}\frac{d}{dx}(4x^2 - 7)$$

$$= -2(4x^2 - 7)^{-3}(8x) = -\frac{16x}{(4x^2 - 7)^3}$$ ◼◼◼◼

EXAMPLE 5   Find the slope of the tangent line to the graph of the function

$$f(x) = \left(\frac{2x + 1}{3x + 2}\right)^3$$

at the point $(0, \frac{1}{8})$.

SOLUTION ✔ The slope of the tangent line to the graph of f at any point is given by $f'(x)$. To compute $f'(x)$, we use the general power rule followed by the quotient rule, obtaining

$$f'(x) = 3\left(\frac{2x+1}{3x+2}\right)^2 \frac{d}{dx}\left(\frac{2x+1}{3x+2}\right)$$

$$= 3\left(\frac{2x+1}{3x+2}\right)^2 \left[\frac{(3x+2)(2) - (2x+1)(3)}{(3x+2)^2}\right]$$

$$= 3\left(\frac{2x+1}{3x+2}\right)^2 \left[\frac{6x+4-6x-3}{(3x+2)^2}\right]$$

$$= \frac{3(2x+1)^2}{(3x+2)^4}$$

In particular, the slope of the tangent line to the graph of f at $(0, \frac{1}{8})$ is given by

$$f'(0) = \frac{3(0+1)^2}{(0+2)^4} = \frac{3}{16}$$

■ ■ ■ ■

Exploring with Technology

Refer to Example 5.

1. Use a graphing utility to plot the graph of the function f using the viewing rectangle $[-2, 1] \times [-1, 2]$. Then draw the tangent line to the graph of f at the point $(0, \frac{1}{8})$.
2. For a better picture, repeat part 1 using the viewing rectangle $[-1, 1] \times [-0.1, 0.3]$.
3. Use the numerical differentiation capability of the graphing utility to verify that the slope of the tangent line at $(0, \frac{1}{8})$ is $\frac{3}{16}$.

APPLICATIONS

  EXAMPLE **6** The membership of The Fitness Center, which opened a few years ago, is approximated by the function

$$N(t) = 100(64 + 4t)^{2/3} \qquad (0 \le t \le 52)$$

where $N(t)$ gives the number of members at the beginning of week t.

a. Find $N'(t)$.
b. How fast was the center's membership increasing initially ($t = 0$)?
c. How fast was the membership increasing at the beginning of the 40th week?
d. What was the membership when the center first opened? At the beginning of the 40th week?

SOLUTION ✔   **a.** Using the general power rule, we obtain

$$N'(t) = \frac{d}{dt}[100(64 + 4t)^{2/3}]$$

$$= 100\frac{d}{dt}(64 + 4t)^{2/3}$$

$$= 100\left(\frac{2}{3}\right)(64 + 4t)^{-1/3}\frac{d}{dt}(64 + 4t)$$

$$= \frac{200}{3}(64 + 4t)^{-1/3}(4)$$

$$= \frac{800}{3(64 + 4t)^{1/3}}$$

b. The rate at which the membership was increasing when the center first opened is given by

$$N'(0) = \frac{800}{3(64)^{1/3}} \approx 66.7$$

or approximately 67 people per week.

c. The rate at which the membership was increasing at the beginning of the 40th week is given by

$$N'(40) = \frac{800}{3(64 + 160)^{1/3}} \approx 43.9$$

or approximately 44 people per week.

d. The membership when the center first opened is given by

$$N(0) = 100(64)^{2/3} = 100(16)$$

or approximately 1600 people. The membership at the beginning of the 40th week is given by

$$N(40) = 100(64 + 160)^{2/3} \approx 3688.3$$

or approximately 3688 people.   ■ ■ ■ ■

Group Discussion

The profit P of a one-product software manufacturer depends on the number of units of its products sold. The manufacturer estimates that it will sell x units of its product per week. Suppose $P = g(x)$ and $x = f(t)$, where g and f are differentiable functions.

1. Write an expression giving the rate of change of the profit with respect to the number of units sold.

2. Write an expression giving the rate of change of the number of units sold per week.

3. Write an expression giving the rate of change of the profit per week.

EXAMPLE **7**

FIGURE 3.6
Cross section of the aorta

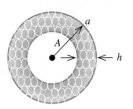

Arteriosclerosis begins during childhood when plaque (soft masses of fatty material) forms in the arterial walls, blocking the flow of blood through the arteries and leading to heart attacks, strokes, and gangrene. Suppose the idealized cross section of the aorta is circular with radius a cm and by year t the thickness of the plaque (assume it is uniform) is $h = f(t)$ cm (Figure 3.6). Then the area of the opening is given by $A = \pi(a - h)^2$ square centimeters (cm²).

Suppose the radius of an individual's artery is 1 cm ($a = 1$) and the thickness of the plaque in year t is given by

$$h = g(t) = 1 - 0.01(10{,}000 - t^2)^{1/2} \text{ cm}$$

Since the area of the arterial opening is given by

$$A = f(h) = \pi(1 - h)^2$$

the rate at which A is changing with respect to time is given by

$$\frac{dA}{dt} = \frac{dA}{dh} \cdot \frac{dh}{dt} = f'(h) \cdot g'(t) \qquad \text{(By the chain rule)}$$

$$= 2\pi(1 - h)(-1)\left[-0.01\left(\frac{1}{2}\right)(10{,}000 - t^2)^{-1/2}(-2t)\right] \qquad \text{(Using the chain rule twice)}$$

$$= -2\pi(1 - h)\left[\frac{0.01t}{(10{,}000 - t^2)^{1/2}}\right]$$

$$= -\frac{0.02\pi(1 - h)t}{\sqrt{10{,}000 - t^2}}$$

For example, when $t = 50$,

$$h = g(50) = 1 - 0.01(10{,}000 - 2500)^{1/2} \approx 0.134$$

so that

$$\frac{dA}{dt} = -\frac{0.02\pi(1 - 0.134)50}{\sqrt{10{,}000 - 2500}} \approx -0.03$$

That is, the area of the arterial opening is decreasing at the rate of 0.03 cm²/year.

■ ■ ■ ■

 Group Discussion
Suppose the population P of a certain bacteria culture is given by $P = f(T)$, where T is the temperature of the medium. Further, suppose the temperature T is a function of time t in seconds—that is, $T = g(t)$. Give an interpretation of each of the following quantities:

1. $\dfrac{dP}{dT}$ **2.** $\dfrac{dT}{dt}$ **3.** $\dfrac{dP}{dt}$ **4.** $(f \circ g)(t)$ **5.** $f'(g(t))g'(t)$

FINDING THE DERIVATIVE OF A COMPOSITE FUNCTION

EXAMPLE 1 Find the rate of change of $f(x) = \sqrt{x}(1 + 0.02x^2)^{3/2}$ when $x = 2.1$.

SOLUTION ✔ Using the numerical derivative operation of a graphing utility, we find

$$f'(2.1) = 0.582146320119$$

or approximately 0.58 unit per unit change in x. ■■■■

EXAMPLE 2 The management of Astro World ("The Amusement Park of the Future") estimates that the total number of visitors (in thousands) to the amusement park t hours after opening time at 9 A.M. is given by

$$N(t) = \frac{30t}{\sqrt{2 + t^2}}$$

What is the rate at which visitors are admitted to the amusement park at 10:30 A.M.?

SOLUTION ✔ Using the numerical derivative operation of a graphing utility, we find

$$N'(1.5) \approx 6.8481$$

or approximately 6848 visitors per hour. ■■■■

Exercises

In Exercises 1–6, use the numerical derivative operation of a graphing utility to find the rate of change of $f(x)$ at the given value of x. Give your answer accurate to four decimal places.

1. $f(x) = \sqrt{x^2 - x^4}$; $x = 0.5$

2. $f(x) = x - \sqrt{1 - x^2}$; $x = 0.4$

3. $f(x) = x\sqrt{1 - x^2}$; $x = 0.2$

4. $f(x) = (x + \sqrt{x^2 + 4})^{3/2}$; $x = 1$

5. $f(x) = \dfrac{\sqrt{1 + x^2}}{x^3 + 2}$; $x = -1$

6. $f(x) = \dfrac{x^3}{1 + (1 + x^2)^{3/2}}$; $x = 3$

7. WORLDWIDE PRODUCTION OF VEHICLES According to *Automotive News,* the worldwide production of vehicles between 1960 and 1990 is given by the function

$$f(t) = 16.5\sqrt{1 + 2.2t} \qquad (0 \le t \le 3)$$

where $f(t)$ is measured in units of a million and t is measured in decades, with $t = 0$ corresponding to the beginning of 1960. What was the rate of change of the worldwide production of vehicles at the beginning of 1970? At the beginning of 1980?
Source: Automotive News

8. ACCUMULATION YEARS People from their mid-40s to their mid-50s are in the prime investing years. Demographic studies of this type are of particular importance to financial institutions. The function

$$N(t) = 34.4(1 + 0.32125t)^{0.15} \qquad (0 \le t \le 12)$$

gives the projected number of people in this age group in the United States (in millions) in year t where $t = 0$ corresponds to the beginning of 1996.
a. How large is this segment of the population projected to be at the beginning of 2005?
b. How fast will this segment of the population be growing at the beginning of 2005?
Source: Census Bureau

1. Find the derivative of

$$f(x) = -\frac{1}{\sqrt{2x^2 - 1}}$$

2. Suppose the life expectancy at birth (in years) of a female in a certain country is described by the function

$$g(t) = 50.02(1 + 1.09t)^{0.1} \qquad (0 \le t \le 150)$$

where t is measured in years and $t = 0$ corresponds to the beginning of 1900.
a. What is the life expectancy at birth of a female born at the beginning of 1980? At the beginning of the year 2000?
b. How fast is the life expectancy at birth of a female born at any time t changing?

Solutions to Self-Check Exercises 3.3 can be found on page 223.

3.3 Exercises

In Exercises 1–46, find the derivative of the given function.

1. $f(x) = (2x - 1)^4$

2. $f(x) = (1 - x)^3$

3. $f(x) = (x^2 + 2)^5$

4. $f(t) = 2(t^3 - 1)^5$

5. $f(x) = (2x - x^2)^3$

6. $f(x) = 3(x^3 - x)^4$

7. $f(x) = (2x + 1)^{-2}$

8. $f(t) = \frac{1}{2}(2t^2 + t)^{-3}$

9. $f(x) = (x^2 - 4)^{3/2}$

10. $f(t) = (3t^2 - 2t + 1)^{3/2}$

11. $f(x) = \sqrt{3x - 2}$

12. $f(t) = \sqrt{3t^2 - t}$

13. $f(x) = \sqrt[3]{1 - x^2}$

14. $f(x) = \sqrt{2x^2 - 2x + 3}$

15. $f(x) = \frac{1}{(2x + 3)^3}$

16. $f(x) = \frac{2}{(x^2 - 1)^4}$

17. $f(t) = \frac{1}{\sqrt{2t - 3}}$

18. $f(x) = \frac{1}{\sqrt{2x^2 - 1}}$

19. $y = \frac{1}{(4x^4 + x)^{3/2}}$

20. $f(t) = \frac{4}{\sqrt[3]{2t^2 + t}}$

21. $f(x) = (3x^2 + 2x + 1)^{-2}$

22. $f(t) = (5t^3 + 2t^2 - t + 4)^{-3}$

23. $f(x) = (x^2 + 1)^3 - (x^3 + 1)^2$

24. $f(t) = (2t - 1)^4 + (2t + 1)^4$

25. $f(t) = (t^{-1} - t^{-2})^3$

26. $f(v) = (v^{-3} + 4v^{-2})^3$

27. $f(x) = \sqrt{x + 1} + \sqrt{x - 1}$

28. $f(u) = (2u + 1)^{3/2} + (u^2 - 1)^{-3/2}$

29. $f(x) = 2x^2(3 - 4x)^4$

30. $h(t) = t^2(3t + 4)^3$

31. $f(x) = (x - 1)^2(2x + 1)^4$

32. $g(u) = (1 + u^2)^5(1 - 2u^2)^8$

33. $f(x) = \left(\frac{x + 3}{x - 2}\right)^3$

34. $f(x) = \left(\frac{x + 1}{x - 1}\right)^5$

35. $s(t) = \left(\frac{t}{2t + 1}\right)^{3/2}$

36. $g(s) = \left(s^2 + \frac{1}{s}\right)^{3/2}$

37. $g(u) = \sqrt{\frac{u + 1}{3u + 2}}$

38. $g(x) = \sqrt{\frac{2x + 1}{2x - 1}}$

39. $f(x) = \frac{x^2}{(x^2 - 1)^4}$

40. $g(u) = \frac{2u^2}{(u^2 + u)^3}$

41. $h(x) = \frac{(3x^2 + 1)^3}{(x^2 - 1)^4}$

42. $g(t) = \frac{(2t - 1)^2}{(3t + 2)^4}$

43. $f(x) = \frac{\sqrt{2x + 1}}{x^2 - 1}$

44. $f(t) = \frac{4t^2}{\sqrt{2t^2 + 2t - 1}}$

45. $g(t) = \frac{\sqrt{t + 1}}{\sqrt{t^2 + 1}}$

46. $f(x) = \frac{\sqrt{x^2 + 1}}{\sqrt{x^2 - 1}}$

In Exercises 47–52, find $\dfrac{dy}{du}$, $\dfrac{du}{dx}$, and $\dfrac{dy}{dx}$.

47. $y = u^{4/3}$ and $u = 3x^2 - 1$

48. $y = \sqrt{u}$ and $u = 7x - 2x^2$

49. $y = u^{-2/3}$ and $u = 2x^3 - x + 1$

50. $y = 2u^2 + 1$ and $u = x^2 + 1$

51. $y = \sqrt{u} + \dfrac{1}{\sqrt{u}}$ and $u = x^3 - x$

52. $y = \dfrac{1}{u}$ and $u = \sqrt{x} + 1$

53. Suppose $F(x) = g(f(x))$ and $f(2) = 3$, $f'(2) = -3$, $g(3) = 5$, and $g'(3) = 4$. find $F'(2)$.

54. Suppose $h = f \circ g$. Find $h'(0)$ given that $f(0) = 6$, $f'(5) = -2$, $g(0) = 5$, and $g'(0) = 3$.

55. Suppose $F(x) = f(x^2 + 1)$. Find $F'(1)$ if $f'(2) = 3$.

56. Let $F(x) = f(f(x))$. Does it follow that $F'(x) = [f'(x)]^2$?
Hint: Let $f(x) = x^2$.

57. Suppose $h = g \circ f$. Does it follow that $h' = g' \circ f'$?
Hint: Let $f(x) = x$ and $g(x) = x^2$.

58. Suppose $h = f \circ g$. Show that $h' = (f' \circ g)g'$.

In Exercises 59–62, find an equation of the tangent line to the graph of the function at the given point.

59. $f(x) = (1 - x)(x^2 - 1)^2$; $(2, -9)$

60. $f(x) = \left(\dfrac{x+1}{x-1}\right)^2$; $(3, 4)$

61. $f(x) = x\sqrt{2x^2 + 7}$; $(3, 15)$

62. $f(x) = \dfrac{8}{\sqrt{x^2 + 6x}}$; $(2, 2)$

63. TELEVISION VIEWING The number of viewers of a television series introduced several years ago is approximated by the function

$$N(x) = (60 + 2x)^{2/3} \qquad (1 \le x \le 26)$$

where $N(x)$ (measured in millions) denotes the number of weekly viewers of the series in the xth week. Find the rate of increase of the weekly audience at the end of week 2 and at the end of week 12. How many viewers were there in week 2 and in week 24?

64. MALE LIFE EXPECTANCY Suppose the life expectancy of a male at birth in a certain country is described by the function

$$f(t) = 46.9(1 + 1.09t)^{0.1} \qquad (0 \le t \le 150)$$

where t is measured in years and $t = 0$ corresponds to the beginning of 1900. How long can a male born at the beginning of the year 2000 in that country expect to live? What is the rate of change of the life expectancy of a male born in that country at the beginning of the year 2000?

65. CONCENTRATION OF CARBON MONOXIDE IN THE AIR According to a joint study conducted by Oxnard's Environmental Management Department and a state government agency, the concentration of carbon monoxide in the air due to automobile exhaust t yr from now is given by

$$C(t) = 0.01(0.2t^2 + 4t + 64)^{2/3}$$

parts per million.
a. Find the rate at which the level of carbon monoxide is changing with respect to time.
b. Find the rate at which the level of carbon monoxide will be changing 5 yr from now.

66. CONTINUING EDUCATION ENROLLMENT The registrar of Kellogg University estimates that the total student enrollment in the Continuing Education division will be given by

$$N(t) = -\dfrac{20{,}000}{\sqrt{1 + 0.2t}} + 21{,}000$$

where $N(t)$ denotes the number of students enrolled in the division t yr from now. Find an expression for $N'(t)$. How fast is the student enrollment increasing currently? How fast will it be increasing 5 yr from now?

67. AIR POLLUTION According to the South Coast Air Quality Management District, the level of nitrogen dioxide, a brown gas that impairs breathing, present in the atmosphere on a certain May day in downtown Los Angeles is approximated by

$$A(t) = 0.03t^3(t - 7)^4 + 60.2 \qquad (0 \le t \le 7)$$

where $A(t)$ is measured in pollutant standard index and t is measured in hours, with $t = 0$ corresponding to 7 A.M.
a. Find $A'(t)$.
b. Find $A'(1)$, $A'(3)$, and $A'(4)$ and interpret your results.

68. EFFECT OF LUXURY TAX ON CONSUMPTION Government economists of a developing country determined that the purchase of imported perfume is related to a proposed "luxury tax" by the formula

$$N(x) = \sqrt{10,000 - 40x - 0.02x^2} \qquad (0 \le x \le 200)$$

where $N(x)$ measures the percentage of normal consumption of perfume when a "luxury tax" of x percent is imposed on it. Find the rate of change of $N(x)$ for taxes of 10%, 100%, and 150%.

69. PULSE RATE OF AN ATHLETE The pulse rate (the number of heartbeats per minute) of a long-distance runner t seconds after leaving the starting line is given by

$$P(t) = \frac{300\sqrt{\frac{1}{2}t^2 + 2t + 25}}{t + 25} \qquad (t \ge 0)$$

Compute $P'(t)$. How fast is the athlete's pulse rate increasing 10 sec, 60 sec, and 2 min into the run? What is her pulse rate 2 min into the run?

70. THURSTONE LEARNING MODEL Psychologist L. L. Thurstone suggested the following relationship between learning time T and the length of a list n:

$$T = f(n) = An\sqrt{n - b}$$

where A and b are constants that depend on the person and the task.
a. Compute dT/dn and interpret your result.
b. For a certain person and a certain task, suppose $A = 4$ and $b = 4$. Compute $f'(13)$ and $f'(29)$ and interpret your results.

71. OIL SPILLS In calm waters, the oil spilling from the ruptured hull of a grounded tanker spreads in all directions. Assuming that the area polluted is a circle and that its radius is increasing at a rate of 2 ft/sec, determine how fast the area is increasing when the radius of the circle is 40 ft.

72. ARTERIOSCLEROSIS Refer to Example 7, page 217. Suppose the radius of an individual's artery is 1 cm and the thickness of the plaque (in centimeters) t yr from now is given by

$$h = g(t) = \frac{0.5t^2}{t^2 + 10} \qquad (0 \le t \le 10)$$

How fast will the arterial opening be decreasing 5 yr from now?

73. TRAFFIC FLOW Opened in the late 1950s, the Central Artery in downtown Boston was designed to move 75,000 vehicles a day. The number of vehicles moved per day is approximated by the function

$$x = f(t) = 6.25t^2 + 19.75t + 74.75 \qquad (0 \le t \le 5)$$

where x is measured in thousands and t in decades, with $t = 0$ corresponding to the beginning of 1959. Suppose

the average speed of traffic flow in mph is given by

$$S = g(x) = -0.00075x^2 + 67.5 \qquad (75 \le x \le 350)$$

where x has the same meaning as before. What was the rate of change of the average speed of traffic flow at the beginning of 1999? What was the average speed of traffic flow at that time?

74. HOTEL OCCUPANCY RATES The occupancy rate of the all-suite Wonderland Hotel, located near an amusement park, is given by the function

$$r(t) = \frac{10}{81}t^3 - \frac{10}{3}t^2 + \frac{200}{9}t + 60 \qquad (0 \le t \le 12)$$

where t is measured in months and $t = 0$ corresponds to the beginning of January. Management has estimated that the monthly revenue (in thousands of dollars per month) is approximated by the function

$$R(r) = -\frac{3}{5000}r^3 + \frac{9}{50}r^2 \qquad (0 \le r \le 100)$$

where r is the occupancy rate.
a. Find an expression that gives the rate of change of Wonderland's occupancy rate with respect to time.
b. Find an expression that gives the rate of change of Wonderland's monthly revenue with respect to the occupancy rate.
c. What is the rate of change of Wonderland's monthly revenue with respect to time at the beginning of January? At the beginning of June?
Hint: Use the chain rule to find $R'(r(0))r'(0)$ and $R'(r(6))r'(6)$.

75. EFFECT OF HOUSING STARTS ON JOBS The president of a major housing construction firm claims that the number of construction jobs created is given by

$$N(x) = 1.42x$$

where x denotes the number of housing starts. Suppose the number of housing starts in the next t mo is expected to be

$$x(t) = \frac{7t^2 + 140t + 700}{3t^2 + 80t + 550}$$

million units/year. Find an expression that gives the rate at which the number of construction jobs will be created t mo from now. At what rate will construction jobs be created 1 yr from now?

76. DEMAND FOR PCS The quantity demanded per month, x, of a certain make of personal computer (PC) is related to the average unit price, p (in dollars), of PCs by the equation

$$x = f(p) = \frac{100}{9}\sqrt{810,000 - p^2}$$

It is estimated that t mo from now, the average price of a PC will be given by

$$p(t) = \frac{400}{1 + \frac{1}{8}\sqrt{t}} + 200 \qquad (0 \le t \le 60)$$

dollars. Find the rate at which the quantity demanded per month of the PCs will be changing 16 mo from now.

77. **CRUISE SHIP BOOKINGS** The management of Cruise World, operators of Caribbean luxury cruises, expects that the percentage of young adults booking passage on their cruises in the years ahead will rise dramatically. They have constructed the following model, which gives the percentage of young adult passengers in year t:

$$p = f(t) = 50\left(\frac{t^2 + 2t + 4}{t^2 + 4t + 8}\right) \qquad (0 \le t \le 5)$$

Young adults normally pick shorter cruises and generally spend less on their passage. The following model gives an approximation of the average amount of money R (in dollars) spent per passenger on a cruise when the percentage of young adults is p:

$$R(p) = 1000\left(\frac{p + 4}{p + 2}\right)$$

Find the rate at which the price of the average passage will be changing 2 yr from now.

In Exercises 78–81, determine whether the statement is true or false. If it is true, explain why it is true. If it is false, give an example to show why it is false.

78. If f and g are differentiable and $h = f \circ g$, then $h'(x) = f'[g(x)]g'(x)$.

79. If f is differentiable and c is a constant, then

$$\frac{d}{dx}[f(cx)] = cf'(cx).$$

80. If f is differentiable, then

$$\frac{d}{dx}\sqrt{f(x)} = \frac{f'(x)}{2\sqrt{f(x)}}$$

81. If f is differentiable, then

$$\frac{d}{dx}\left[f\left(\frac{1}{x}\right)\right] = f'\left(\frac{1}{x}\right)$$

82. In Section 3.1 we proved that

$$\frac{d}{dx}(x^n) = nx^{n-1}$$

for the special case when $n = 2$. Use the chain rule to show that

$$\frac{d}{dx}(x^{1/n}) = \frac{1}{n}x^{1/n-1}$$

for any nonzero integer n, assuming that $f(x) = x^{1/n}$ is differentiable.
Hint: Let $f(x) = x^{1/n}$ so that $[f(x)]^n = x$. Differentiate both sides with respect to x.

83. With the aid of Exercise 82, prove that

$$\frac{d}{dx}(x^r) = rx^{r-1}$$

for any rational number r.
Hint: Let $r = m/n$, where m and n are integers with $n \ne 0$, and write $x^r = (x^m)^{1/n}$.

SOLUTIONS TO SELF-CHECK EXERCISES 3.3

1. Rewriting, we have

$$f(x) = -(2x^2 - 1)^{-1/2}$$

Using the general power rule, we find

$$f'(x) = -\frac{d}{dx}(2x^2 - 1)^{-1/2}$$
$$= -\left(-\frac{1}{2}\right)(2x^2 - 1)^{-3/2}\frac{d}{dx}(2x^2 - 1)$$
$$= \frac{1}{2}(2x^2 - 1)^{-3/2}(4x)$$
$$= \frac{2x}{(2x^2 - 1)^{3/2}}$$

2. a. The life expectancy at birth of a female born at the beginning of 1980 is given by

$$g(80) = 50.02[1 + 1.09(80)]^{0.1} \approx 78.29$$

or approximately 78 yr. Similarly, the life expectancy at birth of a female born at the beginning of the year 2000 is given by

$$g(100) = 50.02[1 + 1.09(100)]^{0.1} \approx 80.04$$

or approximately 80 yr.

b. The rate of change of the life expectancy at birth of a female born at any time t is given by $g'(t)$. Using the general power rule, we have

$$g'(t) = 50.02 \frac{d}{dt}(1 + 1.09t)^{0.1}$$

$$= (50.02)(0.1)(1 + 1.09t)^{-0.9} \frac{d}{dt}(1 + 1.09t)$$

$$= (50.02)(0.1)(1.09)(1 + 1.09t)^{-0.9}$$

$$= 5.45218(1 + 1.09t)^{-0.9}$$

$$= \frac{5.45218}{(1 + 1.09t)^{0.9}}$$

3.4 Marginal Functions in Economics

Marginal analysis is the study of the rate of change of economic quantities. For example, an economist is not merely concerned with the value of an economy's gross domestic product (GDP) at a given time but is equally concerned with the rate at which it is growing or declining. In the same vein, a manufacturer is not only interested in the total cost corresponding to a certain level of production of a commodity but is also interested in the rate of change of the total cost with respect to the level of production, and so on. Let's begin with an example to explain the meaning of the adjective *marginal,* as used by economists.

COST FUNCTIONS

 EXAMPLE 1

Suppose the total cost in dollars incurred per week by the Polaraire Company for manufacturing x refrigerators is given by the total cost function

$$C(x) = 8000 + 200x - 0.2x^2 \qquad (0 \leq x \leq 400)$$

a. What is the actual cost incurred for manufacturing the 251st refrigerator?

b. Find the rate of change of the total cost function with respect to x when $x = 250$.

c. Compare the results obtained in parts (a) and (b).

SOLUTION ✔   **a.** The actual cost incurred in producing the 251st refrigerator is the difference between the total cost incurred in producing the first 251 refrigerators and the total cost of producing the first 250 refrigerators:

$$C(251) - C(250) = [8000 + 200(251) - 0.2(251)^2]$$
$$- [8000 + 200(250) - 0.2(250)^2]$$
$$= 45{,}599.8 - 45{,}500$$
$$= 99.80$$

or $99.80.

b. The rate of change of the total cost function C with respect to x is given by the derivative of C—that is, $C'(x) = 200 - 0.4x$. Thus, when the level of production is 250 refrigerators, the rate of change of the total cost with respect to x is given by

$$C'(250) = 200 - 0.4(250)$$
$$= 100$$

or $100.

c. From the solution to part (a), we know that the actual cost for producing the 251st refrigerator is $99.80. This answer is very closely approximated by the answer to part (b), $100. To see why this is so, observe that the difference $C(251) - C(250)$ may be written in the form

$$\frac{C(251) - C(250)}{1} = \frac{C(250 + 1) - C(250)}{1} = \frac{C(250 + h) - C(250)}{h}$$

where $h = 1$. In other words, the difference $C(251) - C(250)$ is precisely the average rate of change of the total cost function C over the interval $[250, 251]$, or, equivalently, the slope of the secant line through the points $(250, 45{,}500)$ and $(251, 45{,}599.8)$. However, the number $C'(250) = 100$ is the instantaneous rate of change of the total cost function C at $x = 250$, or, equivalently, the slope of the tangent line to the graph of C at $x = 250$.

Now when h is small, the average rate of change of the function C is a good approximation to the instantaneous rate of change of the function C, or, equivalently, the slope of the secant line through the points in question is a good approximation to the slope of the tangent line through the point in question. Thus, we may expect

$$C(251) - C(250) = \frac{C(251) - C(250)}{1} = \frac{C(250 + h) - C(250)}{h}$$
$$\approx \lim_{h \to 0} \frac{C(250 + h) - C(250)}{h} = C'(250)$$

which is precisely the case in this example.   ■ ■ ■ ■

The actual cost incurred in producing an additional unit of a certain commodity given that a plant is already at a certain level of operation is called the **marginal cost.** Knowing this cost is very important to management in their decision-making processes. As we saw in Example 1, the marginal cost is approximated by the rate of change of the total cost function evaluated at the appropriate point. For this reason, economists have defined the **marginal cost function** to be the derivative of the corresponding total cost function. In other words, if C is a total cost function, then the marginal cost function is defined to be its derivative C'. Thus, the adjective *marginal* is synonymous with *derivative of.*

 EXAMPLE 2   A subsidiary of the Elektra Electronics Company manufactures a programmable pocket calculator. Management determined that the daily total cost of producing these calculators (in dollars) is given by

$$C(x) = 0.0001x^3 - 0.08x^2 + 40x + 5000$$

where x stands for the number of calculators produced.

a. Find the marginal cost function.
b. What is the marginal cost when $x = 200$, 300, 400, and 600?
c. Interpret your results.

SOLUTION ✔   **a.** The marginal cost function C' is given by the derivative of the total cost function C. Thus,

$$C'(x) = 0.0003x^2 - 0.16x + 40$$

b. The marginal cost when $x = 200$, 300, 400, and 600 is given by

$$C'(200) = 0.0003(200)^2 - 0.16(200) + 40 = 20$$
$$C'(300) = 0.0003(300)^2 - 0.16(300) + 40 = 19$$
$$C'(400) = 0.0003(400)^2 - 0.16(400) + 40 = 24$$
$$C'(600) = 0.0003(600)^2 - 0.16(600) + 40 = 52$$

or $20, $19, $24, and $52, respectively.

c. From the results of part (b), we see that Elektra's actual cost for producing the 201st calculator is approximately $20. The actual cost incurred for producing one additional calculator when the level of production is already 300 calculators is approximately $19, and so on. Observe that when the level of production is already 600 units, the actual cost of producing one additional unit is approximately $52. The higher cost for producing this additional unit when the level of production is 600 units may be the result of several factors, among them excessive costs incurred because of overtime or higher maintenance, production breakdown caused by greater stress and strain on the equipment, and so on. The graph of the total cost function appears in Figure 3.7.   ■ ■ ■ ■

FIGURE 3.7
The cost of producing x calculators is given by $C(x)$.

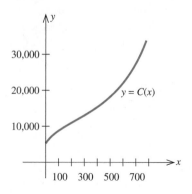

AVERAGE COST FUNCTIONS

Let's now introduce another marginal concept closely related to the marginal cost. Let $C(x)$ denote the total cost incurred in producing x units of a certain commodity. Then the **average cost** of producing x units of the commodity is obtained by dividing the total production cost by the number of units produced. This leads to the following definition.

Average Cost Function

> Suppose $C(x)$ is a total cost function. Then the **average cost function**, denoted by $\overline{C}(x)$ (read "C bar of x"), is
>
> $$\frac{C(x)}{x} \qquad (4)$$

The derivative $\overline{C}'(x)$ of the average cost function, called the **marginal average cost function**, measures the rate of change of the average cost function with respect to the number of units produced.

EXAMPLE 3   The total cost of producing x units of a certain commodity is given by

$$C(x) = 400 + 20x$$

dollars.

a. Find the average cost function $\overline{C}$.
b. Find the marginal average cost function $\overline{C}'$.
c. Interpret the results obtained in parts (a) and (b).

SOLUTION ✔   **a.** The average cost function is given by

$$\overline{C}(x) = \frac{C(x)}{x} = \frac{400 + 20x}{x}$$

$$= 20 + \frac{400}{x}$$

b. The marginal average cost function is

$$\overline{C}'(x) = -\frac{400}{x^2}$$

c. Since the marginal average cost function is negative for all admissible values of x, the rate of change of the average cost function is negative for all $x > 0$; that is, $\overline{C}(x)$ decreases as x increases. However, the graph of $\overline{C}$ always lies

FIGURE 3.8
As the level of production increases, the average cost approaches $20.

above the horizontal line $y = 20$, but it approaches the line since

$$\lim_{x \to \infty} \overline{C}(x) = \lim_{x \to \infty} \left(20 + \frac{400}{x} \right) = 20$$

A sketch of the graph of the function $\overline{C}(x)$ appears in Figure 3.8. This result is fully expected if we consider the economic implications. Note that as the level of production increases, the fixed cost per unit of production, represented by the term $(400/x)$, drops steadily. The average cost approaches the constant unit of production, which is $20 in this case.   ▪ ▪ ▪ ▪

    EXAMPLE 4   Once again consider the subsidiary of the Elektra Electronics Company. The daily total cost for producing its programmable calculators is given by

$$C(x) = 0.0001x^3 - 0.08x^2 + 40x + 5000$$

dollars, where x stands for the number of calculators produced (see Example 2).

a. Find the average cost function $\overline{C}$.
b. Find the marginal average cost function $\overline{C}'$. Compute $\overline{C}'(500)$.
c. Sketch the graph of the function $\overline{C}$ and interpret the results obtained in parts (a) and (b).

SOLUTION ✔   **a.** The average cost function is given by

$$\overline{C}(x) = \frac{C(x)}{x} = 0.0001x^2 - 0.08x + 40 + \frac{5000}{x}$$

b. The marginal average cost function is given by

$$\overline{C}'(x) = 0.0002x - 0.08 - \frac{5000}{x^2}$$

Also,

$$\overline{C}'(500) = 0.0002(500) - 0.08 - \frac{5000}{(500)^2} = 0$$

c. To sketch the graph of the function $\overline{C}$, observe that if x is a small positive number, then $\overline{C}(x) > 0$. Furthermore, $\overline{C}(x)$ becomes arbitrarily large as x approaches zero from the right, since the term $(5000/x)$ becomes arbitrarily

FIGURE 3.9

The average cost reaches a minimum of $35 when 500 calculators are produced.

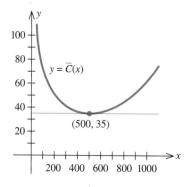

large as x approaches zero. Next, the result $\overline{C}'(500) = 0$ obtained in part (b) tells us that the tangent line to the graph of the function $\overline{C}$ is horizontal at the point (500, 35) on the graph. Finally, plotting the points on the graph corresponding to, say, $x = 100, 200, 300, \ldots, 900$, we obtain the sketch in Figure 3.9. As expected, the average cost drops as the level of production increases. But in this case, as opposed to the case in Example 3, the average cost reaches a minimum value of $35, corresponding to a production level of 500, and *increases* thereafter.

This phenomenon is typical in situations where the marginal cost increases from some point on as production increases, as in Example 2. This situation is in contrast to that of Example 3, in which the marginal cost remains constant at any level of production. ■ ■ ■ ■

Exploring with Technology

Refer to Example 4.

1. Use a graphing utility to plot the graph of the average cost function

$$\overline{C}(x) = 0.0001x^2 - 0.08x + 40 + \frac{5000}{x}$$

using the viewing rectangle [0, 1000] × [0, 100]. Then, using ZOOM and TRACE, show that the lowest point on the graph of $\overline{C}$ is (500, 35).

2. Draw the tangent line to the graph of $\overline{C}$ at (500, 35). What is its slope? Is this expected?

3. Plot the graph of the marginal average cost function

$$\overline{C}'(x) = 0.0002x - 0.08 - \frac{5000}{x^2}$$

using the viewing rectangle [0, 2000] × [−1, 1]. Then use ZOOM and TRACE to show that the zero of the function $\overline{C}'$ occurs at $x = 500$. Verify this result using the root-finding capability of your graphing utility. Is this result compatible with that obtained in part (2)? Explain your answer.

REVENUE FUNCTIONS

Another marginal concept, the *marginal revenue function*, is associated with the revenue function R, given by

$$R(x) = px \tag{5}$$

where x is the number of units sold of a certain commodity and p is the unit selling price. In general, however, the unit selling price p of a commodity is related to the quantity x of the commodity demanded. This relationship, $p = f(x)$, is called a *demand equation* (see Section 2.3). Solving the demand equation for p in terms of x, we obtain the unit price function f, given by

$$p = f(x)$$

Thus, the revenue function R is given by

$$R(x) = px = xf(x)$$

where f is the unit price function. The derivative R' of the function R, called the **marginal revenue function,** measures the rate of change of the revenue function.

EXAMPLE 5   Suppose the relationship between the unit price p in dollars and the quantity demanded x of the Acrosonic model F loudspeaker system is given by the equation

$$p = -0.02x + 400 \qquad (0 \le x \le 20{,}000)$$

a. Find the revenue function R.
b. Find the marginal revenue function R'.
c. Compute $R'(2000)$ and interpret your result.

SOLUTION ✔   **a.** The revenue function R is given by

$$\begin{aligned} R(x) &= px \\ &= x(-0.02x + 400) \\ &= -0.02x^2 + 400x \qquad (0 \le x \le 20{,}000) \end{aligned}$$

b. The marginal revenue function R' is given by

$$R'(x) = -0.04x + 400$$

c. $$R'(2000) = -0.04(2000) + 400 = 320$$

Thus, the actual revenue to be realized from the sale of the 2001st loudspeaker system is approximately $320.   ■ ■ ■ ■

PROFIT FUNCTIONS

Our final example of a marginal function involves the profit function. The profit function P is given by

$$P(x) = R(x) - C(x) \tag{6}$$

where R and C are the revenue and cost functions and x is the number of units of a commodity produced and sold. The **marginal profit function** $P'(x)$ measures the rate of change of the profit function P and provides us with a good approximation of the actual profit or loss realized from the sale of the $(x + 1)$st unit of the commodity (assuming the xth unit has been sold).

EXAMPLE 6

Refer to Example 5. Suppose the cost of producing x units of the Acrosonic model F loudspeaker is

$$C(x) = 100x + 200,000 \text{ dollars}$$

a. Find the profit function P.
b. Find the marginal profit function P'.
c. Compute $P'(2000)$ and interpret your result.
d. Sketch the graph of the profit function P.

SOLUTION ✔

a. From the solution to Example 5(a), we have

$$R(x) = -0.02x^2 + 400x$$

Thus, the required profit function P is given by

$$\begin{aligned} P(x) &= R(x) - C(x) \\ &= (-0.02x^2 + 400x) - (100x + 200,000) \\ &= -0.02x^2 + 300x - 200,000 \end{aligned}$$

b. The marginal profit function P' is given by

$$P'(x) = -0.04x + 300$$

c. $$P'(2000) = -0.04(2000) + 300 = 220$$

Thus, the actual profit realized from the sale of the 2001st loudspeaker system is approximately \$220.

d. The graph of the profit function P appears in Figure 3.10.   ■ ■ ■ ■

FIGURE 3.10
The total profit made when x loudspeakers are produced is given by $P(x)$.

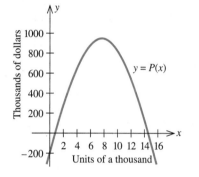

ELASTICITY OF DEMAND

Finally, let us use the marginal concepts introduced in this section to derive an important criterion used by economists to analyze the demand function: *elasticity of demand.*

In what follows, it will be convenient to write the demand function f in the form $x = f(p)$; that is, we will think of the quantity demanded of a certain commodity as a function of its unit price. Since the quantity demanded of a commodity usually decreases as its unit price increases, the function f is typically a decreasing function of p (Figure 3.11a).

FIGURE 3.11

(a) A demand function

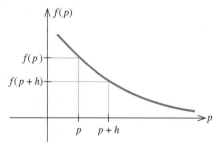

(b) $f(p + h)$ is the quantity demanded when the unit price increases from p to $p + h$ dollars.

Suppose the unit price of a commodity is increased by h dollars from p dollars to $(p + h)$ dollars (Figure 3.11b). Then the quantity demanded drops from $f(p)$ units to $f(p + h)$ units, a change of $[f(p + h) - f(p)]$ units. The percentage change in the unit price is

$$\frac{h}{p}(100) \qquad \left(\frac{\text{Change in unit price}}{\text{Price } p}\right)(100)$$

and the corresponding percentage change in the quantity demanded is

$$100\left[\frac{f(p + h) - f(p)}{f(p)}\right] \qquad \left(\frac{\text{Change in quantity demanded}}{\text{Quantity demanded at price } p}\right)(100)$$

Now, one good way to measure the effect that a percentage change in price has on the percentage change in the quantity demanded is to look at the ratio of the latter to the former. We find

$$\frac{\text{Percentage change in the quantity demanded}}{\text{Percentage change in the unit price}} = \frac{100\left[\dfrac{f(p + h) - f(p)}{f(p)}\right]}{100\left(\dfrac{h}{p}\right)}$$

$$= \frac{\dfrac{f(p + h) - f(p)}{h}}{\dfrac{f(p)}{p}}$$

If f is differentiable at p, then

$$\frac{f(p + h) - f(p)}{h} \approx f'(p)$$

when h is small. Therefore, if h is small, then the ratio is approximately equal to

$$\frac{f'(p)}{\dfrac{f(p)}{p}} = \frac{pf'(p)}{f(p)}$$

Economists call the negative of this quantity the *elasticity of demand*.

Elasticity of Demand

If f is a differentiable demand function defined by $x = f(p)$, then the **elasticity of demand** at price p is given by

$$E(p) = -\frac{pf'(p)}{f(p)} \tag{7}$$

REMARK It will be shown later (Section 4.1) that if f is decreasing on an interval, then $f'(p) < 0$ for p in that interval. In light of this, we see that since both p and $f(p)$ are positive, the quantity $\dfrac{pf'(p)}{f(p)}$ is negative. Because economists would rather work with a positive value, the elasticity of demand $E(p)$ is defined to be the negative of this quantity.   ■ ■ ■

EXAMPLE 7   Consider the demand equation

$$p = -0.02x + 400 \qquad (0 \le x \le 20{,}000)$$

which describes the relationship between the unit price in dollars and the quantity demanded x of the Acrosonic model F loudspeaker systems.

a. Find the elasticity of demand $E(p)$.
b. Compute $E(100)$ and interpret your result.
c. Compute $E(300)$ and interpret your result.

SOLUTION ✔   **a.** Solving the given demand equation for x in terms of p, we find

$$x = f(p) = -50p + 20{,}000$$

from which we see that

$$f'(p) = -50$$

Therefore,

$$E(p) = -\frac{pf'(p)}{f(p)} = -\frac{p(-50)}{-50p + 20{,}000}$$

$$= \frac{p}{400 - p}$$

b. $E(100) = \dfrac{100}{400 - 100} = \dfrac{1}{3}$, which is the elasticity of demand when $p = 100$. To interpret this result, recall that $E(100)$ is the negative of the ratio of the percentage change in the quantity demanded to the percentage change in the unit price when $p = 100$. Therefore, our result tells us that when the unit price p is set at \$100 per speaker, an increase of 1% in the unit price will cause an increase of approximately 0.33% in the quantity demanded.

c. $E(300) = \dfrac{300}{400 - 300} = 3$, which is the elasticity of demand when $p = 300$. It tells us that when the unit price is set at \$300 per speaker, an increase of 1% in the unit price will cause a decrease of approximately 3% in the quantity demanded.   ■ ■ ■ ■

Economists often use the following terminology to describe demand in terms of elasticity.

Elasticity of Demand

The demand is said to be **elastic** if $E(p) > 1$.
The demand is said to be **unitary** if $E(p) = 1$.
The demand is said to be **inelastic** if $E(p) < 1$.

As an illustration, our computations in Example 7 revealed that demand for Acrosonic loudspeakers is elastic when $p = 300$ but inelastic when $p = 100$. These computations confirm that when demand is elastic, a small percentage change in the unit price will result in a greater percentage change in the quantity demanded; and when demand is inelastic, a small percentage change in the unit price will cause a smaller percentage change in the quantity demanded. Finally, when demand is unitary, a small percentage change in the unit price will result in the same percentage change in the quantity demanded.

We can describe the way revenue responds to changes in the unit price using the notion of elasticity. If the quantity demanded of a certain commodity is related to its unit price by the equation $x = f(p)$, then the revenue realized through the sale of x units of the commodity at a price of p dollars each is

$$R(p) = px = pf(p)$$

The rate of change of the revenue with respect to the unit price p is given by

$$R'(p) = f(p) + pf'(p)$$
$$= f(p)\left[1 + \frac{pf'(p)}{f(p)}\right]$$
$$= f(p)[1 - E(p)]$$

Now, suppose demand is elastic when the unit price is set at a dollars. Then $E(a) > 1$, and so $1 - E(a) < 0$. Since $f(p)$ is positive for all values of p, we see that

$$R'(a) = f(a)[1 - E(a)] < 0$$

and so $R(p)$ is decreasing at $p = a$. This implies that a small increase in the unit price when $p = a$ results in a decrease in the revenue, whereas a small decrease in the unit price will result in an increase in the revenue. Similarly, you can show that if the demand is inelastic when the unit price is set at a dollars, then a small increase in the unit price will cause the revenue to increase, and a small decrease in the unit price will cause the revenue to decrease. Finally, if the demand is unitary when the unit price is set at a dollars, then $E(a) = 1$ and $R'(a) = 0$. This implies that a small increase or decrease in the unit price will not result in a change in the revenue. The following statements summarize this discussion.

1. If the demand is elastic at p $(E(p) > 1)$, then an increase in the unit price will cause the revenue to decrease, whereas a decrease in the unit price will cause the revenue to increase.
2. If the demand is inelastic at p $(E(p) < 1)$, then an increase in the unit price will cause the revenue to increase, and a decrease in the unit price will cause the revenue to decrease.
3. If the demand is unitary at p $(E(p) = 1)$, then an increase in the unit price will cause the revenue to stay about the same.

These results are illustrated in Figure 3.12.

FIGURE 3.12
The revenue is increasing on an interval where the demand is inelastic, decreasing on an interval where the demand is elastic, and stationary at the point where the demand is unitary.

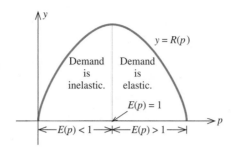

REMARK As an aid to remembering this, note the following:

1. If demand is elastic, then the change in revenue and the change in the unit price move in opposite directions.
2. If demand is inelastic, then they move in the same direction. ■ ■ ■

EXAMPLE 8

Refer to Example 7.

a. Is demand elastic, unitary, or inelastic when $p = 100$? When $p = 300$?
b. If the price is $100, will raising the unit price slightly cause the revenue to increase or decrease?

SOLUTION ✔

a. From the results of Example 7, we see that $E(100) = \frac{1}{3} < 1$ and $E(300) = 3 > 1$. We conclude accordingly that demand is inelastic when $p = 100$ and elastic when $p = 300$.
b. Since demand is inelastic when $p = 100$, raising the unit price slightly will cause the revenue to increase. ■ ■ ■ ■

SELF-CHECK EXERCISES 3.4

 1. The weekly demand for Pulsar VCRs (videocassette recorders) is given by the demand equation

$$p = -0.02x + 300 \qquad (0 \le x \le 15{,}000)$$

where p denotes the wholesale unit price in dollars and x denotes the quantity demanded. The weekly total cost function associated with manufacturing these VCRs is

$$C(x) = 0.000003x^3 - 0.04x^2 + 200x + 70{,}000 \text{ dollars}$$

a. Find the revenue function R and the profit function P.
b. Find the marginal cost function C', the marginal revenue function R', and the marginal profit function P'.
c. Find the marginal average cost function $\overline{C}'$.
d. Compute $C'(3000)$, $R'(3000)$, and $P'(3000)$ and interpret your results.

2. Refer to the preceding exercise. Determine whether the demand is elastic, unitary, or inelastic when $p = 100$ and when $p = 200$.

Solutions to Self-Check Exercises 3.4 can be found on page 239.

3.4 Exercises

1. **PRODUCTION COSTS** The graph of a typical total cost function $C(x)$ associated with the manufacture of x units of a certain commodity is shown in the following figure.
   a. Explain why the function C is always increasing.
   b. As the level of production x increases, the cost per unit drops so that $C(x)$ increases but at a slower pace. However, a level of production is soon reached at which the cost per unit begins to increase dramatically (due to a shortage of raw material, overtime, breakdown of machinery due to excessive stress and strain) so that $C(x)$ continues to increase at a faster pace. Use the graph of C to find the approximate level of production x_0 where this occurs.

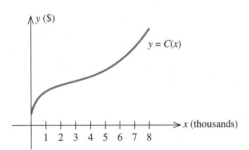

A calculator is recommended for Exercises 2–33.

2. **MARGINAL COST** The total weekly cost (in dollars) incurred by the Lincoln Record Company in pressing x long-playing records is

   $$C(x) = 2000 + 2x - 0.0001x^2 \qquad (0 \le x \le 6000)$$

   a. What is the actual cost incurred in producing the 1001st and the 2001st record?
   b. What is the marginal cost when $x = 1000$ and 2000?

3. **MARGINAL COST** A division of Ditton Industries manufactures the Futura model microwave oven. The daily cost (in dollars) of producing these microwave ovens is

   $$C(x) = 0.0002x^3 - 0.06x^2 + 120x + 5000$$

   where x stands for the number of units produced.
   a. What is the actual cost incurred in manufacturing the 101st oven? The 201st oven? The 301st oven?
   b. What is the marginal cost when $x = 100, 200$, and 300?

4. **MARGINAL AVERAGE COST** The Custom Office Company makes a line of executive desks. It is estimated that the total cost for making x units of their Senior Executive Model is

   $$C(x) = 100x + 200{,}000$$

   dollars per year.
   a. Find the average cost function $\overline{C}$.
   b. Find the marginal average cost function $\overline{C}'$.
   c. What happens to $\overline{C}(x)$ when x is very large? Interpret your results.

5. **MARGINAL AVERAGE COST** The management of the ThermoMaster Company, whose Mexican subsidiary manufactures an indoor–outdoor thermometer, has estimated that the total weekly cost (in dollars) for producing x thermometers is

   $$C(x) = 5000 + 2x$$

   a. Find the average cost function $\overline{C}$.
   b. Find the marginal average cost function $\overline{C}'$.
   c. Interpret your results.

6. Find the average cost function $\overline{C}$ and the marginal average cost function $\overline{C}'$ associated with the total cost function C of Exercise 2.

7. Find the average cost function $\overline{C}$ and the marginal average cost function $\overline{C}'$ associated with the total cost function C of Exercise 3.

8. **MARGINAL REVENUE** The Williams Commuter Air Service realizes a monthly revenue of

   $$R(x) = 8000x - 100x^2$$

   dollars when the price charged per passenger is x dollars.
   a. Find the marginal revenue R'.
   b. Compute $R'(39)$, $R'(40)$, and $R'(41)$. What do your results imply?

9. **MARGINAL REVENUE** The management of the Acrosonic Company plans to market the Electro-Stat, an electrostatic speaker system. The marketing department has determined that the demand for these speakers is

   $$p = -0.04x + 800 \qquad (0 \le x \le 20{,}000)$$

   where p denotes the speaker's unit price (in dollars) and x denotes the quantity demanded.
   a. Find the revenue function R.
   b. Find the marginal revenue function R'.
   c. Compute $R'(5000)$ and interpret your result.

10. **MARGINAL PROFIT** Lynbrook West, an apartment complex, has 100 two-bedroom units. The monthly profit (in dollars) realized from renting x apartments is

$$P(x) = -10x^2 + 1760x - 50,000$$

a. What is the actual profit realized from renting the 51st unit, assuming that 50 units have already been rented?
b. Compute the marginal profit when $x = 50$ and compare your results with that obtained in part (a).

11. **MARGINAL PROFIT** Refer to Exercise 9. Acrosonic's production department estimates that the total cost (in dollars) incurred in manufacturing x Electro-Stat speaker systems in the first year of production will be

$$C(x) = 200x + 300,000$$

a. Find the profit function P.
b. Find the marginal profit function P'.
c. Compute $P'(5000)$ and $P'(8000)$.
d. Sketch the graph of the profit function and interpret your results.

12. **MARGINAL COST, REVENUE, AND PROFIT** The weekly demand for the Pulsar 25 color console television is

$$p = 600 - 0.05x \qquad (0 \le x \le 12,000)$$

where p denotes the wholesale unit price in dollars and x denotes the quantity demanded. The weekly total cost function associated with manufacturing the Pulsar 25 is given by

$$C(x) = 0.000002x^3 - 0.03x^2 + 400x + 80,000$$

where $C(x)$ denotes the total cost incurred in producing x sets.
a. Find the revenue function R and the profit function P.
b. Find the marginal cost function C', the marginal revenue function R', and the marginal profit function P'.
c. Compute $C'(2000)$, $R'(2000)$, and $P'(2000)$ and interpret your results.
d. Sketch the graphs of the functions C, R, and P and interpret parts (b) and (c) using the graphs obtained.

13. **MARGINAL COST, REVENUE, AND PROFIT** The Pulsar Corporation also manufactures a series of 19-inch color television sets. The quantity x of these sets demanded each week is related to the wholesale unit price p by the equation

$$p = -0.006x + 180$$

The weekly total cost incurred by Pulsar for producing x sets is

$$C(x) = 0.000002x^3 - 0.02x^2 + 120x + 60,000$$

dollars. Answer the questions in Exercise 12 for these data.

14. **MARGINAL AVERAGE COST** Find the average cost function $\overline{C}$ associated with the total cost function C of Exercise 12.
a. What is the marginal average cost function $\overline{C}'$?
b. Compute $\overline{C}'(5,000)$ and $\overline{C}'(10,000)$ and interpret your results.
c. Sketch the graph of $\overline{C}$.

15. **MARGINAL AVERAGE COST** Find the average cost function $\overline{C}$ associated with the total cost function C of Exercise 13.
a. What is the marginal average cost function $\overline{C}'$?
b. Compute $\overline{C}'(5,000)$ and $\overline{C}'(10,000)$ and interpret your results.

16. **MARGINAL REVENUE** The quantity of Sicard wristwatches demanded per month is related to the unit price by the equation

$$p = \frac{50}{0.01x^2 + 1} \qquad (0 \le x \le 20)$$

where p is measured in dollars and x in units of a thousand.
a. Find the revenue function R.
b. Find the marginal revenue function R'.
c. Compute $R'(2)$ and interpret your result.

17. **MARGINAL PROPENSITY TO CONSUME** The consumption function of the U.S. economy for 1929 to 1941 is

$$C(x) = 0.712x + 95.05$$

where $C(x)$ is the personal consumption expenditure and x is the personal income, both measured in billions of dollars. Find the rate of change of consumption with respect to income, dC/dx. This quantity is called the *marginal propensity to consume.*

18. **MARGINAL PROPENSITY TO CONSUME** Refer to Exercise 17. Suppose a certain economy's consumption function is

$$C(x) = 0.873x^{1.1} + 20.34$$

where $C(x)$ and x are measured in billions of dollars. Find the marginal propensity to consume when $x = 10$.

19. MARGINAL PROPENSITY TO SAVE Suppose $C(x)$ measures an economy's personal consumption expenditure and x the personal income, both in billions of dollars. Then,

$$S(x) = x - C(x) \qquad \text{(Income minus consumption)}$$

measures the economy's savings corresponding to an income of x billion dollars. Show that

$$\frac{dS}{dx} = 1 - \frac{dC}{dx}$$

The quantity dS/dx is called the *marginal propensity to save*.

20. Refer to Exercise 19. For the consumption function of Exercise 17, find the marginal propensity to save.

21. Refer to Exercise 19. For the consumption function of Exercise 18, find the marginal propensity to save when $x = 10$.

For each demand equation in Exercises 22–27, compute the elasticity of demand and determine whether the demand is elastic, unitary, or inelastic at the indicated price.

22. $x = -\dfrac{3}{2}p + 9; p = 2$

23. $x = -\dfrac{5}{4}p + 20; p = 10$

24. $x + \dfrac{1}{3}p - 20 = 0; p = 30$

25. $0.4x + p - 20 = 0; p = 10$

26. $p = 144 - x^2; p = 96$

27. $p = 169 - x^2; p = 29$

28. ELASTICITY OF DEMAND The management of the Titan Tire Company has determined that the quantity demanded x of their Super Titan tires per week is related to the unit price p by the equation

$$x = \sqrt{144 - p}$$

where p is measured in dollars and x in units of a thousand.
a. Compute the elasticity of demand when $p = 63, 96,$ and 108.
b. Interpret the results obtained in part (a).
c. Is the demand elastic, unitary, or inelastic when $p = 63, 96,$ and 108?

29. ELASTICITY OF DEMAND The demand equation for the Roland portable hair dryer is given by

$$x = \frac{1}{5}(225 - p^2) \qquad (0 \le p \le 15)$$

where x (measured in units of a hundred) is the quantity demanded per week and p is the unit price in dollars.
a. Is the demand elastic or inelastic when $p = 8$ and when $p = 10$?
b. When is the demand unitary?
Hint: Solve $E(p) = 1$ for p.
c. If the unit price is lowered slightly from \$10, will the revenue increase or decrease?
d. If the unit price is increased slightly from \$8, will the revenue increase or decrease?

30. ELASTICITY OF DEMAND The quantity demanded per week x (in units of a hundred) of the Mikado miniature camera is related to the unit price p (in dollars) by the demand equation

$$x = \sqrt{400 - 5p} \qquad (0 \le p \le 80)$$

a. Is the demand elastic or inelastic when $p = 40$? When $p = 60$?
b. When is the demand unitary?
c. If the unit price is lowered slightly from \$60, will the revenue increase or decrease?
d. If the unit price is increased slightly from \$40, will the revenue increase or decrease?

31. ELASTICITY OF DEMAND The proprietor of the Showplace, a video club, has estimated that the rental price p (in dollars) of prerecorded videocassette tapes is related to the quantity x rented per week by the demand equation

$$x = \frac{2}{3}\sqrt{36 - p^2} \qquad (0 \le p \le 6)$$

Currently, the rental price is \$2/tape.
a. Is the demand elastic or inelastic at this rental price?
b. If the rental price is increased, will the revenue increase or decrease?

32. ELASTICITY OF DEMAND The demand function for a certain make of exercise bicycle sold exclusively through cable television is

$$p = \sqrt{9 - 0.02x} \qquad (0 \le x \le 450)$$

where p is the unit price in hundreds of dollars and x is the quantity demanded per week. Compute the elasticity of demand and determine the range of prices corresponding to inelastic, unitary, and elastic demand.
Hint: Solve the equation $E(p) = 1$.

33. ELASTICITY OF DEMAND The demand equation for the Sicard wristwatch is given by

$$x = 10\sqrt{\frac{50-p}{p}} \qquad (0 < p \le 50)$$

where x (measured in units of a thousand) is the quantity demanded per week and p is the unit price in dollars. Compute the elasticity of demand and determine the range of prices corresponding to inelastic, unitary, and elastic demand.

In Exercises 34 and 35, determine whether the statement is true or false. If it is true, explain why it is true. If it is false, give an example to show why it is false.

34. If C is a differentiable total cost function, then the marginal average cost function is

$$\overline{C}'(x) = \frac{xC'(x) - C(x)}{x^2}$$

35. If the marginal profit function is positive at $x = a$, then it makes sense to decrease the level of production.

SOLUTIONS TO SELF-CHECK EXERCISES 3.4

1. a. $R(x) = px$
$$= x(-0.02x + 300)$$
$$= -0.02x^2 + 300x \qquad (0 \le x \le 15,000)$$

$$P(x) = R(x) - C(x)$$
$$= -0.02x^2 + 300x$$
$$\quad -(0.000003x^3 - 0.04x^2 + 200x + 70,000)$$
$$= -0.000003x^3 + 0.02x^2 + 100x - 70,000$$

b. $C'(x) = 0.000009x^2 - 0.08x + 200$
$R'(x) = -0.04x + 300$
$P'(x) = -0.000009x^2 + 0.04x + 100$

c. The average cost function is

$$\overline{C}(x) = \frac{C(x)}{x}$$
$$= \frac{0.000003x^3 - 0.04x^2 + 200x + 70,000}{x}$$
$$= 0.000003x^2 - 0.04x + 200 + \frac{70,000}{x}$$

Therefore, the marginal average cost function is

$$\overline{C}'(x) = 0.000006x - 0.04 - \frac{70,000}{x^2}$$

d. Using the results from part (b), we find

$$C'(3000) = 0.000009(3000)^2 - 0.08(3000) + 200$$
$$= 41$$

That is, when the level of production is already 3000 VCRs, the actual cost of producing one additional VCR is approximately \$41. Next,

$$R'(3000) = -0.04(3000) + 300 = 180$$

That is, the actual revenue to be realized from selling the 3001st VCR is approximately \$180. Finally,

$$P'(3000) = -0.000009(3000)^2 + 0.04(3000) + 100$$
$$= 139$$

That is, the actual profit realized from selling the 3001st VCR is approximately \$139.

2. We first solve the given demand equation for x in terms of p, obtaining

$$x = f(p) = -50p + 15,000$$
$$f'(p) = -50$$

Therefore,

$$E(p) = -\frac{pf'(p)}{f(p)} = -\frac{p}{-50p + 15,000}(-50)$$

$$= \frac{p}{300 - p} \qquad (0 \le p < 300)$$

Next, we compute

$$E(100) = \frac{100}{300 - 100} = \frac{1}{2} < 1$$

and we conclude that demand is inelastic when $p = 100$. Also,

$$E(200) = \frac{200}{300 - 200} = 2 > 1$$

and we see that demand is elastic when $p = 200$.

3.5 Higher-Order Derivatives

HIGHER-ORDER DERIVATIVES

The derivative f' of a function f is also a function. As such, the differentiability of f' may be considered. Thus, the function f' has a derivative f'' at a point x in the domain of f' if the limit of the quotient

$$\frac{f'(x + h) - f'(x)}{h}$$

exists as h approaches zero. In other words, it is the derivative of the first derivative.

The function f'' obtained in this manner is called the **second derivative** of the function f, just as the derivative f' of f is often called the first derivative of f. Continuing in this fashion, we are led to considering the third, fourth, and higher-order derivatives of f whenever they exist. Notations for the first, second, third, and, in general, nth derivatives of a function f at a point x are

$$f'(x), f''(x), f'''(x), \ldots, f^{(n)}(x)$$

or

$$D^1 f(x), D^2 f(x), D^3 f(x), \ldots, D^n f(x)$$

If f is written in the form $y = f(x)$, then the notations for its derivatives are

$$y', y'', y''', \ldots, y^{(n)}$$

$$\frac{dy}{dx}, \frac{d^2 y}{dx^2}, \frac{d^3 y}{dx^3}, \ldots, \frac{d^n y}{dx^n}$$

or

$$D^1 y, D^2 y, D^3 y, \ldots, D^n y$$

respectively.

EXAMPLE 1

Find the derivatives of all orders of the polynomial function

$$f(x) = x^5 - 3x^4 + 4x^3 - 2x^2 + x - 8$$

SOLUTION ✔

We have

$$f'(x) = 5x^4 - 12x^3 + 12x^2 - 4x + 1$$

$$f''(x) = \frac{d}{dx} f'(x) = 20x^3 - 36x^2 + 24x - 4$$

$$f'''(x) = \frac{d}{dx} f''(x) = 60x^2 - 72x + 24$$

$$f^{(4)}(x) = \frac{d}{dx} f'''(x) = 120x - 72$$

$$f^{(5)}(x) = \frac{d}{dx} f^{(4)}(x) = 120$$

and, in general,

$$f^{(n)}(x) = 0 \qquad \text{(for } n > 5\text{)}$$

■ ■ ■ ■

EXAMPLE 2

Find the third derivative of the function f defined by $y = x^{2/3}$. What is its domain?

SOLUTION ✔

We have

$$y' = \frac{2}{3} x^{-1/3}$$

$$y'' = \left(\frac{2}{3}\right)\left(-\frac{1}{3}\right) x^{-4/3} = -\frac{2}{9} x^{-4/3}$$

so the required derivative is

$$y''' = \left(-\frac{2}{9}\right)\left(-\frac{4}{3}\right) x^{-7/3} = \frac{8}{27} x^{-7/3} = \frac{8}{27 x^{7/3}}$$

FIGURE 3.13
The graph of the function $y = x^{2/3}$

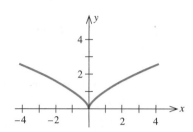

The common domain of the functions f', f'', and f''' is the set of all real numbers except $x = 0$. The domain of $y = x^{2/3}$ is the set of all real numbers. The graph of the function $y = x^{2/3}$ appears in Figure 3.13.   ■ ■ ■ ■

REMARK Always simplify an expression before differentiating it to obtain the next order derivative.   ■ ■ ■

EXAMPLE 3   Find the second derivative of the function $y = (2x^2 + 3)^{3/2}$.

SOLUTION ✔   We have, using the general power rule,

$$y' = \frac{3}{2}(2x^2 + 3)^{1/2}(4x) = 6x(2x^2 + 3)^{1/2}$$

Next, using the product rule and then the chain rule, we find

$$y'' = (6x) \cdot \frac{d}{dx}(2x^2 + 3)^{1/2} + \left[\frac{d}{dx}(6x)\right](2x^2 + 3)^{1/2}$$

$$= (6x)\left(\frac{1}{2}\right)(2x^2 + 3)^{-1/2}(4x) + 6(2x^2 + 3)^{1/2}$$

$$= 12x^2(2x^2 + 3)^{-1/2} + 6(2x^2 + 3)^{1/2}$$

$$= 6(2x^2 + 3)^{-1/2}[2x^2 + (2x^2 + 3)]$$

$$= \frac{6(4x^2 + 3)}{\sqrt{2x^2 + 3}}$$ ■ ■ ■ ■

APPLICATIONS

Just as the derivative of a function f at a point x measures the rate of change of the function f at that point, the second derivative of f (the derivative of f') measures the rate of change of the derivative f' of the function f. The third derivative of the function f, f''', measures the rate of change of f'', and so on.

In Chapter 4 we will discuss applications involving the geometric interpretation of the second derivative of a function. The following example gives an interpretation of the second derivative in a familiar role.

EXAMPLE **4**

Refer to the example on page 111. The distance s (in feet) covered by a maglev moving along a straight track t seconds after starting from rest is given by the function $s = 4t^2$ ($0 \leq t \leq 10$). What is the maglev's acceleration at the end of 30 seconds?

SOLUTION ✔

The velocity of the maglev t seconds from rest is given by

$$v = \frac{ds}{dt} = \frac{d}{dt}(4t^2) = 8t$$

The acceleration of the maglev t seconds from rest is given by the rate of change of the velocity of t—that is,

$$a = \frac{d}{dt}v = \frac{d}{dt}\left(\frac{ds}{dt}\right) = \frac{d^2s}{dt^2} = \frac{d}{dt}(8t) = 8$$

or 8 feet per second per second, normally abbreviated, 8 ft/sec². ■ ■ ■ ■

EXAMPLE **5**

A ball is thrown straight up into the air from the roof of a building. The height of the ball as measured from the ground is given by

$$s = -16t^2 + 24t + 120$$

where s is measured in feet and t in seconds. Find the velocity and acceleration of the ball 3 seconds after it is thrown into the air.

SOLUTION ✔

The velocity v and acceleration a of the ball at any time t are given by

$$v = \frac{ds}{dt} = \frac{d}{dt}(-16t^2 + 24t + 120) = -32t + 24$$

and

$$a = \frac{d^2t}{dt^2} = \frac{d}{dt}\left(\frac{ds}{dt}\right) = \frac{d}{dt}(-32t + 24) = -32$$

Therefore, the velocity of the ball 3 seconds after it is thrown into the air is

$$v = -32(3) + 24 = -72$$

That is, the ball is falling downward at a speed of 72 ft/sec. The acceleration of the ball is 32 ft/sec² downward at any time during the motion. ■ ■ ■ ■

Another interpretation of the second derivative of a function—this time from the field of economics—follows. Suppose the consumer price index (CPI) of an economy between the years a and b is described by the function $I(t)$ ($a \leq t \leq b$) (Figure 3.14). Then, the first derivative of I, $I'(t)$, gives the rate of inflation of the economy at any time t. The second derivative of I, $I''(t)$, gives the *rate of change of the inflation rate* at any time t. Thus, when the economist or politician claims that "inflation is slowing," what he or she is saying is that the rate of inflation is decreasing. Mathematically, this is equivalent to noting that the second derivative $I''(t)$ is negative at the time t under consideration. Observe that $I'(t)$ could be positive at a time when $I''(t)$ is negative (see Example 6). Thus, one may not draw the conclusion from the aforementioned quote that prices of goods and services are about to drop!

FIGURE 3.14

The CPI of a certain economy from year a to year b is given by $I(t)$.

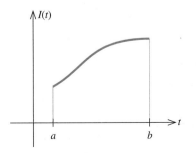

EXAMPLE **6**

An economy's CPI is described by the function

$$I(t) = -0.2t^3 + 3t^2 + 100 \qquad (0 \le t \le 9)$$

where $t = 0$ corresponds to the year 1988. Compute $I'(6)$ and $I''(6)$ and use these results to show that even though the CPI was rising at the beginning of 1994, "inflation was moderating" at that time.

SOLUTION ✔

We find

$$I'(t) = -0.6t^2 + 6t \qquad \text{and} \qquad I''(t) = -1.2t + 6$$

Thus,

$$I'(6) = -0.6(6)^2 + 6(6) = 14.4$$
$$I''(6) = -1.2(6) + 6 = -1.2$$

Our computations reveal that at the beginning of 1994 ($t = 6$), the CPI was increasing at the rate of 14.4 points per year, whereas the rate of the inflation rate was decreasing by 1.2 points per year. Thus, inflation was moderating at that time (Figure 3.15). In Section 4.2, we will see that relief actually began in early 1993.   ■ ■ ■ ■

FIGURE 3.15
The CPI of an economy is given by $I(t)$.

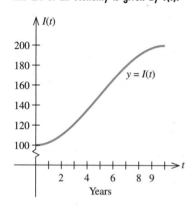

SELF-CHECK EXERCISES 3.5

1. Find the third derivative of

$$f(x) = 2x^5 - 3x^3 + x^2 - 6x + 10$$

2. Let

$$f(x) = \frac{1}{1 + x}$$

Find $f'(x)$, $f''(x)$, and $f'''(x)$.

3. A certain species of turtle faces extinction because dealers collect truckloads of turtle eggs to be sold as aphrodisiacs. After severe conservation measures are implemented, it is hoped that the turtle population will grow according to the rule

$$N(t) = 2t^3 + 3t^2 - 4t + 1000 \qquad (0 \le t \le 10)$$

where $N(t)$ denotes the population at the end of year t. Compute $N''(2)$ and $N''(8)$ and interpret your results.

Solutions to Self-Check Exercises 3.5 can be found on page 248.

3.5 Exercises

In Exercises 1–20, find the first and second derivatives of the given function.

1. $f(x) = 4x^2 - 2x + 1$

2. $f(x) = -0.2x^2 + 0.3x + 4$

3. $f(x) = 2x^3 - 3x^2 + 1$

4. $g(x) = -3x^3 + 24x^2 + 6x - 64$

5. $h(t) = t^4 - 2t^3 + 6t^2 - 3t + 10$

6. $f(x) = x^5 - x^4 + x^3 - x^2 + x - 1$

7. $f(x) = (x^2 + 2)^5$ **8.** $g(t) = t^2(3t + 1)^4$

9. $g(t) = (2t^2 - 1)^2(3t^2)$

10. $h(x) = (x^2 + 1)^2(x - 1)$

11. $f(x) = (2x^2 + 2)^{7/2}$

12. $h(w) = (w^2 + 2w + 4)^{5/2}$

13. $f(x) = x(x^2 + 1)^2$ **14.** $g(u) = u(2u - 1)^3$

15. $f(x) = \dfrac{x}{2x + 1}$ **16.** $g(t) = \dfrac{t^2}{t - 1}$

17. $f(s) = \dfrac{s - 1}{s + 1}$ **18.** $f(u) = \dfrac{u}{u^2 + 1}$

19. $f(u) = \sqrt{4 - 3u}$ **20.** $f(x) = \sqrt{2x - 1}$

In Exercises 21–28, find the third derivative of the given function.

21. $f(x) = 3x^4 - 4x^3$

22. $f(x) = 3x^5 - 6x^4 + 2x^2 - 8x + 12$

23. $f(x) = \dfrac{1}{x}$ **24.** $f(x) = \dfrac{2}{x^2}$

25. $g(s) = \sqrt{3s - 2}$ **26.** $g(t) = \sqrt{2t + 3}$

27. $f(x) = (2x - 3)^4$ **28.** $g(t) = (\tfrac{1}{2}t^2 - 1)^5$

29. ACCELERATION OF A FALLING OBJECT During the construction of an office building, a hammer is accidentally dropped from a height of 256 ft. The distance the hammer falls in t sec is $s = 16t^2$. What is the hammer's velocity when it strikes the ground? What is its acceleration?

30. ACCELERATION OF A CAR The distance s (in feet) covered by a car t sec after starting from rest is given by

$$s = -t^3 + 8t^2 + 20t \qquad (0 \le t \le 6)$$

Find a general expression for the car's acceleration at any time $t (0 \le t \le 6)$. Show that the car is decelerating $2\tfrac{2}{3}$ sec after starting from rest.

31. CRIME RATES The number of major crimes committed in Bronxville between 1988 and 1995 is approximated by the function

$$N(t) = -0.1t^3 + 1.5t^2 + 100 \qquad (0 \le t \le 7)$$

where $N(t)$ denotes the number of crimes committed in year t and $t = 0$ corresponds to the year 1988. Enraged by the dramatic increase in the crime rate, Bronxville's citizens, with the help of the local police, organized "Neighborhood Crime Watch" groups in early 1992 to combat this menace.
a. Verify that the crime rate was increasing from 1988 through 1995.

Hint: Compute $N'(0), N'(1), \ldots, N'(7)$.
b. Show that the Neighborhood Crime Watch program was working by computing $N''(4)$, $N''(5)$, $N''(6)$, and $N''(7)$.

32. GDP OF A DEVELOPING COUNTRY A developing country's gross domestic product (GDP) from 1992 to 2000 is approximated by the function

$$G(t) = -0.2t^3 + 2.4t^2 + 60 \qquad (0 \le t \le 8)$$

where $G(t)$ is measured in billions of dollars and $t = 0$ corresponds to the year 1992.
a. Compute $G'(0), G'(1), \ldots, G'(8)$.
b. Compute $G''(0), G''(1), \ldots, G''(8)$.
c. Using the results obtained in parts (a) and (b), show that after a spectacular growth rate in the early years, the growth of the GDP cooled off.

33. TEST FLIGHT OF A VTOL In a test flight of the McCord Terrier, McCord Aviation's experimental VTOL (vertical takeoff and landing) aircraft, it was determined that t sec after lift-off, when the craft was operated in the
(continued on p. 248)

FINDING THE SECOND DERIVATIVE OF A FUNCTION AT A GIVEN POINT

Some graphing utilities have the capability of numerically computing the second derivative of a function at a point. If your graphing utility has this capability, use it to work through the examples and exercises of this section.

EXAMPLE 1 Use the (second) numerical derivative operation of a graphing utility to find the second derivative of $f(x) = \sqrt{x}$ when $x = 4$.

SOLUTION ✔ Using the (second) numerical derivative operation of a graphing utility, we find

$$f''(4) = \text{der2 } (x\wedge.5, x, 4) = -0.03125 \qquad \blacksquare\blacksquare\blacksquare\blacksquare$$

EXAMPLE 2 The anticipated rise in Alzheimer's patients in the United States is given by

$$f(t) = -0.02765t^4 + 0.3346t^3 - 1.1261t^2$$
$$+ 1.7575t + 3.7745 \qquad (0 \le t \le 6)$$

where $f(t)$ is measured in millions and t is measured in decades, with $t = 0$ corresponding to the beginning of 1990.

a. How fast is the number of Alzheimer's patients in the United States anticipated to be changing at the beginning of 2030?
b. How fast is the rate of change of the number of Alzheimer's patients in the United States anticipated to be changing at the beginning of 2030?
c. Plot the graph of f in the viewing rectangle $[0, 7] \times [0, 12]$.
Source: Alzheimer's Association

SOLUTION ✔ a. Using the numerical derivative operation of a graphing utility, we find that the number of Alzheimer's patients at the beginning of 2030 can be anticipated to be changing at the rate of

$$f'(4) = 1.7311$$

That is, the number is increasing at the rate of approximately 1.7 million patients per decade.
b. Using the (second) numerical derivative operation of a graphing utility, we find that

$$f''(4) = 0.4694$$

That is, the rate of change of the number of Alzheimer's patients is increasing at the rate of approximately 0.5 million patients per decade per decade.

c. The graph is shown in Figure T1.

FIGURE T1
The graph of f in the viewing window $[0, 7] \times [0, 12]$

Exercises

In Exercises 1–8, find the value of the second derivative of f at the given value of x. Express your answer correct to four decimal places.

1. $f(x) = 2x^3 - 3x^2 + 1$; $x = -1$

2. $f(x) = 2.5x^5 - 3x^3 + 1.5x + 4$; $x = 2.1$

3. $f(x) = 2.1x^{3.1} - 4.2x^{1.7} + 4.2$; $x = 1.4$

4. $f(x) = 1.7x^{4.2} - 3.2x^{1.3} + 4.2x - 3.2$; $x = 2.2$

5. $f(x) = \dfrac{x^2 + 2x - 5}{x^3 + 1}$; $x = 2.1$

6. $f(x) = \dfrac{x^3 + x + 2}{2x^2 - 5x + 4}$; $x = 1.2$

7. $f(x) = \dfrac{x^{1/2} + 2x^{3/2} + 1}{2x^{1/2} + 3}$; $x = 0.5$

8. $f(x) = \dfrac{\sqrt{x} - 1}{2x + \sqrt{x} + 4}$; $x = 2.3$

9. RATE OF BANK FAILURES The Federal Deposit Insurance Corporation (FDIC) estimates that the rate at which banks were failing between 1982 and 1994 is given by

$$f(t) = -0.063447t^4 - 1.953283t^3 + 14.632576t^2 - 6.684704t + 47.458874 \qquad (0 \le t \le 12)$$

where $f(t)$ is measured in the number of banks per year and t is measured in years, with $t = 0$ corresponding to the beginning of 1982. Compute $f''(6)$ and interpret your results.
Source: Federal Deposit Insurance Corporation

10. MULTIMEDIA SALES According to the Electronic Industries Association, sales in the multimedia market (hardware and software) are expected to be

$$S(t) = -0.0094t^4 + 0.1204t^3 - 0.0868t^2 + 0.0195t + 3.3325 \qquad (0 \le t \le 10)$$

where $S(t)$ is measured in billions of dollars and t is measured in years, with $t = 0$ corresponding to 1990. Compute $S''(7)$ and interpret your results.
Source: Electronics Industries Association

vertical takeoff mode, its altitude (in feet) was

$$h(t) = \frac{1}{16}t^4 - t^3 + 4t^2 \qquad (0 \le t \le 8)$$

a. Find an expression for the craft's velocity at time t.
b. Find the craft's velocity when $t = 0$ (the initial velocity), $t = 4$, and $t = 8$.
c. Find an expression for the craft's acceleration at time t.
d. Find the craft's acceleration when $t = 0$, 4, and 8.
e. Find the craft's height when $t = 0$, 4, and 8.

34. U.S. CENSUS According to the U.S. Census Bureau, the number of Americans aged 45 to 54 will be approximately

$$N(t) = -0.00233t^4 + 0.00633t^3 - 0.05417t^2$$
$$+ 1.3467t + 25$$

million people in year t, where $t = 0$ corresponds to the beginning of 1990. Compute $N'(10)$ and $N''(10)$ and interpret your results.
Source: U.S. Census Bureau

35. AIR PURIFICATION During testing of a certain brand of air purifier, it was determined that the amount of smoke remaining t min after the start of the test was

$$A(t) = -0.00006t^5 + 0.00468t^4 - 0.1316t^3$$
$$+ 1.915t^2 - 17.63t + 100$$

percent of the original amount. Compute $A'(10)$ and $A''(10)$ and interpret your results.
Source: Consumer Reports

In Exercises 36–39, determine whether the statement is true or false. If it is true, explain why it is true. If it is false, give an example to show why it is false.

36. If the second derivative of f exists at $x = a$, then $f''(a) = [f'(a)]^2$.

37. If $h = fg$ where f and g have second-order derivatives, then

$$h''(x) = f''(x)g(x) + 2f'(x)g'(x) + f(x)g''(x)$$

38. If $f(x)$ is a polynomial function of degree n, then $f^{(n+1)}(x) = 0$.

39. Suppose $P(t)$ represents the population of bacteria at time t and suppose $P'(t) > 0$ and $P''(t) < 0$; then the population is increasing at time t but at a decreasing rate.

40. Let f be the function defined by the rule $f(x) = x^{7/3}$. Show that f has first- and second-order derivatives at all points x, in particular at $x = 0$. Show also that the third derivative of f does *not* exist at $x = 0$.

41. Construct a function f that has derivatives of order up through and including n at a point a but fails to have the $(n + 1)$st derivative there.
Hint: See Exercise 40.

42. Show that a polynomial function has derivatives of all orders.
Hint: Let $P(x) = a_0x^n + a_1x^{n-1} + a_2x^{n-2} + \cdots + a_n$ be a polynomial of degree n, where n is a positive integer and $a_0, a_1, \ldots, a_n$ are constants with $a_0 \ne 0$. Compute $P'(x), P''(x), \ldots$.

SOLUTIONS TO SELF-CHECK EXERCISES 3.5

1. $f'(x) = 10x^4 - 9x^2 + 2x - 6$

$f''(x) = 40x^3 - 18x + 2$

$f'''(x) = 120x^2 - 18$

2. We write $f(x) = (1 + x)^{-1}$ and use the general power rule, obtaining

$$f'(x) = (-1)(1 + x)^{-2}\frac{d}{dx}(1 + x) = -(1 + x)^{-2}(1)$$

$$= -(1 + x)^{-2} = -\frac{1}{(1 + x)^2}$$

Continuing, we find

$$f''(x) = -(-2)(1+x)^{-3}$$
$$= 2(1+x)^{-3} = \frac{2}{(1+x)^3}$$
$$f'''(x) = 2(-3)(1+x)^{-4}$$
$$= -6(1+x)^{-4}$$
$$= -\frac{6}{(1+x)^4}$$

3. $N'(t) = 6t^2 + 6t - 4$

$N''(t) = 12t + 6 = 6(2t+1)$

Therefore, $N''(2) = 30$ and $N''(8) = 102$. The results of our computations reveal that at the end of year 2, the *rate* of growth of the turtle population is increasing at the rate of 30 turtles/year/year. At the end of year 8, the rate is increasing at the rate of 102 turtles/year/year. Clearly, the conservation measures are paying off handsomely.

3.6 Implicit Differentiation and Related Rates

DIFFERENTIATING IMPLICITLY

Up to now we have dealt with functions expressed in the form $y = f(x)$; that is, the dependent variable y is expressed *explicitly* in terms of the independent variable x. However, not all functions are expressed in this form. Consider, for example, the equation

$$x^2y + y - x^2 + 1 = 0 \tag{8}$$

This equation does express y *implicitly* as a function of x. In fact, solving (8) for y in terms of x, we obtain

$$(x^2+1)y = x^2 - 1 \quad \text{(Implicit equation)}$$
$$y = f(x) = \frac{x^2-1}{x^2+1} \quad \text{(Explicit equation)}$$

which gives an explicit representation of f.

Next, consider the equation

$$y^4 - y^3 - y + 2x^3 - x = 8$$

When certain restrictions are placed on x and y, this equation defines y as a function of x. But in this instance, we would be hard pressed to find y explicitly in terms of x. The following question arises naturally: How does one go about computing dy/dx in this case?

As it turns out, thanks to the chain rule, a method *does* exist for computing the derivative of a function directly from the implicit equation defining the function. This method is called **implicit differentiation** and is demonstrated in the next several examples.

EXAMPLE 1 Find $\dfrac{dy}{dx}$ given the equation $y^2 = x$.

SOLUTION ✔ Differentiating both sides of the equation with respect to x, we obtain

$$\frac{d}{dx}(y^2) = \frac{d}{dx}(x)$$

To carry out the differentiation of the term $\dfrac{d}{dx}y^2$, we note that y is a function of x. Writing $y = f(x)$ to remind us of this fact, we find that

$$\frac{d}{dx}(y^2) = \frac{d}{dx}[f(x)]^2 \qquad \text{[Writing } y = f(x)\text{]}$$

$$= 2f(x)f'(x) \qquad \text{(Using the chain rule)}$$

$$= 2y\frac{dy}{dx} \qquad \text{[Returning to using } y \text{ instead of } f(x)\text{]}$$

Therefore, the equation

$$\frac{d}{dx}(y^2) = \frac{d}{dx}(x)$$

is equivalent to

$$2y\frac{dy}{dx} = 1$$

Solving for $\dfrac{dy}{dx}$ yields

$$\frac{dy}{dx} = \frac{1}{2y}$$

■ ■ ■ ■

Before considering other examples, let us summarize the important steps involved in implicit differentiation. (Here we assume that dy/dx exists.)

Finding $\dfrac{dy}{dx}$ by Implicit Differentiation

> **1.** Differentiate both sides of the equation *with respect to x*. (Make sure that the derivative of any term involving y includes the factor dy/dx.)
> **2.** Solve the resulting equation for dy/dx in terms of x and y.

EXAMPLE 2 Find dy/dx given the equation

$$y^3 - y + 2x^3 - x = 8$$

SOLUTION ✔ Differentiating both sides of the given equation with respect to x, we obtain

$$\frac{d}{dx}(y^3 - y + 2x^3 - x) = \frac{d}{dx}(8)$$

$$\frac{d}{dx}(y^3) - \frac{d}{dx}(y) + \frac{d}{dx}(2x^3) - \frac{d}{dx}(x) = 0$$

Now, recalling that y is a function of x, we apply the chain rule to the first two terms on the left. Thus,

$$3y^2 \frac{dy}{dx} - \frac{dy}{dx} + 6x^2 - 1 = 0$$

$$(3y^2 - 1)\frac{dy}{dx} = 1 - 6x^2$$

$$\frac{dy}{dx} = \frac{1 - 6x^2}{3y^2 - 1}$$

EXAMPLE 3   Consider the equation $x^2 + y^2 = 4$.

a. Find dy/dx by implicit differentiation.
b. Find the slope of the tangent line to the graph of the function $y = f(x)$ at the point $(1, \sqrt{3})$.
c. Find an equation of the tangent line of part (b).

SOLUTION ✔   **a.** Differentiating both sides of the equation with respect to x, we obtain

$$\frac{d}{dx}(x^2 + y^2) = \frac{d}{dx}(4)$$

$$\frac{d}{dx}(x^2) + \frac{d}{dx}(y^2) = 0$$

$$2x + 2y\frac{dy}{dx} = 0$$

$$\frac{dy}{dx} = -\frac{x}{y} \qquad (y \neq 0)$$

b. The slope of the tangent line to the graph of the function at the point $(1, \sqrt{3})$ is given by

$$\left.\frac{dy}{dx}\right|_{(1,\sqrt{3})} = \left.-\frac{x}{y}\right|_{(1,\sqrt{3})} = -\frac{1}{\sqrt{3}}$$

FIGURE 3.16
The line $x + \sqrt{3}y - 4 = 0$ is tangent to the graph of the function $y = f(x)$.

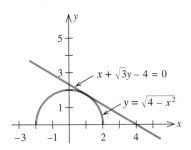

(*Note:* This notation is read "dy/dx evaluated at the point $(1, \sqrt{3})$.")
c. An equation of the tangent line in question is found by using the point-slope form of the equation of a line with the slope $m = -1/\sqrt{3}$ and the point $(1, \sqrt{3})$. Thus,

$$y - \sqrt{3} = -\frac{1}{\sqrt{3}}(x - 1)$$

$$\sqrt{3}y - 3 = -x + 1$$

$$x + \sqrt{3}y - 4 = 0$$

A sketch of this tangent line is shown in Figure 3.16.

We can also solve the equation $x^2 + y^2 = 4$ explicitly for y in terms of x. If we do this, we obtain

$$y = \pm\sqrt{4 - x^2}$$

From this, we see that the equation $x^2 + y^2 = 4$ defines the two functions

$$y = f(x) = \sqrt{4 - x^2}$$
$$y = g(x) = -\sqrt{4 - x^2}$$

Since the point $(1, \sqrt{3})$ does not lie on the graph of $y = g(x)$, we conclude that

$$y = f(x) = \sqrt{4 - x^2}$$

is the required function. The graph of f is the upper semicircle shown in Figure 3.16.   ■ ■ ■ ■

Group Discussion

Refer to Example 3. Yet another function defined implicitly by the equation $x^2 + y^2 = 4$ is the function

$$y = h(x) = \begin{cases} \sqrt{4 - x^2} & \text{if } -2 \le x < 0 \\ -\sqrt{4 - x^2} & \text{if } 0 \le x \le 2 \end{cases}$$

1. Sketch the graph of h.
2. Show that $h'(x) = -x/y$.
3. Find an equation of the tangent line to the graph of h at the point $(1, -\sqrt{3})$.

To find dy/dx at a *specific point* (a, b), differentiate the given equation implicitly with respect to x and then replace x and y by a and b, respectively, *before* solving the equation for dy/dx. This often simplifies the amount of algebra involved.

EXAMPLE 4   Find dy/dx given that x and y are related by the equation

$$x^2y^3 + 6x^2 = y + 12$$

and that $y = 2$ when $x = 1$.

SOLUTION ✔   Differentiating both sides of the given equation with respect to x, we obtain

$$\frac{d}{dx}(x^2y^3) + \frac{d}{dx}(6x^2) = \frac{d}{dx}(y) + \frac{d}{dx}(12)$$

$$x^2 \cdot \frac{d}{dx}(y^3) + y^3 \cdot \frac{d}{dx}(x^2) + 12x = \frac{dy}{dx} \qquad \text{[Using the product rule on } \frac{d}{dx}(x^2y^3)\text{]}$$

$$3x^2y^2\frac{dy}{dx} + 2xy^3 + 12x = \frac{dy}{dx}$$

Substituting $x = 1$ and $y = 2$ into this equation gives

$$3(1)^2(2)^2 \frac{dy}{dx} + 2(1)(2)^3 + 12(1) = \frac{dy}{dx}$$

$$12\frac{dy}{dx} + 16 + 12 = \frac{dy}{dx}$$

and, solving for $\dfrac{dy}{dx}$,

$$\frac{dy}{dx} = -\frac{28}{11}$$

Note that it is not necessary to find an explicit expression for dy/dx. ■ ■ ■ ■

REMARK In Examples 3 and 4, you can verify that the points at which we evaluated dy/dx actually lie on the curve in question by showing that the coordinates of the points satisfy the given equations. ■ ■ ■

EXAMPLE 5

Find dy/dx given that x and y are related by the equation

$$\sqrt{x^2 + y^2} - x^2 = 5$$

SOLUTION ✔ Differentiating both sides of the given equation with respect to x, we obtain

$$\frac{d}{dx}(x^2 + y^2)^{1/2} - \frac{d}{dx}(x^2) = \frac{d}{dx}(5)$$

[Writing $\sqrt{x^2 + y^2} = (x^2 + y^2)^{1/2}$]

$$\frac{1}{2}(x^2 + y^2)^{-1/2}\frac{d}{dx}(x^2 + y^2) - 2x = 0$$

(Using the general power rule on the first term)

$$\frac{1}{2}(x^2 + y^2)^{-1/2}\left(2x + 2y\frac{dy}{dx}\right) - 2x = 0$$

$$2x + 2y\frac{dy}{dx} = 4x(x^2 + y^2)^{1/2}$$

[Transposing $2x$ and multiplying both sides by $2(x^2 + y^2)^{1/2}$]

$$2y\frac{dy}{dx} = 4x(x^2 + y^2)^{1/2} - 2x$$

$$\frac{dy}{dx} = \frac{2x\sqrt{x^2 + y^2} - x}{y}$$

■ ■ ■ ■

RELATED RATES

Implicit differentiation is a useful technique for solving a class of problems known as **related rates** problems. For example, suppose x and y are each functions of a third variable t. Here, x might denote the mortgage rate and y the number of single-family homes sold at any time t. Further, suppose we have an equation that gives the relationship between x and y (the number of houses sold y is related to the mortgage rate x). Differentiating both sides of this equation implicitly with respect to t, we obtain an equation that gives a relationship between dx/dt and dy/dt. In the context of our example, this equation gives us a relationship between the rate of change of the mortgage rate and the rate of change of the number of houses sold, as a function of time. Thus, knowing

$$\frac{dx}{dt} \quad \text{(How fast the mortgage rate is changing at time } t\text{)}$$

we can determine

$$\frac{dy}{dt} \quad \text{(How fast the sale of houses is changing at that instant of time)}$$

EXAMPLE 6 A study prepared for the National Association of Realtors estimates that the number of housing starts in the Southwest, $N(t)$ (in units of a million), over the next 5 years is related to the mortgage rate $r(t)$ (percent per year) by the equation

$$9N^2 + r = 36$$

What is the rate of change of the number of housing starts with respect to time when the mortgage rate is 11% per year and is increasing at the rate of 1.5% per year?

SOLUTION ✔ We are given that

$$r = 11 \quad \text{and} \quad \frac{dr}{dt} = 1.5$$

at a certain instant of time, and we are required to find dN/dt. First, by substituting $r = 11$ into the given equation, we find

$$9N^2 + 11 = 36$$
$$N^2 = \frac{25}{9}$$

or $N = 5/3$ (we reject the negative root). Next, differentiating the given equation implicitly on both sides with respect to t, we obtain

$$\frac{d}{dt}(9N^2) + \frac{d}{dt}(r) = \frac{d}{dt}(36)$$

$$18N\frac{dN}{dt} + \frac{dr}{dt} = 0 \quad \text{(Use the chain rule on the first term.)}$$

Then, substituting $N = 5/3$ and $dr/dt = 1.5$ into this equation gives

$$18\left(\frac{5}{3}\right)\frac{dN}{dt} + 1.5 = 0$$

Solving this equation for dN/dt then gives

$$\frac{dN}{dt} = -\frac{1.5}{30} \approx -0.05$$

Thus, at the instant of time under consideration, the number of housing starts is decreasing at the rate of 50,000 units per year.   ■ ■ ■ ■

EXAMPLE 7   A major audiotape manufacturer is willing to make x thousand ten packs of metal alloy audiocassette tapes per week available in the marketplace when the wholesale price is $\$p$ per ten pack. It is known that the relationship between x and p is governed by the supply equation

$$x^2 - 3xp + p^2 = 5$$

How fast is the supply of tapes changing when the price per ten pack is $11, the quantity supplied is 4000 ten packs, and the wholesale price per ten pack is increasing at the rate of 10 cents per ten pack per week?

SOLUTION ✔   We are given that

$$p = 11, \qquad x = 4, \qquad \frac{dp}{dt} = 0.1$$

at a certain instant of time, and we are required to find dx/dt. Differentiating the given equation on both sides with respect to t, we obtain

$$\frac{d}{dt}(x^2) - \frac{d}{dt}(3xp) + \frac{d}{dt}(p^2) = \frac{d}{dt}(5)$$

$$2x\frac{dx}{dt} - 3\left(p\frac{dx}{dt} + x\frac{dp}{dt}\right) + 2p\frac{dp}{dt} = 0 \qquad \text{(Use the product rule on the second term.)}$$

Substituting the given values of p, x, and dp/dt into the last equation, we have

$$2(4)\frac{dx}{dt} - 3\left[(11)\frac{dx}{dt} + 4(0.1)\right] + 2(11)(0.1) = 0$$

$$8\frac{dx}{dt} - 33\frac{dx}{dt} - 1.2 + 2.2 = 0$$

$$25\frac{dx}{dt} = 1$$

$$\frac{dx}{dt} = 0.04$$

Thus, at the instant of time under consideration the supply of ten pack audio-cassettes is increasing at the rate of (0.04)(1000), or 40, ten packs per week.   ■ ■ ■ ■

In certain related problems, we need to formulate the problem mathematically before analyzing it. The following guidelines can be used to help solve problems of this type.

Solving Related Rates Problems

1. Assign a variable to each quantity. Draw a diagram if needed.
2. Write the *given* values of the variables and their rates of change with respect to t.
3. Find an equation giving the relationship between the variables.
4. Differentiate both sides of this equation implicitly with respect to t.
5. Replace the variables and their derivatives by the numerical data found in step 2 and solve the equation for the required rate of change.

EXAMPLE 8   At a distance of 4000 feet from the launch site, a spectator is observing a rocket being launched. If the rocket lifts off vertically and is rising at a speed of 600 feet/second when it is at an altitude of 3000 feet, how fast is the distance between the rocket and the spectator changing at that instant?

SOLUTION ✔   **Step 1**   Let

$$y = \text{the altitude of the rocket}$$
$$x = \text{the distance between the rocket and the spectator}$$

at any time t (Figure 3.17).

FIGURE 3.17
The rate at which x is changing with respect to time is related to the rate of change of y with respect to time.

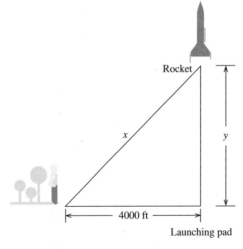

Step 2   We are given that at a certain instant of time

$$y = 3000 \qquad \text{and} \qquad \frac{dy}{dt} = 600$$

and are asked to find dx/dt at that instant.

Step 3   Applying the Pythagorean theorem to the right triangle in Figure 3.17, we find that

$$x^2 = y^2 + 4000^2$$

Therefore, when $y = 3000$,

$$x = \sqrt{3000^2 + 4000^2} = 5000$$

Step 4   Next, we differentiate the equation $x^2 = y^2 + 4000^2$ with respect to t, obtaining

$$2x\frac{dx}{dt} = 2y\frac{dy}{dt}$$

(Remember, both x and y are functions of t.)

Step 5   Substituting $x = 5000$, $y = 3000$, and $dy/dt = 600$, we find

$$2(5000)\frac{dx}{dt} = 2(3000)(600)$$

$$\frac{dx}{dt} = 360$$

Therefore, the distance between the rocket and the spectator is changing at a rate of 360 feet/second.   ■ ■ ■ ■

 Be sure that you do *not* replace the variables in the equation found in step 3 by their numerical values before differentiating the equation.

SELF-CHECK EXERCISES 3.6

1. Given the equation $x^3 + 3xy + y^3 = 4$, find dy/dx by implicit differentiation.

2. Find an equation of the tangent line to the graph of $16x^2 + 9y^2 = 144$ at the point $\left(2, -\frac{4\sqrt{5}}{3}\right)$.

Solutions to Self-Check Exercises 3.6 can be found on page 260.

3.6 Exercises

In Exercises 1–8, find the derivative dy/dx (a) by solving each of the given implicit equations for y explicitly in terms of x and (b) by differentiating each of the given equations implicitly. Show that, in each case, the results are equivalent.

1. $x + 2y = 5$

2. $3x + 4y = 6$

3. $xy = 1$

4. $xy - y - 1 = 0$

5. $x^3 - x^2 - xy = 4$

6. $x^2y - x^2 + y - 1 = 0$

7. $\dfrac{x}{y} - x^2 = 1$

8. $\dfrac{y}{x} - 2x^3 = 4$

In Exercises 9–30, find dy/dx by implicit differentiation.

9. $x^2 + y^2 = 16$

10. $2x^2 + y^2 = 16$

11. $x^2 - 2y^2 = 16$

12. $x^3 + y^3 + y - 4 = 0$

13. $x^2 - 2xy = 6$

14. $x^2 + 5xy + y^2 = 10$

15. $x^2y^2 - xy = 8$

16. $x^2y^3 - 2xy^2 = 5$

17. $x^{1/2} + y^{1/2} = 1$

18. $x^{1/3} + y^{1/3} = 1$

19. $\sqrt{x + y} = x$

20. $(2x + 3y)^{1/3} = x^2$

21. $\dfrac{1}{x^2} + \dfrac{1}{y^2} = 1$

22. $\dfrac{1}{x^3} + \dfrac{1}{y^3} = 5$

23. $\sqrt{xy} = x + y$

24. $\sqrt{xy} = 2x + y^2$

25. $\dfrac{x + y}{x - y} = 3x$

26. $\dfrac{x - y}{2x + 3y} = 2x$

27. $xy^{3/2} = x^2 + y^2$

28. $x^2y^{1/2} = x + 2y^3$

29. $(x + y)^3 + x^3 + y^3 = 0$

30. $(x + y^2)^{10} = x^2 + 25$

In Exercises 31–34, find an equation of the tangent line to the graph of the function f defined by the given equation at the indicated point.

31. $4x^2 + 9y^2 = 36$; $(0, 2)$

32. $y^2 - x^2 = 16$; $(2, 2\sqrt{5})$

33. $x^2y^3 - y^2 + xy - 1 = 0$; $(1, 1)$

34. $(x - y - 1)^3 = x$; $(1, -1)$

In Exercises 35–38, find the second derivative d^2y/dx^2 of each of the functions defined implicitly by the given equation.

35. $xy = 1$ 36. $x^3 + y^3 = 28$

37. $y^2 - xy = 8$ 38. $x^{1/3} + y^{1/3} = 1$

39. The volume of a right-circular cylinder of radius r and height h is $V = \pi r^2 h$. Suppose the radius and height of the cylinder are changing with respect to time t.
   a. Find a relationship between dV/dt, dr/dt, and dh/dt.
   b. At a certain instant of time, the radius and height of the cylinder are 2 and 6 in. and are increasing at the rate of 0.1 and 0.3 in./sec, respectively. How fast is the volume of the cylinder increasing?

40. A car leaves an intersection traveling west. Its position 4 seconds later is 20 ft from the intersection. At the same time, another car leaves the same intersection heading north so that its position 4 sec later is 28 ft from the intersection. If the speed of the cars at that instant of time is 9 ft/sec and 11 ft/sec, respectively, find the rate at which the distance between the two cars is changing.

41. PRICE-DEMAND Suppose the quantity demanded weekly of the Super Titan radial tires is related to its unit price by the equation

$$p + x^2 = 144$$

where p is measured in dollars and x is measured in units of a thousand. How fast is the quantity demanded changing when $x = 9$, $p = 63$, and the price per tire is increasing at the rate of $2/week?

42. PRICE-SUPPLY Suppose the quantity x of Super Titan radial tires made available per week in the marketplace by the Titan Tire Company is related to the unit selling price by the equation

$$p - \frac{1}{2}x^2 = 48$$

where x is measured in units of a thousand and p is in dollars. How fast is the weekly supply of Super Titan radial tires being introduced into the marketplace when

$x = 6$, $p = 66$, and the price per tire is decreasing at the rate of $3/week?

43. PRICE-DEMAND The demand equation for a certain brand of metal alloy audiocassette tape is

$$100x^2 + 9p^2 = 3600$$

where x represents the number (in thousands) of ten packs demanded per week when the unit price is p. How fast is the quantity demanded increasing when the unit price per ten pack is $14 and the selling price is dropping at the rate of $.15 per ten pack per week?
   Hint: To find the value of x when $p = 14$, solve the equation $100x^2 + 9p^2 = 3600$ for x when $p = 14$.

44. EFFECT OF PRICE ON SUPPLY Suppose the wholesale price of a certain brand of medium-size eggs p (in dollars per carton) is related to the weekly supply x (in thousands of cartons) by the equation

$$625p^2 - x^2 = 100$$

If 25,000 cartons of eggs are available at the beginning of a certain week and the price is falling at the rate of 2 cents/carton/week, at what rate is the supply falling?
   Hint: To find the value of p when $x = 25$, solve the supply equation for p when $x = 25$.

45. SUPPLY-DEMAND Refer to Exercise 44. If 25,000 cartons of eggs are available at the beginning of a certain week and the supply is falling at the rate of 1000 cartons/week, at what rate is the wholesale price changing?

46. ELASTICITY OF DEMAND The demand function for a certain make of cartridge typewriter ribbon is

$$p = -0.01x^2 - 0.1x + 6$$

where p is the unit price in dollars and x is the quantity demanded each week, measured in units of a thousand. Compute the elasticity of demand and determine whether the demand is inelastic, unitary, or elastic when $x = 10$.

47. ELASTICITY OF DEMAND The demand function for a certain brand of compact disc is

$$p = -0.01x^2 - 0.2x + 8$$

where p is the wholesale unit price in dollars and x is the quantity demanded each week, measured in units of a thousand. Compute the elasticity of demand and determine whether the demand is inelastic, unitary, or elastic when $x = 15$.

48. The volume V of a cube with sides of length x in. is changing with respect to time. At a certain instant of time, the sides of the cube are 5 in. long and increasing at the rate of 0.1 in./sec. How fast is the volume of the cube changing at that instant of time?

49. OIL SPILLS In calm waters, oil spilling from the ruptured hull of a grounded tanker spreads in all directions. If the area polluted is a circle and its radius is increasing at a rate of 2 ft/sec, determine how fast the area is increasing when the radius of the circle is 40 ft.

50. Two ships leave the same port at noon. Ship A sails north at 15 mph, and ship B sails east at 12 mph. How fast is the distance between them changing at 1 P.M.?

51. A car leaves an intersection traveling east. Its position t sec later is given by $x = t^2 + t$ ft. At the same time, another car leaves the same intersection heading north, traveling $y = t^2 + 3t$ ft in t sec. Find the rate at which the distance between the two cars will be changing 5 sec later.

52. At a distance of 50 ft from the pad, a man observes a helicopter taking off from a heliport. If the helicopter lifts off vertically and is rising at a speed of 44 ft/sec when it is at an altitude of 120 ft, how fast is the distance between the helicopter and the man changing at that instant?

53. A spectator watches a rowing race from the edge of a river bank. The lead boat is moving in a straight line that is 120 ft from the river bank. If the boat is moving at a constant speed of 20 ft/sec, how fast is the boat moving away from the spectator when it is 50 ft past her?

54. A boat is pulled toward a dock by means of a rope which is wound on a drum that is located 4 ft above the bow of the boat. If the rope is being pulled in at the rate of 3 ft/sec, how fast is the boat approaching the dock when it is 25 ft from the dock?

55. Assume that a snowball is in the shape of a sphere. If the snowball melts at a rate that is proportional to its surface area, show that its radius decreases at a constant rate.
Hint: Its volume is $V = (4/3)\pi r^3$, and its surface area is $S = 4\pi r^2$.

56. BLOWING SOAP BUBBLES Carlos is blowing air into a soap bubble at the rate of 8 cm³/sec. Assuming that the bubble is spherical, how fast is its radius changing at the instant of time when the radius is 10 cm? How fast is the surface area of the bubble changing at that instant of time?

57. COAST GUARD PATROL SEARCH MISSION The pilot of a Coast Guard patrol aircraft on a search mission had just spotted a disabled fishing trawler and decided to go in for a closer look. Flying at a constant altitude of 1000 ft and at a steady speed of 264 ft/sec, the aircraft passed directly over the trawler. How fast was the aircraft receding from the trawler when it was 1500 ft from it?

58. A coffee pot in the form of a circular cylinder of radius 4 in. is being filled with water flowing at a constant rate. If the water level is rising at the rate of 0.4 in./sec, what is the rate at which water is flowing into the coffee pot?

59. A 6-ft tall man is walking away from a street light 18 ft high at a speed of 6 ft/sec. How fast is the tip of his shadow moving along the ground?

60. A 20-ft ladder leaning against a wall begins to slide. How fast is the top of the ladder sliding down the wall at the instant of time when the bottom of the ladder is 12 ft from the wall and sliding away from the wall at the rate of 5 ft/sec?
Hint: Refer to the adjacent figure. By the Pythagorean theorem, $x^2 + y^2 = 400$. Find dy/dt when $x = 12$ and $dx/dt = 5$.

61. The base of a 13-ft ladder leaning against a wall begins to slide away from the wall. At the instant of time when the base is 12 ft from the wall, the base is moving at the rate of 8 ft/sec. How fast is the top of the ladder sliding down the wall at that instant of time?

Hint: Refer to the hint in Problem 60.

62. Water flows from a tank of constant cross-sectional area 50 ft² through an orifice of constant cross-sectional area 1.4 ft² located at the bottom of the tank (see the figure).

Initially, the height of the water in the tank was 20 ft and its height t sec later is given by the equation

$$2\sqrt{h} + \frac{1}{25}t - 2\sqrt{20} = 0 \qquad (0 \le t \le 50\sqrt{20})$$

How fast was the height of the water decreasing when its height was 8 ft?

In Exercises 63 and 64, determine whether the statement is true or false. If it is true, explain why it is true. If it is false, give an example to show why it is false.

63. If f and g are differentiable and $f(x)g(y) = 0$, then

$$\frac{dy}{dx} = -\frac{f'(x)g(y)}{f(x)g'(y)} \qquad (f(x) \ne 0 \quad \text{and} \quad g'(y) \ne 0)$$

64. If f and g are differentiable and $f(x) + g(y) = 0$, then

$$\frac{dy}{dx} = -\frac{f'(x)}{g'(y)}$$

SOLUTIONS TO SELF-CHECK EXERCISES 3.6

1. Differentiating both sides of the equation with respect to x, we have

$$3x^2 + 3y + 3xy' + 3y^2y' = 0$$
$$(x^2 + y) + (x + y^2)y' = 0$$
$$y' = -\frac{x^2 + y}{x + y^2}$$

2. To find the slope of the tangent line to the graph of the function at any point, we differentiate the equation implicitly with respect to x, obtaining

$$32x + 18yy' = 0$$
$$y' = -\frac{16x}{9y}$$

In particular, the slope of the tangent line at $\left(2, -\frac{4\sqrt{5}}{3}\right)$ is

$$m = -\frac{16(2)}{9\left(-\frac{4\sqrt{5}}{3}\right)} = \frac{8}{3\sqrt{5}}$$

Using the point-slope form of the equation of a line, we find

$$y - \left(-\frac{4\sqrt{5}}{3}\right) = \frac{8}{3\sqrt{5}}(x - 2)$$

$$y = \frac{8\sqrt{5}}{15}x - \frac{36\sqrt{5}}{15} = \frac{8\sqrt{5}}{15}x - \frac{12\sqrt{5}}{5}$$

3.7 Differentials

The Millers are planning to buy a house in the near future and estimate that they will need a 30-year fixed-rate mortgage for $120,000. If the interest rate increases from the present rate of 9% per year to 9.4% per year between now and the time the Millers decide to secure the loan, approximately how much more per month will their mortgage be? (You will be asked to answer this question in Exercise 44, page 272.)

Questions such as this, in which one wishes to *estimate* the change in the dependent variable (monthly mortgage payment) corresponding to a small change in the independent variable (interest rate per year), occur in many real-life applications. For example:

■ An economist would like to know how a small increase in a country's capital expenditure will affect the country's gross domestic output.

■ A sociologist would like to know how a small increase in the amount of capital investment in a housing project will affect the crime rate.

■ A businesswoman would like to know how raising a product's unit price by a small amount will affect her profit.

■ A bacteriologist would like to know how a small increase in the amount of a bactericide will affect a population of bacteria.

To calculate these changes and estimate their effects, we use the *differential* of a function, a concept that will be introduced shortly.

INCREMENTS

Let x denote a variable quantity and suppose x changes from x_1 to x_2. This change in x is called the **increment in x** and is denoted by the symbol Δx (read "delta x"). Thus,

$$\Delta x = x_2 - x_1 \qquad \text{(Final value minus initial value)} \qquad (9)$$

EXAMPLE 1 Find the increment in x:

a. As x changes from 3 to 3.2   **b.** As x changes from 3 to 2.7

SOLUTION ✔ **a.** Here, $x_1 = 3$ and $x_2 = 3.2$, so

$$\Delta x = x_2 - x_1 = 3.2 - 3 = 0.2$$

b. Here, $x_1 = 3$ and $x_2 = 2.7$. Therefore,

$$\Delta x = x_2 - x_1 = 2.7 - 3 = -0.3 \qquad \blacksquare\blacksquare\blacksquare\blacksquare$$

Observe that Δx plays the same role that h played in Section 2.4.

Now, suppose two quantities, x and y, are related by an equation $y = f(x)$, where f is a function. If x changes from x to $x + \Delta x$, then the

Portfolio

JOHN DECKER

TITLE: Mortgage Counselor
INSTITUTION: A major bank

John Decker stresses that he and his colleagues "strive to grant loans. That's our job." But before allowing someone to file a formal application, Decker takes the person over several "prequalification hurdles" to gauge his or her ability to handle a mortgage.

To start, Decker relies on a two-tiered, debt-to-income ratio to see whether a person has sufficient gross monthly income to make payments. Under the first tier, the proposed monthly payment (principal and interest, property tax, homeowner's insurance, and, when applicable, a condo fee) cannot exceed 28% of an individual's gross monthly income. If Decker's initial calculations are positive, he then must determine the individual's ability to meet monthly mortgage payments while also repaying other debts, such as car and student loans, credit cards, alimony, and so on. These combined payments cannot exceed 36% percent of gross monthly income. A typical person with an $800 mortgage obligation and $550 in other payments would have to earn $3750 per month to clear these first two hurdles.

Contrary to popular belief, bankers *want* to lend money. "The idea is to grant mortgages," says Decker, which contribute substantially to a bank's profitability. But lending money means making sensible decisions about how much to lend as well as a person's ability to repay the loan.

Using a loan-to-value formula, banks might lend 80% of a property's value. In such cases, the applicant puts 20% down to make the purchase. Or the bank may decide to lend up to 95% or as little as 75% of the appraised value.

Understandably, banks don't like to see bankruptcies, late payments, or liens on a personal credit history. Decker notes, however, that even this hurdle doesn't necessarily mean failure in securing a mortgage. He works closely with each individual to overcome any stigma that might prompt a rejection.

Once Decker has put together a successful mortgage application, it is reviewed internally. Then, even though a mortgage is granted, it might come through at a slightly higher interest rate—10.4% instead of the expected 10% rate. Differentials would be used to compute the change in monthly payments, for although this might seem like a small change, on a large mortgage it can be enough of a variable to affect the new customer's ability to make monthly payments.

Using the debt-to-income ratio, Decker plugs the new variable into his formulas to determine whether a problem exists. These formulas used to compute the monthly payments involve the use of exponential functions. On a $100,000, 30-year, fixed-rate mortgage, payments will increase about $30 per month. Decker notes that such a small increase doesn't usually pose a significant problem.

If a customer can't make the new payment, however, Decker explores alternatives until he finds a solution. For Decker, it comes down to this: "If there is any possible way to give a loan, we're going to make it work."

Decker's job is simple: to lend money to people who want to buy a home. The hard part is deciding whether applicants qualify for one of the bank's 40 different mortgage plans. Sifting through income figures, current indebtedness, and credit history helps Decker determine which individuals make the best mortgage candidates.

FIGURE 3.18
An increment of Δx in x induces an increment of $\Delta y = f(x + \Delta x) - f(x)$ in y.

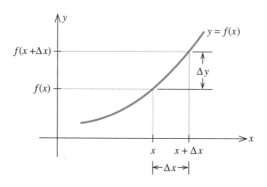

corresponding change in y is called the **increment in y**. It is denoted by Δy and is defined in Figure 3.18 by

$$\Delta y = f(x + \Delta x) - f(x) \tag{10}$$

 EXAMPLE **2**   Let $y = x^3$. Find Δx and Δy:

a. When x changes from 2 to 2.01   **b.** When x changes from 2 to 1.98

SOLUTION ✔   Let $f(x) = x^3$.

a. Here, $\Delta x = 2.01 - 2 = 0.01$. Next,

$$\Delta y = f(x + \Delta x) - f(x) = f(2.01) - f(2)$$
$$= (2.01)^3 - 2^3 = 8.120601 - 8 = 0.120601$$

b. Here, $\Delta x = 1.98 - 2 = -0.02$. Next,

$$\Delta y = f(x + \Delta x) - f(x) = f(1.98) - f(2)$$
$$= (1.98)^3 - 2^3 = 7.762392 - 8 = -0.237608 \qquad ■ ■ ■ ■$$

DIFFERENTIALS

We can obtain a relatively quick and simple way of approximating Δy, the change in y due to a small change Δx, by examining the graph of the function f shown in Figure 3.19.

FIGURE 3.19
If Δx is small, *dy* is a good approximation of Δy.

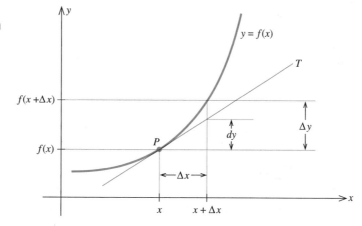

Observe that near the point of tangency P, the tangent line T is close to the graph of f. Therefore, if Δx is small, then dy is a good approximation of Δy. We can find an expression for dy as follows: Notice that the slope of T is given by

$$\frac{dy}{\Delta x} \quad \text{(Rise divided by run)}$$

However, the slope of T is given by $f'(x)$. Therefore, we have

$$\frac{dy}{\Delta x} = f'(x)$$

or $dy = f'(x)\Delta x$. Thus, we have the approximation

$$\Delta y \approx dy = f'(x)\Delta x$$

in terms of the derivative of f at x. The quantity dy is called the *differential of y*.

The Differential

Let $y = f(x)$ define a differentiable function of x. Then,

1. The differential dx of the independent variable x is $dx = \Delta x$.
2. The differential dy of the dependent variable y is

$$dy = f'(x)\Delta x = f'(x)dx \tag{11}$$

REMARKS

1. For the independent variable x: There is no difference between Δx and dx—both measure the change in x from x to $x + \Delta x$.
2. For the dependent variable y: Δy measures the *actual* change in y as x changes from x to $x + \Delta x$, whereas dy measures the *approximate* change in y corresponding to the same change in x.
3. The differential dy depends on both x and dx, but for fixed x, dy is a linear function of dx. ■■■

EXAMPLE 3   Let $y = x^3$.

a. Find the differential dy of y.
b. Use dy to approximate Δy when x changes from 2 to 2.01.
c. Use dy to approximate Δy when x changes from 2 to 1.98.
d. Compare the results of part (b) with those of Example 2.

SOLUTION ✔   **a.** Let $f(x) = x^3$. Then,

$$dy = f'(x) \, dx = 3x^2 dx$$

b. Here, $x = 2$ and $dx = 2.01 - 2 = 0.01$. Therefore,

$$dy = 3x^2 \, dx = 3(2)^2(0.01) = 0.12$$

c. Here, $x = 2$ and $dx = 1.98 - 2 = -0.02$. Therefore,

$$dy = 3x^2 \, dx = 3(2)^2(-0.02) = -0.24$$

d. As you can see, both approximations 0.12 and -0.24 are quite close to the actual changes of Δy obtained in Example 2: 0.120601 and -0.237608.

■ ■ ■ ■

Observe how much easier it is to find an approximation to the exact change in a function with the help of the differential, rather than calculating the exact change in the function itself. In the following examples, we take advantage of this fact.

EXAMPLE **4**

Approximate the value of $\sqrt{26.5}$ using differentials. Verify your result using the $\boxed{\sqrt{}}$ key on your calculator.

SOLUTION ✔

Since we want to compute the square root of a number, let's consider the function $y = f(x) = \sqrt{x}$. Since 25 is the number nearest 26.5 whose square root is readily recognized, let's take $x = 25$. We want to know the change in y, Δy, as x changes from $x = 25$ to $x = 26.5$, an increase of $\Delta x = 1.5$ units. Using Equation (11), we find

$$\Delta y \approx dy = f'(x)\Delta x$$

$$= \left[\frac{1}{2\sqrt{x}} \bigg|_{x=25} \right] \cdot (1.5) = \left(\frac{1}{10} \right)(1.5) = 0.15$$

Therefore,

$$\sqrt{26.5} - \sqrt{25} = \Delta y \approx 0.15$$
$$\sqrt{26.5} \approx \sqrt{25} + 0.15 = 5.15$$

The exact value of $\sqrt{26.5}$, rounded off to five decimal places, is 5.14782. Thus, the error incurred in the approximation is 0.00218.

■ ■ ■ ■

APPLICATIONS

EXAMPLE **5**

The total cost incurred in operating a certain type of truck on a 500-mile trip, traveling at an average speed of v mph, is estimated to be

$$C(v) = 125 + v + \frac{4500}{v}$$

dollars. Find the approximate change in the total operating cost when the average speed is increased from 55 mph to 58 mph.

SOLUTION ✔   With $v = 55$ and $\Delta v = dv = 3$, we find

$$\Delta C \approx dC = C'(v)dv = \left(1 - \frac{4500}{v^2}\right)\bigg|_{v=55} \cdot 3$$

$$= \left(1 - \frac{4500}{3025}\right)(3) \approx -1.46$$

so the total operating cost is found to decrease by $1.46. This might explain why so many independent truckers often exceed the 55 mph speed limit.

■ ■ ■ ■

EXAMPLE **6**   The relationship between the amount of money x spent by Cannon Precision Instruments on advertising and Cannon's total sales $S(x)$ is given by the function

$$S(x) = -0.002x^3 + 0.6x^2 + x + 500 \qquad (0 \le x \le 200)$$

where x is measured in thousands of dollars. Use differentials to estimate the change in Cannon's total sales if advertising expenditures are increased from $100,000 ($x = 100$) to $105,000 ($x = 105$).

SOLUTION ✔   The required change in sales is given by

$$\Delta S \approx dS = S'(100)dx$$

$$= -0.006x^2 + 1.2x + 1\big|_{x=100} \cdot (5) \qquad (dx = 105 - 100 = 5)$$

$$= (-60 + 120 + 1)(5) = 305$$

—that is, an increase of $305,000.

■ ■ ■ ■

EXAMPLE **7**

The Rings of Neptune

a. A ring has an inner radius of r units and an outer radius of R units, where $(R - r)$ is small in comparison to r (Figure 3.20a). Use differentials to estimate the area of the ring.

b. Recent observations, including those of Voyager I and II, showed that Neptune's ring system is considerably more complex than had been believed. For one thing, it is made up of a large number of distinguishable rings rather than one continuous great ring as previously thought (Figure 3.20b). The

FIGURE **3.20**

(a) The area of the ring is the circumference of the inner circle times the thickness.

(b) Neptune and its rings.

outermost ring, 1989N1R, has an inner radius of approximately 62,900 kilometers (measured from the center of the planet), and a radial width of approximately 50 kilometers. Using these data, estimate the area of the ring.

SOLUTION ✔   **a.** Using the fact that the area of a circle of radius x is $A = f(x) = \pi x^2$, we find

$$\pi R^2 - \pi r^2 = f(R) - f(r)$$
$$= \Delta A \qquad \text{(Remember, } \Delta A = \text{change in } f \text{ when}$$
$$\approx dA \qquad \qquad x \text{ changes from } x = r \text{ to } x = R.)$$
$$= f'(r)dr$$

where $dr = R - r$. So, we see that the area of the ring is approximately $2\pi r(R - r)$ square units. In words, the area of the ring is approximately equal to

Circumference of the inner circle $\times$ Thickness of the ring

b. Applying the results of part (a) with $r = 62,900$ and $dr = 50$, we find that the area of the ring is approximately $2\pi(62,900)(50)$, or 19,760,618 square kilometers, which is roughly 4% of Earth's surface.   ▪ ▪ ▪ ▪

Before looking at the next example, we need to familiarize ourselves with some terminology. If a quantity with exact value q is measured or calculated with an error of Δq, then the quantity $\Delta q/q$ is called the *relative error* in the measurement or calculation of q. If the quantity $\Delta q/q$ is expressed as a percentage, it is then called the *percentage error*. Because Δq is approximated by dq, we normally approximate the relative error $\Delta q/q$ by dq/q.

EXAMPLE **8**   Suppose the radius of a ball-bearing is measured to be 0.5 inch, with a maximum error of ± 0.0002 inch. Then, the relative error in r is

$$\frac{dr}{r} = \frac{\pm 0.0002}{0.5} = \pm 0.0004$$

and the percentage error is $\pm 0.04\%$.   ▪ ▪ ▪ ▪

EXAMPLE **9**   Suppose the side of a cube is measured with a maximum percentage error of 2%. Use differentials to estimate the maximum percentage error in the calculated volume of the cube.

SOLUTION ✔   Suppose the side of the cube is x, so its volume is

$$V = x^3$$

We are given that $\left| \dfrac{dx}{x} \right| \leq 0.02$. Now,

$$dV = 3x^2 dx$$

and so

$$\frac{dV}{V} = \frac{3x^2 dx}{x^3} = 3\frac{dx}{x}$$

Therefore,

$$\left|\frac{dV}{V}\right| = 3\left|\frac{dx}{x}\right| \le 3(0.02) = 0.06$$

and we see that the maximum percentage error in the measurement of the volume of the cube is 6%.   ■ ■ ■ ■

Finally, we want to point out that if at some point in reading this section you have a sense of déjà vu, do not be surprised, because the notion of the differential was first used in Section 3.4 (see Example 1). There we took $\Delta x = 1$ since we were interested in finding the marginal cost when the level of production was increased from $x = 250$ to $x = 251$. If we had used differentials, we would have found

$$C(251) - C(250) \approx C'(250)dx$$

so that taking $dx = \Delta x = 1$, we have $C(251) - C(250) \approx C'(250)$, which agrees with the result obtained in Example 1. Thus, in Section 3.4, we touched upon the notion of the differential, albeit in the special case in which $dx = 1$.

SELF-CHECK EXERCISES 3.7

1. Find the differential of $f(x) = \sqrt{x} + 1$.

2. A certain country's government economists have determined that the demand equation for corn in that country is given by

$$p = f(x) = \frac{125}{x^2 + 1}$$

where p is expressed in dollars per bushel and x, the quantity demanded per year, is measured in billions of bushels. The economists are forecasting a harvest of 6 billion bushels for the year. If the actual production of corn were 6.2 billion bushels for the year instead, what would be the approximate drop in the predicted price of corn per bushel?

Solutions to Self-Check Exercises 3.7 can be found on page 273.

3.7 Exercises

In Exercises 1–14, find the differential of the given function.

1. $f(x) = 2x^2$
2. $f(x) = 3x^2 + 1$
3. $f(x) = x^3 - x$
4. $f(x) = 2x^3 + x$
5. $f(x) = \sqrt{x + 1}$
6. $f(x) = \frac{3}{\sqrt{x}}$
7. $f(x) = 2x^{3/2} + x^{1/2}$
8. $f(x) = 3x^{5/6} + 7x^{2/3}$
9. $f(x) = x + \frac{2}{x}$
10. $f(x) = \frac{3}{x - 1}$
11. $f(x) = \frac{x - 1}{x^2 + 1}$
12. $f(x) = \frac{2x^2 + 1}{x + 1}$
13. $f(x) = \sqrt{3x^2 - x}$
14. $f(x) = (2x^2 + 3)^{1/3}$

15. Let f be a function defined by

$$y = f(x) = x^2 - 1$$

a. Find the differential of f.
b. Use your result from part (a) to find the approximate change in y if x changes from 1 to 1.02.
c. Find the actual change in y if x changes from 1 to 1.02 and compare your result with that obtained in part (b).

16. Let f be a function defined by

$$y = f(x) = 3x^2 - 2x + 6$$

a. Find the differential of f.
b. Use your result from part (a) to find the approximate change in y if x changes from 2 to 1.97.
c. Find the actual change in y if x changes from 2 to 1.97 and compare your result with that obtained in part (b).

17. Let f be a function defined by

$$y = f(x) = \frac{1}{x}$$

a. Find the differential of f.
b. Use your result from part (a) to find the approximate change in y if x changes from -1 to -0.95.
c. Find the actual change in y if x changes from -1 to -0.95 and compare your result with that obtained in part (b).

18. Let f be a function defined by

$$y = f(x) = \sqrt{2x + 1}$$

a. Find the differential of f.
b. Use your result from part (a) to find the approximate change in y if x changes from 4 to 4.1.
c. Find the actual change in y if x changes from 4 to 4.1 and compare your result with that obtained in part (b).

In Exercises 19–26, use differentials to approximate the given quantity.

19. $\sqrt{10}$ **20.** $\sqrt{17}$ **21.** $\sqrt{49.5}$

22. $\sqrt{99.7}$ **23.** $\sqrt[3]{7.8}$ **24.** $\sqrt[4]{81.6}$

25. $\sqrt{0.089}$ **26.** $\sqrt[3]{0.00096}$

27. Use a differential to approximate $\sqrt{4.02} + \dfrac{1}{\sqrt{4.02}}$.

Hint: Let $f(x) = \sqrt{x} + \dfrac{1}{\sqrt{x}}$ and compute dy with $x = 4$ and $dx = 0.02$.

28. Use a differential to approximate $\dfrac{2(4.98)}{(4.98)^2 + 1}$.

Hint: Study the hint for Exercise 27.

▭ **A calculator is recommended for the remainder of this exercise set.**

29. ERROR ESTIMATION The length of each edge of a cube is 12 cm, with a possible error in measurement of 0.02 cm. Use differentials to estimate the error that might occur when the volume of the cube is calculated.

30. ESTIMATING THE AMOUNT OF PAINT REQUIRED A coat of paint of thickness 0.05 cm is to be applied uniformly to the faces of a cube of edge 30 cm. Use differentials to find the approximate amount of paint required for the job.

31. ERROR ESTIMATION A hemisphere-shaped dome of radius 60 ft is to be coated with a layer of rust-proofer before painting. Use differentials to estimate the amount of rust-proofer needed if the coat is to be 0.01 in. thick.

Hint: The volume of a hemisphere of radius r is $V = \frac{2}{3}\pi r^3$.

32. GROWTH OF A CANCEROUS TUMOR The volume of a spherical cancer tumor is given by

$$V(r) = \frac{4}{3}\pi r^3$$

If the radius of a tumor is estimated at 1.1 cm, with a maximum error in measurement of 0.005 cm, determine the error that might occur when the volume of the tumor is calculated.

33. UNCLOGGING ARTERIES Research done in the 1930s by the French physiologist Jean Poiseuille showed that the resistance R of a blood vessel of length l and radius r is $R = kl/r^4$, where k is a constant. Suppose a dose of the drug TPA increases r by 10%. How will this affect the resistance R? Assume that l is constant.

34. GROSS DOMESTIC PRODUCT An economist has determined that a certain country's gross domestic product (GDP) is approximated by the function $f(x) = 640x^{1/5}$, where $f(x)$ is measured in billions of dollars and x is the capital outlay in billions of dollars. Use differentials to estimate the change in the country's GDP if the country's capital expenditure changes from \$243 billion to \$248 billion.

35. LEARNING CURVES The length of time (in seconds) a certain individual takes to learn a list of n items is approximated by

$$f(n) = 4n\sqrt{n - 4}$$

Use differentials to approximate the additional time it takes the individual to learn the items on a list when n is increased from 85 to 90 items.

(continued on p. 272)

FINDING THE DIFFERENTIAL OF A FUNCTION

The calculation of the differential of f at a given value of x involves the evaluation of the derivative of f at that point and can be facilitated through the use of the numerical derivative function.

EXAMPLE 1

Use dy to approximate Δy if $y = x^2(2x^2 + x + 1)^{2/3}$ and x changes from 2 to 1.98.

SOLUTION ✔

Let $f(x) = x^2(2x^2 + x + 1)^{2/3}$. Since $dx = 1.98 - 2 = -0.02$, we find the required approximation to be

$$dy = f'(2) \cdot (-0.02)$$

But using the numerical derivative operation, we find

$$f'(2) = 30.5758132855$$

and so

$$dy = (-0.02)(30.5758132855) = -0.611516266 \quad \blacksquare\blacksquare\blacksquare\blacksquare$$

EXAMPLE 2

The Meyers are considering the purchase of a house in the near future and estimate that they will need a loan of \$120,000. Based on a 30-year conventional mortgage with an interest rate of r per year, their monthly repayment will be

$$P = \frac{10{,}000r}{1 - \left(1 + \dfrac{r}{12}\right)^{-360}}$$

dollars. If the interest rate increases from the present rate of 10%/year to 10.2% per year between now and the time the Meyers decide to secure the loan, approximately how much more per month will their mortgage payment be?

SOLUTION ✔

Let's write

$$P = f(r) = \frac{10{,}000r}{1 - \left(1 + \dfrac{r}{12}\right)^{-360}}$$

Then the increase in the mortgage payment will be approximately

$$dP = f'(0.1)dr = f'(0.1)(0.002) \qquad \text{(Since } dr = 0.102 - 0.1\text{)}$$
$$= (8867.59947979)(0.002) \approx 17.7352 \qquad \text{(Using the numerical derivative operation)}$$

or approximately \$17.74 per month. $\quad \blacksquare\blacksquare\blacksquare\blacksquare$

In Exercises 1–6, use dy to approximate Δy for the function $y = f(x)$ when x changes from $x = a$ to $x = b$.

1. $f(x) = 0.21x^7 - 3.22x^4 + 5.43x^2 + 1.42x + 12.42$; $a = 3$, $b = 3.01$

2. $f(x) = \dfrac{0.2x^2 + 3.1}{1.2x + 1.3}$; $a = 2$, $b = 1.96$

3. $f(x) = \sqrt{2.2x^2 + 1.3x + 4}$; $a = 1$, $b = 1.03$

4. $f(x) = x\sqrt{2x^3 - x + 4}$; $a = 2$, $b = 1.98$

5. $f(x) = \dfrac{\sqrt{x^2 + 4}}{x - 1}$; $a = 4$, $b = 4.1$

6. $f(x) = 2.1x^2 + \dfrac{3}{\sqrt{x}} + 5$; $a = 3$, $b = 2.95$

7. CALCULATING MORTGAGE PAYMENTS Refer to Example 2. How much more per month will the Meyers' mortgage payment be if the interest rate increases from 10% to 10.3%/year? To 10.4%/year? To 10.5%/year?

8. ESTIMATING THE AREA OF A RING OF NEPTUNE The ring 1989N2R of the planet Neptune has an inner radius of approximately 53,200 km (measured from the center of the planet) and a radial width of 15 km. Use differentials to estimate the area of the ring.

9. EFFECT OF PRICE INCREASE ON QUANTITY DEMANDED The quantity demanded per week of the Alpha Sports Watch, x (in thousands), is related to its unit price of p dollars by the equation

$$x = f(p) = 10\sqrt{\frac{50 - p}{p}} \qquad (0 \le p \le 50)$$

Use differentials to find the decrease in the quantity of the watches demanded per week if the unit price is increased from \$40 to \$42.

36. **EFFECT OF ADVERTISING ON PROFITS** The relationship between Cunningham Realty's quarterly profits, $P(x)$, and the amount of money x spent on advertising per quarter is described by the function

$$P(x) = -\frac{1}{8}x^2 + 7x + 30 \qquad (0 \le x \le 50)$$

where both $P(x)$ and x are measured in thousands of dollars. Use differentials to estimate the increase in profits when advertising expenditure each quarter is increased from $24,000 to $26,000.

37. **EFFECT OF MORTGAGE RATES ON HOUSING STARTS** A study prepared for the National Association of Realtors estimates that the number of housing starts per year over the next 5 yr will be

$$N(r) = \frac{7}{1 + 0.02r^2}$$

million units, where r (percent) is the mortgage rate. Use differentials to estimate the decrease in the number of housing starts when the mortgage rate is increased from 12% to 12.5%.

38. **SUPPLY-PRICE** The supply equation for a certain brand of transistor radio is given by

$$p = s(x) = 0.3 \sqrt{x} + 10$$

where x is the quantity supplied and p is the unit price in dollars. Use differentials to approximate the change in price when the quantity supplied is increased from 10,000 units to 10,500 units.

39. **DEMAND-PRICE** The demand function for the Sentinel smoke alarm is given by

$$p = d(x) = \frac{30}{0.02x^2 + 1}$$

where x is the quantity demanded (in units of a thousand) and p is the unit price in dollars. Use differentials to estimate the change in the price p when the quantity demanded changes from 5000 to 5500 units/wk.

40. **SURFACE AREA OF AN ANIMAL** Animal physiologists use the formula

$$S = kW^{2/3}$$

to calculate an animal's surface area (in square meters) from its weight W (in kilograms), where k is a constant that depends on the animal under consideration. Sup-

pose a physiologist calculates the surface area of a horse ($k = 0.1$). If the horse's weight is estimated at 300 kg, with a maximum error in measurement of 0.6 kg, determine the percentage error in the calculation of the horse's surface area.

41. **FORECASTING PROFITS** The management of Trappee and Sons, Inc., forecast that they will sell 200,000 cases of their Texa-Pep hot sauce next year. Their annual profit is described by

$$P(x) = -0.000032x^3 + 6x - 100$$

thousand dollars, where x is measured in thousands of cases. If the maximum error in the forecast is 15%, determine the corresponding error in Trappee's profits.

42. **FORECASTING COMMODITY PRICES** A certain country's government economists have determined that the demand equation for soybeans in that country is given by

$$p = f(x) = \frac{55}{2x^2 + 1}$$

where p is expressed in dollars per bushel and x, the quantity demanded per year, is measured in billions of bushels. The economists are forecasting a harvest of 1.8 billion bushels for the year, with a maximum error of 15% in their forecast. Determine the corresponding maximum error in the predicted price per bushel of soybeans.

43. **CRIME STUDIES** A sociologist has found that the number of serious crimes in a certain city per year is described by the function

$$N(x) = \frac{500(400 + 20x)^{1/2}}{(5 + 0.2x)^2}$$

where x (in cents per dollar deposited) is the level of reinvestment in the area in conventional mortgages by the city's ten largest banks. Use differentials to estimate the change in the number of crimes if the level of reinvestment changes from 20 cents/dollar deposited to 22 cents/dollar deposited.

44. **FINANCING A HOME** The Millers are planning to buy a home in the near future and estimate that they will need a 30-yr fixed-rate mortgage for $120,000. Their monthly payment P (in dollars) can be computed using the formula

$$P = \frac{10{,}000r}{1 - \left(1 + \dfrac{r}{12}\right)^{-360}}$$

where r is the interest rate per year.

a. Find the differential of P.

b. If the interest rate increases from the present rate of 9%/year to 9.2%/year between now and the time the Millers decide to secure the loan, approximately how much more will their monthly mortgage payment be? How much more will it be if the interest rate increases to 9.3%/year? To 9.4%/year? To 9.5%/year?

In Exercises 45 and 46, determine whether the statement is true or false. If it is true, explain why it is true. If it is false, give an example to show why it is false.

45. If $y = ax + b$ where a and b are constants, then $\Delta y = dy$.

46. If $A = f(x)$, then the percentage change in A is

$$\frac{100 f'(x)}{f(x)} dx$$

SOLUTIONS TO SELF-CHECK EXERCISES 3.7

1. We find

$$f'(x) = \frac{1}{2} x^{-1/2} = \frac{1}{2\sqrt{x}}$$

Therefore, the required differential of f is

$$dy = \frac{1}{2\sqrt{x}} dx$$

2. We first compute the differential

$$dp = -\frac{250x}{(x^2 + 1)^2} dx$$

Next, using Equation (11) with $x = 6$ and $dx = 0.2$, we find

$$\Delta p \approx dp = -\frac{250(6)}{(36 + 1)^2}(0.2) = -0.22$$

or a drop in price of 22 cents/bushel.

CHAPTER 3 Summary of Principal Formulas and Terms

Formulas

1. Derivative of a constant	$\dfrac{d}{dx}(c) = 0$, c a constant
2. Power rule	$\dfrac{d}{dx}(x^n) = nx^{n-1}$
3. Constant multiple rule	$\dfrac{d}{dx}(cu) = c\dfrac{du}{dx}$, c a constant
4. Sum rule	$\dfrac{d}{dx}(u \pm v) = \dfrac{du}{dx} \pm \dfrac{dv}{dx}$
5. Product rule	$\dfrac{d}{dx}(uv) = u\dfrac{dv}{dx} + v\dfrac{du}{dx}$

6. Quotient rule $\dfrac{d}{dx}\left(\dfrac{u}{v}\right)=\dfrac{v\dfrac{du}{dx}-u\dfrac{dv}{dx}}{v^2}$

7. Chain rule $\dfrac{dy}{dx}=\dfrac{dy}{du}\cdot\dfrac{du}{dx}$

8. General power rule $\dfrac{d}{dx}(u^n)=nu^{n-1}\dfrac{du}{dx}$

9. Average cost function $\overline{C}(x)=\dfrac{C(x)}{x}$

10. Revenue function $R(x)=px$

11. Profit function $P(x)=R(x)-C(x)$

12. Elasticity of demand $E(p)=-\dfrac{pf'(p)}{f(p)}$

13. Differential of y $dy=f'(x)dx$

Terms

marginal cost function

marginal average cost function

marginal revenue function

marginal profit function

elastic demand

unitary demand

inelastic demand

second derivative of f

implicit differentiation

related rates

CHAPTER 3 REVIEW EXERCISES

In Exercises 1–30, find the derivative of the given function.

1. $f(x)=3x^5-2x^4+3x^2-2x+1$

2. $f(x)=4x^6+2x^4+3x^2-2$

3. $g(x)=-2x^{-3}+3x^{-1}+2$

4. $f(t)=2t^2-3t^3-t^{-1/2}$

5. $g(t)=2t^{-1/2}+4t^{-3/2}+2$

6. $h(x)=x^2+\dfrac{2}{x}$ **7.** $f(t)=t+\dfrac{2}{t}+\dfrac{3}{t^2}$

8. $g(s)=2s^2-\dfrac{4}{s}+\dfrac{2}{\sqrt{s}}$ **9.** $h(x)=x^2-\dfrac{2}{x^{3/2}}$

10. $f(x)=\dfrac{x+1}{2x-1}$ **11.** $g(t)=\dfrac{t^2}{2t^2+1}$

12. $h(t)=\dfrac{\sqrt{t}}{\sqrt{t}+1}$

13. $f(x)=\dfrac{\sqrt{x}-1}{\sqrt{x}+1}$

14. $f(t)=\dfrac{t}{2t^2+1}$

15. $f(x)=\dfrac{x^2(x^2+1)}{x^2-1}$

16. $f(x)=(2x^2+x)^3$

17. $f(x)=(3x^3-2)^8$

18. $h(x)=(\sqrt{x}+2)^5$

19. $f(t)=\sqrt{2t^2+1}$

20. $g(t)=\sqrt[3]{1-2t^3}$

21. $s(t)=(3t^2-2t+5)^{-2}$

22. $f(x)=(2x^3-3x^2+1)^{-3/2}$

23. $h(x)=\left(x+\dfrac{1}{x}\right)^2$

24. $h(x)=\dfrac{1+x}{(2x^2+1)^2}$

25. $h(t)=(t^2+t)^4(2t^2)$

26. $f(x)=(2x+1)^3(x^2+x)^2$

27. $g(x)=\sqrt{x}(x^2-1)^3$

28. $f(x)=\dfrac{x}{\sqrt{x^3+2}}$

29. $h(x)=\dfrac{\sqrt{3x+2}}{4x-3}$

30. $f(t)=\dfrac{\sqrt{2t+1}}{(t+1)^3}$

In Exercises 31–36, find the second derivative of the given function.

31. $f(x) = 2x^4 - 3x^3 + 2x^2 + x + 4$

32. $g(x) = \sqrt{x} + \dfrac{1}{\sqrt{x}}$

33. $h(t) = \dfrac{t}{t^2 + 4}$

34. $f(x) = (x^3 + x + 1)^2$

35. $f(x) = \sqrt{2x^2 + 1}$

36. $f(t) = t(t^2 + 1)^3$

In Exercises 37–42, find dy/dx by implicit differentiation.

37. $6x^2 - 3y^2 = 9$

38. $2x^3 - 3xy = 4$

39. $y^3 + 3x^2 = 3y$

40. $x^2 + 2x^2y^2 + y^2 = 10$

41. $x^2 - 4xy - y^2 = 12$

42. $3x^2y - 4xy + x - 2y = 6$

43. Let $f(x) = 2x^3 - 3x^2 - 16x + 3$.
   a. Find the points on the graph of f at which the slope of the tangent line is equal to -4.
   b. Find the equation(s) of the tangent line(s) of part (a).

44. Let $f(x) = \frac{1}{3}x^3 + \frac{1}{2}x^2 - 4x + 1$.
   a. Find the points on the graph of f at which the slope of the tangent line is equal to -2.
   b. Find the equation(s) of the tangent line(s) of part (a).

45. Find an equation of the tangent line to the graph of $y = \sqrt{4 - x^2}$ at the point $(1, \sqrt{3})$.

46. Find an equation of the tangent line to the graph of $y = x(x + 1)^5$ at the point $(1, 32)$.

47. Find the third derivative of the function

$$f(x) = \frac{1}{2x - 1}$$

What is its domain?

48. WORLDWIDE NETWORKED PCS The number of worldwide networked PCs (in millions) is given by

$$N(t) = 3.136x^2 + 3.954x + 116.468 \qquad (0 \le t \le 9)$$

where t is measured in years, with $t = 0$ corresponding to the beginning of 1991.
   a. How many worldwide networked PCs were there at the beginning of 1997?
   b. How fast was the number of worldwide networked PCs changing at the beginning of 1997?

49. The number of subscribers to CNC Cable Television in the town of Randolph is approximated by the function

$$N(x) = 1000(1 + 2x)^{1/2} \qquad (1 \le x \le 30)$$

where $N(x)$ denotes the number of subscribers to the service in the xth week. Find the rate of increase in the number of subscribers at the end of the 12th week.

50. The total weekly cost in dollars incurred by the Herald Record Company in pressing x video discs is given by the total cost function

$$C(x) = 2500 + 2.2x \qquad (0 \le x \le 8000)$$

   a. What is the marginal cost when $x = 1000$? When $x = 2000$?
   b. Find the average cost function $\overline{C}$ and the marginal average cost function $\overline{C}'$.
   c. Using the results from part (b), show that the average cost incurred by Herald in pressing a video disc approaches $2.20/disc when the level of production is high enough.

51. The marketing department of Telecon Corporation has determined that the demand for their cordless phones obeys the relationship

$$p = -0.02x + 600 \qquad (0 \le x \le 30,000)$$

where p denotes the phone's unit price (in dollars) and x denotes the quantity demanded.
   a. Find the revenue function R.
   b. Find the marginal revenue function R'.
   c. Compute $R'(10,000)$ and interpret your result.

52. The weekly demand for the Lectro-Copy photocopying machine is given by the demand equation

$$p = 2000 - 0.04x \qquad (0 \le x \le 50,000)$$

where p denotes the wholesale unit price in dollars and x denotes the quantity demanded. The weekly total cost function for manufacturing these copiers is given by

$$C(x) = 0.000002x^3 - 0.02x^2 + 1000x + 120,000$$

where $C(x)$ denotes the total cost incurred in producing x units.
   a. Find the revenue function R, the profit function P, and the average cost function $\overline{C}$.
   b. Find the marginal cost function C', the marginal revenue function R', the marginal profit function P', and the marginal average cost function $\overline{C}'$.
   c. Compute $C'(3000)$, $R'(3000)$, and $P'(3000)$.
   d. Compute $\overline{C}'(5000)$ and $\overline{C}'(8000)$ and interpret your results.

APPLICATIONS
OF THE DERIVATIVE

4

This chapter further explores the power of the derivative, which we use to help analyze the properties of functions. The information obtained can then be used to accurately sketch graphs of functions. We also see how the derivative is used in solving a large class of optimization problems, including finding what level of production will yield a maximum profit for a company, finding what level of production will result in minimal cost to a company, finding the maximum height attained by a rocket, finding the maximum velocity at which air is expelled when a person coughs, and a host of other problems.

What is the maximum altitude and the maximum velocity attained by the rocket? In Example 7, page 348, you will see how the techniques of calculus can be used to help answer these questions.

4.1 Applications of the First Derivative

DETERMINING THE INTERVALS WHERE A FUNCTION IS INCREASING OR DECREASING

According to a study by the U.S. Department of Energy and the Shell Development Company, a typical car's fuel economy as a function of its speed is described by the graph shown in Figure 4.1. Observe that the fuel economy $f(x)$ in miles per gallon (mpg) improves as x, the vehicle's speed in miles per hour (mph), increases from 0 to 42, and then drops as the speed increases beyond 42 mph. We use the terms *increasing* and *decreasing* to describe the behavior of a function as we move from left to right along its graph.

FIGURE 4.1
A typical car's fuel economy improves as the speed at which it is driven increases from 0 mph to 42 mph and drops at speeds greater than 42 mph.

Source: U.S. Department of Energy and Shell Development Co.

More precisely, we have the following definitions.

Increasing and Decreasing Functions

A function f is **increasing** on an interval (a, b) if for any two numbers x_1 and x_2 in (a, b), $f(x_1) < f(x_2)$ whenever $x_1 < x_2$ (Figure 4.2a).
A function f is **decreasing** on an interval (a, b) if for any two numbers x_1 and x_2 in (a, b), $f(x_1) > f(x_2)$ whenever $x_1 < x_2$ (Figure 4.2b).

FIGURE 4.2

(a) f is increasing on (a, b). **(b)** f is decreasing on (a, b).

We say that f is *increasing at a point* c if there exists an interval (a, b) containing c such that f is increasing on (a, b). Similarly, we say that f is *decreasing at a point* c if there exists an interval (a, b) containing c such that f is decreasing on (a, b).

Since the rate of change of a function at a point $x = c$ is given by the derivative of the function at that point, the derivative lends itself naturally to being a tool for determining the intervals where a differentiable function is increasing or decreasing. Indeed, as we saw in Chapter 2, the derivative of a function at a point measures both the slope of the tangent line to the graph of the function at that point and the rate of change of the function at the same point. In fact, at a point where the derivative is positive, the slope of the tangent line to the graph is positive, and the function is increasing. At a point where the derivative is negative, the slope of the tangent line to the graph is negative, and the function is decreasing (Figure 4.3).

FIGURE 4.3

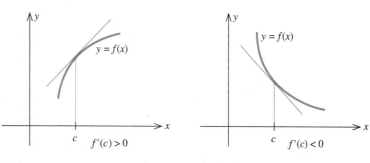

(a) f is increasing at $x = c$.   **(b)** f is decreasing at $x = c$.

These observations lead to the following important theorem, which we state without proof.

THEOREM 1

a. If $f'(x) > 0$ for each value of x in an interval (a, b), then f is increasing on (a, b).

b. If $f'(x) < 0$ for each value of x in an interval (a, b), then f is decreasing on (a, b).

c. If $f'(x) = 0$ for each value of x in an interval (a, b), then f is constant on (a, b).

EXAMPLE 1   Find the interval where the function $f(x) = x^2$ is increasing and the interval where it is decreasing.

SOLUTION ✔   The derivative of $f(x) = x^2$ is $f'(x) = 2x$. Since

$$f'(x) = 2x > 0 \quad \text{if } x > 0$$
$$f'(x) = 2x < 0 \quad \text{if } x < 0$$

f is increasing on the interval $(0, \infty)$ and decreasing on the interval $(-\infty, 0)$ (Figure 4.4).

FIGURE 4.4
The graph of f falls on $(-\infty, 0)$ where $f'(x) < 0$ and rises on $(0, \infty)$ where $f'(x) > 0$.

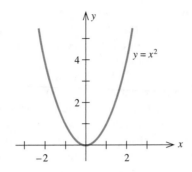

Recall that the graph of a continuous function cannot have any breaks. As a consequence, a continuous function cannot change sign unless it equals zero for some value of x. (See Theorem 5, page 143.) This observation suggests the following procedure for determining the sign of the derivative f' of a function f, and hence the intervals where the function f is increasing and where it is decreasing.

Determining the Intervals Where a Function Is Increasing or Decreasing

1. Find all values of x for which $f'(x) = 0$ or f' is discontinuous and identify the open intervals determined by these points.
2. Select a test point c in each interval found in step 1 and determine the sign of $f'(c)$ in that interval.
   a. If $f'(c) > 0$, f is increasing on that interval.
   b. If $f'(c) < 0$, f is decreasing on that interval.

EXAMPLE 2 Determine the intervals where the function $f(x) = x^3 - 3x^2 - 24x + 32$ is increasing and where it is decreasing.

SOLUTION ✔ 1. The derivative of f is

$$f'(x) = 3x^2 - 6x - 24 = 3(x + 2)(x - 4)$$

and it is continuous everywhere. The zeros of $f'(x)$ are $x = -2$ and $x = 4$, and these points divide the real line into the intervals $(-\infty, -2)$, $(-2, 4)$, and $(4, \infty)$.

2. To determine the sign of $f'(x)$ in the intervals $(-\infty, -2)$, $(-2, 4)$, and $(4, \infty)$, compute $f'(x)$ at a convenient test point in each interval. The results are shown in the following table:

Interval	Test Point c	$f'(c)$	Sign of $f'(x)$
$(-\infty, -2)$	-3	21	$+$
$(-2, 4)$	0	-24	$-$
$(4, \infty)$	5	21	$+$

FIGURE 4.5
Sign diagram for f'

Using these results, we obtain the sign diagram shown in Figure 4.5. We conclude that f is increasing on the intervals $(-\infty, -2)$ and $(4, \infty)$ and is decreasing on the interval $(-2, 4)$. Figure 4.6 shows the graph of f.

FIGURE 4.6
The graph of f rises on $(-\infty, -2)$, falls on $(-2, 4)$, and rises again on $(4, \infty)$.

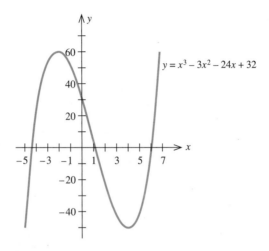

■ ■ ■ ■

Exploring with Technology

Refer to Example 2.

1. Use a graphing utility to plot the graphs of

$$f(x) = x^3 - 3x^2 - 24x + 32$$

and its derivative function

$$f'(x) = 3x^2 - 6x - 24$$

using the viewing rectangle $[-10, 10] \times [-50, 70]$.
2. By looking at the graph of f', determine the intervals where $f'(x) > 0$ and the intervals where $f'(x) < 0$. Next, look at the graph of f and determine the intervals where it is increasing and the intervals where it is decreasing. Describe the relationship. Is it what you expected?

REMARK Do not be concerned with how the graphs in this section are obtained. We will learn how to sketch these graphs later. However, if you are familiar with the use of a graphing utility, you may go ahead and verify each graph.

■ ■ ■

EXAMPLE 3 Find the interval where the function $f(x) = x^{2/3}$ is increasing and the interval where it is decreasing.

SOLUTION ✔ **1.** The derivative of f is

$$f'(x) = \frac{2}{3}x^{-1/3} = \frac{2}{3x^{1/3}}$$

The function f' is not defined at $x = 0$, so f' is discontinuous there. It is continuous everywhere else. Furthermore, f' is not equal to zero anywhere. The point $x = 0$ divides the real line (the domain of f) into the intervals $(-\infty, 0)$ and $(0, \infty)$.

2. Pick a test point (say, $x = -1$) in the interval $(-\infty, 0)$ and compute

$$f'(-1) = -\frac{2}{3}$$

Since $f'(-1) < 0$, we see that $f'(x) < 0$ on $(-\infty, 0)$. Next, we pick a test point (say, $x = 1$) in the interval $(0, \infty)$ and compute

$$f'(1) = \frac{2}{3}$$

Since $f'(1) > 0$, we see that $f'(x) > 0$ on $(0, \infty)$. Figure 4.7 shows these results in the form of a sign diagram.

> **Group Discussion**
>
> True or false? If f is continuous at c and f is increasing at c, then $f'(c) \neq 0$. Explain your answer.
> Hint: Consider $f(x) = x^3$ and $c = 0$.

FIGURE 4.7
Sign diagram for f'

f' not defined at $x = 0$

$- - - - - - - -\downarrow+ + + + + + +$

$\begin{array}{ccc} & | & \\ -1 & 0 & 1 \end{array}$ → x

We conclude that f is decreasing on the interval $(-\infty, 0)$ and increasing on the interval $(0, \infty)$. The graph of f, shown in Figure 4.8, confirms these results.

FIGURE 4.8
f decreases on $(-\infty, 0)$ and increases on $(0, \infty)$.

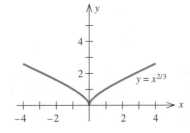

$y = x^{2/3}$

■ ■ ■ ■

EXAMPLE **4**   Find the intervals where the function $f(x) = x + 1/x$ is increasing and where it is decreasing.

SOLUTION ✔   **1.** The derivative of f is

$$f'(x) = 1 - \frac{1}{x^2} = \frac{x^2 - 1}{x^2}$$

Since f' is not defined at $x = 0$, it is discontinuous there. Furthermore, $f'(x)$ is equal to zero when $x^2 - 1 = 0$ or $x = \pm 1$. These values of x partition the domain of f' into the open intervals $(-\infty, -1)$, $(-1, 0)$, $(0, 1)$, and $(1, \infty)$, where the sign of f' is different from zero.

2. To determine the sign of f' in each of these intervals, we compute $f'(x)$ at the test points $x = -2, -\frac{1}{2}, \frac{1}{2}$, and 2, respectively, obtaining $f'(-2) = \frac{3}{4}$, $f'(-\frac{1}{2}) = -3$, $f'(\frac{1}{2}) = -3$, and $f'(2) = \frac{3}{4}$. From the sign diagram for f' (Figure 4.9), we conclude that f is increasing on $(-\infty, -1)$ and $(1, \infty)$ and decreasing on $(-1, 0)$ and $(0, 1)$.

FIGURE 4.9
f' does not change sign as we move across $x = 0$.

The graph of f appears in Figure 4.10. Note that f' does not change sign as we move across the point of discontinuity, $x = 0$. (Compare this with Example 3.)

FIGURE 4.10
The graph of f rises on $(-\infty, -1)$, falls on $(-1, 0)$ and $(0, 1)$, and rises again on $(1, \infty)$.

   Example 4 reminds us that we must not automatically conclude that the derivative f' must change sign when we move across a point of discontinuity or a zero of f'.

Group Discussion

Consider the profit function P associated with a certain commodity defined by $\quad\quad P(x) = R(x) - C(x) \quad\quad (x \geq 0)$

where R is the revenue function, C is the total cost function, and x is the number of units of the product produced and sold.
1. Find an expression for $P'(x)$.
2. Find relationships in terms of the derivatives of R and C so that
   a. P is increasing at $x = a$.
   b. P is decreasing at $x = a$.
   c. P is neither increasing nor decreasing at $x = a$.
   Hint: Recall that the derivative of a function at $x = a$ measures the rate of change of the function at that point.
3. Explain the results of part 2 in economic terms.

Exploring with Technology

1. Use a graphing utility to sketch the graphs of $f(x) = x^3 - ax$ for $a = -2, -1, 0, 1, 2$, using the viewing rectangle $[-2, 2] \times [-2, 2]$.
2. Use the results of part 1 to guess at the values of a so that f is increasing on $(-\infty, \infty)$.
3. Prove your conjecture analytically.

RELATIVE EXTREMA

Besides helping us determine where the graph of a function is increasing and decreasing, the first derivative may be used to help us locate certain "high points" and "low points" on the graph of f. Knowing these points is invaluable in sketching the graphs of functions and solving optimization problems. These "high points" and "low points" correspond to the *relative (local) maxima* and *relative minima* of a function. They are so called because they are the highest or the lowest points when compared with points nearby.

The graph shown in Figure 4.11 gives the U.S. budget deficit from 1980

FIGURE 4.11
U.S. budget deficit from 1980 to 1991

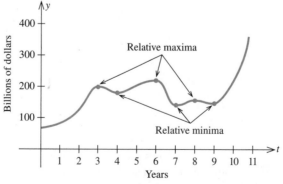

Source: Office of Management and Budget

($t = 0$) through 1991. The relative maxima and the relative minima of the function f are indicated on the graph.

More generally, we have the following definition:

Relative Maximum

A function f has a relative maximum at $x = c$ if there exists an open interval (a, b) containing c such that $f(x) \leq f(c)$ for all x in (a, b).

Geometrically, this means that there is *some* interval containing $x = c$ such that no point on the graph of f with its x-coordinate in that interval can lie above the point $(c, f(c))$; that is, $f(c)$ is the largest value of $f(x)$ in some interval around $x = c$. Figure 4.12 depicts the graph of a function f that has a relative maximum at $x = x_1$ and another at $x = x_3$.

Observe that all the points on the graph of f with x-coordinates in the interval I_1 containing x_1 (shown in blue) lie on or below the point $(x_1, f(x_1))$.

FIGURE 4.12
f has a relative maximum at $x = x_1$ and at $x = x_3$.

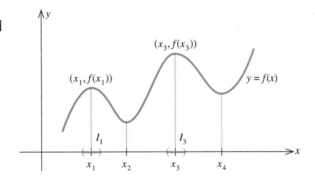

This is also true for the point $(x_3, f(x_3))$ and the interval I_3. Thus, even though there are points on the graph of f that are "higher" than the points $(x_1, f(x_1))$ and $(x_3, f(x_3))$, the latter points are "highest" relative to points in their respective neighborhoods (intervals). Points on the graph of a function f that are "highest" and "lowest" with respect to *all* points in the domain of f will be studied in Section 4.4.

The definition of the relative minimum of a function parallels that of the relative maximum of a function.

Relative Minimum

A function f has a relative minimum at $x = c$ if there exists an open interval (a, b) containing c such that $f(x) \geq f(c)$ for all x in (a, b).

The graph of the function f, depicted in Figure 4.12, has a relative minimum at $x = x_2$ and another at $x = x_4$.

FINDING THE RELATIVE EXTREMA

We refer to the relative maximum and relative minimum of a function as the **relative extrema** of that function. As a first step in our quest to find the relative extrema of a function, we consider functions that have derivatives at such points. Suppose that f is a function that is differentiable in some interval (a, b) that contains a point $x = c$ and that f has a relative maximum at $x = c$ (Figure 4.13a).

FIGURE 4.13

(a) f has a relative maximum at $x = c$.

(b) f has a relative minimum at $x = c$.

Observe that the slope of the tangent line to the graph of f must change from positive to negative as we move across the point $x = c$ from left to right. Therefore, the tangent line to the graph of f at the point $(c, f(c))$ must be horizontal; that is, $f'(c) = 0$ (Figure 4.13a).

Using a similar argument, it may be shown that the derivative f' of a differentiable function f must also be equal to zero at a point $x = c$, where f has a relative minimum (Figure 4.13b).

This analysis reveals an important characteristic of the relative extrema of a differentiable function f: *At any point c where f has a relative extremum,* $f'(c) = 0$.

FIGURE 4.14

$f'(0) = 0$, but f does not have a relative extremum at $(0, 0)$.

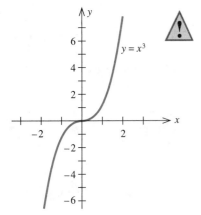

Before we develop a procedure for finding such points, a few words of caution are in order. First, this result tells us that if a differentiable function f has a relative extremum at a point $x = c$, then $f'(c) = 0$. The converse of this statement—if $f'(c) = 0$ at some point $x = c$, then f must have a relative extremum at that point—is not true. Consider, for example, the function $f(x) = x^3$. Here, $f'(x) = 3x^2$, so $f'(0) = 0$. Yet, f has neither a relative maximum nor a relative minimum at $x = 0$ (Figure 4.14).

Second, our result assumes that the function is differentiable and thus has a derivative at a point that gives rise to a relative extremum. The functions $f(x) = |x|$ and $g(x) = x^{2/3}$ demonstrate that a relative extremum of a function may exist at a point at which the derivative does not exist. Both these functions fail to be differentiable at $x = 0$, but each has a relative minimum there. Figure 4.15 shows the graphs of these functions. Note that the slopes of the

FIGURE 4.15
Each of these functions has a relative extremum at (0, 0), but the derivative does not exist there.

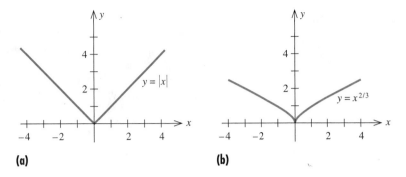

(a)   **(b)**

tangent lines change from negative to positive as we move across $x = 0$, just as in the case of a function that is differentiable at a value of x that gives rise to a relative minimum.

We refer to a point in the domain of f that *may* give rise to a relative extremum as a critical point.

Critical Point of f

> A **critical point** of a function f is any point x in the domain of f such that $f'(x) = 0$ or $f'(x)$ does not exist.

Figure 4.16 depicts the graph of a function that has critical points at $x = a$, b, c, d, and e. Observe that $f'(x) = 0$ at $x = a$, b, and c. Next, since there is a corner at $x = d$, $f'(x)$ does not exist there. Finally, $f'(x)$ does not exist at $x = e$ because the tangent line there is vertical. Also, observe that the critical points $x = a$, b, and d give rise to relative extrema of f, whereas the critical points $x = c$ and $x = e$ do not.

FIGURE 4.16
Critical points of f

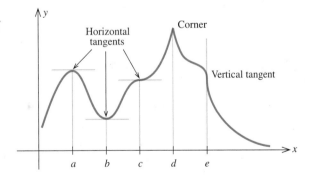

Having defined what a critical point is, we can now state a formal procedure for finding the relative extrema of a continuous function that is differentiable everywhere except at isolated values of x. Incorporated into the procedure is the so-called **first derivative test,** which helps us determine whether a point gives rise to a relative maximum or a relative minimum of the function f.

Procedure for Finding Relative Extrema (the First Derivative Test)

1. Determine the critical points of f.

2. Determine the sign of $f'(x)$ to the left and right of each critical point.

   a. If $f'(x)$ changes sign from *positive* to *negative* as we move across a critical point $x = c$, then $f(c)$ is a relative maximum.

   b. If $f'(x)$ changes sign from *negative* to *positive* as we move across a critical point $x = c$, then $f(c)$ is a relative minimum.

   c. If $f'(x)$ does not change sign as we move across a critical point $x = c$, then $f(c)$ is not a relative extremum.

EXAMPLE 5   Find the relative maxima and relative minima of the function $f(x) = x^2$.

SOLUTION ✔   The derivative of $f(x) = x^2$ is given by $f'(x) = 2x$. Setting $f'(x) = 0$ yields $x = 0$ as the only critical point of f. Since

$$f'(x) < 0 \quad \text{if } x < 0$$
$$f'(x) > 0 \quad \text{if } x > 0$$

we see that $f'(x)$ changes sign from negative to positive as we move across the critical point $x = 0$. Thus, we conclude that $f(0) = 0$ is a relative minimum of f (Figure 4.17).

FIGURE 4.17
f has a relative minimum at $x = 0$.

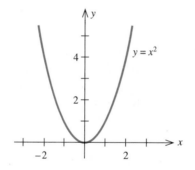

■ ■ ■ ■

EXAMPLE 6   Find the relative maxima and relative minima of the function $f(x) = x^{2/3}$ (see Example 3).

SOLUTION ✔   The derivative of f is $f'(x) = (2/3)x^{-1/3}$. As noted in Example 3, f' is not defined at $x = 0$, is continuous everywhere else, and is not equal to zero in its domain. Thus, $x = 0$ is the only critical point of the function f.

The sign diagram obtained in Example 3 is reproduced in Figure 4.18. We can see that the sign of $f'(x)$ changes from negative to positive as we

FIGURE 4.18
Sign diagram for f'

move across $x = 0$ from left to right. Thus, an application of the first derivative test tells us that $f(0) = 0$ is a relative minimum of f (Figure 4.19).

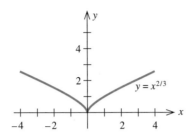

FIGURE 4.19
f has a relative minimum at $x = 0$.

$y = x^{2/3}$

Group Discussion

Recall that the average cost function $\overline{C}$ is defined by

$$\overline{C} = \frac{C(x)}{x}$$

where $C(x)$ is the total cost function and x is the number of units of a commodity manufactured (see Section 3.4).

1. Show that

$$\overline{C}'(x) = \frac{C'(x) - \overline{C}(x)}{x} \qquad (x > 0)$$

2. Use the result of part 1 to conclude that $\overline{C}$ is decreasing for values of x at which $C'(x) < \overline{C}(x)$. Find similar conditions for which $\overline{C}$ is increasing and for which $\overline{C}$ is constant.

3. Explain the results of part 2 in economic terms.

EXAMPLE 7   Find the relative maxima and relative minima of the function

$$f(x) = x^3 - 3x^2 - 24x + 32$$

SOLUTION ✔   The derivative of f is

$$f'(x) = 3x^2 - 6x - 24 = 3(x + 2)(x - 4)$$

and it is continuous everywhere. The zeros of $f'(x)$, $x = -2$ and $x = 4$, are the only critical points of the function f. The sign diagram for f' is shown in Figure 4.20. Examine the two critical points $x = -2$ and $x = 4$ for a relative extremum using the first derivative test and the sign diagram for f':

FIGURE 4.20
Sign diagram for f'

1. *The critical point $x = -2$:* Since the function $f'(x)$ changes sign from positive to negative as we move across $x = -2$ from left to right, we conclude that a relative maximum of f occurs at $x = -2$. The value of $f(x)$ when $x = -2$ is

$$f(-2) = (-2)^3 - 3(-2)^2 - 24(-2) + 32 = 60$$

2. *The critical point* $x = 4$: $f'(x)$ changes sign from negative to positive as we move across $x = 4$ from left to right, so $f(4) = -48$ is a relative minimum of f. The graph of f appears in Figure 4.21.

FIGURE 4.21
f has a relative maximum at $x = -2$ and a relative minimum at $x = 4$.

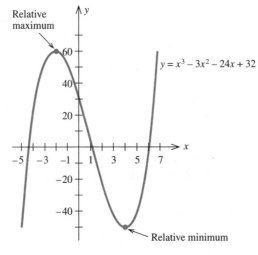

EXAMPLE 8   Find the relative maxima and the relative minima of the function

$$f(x) = x + \frac{1}{x}$$

SOLUTION ✔   The derivative of f is

$$f'(x) = 1 - \frac{1}{x^2} = \frac{x^2 - 1}{x^2} = \frac{(x + 1)(x - 1)}{x^2}$$

FIGURE 4.22
$x = 0$ is not a critical point because f is not defined at $x = 0$.

Since f' is equal to zero at $x = -1$ and $x = 1$, these are critical points for the function f. Next, observe that f' is discontinuous at $x = 0$. However, because *f is not defined at that point*, the point $x = 0$ does not qualify as a critical point of f. Figure 4.22 shows the sign diagram for f'.

Since $f'(x)$ changes sign from positive to negative as we move across $x = -1$ from left to right, the first derivative test implies that $f(-1) = -2$ is a relative maximum of the function f. Next, $f'(x)$ changes sign from negative to positive as we move across $x = 1$ from left to right, so $f(1) = 2$ is a relative minimum of the function f. The graph of f appears in Figure 4.23. Note that this function has a relative maximum that lies below its relative minimum.

FIGURE 4.23

$$f(x) = x + \frac{1}{x}$$

Refer to Example 8.

1. Use a graphing utility to plot the graphs of $f(x) = x + 1/x$ and its derivative function $f'(x) = 1 - 1/x^2$, using the viewing rectangle $[-4, 4] \times [-8, 8]$.
2. By studying the graph of f', determine the critical points of f. Next, note the sign of $f'(x)$ immediately to the left and to the right of each critical point. What can you conclude about each critical point? Are your conclusions borne out by the graph of f?

APPLICATIONS

EXAMPLE **9**

The profit function of the Acrosonic Company is given by

$$P(x) = -0.02x^2 + 300x - 200{,}000$$

dollars, where x is the number of Acrosonic model F loudspeaker systems produced. Find where the function P is increasing and where it is decreasing.

SOLUTION ✔

The derivative P' of the function P is

$$P'(x) = -0.04x + 300 = -0.04(x - 7500)$$

Thus, $P'(x) = 0$ when $x = 7500$. Furthermore, $P'(x) > 0$ for x in the interval $(0, 7500)$, and $P'(x) < 0$ for x in the interval $(7500, \infty)$. This means that the profit function P is increasing on $(0, 7500)$ and decreasing on $(7500, \infty)$ (Figure 4.24).

FIGURE 4.24
The profit function is increasing on $(0, 7500)$ and decreasing on $(7500, \infty)$.

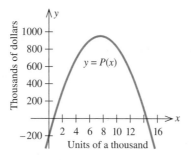

■ ■ ■ ■

EXAMPLE **10**

The number of major crimes committed in the city of Bronxville from 1993 to 2000 is approximated by the function

$$N(t) = -0.1t^3 + 1.5t^2 + 100 \qquad (0 \le t \le 7)$$

where $N(t)$ denotes the number of crimes committed in year t and $t = 0$ corresponds to the beginning of 1993. Find where the function N is increasing and where it is decreasing.

SOLUTION ✔   The derivative N' of the function N is

$$N'(t) = -0.3t^2 + 3t = -0.3t(t - 10)$$

Since $N'(t) > 0$ for t in the interval $(0, 7)$, the function N is increasing throughout that interval (Figure 4.25).

FIGURE 4.25
The number of crimes, $N(t)$, is increasing over the 7-year interval.

SELF-CHECK EXERCISES **4.1**

1. Find the intervals where the function $f(x) = \frac{2}{3}x^3 - x^2 - 12x + 3$ is increasing and the intervals where it is decreasing.

2. Find the relative extrema of $f(x) = \dfrac{x^2}{1 - x^2}$.

Solutions to Self-Check Exercises 4.1 can be found on page 301.

4.1 Exercises

In Exercises 1–8, you are given the graph of a function f. Determine the intervals where f is increasing, constant, or decreasing.

1.

2.

3.

4.

5.

6.

7.

8.

9. THE BOSTON MARATHON The graph of the function *f* shown in the accompanying figure gives the elevation of that part of the Boston Marathon course that includes the notorious Heartbreak Hill. Determine the intervals (stretches of the course) where the function *f* is increasing (the runner is laboring), where it is constant (the runner is taking a breather), and where it is decreasing (the runner is coasting).

In Exercises 10–35, find the interval(s) where each function is increasing and the interval(s) where it is decreasing.

10. $f(x) = 4 - 5x$

11. $f(x) = 3x + 5$

12. $f(x) = 2x^2 + x + 1$

13. $f(x) = x^2 - 3x$

14. $f(x) = x^3 - 3x^2$

15. $g(x) = x - x^3$

16. $f(x) = x^3 - 3x + 4$

17. $g(x) = x^3 + 3x^2 + 1$

18. $f(x) = \frac{2}{3}x^3 - 2x^2 - 6x - 2$

19. $f(x) = \frac{1}{3}x^3 - 3x^2 + 9x + 20$

20. $g(x) = x^4 - 2x^2 + 4$

21. $h(x) = x^4 - 4x^3 + 10$

22. $h(x) = \frac{1}{2x + 3}$

23. $f(x) = \frac{1}{x - 2}$

24. $g(t) = \frac{2t}{t^2 + 1}$

25. $h(t) = \frac{t}{t - 1}$

26. $f(x) = x^{2/3} + 5$

27. $f(x) = x^{3/5}$

28. $f(x) = (x - 5)^{2/3}$

29. $f(x) = \sqrt{x + 1}$

30. $g(x) = x\sqrt{x + 1}$

31. $f(x) = \sqrt{16 - x^2}$

32. $h(x) = \frac{x^2}{x - 1}$

33. $f(x) = \frac{x^2 - 1}{x}$

34. $g(x) = \frac{x}{(x + 1)^2}$

35. $f(x) = \frac{1}{(x - 1)^2}$

In Exercises 36–43, you are given the graph of a function *f*. Determine the relative maxima and relative minima, if any.

36.

37.

38.

39.

40.

41.

42.

43.

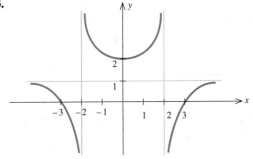

In Exercises 44–47, match the graph of the function with the graph of its derivative in (a)–(d).

44.

45.

46.

47.

(a)

(b)

(c)

(d)

In Exercises 48–71, find the relative maxima and relative minima, if any, of each function.

48. $g(x) = x^2 + 3x + 8$ **49.** $f(x) = x^2 - 4x$

50. $h(t) = -t^2 + 6t + 6$ **51.** $f(x) = \frac{1}{2}x^2 - 2x + 4$

52. $f(x) = x^{5/3}$ **53.** $f(x) = x^{2/3} + 2$

54. $f(x) = x^3 - 3x + 6$ **55.** $g(x) = x^3 - 3x^2 + 4$

56. $f(x) = \frac{1}{2}x^4 - x^2$

57. $F(x) = \frac{1}{3}x^3 - x^2 - 3x + 4$

58. $h(x) = \frac{1}{2}x^4 - 3x^2 + 4x - 8$

59. $g(x) = x^4 - 4x^3 + 8$

60. $F(t) = 3t^5 - 20t^3 + 20$

61. $f(x) = 3x^4 - 2x^3 + 4$

62. $h(x) = \dfrac{x}{x+1}$ **63.** $g(x) = \dfrac{x+1}{x}$

64. $g(x) = 2x^2 + \dfrac{4000}{x} + 10$

65. $f(x) = x + \dfrac{9}{x} + 2$

66. $g(x) = \dfrac{x}{x^2 - 1}$ **67.** $f(x) = \dfrac{x}{1 + x^2}$

68. $g(t) = \dfrac{t^2}{1 + t^2}$ **69.** $f(x) = \dfrac{x^2}{x^2 - 4}$

70. $g(x) = x\sqrt{x - 4}$ **71.** $f(x) = (x - 1)^{2/3}$

72. A stone is thrown straight up from the roof of an 80-ft building. The distance (in feet) of the stone from the ground at any time t (in seconds) is given by

$$h(t) = -16t^2 + 64t + 80$$

When is the stone rising, and when is it falling? If the stone were to miss the building, when would it hit the ground? Sketch the graph of h.

Hint: The stone is on the ground when $h(t) = 0$.

73. PROFIT FUNCTIONS The Mexican subsidiary of the Thermo-Master Company manufactures an indoor–outdoor thermometer. Management estimates that the profit (in dollars) realizable by the company for the manufacture and sale of x units of thermometers per week is

$$P(x) = -0.001x^2 + 8x - 5000$$

Find the intervals where the profit function P is increasing and the intervals where P is decreasing.

74. **FLIGHT OF A ROCKET** The height (in feet) attained by a rocket t sec into flight is given by the function

$$h(t) = -\frac{1}{3}t^3 + 16t^2 + 33t + 10$$

When is the rocket rising, and when is it descending?

75. **ENVIRONMENT OF FORESTS** Following the lead of the National Wildlife Federation, the Department of the Interior of a South American country began to record an index of environmental quality that measured progress and decline in the environmental quality of its forests. The index for the years 1984 through 1994 is approximated by the function

$$I(t) = \frac{1}{3}t^3 - \frac{5}{2}t^2 + 80 \qquad (0 \le t \le 10)$$

where $t = 0$ corresponds to the year 1984. Find the intervals where the function I is increasing and the intervals where it is decreasing. Interpret your results.
Source: World Almanac

76. **AVERAGE SPEED OF A HIGHWAY VEHICLE** The average speed of a vehicle on a stretch of Route 134 between 6 A.M. and 10 A.M. on a typical weekday is approximated by the function

$$f(t) = 20t - 40\sqrt{t} + 50 \qquad (0 \le t \le 4)$$

where $f(t)$ is measured in miles per hour and t is measured in hours, with $t = 0$ corresponding to 6 A.M. Find the interval where f is increasing and the interval where f is decreasing and interpret your results.

77. **AVERAGE COST** The average cost (in dollars) incurred by the Lincoln Record Company per week in pressing x compact discs is given by

$$\overline{C}(x) = -0.0001x + 2 + \frac{2000}{x} \qquad (0 < x \le 6000)$$

Show that $\overline{C}(x)$ is always decreasing over the interval (0, 6000).

78. **AIR POLLUTION** According to the South Coast Air Quality Management District, the level of nitrogen dioxide, a brown gas that impairs breathing, present in the atmosphere on a certain May day in downtown Los Angeles is approximated by

$$A(t) = 0.03t^3(t - 7)^4 + 60.2 \qquad (0 \le t \le 7)$$

where $A(t)$ is measured in pollutant standard index (PSI) and t is measured in hours with $t = 0$ corresponding to 7 A.M. At what time of day is the air pollution increasing, and at what time is it decreasing?

79. **PROJECTED RETIREMENT FUNDS** Based on data from the Central Provident Fund of a certain country (a government agency similar to the Social Security Administration), the estimated cash in the fund in 1995 is given by

$$A(t) = -96.6t^4 + 403.6t^3$$
$$+ 660.9t^2 + 250 \qquad (0 \le t \le 5)$$

where $A(t)$ is measured in billions of dollars and t is measured in decades, with $t = 0$ corresponding to the year 1995. Find the interval where A is increasing and the interval where A is decreasing and interpret your results.
Hint: Use the quadratic formula.

80. **LEARNING CURVES** The Emory Secretarial School finds from experience that the average student taking Advanced Typing will progress according to the rule

$$N(t) = \frac{60t + 180}{t + 6} \qquad (t \ge 0)$$

where $N(t)$ measures the number of words per minute the student can type after t weeks in the course. Compute $N'(t)$ and use this result to show that the function N is increasing on the interval $(0, \infty)$.

81. **DRUG CONCENTRATION IN THE BLOOD** The concentration (in milligrams per cubic centimeter) of a certain drug in a patient's body t hr after injection is given by

$$C(t) = \frac{t^2}{2t^3 + 1} \qquad (0 \le t \le 4)$$

When is the concentration of the drug increasing, and when is it decreasing?

82. **AGE OF DRIVERS IN CRASH FATALITIES** The number of crash fatalities per 100,000 vehicle miles of travel (based on 1994 data) is approximated by the model

$$f(x) = \frac{15}{0.08333x^2 + 1.91667x + 1} \qquad (0 \le x \le 11)$$

where x is the age of the driver in years with $x = 0$ corresponding to age 16. Show that f is decreasing on (0, 11) and interpret your result.
Source: National Highway Traffic Safety Administration

83. **AIR POLLUTION** The amount of nitrogen dioxide, a brown gas that impairs breathing, present in the atmosphere on a certain May day in the city of Long Beach is approximated by

$$A(t) = \frac{136}{1 + 0.25(t - 4.5)^2} + 28 \qquad (0 \le t \le 11)$$

where $A(t)$ is measured in pollutant standard index (PSI) and t is measured in hours, with $t = 0$ corresponding to

(continued on p. 300)

USING THE FIRST DERIVATIVE TO ANALYZE A FUNCTION

A graphing utility is an effective tool for analyzing the properties of functions. This is especially true when we also bring into play the power of calculus, as the following examples show.

EXAMPLE 1 Let $f(x) = 2.4x^4 - 8.2x^3 + 2.7x^2 + 4x + 1$.

a. Use a graphing utility to plot the graph of f.
b. Find the intervals where f is increasing and the intervals where f is decreasing.
c. Find the relative extrema of f.

SOLUTION ✔ **a.** The graph of f in the viewing rectangle $[-2, 4] \times [-10, 10]$ is shown in Figure T1.

FIGURE T1
The graph of f in the viewing rectangle $[-2, 4] \times [-10, 10]$

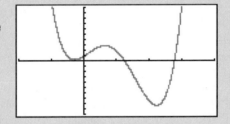

b. We compute

$$f'(x) = 9.6x^3 - 24.6x^2 + 5.4x + 4$$

and observe that f' is continuous everywhere, so the critical points of f occur at values of x where $f'(x) = 0$. To solve this last equation, observe that $f'(x)$ is a *polynomial function* of degree 3. The easiest way to solve the polynomial equation

$$9.6x^3 - 24.6x^2 + 5.4x + 4 = 0$$

is to use the function on a graphing utility for solving polynomial equations. (Not all graphing utilities have this function.) You can also use TRACE and ZOOM, but this will not give the same accuracy without a much greater effort.
    We find

$$x_1 \approx 2.22564943249, \qquad x_2 \approx 0.63272944121, \qquad x_3 \approx -0.295878873696$$

Referring to Figure T1, we conclude that f is decreasing on $(-\infty, -0.2959)$ and $(0.6327, 2.2256)$ (correct to four decimal places) and f is increasing on $(-0.2959, 0.6327)$ and $(2.2256, \infty)$.

c. Using the evaluation function of a graphing utility, we find the value of f at each of the critical points found in part (b). Upon referring to Figure T1 once again, we see that $f(x_3) \approx 0.2836$ and $f(x_1) \approx -8.2366$ are relative minimum values of f and $f(x_2) \approx 2.9194$ is a relative maximum value of f.

∎∎∎∎

REMARK The equation $f'(x) = 0$ in Example 1 is a polynomial equation, and so it is easily solved using the function for solving polynomial equations. We could also solve the equation using the function for finding the roots of equations, but that would require much more work. For equations that are *not* polynomial equations, however, our only choice is to use the function for finding the roots of equations.

If the derivative of a function is difficult to compute or simplify and we do not require great precision in the solution, we can find the relative extrema of the function using a combination of ZOOM and TRACE. This technique, which does not require the use of the derivative of f, is illustrated in the following example.

EXAMPLE 2 Let $f(x) = x^{1/3}(x^2 + 1)^{-3/2}3^{-x}$.

a. Use a graphing utility to plot the graph of f.*
b. Find the relative extrema of f.

SOLUTION ✔ a. The graph of f in the viewing rectangle $[-4, 2] \times [-2, 1]$ is shown in Figure T2.

FIGURE T2
The graph of f in the viewing rectangle $[-4, 2] \times [-2, 1]$

b. From the graph of f in Figure T2, we see that f has relative maxima when $x \approx -2$ and $x \approx 0.25$ and a relative minimum when $x \approx -0.75$. To obtain a better approximation of the first relative maximum, we zoom-in with the

---

* Functions of the form $f(x) = 3^{-x}$ are called exponential functions, and we will study them in greater detail in Chapter 5.

cursor at approximately the point on the graph corresponding to $x \approx -2$. Then, using **TRACE,** we see that a relative maximum occurs when $x \approx -1.76$ with value $y \approx -1.01$. Similarly, we find the other relative maximum where $x \approx 0.20$ with value $y \approx 0.44$. Repeating the procedure, we find the relative minimum at the point where $x \approx -0.86$ and $y \approx -1.07$. ■ ■ ■ ■

Finally, we comment that if you have access to a computer and software such as Derive, Maple, or Mathematica, then symbolic differentiation will yield the derivative $f'(x)$ of any differentiable function. This software will also solve the equation $f'(x) = 0$ with ease. Thus, the use of a computer will simplify even more greatly the analysis of functions.

Exercises

In Exercises 1–4, use a graphing utility to find (a) the intervals where *f* is increasing and the intervals where *f* is decreasing and (b) the relative extrema of *f*. Express your answers accurate to four decimal places.

1. $f(x) = 3.4x^4 - 6.2x^3 + 1.8x^2 + 3x - 2$

2. $f(x) = 1.8x^4 - 9.1x^3 + 5x - 4$

3. $f(x) = 2x^5 - 5x^3 + 8x^2 - 3x + 2$

4. $f(x) = 3x^5 - 4x^2 + 3x - 1$

In Exercises 5–8, use the ZOOM and TRACE features of a graphing utility to find (a) the intervals where *f* is increasing and the intervals where *f* is decreasing and (b) the relative extrema of *f*. Express your answers accurate to two decimal places.

5. $f(x) = (2x + 1)^{1/3}(x^2 + 1)^{-2/3}$

6. $f(x) = [x^2(x^3 - 1)]^{1/3} + \dfrac{1}{x}$

7. $f(x) = x - \sqrt{1 - x^2}$

8. $f(x) = \dfrac{\sqrt{x}(x^2 - 1)^2}{x - 2}$

9. **RATE OF BANK FAILURES** The Federal Deposit Insurance Company (FDIC) estimates that the rate at which banks were failing between 1982 and 1994 is given by

$$f(t) = 0.063447t^4 - 1.953283t^3 + 14.632576t^2$$
$$- 6.684704t + 47.458874 \qquad (0 \le t \le 12)$$

where $f(t)$ is measured in the number of banks per year and t is measured in years, with $t = 0$ corresponding to the beginning of 1982.
a. Use a graphing utility to plot the graph of f in the viewing rectangle $[0, 13] \times [0, 220]$.
b. Determine the intervals where f is increasing and where f is decreasing and interpret your result.
c. Find the relative maximum of f and interpret your result.
Source: Federal Deposit Insurance Corporation

10. **MANUFACTURING CAPACITY** Data obtained from the Federal Reserve show that the annual increase in manufacturing capacity between 1988 and 1994 is given by

$$f(t) = 0.0388889t^3 - 0.283333t^2 + 0.477778t$$
$$+ 2.04286 \qquad (0 \le t \le 6)$$

where $f(t)$ is a percentage and t is measured in years, with $t = 0$ corresponding to the beginning of 1988.

a. Use a graphing utility to plot the graph of f in the viewing rectangle $[0, 8] \times [0, 4]$.

b. Determine the intervals where f is increasing and where f is decreasing and interpret your result.

Source: Federal Reserve

11. **HOME SALES** According to the Greater Boston Real Estate Board—Multiple Listing Service, the average number of days a single-family home remains for sale from listing to accepted offer is approximated by the function

$$f(t) = 0.0171911t^4 - 0.662121t^3 + 6.18083t^2$$
$$- 8.97086t + 53.3357 \qquad (0 \le t \le 10)$$

where t is measured in years, with $t = 0$ corresponding to the beginning of 1984.

a. Use a graphing utility to plot the graph of f in the viewing rectangle $[0, 12] \times [0, 120]$.

b. Use the graph of f to find, approximately, the intervals where f is increasing and the intervals where f is decreasing. What does this result tell us about the sales of single-family homes in the greater Boston area from 1984 through 1994?

Source: Greater Boston Real Estate Board—Multiple Listing Service

12. **MORNING TRAFFIC RUSH** The speed of traffic flow on a certain stretch of Route 123 between 6 A.M. and 10 A.M. on a typical weekday is approximated by the function

$$f(t) = 20t - 40\sqrt{t} + 52 \qquad (0 \le t \le 4)$$

where $f(t)$ is measured in miles per hour and t is measured in hours, with $t = 0$ corresponding to 6 A.M. Find the interval where f is increasing, the interval where f is decreasing, and the relative extrema of f. Interpret your results.

13. **AIR POLLUTION** The amount of nitrogen dioxide, a brown gas that impairs breathing, present in the atmosphere on a certain May day in the city of Long Beach, is approximated by

$$A(t) = \frac{136}{1 + 0.25(t - 4.5)^2} + 28 \qquad (0 \le t \le 11)$$

where $A(t)$ is measured in pollutant standard index (PSI) and t is measured in hours, with $t = 0$ corresponding to 7 A.M. When is the PSI increasing and when is it decreasing? At what time is the PSI highest, and what is its value at that time?

7 A.M. Find the intervals where A is increasing and where A is decreasing and interpret your results.

84. **PRISON OVERCROWDING**   The 1980s saw a trend toward old-fashioned punitive deterrence as opposed to the more liberal penal policies and community-based corrections popular in the 1960s and early 1970s. As a result, prisons became more crowded, and the gap between the number of people in prison and the prison capacity widened. Based on figures from the U.S. Department of Justice, the number of prisoners (in thousands) in federal and state prisons is approximated by the function

$$N(t) = 3.5t^2 + 26.7t + 436.2 \qquad (0 \le t \le 10)$$

where t is measured in years and $t = 0$ corresponds to 1984. The number of inmates for which prisons were designed is given by

$$C(t) = 24.3t + 365 \qquad (0 \le t \le 10)$$

where $C(t)$ is measured in thousands and t has the same meaning as before. Show that the gap between the number of prisoners and the number for which the prisons were designed has been widening at any time t.
Source: U.S. Dept. of Justice

Hint: First, write a function G that gives the gap between the number of prisoners and the number for which the prisons were designed at any time t. Then show that $G'(t) > 0$ for all values of t in the interval $(0, 10)$.

85. Using Theorem 1, verify that the linear function $f(x) = mx + b$ is (a) increasing everywhere if $m > 0$, (b) decreasing everywhere if $m < 0$, and (c) constant if $m = 0$.

86. In what interval is the quadratic function

$$f(x) = ax^2 + bx + c \qquad (a \ne 0)$$

increasing? In what interval is f decreasing?

In Exercises 87–92, determine whether the statement is true or false. If it is true, explain why it is true. If it is false, give an example to show why it is false.

87. If f is decreasing on (a, b), then $f'(x) < 0$ for each x in (a, b).

88. If f and g are both increasing on (a, b), then $f + g$ is increasing on (a, b).

89. If f and g are both decreasing on (a, b), then $f - g$ is decreasing on (a, b).

90. If $f(x)$ and $g(x)$ are positive on (a, b) and both f and g are increasing on (a, b), then fg is increasing on (a, b).

91. If $f'(c) = 0$, then f has a relative maximum or a relative minimum at $x = c$.

92. If f has a relative minimum at $x = c$, then $f'(c) = 0$.

93. Let

$$f(x) = \begin{cases} -3x & \text{if } x < 0 \\ 2x + 4 & \text{if } x \ge 0 \end{cases}$$

   a. Compute $f'(x)$ and show that it changes sign from negative to positive as we move across $x = 0$.
   b. Show that f does not have a relative minimum at $x = 0$. Does this contradict the first derivative test? Explain your answer.

94. Let

$$f(x) = \begin{cases} -x^2 + 3 & \text{if } x \ne 0 \\ 2 & \text{if } x = 0 \end{cases}$$

   a. Compute $f'(x)$ and show that it changes sign from positive to negative as we move across $x = 0$.
   b. Show that f does not have a relative maximum at $x = 0$. Does this contradict the first derivative test? Explain your answer.

95. Show that the quadratic function

$$f(x) = ax^2 + bx + c \qquad (a \ne 0)$$

has a relative extremum when $x = -b/2a$. Also, show that the relative extremum is a relative maximum if $a < 0$ and a relative minimum if $a > 0$.

96. Show that the cubic function

$$f(x) = ax^3 + bx^2 + cx + d \qquad (a \ne 0)$$

has no relative extremum if and only if $b^2 - 3ac \le 0$.

97. Refer to Example 6, page 144.
   a. Show that f is increasing on the interval $(0, 1)$.
   b. Show that $f(0) = -1$ and $f(1) = 1$ and use the result of part (a) together with the intermediate value theorem to conclude that there is exactly one root of $f(x) = 0$ in $(0, 1)$.

98. Show that the function

$$f(x) = \frac{ax + b}{cx + d}$$

does not have a relative extremum if $ad - bc \ne 0$. What can you say about f if $ad - bc = 0$?

SOLUTIONS TO SELF-CHECK EXERCISES 4.1

1. The derivative of f is

$$f'(x) = 2x^2 - 2x - 12 = 2(x + 2)(x - 3)$$

and it is continuous everywhere. The zeros of $f'(x)$ are $x = -2$ and $x = 3$. The sign diagram of f' is shown in the accompanying figure. We conclude that f is increasing on the intervals $(-\infty, -2)$ and $(3, \infty)$ and decreasing on the interval $(-2, 3)$.

2. The derivative of f is

$$f'(x) = \frac{(1 - x^2)\dfrac{d}{dx}(x^2) - x^2\dfrac{d}{dx}(1 - x^2)}{(1 - x^2)^2}$$

$$= \frac{(1 - x^2)(2x) - x^2(-2x)}{(1 - x^2)^2} = \frac{2x}{(1 - x^2)^2}$$

and it is continuous everywhere except at $x = \pm 1$. Since $f'(x)$ is equal to zero at $x = 0$, $x = 0$ is a critical point of f. Next, observe that $f'(x)$ is discontinuous at $x = \pm 1$, but since these points are not in the domain of f, they do not qualify as critical points of f. Finally, from the sign diagram of f' shown in the accompanying figure, we conclude that $f(0) = 0$ is a relative minimum of f.

4.2 Applications of the Second Derivative

DETERMINING THE INTERVALS OF CONCAVITY

Consider the graphs shown in Figure 4.26, which give the estimated population of the world and of the United States through the year 2000. Both graphs are rising, indicating that both the U.S. population and the world population will continue to increase through the year 2000. But observe that the graph in Figure 4.26a opens upward, whereas the graph in Figure 4.26b opens downward. What is the significance of this? To answer this question, let us look at the slopes of the tangent lines to various points on each graph (Figure 4.27).

FIGURE 4.26

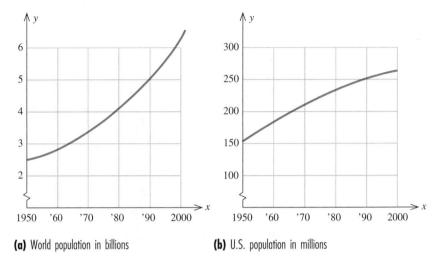

(a) World population in billions   **(b)** U.S. population in millions

Source: U.S. Dept. of Commerce and Worldwatch Institute

In Figure 4.27a we see that the slopes of the tangent lines to the graph are increasing as we move from left to right. Since the slope of the tangent line to the graph at a point on the graph measures the rate of change of the function at that point, we conclude that the world population is not only increasing through the year 2000 but is increasing at an *increasing* pace. A similar analysis of Figure 4.27b reveals that the U.S. population is increasing, but at a *decreasing* pace.

FIGURE 4.27

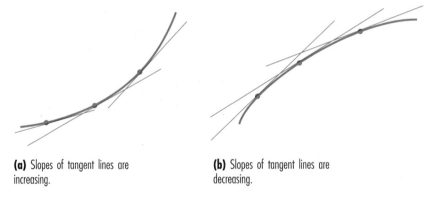

(a) Slopes of tangent lines are increasing.   **(b)** Slopes of tangent lines are decreasing.

The shape of a curve can be described using the notion of concavity.

Concavity of a Function *f*

Let the function *f* be differentiable on an interval (a, b). Then,

1. *f* is **concave upward** on (a, b) if f' is increasing on (a, b).

2. *f* is **concave downward** on (a, b) if f' is decreasing on (a, b).

FIGURE 4.28

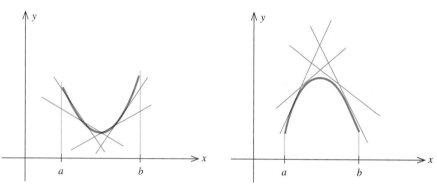

(a) f is concave upward on (a, b).

(b) f is concave downward on (a, b).

Geometrically, a curve is concave upward if it lies above its tangent lines (Figure 4.28a). Similarly, a curve is concave downward if it lies below its tangent lines (Figure 4.28b).

We also say that f is *concave upward at a point* $x = c$ if there exists an interval (a, b) containing c in which f is concave upward. Similarly, we say that f is *concave downward at a point* $x = c$ if there exists an interval (a, b) containing c in which f is concave downward.

If a function f has a second derivative f'', we can use f'' to determine the intervals of concavity of the function. Recall that $f''(x)$ measures the rate of change of the slope $f'(x)$ of the tangent line to the graph of f at the point $(x, f(x))$. Thus, if $f''(x) > 0$ on an interval (a, b), then the slopes of the tangent lines to the graph of f are increasing on (a, b) and so f is concave upward on (a, b). Similarly, if $f''(x) < 0$ on (a, b), then f is concave downward on (a, b). These observations suggest the following theorem.

THEOREM 2

a. If $f''(x) > 0$ for each value of x in (a, b), then f is concave upward on (a, b).

b. If $f''(x) < 0$ for each value of x in (a, b), then f is concave downward on (a, b).

The following procedure, based on the conclusions of Theorem 2, may be used to determine the intervals of concavity of a function.

**Determining the
Intervals of
Concavity of f**

1. Determine the values of x for which f'' is zero or where f'' is not defined, and identify the open intervals determined by these points.
2. Determine the sign of f'' in each interval found in step 1. To do this, compute $f''(c)$, where c is any conveniently chosen test point in the interval.
   a. If $f''(c) > 0$, f is concave upward on that interval.
   b. If $f''(c) < 0$, f is concave downward on that interval.

EXAMPLE **1** Determine where the function $f(x) = x^3 - 3x^2 - 24x + 32$ is concave upward and where it is concave downward.

SOLUTION ✔ Here,

$$f'(x) = 3x^2 - 6x - 24$$

$$f''(x) = 6x - 6 = 6(x - 1)$$

FIGURE 4.29
Sign diagram for f''

and f'' is defined everywhere. Setting $f''(x) = 0$ gives $x = 1$. The sign diagram of f'' appears in Figure 4.29. We conclude that f is concave downward on the interval $(-\infty, 1)$ and is concave upward on the interval $(1, \infty)$. Figure 4.30 shows the graph of f.

FIGURE 4.30
f is concave downward on $(-\infty, 1)$ and concave upward on $(1, \infty)$.

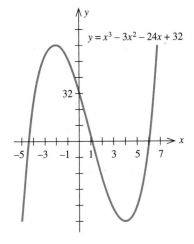

$y = x^3 - 3x^2 - 24x + 32$

EXAMPLE **2** Determine the intervals where the function $f(x) = x + 1/x$ is concave upward and where it is concave downward.

SOLUTION ✔ We have

$$f'(x) = 1 - \frac{1}{x^2}$$

$$f''(x) = \frac{2}{x^3}$$

We deduce from the sign diagram for f'' (Figure 4.31) that the function f is concave downward on the interval $(-\infty, 0)$ and concave upward on the interval $(0, \infty)$. The graph of f is sketched in Figure 4.32.

FIGURE 4.31
The sign diagram for f''

f'' is not defined at $x = 0$

FIGURE 4.32
f is concave downward on $(-\infty, 0)$ and concave upward on $(0, \infty)$.

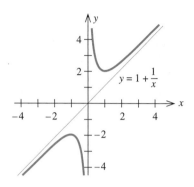

$$y = 1 + \frac{1}{x}$$

Exploring with Technology

Refer to Example 1.

1. Use a graphing utility to plot the graph of

$$f(x) = x^3 - 3x^2 - 24x + 32$$

and its second derivative

$$f''(x) = 6x - 6$$

using the viewing rectangle $[-10, 10] \times [-80, 90]$.

2. By studying the graph of f'', determine the intervals where $f''(x) > 0$ and the intervals where $f''(x) < 0$. Next, look at the graph of f and determine the intervals where the graph of f is concave upward and the intervals where the graph of f is concave downward. Are these observations what you might have expected?

INFLECTION POINTS

FIGURE 4.33
The graph of *S* has a point of inflection at (50, 2700).

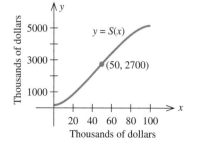

Figure 4.33 shows the total sales S of a manufacturer of automobile air conditioners versus the amount of money x that the company spends on advertising its product. Notice that the graph of the continuous function $y = S(x)$ changes concavity—from upward to downward—at the point (50, 2700). This point is called an *inflection point* of S. To understand the significance of this inflection point, observe that the total sales increase rather slowly at first, but as more money is spent on advertising, the total sales increase rapidly. This rapid increase reflects the effectiveness of the company's ads. However, a point is soon reached after which any additional advertising expenditure results in increased sales but at a slower rate of increase. This point, commonly known as the *point of diminishing returns,* is the point of inflection of the function S. We will return to this example later.

Let's now state formally the definition of an inflection point.

Inflection Point

A point on the graph of a differentiable function f at which the concavity changes is called an **inflection point.**

Observe that the graph of a function crosses its tangent line at a point of inflection (Figure 4.34).

FIGURE 4.34
At each point of inflection, the graph of a function crosses its tangent line.

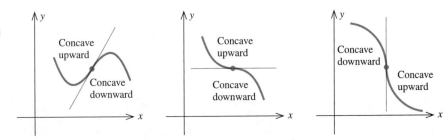

The following procedure may be used to find inflection points.

Finding Inflection Points

1. Compute $f''(x)$.
2. Determine the points in the domain of f for which $f''(x) = 0$ or $f''(x)$ does not exist.
3. Determine the sign of $f''(x)$ to the left and right of each point $x = c$ found in step 2. If there is a change in the sign of $f''(x)$ as we move across the point $x = c$, then $(c, f(c))$ is an inflection point of f.

 The points determined in step 2 are only *candidates* for the inflection points of f. For example, you can easily verify that $f''(0) = 0$ if $f(x) = x^4$, but a sketch of the graph of f will show that $(0, 0)$ is not an inflection point of f.

EXAMPLE 3   Find the points of inflection of the function $f(x) = x^3$.

SOLUTION ✔
$$f'(x) = 3x^2$$
$$f''(x) = 6x$$

Observe that f'' is continuous everywhere and is zero if $x = 0$. The sign diagram of f'' is shown in Figure 4.35. From this diagram, we see that $f''(x)$

FIGURE 4.35
Sign diagram for f''

FIGURE 4.36
f has an inflection point at (0, 0).

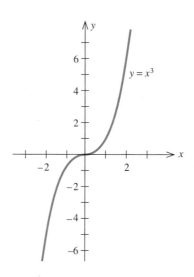

changes sign as we move across $x = 0$. Thus, the point $(0, 0)$ is an inflection point of the function f (Figure 4.36). ■ ■ ■ ■

EXAMPLE **4**   Determine the intervals where the function $f(x) = (x - 1)^{5/3}$ is concave upward and where it is concave downward and find the inflection points of f.

SOLUTION ✔   The first derivative of f is

$$f'(x) = \frac{5}{3}(x - 1)^{2/3}$$

FIGURE 4.37
Sign diagram for f''

and the second derivative of f is

$$f''(x) = \frac{10}{9}(x - 1)^{-1/3} = \frac{10}{9(x - 1)^{1/3}}$$

We see that f'' is not defined at $x = 1$. Furthermore, $f''(x)$ is not equal to zero anywhere. The sign diagram of f'' is shown in Figure 4.37. From the sign diagram, we see that f is concave downward on $(-\infty, 1)$ and concave upward on $(1, \infty)$. Next, since $x = 1$ does lie in the domain of f, our computations also reveal that the point $(1, 0)$ is an inflection point of f (Figure 4.38).

FIGURE 4.38
f has an inflection point at (1, 0).

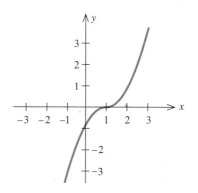

■ ■ ■ ■

EXAMPLE 5   Determine the intervals where the function

$$f(x) = \frac{1}{x^2 + 1}$$

is concave upward and where it is concave downward and find the inflection points of f.

SOLUTION ✔   The first derivative of f is

$$f'(x) = \frac{d}{dx}(x^2 + 1)^{-1} = -2x(x^2 + 1)^{-2} \qquad \text{(Using the general power rule)}$$

$$= -\frac{2x}{(x^2 + 1)^2}$$

Next, using the quotient rule, we find

$$f''(x) = \frac{(x^2 + 1)^2(-2) + (2x)2(x^2 + 1)(2x)}{(x^2 + 1)^4}$$

$$= \frac{(x^2 + 1)[-2(x^2 + 1) + 8x^2]}{(x^2 + 1)^4} = \frac{(x^2 + 1)(6x^2 - 2)}{(x^2 + 1)^4}$$

$$= \frac{2(3x^2 - 1)}{(x^2 + 1)^3} \qquad \text{(Canceling the common factors)}$$

FIGURE 4.39
Sign diagram for f''

Observe that f'' is continuous everywhere and is zero if

$$3x^2 - 1 = 0$$

$$x^2 = \frac{1}{3}$$

FIGURE 4.40
The graph of $f(x) = \dfrac{1}{x^2 + 1}$
is concave upward on
$(-\infty, -\sqrt{3}/3) \cup (\sqrt{3}/3, \infty)$
and concave downward on
$(-\sqrt{3}/3, \sqrt{3}/3)$.

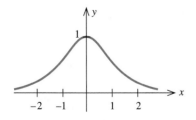

or $x = \pm\sqrt{3}/3$. The sign diagram for f'' is shown in Figure 4.39. From the sign diagram for f'', we see that f is concave upward on $(-\infty, -\sqrt{3}/3) \cup (\sqrt{3}/3, \infty)$ and concave downward on $(-\sqrt{3}/3, \sqrt{3}/3)$. Also, observe that $f''(x)$ changes sign as we move across the points $x = -\sqrt{3}/3$ and $x = \sqrt{3}/3$. Since

$$f\left(-\frac{\sqrt{3}}{3}\right) = \frac{1}{1/3 + 1} = \frac{3}{4} \qquad \text{and} \qquad f\left(\frac{\sqrt{3}}{3}\right) = \frac{3}{4}$$

we see that the points $(-\sqrt{3}/3, 3/4)$ and $(\sqrt{3}/3, 3/4)$ are inflection points of f. The graph of f is shown in Figure 4.40.   ■ ■ ■ ■

Group Discussion

1. Suppose $(c, f(c))$ is an inflection point of f. Can you conclude that f has a relative extremum at $x = c$? Explain your answer.

2. True or false: A polynomial function of degree 3 has exactly one inflection point.

Hint: Study the function $f(x) = ax^3 + bx^2 + cx + d \ (a \neq 0)$.

APPLICATIONS

Examples 6 and 7 illustrate familiar interpretations of the significance of the inflection point of a function.

EXAMPLE 6

The total sales S (in thousands of dollars) of the Arctic Air Corporation, a manufacturer of automobile air conditioners, is related to the amount of money x (in thousands of dollars) the company spends on advertising its products by the formula

$$S = -0.01x^3 + 1.5x^2 + 200 \qquad (0 \leq x \leq 100)$$

Find the inflection point of the function S.

SOLUTION ✔

The first two derivatives of S are given by

$$S' = -0.03x^2 + 3x$$
$$S'' = -0.06x + 3$$

Setting $S'' = 0$ gives $x = 50$ as the only candidate for an inflection point of S. Moreover, since

$$S'' > 0 \quad \text{for} \quad x < 50$$
$$S'' < 0 \quad \text{for} \quad x > 50$$

the point $(50, 2700)$ is an inflection point of the function S. The graph of S appears in Figure 4.41. Notice that this is the graph of the function we discussed earlier.

FIGURE 4.41
The graph of $S(x)$ has a point of inflection at (50, 2700).

EXAMPLE **7**   An economy's consumer price index (CPI) is described by the function

$$I(t) = -0.2t^3 + 3t^2 + 100 \qquad (0 \le t \le 9)$$

where $t = 0$ corresponds to the year 1991. Find the point of inflection of the function I and discuss its significance.

SOLUTION ✔   The first two derivatives of I are given by

$$I'(t) = -0.6t^2 + 6t$$
$$I''(t) = -1.2t + 6 = -1.2(t - 5)$$

Setting $I''(t) = 0$ gives $t = 5$ as the only candidate for an inflection point of I. Next, we observe that

$$I'' > 0 \quad \text{for} \quad t < 5$$
$$I'' < 0 \quad \text{for} \quad t > 5$$

so the point $(5, 150)$ is an inflection point of I. The graph of I is sketched in Figure 4.42.

Since the second derivative of I measures the rate of change of the inflation rate, our computations reveal that the rate of inflation had in fact peaked at $t = 5$. Thus, relief actually began at the beginning of 1996. ■ ■ ■ ■

FIGURE 4.42
The graph of $I(t)$ has a point of inflection at $(5, 150)$.

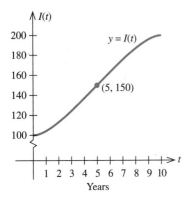

THE SECOND DERIVATIVE TEST

We now show how the second derivative f'' of a function f can be used to help us determine whether a critical point of f is a relative extremum of f. Figure 4.43a shows the graph of a function that has a relative maximum at $x = c$. Observe that f is concave downward at that point. Similarly, Figure 4.43b shows that at a relative minimum of f the graph is concave upward. But from our previous work we know that f is concave downward at $x = c$ if $f''(c) < 0$ and f is concave upward at $x = c$ if $f''(c) > 0$. These observations suggest the following alternative procedure for determining whether a critical point of f gives rise to a relative extremum of f. This result is called the **second derivative test** and is applicable when f'' exists.

FIGURE 4.43

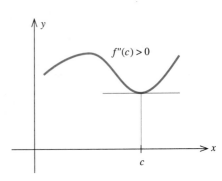

(a) f has a relative maximum at $x = c$.

(b) f has a relative minimum at $x = c$.

The Second Derivative Test

1. Compute $f'(x)$ and $f''(x)$.
2. Find all the critical points of f at which $f'(x) = 0$.
3. Compute $f''(c)$ for each such critical point c.
    a. If $f''(c) < 0$, then f has a relative maximum at c.
    b. If $f''(c) > 0$, then f has a relative minimum at c.
    c. If $f''(c) = 0$, the test fails; that is, it is inconclusive.

REMARK As stated in step 3c, the second derivative test does not yield a conclusion if $f''(c) = 0$ or if $f''(c)$ does not exist. In other words, $x = c$ may give rise to a relative extremum or an inflection point (see Exercise 91, page 320). In such cases, you should revert to the first derivative test.   ■ ■ ■

EXAMPLE 8 Determine the relative extrema of the function

$$f(x) = x^3 - 3x^2 - 24x + 32$$

using the second derivative test. (See Example 7, Section 4.1.)

SOLUTION ✔ We have

$$f'(x) = 3x^2 - 6x - 24 = 3(x + 2)(x - 4)$$

so $f'(x) = 0$ gives $x = -2$ and $x = 4$, the critical points of f, as in Example 7. Next, we compute

$$f''(x) = 6x - 6 = 6(x - 1)$$

Since

$$f''(-2) = 6(-2 - 1) = -18 < 0$$

the second derivative test implies that $f(-2) = 60$ is a relative maximum of f. Also,

$$f''(4) = 6(4 - 1) = 18 > 0$$

and the second derivative test implies that $f(4) = -48$ is a relative minimum of f, which confirms the results obtained earlier.   ■ ■ ■ ■

Group Discussion
Suppose a function f has the following properties:

1. $f'(x) > 0$ for all x in an interval (a, b).
2. There is a point c between a and b such that $f'(c) = 0$.

What special property can you ascribe to the point $(c, f(c))$? Answer the question if Property 1 is replaced by the property that $f''(x) < 0$ for all x in (a, b).

COMPARING THE FIRST AND SECOND DERIVATIVE TESTS

Notice that both the first derivative test and the second derivative test are used to classify the critical points of f. What are the pros and cons of the two tests? Since the second derivative test is applicable only when f'' exists, it is less versatile than the first derivative test. For example, it cannot be used to locate the relative minimum $f(0) = 0$ of the function $f(x) = x^{2/3}$.

Furthermore, the second derivative test is inconclusive when f'' is equal to zero at a critical point of f, whereas the first derivative test always yields positive conclusions. The second derivative test is also inconvenient to use when f'' is difficult to compute. On the plus side, if f'' is computed easily, then we use the second derivative test since it involves just the evaluation of f'' at the critical point(s) of f. Also, the conclusions of the second derivative test are important in theoretical work.

We close this section by summarizing the different roles played by the first derivative f' and the second derivative f'' of a function f in determining the properties of the graph of f. The first derivative f' tells us where f is increasing and where f is decreasing, whereas the second derivative f'' tells us where f is concave upward and where f is concave downward. These different properties of f are reflected by the signs of f' and f'' in the interval of interest. The following table shows the general characteristics of the function f for various possible combinations of the signs of f' and f'' in the interval (a, b).

Signs of f' and f'	Properties of the Graph of f	General Shape of the Graph of f
$f'(x) > 0$	f increasing	
$f''(x) > 0$	f concave upward	
$f'(x) > 0$	f increasing	
$f''(x) < 0$	f concave downward	
$f'(x) < 0$	f decreasing	
$f''(x) > 0$	f concave upward	
$f'(x) < 0$	f decreasing	
$f''(x) < 0$	f concave downward	

SELF-CHECK EXERCISES 4.2

1. Determine where the function $f(x) = 4x^3 - 3x^2 + 6$ is concave upward and where it is concave downward.

2. Using the second derivative test, if applicable, find the relative extrema of the function $f(x) = 2x^3 - \frac{1}{2}x^2 - 12x - 10$.

3. A certain country's gross domestic product (GDP) (in millions of dollars) in year t is described by the function

$$G(t) = -2t^3 + 45t^2 + 20t + 6000 \qquad (0 \le t \le 11)$$

where $t = 0$ corresponds to the beginning of the year 1989. Find the inflection point of the function G and discuss its significance.

Solutions to Self-Check Exercises 4.2 can be found on page 320.

4.2 Exercises

In Exercises 1–8, you are given the graph of a function f. Determine the intervals where f is concave upward and where it is concave downward. Also, find all inflection points of f, if any.

1.

2.

3.

4.

5.

6.

7.

8.

(b)

In Exercises 9–12, determine which graph—a, b, or c—is the graph of the function f with the specified properties.

(c)

9. $f(2) = 1$, $f'(2) > 0$, and $f''(2) < 0$

(a)

(b)

(c)

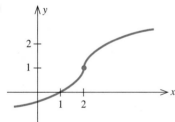

11. $f'(0)$ is undefined, f is decreasing on $(-\infty, 0)$, f is concave downward on $(0, 3)$, and f has an inflection point at $x = 3$.

(a)

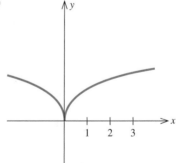

10. $f(1) = 2$, $f'(x) > 0$ on $(-\infty, 1) \cup (1, \infty)$, and $f''(1) = 0$

(a)

(b)

(c)

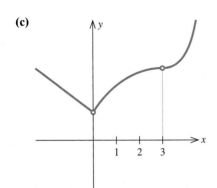

12. f is decreasing on $(-\infty, 2)$ and increasing on $(2, \infty)$, f is concave upward on $(1, \infty)$, and f has inflection points at $x = 0$ and $x = 1$.

(a)

(b)

(c)

13. EFFECT OF ADVERTISING ON BANK DEPOSITS The following graphs were used by the CEO of the Madison Savings Bank to illustrate what effect a projected promotional campaign would have on its deposits over the next year. The functions D_1 and D_2 give the projected amount of money on deposit with the bank over the next 12 mo with and without the proposed promotional campaign, respectively.
a. Determine the signs of $D_1'(t)$, $D_2'(t)$, $D_1''(t)$, and $D_2''(t)$ on the interval $(0, 12)$.
b. Explain the significance of your results in the context of the problem.

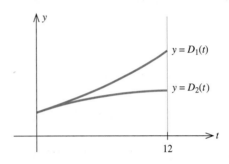

14. ASSEMBLY TIME OF A WORKER In the following graph, $N(t)$ gives the number of transistor radios assembled by the average worker by the tth hour, where $t = 0$ corresponds to 8 A.M. and $0 \leq t \leq 4$. Explain the significance of the inflection point P shown on the graph.

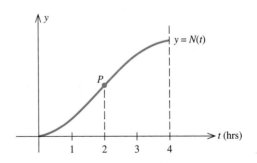

15. WATER POLLUTION When organic waste is dumped into a pond, the oxidation process that takes place reduces the pond's oxygen content. However, given time, nature will restore the oxygen content to its natural level. In the graph on page 316, $P(t)$ gives the oxygen content (as a percentage of its normal level) t days after organic waste has been dumped into the pond. Explain the significance of the inflection point Q.

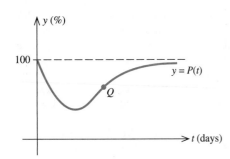

In Exercises 16–21 show that the function is concave upward wherever it is defined.

16. $f(x) = 4x^2 - 12x + 7$

17. $g(x) = x^4 + \dfrac{1}{2}x^2 + 6x + 10$

18. $h(x) = \dfrac{1}{x^2}$ **19.** $f(x) = \dfrac{1}{x^4}$

20. $h(x) = \sqrt{x^2 + 4}$ **21.** $g(x) = -\sqrt{4 - x^2}$

In Exercises 22–43, determine where the function is concave upward and where it is concave downward.

22. $g(x) = -x^2 + 3x + 4$ **23.** $f(x) = 2x^2 - 3x + 4$

24. $g(x) = x^3 - x$ **25.** $f(x) = x^3 - 1$

26. $f(x) = 3x^4 - 6x^3 + x - 8$

27. $f(x) = x^4 - 6x^3 + 2x + 8$

28. $f(x) = \sqrt[3]{x}$ **29.** $f(x) = x^{4/7}$

30. $g(x) = \sqrt{x - 2}$ **31.** $f(x) = \sqrt{4 - x}$

32. $g(x) = \dfrac{x}{x + 1}$ **33.** $f(x) = \dfrac{1}{x - 2}$

34. $g(x) = \dfrac{x}{1 + x^2}$ **35.** $f(x) = \dfrac{1}{2 + x^2}$

36. $f(x) = \dfrac{x + 1}{x - 1}$ **37.** $h(t) = \dfrac{t^2}{t - 1}$

38. $h(r) = -\dfrac{1}{(r - 2)^2}$ **39.** $g(x) = x + \dfrac{1}{x^2}$

40. $f(x) = (x - 2)^{2/3}$ **41.** $g(t) = (2t - 4)^{1/3}$

42. $f(x) = \dfrac{x^2 - 2}{x^3}$ **43.** $f(x) = \dfrac{x^2}{x^2 - 1}$

In Exercises 44–55, find the inflection points, if any, of each function.

44. $g(x) = x^3 - 6x$ **45.** $f(x) = x^3 - 2$

46. $g(x) = 2x^3 - 3x^2 + 18x - 8$

47. $f(x) = 6x^3 - 18x^2 + 12x - 15$

48. $f(x) = x^4 - 2x^3 + 6$ **49.** $f(x) = 3x^4 - 4x^3 + 1$

50. $f(x) = \sqrt[5]{x}$ **51.** $g(t) = \sqrt[3]{t}$

52. $f(x) = (x - 2)^{4/3}$ **53.** $f(x) = (x - 1)^3 + 2$

54. $f(x) = 2 + \dfrac{3}{x}$ **55.** $f(x) = \dfrac{2}{1 + x^2}$

In Exercises 56–73, find the relative extrema, if any, of each function. Use the second derivative test, if applicable.

56. $g(x) = 2x^2 + 3x + 7$ **57.** $f(x) = -x^2 + 2x + 4$

58. $g(x) = x^3 - 6x$ **59.** $f(x) = 2x^3 + 1$

60. $f(x) = 2x^3 + 3x^2 - 12x - 4$

61. $f(x) = \dfrac{1}{3}x^3 - 2x^2 - 5x - 10$

62. $f(t) = 2t + \dfrac{3}{t}$ **63.** $g(t) = t + \dfrac{9}{t}$

64. $f(x) = \dfrac{2x}{x^2 + 1}$ **65.** $f(x) = \dfrac{x}{1 - x}$

66. $g(x) = x^2 + \dfrac{2}{x}$ **67.** $f(t) = t^2 - \dfrac{16}{t}$

68. $g(x) = \dfrac{1}{1 + x^2}$ **69.** $g(s) = \dfrac{s}{1 + s^2}$

70. $f(x) = \dfrac{x^2}{x^2 + 1}$ **71.** $f(x) = \dfrac{x^4}{x - 1}$

72. $f(x) = \dfrac{x^2 + 4}{x^2 - 1}$ **73.** $g(x) = \dfrac{2 - x}{(x + 2)^3}$

74. EFFECT OF BUDGET CUTS ON DRUG-RELATED CRIMES The following graphs were used by a police commissioner to illustrate what effect a budget cut would have on crime in the city. The number $N_1(t)$ gives the projected number

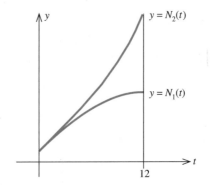

of drug-related crimes in the next 12 mo. The number $N_2(t)$ gives the projected number of drug-related crimes in the same time frame if next year's budget is cut.
a. Explain why $N_1'(t)$ and $N_2'(t)$ are both positive on the interval $(0, 12)$.
b. What are the signs of $N_1''(t)$ and $N_2''(t)$ on the interval $(0, 12)$?
c. Interpret the results of part (b).

75. **DEMAND FOR RNs** The following graph gives the total number of help-wanted ads for RNs (registered nurses) in 22 cities over the last 12 mo as a function of time t (t measured in months).
a. Explain why $N'(t)$ is positive on the interval $(0, 12)$.
b. Determine the signs of $N''(t)$ on the interval $(0, 6)$ and the interval $(6, 12)$.
c. Interpret the results of part (b).

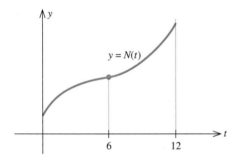

76. In the following figure, water is poured into the vase at a constant rate (in appropriate units), and the water level rises to a height of $f(t)$ units at time t as measured from the base of the vase. The graph of f follows. Explain the shape of the curve in terms of its concavity. What is the significance of the inflection point?

77. In the following figure, water is poured into an urn at a constant rate (in appropriate units), and the water level rises to a height of $f(t)$ units at time t as measured from the base of the urn. Sketch the graph of f and explain its shape, indicating where it is concave upward and concave downward. Indicate the inflection point on the graph and explain its significance.
Hint: Study Exercise 76.

78. **EFFECT OF ADVERTISING ON HOTEL REVENUE** The total annual revenue R of the Miramar Resorts Hotel is related to the amount of money x the hotel spends on advertising its services by the function

$$R(x) = -0.003x^3 + 1.35x^2 + 2x + 8000$$
$$(0 \le x \le 400)$$

where both R and x are measured in thousands of dollars. Find the inflection point of R and discuss its significance.

79. **EFFECT OF ADVERTISING ON SALES** The total sales S of the Cannon Precision Instruments Corporation is related to the amount of money x that Cannon spends on advertising its products by the function

$$S(x) = -0.002x^3 + 0.6x^2 + x + 500$$
$$(0 \le x \le 200)$$

where S and x are measured in thousands of dollars. Find the inflection point of the function S and discuss its significance.

80. **FORECASTING PROFITS** As a result of increasing energy costs, the growth rate of the profit of the 4-yr-old Venice Glassblowing Company has begun to decline. Venice's management, after consulting with energy experts, decides to implement certain energy-conservation measures aimed at cutting energy bills. The general manager reports that, according to his calculations, the growth rate of Venice's profit should be on the increase again within 4 yr. If Venice's profit (in hundreds of dollars) x years from now is given by the function

$$P(x) = x^3 - 9x^2 + 40x + 50 \qquad (0 \le x \le 8)$$

determine whether the general manager's forecast will be accurate.
Hint: Find the inflection point of the function P and study the concavity of P.

81. **WORKER EFFICIENCY** An efficiency study conducted for the Elektra Electronics Company showed that the number of Space Commander walkie-talkies assembled by the average worker t hours after starting work at 8 A.M. is given by

$$N(t) = -t^3 + 6t^2 + 15t \qquad (0 \le t \le 4)$$

At what time during the morning shift is the average worker performing at peak efficiency?

(continued on p. 320)

FINDING THE INFLECTION POINTS OF A FUNCTION

A graphing utility can be used to find the inflection points of a function and hence the intervals where the graph of the function is concave upward and the intervals where it is concave downward. Some graphing utilities have an operation for finding inflection points directly. If your graphing utility has this capability, use it to work through the example and exercises in this section.

EXAMPLE 1

Let $f(x) = 2.5x^5 - 12.4x^3 + 4.2x^2 - 5.2x + 4$.

a. Use a graphing utility to plot the graph of f.
b. Find the inflection points of f.
c. Find the intervals where f is concave upward and where it is concave downward.

SOLUTION ✔

a. The graph of f using the viewing rectangle $[-3, 3] \times [-25, 60]$ is shown in Figure T1.

FIGURE T1
The graph of f in the viewing rectangle $[-3, 3] \times [-25, 60]$

b. From Figure T1 we see that f has three inflection points—one occurring at the point where the x-coordinate is approximately -1, another at the point where $x \approx 0$, and the third at the point where $x \approx 1$. To find the first inflection point, we use the inflection operation, moving the cursor to the point on the graph of f where $x \approx -1$. We obtain the point $(-1.2728, 34.6395)$ (accurate to four decimal places). Next, setting the cursor near the point $x = 0$ yields the inflection point $(0.1139, 3.4440)$. Finally, with the cursor set at $x = 1$, we obtain the third inflection point $(1.1589, -10.4594)$.
c. From the results of part (b), we see that f is concave upward on the intervals $(-1.2728, 0.1139)$ and $(1.1589, \infty)$ and concave downward on $(-\infty, -1.2728)$ and $(0.1139, 1.1589)$. ∎

In Exercises 1–8, use a graphing utility to find (a) the intervals where *f* is concave upward and the intervals where *f* is concave downward and (b) the inflection points of *f*. Express your answers accurate to four decimal places.

1. $f(x) = 1.8x^4 - 4.2x^3 + 2.1x + 2$

2. $f(x) = -2.1x^4 + 3.1x^3 + 2x^2 - x + 1.2$

3. $f(x) = 1.2x^5 - 2x^4 + 3.2x^3 - 4x + 2$

4. $f(x) = -2.1x^5 + 3.2x^3 - 2.2x^2 + 4.2x - 4$

5. $f(x) = x^3(x^2 + 1)^{-1/3}$

6. $f(x) = x^2(x^3 - 1)^3$

7. $f(x) = \dfrac{x^2 - 1}{x^3}$

8. $f(x) = \dfrac{x + 1}{\sqrt{x}}$

9. GROWTH OF HMOs Based on data compiled by the Group Health Association of America, the number of people receiving their care in an HMO (Health Maintenance Organization) from the beginning of 1984 through 1994 is approximated by the function

$$f(t) = 0.0514t^3 - 0.853t^2 + 6.8147t + 15.6524$$
$$(0 < t \le 11)$$

where $f(t)$ gives the number of people in millions and t is measured in years, with $t = 0$ corresponding to the beginning of 1984.
a. Use a graphing utility to plot the graph of f in the viewing rectangle $[0, 12] \times [0, 120]$.
b. Find the points of inflection of f.
c. At what time in the given time interval was the number of people receiving their care at an HMO increasing fastest?
Source: Group Health Association of America

10. MANUFACTURING CAPACITY Data obtained from the Federal Reserve show that the annual increase in manufac-

turing capacity between 1988 and 1994 is given by

$$f(t) = 0.0388889t^3 - 0.283333t^2 + 0.477778t$$
$$+ 2.04286 \qquad (0 \le t \le 6)$$

where $f(t)$ is a percentage and t is measured in years, with $t = 0$ corresponding to the beginning of 1988.
a. Use a graphing utility to plot the graph of f in the viewing rectangle $[0, 8] \times [0, 4]$.
b. Find the point of inflection and interpret your result.
Source: Federal Reserve

11. Time on the Market According to the Greater Boston Real Estate Board—Multiple Listing Service, the average number of days a single-family home remains for sale from listing to accepted offer is approximated by the function

$$f(t) = 0.0171911t^4 - 0.662121t^3 + 6.18083t^2$$
$$- 8.97086t + 53.3357 \qquad (0 \le t \le 10)$$

where t is measured in years, with $t = 0$ corresponding to the beginning of 1984.
a. Use a graphing utility to plot the graph of f in the viewing rectangle $[0, 12] \times [0, 120]$.
b. Find the points of inflection and interpret your result.
Source: Greater Boston Real Estate Board—Multiple Listing Service

12. MULTIMEDIA SALES According to the Electronic Industries Association, sales in the multimedia market (hardware and software) are expected to be

$$S(t) = -0.0094t^4 + 0.1204t^3 - 0.0868t^2$$
$$+ 0.0195t + 3.3325 \qquad (0 \le t \le 10)$$

where $S(t)$ is measured in billions of dollars and t is measured in years, with $t = 0$ corresponding to 1990.
a. Plot the graph of S in the viewing rectangle $[0, 12] \times [0, 25]$.
b. Find the inflection point of S and interpret your result.
Source: Electronic Industries Association

82. COST OF PRODUCING CALCULATORS A subsidiary of Elektra Electronics manufactures programmable calculators. Management determines that the daily cost $C(x)$ (in dollars) of producing these calculators is

$$C(x) = 0.0001x^3 - 0.08x^2 + 40x + 5000$$

where x is the number of calculators produced. Find the inflection point of the function C and interpret your result.

83. FLIGHT OF A ROCKET The altitude (in feet) of a rocket t sec into flight is given by

$$s = f(t) = -t^3 + 54t^2 + 480t + 6$$

Find the point of inflection of the function f and interpret your result. What is the maximum velocity attained by the rocket?

84. AIR POLLUTION The level of ozone, an invisible gas that irritates and impairs breathing, present in the atmosphere on a certain May day in the city of Riverside was approximated by

$$A(t) = 1.0974t^3 - 0.0915t^4 \qquad (0 \le t \le 11)$$

where $A(t)$ is measured in pollutant standard index (PSI) and t is measured in hours, with $t = 0$ corresponding to 7 A.M. Use the second derivative test to show that the function A has a relative maximum at approximately $t = 9$. Interpret your results.

In Exercises 85–87, determine whether the statement is true or false. If it is true, explain why it is true. If it is false, give an example to show why it is false.

85. If the graph of f is concave upward on (a, b) then the graph of $-f$ is concave downward on (a, b).

86. If the graph of f is concave upward on (a, c) and concave downward on (c, b), where $a < c < b$, then f has an inflection point at $x = c$.

87. If $x = c$ is a critical point of f where $a < c < b$ and $f''(x) < 0$ on (a, b), then f has a relative maximum at $x = c$.

88. Show that the quadratic function

$$f(x) = ax^2 + bx + c \qquad (a \ne 0)$$

is concave upward if $a > 0$ and concave downward if $a < 0$. Thus, by examining the sign of the coefficient of x^2, one can tell immediately whether the parabola opens upward or downward.

89. Suppose f has an inflection point at $(a, f(a))$. Must the function f' have a relative extremum at $x = a$? Explain your answer.

90. Show that the cubic function

$$f(x) = ax^3 + bx^2 + cx + d \qquad (a \ne 0)$$

has one and only one inflection point. Find the coordinates of this point.

91. Consider the functions $f(x) = x^3$, $g(x) = x^4$, and $h(x) = -x^4$.
a. Show that $x = 0$ is a critical point of each of the functions f, g, and h.
b. Show that the second derivative of each of the functions f, g, and h equals zero at $x = 0$.
c. Show that f has neither a relative maximum nor a relative minimum at $x = 0$, that g has a relative minimum at $x = 0$, and that h has a relative maximum at $x = 0$.

SOLUTIONS TO SELF-CHECK EXERCISES 4.2

1. We first compute

$$f'(x) = 12x^2 - 6x$$
$$f''(x) = 24x - 6 = 6(4x - 1)$$

Observe that f'' is continuous everywhere and has a zero at $x = \frac{1}{4}$. The sign diagram of f'' is shown in the accompanying figure.

From the sign diagram for f'', we see that f is concave upward on $(\frac{1}{4}, \infty)$ and concave downward on $(-\infty, \frac{1}{4})$.

2. First, we find the critical points of f by solving the equation

$$f'(x) = 6x^2 - x - 12 = 0$$

That is,

$$(3x + 4)(2x - 3) = 0$$

giving $x = -\frac{4}{3}$ and $x = \frac{3}{2}$. Next, we compute

$$f''(x) = 12x - 1$$

Since

$$f''\left(-\frac{4}{3}\right) = 12\left(-\frac{4}{3}\right) - 1 = -17 < 0$$

the second derivative test implies that $f\left(-\frac{4}{3}\right) = \frac{10}{27}$ is a relative maximum of f. Also,

$$f''\left(\frac{3}{2}\right) = 12\left(\frac{3}{2}\right) - 1 = 17 > 0$$

and we see that $f(\frac{3}{2}) = -\frac{179}{8}$ is a relative minimum.

3. We compute the second derivative of G. Thus,

$$G'(t) = -6t^2 + 90t + 20$$
$$G''(t) = -12t + 90$$

Now, G'' is continuous everywhere, and $G''(t) = 0$, where $t = \frac{15}{2}$, giving $t = \frac{15}{2}$ as the only candidate for an inflection point of G. Since $G''(t) > 0$ for $t < \frac{15}{2}$ and $G''(t) < 0$ for $t > \frac{15}{2}$, we see that the point $(\frac{15}{2}, \frac{15,675}{2})$ is an inflection point of G. The results of our computations tell us that the country's GDP was increasing most rapidly at the beginning of July 1996.

4.3 Curve Sketching

A REAL-LIFE EXAMPLE

As we have seen on numerous occasions, the graph of a function is a useful aid for visualizing the function's properties. From a practical point of view, the graph of a function also gives, at one glance, a complete summary of all the information captured by the function.

Consider, for example, the graph of the function giving the Dow-Jones Industrial Average (DJIA) on Black Monday, October 19, 1987 (Figure 4.44). Here, $t = 0$ corresponds to 8:30 A.M., when the market was open for business, and $t = 7.5$ corresponds to 4 P.M., the closing time. The following information may be gleaned from studying the graph.

The graph is *decreasing* rapidly from $t = 0$ to $t = 1$, reflecting the sharp drop in the index in the first hour of trading. The point (1, 2047) is a *relative minimum* point of the function, and this turning point coincides with the start

FIGURE 4.44
The Dow-Jones Industrial Average on Black Monday

Source: *Wall Street Journal*

of an aborted recovery. The short-lived rally, represented by the portion of the graph that is *increasing* on the interval (1, 2), quickly fizzled out at $t = 2$ (10:30 A.M.). The *relative maximum* point (2, 2150) marks the highest point of the recovery. The function is decreasing in the rest of the interval. The point (4, 2006) is an *inflection point* of the function; it shows that there was a temporary respite at $t = 4$ (12:30 P.M.). However, selling pressure continued unabated, and the DJIA continued to fall until the closing bell. Finally, the graph also shows that the index opened at the high of the day [$f(0) = 2247$ is the *absolute maximum* of the function] and closed at the low of the day [$f(15/2) = 1739$ is the *absolute minimum* of the function], a drop of 508 points!*

Before we turn our attention to the actual task of sketching the graph of a function, let's look at some properties of graphs that will be helpful in this connection.

FIGURE 4.45
The graph of *f* has a vertical asymptote at $x = 1$.

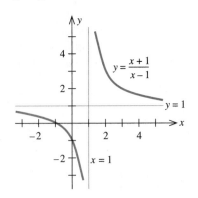

VERTICAL ASYMPTOTES

Before going on, you might want to review the material on one-sided limits and the limit at infinity of a function (Sections 2.4 and 2.5).

Consider the graph of the function

$$f(x) = \frac{x + 1}{x - 1}$$

shown in Figure 4.45. Observe that $f(x)$ increases without bound (tends to infinity) as x approaches $x = 1$ from the right; that is,

$$\lim_{x \to 1^+} \frac{x + 1}{x - 1} = \infty$$

---

* Absolute maxima and absolute minima of functions are covered in Section 4.4.

You can verify this by taking a sequence of values of x approaching $x = 1$ from the right and looking at the corresponding values of $f(x)$.

Here is another way of looking at the situation: Observe that if x is a number that is a little larger than 1, then both $(x + 1)$ and $(x - 1)$ are positive, so $(x + 1)/(x - 1)$ is also positive. As x approaches $x = 1$, the numerator $(x + 1)$ approaches the number 2, but the denominator $(x - 1)$ approaches zero, so the quotient $(x + 1)/(x - 1)$ approaches infinity, as observed earlier. The line $x = 1$ is called a *vertical asymptote* of the graph of f.

For the function $f(x) = (x + 1)/(x - 1)$, you can show that

$$\lim_{x \to 1^-} \frac{x + 1}{x - 1} = -\infty$$

and this tells us how $f(x)$ approaches the asymptote $x = 1$ from the left.

More generally, we have the following definition:

Vertical Asymptote

The line $x = a$ is a **vertical asymptote** of the graph of a function f if either

$$\lim_{x \to a^+} f(x) = \infty \quad \text{or} \quad -\infty$$

or

$$\lim_{x \to a^-} f(x) = \infty \quad \text{or} \quad -\infty$$

REMARK Although a vertical asymptote of a graph is not part of the graph, it serves as a useful aid for sketching the graph. ■ ■ ■

For rational functions

$$f(x) = \frac{P(x)}{Q(x)}$$

there is a simple criterion for determining whether the graph of f has any vertical asymptotes.

Finding Vertical Asymptotes of Rational Functions

Suppose f is a rational function

$$f(x) = \frac{P(x)}{Q(x)}$$

where P and Q are polynomial functions. Then, the line $x = a$ is a vertical asymptote of the graph of f if $Q(a) = 0$ but $P(a) \neq 0$.

For the function

$$f(x) = \frac{x + 1}{x - 1}$$

considered earlier, $P(x) = x + 1$ and $Q(x) = x - 1$. Observe that $Q(1) = 0$ but $P(1) = 2 \neq 0$, so $x = 1$ is a vertical asymptote of the graph of f.

EXAMPLE 1   Find the vertical asymptotes of the graph of the function

$$f(x) = \frac{x^2}{4 - x^2}$$

SOLUTION ✔   The function f is a rational function with $P(x) = x^2$ and $Q(x) = 4 - x^2$. The zeros of Q are found by solving

$$4 - x^2 = 0$$

—that is,

$$(2 - x)(2 + x) = 0$$

giving $x = -2$ and $x = 2$. These are candidates for the vertical asymptotes of the graph of f. Examining $x = -2$, we compute $P(-2) = (-2)^2 = 4 \neq 0$, and we see that $x = -2$ is indeed a vertical asymptote of the graph of f. Similarly, we find $P(2) = 2^2 = 4 \neq 0$, and so $x = 2$ is also a vertical asymptote of the graph of f. The graph of f sketched in Figure 4.46 confirms these results.

FIGURE 4.46
$x = -2$ and $x = 2$ are vertical asymptotes of the graph of f.

 Recall that in order for the line $x = a$ to be a vertical asymptote of the graph of a rational function f, *only* the denominator of $f(x)$ must be equal to zero at $x = 0$. If *both* $P(a)$ and $Q(a)$ are equal to zero, then $x = a$ need *not* be a vertical asymptote. For example, look at the function

$$f(x) = \frac{4(x^2 - 4)}{x - 2}$$

whose graph appears in Figure 2.24a, page 113.

HORIZONTAL ASYMPTOTES

Let's return to the function f defined by

$$f(x) = \frac{x + 1}{x - 1}$$

(Figure 4.47).

FIGURE 4.47
The graph of f has a horizontal asymptote at $y = 1$.

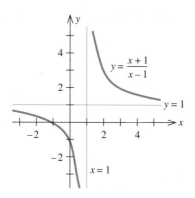

Observe that $f(x)$ approaches the horizontal line $y = 1$ as x approaches infinity, and, in this case, $f(x)$ approaches $y = 1$ as x approaches minus infinity as well. The line $y = 1$ is called a horizontal asymptote of the graph of f. More generally, we have the following definition.

Horizontal Asymptote

> The line $y = b$ is a **horizontal asymptote** of the graph of a function f if either
> $$\lim_{x \to \infty} f(x) = b \qquad \text{or} \qquad \lim_{x \to -\infty} f(x) = b$$

For the function

$$f(x) = \frac{x + 1}{x - 1}$$

we see that

$$\lim_{x \to \infty} \frac{x + 1}{x - 1} = \lim_{x \to \infty} \frac{1 + \dfrac{1}{x}}{1 - \dfrac{1}{x}} \qquad \text{(Divide numerator and denominator by } x.\text{)}$$

$$= 1$$

Also,

$$\lim_{x \to -\infty} \frac{x + 1}{x - 1} = \lim_{x \to -\infty} \frac{1 + \dfrac{1}{x}}{1 - \dfrac{1}{x}}$$

$$= 1$$

In either case, we conclude that $y = 1$ is a horizontal asymptote of the graph of f, as observed earlier.

EXAMPLE **2**　Find the horizontal asymptotes of the graph of the function

$$f(x) = \frac{x^2}{4 - x^2}$$

SOLUTION ✔　We compute

$$\lim_{x \to \infty} \frac{x^2}{4 - x^2} = \lim_{x \to \infty} \frac{1}{\dfrac{4}{x^2} - 1} \qquad \text{(Dividing numerator and denominator by } x^2\text{)}$$

$$= -1$$

and so $y = -1$ is a horizontal asymptote, as before. The graph of f sketched in Figure 4.48 confirms this result.

FIGURE 4.48
The graph of f has a horizontal asymptote at $y = -1$.

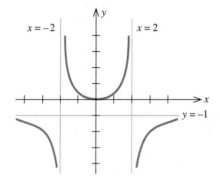

We next state an important property of polynomial functions.

> A polynomial function has no vertical or horizontal asymptotes.

To see this, note that a polynomial function $P(x)$ can be written as a rational function with denominator equal to 1. Thus,

$$P(x) = \frac{P(x)}{1}$$

Since the denominator is never equal to zero, P has no vertical asymptotes. Next, if P is a polynomial of degree greater than or equal to 1, then

$$\lim_{x \to \infty} P(x) \qquad \text{and} \qquad \lim_{x \to -\infty} P(x)$$

are either infinity or minus infinity; that is, they do not exist. Therefore, P has no horizontal asymptotes.

In the last two sections, we saw how the first and second derivatives of a function are used to reveal various properties of the graph of a function f. We now show how this information can be used to help us sketch the graph of f. We begin by giving a general procedure for curve sketching.

A Guide to Curve Sketching

1. Determine the domain of f.
2. Find the x- and y-intercepts of f.*
3. Determine the behavior of f for large absolute values of x.
4. Find all horizontal and vertical asymptotes of f.
5. Determine the intervals where f is increasing and where f is decreasing.
6. Find the relative extrema of f.
7. Determine the concavity of f.
8. Find the inflection points of f.
9. Plot a few additional points to help further identify the shape of the graph of f and sketch the graph.

We now illustrate the techniques of curve sketching with several examples.

TWO STEP-BY-STEP EXAMPLES

EXAMPLE 3   Sketch the graph of the function

$$y = f(x) = x^3 - 6x^2 + 9x + 2$$

SOLUTION ✔   Obtain the following information on the graph of f.

1. The domain of f is the interval $(-\infty, \infty)$.
2. By setting $x = 0$, we find that the y-intercept is 2. The x-intercept is found by setting $y = 0$, which in this case leads to a cubic equation. Since the solution is not readily found, we will not use this information.
3. Since

$$\lim_{x \to -\infty} f(x) = \lim_{x \to -\infty} (x^3 - 6x^2 + 9x + 2) = -\infty$$

$$\lim_{x \to \infty} f(x) = \lim_{x \to \infty} (x^3 - 6x^2 + 9x + 2) = \infty$$

we see that f decreases without bound as x decreases without bound and that f increases without bound as x increases without bound.
4. Since f is a polynomial function, there are no asymptotes.

$$f'(x) = 3x^2 - 12x + 9 = 3(x^2 - 4x + 3)$$
$$= 3(x - 3)(x - 1)$$

5. Setting $f'(x) = 0$ gives $x = 1$ or $x = 3$. The sign diagram for f' shows that f is increasing on the intervals $(-\infty, 1)$ and $(3, \infty)$ and decreasing on the interval $(1, 3)$ (Figure 4.49).

FIGURE 4.49
Sign diagram for f'

---

* The equation $f(x) = 0$ may be difficult to solve, in which case one may decide against finding the x-intercepts or use technology, if available, for assistance.

6. From the results of step 5, we see that $x = 1$ and $x = 3$ are critical points of f. Furthermore, f' changes sign from positive to negative as we move across $x = 1$, so a relative maximum of f occurs at $x = 1$. Similarly, we see that a relative minimum of f occurs at $x = 3$. Now,

$$f(1) = 1 - 6 + 9 + 2 = 6$$
$$f(3) = 3^3 - 6(3)^2 + 9(3) + 2 = 2$$

so $f(1) = 6$ is a relative maximum of f and $f(3) = 2$ is a relative minimum of f.

7.
$$f''(x) = 6x - 12 = 6(x - 2)$$

which is equal to zero when $x = 2$. The sign diagram of f'' shows that f is concave downward on the interval $(-\infty, 2)$ and concave upward on the interval $(2, \infty)$ (Figure 4.50).

8. From the results of step 7, we see that f'' changes sign as we move across the point $x = 2$. Next,

$$f(2) = 2^3 - 6(2)^2 + 9(2) + 2 = 4$$

and so the required inflection point of f is $(2, 4)$.

9. Summarizing, we have

FIGURE 4.50
Sign diagram for f''

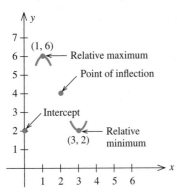

Domain	$(-\infty, \infty)$
Intercept	$(0, 2)$
$\lim\limits_{x \to -\infty} f(x); \lim\limits_{x \to \infty} f(x)$	$-\infty; \infty$
Asymptotes	None
Intervals where f is ↗ or ↘	↗ on $(-\infty, 1) \cup (3, \infty)$; ↘ on $(1, 3)$
Relative extrema	Rel. max. at $(1, 6)$; rel. min. at $(3, 2)$
Concavity	Downward on $(-\infty, 2)$; upward on $(2, \infty)$
Point of inflection	$(2, 4)$

In general, it is a good idea to start graphing by plotting the intercept, relative extrema, and inflection point (Figure 4.51). Then, using the rest of the information, we complete the graph of f, as sketched in Figure 4.52.

FIGURE 4.51
We first plot the intercept, the relative extrema, and the inflection point.

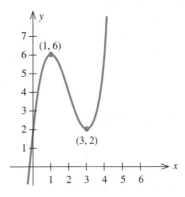

FIGURE 4.52
The graph of $y = x^3 - 6x^2 + 9x + 2$

> **Group Discussion**
>
> The average price of gasoline at the pump over a 3-month period, during which there was a temporary shortage of oil, is described by the function f defined on the interval $[0, 3]$. During the first month, the price was increasing at an increasing rate. Starting with the second month, the good news was that the rate of increase was slowing down, although the price of gas was still increasing. This pattern continued until the end of the second month. The price of gas peaked at the end of $t = 2$ and began to fall at an increasing rate until $t = 3$.
> **1.** Describe the signs of $f'(t)$ and $f''(t)$ over each of the intervals $(0, 1)$, $(1, 2)$, and $(2, 3)$.
> **2.** Make a sketch showing a plausible graph of f over $[0, 3]$.

EXAMPLE 4   Sketch the graph of the function

$$y = f(x) = \frac{x + 1}{x - 1}$$

SOLUTION ✔   Obtain the following information:

1. f is undefined when $x = 1$, so the domain of f is the set of all real numbers other than $x = 1$.

2. Setting $y = 0$ gives -1, the x-intercept of f. Next, setting $x = 0$ gives -1 as the y-intercept of f.

3. Earlier we found that

$$\lim_{x \to \infty} \frac{x + 1}{x - 1} = 1 \quad \text{and} \quad \lim_{x \to -\infty} \frac{x + 1}{x - 1} = 1$$

(see page 325). Consequently, we see that $f(x)$ approaches the line $y = 1$ as $|x|$ becomes arbitrarily large. For $x > 1$, $f(x) > 1$ and $f(x)$ approaches the line $y = 1$ from above. For $x < 1$, $f(x) < 1$, so $f(x)$ approaches the line $y = 1$ from below.

4. The straight line $x = 1$ is a vertical asymptote of the graph of f. Also, from the results of step 3, we conclude that $y = 1$ is a horizontal asymptote of the graph of f.

5.
$$f'(x) = \frac{(x - 1)(1) - (x + 1)(1)}{(x - 1)^2} = -\frac{2}{(x - 1)^2}$$

and is discontinuous at $x = 1$. The sign diagram of f' shows that $f'(x) < 0$ whenever it is defined. Thus, f is decreasing on the intervals $(-\infty, 1)$ and $(1, \infty)$ (Figure 4.53).

6. From the results of step 5, we see that there are no critical points of f since $f'(x)$ is never equal to zero for any value of x in the domain of f.

FIGURE 4.53
The sign diagram for f'

f' is not defined here

FIGURE 4.54
The sign diagram for f''

f'' is not defined here

FIGURE 4.55
The graph of f has a horizontal asymptote at $y = 1$ and a vertical asymptote at $x = 1$.

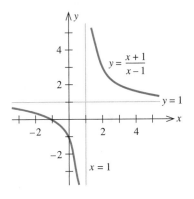

7.
$$f''(x) = \frac{d}{dx}[-2(x-1)^{-2}] = 4(x-1)^{-3} = \frac{4}{(x-1)^3}$$

The sign diagram of f'' shows immediately that f is concave downward on the interval $(-\infty, 1)$ and concave upward on the interval $(1, \infty)$ (Figure 4.54).

8. From the results of step 7, we see that there are no candidates for inflection points of f since $f''(x)$ is never equal to zero for any value of x in the domain of f. Hence, f has no inflection points.

9. Summarizing, we have

Domain	$(-\infty, 1) \cup (1, \infty)$
Intercepts	$(0, -1)$; $(-1, 0)$
$\lim_{x \to -\infty} f(x)$; $\lim_{x \to \infty} f(x)$	1; 1
Asymptotes	$x = 1$ is a vertical asymptote; $y = 1$ is a horizontal asymptote
Intervals where f is ↗ or ↘	↘ on $(-\infty, 1) \cup (1, \infty)$
Relative extrema	None
Concavity	Downward on $(-\infty, 1)$; upward on $(1, \infty)$
Points of inflection	None

The graph of f is sketched in Figure 4.55.   ■ ■ ■ ■

SELF-CHECK EXERCISES 4.3

1. Find the horizontal and vertical asymptotes of the graph of the function

$$f(x) = \frac{2x^2}{x^2 - 1}$$

2. Sketch the graph of the function

$$f(x) = \frac{2}{3}x^3 - 2x^2 - 6x + 4$$

Solutions to Self-Check Exercises 4.3 can be found on page 339.

4.3 Exercises

In Exercises 1–10, find the horizontal and vertical asymptotes of the graph.

1.

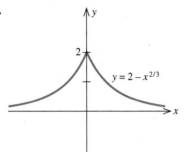

$y = 2 - x^{2/3}$

2.

$y = \dfrac{1}{(x+1)^2}$

3.

$y = \dfrac{1}{x^3}$

4.

$y = \dfrac{1}{x^2+1}$

5.

$y = \dfrac{x}{x^2-1}$

6.

$y = \dfrac{x}{x^2+1}$

7.

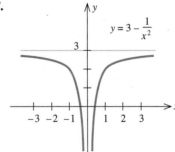

$y = 3 - \dfrac{1}{x^2}$

8.

$y = \dfrac{1}{(x+2)^2}$

9.

10.

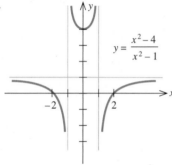

In Exercises 11–28, find the horizontal and vertical asymptotes of the graph of the function. (You need not sketch the graph.)

11. $f(x) = \dfrac{1}{x}$

12. $f(x) = \dfrac{1}{x + 2}$

13. $f(x) = -\dfrac{2}{x^2}$

14. $g(x) = \dfrac{1}{1 + 2x^2}$

15. $f(x) = \dfrac{x - 1}{x + 1}$

16. $g(t) = \dfrac{t + 1}{2t - 1}$

17. $h(x) = x^3 - 3x^2 + x + 1$

18. $g(x) = 2x^3 + x^2 + 1$

19. $f(t) = \dfrac{t^2}{t^2 - 9}$

20. $g(x) = \dfrac{x^3}{x^2 - 4}$

21. $f(x) = \dfrac{3x}{x^2 - x - 6}$

22. $g(x) = \dfrac{2x}{x^2 + x - 2}$

23. $g(t) = 2 + \dfrac{5}{(t - 2)^2}$

24. $f(x) = 1 + \dfrac{2}{x - 3}$

25. $f(x) = \dfrac{x^2 - 2}{x^2 - 4}$

26. $h(x) = \dfrac{2 - x^2}{x^2 + x}$

27. $g(x) = \dfrac{x^3 - x}{x(x + 1)}$

28. $f(x) = \dfrac{x^4 - x^2}{x(x - 1)(x + 2)}$

In Exercises 29 and 30, you are given the graphs of two functions f and g. One function is the derivative function of the other. Identify each of them.

29.

30.

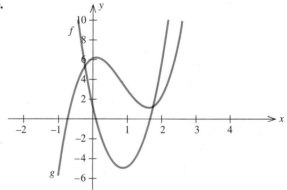

31. TERMINAL VELOCITY A skydiver leaps from the gondola of a hot-air balloon. As she free-falls, air resistance, which is proportional to her velocity, builds up to a point where it balances the force due to gravity. The resulting motion may be described in terms of her velocity as follows: Starting at rest (zero velocity), her velocity increases and approaches a constant velocity, called the *terminal velocity*. Sketch a graph of her velocity v versus time t.

In Exercises 32–35, use the information summarized in the table to sketch the graph of *f*.

32. $f(x) = x^3 - 3x^2 + 1$

Domain	$(-\infty, \infty)$
Intercept	y-intercept: 1
Asymptotes	None
Intervals where f is ↗ and ↘	↗ on $(-\infty, 0) \cup (2, \infty)$; ↘ on $(0, 2)$
Relative extrema	Rel. max. at $(0, 1)$; rel. min. at $(2, -3)$
Concavity	Downward on $(-\infty, 1)$; upward on $(1, \infty)$
Point of inflection	$(1, -1)$

33. $f(x) = \dfrac{1}{9}(x^4 - 4x^3)$

Domain	$(-\infty, \infty)$
Intercepts	x-intercepts: 0, 4; y-intercept: 0
Asymptotes	None
Intervals where f is ↗ and ↘	↗ on $(3, \infty)$; ↘ on $(-\infty, 3)$
Relative extrema	Rel. min. at $(3, -3)$
Concavity	Downward on $(0, 2)$; upward on $(-\infty, 0) \cup (2, \infty)$
Points of inflection	$(0, 0)$ and $(2, -16/9)$

34. $f(x) = \dfrac{4x - 4}{x^2}$

Domain	$(-\infty, 0) \cup (0, \infty)$
Intercept	x-intercept: 1
Asymptotes	x-axis and y-axis
Intervals where f is ↗ and ↘	↗ on $(0, 2)$; ↘ on $(-\infty, 0) \cup (2, \infty)$
Relative extrema	Rel. max. at $(2, 1)$
Concavity	Downward on $(-\infty, 0) \cup (0, 3)$; upward on $(3, \infty)$
Points of inflection	$(3, 8/9)$

35. $f(x) = x - 3x^{1/3}$

Domain	$(-\infty, \infty)$
Intercepts	x-intercepts: $\pm 3\sqrt{3}$
Asymptotes	None
Intervals where f is ↗ and ↘	↗ on $(-\infty, -1) \cup (1, \infty)$; ↘ on $(-1, 1)$
Relative extrema	Rel. max. at $(-1, 2)$; rel. min. at $(1, -2)$
Concavity	Downward on $(-\infty, 0)$; upward on $(0, \infty)$
Points of inflection	$(0, 0)$

In Exercises 36–59, sketch the graph of the function, using the curve-sketching guide of this section.

36. $f(x) = x^2 - 2x + 3$

37. $g(x) = 4 - 3x - 2x^3$

38. $f(x) = 2x^3 + 1$

39. $h(x) = x^3 - 3x + 1$

40. $f(t) = 2t^3 - 15t^2 + 36t - 20$

41. $f(x) = -2x^3 + 3x^2 + 12x + 2$

42. $f(t) = 3t^4 + 4t^3$

43. $h(x) = \dfrac{3}{2}x^4 - 2x^3 - 6x^2 + 8$

44. $f(x) = \sqrt{x^2 + 5}$

45. $f(t) = \sqrt{t^2 - 4}$

46. $f(x) = \sqrt[3]{x^2}$

47. $g(x) = \dfrac{1}{2}x - \sqrt{x}$

48. $f(x) = \dfrac{1}{x + 1}$

49. $g(x) = \dfrac{2}{x - 1}$

50. $g(x) = \dfrac{x}{x - 1}$

51. $h(x) = \dfrac{x + 2}{x - 2}$

52. $g(x) = \dfrac{x}{x^2 - 4}$

53. $f(t) = \dfrac{t^2}{1 + t^2}$

54. $f(x) = \dfrac{x^2 - 9}{x^2 - 4}$

55. $g(t) = -\dfrac{t^2 - 2}{t - 1}$

56. $h(x) = \dfrac{1}{x^2 - x - 2}$

57. $g(t) = \dfrac{t + 1}{t^2 - 2t - 1}$

58. $g(x) = (x + 2)^{3/2} + 1$

59. $h(x) = (x - 1)^{2/3} + 1$

60. COST OF REMOVING TOXIC POLLUTANTS A city's main well was recently found to be contaminated with trichloroethylene (a cancer-causing chemical) as a result of an abandoned chemical dump leaching chemicals into the water. A proposal submitted to the city council indicated that the cost, measured in millions of dollars, of removing $x\%$ of the toxic pollutants is given by

$$C(x) = \frac{0.5x}{100 - x}$$

a. Find the vertical asymptote of $C(x)$.
b. Is it possible to remove 100% of the toxic pollutant from the water?

61. AVERAGE COST OF PRODUCING VIDEO DISCS The average cost per disc (in dollars) incurred by the Herald Record Company in pressing x video discs is given by the average cost function

$$\overline{C}(x) = 2.2 + \frac{2500}{x}$$

a. Find the horizontal asymptote of $\overline{C}(x)$.
b. What is the limiting value of the average cost?

(continued on p. 338)

ANALYZING THE PROPERTIES OF A FUNCTION

One of the main purposes of studying Section 4.3 is to see how the many concepts of calculus come together to paint a picture of a function. The techniques of graphing also play a very practical role. For example, using the techniques of graphing developed in Section 4.3, you can tell if the graph of a function generated by a graphing utility is reasonably complete. Furthermore, these techniques can often reveal details that are missing from a graph.

EXAMPLE 1

Consider the function $f(x) = 2x^3 - 3.5x^2 + x - 10$. A plot of the graph of f in the standard viewing rectangle is shown in Figure T1. Since the domain of f is the interval $(-\infty, \infty)$, we see that Figure T1 does not reveal the part of the graph to the left of the y-axis. This suggests that we enlarge the viewing rectangle accordingly. Figure T2 shows the graph of f in the viewing rectangle $[-10, 10] \times [-20, 10]$.

FIGURE T1

The graph of f in the standard viewing rectangle

FIGURE T2

The graph of f in the viewing rectangle $[-10, 10] \times [-20, 10]$

The behavior of f for large values of f [$\lim_{x \to -\infty} f(x) = -\infty$ and $\lim_{x \to \infty} f(x) = \infty$] suggests that this viewing rectangle has captured a sufficiently complete picture of f. Next, an analysis of the first derivative of f,

$$f'(x) = 6x^2 - 7x + 1 = (6x - 1)(x - 1)$$

reveals that f has critical values at $x = 1/6$ and $x = 1$. In fact, a sign diagram of f' shows that f has a relative maximum at $x = 1/6$ and a relative minimum at $x = 1$, details that are not revealed in the graph of f shown in Figure T2. To examine this portion of the graph of f, we use, say, the viewing rectangle $[-1, 2] \times [-11, -9]$. The resulting graph of f is shown in Figure T3, which certainly reveals the hitherto missing details! Thus, through an interaction of calculus and a graphing utility, we are able to obtain a good picture of the properties of f.

FINDING x-INTERCEPTS

As noted in Section 4.3, it is not always easy to find the x-intercepts of the graph of a function. But this information is very important in applications. By using the function for solving polynomial equations or the function for finding the roots of an equation, we can solve the equation $f(x) = 0$ quite easily and hence yield the x-intercepts of the graph of a function.

EXAMPLE 2

Let $f(x) = x^3 - 3x^2 + x + 1.5$.

a. Use the function for solving polynomial equations on a graphing utility to find the x-intercepts of the graph of f.
b. Use the function for finding the roots of an equation on a graphing utility to find the x-intercepts of the graph of f.

SOLUTION ✔

a. Observe that f is a polynomial function of degree 3, and so we may use the function for solving polynomial equations to solve the equation $x^3 - 3x^2 + x + 1.5 = 0$ $[f(x) = 0]$. We find that the solutions (x-intercepts) are

$$x_1 \approx -0.525687120865, \qquad x_2 \approx 1.2586520225, \qquad x_3 \approx 2.26703509836$$

FIGURE T4
The graph of
$f(x) = x^3 - 3x^2 + x + 1.5$

b. Using the graph of f (Figure T4), we see that $x_1 \approx -0.5$, $x_2 \approx 1$, and $x_3 \approx 2$. Using the function for finding the roots of an equation on a graphing utility, and these values of x as initial guesses, we find

$$x_1 \approx -0.5256871209, \qquad x_2 \approx 1.2586520225, \qquad x_3 \approx 2.2670350984$$

REMARK The function for solving polynomial equations on a graphing utility will solve a polynomial equation $f(x) = 0$, where f is a polynomial function. The function for finding the roots of a polynomial, however, will solve equations $f(x) = 0$ even if f is not a polynomial.

EXAMPLE 3

Unless payroll taxes are increased significantly and/or benefits are scaled back drastically, it is a matter of time before the current Social Security system goes broke. Based on data from the Board of Trustees of the Social Security Administration, the assets of the system—the Social Security "trust fund"—may be approximated by

$$f(t) = -0.0129t^4 + 0.3087t^3 + 2.1760t^2 + 62.8466t + 506.2955 \qquad (0 \le t \le 35)$$

where $f(t)$ is measured in millions of dollars and t is measured in years, with $t = 0$ corresponding to 1995.

a. Use a graphing calculator to sketch the graph of f.
b. Based on this model, when can the Social Security system be expected to go broke?
Source: Social Security Administration

SOLUTION ✔ **a.** The graph of f in the window $[0, 35] \times [-1000, 3500]$ is shown in Figure T5.
b. Using the function for finding the roots on a graphing utility, we find that $y = 0$ when $t \approx 34.1$, and this tells us that the system is expected to go broke around 2029.

■ ■ ■ ■

Exercises

In Exercises 1–4, use the method of Example 1 to analyze the function. (*Note:* Your answers will not be unique.)

1. $f(x) = 4x^3 - 4x^2 + x + 10$

2. $f(x) = x^3 + 2x^2 + x - 12$

3. $f(x) = \frac{1}{2}x^4 + x^3 + \frac{1}{2}x^2 - 10$

4. $f(x) = 2.25x^4 - 4x^3 + 2x^2 + 2$

In Exercises 5–10, find the *x*-intercepts of the graph of *f*. Give your answer accurate to four decimal places.

5. $f(x) = 0.2x^3 - 1.2x^2 + 0.8x + 2.1$

6. $f(x) = -0.5x^3 + 1.7x^2 - 1.2$

7. $f(x) = 0.3x^4 - 1.2x^3 + 0.8x^2 + 1.1x - 2$

8. $f(x) = -0.2x^4 + 0.8x^3 - 2.1x + 1.2$

9. $f(x) = 2x^2 - \sqrt{x + 1} - 3$

10. $f(x) = x - \sqrt{1 - x^2}$

62. **CONCENTRATION OF A DRUG IN THE BLOODSTREAM**   The concentration (in milligrams per cubic centimeter) of a certain drug in a patient's bloodstream t hr after injection is given by

$$C(t) = \frac{0.2t}{t^2 + 1}$$

a. Find the horizontal asymptote of $C(t)$.
b. Interpret your result.

63. **EFFECT OF ENZYMES ON CHEMICAL REACTIONS**   Certain proteins, known as enzymes, serve as catalysts for chemical reactions in living things. In 1913 Leonor Michaelis and L. M. Menten discovered the following formula giving the initial speed V (in moles per liter per second) at which the reaction begins in terms of the amount of substrate x (the substance that is being acted upon, measured in moles per liter):

$$V = \frac{ax}{x + b}$$

where a and b are positive constants.
a. Find the horizontal asymptote of V.
b. Interpret your result.

64. **WORKER EFFICIENCY**   An efficiency study showed that the total number of cordless telephones assembled by an average worker at Delphi Electronics t hr after starting work at 8 A.M. is given by

$$N(t) = -\frac{1}{2}t^3 + 3t^2 + 10t \qquad (0 \le t \le 4)$$

Sketch the graph of the function N and interpret your results.

65. **GDP OF A DEVELOPING COUNTRY**   A developing country's gross domestic product (GDP) from 1992 to 2000 is approximated by the function

$$G(t) = -0.2t^3 + 2.4t^2 + 60 \qquad (0 \le t \le 8)$$

where $G(t)$ is measured in billions of dollars and $t = 0$ corresponds to the year 1992. Sketch the graph of the function G and interpret your results.

66. **CONCENTRATION OF A DRUG IN THE BLOODSTREAM**   The concentration (in millimeters per cubic centimeter) of a certain drug in a patient's bloodstream t hours after injection is given by

$$C(t) = \frac{0.2t}{t^2 + 1}$$

Sketch the graph of the function C and interpret your results.

67. **OXYGEN CONTENT OF A POND**   When organic waste is dumped into a pond, the oxidation process that takes place reduces the pond's oxygen content. However, given time, nature will restore the oxygen content to its natural level. Suppose the oxygen content t days after organic waste has been dumped into the pond is given by

$$f(t) = 100 \left(\frac{t^2 - 4t + 4}{t^2 + 4} \right) \qquad (0 \le t < \infty)$$

percent of its normal level. Sketch the graph of the function f and interpret your results.

68. **BOX-OFFICE RECEIPTS**   The total worldwide box-office receipts for a long-running movie are approximated by the function

$$T(x) = \frac{120x^2}{x^2 + 4}$$

where $T(x)$ is measured in millions of dollars and x is the number of years since the movie's release. Sketch the graph of the function T and interpret your results.

69. **COST OF REMOVING TOXIC POLLUTANTS**   A city's main well was recently found to be contaminated with trichloroethylene, a cancer-causing chemical, as a result of an abandoned chemical dump leaching chemicals into the water. A proposal submitted to the city council indicates that the cost, measured in millions of dollars, of removing $x\%$ of the toxic pollutant is given by

$$C(x) = \frac{0.5x}{100 - x}$$

Sketch the graph of the function C and interpret your results.

SOLUTIONS TO SELF-CHECK EXERCISES 4.3

1. Since

$$\lim_{x \to \infty} \frac{2x^2}{x^2 - 1} = \lim_{x \to \infty} \frac{2}{1 - \dfrac{1}{x^2}} \qquad \text{(Divide the numerator and denominator by } x^2.\text{)}$$

$$= 2$$

we see that $y = 2$ is a horizontal asymptote. Next, since

$$x^2 - 1 = (x + 1)(x - 1) = 0$$

implies $x = -1$ or $x = 1$, these are candidates for the vertical asymptotes of f. Since the numerator of f is not equal to zero for $x = -1$ or $x = 1$, we conclude that $x = -1$ and $x = 1$ are vertical asymptotes of the graph of f.

2. We obtain the following information on the graph of f.
(1) The domain of f is the interval $(-\infty, \infty)$.
(2) By setting $x = 0$, we find the y-intercept is 4.
(3) Since

$$\lim_{x \to -\infty} f(x) = \lim_{x \to -\infty} \left(\frac{2}{3}x^3 - 2x^2 - 6x + 4 \right) = -\infty$$

$$\lim_{x \to \infty} f(x) = \lim_{x \to \infty} \left(\frac{2}{3}x^3 - 2x^2 - 6x + 4 \right) = \infty$$

we see that $f(x)$ decreases without bound as x decreases without bound and that $f(x)$ increases without bound as x increases without bound.
(4) Since f is a polynomial function, there are no asymptotes.

$$f'(x) = 2x^2 - 4x - 6 = 2(x^2 - 2x - 3)$$
$$= 2(x + 1)(x - 3)$$

(5) Setting $f'(x) = 0$ gives $x = -1$ or $x = 3$. The accompanying sign diagram for f' shows that f is increasing on the intervals $(-\infty, -1)$ and $(3, \infty)$ and decreasing on $(-1, 3)$.

Sign diagram for f'

(6) From the results of step 5, we see that $x = -1$ and $x = 3$ are critical points of f. Furthermore, the sign diagram of f' tells us that $x = -1$ gives rise to a relative maximum of f and $x = 3$ gives rise to a relative minimum of f. Now,

$$f(-1) = \frac{2}{3}(-1)^3 - 2(-1)^2 - 6(-1) + 4 = \frac{22}{3}$$

$$f(3) = \frac{2}{3}(3)^3 - 2(3)^2 - 6(3) + 4 = -14$$

so $f(-1) = 22/3$ is a relative maximum of f and $f(3) = -14$ is a relative minimum of f.

(7) $$f''(x) = 4x - 4 = 4(x - 1)$$

which is equal to zero when $x = 1$. The accompanying sign diagram of f'' shows that f is concave downward on the interval $(-\infty, 1)$ and concave upward on the interval $(1, \infty)$.

Sign diagram for f''

(8) From the results of step 7, we see that $x = 1$ is the only candidate for an inflection point of f. Since $f''(x)$ changes sign as we move across the point $x = 1$ and

$$f(1) = \frac{2}{3}(1)^3 - 2(1)^2 - 6(1) + 4 = -\frac{10}{3}$$

we see that the required inflection point is $(1, -10/3)$.
(9) Summarizing this information, we have

Domain	$(-\infty, \infty)$
Intercept	$(0, 4)$
Intervals where f is $\nearrow$ or $\searrow$	$\nearrow$ on $(-\infty, -1) \cup (3, \infty)$; $\searrow$ on $(-1, 3)$
Relative extrema	Rel. max. at $(-1, 22/3)$; rel. min. at $(3, -14)$
Concavity	Downward on $(-\infty, 1)$; upward on $(1, \infty)$
Point of inflection	$(1, -10/3)$
$\lim_{x \to -\infty} f(x); \lim_{x \to \infty} f(x)$	$-\infty$; ∞
Asymptotes	None

The graph of f is sketched in the accompanying figure.

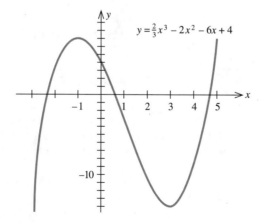

4.4 Optimization I

ABSOLUTE EXTREMA

The graph of the function f in Figure 4.56 shows the average age of cars in use in the United States from the beginning of 1946 ($t = 0$) to the beginning of 1990 ($t = 44$). Observe that the highest average age of cars in use during this period is 9 years, whereas the lowest average age of cars in use during the same period is $5\frac{1}{2}$ years. The number 9, the largest value of $f(t)$ for all values of t in the interval [0, 44] (the domain of f) is called the *absolute maximum value of f* on that interval. The number $5\frac{1}{2}$, the smallest value of $f(t)$ for all values of t in [0, 44], is called the *absolute minimum value of f* on that interval. Notice, too, that the absolute maximum value of f is attained at the end point $t = 0$ of the interval, whereas the absolute minimum value of f is attained at the two interior points $t = 12$ (corresponding to 1958) and $t = 23$ (corresponding to 1969).

FIGURE 4.56
$f(t)$ gives the average age of cars in use in year t, t in [0, 44].

Source: American Automobile Association

A precise definition of the **absolute extrema** (absolute maximum or absolute minimum) of a function follows.

The Absolute Extrema of a Function f

If $f(x) \leq f(c)$ for all x in the domain of f, then $f(c)$ is called the **absolute maximum value** of f.

If $f(x) \geq f(c)$ for all x in the domain of f, then $f(c)$ is called the **absolute minimum value** of f.

Figure 4.57 shows the graphs of several functions and gives the absolute maximum and absolute minimum of each function, if they exist.

FIGURE 4.57

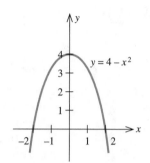

(a) $f(0) = 0$ is the absolute minimum of f; f has no absolute maximum.

(b) $f(0) = 4$ is the absolute maximum of f; f has no absolute minimum.

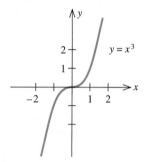

(c) $f(\sqrt{2}/2) = 1/2$ is the absolute maximum of f. $f(-\sqrt{2}/2) = -1/2$ is the absolute minimum of f.

(d) f has no absolute extrema.

ABSOLUTE EXTREMA ON A CLOSED INTERVAL

As the preceding examples show, a continuous function defined on an arbitrary interval does not always have an absolute maximum or an absolute minimum. But an important case arises often in practical applications in which both the absolute maximum and the absolute minimum of a function are guaranteed to exist. This occurs when a continuous function is defined on a *closed* interval. Let's state this important result in the form of a theorem, whose proof we will omit.

THEOREM 3

If a function f is continuous on a closed interval $[a, b]$, then f has both an absolute maximum value and an absolute minimum value on $[a, b]$.

Observe that if an absolute extremum of a continuous function f occurs at a point in an open interval (a, b), then it must be a relative extremum of f and hence its x-coordinate must be a critical point of f. Otherwise, the

FIGURE 4.58
The relative minimum of f at x_3 is the absolute minimum of f. The right end point b gives rise to the absolute maximum value $f(b)$ of f.

FIGURE 4.58
The relative minimum of f at x_3 is the absolute minimum of f. The right end point b gives rise to the absolute maximum value $f(b)$ of f.

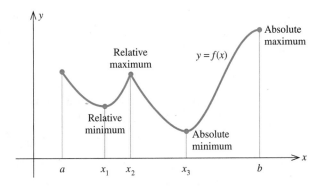

absolute extremum of f must occur at one or both of the end points of the interval $[a, b]$. A typical situation is illustrated in Figure 4.58.

Here x_1, x_2, and x_3 are critical points of f. The absolute minimum of f occurs at x_3, which lies in the open interval (a, b) and is a critical point of f. The absolute maximum of f occurs at b, an end point. This observation suggests the following procedure for finding the absolute extrema of a continuous function on a closed interval.

Finding the Absolute Extrema of f on a Closed Interval

1. Find the critical points of f that lie in (a, b).
2. Compute the value of f at each critical point found in step 1 and compute $f(a)$ and $f(b)$.
3. The absolute maximum value and absolute minimum value of f will correspond to the largest and smallest numbers, respectively, found in step 2.

EXAMPLE 1 Find the absolute extrema of the function $F(x) = x^2$ defined on the interval $[-1, 2]$.

SOLUTION ✔ The function F is continuous on the closed interval $[-1, 2]$ and differentiable on the open interval $(-1, 2)$. The derivative of F is

$$F'(x) = 2x$$

so $x = 0$ is the only critical point of F. Next, evaluate $F(x)$ at $x = -1$, $x = 0$, and $x = 2$. Thus,

$$F(-1) = 1, \qquad F(0) = 0, \qquad F(2) = 4$$

It follows that 0 is the absolute minimum value of F and 4 is the absolute maximum value of F. The graph of F, in Figure 4.59, confirms our results.

FIGURE 4.59
F has an absolute minimum value of 0 and an absolute maximum value of 4.

EXAMPLE **2**

Find the absolute extrema of the function

$$f(x) = x^3 - 2x^2 - 4x + 4$$

defined on the interval $[0, 3]$.

SOLUTION ✔

The function f is continuous on the closed interval $[0, 3]$ and differentiable on the open interval $(0, 3)$. The derivative of f is

$$f'(x) = 3x^2 - 4x - 4 = (3x + 2)(x - 2)$$

and it is equal to zero when $x = -\frac{2}{3}$ and $x = 2$. Since the point $x = -\frac{2}{3}$ lies outside the interval $[0, 3]$, it is dropped from further consideration, and $x = 2$ is seen to be the sole critical point of f. Next, we evaluate $f(x)$ at the critical point of f as well as the end points of f, obtaining

$$f(0) = 4, \qquad f(2) = -4, \qquad f(3) = 1$$

From these results, we conclude that -4 is the absolute minimum value of f and 4 is the absolute maximum value of f. The graph of f, which appears in Figure 4.60, confirms our results. Observe that the absolute maximum of f occurs at the end point $x = 0$ of the interval $[0, 3]$, while the absolute minimum of f occurs at $x = 2$, which is a point in the interval $(0, 3)$.

FIGURE 4.60
f has an absolute maximum value of 4 and an absolute minimum value of -4.

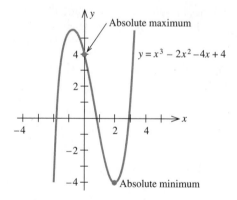

Absolute maximum

$y = x^3 - 2x^2 - 4x + 4$

Absolute minimum

Exploring with Technology

Let $f(x) = x^3 - 2x^2 - 4x + 4$. (This is the function of Example 2.)

1. Use a graphing utility to plot the graph of f, using the viewing rectangle $[0, 3] \times [-5, 5]$. Use **TRACE** to find the absolute extrema of f on the interval $[0, 3]$ and thus verify the results obtained analytically in Example 2.

2. Plot the graph of f using the viewing rectangle $[-2, 1] \times [-5, 6]$. Use **ZOOM** and **TRACE** to find the absolute extrema of f on the interval $[-2, 1]$. Verify your results analytically.

EXAMPLE 3

Find the absolute maximum and absolute minimum values of the function $f(x) = x^{2/3}$ on the interval $[-1, 8]$.

SOLUTION ✔

The derivative of f is

$$f'(x) = \frac{2}{3}x^{-1/3} = \frac{2}{3x^{1/3}}$$

FIGURE 4.61
f has an absolute minimum value of $f(0) = 0$ and an absolute maximum value of $f(8) = 4$.

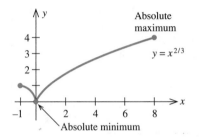

Note that f' is not defined at $x = 0$, is continuous everywhere else, and does not equal zero for all x. Therefore, $x = 0$ is the only critical point of f. Evaluating $f(x)$ at $x = -1$, 0, and 8, we obtain

$$f(-1) = 1, \quad f(0) = 0, \quad f(8) = 4$$

We conclude that the absolute minimum value of f is 0, attained at $x = 0$, and the absolute maximum value of f is 4, attained at $x = 8$ (Figure 4.61).

■ ■ ■ ■

APPLICATIONS

Many real-world applications call for finding the absolute maximum value or the absolute minimum value of a given function. For example, management is interested in finding what level of production will yield the maximum profit for a company; a farmer is interested in finding the right amount of fertilizer to maximize crop yield; a doctor is interested in finding the maximum concentration of a drug in a patient's body and the time at which it occurs; and an engineer is interested in finding the dimension of a container with a specified shape and volume that can be constructed at a minimum cost.

EXAMPLE 4

The Acrosonic Company's total profit (in dollars) from manufacturing and selling x units of their model F loudspeaker systems is given by

$$P(x) = -0.02x^2 + 300x - 200{,}000 \qquad (0 \le x \le 20{,}000)$$

How many units of the loudspeaker system must Acrosonic produce to maximize its profits?

SOLUTION ✔

To find the absolute maximum of P on $[0, 20{,}000]$, first find the critical points of P on the interval $(0, 20{,}000)$. To do this, compute

$$P'(x) = -0.04x + 300$$

Solving the equation $P'(x) = 0$ gives $x = 7500$. Next, evaluate $P(x)$ at $x = 7500$ as well as the end points $x = 0$ and $x = 20{,}000$ of the interval $[0, 20{,}000]$, obtaining

$$P(0) = -200{,}000$$
$$P(7500) = 925{,}000$$
$$P(20{,}000) = -2{,}200{,}000$$

FIGURE 4.62

P has an absolute maximum at (7500, 925,000).

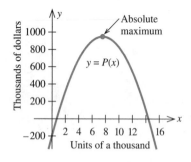

From these computations we see that the absolute maximum value of the function P is 925,000. Thus, by producing 7500 units, Acrosonic will realize a maximum profit of $925,000. The graph of P is sketched in Figure 4.62.

■ ■ ■ ■

> **Group Discussion**
>
> Recall that the total profit function P is defined as $P(x) = R(x) - C(x)$, where R is the total revenue function, C is the total cost function, and x is the number of units of a product produced and sold. (Assume all derivatives exist.)
>
> **1.** Show that at the level of production x_0 that yields the maximum profit for the company, the following two conditions are satisfied:
>
> $$R'(x_0) = C'(x_0) \qquad \text{and} \qquad R''(x_0) < C''(x_0)$$
>
> **2.** Interpret the two conditions in part 1 in economic terms and explain why they make sense.

EXAMPLE 5

When a person coughs, the trachea (windpipe) contracts, allowing air to be expelled at a maximum velocity. It can be shown that during a cough the velocity v of airflow is given by the function

$$v = f(r) = kr^2(R - r)$$

where r is the trachea's radius (in centimeters) during a cough, R is the trachea's normal radius (in centimeters), and k is a positive constant that depends on the length of the trachea. Find the radius r for which the velocity of airflow is greatest.

SOLUTION ✔

To find the absolute maximum of f on $[0, R]$, first find the critical points of f on the interval $(0, R)$. We compute

$$f'(r) = 2kr(R - r) - kr^2 \qquad \text{(Using the product rule)}$$
$$= -3kr^2 + 2kRr = kr(-3r + 2R)$$

FIGURE 4.63

The velocity of airflow is greatest when the radius of the contracted trachea is $\frac{2}{3}R$.

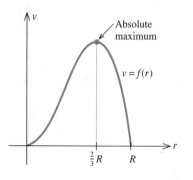

Setting $f'(r) = 0$ gives $r = 0$ or $r = \frac{2}{3}R$, and so $r = \frac{2}{3}R$ is the sole critical point of f ($r = 0$ is an end point). Evaluating $f(r)$ at $r = \frac{2}{3}R$, as well as at the end points $r = 0$ and $r = R$, we obtain

$$f(0) = 0$$
$$f\left(\frac{2}{3}R\right) = \frac{4k}{27}R^3$$
$$f(R) = 0$$

from which we deduce that the velocity of airflow is greatest when the radius of the contracted trachea is $\frac{2}{3}R$—that is, when the radius is contracted by approximately 33%. The graph of the function f is shown in Figure 4.63.

■ ■ ■ ■

Group Discussion

Prove that if a cost function $C(x)$ is concave upward $[C''(x) > 0]$, then the level of production that will result in the smallest average production cost occurs when

$$\overline{C}(x) = C'(x)$$

—that is, when the average cost $\overline{C}(x)$ is equal to the marginal cost $C'(x)$.

Hints:

1. Show that

$$\overline{C}'(x) = \frac{xC'(x) - C(x)}{x^2}$$

so that the critical point of the function $\overline{C}$ occurs when

$$xC'(x) - C(x) = 0$$

2. Show that at a critical point of $\overline{C}$

$$\overline{C}''(x) = \frac{C''(x)}{x}$$

Use the second derivative test to reach the desired conclusion.

EXAMPLE **6**

The daily average cost function (in dollars per unit) of the Elektra Electronics Company is given by

$$\overline{C}(x) = 0.0001x^2 - 0.08x + 40 + \frac{5000}{x} \qquad (x > 0)$$

where x stands for the number of programmable calculators Elektra produces. Show that a production level of 500 units per day results in a minimum average cost for the company.

SOLUTION ✔

The domain of the function $\overline{C}$ is the interval $(0, \infty)$, which is not closed. To solve the problem, we resort to the graphic method. Using the techniques of graphing from the last section, we sketch the graph of $\overline{C}$ (Figure 4.64).

FIGURE 4.64
The minimum average cost is $35 per unit.

Now,

$$\overline{C}'(x) = 0.0002x - 0.08 - \frac{5000}{x^2}$$

Substituting the given value of x, 500, into $\overline{C}'(x)$ gives $\overline{C}'(500) = 0$, so $x = 500$ is a critical point of $\overline{C}$. Next,

$$\overline{C}''(x) = 0.0002 + \frac{10,000}{x^3}$$

Thus,

$$\overline{C}''(500) = 0.0002 + \frac{10,000}{(500)^3} > 0$$

and by the second derivative test, a relative minimum of the function $\overline{C}$ occurs at the point $x = 500$. Furthermore, $\overline{C}''(x) > 0$ for $x > 0$, which implies that the graph of $\overline{C}$ is concave upward everywhere, so the relative minimum of $\overline{C}$ must be the absolute minimum of $\overline{C}$. The minimum average cost is given by

$$\overline{C}(500) = 0.0001(500)^2 - 0.08(500) + 40 + \frac{5000}{500}$$

$$= 35$$

or \$35 per unit. ■ ■ ■ ■

Exploring with Technology

Refer to the preceding Group Discussion and Example 6.

1. Using a graphing utility, plot the graphs of

$$\overline{C}(x) = 0.0001x^2 - 0.08x + 40 + \frac{5000}{x}$$

$$C'(x) = 0.0003x^2 - 0.16x + 40$$

using the viewing rectangle $[0, 1000] \times [0, 150]$.
Note: $C(x) = 0.0001x^3 - 0.08x^2 + 40x + 5000$ (Why?)
2. Find the point of intersection of the graphs of $\overline{C}$ and C' and thus verify the assertion in the Group Discussion for the special case studied in Example 6.

EXAMPLE **7**   The altitude (in feet) of a rocket t seconds into flight is given by

$$s = f(t) = -t^3 + 96t^2 + 195t + 5 \qquad (t \geq 0)$$

a. Find the maximum altitude attained by the rocket.
b. Find the maximum velocity attained by the rocket.

SOLUTION ✔   **a.** The maximum altitude attained by the rocket is given by the largest value of the function f in the closed interval $[0, T]$, where T denotes the time the rocket impacts Earth. We know that such a number exists because the dominant term in the expression for the continuous function f is $-t^3$. So for t large enough, the value of $f(t)$ must change from positive to negative and, in particular, it must attain the value 0 for some T.

FIGURE 4.65
The maximum altitude of the rocket is 143,655 feet.

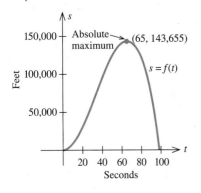

To find the absolute maximum of f, compute

$$f'(t) = -3t^2 + 192t + 195$$
$$= -3(t - 65)(t + 1)$$

and solve the equation $f'(t) = 0$, obtaining $t = -1$ and $t = 65$. Ignore $t = -1$ since it lies outside the interval $[0, T]$. This leaves the critical point $t = 65$ of f. Continuing, we compute

$$f(0) = 5, \qquad f(65) = 143,655, \qquad f(T) = 0$$

and conclude, accordingly, that the absolute maximum value of f is 143,655. Thus, the maximum altitude of the rocket is 143,655 feet, attained 65 seconds into flight. The graph of f is sketched in Figure 4.65.

b. To find the maximum velocity attained by the rocket, find the largest value of the function that describes the rocket's velocity at any time t—namely,

$$v = f'(t) = -3t^2 + 192t + 195 \qquad (t \geq 0)$$

We find the critical point of v by setting $v' = 0$. But

$$v' = -6t + 192$$

and the critical point of v is $t = 32$. Since

$$v'' = -6 < 0$$

FIGURE 4.66
The maximum velocity of the rocket is 3267 feet per second.

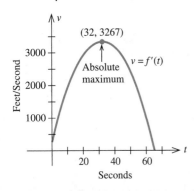

the second derivative test implies that a relative maximum of v occurs at $t = 32$. Our computation has in fact clarified the property of the "velocity curve." Since $v'' < 0$ everywhere, the velocity curve is concave downward everywhere. With this observation, we assert that the relative maximum must in fact be the absolute maximum of v. The maximum velocity of the rocket is given by evaluating v at $t = 32$,

$$f'(32) = -3(32)^2 + 192(32) + 195$$

or 3267 feet per second. The graph of the velocity function v is sketched in Figure 4.66. ◾◾◾◾

SELF-CHECK EXERCISES 4.4

1. Let $f(x) = x - 2\sqrt{x}$.
   a. Find the absolute extrema of f on the interval $[0, 9]$.
   b. Find the absolute extrema of f.

2. Find the absolute extrema of $f(x) = 3x^4 + 4x^3 + 1$ on $[-2, 1]$.

3. The operating rate (expressed as a percentage) of factories, mines, and utilities in a certain region of the country on the tth day of the year 2000 is given by the function

$$f(t) = 80 + \frac{1200t}{t^2 + 40,000} \qquad (0 \leq t \leq 250)$$

On which day of the first 250 days of 2000 was the manufacturing capacity operating rate highest?

Solutions to Self-Check Exercises 4.4 can be found on page 356.

4.4 Exercises

In Exercises 1–8, you are given the graph of some function _f_ defined on the indicated interval. Find the absolute maximum and the absolute minimum of _f_, if they exist.

1.

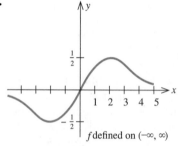

_f_ defined on $(-\infty, \infty)$

2.

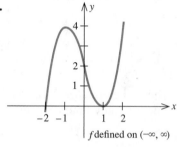

_f_ defined on $(-\infty, \infty)$

3.

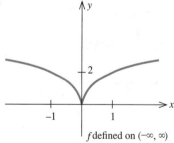

_f_ defined on $(-\infty, \infty)$

4.

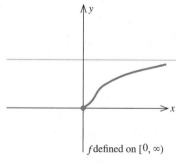

_f_ defined on $[0, \infty)$

5.

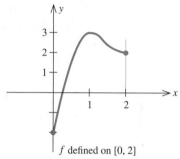

_f_ defined on $[0, 2]$

6.

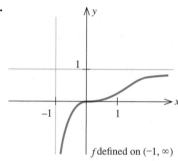

_f_ defined on $(-1, \infty)$

7.

$(\frac{3}{2}, -\frac{27}{16})$

_f_ defined on $[-1, 2]$

8.

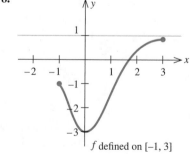

_f_ defined on $[-1, 3]$

In Exercises 9–38, find the absolute maximum value and the absolute minimum value, if any, of the given function.

9. $f(x) = 2x^2 + 3x - 4$

10. $g(x) = -x^2 + 4x + 3$

11. $h(x) = x^{1/3}$

12. $f(x) = x^{2/3}$

13. $f(x) = \dfrac{1}{1 + x^2}$

14. $f(x) = \dfrac{x}{1 + x^2}$

15. $f(x) = x^2 - 2x - 3$ on $[-2, 3]$

16. $g(x) = x^2 - 2x - 3$ on $[0, 4]$

17. $f(x) = -x^2 + 4x + 6$ on $[0, 5]$

18. $f(x) = -x^2 + 4x + 6$ on $[3, 6]$

19. $f(x) = x^3 + 3x^2 - 1$ on $[-3, 2]$

20. $g(x) = x^3 + 3x^2 - 1$ on $[-3, 1]$

21. $g(x) = 3x^4 + 4x^3$ on $[-2, 1]$

22. $f(x) = \dfrac{1}{2}x^4 - \dfrac{2}{3}x^3 - 2x^2 + 3$ on $[-2, 3]$

23. $f(x) = \dfrac{x + 1}{x - 1}$ on $[2, 4]$

24. $g(t) = \dfrac{t}{t - 1}$ on $[2, 4]$

25. $f(x) = 4x + \dfrac{1}{x}$ on $[1, 3]$

26. $f(x) = 9x - \dfrac{1}{x}$ on $[1, 3]$

27. $f(x) = \dfrac{1}{2}x^2 - 2\sqrt{x}$ on $[0, 3]$

28. $g(x) = \dfrac{1}{8}x^2 - 4\sqrt{x}$ on $[0, 9]$

29. $f(x) = \dfrac{1}{x}$ on $(0, \infty)$

30. $g(x) = \dfrac{1}{x + 1}$ on $(0, \infty)$

31. $f(x) = 3x^{2/3} - 2x$ on $[0, 3]$

32. $g(x) = x^2 + 2x^{2/3}$ on $[-2, 2]$

33. $f(x) = x^{2/3}(x^2 - 4)$ on $[-1, 2]$

34. $f(x) = x^{2/3}(x^2 - 4)$ on $[-1, 3]$

35. $f(x) = \dfrac{x}{x^2 + 2}$ on $[-1, 2]$

36. $f(x) = \dfrac{1}{x^2 + 2x + 5}$ on $[-2, 1]$

37. $f(x) = \dfrac{x}{\sqrt{x^2 + 1}}$ on $[-1, 1]$

38. $g(x) = x\sqrt{4 - x^2}$ on $[0, 2]$

39. A stone is thrown straight up from the roof of an 80-ft building. The height (in feet) of the stone at any time t (in seconds), measured from the ground, is given by

$$h(t) = -16t^2 + 64t + 80$$

What is the maximum height the stone reaches?

40. **MAXIMIZING PROFITS** Lynbrook West, an apartment complex, has 100 two-bedroom units. The monthly profit (in dollars) realized from renting out x apartments is given by

$$P(x) = -10x^2 + 1760x - 50,000$$

How many units should be rented out in order to maximize the monthly rental profit? What is the maximum monthly profit realizable?

41. **MAXIMIZING PROFITS** The estimated monthly profit (in dollars) realizable by the Cannon Precision Instruments Corporation for manufacturing and selling x units of its model M1 camera is

$$P(x) = -0.04x^2 + 240x - 10,000$$

How many cameras should Cannon produce per month in order to maximize its profits?

42. **FLIGHT OF A ROCKET** The altitude (in feet) attained by a model rocket t sec into flight is given by the function

$$h(t) = -\dfrac{1}{3}t^3 + 4t^2 + 20t + 2$$

Find the maximum altitude attained by the rocket.

43. **FEMALE SELF-EMPLOYED WORKFORCE** Based on data obtained from the U.S. Department of Labor, the number of nonfarm, full-time, self-employed women can be approximated by

$$N(t) = 0.81t - 1.14\sqrt{t} + 1.53 \qquad (0 \le t \le 6)$$

where $N(t)$ is measured in millions and t is measured in 5-yr intervals, with $t = 0$ corresponding to the beginning of 1963. Determine the absolute extrema of the function N on the interval $[0, 6]$. Interpret your results.
Source: U.S. Department of Labor

44. **MAXIMIZING PROFITS** The management of Trappee and Sons, Inc., producers of the famous Texa-Pep hot sauce, estimate that their profit (in dollars) from the daily production and sale of x cases (each case consisting of 24 bottles) of the hot sauce is given by

$$P(x) = -0.000002x^3 + 6x - 400$$

What is the largest possible profit Trappee can make in 1 day?

45. MAXIMIZING PROFITS The quantity demanded per month of the Walter Serkin recording of Beethoven's *Moonlight Sonata,* manufactured by Phonola Record Industries, is related to the price per compact disc. The equation

$$p = -0.00042x + 6 \qquad (0 \le x \le 12{,}000)$$

where p denotes the unit price in dollars and x is the number of discs demanded, relates the demand to the price. The total monthly cost (in dollars) for pressing and packaging x copies of this classical recording is given by

$$C(x) = 600 + 2x - 0.00002x^2 \qquad (0 \le x \le 20{,}000)$$

How many copies should Phonola produce per month in order to maximize its profits?

Hint: The revenue is $R(x) = px$, and the profit is $P(x) = R(x) - C(x)$.

46. MAXIMIZING PROFIT A manufacturer of tennis rackets finds that the total cost $C(x)$ (in dollars) of manufacturing x rackets/day is given by $C(x) = 400 + 4x + 0.0001x^2$. Each racket can be sold at a price of p dollars, where p is related to x by the demand equation $p = 10 - 0.0004x$. If all rackets that are manufactured can be sold, find the daily level of production that will yield a maximum profit for the manufacturer.

47. MAXIMIZING PROFIT The weekly demand for the Pulsar 25-in. color console television is given by the demand equation

$$p = -0.05x + 600 \qquad (0 \le x \le 12{,}000)$$

where p denotes the wholesale unit price in dollars and x denotes the quantity demanded. The weekly total cost function associated with manufacturing these sets is given by

$$C(x) = 0.000002x^3 - 0.03x^2 + 400x + 80{,}000$$

where $C(x)$ denotes the total cost incurred in producing x sets. Find the level of production that will yield a maximum profit for the manufacturer.

Hint: Use the quadratic formula.

48. MINIMIZING AVERAGE COSTS Suppose the total cost function for manufacturing a certain product is $C(x) = 0.2(0.01x^2 + 120)$ dollars, where x represents the number of units produced. Find the level of production that will minimize the average cost.

49. MINIMIZING PRODUCTION COSTS The total monthly cost (in dollars) incurred by Cannon Precision Instruments Corporation for manufacturing x units of the model M1

camera is given by the function

$$C(x) = 0.0025x^2 + 80x + 10{,}000$$

a. Find the average cost function $\overline{C}$.
b. Find the level of production that results in the smallest average production cost.
c. Find the level of production for which the average cost is equal to the marginal cost.
d. Compare the result of part (c) with that of part (b).

50. MINIMIZING PRODUCTION COSTS The daily total cost (in dollars) incurred by Trappee and Sons, Inc., for producing x cases of Texa-Pep hot sauce is given by the function

$$C(x) = 0.000002x^3 + 5x + 400$$

Using this function, answer the questions posed in Exercise 49.

51. MAXIMIZING REVENUE Suppose the quantity demanded per week of a certain dress is related to the unit price p by the demand equation $p = \sqrt{800 - x}$, where p is in dollars and x is the number of dresses made. How many dresses should be made and sold per week in order to maximize the revenue?

Hint: $R(x) = px$.

52. MAXIMIZING REVENUE The quantity demanded per month of the Sicard wristwatch is related to the unit price by the equation

$$p = \frac{50}{0.01x^2 + 1} \qquad (0 \le x \le 20)$$

where p is measured in dollars and x is measured in units of a thousand. How many watches must be sold to yield a maximum revenue?

53. OXYGEN CONTENT OF A POND When organic waste is dumped into a pond, the oxidation process that takes place reduces the pond's oxygen content. However, given time, nature will restore the oxygen content to its natural level. Suppose the oxygen content t days after organic waste has been dumped into the pond is given by

$$f(t) = 100 \left[\frac{t^2 - 4t + 4}{t^2 + 4} \right] \qquad (0 \le t < \infty)$$

percent of its normal level.
a. When is the level of oxygen content lowest?
b. When is the rate of oxygen regeneration greatest?

54. AIR POLLUTION The amount of nitrogen dioxide, a brown gas that impairs breathing, present in the atmosphere on a certain May day in the city of Long Beach is approximated by

$$A(t) = \frac{136}{1 + 0.25(t - 4.5)^2} + 28 \qquad (0 \le t \le 11)$$

where $A(t)$ is measured in pollutant standard index (PSI) and t is measured in hours, with $t = 0$ corresponding to 7 A.M. Determine the time of day when the pollution is at its highest level.

55. **MAXIMIZING REVENUE** The average revenue is defined as the function

$$\overline{R}(x) = \frac{R(x)}{x} \qquad (x > 0)$$

Prove that if a revenue function $R(x)$ is concave downward [$R''(x) < 0$], then the level of sales that will result in the largest average revenue occurs when $\overline{R}(x) = R'(x)$.

56. **VELOCITY OF BLOOD** According to a law discovered by the nineteenth-century physician Jean Louis Marie Poiseuille, the velocity (in centimeters per second) of blood r cm from the central axis of an artery is given by

$$v(r) = k(R^2 - r^2)$$

where k is a constant and R is the radius of the artery. Show that the velocity of blood is greatest along the central axis.

57. **GDP OF A DEVELOPING COUNTRY** A developing country's gross domestic product (GDP) from 1988 to 1996 is approximated by the function

$$G(t) = -0.2t^3 + 2.4t^2 + 60 \qquad (0 \le t \le 8)$$

where $G(t)$ is measured in billions of dollars and $t = 0$ corresponds to the year 1988. Show that the growth rate of the country's GDP was maximal in 1992.

58. **CRIME RATES** The number of major crimes committed in the city of Bronxville between 1987 and 1994 is approximated by the function

$$N(t) = -0.1t^3 + 1.5t^2 + 100 \qquad (0 \le t \le 7)$$

where $N(t)$ denotes the number of crimes committed in year t ($t = 0$ corresponds to the year 1987). Enraged by the dramatic increase in the crime rate, the citizens of Bronxville, with the help of the local police, organized "Neighborhood Crime Watch" groups in early 1991 to combat this menace. Show that the growth in the crime rate was maximal in 1992, giving credence to the claim that the Neighborhood Crime Watch program was working.

59. **SOCIAL SECURITY SURPLUS** Based on data from the Social Security Administration, the estimated cash in the Social Security retirement and disability trust funds may be approximated by

$$f(t) = -0.0129t^4 + 0.3087t^3 + 2.1760t^2 + 62.8466t$$
$$+ 506.2955 \qquad (0 \le t \le 35)$$

where $f(t)$ is measured in billions of dollars and t is measured in years, with $t = 0$ corresponding to the year 1995. Show that the Social Security surplus will be at its highest level at approximately the middle of the year 2018.

Hint: Show that $t = 23.6811$ is an approximate critical point of $f'(t)$.

Source: Social Security Administration

60. **ENERGY EXPENDED BY A FISH** It has been conjectured that a fish swimming a distance of L ft at a speed of v ft/sec relative to the water and against a current flowing at the rate of u ft/sec ($u < v$) expends a total energy given by

$$E(v) = \frac{aLv^3}{v - u}$$

where E is measured in foot-pounds (ft-lb) and a is a constant. Find the speed v at which the fish must swim in order to minimize the total energy expended. (*Note:* This result has been verified by biologists.)

61. **REACTION TO A DRUG** The strength of a human body's reaction R to a dosage D of a certain drug is given by

$$R = D^2 \left(\frac{k}{2} - \frac{D}{3} \right)$$

where k is a positive constant. Show that the maximum reaction is achieved if the dosage is k units.

62. Refer to Exercise 61. Show that the rate of change in the reaction R with respect to the dosage D is maximal if $D = k/2$.

In Exercises 63–66, determine whether the statement is true or false. If it is true, explain why it is true. If it is false, give an example to show why it is false.

63. If f is defined on a closed interval $[a, b]$, then f has an absolute maximum value.

64. If f is continuous on an open interval (a, b), then f does not have an absolute minimum value.

65. If f is not continuous on the closed interval $[a, b]$, then f cannot have an absolute maximum value.

(continued on p. 356)

FINDING THE ABSOLUTE EXTREMA OF A FUNCTION

Some graphing utilities have a function for finding the absolute maximum and the absolute minimum values of a continuous function on a closed interval. If your graphing utility has this capability, use it to work through the example and exercises of this section.

EXAMPLE **1**   Let $f(x) = \dfrac{2x + 4}{(x^2 + 1)^{3/2}}$.

a. Use a graphing utility to plot the graph of f in the viewing rectangle $[-3, 3] \times [-1, 5]$.

b. Find the absolute maximum and absolute minimum values of f on the interval $[-3, 3]$. Express your answers accurate to four decimal places.

SOLUTION ✓   **a.** The graph of f is shown in Figure T1.

FIGURE T1
The graph of f in the viewing rectangle $[-3, 3] \times [-1, 5]$

b. Using the function on a graphing utility for finding the absolute minimum value of a continuous function on a closed interval, we find the absolute minimum value of f to be -0.0632. Similarly, using the function for finding the absolute maximum value, we find the absolute maximum value to be 4.1593.

■ ■ ■ ■

REMARK   Some graphing utilities will enable you to find the absolute minimum and absolute maximum values of a continuous function on a closed interval without having to graph the function.

In Exercises 1–6, use a graphing utility to find the absolute maximum and the absolute minimum values of *f* in the given interval using the method of Example 1. Express your answers accurate to four decimal places.

1. $f(x) = 3x^4 - 4.2x^3 + 6.1x - 2; [-2, 3]$

2. $f(x) = 2.1x^4 - 3.2x^3 + 4.1x^2 + 3x - 4; [-1, 2]$

3. $f(x) = \dfrac{2x^3 - 3x^2 + 1}{x^2 + 2x - 8}; [-3, 1]$

4. $f(x) = \sqrt{x}(x^3 - 4)^2; [0.5, 1]$

5. $f(x) = \dfrac{x^3 - 1}{x^2}; [1, 3]$

6. $f(x) = \dfrac{x^3 - x^2 + 1}{x - 2}; [1, 3]$

7. RATE OF BANK FAILURES The Federal Deposit Insurance Company (FDIC) estimates that the rate at which banks were failing between 1982 and 1994 is given by

$$f(t) = 0.063447t^4 - 1.953283t^3 + 14.632576t^2$$
$$- 6.684704t + 47.458874 \qquad (0 \le t \le 12)$$

where $f(t)$ is the number of banks per year and t is measured in years, with $t = 0$ corresponding to the beginning of 1982.
a. Use a graphing utility to plot the graph of f in the viewing rectangle $[0, 12] \times [0, 220]$.
b. What is the highest rate of bank failures during the period in question?
Source: Federal Deposit Insurance Corporation

8. BOSTON'S DAILY TEMPERATURE FOR 1993 The average daily temperature in Boston for 1993 is given by

$$f(t) = 0.0434841t^4 - 1.18523t^3 + 9.38548t^2$$
$$- 17.7553t + 38.9272 \qquad (0 \le t \le 12)$$

where $f(t)$ is measured in degrees Fahrenheit and t is in months, with $t = 0$ corresponding to the beginning of

1993. Find the absolute extrema for f and interpret your results.
Source: Robert Lautzenheiser, Climatologist

9. TIME ON THE MARKET According to the Greater Boston Real Estate Board—Multiple Listing Service, the average number of days a single-family home remains for sale from listing to accepted offer is approximated by the function

$$f(t) = 0.0171911t^4 - 0.662121t^3 + 6.18083t^2$$
$$- 8.97086t + 53.3357 \qquad (0 \le t \le 10)$$

where t is measured in years, with $t = 0$ corresponding to the beginning of 1984.
a. Use a graphing utility to plot the graph of f in the viewing rectangle $[0, 12] \times [0, 120]$.
b. Find the absolute maximum value and the absolute minimum value of f in the interval $[0, 12]$. Interpret your results.
Source: Greater Boston Real Estate Board—Multiple Listing Service

10. WHY SSS BENEFITS MAY EXCEED PAYROLL TAXES Unless payroll taxes are increased significantly and/or benefits are scaled back drastically, it is a matter of time before the current Social Security system goes broke. Based on data from the Board of Trustees of the Social Security Administration, the assets of the system—the Social Security "trust fund"—may be approximated by

$$f(t) = -0.0129t^4 + 0.3087t^3 + 2.1760t^2$$
$$+ 62.8466t + 506.2955 \qquad (0 \le t \le 35)$$

where $f(t)$ is measured in millions of dollars and t is measured in years, with $t = 0$ corresponding to 1995.
a. Use a graphing calculator to sketch the graph of f.
b. Based on this model, when will the Social Security system start to pay out more benefits than it gets in payroll taxes?
Source: Social Security Administration

355

66. If $f''(x) < 0$ on (a, b) and $f'(c) = 0$ where $a < c < b$, then $f(c)$ is the absolute maximum value of f on $[a, b]$.

67. Let f be a constant function—that is, let $f(x) = c$, where c is some real number. Show that every point $x = a$ is an absolute maximum and, at the same time, an absolute minimum of f.

68. Show that a polynomial function defined on the interval $(-\infty, \infty)$ cannot have both an absolute maximum and an absolute minimum unless it is a constant function.

69. One condition that must be satisfied before Theorem 3

(page 342) is applicable is that the function f must be continuous on the closed interval $[a, b]$. Define a function f on the closed interval $[-1, 1]$ by

$$f(x) = \begin{cases} \dfrac{1}{x} & \text{if } x \in [-1, 1] \\ 0 & \text{if } x = 0 \end{cases} \qquad (x \neq 0)$$

a. Show that f is not continuous at $x = 0$.
b. Show that $f(x)$ does not attain an absolute maximum or an absolute minimum on the interval $[-1, 1]$.
c. Confirm your results by sketching the function f.

SOLUTIONS TO SELF-CHECK EXERCISES 4.4

1. a. The function f is continuous in its domain and differentiable in the interval $(0, 9)$. The derivative of f is

$$f'(x) = 1 - x^{-1/2} = \frac{x^{1/2} - 1}{x^{1/2}}$$

and it is equal to zero when $x = 1$. Evaluating $f(x)$ at the end points $x = 0$ and $x = 9$ and at the critical point $x = 1$ of f, we have

$$f(0) = 0, \qquad f(1) = -1, \qquad f(9) = 3$$

From these results, we see that -1 is the absolute minimum value of f and 3 is the absolute maximum value of f.

b. In this case, the domain of f is the interval $[0, \infty)$, which is not closed. Therefore, we resort to the graphic method. Using the techniques of graphing, we sketch in the accompanying figure the graph of f.

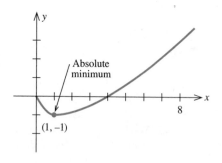

The graph of f shows that -1 is the absolute minimum value of f, but f has no absolute maximum since $f(x)$ increases without bound as x increases without bound.

2. The function f is continuous on the interval $[-2, 1]$. It is also differentiable on the open interval $(-2, 1)$. The derivative of f is

$$f'(x) = 12x^3 + 12x^2 = 12x^2(x + 1)$$

and it is continuous on $(-2, 1)$. Setting $f'(x) = 0$ gives $x = -1$ and $x = 0$ as critical points of f. Evaluating $f(x)$ at these critical points of f as well as at the end points of the interval $[-2, 1]$, we obtain

$$f(-2) = 17, \quad f(-1) = 0, \quad f(0) = 1, \quad f(1) = 8$$

From these results, we see that 0 is the absolute minimum value of f and 17 is the absolute maximum value of f.

3. The problem is solved by finding the absolute maximum of the function f on $[0, 250]$. Differentiating $f(t)$, we obtain

$$f'(t) = \frac{(t^2 + 40,000)(1200) - 1200t(2t)}{(t^2 + 40,000)^2}$$
$$= \frac{-1200(t^2 - 40,000)}{(t^2 + 40,000)^2}$$

Upon setting $f'(t) = 0$ and solving the resulting equation, we obtain $t = -200$ or 200. Since -200 lies outside the interval $[0, 250]$, we are interested only in the critical point $t = 200$ of f. Evaluating $f(t)$ at $t = 0$, $t = 200$, and $t = 250$, we find

$$f(0) = 80, \quad f(200) = 83, \quad f(250) = 82.93$$

We conclude that the manufacturing capacity operating rate was the highest on the 200th day of 2000—that is, a little past the middle of July 2000.

4.5 Optimization II

Section 4.4 outlined how to find the solution to certain optimization problems in which the objective function is given. In this section we consider problems in which we are required to first find the appropriate function to be optimized. The following guidelines will be useful for solving these problems.

Guidelines for Solving Optimization Problems

1. Assign a letter to each variable mentioned in the problem. If appropriate, draw and label a figure.
2. Find an expression for the quantity to be optimized.
3. Use the conditions given in the problem to write the quantity to be optimized as a function f of *one* variable. Note any restrictions to be placed on the domain of f from physical considerations of the problem.
4. Optimize the function f over its domain using the methods of Section 4.4.

REMARK In carrying out step 4, remember that if the function f to be optimized is continuous on a closed interval, then the absolute maximum and absolute minimum of f are, respectively, the largest and smallest values of $f(x)$ on the set composed of the critical points of f and the end points of the interval. If the domain of f is not a closed interval, then we resort to the graphic method. ■■■

MAXIMIZATION PROBLEMS

EXAMPLE 1

A man wishes to have a rectangular-shaped garden in his backyard. He has 50 feet of fencing material with which to enclose his garden. Find the dimensions for the largest garden he can have if he uses all of the fencing material.

SOLUTION ✔

Step 1 Let x and y denote the dimensions (in feet) of two adjacent sides of the garden (Figure 4.67) and let A denote its area.

Step 2 The area of the garden,

$$A = xy \tag{1}$$

is the quantity to be maximized.

Step 3 The perimeter of the rectangle, $(2x + 2y)$ feet, must equal 50 feet. Therefore, we have the equation

$$2x + 2y = 50$$

Next, solving this equation for y in terms of x yields

$$y = 25 - x \tag{2}$$

which, when substituted into Equation (1), gives

$$A = x(25 - x)$$
$$= -x^2 + 25x$$

(Remember, the function to be optimized must involve just one variable.) Since the sides of the rectangle must be nonnegative, we must have $x \geq 0$ and $y = 25 - x \geq 0$; that is, we must have $0 \leq x \leq 25$. Thus, the problem is reduced to that of finding the absolute maximum of $A = f(x) = -x^2 + 25x$ on the closed interval $[0, 25]$.

Step 4 Observe that f is continuous on $[0, 25]$, so the absolute maximum value of f must occur at the end point(s) or at the critical point(s) of f. The derivative of the function A is given by

$$A' = f'(x) = -2x + 25$$

Setting $A' = 0$ gives

$$-2x + 25 = 0$$

or $x = 12.5$, as the critical point of A. Next, we evaluate the function $A = f(x)$ at $x = 12.5$ and at the end points $x = 0$ and $x = 25$ of the interval $[0, 25]$, obtaining

$$f(0) = 0, \qquad f(12.5) = 156.25, \qquad f(25) = 0$$

We see that the absolute maximum value of the function f is 156.25. From Equation (2) we see that when $x = 12.5$, the value of y is given by $y = 12.5$. Thus, the garden would be of maximum area (156.25 square feet) if it were in the form of a square with sides 12.5 feet long.   ■■■■

FIGURE 4.67

What is the maximum rectangular area that can be enclosed with 50 feet of fencing?

EXAMPLE 2   By cutting away identical squares from each corner of a rectangular piece of cardboard and folding up the resulting flaps, the cardboard may be turned into an open box. If the cardboard is 16 inches long and 10 inches wide, find the dimensions of the box that will yield the maximum volume.

SOLUTION ✔   **Step 1**   Let x denote the length (in inches) of one side of each of the identical squares to be cut out of the cardboard (Figure 4.68) and let V denote the volume of the resulting box.

FIGURE 4.68
The dimensions of the open box are $(16 - 2x)$ by $(10 - 2x)$ by x inches.

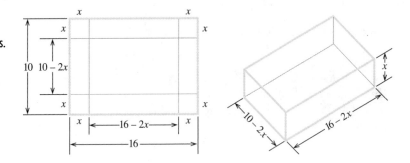

Step 2   The dimensions of the box are $(16 - 2x)$ inches long, $(10 - 2x)$ inches wide, and x inches high. Therefore, its volume (in cubic inches),

$$V = (16 - 2x)(10 - 2x)x$$
$$= 4(x^3 - 13x^2 + 40x) \qquad \text{(Expanding the expression)}$$

is the quantity to be maximized.

Step 3   Since each side of the box must be nonnegative, x must satisfy the inequalities $x \geq 0$, $16 - 2x \geq 0$, and $10 - 2x \geq 0$. This set of inequalities is satisfied if $0 \leq x \leq 5$. Thus, the problem at hand is equivalent to that of finding the absolute maximum of

$$V = f(x) = 4(x^3 - 13x^2 + 40x)$$

on the closed interval $[0, 5]$.

Step 4   Observe that f is continuous on $[0, 5]$, so the absolute maximum value of f must be attained at the end point(s) or at the critical point(s) of f.

Differentiating $f(x)$, we obtain

$$f'(x) = 4(3x^2 - 26x + 40)$$
$$= 4(3x - 20)(x - 2)$$

Upon setting $f'(x) = 0$ and solving the resulting equation for x, we obtain $x = 20/3$ or $x = 2$. Since $20/3$ lies outside the interval $[0, 5]$, it is no longer considered, and we are interested only in the critical point $x = 2$ of f. Next, evaluating $f(x)$ at $x = 0$, $x = 5$ (the end points of the interval $[0, 5]$), and $x = 2$, we obtain

$$f(0) = 0, \qquad f(2) = 144, \qquad f(5) = 0$$

Thus, the volume of the box is maximized by taking $x = 2$. The dimensions of the box are $12 \times 6 \times 2$ inches, and the volume is 144 cubic inches.   ■ ■ ■ ■

Exploring with Technology

Refer to Example 2.

1. Use a graphing utility to plot the graph of

$$f(x) = 4(x^3 - 13x^2 + 40x)$$

using the viewing rectangle $[0, 5] \times [0, 150]$. Explain what happens to $f(x)$ as x increases from $x = 0$ to $x = 5$ and give a physical interpretation.

2. Using ZOOM and TRACE, find the absolute maximum of f on the interval $[0, 5]$ and thus verify the solution for Example 2 obtained analytically.

EXAMPLE **3**   A city's Metropolitan Transit Authority (MTA) operates a subway line for commuters from a certain suburb to the downtown metropolitan area. Currently, an average of 6000 passengers a day take the trains, paying a fare of $1.50 per ride. The board of the MTA, contemplating raising the fare to $1.75 per ride in order to generate a larger revenue, engages the services of a consulting firm. The firm's study reveals that for each $.25 increase in fare, the ridership will be reduced by an average of 1000 passengers a day. Thus, the consulting firm recommends that MTA stick to the current fare of $1.50 per ride, which already yields a maximum revenue. Show that the consultants are correct.

SOLUTION ✔   **Step 1**   Let x denote the number of passengers per day, p denote the fare per ride, and R be MTA's revenue.

Step 2   To find a relationship between x and p, observe that the given data imply that when $x = 6000$, $p = 1.5$, and when $x = 5000$, $p = 1.75$. Therefore, the points $(6000, 1.5)$ and $(5000, 1.75)$ lie on a straight line. (Why?) To find the linear relationship between p and x, use the point-slope form of the equation of a straight line. Now, the slope of the line is

$$m = \frac{1.75 - 1.50}{5000 - 6000} = -0.00025$$

Therefore, the required equation is

$$p - 1.5 = -0.00025(x - 6000)$$
$$= -0.00025x + 1.5$$
$$p = -0.00025x + 3$$

Therefore, the revenue,

$$R = f(x) = xp = -0.00025x^2 + 3x$$ (Number of riders times unit fare)

is the quantity to be maximized.

Step 3 Since both p and x must be nonnegative, we see that $0 \le x \le 12{,}000$, and the problem is that of finding the absolute maximum of the function f on the closed interval $[0, 12{,}000]$.

Step 4 Observe that f is continuous on $[0, 12{,}000]$. To find the critical point of R, we compute

$$f'(x) = -0.0005x + 3$$

and set it equal to zero, giving $x = 6000$. Evaluating the function f at $x = 6000$, as well as at the end points $x = 0$ and $x = 12{,}000$, yields

$$f(0) = 0$$
$$f(6000) = 9000$$
$$f(12{,}000) = 0$$

We conclude that a maximum revenue of $9000 per day is realized when the ridership is 6000 per day. The optimum price of the fare per ride is therefore $1.50, as recommended by the consultants. The graph of the revenue function R is shown in Figure 4.69. ■ ■ ■ ■

FIGURE 4.69
f has an absolute maximum of 9000 when $x = 6000$.

MINIMIZATION PROBLEMS

EXAMPLE 4

The Betty Moore Company requires that its corned beef hash containers have a capacity of 54 cubic inches, have the shape of right-circular cylinders, and be made of tin. Determine the radius and height of the container that requires the least amount of metal.

SOLUTION ✔

Step 1 Let the radius and height of the container be r and h inches, respectively, and let S denote the surface area of the container (Figure 4.70).

Step 2 The amount of tin used to construct the container is given by the total surface area of the cylinder. Now, the area of the base and the top of the cylinder are each πr^2 square inches and the area of the side is $2\pi rh$ square inches. Therefore,

$$S = 2\pi r^2 + 2\pi rh$$ (3)

is the quantity to be minimized.

Step 3 The requirement that the volume of a container be 54 cubic inches implies that

$$\pi r^2 h = 54$$ (4)

Solving Equation (4) for h, we obtain

$$h = \frac{54}{\pi r^2}$$ (5)

FIGURE 4.70
We want to minimize the amount of material used to construct the container.

which, when substituted into (3), yields

$$S = 2\pi r^2 + 2\pi r \left(\frac{54}{\pi r^2}\right)$$

$$= 2\pi r^2 + \frac{108}{r}$$

Clearly, the radius r of the container must satisfy the inequality $r > 0$. The problem now is reduced to finding the absolute minimum of the function $S = f(r)$ on the interval $(0, \infty)$.

Step 4   Using the curve-sketching techniques of Section 4.3, we obtain the graph of f in Figure 4.71.

To find the critical point of f, we compute

$$S' = 4\pi r - \frac{108}{r^2}$$

and solve the equation $S' = 0$ for r:

$$4\pi r - \frac{108}{r^2} = 0$$

$$4\pi r^3 - 108 = 0$$

$$r^3 = \frac{27}{\pi}$$

$$r = \frac{3}{\sqrt[3]{\pi}} \approx 2 \qquad \textbf{(6)}$$

Next, let's show that this value of r gives rise to the absolute minimum of f. To show this, we first compute

$$S'' = 4\pi + \frac{216}{r^3}$$

Since $S'' > 0$ for $r = 3/\sqrt[3]{\pi}$, the second derivative test implies that the value of r in Equation (6) gives rise to a relative minimum of f. Finally, this relative minimum of f is also the absolute minimum of f since f is always concave upward ($S'' > 0$ for all $r > 0$). To find the height of the given container, we substitute the value of r given in (6) into (5). Thus,

$$h = \frac{54}{\pi r^2} = \frac{54}{\pi \left(\frac{3}{\pi^{1/3}}\right)^2}$$

$$= \frac{54\pi^{2/3}}{(\pi)9}$$

$$= \frac{6}{\pi^{1/3}} = \frac{6}{\sqrt[3]{\pi}}$$

$$= 2r$$

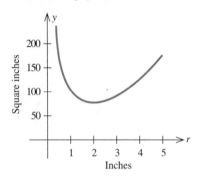

FIGURE 4.71
The total surface area of the right cylindrical container is graphed as a function of r.

Square inches

200
150
100
50

1   2   3   4   5
Inches

We conclude that the required container has a radius of approximately 2 inches and a height of approximately 4 inches, or twice the size of the radius.   ■ ■ ■ ■

AN INVENTORY PROBLEM

One problem faced by many companies is that of controlling the inventory of goods carried. Ideally, the manager must ensure that the company has sufficient stock to meet customer demand at all times. At the same time, she must make sure that this is accomplished without overstocking (incurring unnecessary storage costs) and also without having to place orders too frequently (incurring reordering costs).

EXAMPLE 5

The Dixie Import-Export Company is the sole agent for the Excalibur 250-cc motorcycle. Management estimates that the demand for these motorcycles is 10,000 per year and that they will sell at a uniform rate throughout the year. The cost incurred in ordering each shipment of motorcycles is $10,000, and the cost per year of storing each motorcycle is $200.

Dixie's management faces the following problem: Ordering too many motorcycles at one time ties up valuable storage space and increases the storage cost. On the other hand, placing orders too frequently increases the ordering costs. How large should each order be, and how often should orders be placed, to minimize ordering and storage costs?

SOLUTION ✔

Let x denote the number of motorcycles in each order (the lot size). Then, assuming that each shipment arrives just as the previous shipment has been sold, the average number of motorcycles in storage during the year is $x/2$. You can see that this is the case by examining Figure 4.72. Thus, Dixie's storage cost for the year is given by $200(x/2)$, or $100x$ dollars.

FIGURE 4.72
As each lot is depleted, the new lot arrives. The average inventory level is $x/2$ if x is the lot size.

Next, since the company requires 10,000 motorcycles for the year and since each order is for x motorcycles, the number of orders required is

$$\frac{10,000}{x}$$

This gives an ordering cost of

$$10,000\left(\frac{10,000}{x}\right) = \frac{100,000,000}{x}$$

dollars for the year. Thus, the total yearly cost incurred by Dixie, which includes the ordering and storage costs attributed to the sale of these motorcycles, is given by

$$C(x) = 100x + \frac{100{,}000{,}000}{x}$$

The problem is reduced to finding the absolute minimum of the function C in the interval $(0, 10{,}000]$. To accomplish this, we compute

$$C'(x) = 100 - \frac{100{,}000{,}000}{x^2}$$

Setting $C'(x) = 0$ and solving the resulting equation, we obtain $x = \pm 1000$. Since the number -1000 is outside the domain of the function C, it is rejected, leaving $x = 1000$ as the only critical point of C. Next, we find

$$C''(x) = \frac{200{,}000{,}000}{x^3}$$

Since $C''(1000) > 0$, the second derivative test implies that the critical point $x = 1000$ is a relative minimum of the function C (Figure 4.73). Also, since $C''(x) > 0$ for all x in $(0, 10{,}000]$, the function C is concave upward everywhere so that the point $x = 1000$ also gives the absolute minimum of C. Thus, to minimize the ordering and storage costs, Dixie should place $10{,}000/1000$, or 10, orders a year, each for a shipment of 1000 motorcycles.

FIGURE 4.73
C has an absolute minimum at
$(1000, 200{,}000)$.

SELF-CHECK EXERCISES 4.5

1. A man wishes to have an enclosed vegetable garden in his backyard. If the garden is to be a rectangular area of 300 ft², find the dimensions of the garden that will minimize the amount of fencing material needed.

2. The demand for the Super Titan tires is 1,000,000/year. The setup cost for each production run is $4000, and the manufacturing cost is $20/tire. The cost of storing each tire over the year is $2. Assuming uniformity of demand throughout the year and instantaneous production, determine how many tires should be manufactured per production run in order to keep the production cost to a minimum.

Solutions to Self-Check Exercises 4.5 can be found on page 368.

4.5 Exercises

1. **ENCLOSING THE LARGEST AREA** The owner of the Rancho Los Feliz has 3000 yd of fencing material with which to enclose a rectangular piece of grazing land along the straight portion of a river. If fencing is not required along the river, what are the dimensions of the largest area he can enclose? What is this area?

2. **ENCLOSING THE LARGEST AREA** Refer to Exercise 1. As an alternative plan, the owner of the Rancho Los Feliz might use the 3000 yd of fencing material to enclose the rectangular piece of grazing land along the straight portion of the river and then subdivide it by means of a fence running parallel to the sides. Again, no fencing is required along the river. What are the dimensions of the largest area that can be enclosed? What is this area? (See the accompanying figure.)

3. **MINIMIZING CONSTRUCTION COSTS** The management of the UNICO department store has decided to enclose an 800-ft^2 area outside the building for displaying potted plants and flowers. One side will be formed by the external wall of the store, two sides will be constructed of pine boards, and the fourth side will be made of galvanized steel fencing material. If the pine board fencing costs \$6/running foot and the steel fencing costs \$3/running foot, determine the dimensions of the enclosure that can be erected at minimum cost.

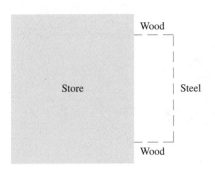

4. **PACKAGING** By cutting away identical squares from each corner of a rectangular piece of cardboard and folding up the resulting flaps, an open box may be made. If the cardboard is 15 in. long and 8 in. wide, find the dimensions of the box that will yield the maximum volume.

5. **METAL FABRICATION** If an open box is made from a tin sheet 8 in. square by cutting out identical squares from each corner and bending up the resulting flaps, determine the dimensions of the largest box that can be made.

6. **MINIMIZING CONSTRUCTION COSTS** If an open box has a square base and a volume of 108 in.^{3}, and is constructed from a tin sheet, find the dimensions of the box, assuming a minimum amount of material is used in its construction.

7. **MINIMIZING CONSTRUCTION COSTS** What are the dimensions of a closed rectangular box that has a square cross section, a capacity of 128 in.^{3}, and is constructed using the least amount of material?

8. **MINIMIZING CONSTRUCTION COSTS** A rectangular box is to have a square base and a volume of 20 ft^3. If the material for the base costs 30 cents/square foot, the material for the sides costs 10 cents/square foot, and the material for the top costs 20 cents/square foot, determine the dimensions of the box that can be constructed at minimum cost.

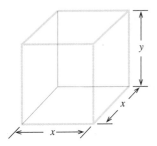

9. **PARCEL POST REGULATIONS** Postal regulations specify that a parcel sent by parcel post may have a combined length and girth of no more than 108 in. Find the dimensions of a rectangular package that has a square cross section

and the largest volume that may be sent through the mail. What is the volume of such a package?

Hint: The length plus the girth is $4x + h$ (see the accompanying figure).

10. **BOOK DESIGN** A production editor at Saunders-Roe Publishing decided that the pages of a book should have 1-in. margins at the top and bottom and $\frac{1}{2}$-in. margins on the sides. She further stipulated that each page should have an area of 50 in.^{2} (see the accompanying figure). Determine the page dimensions that will result in the maximum printed area on the page.

11. **PARCEL POST REGULATIONS** Postal regulations specify that a parcel sent by parcel post may have a combined length and girth of no more than 108 in. Find the dimensions of the cylindrical package of greatest volume that may be sent through the mail. What is the volume of such a package? Compare with Exercise 9.

Hint: The length plus the girth is $2\pi r + l$.

12. **MINIMIZING COSTS** For its beef stew, the Betty Moore Company uses tin containers that have the form of right circular cylinders. Find the radius and height of a container if it has a capacity of 36 in.^{3} and is constructed using the least amount of metal.

13. **PRODUCT DESIGN** The cabinet that will enclose the Acrosonic model D loudspeaker system will be rectangular and will have an internal volume of 2.4 ft^3. For aesthetic reasons, it has been decided that the height of the cabinet is to be 1.5 times its width. If the top, bottom, and sides of the cabinet are constructed of veneer costing 40 cents/square foot and the front (ignore the cutouts in the baffle) and rear are constructed of particle board costing

20 cents/square foot, what are the dimensions of the enclosure that can be constructed at a minimum cost?

14. **DESIGNING A NORMAN WINDOW** A Norman window has the shape of a rectangle surmounted by a semicircle (see the accompanying figure). If a Norman window is to have a perimeter of 28 ft, what should its dimensions be in order to allow the maximum amount of light through the window?

15. **OPTIMAL CHARTER-FLIGHT FARE** If exactly 200 people sign up for a charter flight, the Leisure World Travel Agency charges $300/person. However, if more than 200 people sign up for the flight (assume this is the case), then each fare is reduced by $1 for each additional person. Determine how many passengers will result in a maximum revenue for the travel agency. What is the maximum revenue? What would be the fare per passenger in this case?

Hint: Let x denote the number of passengers above 200. Show that the revenue function R is given by $R(x) = (200 + x) \cdot (300 - x)$.

16. **MAXIMIZING YIELD** An apple orchard has an average yield of 36 bushels of apples/tree if tree density is 22 trees/acre. For each unit increase in tree density, the yield decreases by 2 bushels. How many trees should be planted in order to maximize the yield?

17. **STRENGTH OF A BEAM** A wooden beam has a rectangular cross section of height h in. and width w in. (see the accompanying figure). The strength S of the beam is directly proportional to its width and the square of its height. What are the dimensions of the cross section of the strongest beam that can be cut from a round log of diameter 24 in.?

Hint: $S = kh^2w$, where k is a constant of proportionality.

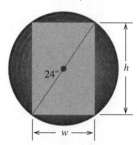

18. **DESIGNING A GRAIN SILO** A grain silo has the shape of a right circular cylinder surmounted by a hemisphere (see the accompanying figure). If the silo is to have a capacity of 504π ft^3, find the radius and height of the silo that requires the least amount of material to construct.

    Hint: The volume of the silo is $\pi r^2 h + \frac{2}{3}\pi r^3$, and the surface area (including the floor) is $\pi(3r^2 + 2rh)$.

19. **MINIMIZING CONSTRUCTION COSTS** In the following diagram, S represents the position of a power relay station located on a straight coast, and E shows the location of a marine biology experimental station on an island. A cable is to be laid connecting the relay station with the experimental station. If the cost of running the cable on land is \$1/running foot and the cost of running the cable under water is \$3/running foot, locate the point P that will result in a minimum cost (solve for x).

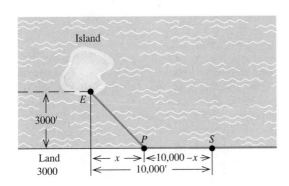

20. **FLIGHTS OF BIRDS** During daylight hours, some birds fly more slowly over water than over land because some of their energy is expended in overcoming the downdrafts of air over open bodies of water. Suppose a bird that flies at a constant speed of 4 mph over water and 6 mph over land starts its journey at the point E on an island and ends at its nest N on the shore of the mainland, as shown in the accompanying figure. Find the location of the point P that allows the bird to complete its journey in the minimum time (solve for x).

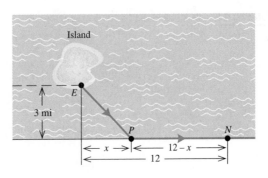

21. **OPTIMAL SPEED OF A TRUCK** A truck gets $400/x$ mpg when driven at a constant speed of x mph (between 50 and 70 mph). If the price of fuel is \$1/gallon and the driver is paid \$8/hour, at what speed between 50 and 70 mph is it most economical to drive?

22. **INVENTORY CONTROL AND PLANNING** The demand for motorcycle tires imported by the Dixie Import-Export Company is 40,000/year and may be assumed to be uniform throughout the year. The cost of ordering a shipment of tires is \$400, and the cost of storing each tire for a year is \$2. Determine how many tires should be in each shipment if the ordering and storage costs are to be minimized. (Assume that each shipment arrives just as the previous one has been sold.)

23. **INVENTORY CONTROL AND PLANNING** The McDuff Preserves Company expects to bottle and sell 2,000,000 32-oz jars of jam. The company orders its containers from the Consolidated Bottle Company. The cost of ordering a shipment of bottles is \$200, and the cost of storing each empty bottle for a year is \$.40. How many orders should McDuff place per year and how many bottles should be in each shipment if the ordering and storage costs are to be minimized? (Assume that each shipment of bottles is used up before the next shipment arrives.)

24. **INVENTORY CONTROL AND PLANNING** The Neilsen Cookie Company sells its assorted butter cookies in containers that have a net content of 1 lb. The estimated demand for the cookies is 1,000,000 units. The setup cost for each production run is \$500, and the manufacturing cost is \$.50 for each container of cookies. The cost of storing each container of cookies over the year is \$.40. Assuming uniformity of demand throughout the year and instantaneous production, how many containers of cookies should Neilsen produce per production run in order to minimize the production cost?

    Hint: Following the method of Example 5, show that the total production cost is given by the function

    $$C(x) = \frac{500,000,000}{x} + 0.2x + 500,000$$

    Then minimize the function C on the interval $(0, 1,000,000)$.

SOLUTIONS TO SELF-CHECK EXERCISES 4.5

1. Let x and y (measured in feet) denote the length and width of the rectangular garden. Since the area is to be 300 ft², we have

$$xy = 300$$

Next, the amount of fencing to be used is given by the perimeter, and this quantity is to be minimized. Thus, we want to minimize

$$2x + 2y$$

or, since $y = 300/x$ (obtained by solving for y in the first equation), we see that the expression to be minimized is

$$f(x) = 2x + 2\left(\frac{300}{x}\right)$$

$$= 2x + \frac{600}{x}$$

for positive values of x. Now,

$$f'(x) = 2 - \frac{600}{x^2}$$

Setting $f'(x) = 0$ yields $x = -\sqrt{300}$ or $x = \sqrt{300}$. We consider only the critical point $x = \sqrt{300}$ since $-\sqrt{300}$ lies outside the interval $(0, \infty)$. We then compute

$$f''(x) = \frac{1200}{x^3}$$

Since

$$f''(\sqrt{300}) > 0$$

the second derivative test implies that a relative minimum of f occurs at $x = \sqrt{300}$. In fact, since $f''(x) > 0$ for all x in $(0, \infty)$, we conclude that $x = \sqrt{300}$ gives rise to the absolute minimum of f. The corresponding value of y, obtained by substituting this value of x into the equation $xy = 300$, is $y = \sqrt{300}$. Therefore, the required dimensions of the vegetable garden are approximately 17.3 ft $\times$ 17.3 ft.

2. Let x denote the number of tires in each production run. Then, the average number of tires in storage is $x/2$, so the storage cost incurred by the company is $2(x/2)$, or x dollars. Next, since the company needs to manufacture 1,000,000 tires for the year in order to meet the demand, the number of production runs is $1,000,000/x$. This gives setup costs amounting to

$$4000\left(\frac{1,000,000}{x}\right) = \frac{4,000,000,000}{x}$$

dollars for the year. The total manufacturing cost is $20,000,000. Thus, the total yearly cost incurred by the company is given by

$$C(x) = x + \frac{4,000,000,000}{x} + 20,000,000$$

Differentiating $C(x)$, we find

$$C'(x) = 1 - \frac{4,000,000,000}{x^2}$$

Setting $C'(x) = 0$ gives $x = 63,246$ as the critical point in the interval $(0, 1,000,000)$. Next, we find

$$C''(x) = \frac{8,000,000,000}{x^3}$$

Since $C''(x) > 0$ for all $x > 0$, we see that C is concave upward for all $x > 0$. Furthermore, $C''(63,246) > 0$ implies that $x = 63,246$ gives rise to a relative minimum of C (by the second derivative test). Since C is always concave upward for $x > 0$, $x = 63,246$ gives the absolute minimum of C. Therefore, the company should manufacture 63,246 tires in each production run.

CHAPTER 4 Summary of Principal Terms

Terms

increasing function

decreasing function

relative maximum

relative minimum

relative extrema

critical point

first derivative test

concave upward

concave downward

inflection point

second derivative test

vertical asymptote

horizontal asymptote

absolute extrema

absolute maximum value

absolute minimum value

CHAPTER 4 REVIEW EXERCISES

In Exercises 1–10, (a) find the intervals where the given function f is increasing and where it is decreasing, (b) find the relative extrema of f, (c) find the intervals where f is concave upward and where it is concave downward, and (d) find the inflection points, if any, of f.

1. $f(x) = \frac{1}{3}x^3 - x^2 + x - 6$

2. $f(x) = (x - 2)^3$

3. $f(x) = x^4 - 2x^2$

4. $f(x) = x + \frac{4}{x}$

5. $f(x) = \frac{x^2}{x - 1}$

6. $f(x) = \sqrt{x - 1}$

7. $f(x) = (1 - x)^{1/3}$

8. $f(x) = x\sqrt{x - 1}$

9. $f(x) = \frac{2x}{x + 1}$

10. $f(x) = \frac{-1}{1 + x^2}$

In Exercises 11–18, obtain as much information as possible on each of the given functions. Then use this information to sketch the graph of the function.

11. $f(x) = x^2 - 5x + 5$

12. $f(x) = -2x^2 - x + 1$

13. $g(x) = 2x^3 - 6x^2 + 6x + 1$

14. $g(x) = \dfrac{1}{3}x^3 - x^2 + x - 3$

15. $h(x) = x\sqrt{x-2}$ **16.** $h(x) = \dfrac{2x}{1+x^2}$

17. $f(x) = \dfrac{x-2}{x+2}$ **18.** $f(x) = x - \dfrac{1}{x}$

In Exercises 19–22, find the horizontal and vertical asymptotes of the graphs of the given functions. Do not sketch the graphs.

19. $f(x) = \dfrac{1}{2x+3}$ **20.** $f(x) = \dfrac{2x}{x+1}$

21. $f(x) = \dfrac{5x}{x^2 - 2x - 8}$ **22.** $f(x) = \dfrac{x^2 + x}{x(x-1)}$

In Exercises 23–32, find the absolute maximum value and the absolute minimum value, if any, of the given function.

23. $f(x) = 2x^2 + 3x - 2$ **24.** $g(x) = x^{2/3}$

25. $g(t) = \sqrt{25 - t^2}$

26. $f(x) = \dfrac{1}{3}x^3 - x^2 + x + 1$ on $[0, 2]$

27. $h(t) = t^3 - 6t^2$ on $[2, 5]$

28. $g(x) = \dfrac{x}{x^2 + 1}$ on $[0, 5]$

29. $f(x) = x - \dfrac{1}{x}$ on $[1, 3]$

30. $h(t) = 8t - \dfrac{1}{t^2}$ on $[1, 3]$

31. $f(s) = s\sqrt{1 - s^2}$ on $[-1, 1]$

32. $f(x) = \dfrac{x^2}{x-1}$ on $[-1, 3]$

33. Odyssey Travel Agency's monthly profit (in thousands of dollars) depends on the amount of money x (in thousands of dollars) spent on advertising per month according to the rule

$$P(x) = -x^2 + 8x + 20$$

What should Odyssey's monthly advertising budget be in order to maximize its monthly profits?

34. The Department of the Interior of an African country began to record an index of environmental quality to measure progress or decline in the environmental quality of its wildlife. The index for the years 1984 through 1994 is approximated by the function

$$I(t) = \frac{50t^2 + 600}{t^2 + 10} \qquad (0 \le t \le 10)$$

a. Compute $I'(t)$ and show that $I(t)$ is decreasing on the interval $(0, 10)$.
b. Compute $I''(t)$. Study the concavity of the graph of I.
c. Sketch the graph of I.
d. Interpret your results.

35. The weekly demand for video discs manufactured by the Herald Record Company is given by

$$p = -0.0005x^2 + 60$$

where p denotes the unit price in dollars and x denotes the quantity demanded. The weekly total cost function associated with producing these discs is given by

$$C(x) = -0.001x^2 + 18x + 4000$$

where $C(x)$ denotes the total cost incurred in pressing x discs. Find the production level that will yield a maximum profit for the manufacturer.
Hint: Use the quadratic formula.

36. The total monthly cost (in dollars) incurred by the Carlota Music Company in manufacturing x units of its Professional Series guitars is given by the function

$$C(x) = 0.001x^2 + 100x + 4000$$

a. Find the average cost function $\overline{C}$.
b. Determine the production level that will result in the smallest average production cost.

37. The average worker at Wakefield Avionics, Inc., can assemble

$$N(t) = -2t^3 + 12t^2 + 2t \qquad (0 \le t \le 4)$$

ready-to-fly radio-controlled model airplanes t hr into the 8 A.M. to 12 noon morning shift. At what time during this shift is the average worker performing at peak efficiency?

38. You wish to construct a closed rectangular box that has a volume of 4 ft³. The length of the base of the box will be twice as long as its width. The material for the top and bottom of the box costs 30 cents/square foot. The material for the sides of the box costs 20 cents/square foot. Find the dimensions of the least expensive box that can be constructed.

39. The Lehen Vinters Company imports a certain brand of beer. The demand, which may be assumed to be uniform, is 800,000 cases/year. The cost of ordering a shipment of beer is $500, and the cost of storing each case of beer for a year is $2. Determine how many cases of beer should be in each shipment if the ordering and storage costs are to be kept to a minimum. (Assume that each shipment of beer arrives just as the previous one has been sold.)

40. Let

$$f(x) = \begin{cases} x^3 + 1 & \text{if } x \neq 0 \\ 2 & \text{if } x = 0 \end{cases}$$

a. Compute $f'(x)$ and show that it does not change sign as we move across $x = 0$.

b. Show that f has a relative maximum at $x = 0$. Does this contradict the first derivative test? Explain your answer.

5
EXPONENTIAL AND LOGARITHMIC FUNCTIONS

The exponential function is, without doubt, the most important function in mathematics and its applications. After a brief introduction to the exponential function and its *inverse*, the logarithmic function, we learn how to differentiate such functions. This lays the foundation for exploring the many applications involving exponential functions. For example, we look at the role played by exponential functions in computing earned interest on a bank account, in studying the growth of a bacteria population in the laboratory, in studying the way radioactive matter decays, in studying the rate at which a factory worker learns a certain process, and in studying the rate at which a communicable disease is spread over time.

How many bacteria will there be in a culture at the end of a certain period of time? How fast will the bacteria population be growing at the end of that time? Example 1, page 426, answers these questions.

5.1 Exponential Functions

EXPONENTIAL FUNCTIONS AND THEIR GRAPHS

Suppose you deposit a sum of $1000 in an account earning interest at the rate of 10% per year *compounded continuously* (the way most financial institutions compute interest). Then, the accumulated amount at the end of t years ($0 \le t \le 20$) is described by the function f, whose graph appears in Figure 5.1.* Such a function is called an *exponential function.* Observe that the graph of f rises rather slowly at first but very rapidly as time goes by. For purposes of comparison, we have also shown the graph of the function $y = g(t) = 1000(1 + 0.10t)$, giving the accumulated amount for the same principal ($1000) but earning *simple* interest at the rate of 10% per year. The moral of the story: It is never too early to save.

Exponential functions play an important role in many real-world applications, as you will see throughout this chapter.

FIGURE 5.1
Under continuous compounding, a sum of money grows exponentially.

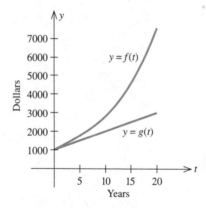

Observe that whenever b is a positive number and n is any real number, the expression b^n is a real number. This enables us to define an exponential function as follows:

Exponential Function

> The function defined by
>
> $$f(x) = b^x \qquad (b > 0, b \ne 1)$$
>
> is called an **exponential function with base b and exponent x.** The domain of f is the set of all real numbers.

---

* We will derive the rule for f in Section 5.3.

For example, the exponential function with base 2 is the function

$$f(x) = 2^x$$

with domain $(-\infty, \infty)$. The values of $f(x)$ for selected values of x follow:

$$f(3) = 2^3 = 8, \qquad f\left(\frac{3}{2}\right) = 2^{3/2} = 2 \cdot 2^{1/2} = 2\sqrt{2}, \qquad f(0) = 2^0 = 1,$$

$$f(-1) = 2^{-1} = \frac{1}{2}, \qquad f\left(-\frac{2}{3}\right) = 2^{-2/3} = \frac{1}{2^{2/3}} = \frac{1}{\sqrt[3]{4}}$$

Computations involving exponentials are facilitated by the laws of exponents. These laws were stated in Section 1.1, and you might want to review the material there. For convenience, however, we will restate these laws.

Laws of Exponents

Let a and b be positive numbers and let x and y be real numbers. Then,

1. $b^x \cdot b^y = b^{x+y}$

2. $\dfrac{b^x}{b^y} = b^{x-y}$

3. $(b^x)^y = b^{xy}$

4. $(ab)^x = a^x b^x$

5. $\left(\dfrac{a}{b}\right)^x = \dfrac{a^x}{b^x}$

The use of the laws of exponents is illustrated in the next example.

EXAMPLE 1

a. $16^{7/4} \cdot 16^{-1/2} = 16^{7/4-1/2} = 16^{5/4} = 2^5 = 32$ (Law 1)

b. $\dfrac{8^{5/3}}{8^{-1/3}} = 8^{5/3-(-1/3)} = 8^2 = 64$ (Law 2)

c. $(64^{4/3})^{-1/2} = 64^{(4/3)(-1/2)} = 64^{-2/3}$

$$= \frac{1}{64^{2/3}} = \frac{1}{(64^{1/3})^2} = \frac{1}{4^2} = \frac{1}{16} \qquad \text{(Law 3)}$$

d. $(16 \cdot 81)^{-1/4} = 16^{-1/4} \cdot 81^{-1/4} = \dfrac{1}{16^{1/4}} \cdot \dfrac{1}{81^{1/4}} = \dfrac{1}{2} \cdot \dfrac{1}{3} = \dfrac{1}{6}$ (Law 4)

e. $\left(\dfrac{3^{1/2}}{2^{1/3}}\right)^4 = \dfrac{3^{4/2}}{2^{4/3}} = \dfrac{9}{2^{4/3}}$ (Law 5)   ■■■■

EXAMPLE 2

Let $f(x) = 2^{2x-1}$. Find the value of x for which $f(x) = 16$.

SOLUTION ✔

We want to solve the equation

$$2^{2x-1} = 16 = 2^4$$

But this equation holds if and only if

$$2x - 1 = 4 \qquad (b^m = b^n \Rightarrow m = n)$$

giving $x = \frac{5}{2}$.   ■■■■

Exponential functions play an important role in mathematical analysis. Because of their special characteristics, they are some of the most useful functions and are found in virtually every field where mathematics is applied. To mention a few examples: Under ideal conditions the number of bacteria present at any time t in a culture may be described by an exponential function of t; radioactive substances decay over time in accordance with an "exponential" law of decay; money left on fixed deposit and earning compound interest grows exponentially; and some of the most important distribution functions encountered in statistics are exponential.

Let's begin our investigation into the properties of exponential functions by studying their graphs.

EXAMPLE 3

Sketch the graph of the exponential function $y = 2^x$.

SOLUTION ✔

First, as discussed earlier, the domain of the exponential function $y = f(x) = 2^x$ is the set of real numbers. Next, putting $x = 0$ gives $y = 2^0 = 1$, the y-intercept of f. There is no x-intercept since there is no value of x for which $y = 0$. To find the range of f, consider the following table of values:

FIGURE 5.2
The graph of $y = 2^x$

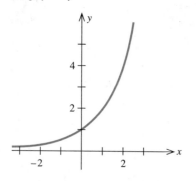

x	-5	-4	-3	-2	-1	0	1	2	3	4	5
y	1/32	1/16	1/8	1/4	1/2	1	2	4	8	16	32

We see from these computations that 2^x decreases and approaches zero as x decreases without bound and that 2^x increases without bound as x increases without bound. Thus, the range of f is the interval $(0, \infty)$—that is, the set of positive real numbers. Finally, we sketch the graph of $y = f(x) = 2^x$ in Figure 5.2.   ■■■■

EXAMPLE 4

Sketch the graph of the exponential function $y = (1/2)^x$.

SOLUTION ✔

The domain of the exponential function $y = (1/2)^x$ is the set of all real numbers. The y-intercept is $(1/2)^0 = 1$; there is no x-intercept since there is no value of x for which $y = 0$. From the following table of values

FIGURE 5.3
The graph of $y = (1/2)^x$

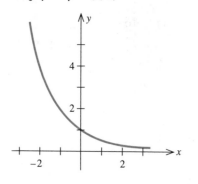

x	-5	-4	-3	-2	-1	0	1	2	3	4	5
y	32	16	8	4	2	1	1/2	1/4	1/8	1/16	1/32

we deduce that $(1/2)^x = 1/2^x$ increases without bound as x decreases without bound and that $(1/2)^x$ decreases and approaches zero as x increases without bound. Thus, the range of f is the interval $(0, \infty)$. The graph of $y = f(x) = (1/2)^x$ is sketched in Figure 5.3.   ■■■■

The functions $y = 2^x$ and $y = (1/2)^x$, whose graphs you studied in Examples 3 and 4, are special cases of the exponential function $y = f(x) = b^x$, obtained by setting $b = 2$ and $b = 1/2$, respectively. In general, the exponential function $y = b^x$ with $b > 1$ has a graph similar to $y = 2^x$, whereas the graph of $y = b^x$ for $0 < b < 1$ is similar to that of $y = (1/2)^x$ (Exercises 27 and 28). When $b = 1$, the function $y = b^x$ reduces to the constant function $y = 1$. For comparison, the graphs of all three functions are sketched in Figure 5.4.

FIGURE 5.4
$y = b^x$ is an increasing function of x if $b > 1$, a constant function if $b = 1$, and a decreasing function if $0 < b < 1$.

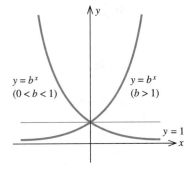

Properties of the Exponential Function

The exponential function $y = b^x$ ($b > 0$, $b \neq 1$) has the following properties:

1. Its domain is $(-\infty, \infty)$.
2. Its range is $(0, \infty)$.
3. Its graph passes through the point $(0, 1)$.
4. It is continuous on $(-\infty, \infty)$.
5. It is increasing on $(-\infty, \infty)$ if $b > 1$ and decreasing on $(-\infty, \infty)$ if $b < 1$.

THE BASE e

Exponential functions to the base e, where e is an irrational number whose value is $2.7182818\ldots$, play an important role in both theoretical and applied problems. It can be shown, although we will not do so here, that

$$e = \lim_{m \to \infty} \left(1 + \frac{1}{m}\right)^m \tag{1}$$

However, you may convince yourself of the plausibility of this definition of the number e by examining Table 5.1, which may be constructed with the help of a calculator.

Table 5.1

m	10	100	1000	10,000	100,000	1,000,000
$\left(1 + \dfrac{1}{m}\right)^m$	2.59374	2.70481	2.71692	2.71815	2.71827	2.71828

Exploring with Technology

To obtain a visual confirmation of the fact that the expression $(1 + 1/m)^m$ approaches the number $e = 2.71828\ldots$ as m increases without bound, plot the graph of $f(x) = (1 + 1/x)^x$ in a suitable viewing rectangle and observe that $f(x)$ approaches $2.71828\ldots$ as x increases without bound. Use ZOOM and TRACE to find the value of $f(x)$ for large values of x.

    EXAMPLE **5**   Sketch the graph of the function $y = e^x$.

SOLUTION ✔   Since $e > 1$, it follows from our previous discussion that the graph of $y = e^x$ is similar to the graph of $y = 2^x$ (see Figure 5.2). With the aid of a calculator, we obtain the following table:

x	−3	−2	−1	0	1	2	3
y	0.05	0.14	0.37	1	2.72	7.39	20.09

The graph of $y = e^x$ is sketched in Figure 5.5.

FIGURE 5.5
The graph of $y = e^x$

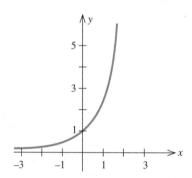

Next, we consider another exponential function to the base e that is closely related to the previous function and is particularly useful in constructing models that describe "exponential decay."

    EXAMPLE **6**   Sketch the graph of the function $y = e^{-x}$.

SOLUTION ✔   Since $e > 1$, it follows that $0 < 1/e < 1$, so $f(x) = e^{-x} = 1/e^x = (1/e)^x$ is an exponential function with base less than 1. Therefore, it has a graph similar to that of the exponential function $y = (1/2)^x$. As before, we construct the following table of values of $y = e^{-x}$ for selected values of x:

x	−3	−2	−1	0	1	2	3
y	20.09	7.39	2.72	1	0.37	0.14	0.05

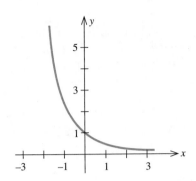

FIGURE **5.6**
The graph of $y = e^{-x}$

Using this table, we sketch the graph of $y = e^{-x}$ in Figure 5.6.   ■ ■ ■ ■

SELF-CHECK EXERCISES 5.1

1. Solve the equation $2^{2x+1} \cdot 2^{-3} = 2^{x-1}$.
2. Sketch the graph of $y = e^{0.4x}$.

Solutions to Self-Check Exercises 5.1 can be found on page 382.

5.1 Exercises

In Exercises 1–8, evaluate the expression.

1. a. $4^{-3} \cdot 4^5$ **b.** $3^{-3} \cdot 3^6$

2. a. $(2^{-1})^3$ **b.** $(3^{-2})^3$

3. a. $9(9)^{-1/2}$ **b.** $5(5)^{-1/2}$

4. a. $\left[\left(-\dfrac{1}{2}\right)^3\right]^{-2}$ **b.** $\left[\left(-\dfrac{1}{3}\right)^2\right]^{-3}$

5. a. $\dfrac{(-3)^4(-3)^5}{(-3)^8}$ **b.** $\dfrac{(2^{-4})(2^6)}{2^{-1}}$

6. a. $3^{1/4} \cdot (9)^{-5/8}$ **b.** $2^{3/4} \cdot (4)^{-3/2}$

7. a. $\dfrac{5^{3.3} \cdot 5^{-1.6}}{5^{-0.3}}$ **b.** $\dfrac{4^{2.7} \cdot 4^{-1.3}}{4^{-0.4}}$

8. a. $\left(\dfrac{1}{16}\right)^{-1/4}\left(\dfrac{27}{64}\right)^{-1/3}$ **b.** $\left(\dfrac{8}{27}\right)^{-1/3}\left(\dfrac{81}{256}\right)^{-1/4}$

In Exercises 9–16, simplify the expression.

9. a. $(64x^9)^{1/3}$ **b.** $(25x^3y^4)^{1/2}$

10. a. $(2x^3)(-4x^{-2})$ **b.** $(4x^{-2})(-3x^5)$

11. a. $\dfrac{6a^{-5}}{3a^{-3}}$ **b.** $\dfrac{4b^{-4}}{12b^{-6}}$

12. a. $y^{-3/2}y^{5/3}$ **b.** $x^{-3/5}x^{8/3}$

13. a. $(2x^3y^2)^3$ **b.** $(4x^2y^2z^3)^2$

14. a. $(x^{r/s})^{s/r}$ **b.** $(x^{-b/a})^{-a/b}$

15. a. $\dfrac{5^0}{(2^{-3}x^{-3}y^2)^2}$ **b.** $\dfrac{(x+y)(x-y)}{(x-y)^0}$

16. a. $\dfrac{(a^m \cdot a^{-n})^{-2}}{(a^{m+n})^2}$ **b.** $\left(\dfrac{x^{2n-2}y^{2n}}{x^{5n+1}y^{-n}}\right)^{1/3}$

In Exercises 17–26, solve the equation for x.

17. $6^{2x} = 6^4$ **18.** $5^{-x} = 5^3$

19. $3^{3x-4} = 3^5$ **20.** $10^{2x-1} = 10^{x+3}$

21. $(2.1)^{x+2} = (2.1)^5$ **22.** $(-1.3)^{x-2} = (-1.3)^{2x+1}$

23. $8^x = \left(\dfrac{1}{32}\right)^{x-2}$ **24.** $3^{x-x^2} = \dfrac{1}{9^x}$

25. $3^{2x} - 12 \cdot 3^x + 27 = 0$

26. $2^{2x} - 4 \cdot 2^x + 4 = 0$

(continued on p. 382)

Although the proof is outside the scope of this book, it can be proved that an exponential function of the form $f(x) = b^x$, where $b > 1$, will ultimately grow faster than the power function $g(x) = x^n$ for *any* positive real number n. To give a visual demonstration of this result for the special case of the exponential function $f(x) = e^x$, we can use a graphing calculator to plot the graphs of both f and g (for selected values of n) on the same set of axes in an appropriate viewing rectangle and observe that the graph of f ultimately lies above that of g.

EXAMPLE 1 Use a graphing utility to plot the graphs of (a) $f(x) = e^x$ and $g(x) = x^3$ on the same set of axes in the viewing rectangle $[0, 6] \times [0, 250]$ and (b) $f(x) = e^x$ and $g(x) = x^5$ in the viewing rectangle $[0, 20] \times [0, 1{,}000{,}000]$.

SOLUTION ✔
a. The graphs of $f(x) = e^x$ and $g(x) = x^3$ in the viewing rectangle $[0, 6] \times [0, 250]$ are shown in Figure T1a.
b. The graphs of $f(x) = e^x$ and $g(x) = x^5$ in the viewing rectangle $[0, 20] \times [0, 1{,}000{,}000]$ are shown in Figure T1b.

FIGURE T1

(a) The graphs of $f(x) = e^x$ and $g(x) = x^3$ in the viewing rectangle $[0, 6] \times [0, 250]$

(b) The graphs of $f(x) = e^x$ and $g(x) = x^5$ in the viewing rectangle $[0, 20] \times [0, 1{,}000{,}000]$

In the exercises that follow, you are asked to use a graphing utility to reveal the properties of exponential functions.

Exercises

In Exercises 1 and 2, use a graphing utility to plot the graphs of the functions *f* and *g* on the same set of axes in the specified viewing rectangle.

1. $f(x) = e^x$ and $g(x) = x^2$; $[0, 4] \times [0, 30]$

2. $f(x) = e^x$ and $g(x) = x^4$; $[0, 15] \times [0, 20{,}000]$

In Exercises 3 and 4, use a graphing utility to plot the graphs of the functions *f* and *g* on the same set of axes in an appropriate viewing rectangle to demonstrate that *f* ultimately grows faster than *g*. (*Note:* Your answer will not be unique.)

3. $f(x) = 2^x$ and $g(x) = x^{2.5}$

4. $f(x) = 3^x$ and $g(x) = x^3$

5. Use a graphing utility to plot the graphs of $f(x) = 2^x$, $g(x) = 3^x$, and $h(x) = 4^x$ on the same set of axes in the viewing rectangle $[0, 5] \times [0, 100]$. Comment on the relationship between the base *b* and the growth of the function $f(x) = b^x$.

6. Use a graphing utility to plot the graphs of $f(x) = (1/2)^x$, $g(x) = (1/3)^x$, and $h(x) = (1/4)^x$ on the same set of axes in the viewing rectangle $[0, 4] \times [0, 1]$. Comment on the relationship between the base *b* and the growth of the function $f(x) = b^x$.

7. Use a graphing utility to plot the graphs of $f(x) = e^x$, $g(x) = 2e^x$, and $h(x) = 3e^x$ on the same set of axes in the viewing rectangle $[-3, 3] \times [0, 10]$. Comment on the role played by the constant *k* in the graph of $f(x) = ke^x$.

8. Use a graphing utility to plot the graphs of $f(x) = -e^x$, $g(x) = -2e^x$, and $h(x) = -3e^x$ on the same set of axes in the viewing rectangle $[-3, 3] \times [-10, 0]$. Comment on the role played by the constant *k* in the graph of $f(x) = ke^x$.

9. Use a graphing utility to plot the graphs of $f(x) = e^{0.5x}$, $g(x) = e^x$, and $h(x) = e^{1.5x}$ on the same set of axes in the viewing rectangle $[-2, 2] \times [0, 4]$. Comment on the role played by the constant *k* in the graph of $f(x) = e^{kx}$.

10. Use a graphing utility to plot the graphs of $f(x) = e^{-0.5x}$, $g(x) = e^{-x}$, and $h(x) = e^{-1.5x}$ on the same set of axes in the viewing rectangle $[-2, 2] \times [0, 4]$. Comment on the role played by the constant *k* in the graph of $f(x) = e^{kx}$.

In Exercises 27–35, sketch the graphs of the given functions on the same axes. A calculator is recommended for these exercises.

27. $y = 2^x$, $y = 3^x$, and $y = 4^x$

28. $y = \left(\frac{1}{2}\right)^x$, $y = \left(\frac{1}{3}\right)^x$, and $y = \left(\frac{1}{4}\right)^x$

29. $y = 2^{-x}$, $y = 3^{-x}$, and $y = 4^{-x}$

30. $y = 4^{0.5x}$ and $y = 4^{-0.5x}$

31. $y = 4^{0.5x}$, $y = 4^x$, and $y = 4^{2x}$

32. $y = e^x$, $y = 2e^x$, and $y = 3e^x$

33. $y = e^{0.5x}$, $y = e^x$, and $y = e^{1.5x}$

34. $y = e^{-0.5x}$, $y = e^{-x}$, and $y = e^{-1.5x}$

35. $y = 0.5e^{-x}$, $y = e^{-x}$, and $y = 2e^{-x}$

In Exercises 36–39, determine whether the statement is true or false. If it is true, explain why it is true. If it is false, give an example to show why it is false.

36. $(x^2 + 1)^3 = x^6 + 1$

37. $e^{xy} = e^x e^y$

38. If $x < y$, then $e^x < e^y$.

39. If $0 < b < 1$ and $x < y$, then $b^x > b^y$.

SOLUTIONS TO SELF-CHECK EXERCISES 5.1

1. $2^{2x+1} \cdot 2^{-3} = 2^{x-1}$

$\frac{2^{2x+1}}{2^{x-1}} \cdot 2^{-3} = 1$ (Dividing both sides by 2^{x-1})

$2^{(2x+1)-(x-1)-3} = 1$

$2^{x-1} = 1$

This is true if and only if $x - 1 = 0$ or $x = 1$.

2. We first construct the following table of values:

x	−3	−2	−1	0	1	2	3	4
$y = e^{0.4x}$	0.3	0.4	0.7	1	1.5	2.2	3.3	5

Next, we plot these points and join them by a smooth curve to obtain the graph of f shown in the accompanying figure.

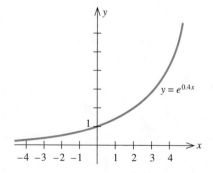

5.2 Logarithmic Functions

LOGARITHMS

You are already familiar with exponential equations of the form

$$b^y = x \qquad (b > 0, b \neq 1)$$

where the variable x is expressed in terms of a real number b and a variable y. But what about solving this same equation for y? You may recall from your study of algebra that the number y is called the **logarithm of x to the base b** and is denoted by $\log_b x$. It is the exponent to which the base b must be raised in order to obtain the number x.

Logarithm of x to the Base b

$$y = \log_b x \qquad \text{if and only if} \qquad x = b^y \qquad (x > 0)$$

Observe that the logarithm $\log_b x$ is defined only for positive values of x.

EXAMPLE 1

a. $\log_{10} 100 = 2$ since $100 = 10^2$
b. $\log_5 125 = 3$ since $125 = 5^3$
c. $\log_3 \dfrac{1}{27} = -3$ since $\dfrac{1}{27} = \dfrac{1}{3^3} = 3^{-3}$
d. $\log_{20} 20 = 1$ since $20 = 20^1$ ■■■■

EXAMPLE 2

Solve each of the following equations for x.

a. $\log_3 x = 4$ **b.** $\log_{16} 4 = x$ **c.** $\log_x 8 = 3$

SOLUTION ✔

a. By definition, $\log_3 x = 4$ implies $x = 3^4 = 81$.
b. $\log_{16} 4 = x$ is equivalent to $4 = 16^x = (4^2)^x = 4^{2x}$, or $4^1 = 4^{2x}$, from which we deduce that

$$2x = 1 \qquad (b^m = b^n \Rightarrow m = n)$$

$$x = \frac{1}{2}$$

c. Referring once again to the definition, we see that the equation $\log_x 8 = 3$ is equivalent to

$$8 = (2^3) = x^3$$

$$x = 2 \qquad (a^m = b^m \Rightarrow a = b)$$ ■■■■

The two widely used systems of logarithms are the system of **common logarithms**, which uses the number 10 as the base, and the system of **natural logarithms**, which uses the irrational number $e = 2.71828\ldots$ as the base. Also, it is standard practice to write **log** for $\log_{10}$ and **ln** for $\log_e$.

Logarithmic Notation

$$\log x = \log_{10} x \quad \text{(Common logarithm)}$$
$$\ln x = \log_e x \quad \text{(Natural logarithm)}$$

The system of natural logarithms is widely used in theoretical work. Using natural logarithms rather than logarithms to other bases often leads to simpler expressions.

LAWS OF LOGARITHMS

Computations involving logarithms are facilitated by the following **laws of logarithms.**

Laws of Logarithms

If m and n are positive numbers, then

1. $\log_b mn = \log_b m + \log_b n$

2. $\log_b \dfrac{m}{n} = \log_b m - \log_b n$

3. $\log_b m^n = n \log_b m$

4. $\log_b 1 = 0$

5. $\log_b b = 1$

Do not confuse the expression $\log m/n$ (Law 2) with the expression $\log m/\log n$. For example,

$$\log \frac{100}{10} = \log 100 - \log 10 = 2 - 1 = 1 \neq \frac{\log 100}{\log 10} = \frac{2}{1} = 2$$

You will be asked to prove these laws in Exercises 53–55. Their derivations are based on the definition of a logarithm and the corresponding laws of exponents. The following examples illustrate the properties of logarithms.

EXAMPLE 3

a. $\log(2 \cdot 3) = \log 2 + \log 3$ **b.** $\ln \dfrac{5}{3} = \ln 5 - \ln 3$

c. $\log \sqrt{7} = \log 7^{1/2} = \dfrac{1}{2} \log 7$ **d.** $\log_5 1 = 0$

e. $\log_{45} 45 = 1$

EXAMPLE 4

Given that $\log 2 \approx 0.3010$, $\log 3 \approx 0.4771$, and $\log 5 \approx 0.6990$, use the laws of logarithms to find

a. $\log 15$ **b.** $\log 7.5$ **c.** $\log 81$ **d.** $\log 50$

SOLUTION ✔   **a.** Note that $15 = 3 \cdot 5$, so by Law 1 for logarithms,

$$\begin{aligned} \log 15 &= \log 3 \cdot 5 \\ &= \log 3 + \log 5 \\ &\approx 0.4771 + 0.6990 \\ &= 1.1761 \end{aligned}$$

b. Observing that $7.5 = 15/2 = (3 \cdot 5)/2$, we apply Laws 1 and 2, obtaining

$$\begin{aligned} \log 7.5 &= \log \frac{(3)(5)}{2} \\ &= \log 3 + \log 5 - \log 2 \\ &\approx 0.4771 + 0.6990 - 0.3010 \\ &= 0.8751 \end{aligned}$$

c. Since $81 = 3^4$, we apply Law 3 to obtain

$$\begin{aligned} \log 81 &= \log 3^4 \\ &= 4 \log 3 \\ &\approx 4(0.4771) \\ &= 1.9084 \end{aligned}$$

d. We write $50 = 5 \cdot 10$ and find

$$\begin{aligned} \log 50 &= \log(5)(10) \\ &= \log 5 + \log 10 \\ &\approx 0.6990 + 1 \qquad \text{(Using Law 5)} \\ &= 1.6990 \end{aligned}$$

■ ■ ■ ■

EXAMPLE 5   Expand and simplify the following expressions:

a. $\log_3 x^2 y^3$ **b.** $\log_2 \dfrac{x^2 + 1}{2^x}$ **c.** $\ln \dfrac{x^2 \sqrt{x^2 - 1}}{e^x}$

SOLUTION ✔   **a.** $\begin{aligned}[t] \log_3 x^2 y^3 &= \log_3 x^2 + \log_3 y^3 \qquad \text{(Law 1)} \\ &= 2 \log_3 x + 3 \log_3 y \qquad \text{(Law 3)} \end{aligned}$

b. $\begin{aligned}[t] \log_2 \frac{x^2 + 1}{2^x} &= \log_2(x^2 + 1) - \log_2 2^x \qquad \text{(Law 2)} \\ &= \log_2(x^2 + 1) - x \log_2 2 \qquad \text{(Law 3)} \\ &= \log_2(x^2 + 1) - x \qquad \text{(Law 5)} \end{aligned}$

c. $\begin{aligned}[t] \ln \frac{x^2 \sqrt{x^2 - 1}}{e^x} &= \ln \frac{x^2(x^2 - 1)^{1/2}}{e^x} \qquad \text{(Rewriting)} \\ &= \ln x^2 + \ln(x^2 - 1)^{1/2} - \ln e^x \qquad \text{(Laws 1 and 2)} \\ &= 2 \ln x + \frac{1}{2} \ln(x^2 - 1) - x \ln e \qquad \text{(Law 3)} \\ &= 2 \ln x + \frac{1}{2} \ln(x^2 - 1) - x \qquad \text{(Law 5)} \end{aligned}$

■ ■ ■ ■

LOGARITHMIC FUNCTIONS AND THEIR GRAPHS

The definition of the logarithm implies that if b and n are positive numbers and b is different from 1, then the expression $\log_b n$ is a real number. This enables us to define a logarithmic function as follows:

Logarithmic Function

> The function defined by
>
> $$f(x) = \log_b x \qquad (b > 0, b \neq 1)$$
>
> is called the **logarithmic function with base b.** The domain of f is the set of all positive numbers.

FIGURE 5.7

The points (u, v) and (v, u) are mirror reflections of each other.

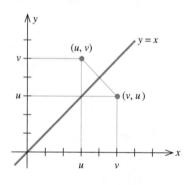

One easy way to obtain the graph of the logarithmic function $y = \log_b x$ is to construct a table of values of the logarithm (base b). However, another method—and a more instructive one—is based on exploiting the intimate relationship between logarithmic and exponential functions.

If a point (u, v) lies on the graph of $y = \log_b x$, then

$$v = \log_b u$$

But we can also write this equation in exponential form as

$$u = b^v$$

So the point (v, u) also lies on the graph of the function $y = b^x$. Let's look at the relationship between the points (u, v) and (v, u) and the line $y = x$ (Figure 5.7). If we think of the line $y = x$ as a mirror, then the point (v, u) is the mirror reflection of the point (u, v). Similarly, the point (u, v) is a mirror reflection of the point (v, u). We can take advantage of this relationship to help us draw the graph of logarithmic functions. For example, if we wish to draw the graph of $y = \log_b x$, where $b > 1$, then we need only draw the mirror reflection of the graph of $y = b^x$ with respect to the line $y = x$ (Figure 5.8).

FIGURE 5.8

The graphs of $y = b^x$ and $y = \log_b x$ are mirror reflections of each other.

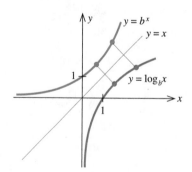

You may discover the following properties of the logarithmic function by taking the reflection of the graph of an appropriate exponential function (Exercises 31 and 32).

Properties of the Logarithmic Function

The logarithmic function $y = \log_b x$ $(b > 0, b \neq 1)$ has the following properties:

1. Its domain is $(0, \infty)$.
2. Its range is $(-\infty, \infty)$.
3. Its graph passes through the point $(1, 0)$.
4. It is continuous on $(0, \infty)$.
5. It is increasing on $(0, \infty)$ if $b > 1$ and decreasing on $(0, \infty)$ if $b < 1$.

EXAMPLE **6**   Sketch the graph of the function $y = \ln x$.

SOLUTION ✔   We first sketch the graph of $y = e^x$. Then, the required graph is obtained by tracing the mirror reflection of the graph of $y = e^x$ with respect to the line $y = x$ (Figure 5.9).

FIGURE 5.9
The graph of $y = \ln x$ is the mirror reflection of the graph of $y = e^x$.

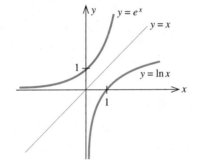

PROPERTIES RELATING THE EXPONENTIAL AND LOGARITHMIC FUNCTIONS

We made use of the relationship that exists between the exponential function $f(x) = e^x$ and the logarithmic function $g(x) = \ln x$ when we sketched the graph of g in Example 6. This relationship is further described by the following properties, which are an immediate consequence of the definition of the logarithm of a number.

Properties Relating e^x and $\ln x$

$$e^{\ln x} = x \qquad (x > 0) \qquad\qquad (2)$$
$$\ln e^x = x \qquad (\text{for any real number } x) \qquad (3)$$

(Try to verify these properties.)

From Properties 2 and 3, we conclude that the composite function

$$(f \circ g)(x) = f[g(x)]$$
$$= e^{\ln x} = x$$
$$(g \circ f)(x) = g[f(x)]$$
$$= \ln e^x = x$$

Thus,

$$f[g(x)] = g[f(x)]$$
$$= x$$

Any two functions f and g that satisfy this relationship are said to be **inverses** of each other. Note that the function f undoes what the function g does, and vice versa, so the composition of the two functions in any order results in the identity function $F(x) = x$.

The relationships expressed in Equations (2) and (3) are useful in solving equations that involve exponentials and logarithms.

Exploring with Technology

You can demonstrate the validity of Properties 2 and 3, which state that the exponential function $f(x) = e^x$ and the logarithmic function $g(x) = \ln x$ are inverses of each other as follows:

1. Sketch the graph of $(f \circ g)(x) = e^{\ln x}$, using the viewing rectangle $[0, 10] \times [0, 10]$. Interpret the result.
2. Sketch the graph of $(g \circ f)(x) = \ln e^x$, using the standard viewing rectangle. Interpret the result.

 EXAMPLE 7   Solve the equation $2e^{x+2} = 5$.

SOLUTION ✔   We first divide both sides of the equation by 2 to obtain

$$e^{x+2} = \frac{5}{2} = 2.5$$

Next, taking the natural logarithm of each side of the equation and using Equation (3), we have

$$\ln e^{x+2} = \ln 2.5$$
$$x + 2 = \ln 2.5$$
$$x = -2 + \ln 2.5$$
$$\approx -1.08$$

■ ■ ■ ■

 EXAMPLE 8   Solve the equation $5 \ln x + 3 = 0$.

SOLUTION ✔   Adding -3 to both sides of the equation leads to

$$5 \ln x = -3$$
$$\ln x = -\frac{3}{5} = -0.6$$

and so

$$e^{\ln x} = e^{-0.6}$$

Using equation (2), we conclude that

$$x = e^{-0.6}$$
$$\approx 0.55$$

■ ■ ■ ■

> **Group Discussion**
> Consider the equation $y = y_0 b^{kx}$, where y_0 and k are positive constants and $b > 0$, $b \neq 1$. Suppose we want to express y in the form $y = y_0 e^{px}$. Use the laws of logarithms to show that $p = k \ln b$ and hence that $y = y_0 e^{(k \ln b)x}$ is an alternative form of $y = y_0 b^{kx}$ using the base e.

SELF-CHECK EXERCISES 5.2

1. Sketch the graph of $y = 3^x$ and $y = \log_3 x$ on the same set of axes.
2. Solve the equation $3e^{x+1} - 2 = 4$.

Solutions to Self-Check Exercises 5.2 can be found on page 391.

5.2 Exercises

In Exercises 1–10, express the given equation in logarithmic form.

1. $2^6 = 64$

2. $3^5 = 243$

3. $3^{-2} = \dfrac{1}{9}$

4. $5^{-3} = \dfrac{1}{125}$

5. $\left(\dfrac{1}{3}\right)^1 = \dfrac{1}{3}$

6. $\left(\dfrac{1}{2}\right)^{-4} = 16$

7. $32^{3/5} = 8$

8. $81^{3/4} = 27$

9. $10^{-3} = 0.001$

10. $16^{-1/4} = 0.5$

In Exercises 11–16, use the facts that log 3 = 0.4771 and log 4 = 0.6021 to find the value of the given logarithm.

11. $\log 12$

12. $\log \dfrac{3}{4}$

13. $\log 16$

14. $\log \sqrt{3}$

15. $\log 48$

16. $\log \dfrac{1}{300}$

In Exercises 17–26, use the laws of logarithms to simplify the given expression.

17. $\log x(x+1)^4$

18. $\log x(x^2+1)^{-1/2}$

19. $\log \dfrac{\sqrt{x+1}}{x^2+1}$

20. $\ln \dfrac{e^x}{1+e^x}$

21. $\ln xe^{-x^2}$

22. $\ln x(x+1)(x+2)$

23. $\ln \dfrac{x^{1/2}}{x^2\sqrt{1+x^2}}$

24. $\ln \dfrac{x^2}{\sqrt{x}(1+x)^2}$

25. $\ln x^x$

26. $\ln x^{x^2+1}$

In Exercises 27–30, sketch the graph of the given equation.

27. $y = \log_3 x$

28. $y = \log_{1/3} x$

29. $y = \ln 2x$

30. $y = \ln \dfrac{1}{2} x$

In Exercises 31 and 32, sketch the graphs of the given equations on the same coordinate axes.

31. $y = 2^x$ and $y = \log_2 x$

32. $y = e^{3x}$ and $y = \ln 3x$

In Exercises 33–42, use logarithms to solve the given equation for t.

33. $e^{0.4t} = 8$

34. $\dfrac{1}{3} e^{-3t} = 0.9$

35. $5e^{-2t} = 6$

36. $4e^{t-1} = 4$

37. $2e^{-0.2t} - 4 = 6$

38. $12 - e^{0.4t} = 3$

39. $\dfrac{50}{1 + 4e^{0.2t}} = 20$

40. $\dfrac{200}{1 + 3e^{-0.3t}} = 100$

41. $A = Be^{-t/2}$

42. $\dfrac{A}{1 + Be^{t/2}} = C$

43. BLOOD PRESSURE A normal child's systolic blood pressure may be approximated by the function

$$p(x) = m(\ln x) + b$$

where $p(x)$ is measured in millimeters of mercury, x is measured in pounds, and m and b are constants. Given that $m = 19.4$ and $b = 18$, determine the systolic blood pressure of a child who weighs 92 lb.

44. MAGNITUDE OF EARTHQUAKES On the Richter scale, the magnitude R of an earthquake is given by the formula

$$R = \log \frac{I}{I_0}$$

where I is the intensity of the earthquake being measured and I_0 is the standard reference intensity.
a. Express the intensity I of an earthquake of magnitude $R = 5$ in terms of the standard intensity I_0.
b. Express the intensity I of an earthquake of magnitude $R = 8$ in terms of the standard intensity I_0. How many times greater is the intensity of an earthquake of magnitude 8 than one of magnitude 5?
c. In modern times the greatest loss of life attributable to an earthquake occurred in eastern China in 1976. Known as the Tangshan earthquake, it registered 8.2 on the Richter scale. How does the intensity of this earthquake compare with the intensity of an earthquake of magnitude $R = 5$?

45. SOUND INTENSITY The relative loudness of a sound D of intensity I is measured in decibels (db), where

$$D = 10 \log \frac{I}{I_0}$$

and I_0 is the standard threshold of audibility.
a. Express the intensity I of a 30-db sound (the sound level of normal conversation) in terms of I_0.
b. Determine how many times greater the intensity of an 80-db sound (rock music) is than that of a 30-db sound.
c. Prolonged noise above 150 db causes immediate and permanent deafness. How does the intensity of a 150-db sound compare with the intensity of an 80-db sound?

46. BAROMETRIC PRESSURE Halley's law states that the barometric pressure (in inches of mercury) at an altitude of x mi above sea level is approximately given by the equation

$$p(x) = 29.92e^{-0.2x} \qquad (x \geq 0)$$

If the barometric pressure as measured by a hot-air balloonist is 20 in. of mercury, what is the balloonist's altitude?

47. FORENSIC SCIENCE Forensic scientists use the following law to determine the time of death of accident or murder victims. If T denotes the temperature of a body t hr after death, then

$$T = T_0 + (T_1 - T_0)(0.97)^t$$

where T_0 is the air temperature and T_1 is the body temperature at the time of death. John Doe was found murdered at midnight in his house, when the room temperature was 70°F and his body temperature was 80°F. When was he killed? Assume that the normal body temperature is 98.6°F.

In Exercises 48–51, determine whether the statement is true or false. If it is true, explain why it is true. If it is false, give an example to show why it is false.

48. $(\ln x)^3 = 3 \ln x$ for all x in $(0, \infty)$.

49. $\ln a - \ln b = \ln(a - b)$ for all positive real numbers a and b.

50. The function $f(x) = 1/\ln x$ is continuous on $(1, \infty)$.

51. The function $f(x) = \ln |x|$ is continuous for all $x \neq 0$.

52. a. Given that $2^x = e^{kx}$, find k.
   b. Show that, in general, if b is a nonnegative real number, then any equation of the form $y = b^x$ may be written in the form $y = e^{kx}$, for some real number k.

53. Use the definition of a logarithm to prove:
   a. $\log_b mn = \log_b m + \log_b n$
   b. $\log_b \dfrac{m}{n} = \log_b m - \log_b n$

   Hint: Let $\log_b m = p$ and $\log_b n = q$. Then, $b^p = m$ and $b^q = n$.

54. Use the definition of a logarithm to prove

$$\log_b m^n = n \log_b m$$

55. Use the definition of a logarithm to prove:
   a. $\log_b 1 = 0$
   b. $\log_b b = 1$

SOLUTIONS TO SELF-CHECK EXERCISES 5.2

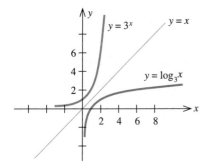

1. First, sketch the graph of $y = 3^x$ with the help of the following table of values:

x	-3	-2	-1	0	1	2	3
$y = 3^x$	1/27	1/9	1/3	0	3	9	27

Next, take the mirror reflection of this graph with respect to the line $y = x$ to obtain the graph of $y = \log_3 x$.

2.
$$3e^{x+1} - 2 = 4$$
$$3e^{x+1} = 6$$
$$e^{x+1} = 2$$
$$\ln e^{x+1} = \ln 2$$
$$(x + 1)\ln e = \ln 2 \quad \text{(Law 3)}$$
$$x + 1 = \ln 2 \quad \text{(Law 5)}$$
$$x = \ln 2 - 1$$
$$\approx -0.3069$$

5.3 Compound Interest

COMPOUND INTEREST

Compound interest is a natural application of the exponential function to the business world. (Albert Einstein called compound interest the greatest invention of mankind.) We begin by recalling that simple interest is interest that is computed only on the original principal. Thus, if I denotes the interest on a principal P (in dollars) at an interest rate of r per year for t years, then we have

$$I = Prt$$

The **accumulated amount** A, the sum of the principal and interest after t years, is given by

$$A = P + I = P + Prt$$
$$= P(1 + rt) \qquad \text{(Simple interest formula)} \qquad \textbf{(4)}$$

Frequently, interest earned is periodically added to the principal and thereafter earns interest itself at the same rate. This is called **compound interest.** To find a formula for the accumulated amount, let's consider a numerical example. Suppose \$1000 (the principal) is deposited in a bank for a **term** of 3 years, earning interest at the rate of 8% per year (called the **nominal,** or **stated, rate**) compounded annually. Then, using Formula (4) with $P = 1000$, $r = 0.08$, and $t = 1$, we see that the accumulated amount at the end of the first year is

$$A_1 = P(1 + rt)$$
$$= 1000[1 + 0.08(1)] = 1000(1.08) = 1080$$

or \$1080.

To find the accumulated amount A_2 at the end of the second year, we use (4) once again, this time with $P = A_1$. (Remember, the principal *and* interest now earn interest over the second year.) We obtain

$$A_2 = P(1 + rt) = A_1(1 + rt)$$
$$= 1000[1 + 0.08(1)][1 + 0.08(1)]$$
$$= 1000[1 + 0.08]^2 = 1000(1.08)^2 \approx 1166.40$$

or approximately \$1166.40.

Finally, the accumulated amount A_3 at the end of the third year is found using (4) with $P = A_2$, giving

$$A_3 = P(1 + rt) = A_2(1 + rt)$$
$$= 1000[1 + 0.08(1)]^2[1 + 0.08(1)]$$
$$= 1000[1 + 0.08]^3 = 1000(1.08)^3 \approx 1259.71$$

or approximately \$1259.71.

If you reexamine our calculations in this example, you will see that the accumulated amounts at the end of each year have the following form:

First year:	$A_1 = 1000(1 + 0.08)$	or	$A_1 = P(1 + r)$
Second year:	$A_2 = 1000(1 + 0.08)^2$	or	$A_2 = P(1 + r)^2$
Third year:	$A_3 = 1000(1 + 0.08)^3$	or	$A_3 = P(1 + r)^3$

These observations suggest the following general result: If P dollars are invested over a term of t years earning interest at the rate of r per year compounded annually, then the accumulated amount is

$$A = P(1 + r)^t \qquad \textbf{(5)}$$

Formula (5) was derived under the assumption that interest was compounded *annually*. In practice, however, interest is usually compounded more than once a year. The interval of time between successive interest calculations is called the **conversion period.**

If interest at a nominal rate of r per year is compounded m times a year on a principal of P dollars, then the simple interest rate per conversion period is

$$i = \frac{r}{m} \qquad \frac{\text{(Annual interest rate)}}{\text{(Periods per year)}}$$

For example, if the nominal interest rate is 8% per year ($r = 0.08$) and interest is compounded quarterly ($m = 4$), then

$$i = \frac{r}{m} = \frac{0.08}{4} = 0.02$$

or 2% per period.

To find a general formula for the accumulated amount when a principal of P dollars is deposited in a bank for a term of t years and earns interest at the (nominal) rate of r per year compounded m times per year, we proceed as before using Formula (5) repeatedly with the interest rate $i = r/m$. We see that the accumulated amount at the end of each period is as follows:

First period: $A_1 = P(1 + i)$
Second period: $A_2 = A_1(1 + i) = [P(1 + i)](1 + i) = P(1 + i)^2$
Third period: $A_3 = A_2(1 + i) = [P(1 + i)^2](1 + i) = P(1 + i)^3$
$$\vdots \qquad\qquad\qquad \vdots$$
nth period: $A_n = A_{n-1}(1 + i) = [P(1 + i)^{n-1}](1 + i) = P(1 + i)^n$

But there are $n = mt$ periods in t years (number of conversion periods times the term). Therefore, the accumulated amount at the end of t years is given by

$$A = P(1 + i)^n$$

Compound Interest Formula

$$A = P(1 + i)^n \qquad\qquad (6)$$

where $i = r/m$, $n = mt$, and

A = Accumulated amount at the end of n conversion periods
P = Principal
r = Nominal interest rate per year
m = Number of conversion periods per year
t = Term (number of years)

 EXAMPLE 1   Find the accumulated amount after 3 years if $1000 is invested at 8% per year compounded (a) annually, (b) semiannually, (c) quarterly, and (d) monthly.

SOLUTION ✔ **a.** Here, $P = 1000$, $r = 0.08$, and $m = 1$. Thus, $i = r = 0.08$ and $n = 3$, so Formula (6) gives

$$A = 1000(1.08)^3$$
$$= 1259.71$$

or $1259.71.

b. Here, $P = 1000$, $r = 0.08$, and $m = 2$. Thus, $i = 0.08/2 = 0.04$ and $n = (3)(2) = 6$, so that (6) gives

$$A = 1000(1.04)^6$$
$$= 1265.32$$

or $1265.32.

c. In this case, $P = 1000$, $r = 0.08$, and $m = 4$. Thus, $i = 0.08/4 = 0.02$ and $n = (3)(4) = 12$, so (6) gives

$$A = 1000(1.02)^{12}$$
$$= 1268.24$$

or $1268.24.

d. Here, $P = 1000$, $r = 0.08$, and $m = 12$. Thus, $i = 0.08/12 = 0.0067$ and $n = (3)(12) = 36$, so (6) gives

$$A = 1000(1.0067)^{36}$$
$$= 1271.75$$

or $1271.75. These results are summarized in Table 5.2.

Table 5.2				
Nominal Rate, r	Conversion Period	Interest Rate/ Conversion Period	Initial Investment	Accumulated Amount
8%	Annual ($m = 1$)	8%	$1000	$1259.71
8	Semiannual ($m = 2$)	4	1000	1265.32
8	Quarterly ($m = 4$)	2	1000	1268.24
8	Monthly ($m = 12$)	2/3	1000	1271.75

▪ ▪ ▪ ▪

EFFECTIVE RATE OF INTEREST

In the last example we saw that the interest actually earned on an investment depends on the frequency with which the interest is compounded. Thus, the stated, or nominal, rate of 8% per year does not reflect the actual rate at which interest is earned. This suggests that we need to find a common basis for comparing interest rates. One such way of comparing interest rates is provided by using the effective rate. The effective rate is the *simple* interest rate that would produce the same accumulated amount in 1 year as the nominal rate compounded m times a year. The effective rate is also called the **true rate.**

To derive a relation between the nominal interest rate, r per year compounded m times, and its corresponding effective rate, r_{eff} per year, let's assume an initial investment of P dollars. Then, the accumulated amount after 1 year at a simple interest rate of r_{eff} per year is

$$A = P(1 + r_{\text{eff}})$$

Also, the accumulated amount after 1 year at an interest rate of r per year compounded m times a year is

$$A = P(1 + i)^n = P\left(1 + \frac{r}{m}\right)^m \qquad \text{(Since } i = r/m\text{)}$$

Equating the two expressions gives

$$P(1 + r_{\text{eff}}) = P\left(1 + \frac{r}{m}\right)^m$$

$$1 + r_{\text{eff}} = \left(1 + \frac{r}{m}\right)^m \qquad \text{(Dividing both sides by } P\text{)}$$

or, upon solving for r_{eff}, we obtain the formula for computing the effective rate of interest:

Effective Rate of Interest Formula

$$r_{\text{eff}} = \left(1 + \frac{r}{m}\right)^m - 1 \qquad\qquad\qquad (7)$$

where

r_{eff} = Effective rate of interest
r = Nominal interest rate per year
m = Number of conversion periods per year

 EXAMPLE 2

Find the effective rate of interest corresponding to a nominal rate of 8% per year compounded (a) annually, (b) semiannually, (c) quarterly, and (d) monthly.

SOLUTION ✔

a. The effective rate of interest corresponding to a nominal rate of 8% per year compounded annually is of course given by 8% per year. This result is also confirmed by using Formula (7) with $r = 0.08$ and $m = 1$. Thus,

$$r_{\text{eff}} = (1 + 0.08) - 1 = 0.08$$

b. Let $r = 0.08$ and $m = 2$. Then, (7) yields

$$r_{\text{eff}} = \left(1 + \frac{0.08}{2}\right)^2 - 1$$
$$= (1.04)^2 - 1$$
$$= 0.0816$$

so the required effective rate is 8.16% per year.

c. Let $r = 0.08$ and $m = 4$. Then, (7) yields

$$r_{\text{eff}} = \left(1 + \frac{0.08}{4}\right)^4 - 1$$
$$= (1.02)^4 - 1$$
$$= 0.08243$$

so the corresponding effective rate in this case is 8.243% per year.

d. Let $r = 0.08$ and $m = 12$. Then, (7) yields

$$r_{\text{eff}} = \left(1 + \frac{0.08}{12}\right)^{12} - 1$$
$$= (1.0067)^{12} - 1$$
$$= 0.08343$$

so the corresponding effective rate in this case is 8.343% per year.   ■ ■ ■ ■

Now, if the effective rate of interest r_{eff} is known, then the accumulated amount after t years on an investment of P dollars may be more readily computed by using the formula

$$A = P(1 + r_{\text{eff}})^t$$

The 1968 Truth in Lending Act passed by Congress requires that the effective rate of interest be disclosed in all contracts involving interest charges. The passage of this act has benefited consumers because they now have a common basis for comparing the various nominal rates quoted by different financial institutions. Furthermore, knowing the effective rate enables consumers to compute the actual charges involved in a transaction. Thus, if the effective rates of interest found in Example 2 were known, the accumulated values of Example 1, shown in Table 5.3, could have been readily found.

Table 5.3

Nominal Rate	Frequency of Interest Payment	Effective Rate	Initial Investment	Accumulated Amount After 3 Years	
8%	Annually	8%	$1000	$1000(1 + 0.08)^3$	= $1259.71
8	Semiannually	8.16	1000	$1000(1 + 0.0816)^3$	= 1265.32
8	Quarterly	8.243	1000	$1000(1 + 0.08243)^3$	= 1268.23
8	Monthly	8.343	1000	$1000(1 + 0.08343)^3$	= 1271.75

PRESENT VALUE

Let's return to the compound interest Formula (6), which expresses the accumulated amount at the end of n periods when interest at the rate of r is compounded m times a year. The principal P in (6) is often referred to as the **present value,** and the accumulated value A is called the **future value** since it is realized at a future date. In certain instances an investor may wish to determine how much money he should invest now, at a fixed rate of interest, so that he will realize a certain sum at some future date. This problem may be solved by expressing P in terms of A. Thus, from (6) we find

$$P = A(1 + i)^{-n}$$

Here, as before, $i = r/m$, where m is the number of conversion periods per year.

Present Value Formula for Compound Interest

$$P = A(1 + i)^{-n} \qquad \qquad (8)$$

 EXAMPLE 3   Find how much money should be deposited in a bank paying interest at the rate of 6% per year compounded monthly so that at the end of 3 years the accumulated amount will be $20,000.

SOLUTION ✔   Here, $r = 0.06$ and $m = 12$, so $i = 0.06/12 = 0.005$ and $n = (3)(12) = 36$. Thus, the problem is to determine P given that $A = 20,000$. Using Formula (8), we obtain

$$P = 20{,}000(1.005)^{-36}$$
$$\approx 16{,}713$$

or $16,713.   ■ ■ ■ ■

 EXAMPLE 4   Find the present value of $49,158.60 due in 5 years at an interest rate of 10% per year compounded quarterly.

SOLUTION ✔   Using Formula (8) with $r = 0.1$ and $m = 4$, so that $i = 0.1/4 = 0.025$, $n = (4)(5) = 20$, and $A = 49,158.6$, we obtain

$$P = (49{,}158.6)(1.025)^{-20} \approx 30{,}000$$

or $30,000.   ■ ■ ■ ■

CONTINUOUS COMPOUNDING OF INTEREST

One question that arises naturally in the study of compound interest is: What happens to the accumulated amount over a fixed period of time if the interest is computed more and more frequently?

Intuition suggests that the more often interest is compounded, the larger the accumulated amount will be. This is confirmed by the results of Example 1, where we found that the accumulated amounts did in fact increase when we increased the number of conversion periods per year.

This leads us to another question: Does the accumulated amount approach a limit when the interest is computed more and more frequently over a fixed period of time?

To answer this question, let's look again at the compound interest formula:

$$A = P\left(1 + \frac{r}{m}\right)^{mt} \tag{9}$$

Recall that m is the number of conversion periods per year. So to find an answer to our problem, we should let m approach infinity (get larger and larger) in (9). But first we will rewrite this equation in the form

$$A = P\left[\left(1 + \frac{r}{m}\right)^m\right]^t \qquad [\text{Since } b^{xy} = (b^x)^y]$$

Now, letting $m \to \infty$, we find that

$$\lim_{m \to \infty}\left[P\left(1 + \frac{r}{m}\right)^m\right]^t = P\left[\lim_{m \to \infty}\left(1 + \frac{r}{m}\right)^m\right]^t \qquad \text{(Why?)}$$

Next, upon making the substitution $u = m/r$ and observing that $u \to \infty$ as $m \to \infty$, the foregoing expression reduces to

$$P\left[\lim_{u \to \infty}\left(1 + \frac{1}{u}\right)^{ur}\right]^t = P\left[\lim_{u \to \infty}\left(1 + \frac{1}{u}\right)^u\right]^{rt}$$

But

$$\lim_{u \to \infty}\left(1 + \frac{1}{u}\right)^u = e \qquad [\text{Using (1)}]$$

so

$$\lim_{m \to \infty} P\left[\left(1 + \frac{r}{m}\right)^m\right]^t = Pe^{rt}$$

Our computations tell us that as the frequency with which interest is compounded increases without bound, the accumulated amount approaches Pe^{rt}. In this situation, we say that interest is *compounded continuously*. Let's summarize this important result.

Continuous Compound Interest Formula

$$A = Pe^{rt} \tag{10}$$

where

P = Principal
r = Annual interest rate compounded continuously
t = Time in years
A = Accumulated amount at the end of t years

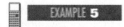 **EXAMPLE 5**   Find the accumulated amount after 3 years if $1000 is invested at 8% per year compounded (a) daily (take the number of days in a year to be 365) and (b) continuously.

SOLUTION ✔ **a.** Using Formula (6) with $P = 1000$, $r = 0.08$, $m = 365$, and $n = (365)(3) = 1095$, we find

$$A = 1000\left(1 + \frac{0.08}{365}\right)^{1095} \approx 1271.22$$

or $1271.22.

b. Here we use Formula (10) with $P = 1000$, $r = 0.08$, and $t = 3$, obtaining

$$A = 1000e^{(0.08)(3)}$$
$$\approx 1271.25$$

or $1271.25.   ■ ■ ■ ■

Exploring with Technology

In the opening paragraph of Section 5.1, we pointed out that the accumulated amount of an account earning interest *compounded continuously* will eventually outgrow by far the accumulated amount of an account earning interest at the same nominal rate but earning simple interest. Illustrate this fact using the following example.

Suppose you deposit $1000 in account I, earning interest at the rate of 10% per year compounded continuously so that the accumulated amount at the end of t years is $A_1(t) = 1000e^{0.1t}$. Suppose you also deposit $1000 in account II, earning simple interest at the rate of 10% per year so that the accumulated amount at the end of t years is $A_2(t) = 1000(1 + 0.1t)$. Use a graphing utility to sketch the graphs of the functions A_1 and A_2 in the viewing rectangle $[0, 20] \times [0, 10,000]$ to see the accumulated amounts $A_1(t)$ and $A_2(t)$ over a 20-year period.

Observe that the accumulated amounts corresponding to interest compounded daily and interest compounded continuously differ by very little. The continuous compound interest formula is a very important tool in theoretical work in financial analysis.

If we solve Formula (10) for P, we obtain

$$P = Ae^{-rt} \qquad \textbf{(11)}$$

which gives the present value in terms of the future (accumulated) value for the case of continuous compounding.

EXAMPLE 6   The Blakely Investment Company owns an office building located in the commercial district of a city. As a result of the continued success of an urban renewal program, local business is enjoying a miniboom. The market value of Blakely's property is

$$V(t) = 300,000e^{\sqrt{t}/2}$$

where $V(t)$ is measured in dollars and t is the time in years from the present. If the expected rate of inflation is 9% compounded continuously for the next 10 years, find an expression for the present value $P(t)$ of the market price of the property valid for the next 10 years. Compute $P(7)$, $P(8)$, and $P(9)$, and interpret your results.

SOLUTION ✔ Using Formula (11) with $A = V(t)$ and $r = 0.09$, we find that the present value of the market price of the property t years from now is

$$P(t) = V(t)e^{-0.09t}$$
$$= 300,000e^{-0.09t+\sqrt{t}/2} \qquad (0 \le t \le 10)$$

Letting $t = 7, 8$, and 9, respectively, we find that

$$P(7) = 300,000e^{-0.09(7)+\sqrt{7}/2} \approx 599,837, \text{ or } \$599,837$$
$$P(8) = 300,000e^{-0.09(8)+\sqrt{8}/2} \approx 600,640, \text{ or } \$600,640$$
$$P(9) = 300,000e^{-0.09(9)+\sqrt{9}/2} \approx 598,115, \text{ or } \$598,115$$

From the results of these computations, we see that the present value of the property's market price seems to decrease after a certain period of growth. This suggests that there is an optimal time for the owners to sell. Later we will show that the highest present value of the property's market price is $600,779, which occurs at time $t = 7.72$ years.   ■ ■ ■ ■

Exploring with Technology

The effective rate of interest is given by

$$r_{\text{eff}} = \left(1 + \frac{r}{m}\right)^m - 1$$

where the number of conversion periods per year is m. In Exercise 27 you will be asked to show that the effective rate of interest r_{eff} corresponding to a nominal interest rate r per year compounded continuously is given by

$$\hat{r}_{\text{eff}} = e^r - 1$$

To obtain a visual confirmation of this result, consider the special case where $r = 0.1$ (10% per year).
1. Use a graphing utility to plot the graph of both

$$y_1 = \left(1 + \frac{0.1}{x}\right)^x - 1 \qquad \text{and} \qquad y_2 = e^{0.1} - 1$$

in the viewing rectangle $[0, 3] \times [0, 0.12]$.
2. Does your result seem to imply that

$$\left(1 + \frac{r}{m}\right)^m - 1$$

approaches

$$\hat{r}_{\text{eff}} = e^r - 1$$

as m increases without bound for the special case $r = 0.1$?

1. Find the present value of $20,000 due in 3 yr at an interest rate of 12%/year compounded monthly.

2. Glen is a retiree living on Social Security and the income from his investment. Currently, his $100,000 investment in a 1-yr CD is yielding 11.6% interest compounded daily. If he reinvests the principal ($100,000) on the due date of the CD in another 1-yr CD paying 9.2% interest compounded daily, find the net decrease in his yearly income from his investment.

3. **a.** What is the accumulated amount after 5 yr if $10,000 is invested at 10%/year compounded continuously?
   b. Find the present value of $10,000 due in 5 yr at an interest rate of 10%/year compounded continuously.

Solutions to Self-Check Exercises 5.3 can be found on page 404.

5.3 Exercises

A calculator is recommended for these exercises.

In Exercises 1–4, find the accumulated amount A if the principal P is invested at an interest rate of r per year for t years.

1. $P = $2500, r = 7\%, t = 10$, compounded semiannually

2. $P = $12,000, r = 8\%, t = 10$, compounded quarterly

3. $P = $150,000, r = 10\%, t = 4$, compounded monthly

4. $P = $150,000, r = 9\%, t = 3$, compounded daily

In Exercises 5 and 6, find the effective rate corresponding to the given nominal rate.

5. **a.** 10%/year compounded semiannually
   b. 9%/year compounded quarterly

6. **a.** 8%/year compounded monthly
   b. 8%/year compounded daily

In Exercises 7 and 8, find the present value of $40,000 due in 4 years at the given rate of interest.

7. **a.** 8%/year compounded semiannually
   b. 8%/year compounded quarterly

8. **a.** 7%/year compounded monthly
   b. 9%/year compounded daily

9. Find the accumulated amount after 4 yr if $5000 is invested at 8%/year compounded continuously.

10. An amount of $25,000 is deposited in a bank that pays interest at the rate of 7%/year, compounded annually. What is the total amount on deposit at the end of 6 yr, assuming there are no deposits or withdrawals during those 6 yr? What is the interest earned in that period of time?

11. **HOUSING PRICES** The Estradas are planning to buy a house 4 yr from now. Housing experts in their area have estimated that the cost of a home will increase at a rate of 9%/year during that 4-yr period. If this economic prediction holds true, how much can they expect to pay for a house that currently costs $80,000?

12. **ENERGY CONSUMPTION** A metropolitan utility company in a western city of the United States expects the consumption of electricity to increase by 8%/year during the next decade, due mainly to the expected population increase. If consumption does increase at this rate, find the amount by which the utility company will have to increase its generating capacity in order to meet the area's needs at the end of the decade.

13. **PENSION FUNDS** The managers of a pension fund have invested $1.5 million in U.S. government certificates of deposit (CDs) that pay interest at the rate of 9.5%/year compounded semiannually over a period of 10 yr. At the end of this period, how much will the investment be worth?

14. **SAVINGS ACCOUNTS**   Bernie invested a sum of money 5 yr ago in a savings account, which has since paid interest at the rate of 8%/yr compounded quarterly. His investment is now worth $22,289.22. How much did he originally invest?

15. **LOAN CONSOLIDATION**   The proprietors of the Coachmen Inn secured two loans from the Union Bank: one for $8000 due in 3 yr and one for $15,000 due in 6 yr, both at an interest rate of 10%/yr compounded semiannually. The bank agreed to allow the two loans to be consolidated into one loan payable in 5 yr at the same interest rate. How much will the proprietors have to pay the bank at the end of 5 yr?

16. **TAX-DEFERRED ANNUITIES**   Kate is in the 28% tax bracket and has $25,000 available for investment during her current tax year. Assume that she remains in the same tax bracket over the next 10 yr and determine the accumulated amount of her investment if she puts the $25,000 into a:
    a. Tax-deferred annuity that pays 12%/year, tax-deferred for 10 yr.
    b. Taxable instrument that pays 12%/year for 10 yr.
    Hint: In this case the yield after taxes is 8.64%/year.

17. **CONSUMER PRICE INDEX**   At an annual inflation rate of 7.5%, how long will it take the Consumer Price Index (CPI) to double?

18. **INVESTMENT RETURNS**   Zoe purchased a house in 1993 for $80,000. In 1999 she sold the house and made a net profit of $28,000. Find the effective annual rate of return on her investment over the 6-yr period.

19. **INVESTMENT RETURNS**   Julio purchased 1000 shares of a certain stock for $25,250 (including commissions). He sold the shares 2 yr later and received $32,100 after deducting commissions. Find the effective annual rate of return on his investment over the 2-yr period.

20. **INVESTMENT OPTIONS**   Investment A offers a 10% return compounded semiannually, and investment B offers a 9.75% return compounded continuously. Which investment has a higher rate of return over a 4-yr period?

21. **PRESENT VALUE**   Find the present value of $59,673 due in 5 yr at an interest rate of 8%/year compounded continuously.

22. **REAL ESTATE INVESTMENTS**   A condominium complex was purchased by a group of private investors for $1.4 million and sold 6 yr later for $3.6 million. Find the annual rate of return (compounded continuously) on their investment.

23. **SAVING FOR COLLEGE**   Having received a large inheritance, a child's parents wish to establish a trust for the child's college education. If 7 yr from now they need an estimated $70,000, how much should they set aside in trust now, if they invest the money at 10.5% compounded (a) quarterly? (b) Continuously?

24. **EFFECT OF INFLATION ON SALARIES**   Omar's current annual salary is $35,000. How much will he need to earn 10 yr from now in order to retain his present purchasing power if the rate of inflation over that period is 6%/year? Assume that inflation is continuously compounded.

25. **PENSIONS**   Eleni, who is now 50 years old, is employed by a firm that guarantees her a pension of $40,000/year at age 65. What is the present value of her first year's pension if inflation over the next 15 yr is (a) 6%? (b) 8%? (c) 12%? Assume that inflation is continuously compounded.

26. **REAL ESTATE INVESTMENTS**   An investor purchased a piece of waterfront property. Because of the development of a marina in the vicinity, the market value of the property is expected to increase according to the rule

$$V(t) = 80,000e^{\sqrt{t}/2}$$

where $V(t)$ is measured in dollars and t is the time in years from the present. If the rate of inflation is expected to be 9% compounded continuously for the next 8 yr, find an expression for the present value $P(t)$ of the property's market price valid for the next 8 yr. What is $P(t)$ expected to be in 4 yr?

27. Show that the effective rate of interest $\hat{r}_{eff}$ that corresponds to a nominal interest rate r per year compounded continuously is given by

$$\hat{r}_{eff} = e^r - 1$$

Hint: From Formula (7) we see that the effective rate $\hat{r}_{eff}$ corresponding to a nominal interest rate r per year compounded m times a year is given by

$$\hat{r}_{eff} = \left(1 + \frac{r}{m}\right)^m - 1$$

Let m tend to infinity in this expression.

28. Refer to Exercise 27. Find the effective rate of interest that corresponds to a nominal rate of 10%/year compounded (a) quarterly, (b) monthly, and (c) continuously.

29. **INVESTMENT ANALYSIS**   Refer to Exercise 27. Bank A pays interest on deposits at a 7% annual rate compounded quarterly, and Bank B pays interest on deposits at a $7\frac{1}{8}$% annual rate compounded continuously. Which bank has the higher effective rate of interest?

30. **INVESTMENT ANALYSIS**   Find the nominal rate of interest that, when compounded monthly, yields an effective rate of interest of 10%/year.
    Hint: Use Equation (7).

31. **INVESTMENT ANALYSIS**   Find the nominal rate of interest that, when compounded continuously, yields an effective rate of interest of 10%/year.
    Hint: See Exercise 27.

MISATO NAKAZAKI

TITLE: Assistant Vice President
INSTITUTION: A large investment corporation

In the securities industry, buying and selling stocks and bonds has always required a mastery of concepts and formulas that outsiders find confusing. As a bond seller, Misato Nakazaki routinely uses terms such as *issue, maturity, current yield, callable* and *convertible bonds,* and so on.

These terms, however, are easily defined. When corporations issue bonds, they are borrowing money at a fixed rate of interest. The bonds are scheduled to mature—to be paid back—on a specific date as much as 30 years into the future. Callable bonds allow the issuer to pay off the loans prior to their expected maturity, reducing overall interest payments. In its simplest terms, current yield is the price of a bond multiplied by the interest rate at which the bond is issued. For example, a bond with a face value of $1000 and an interest rate of 10% yields $100 per year in interest payments. When that same bond is resold at a premium on the secondary market for $1200, its current yield nets only an 8.3% rate of return based on the higher purchase price.

Bonds attract investors for many reasons. A key variable is the sensitivity of the bond's price to future changes in interest rates. If investors get locked into a low-paying bond when future bonds pay higher yields, they lose money. Nakazaki stresses that "no one knows for sure what rates will be over time." Employing differentials allows her to calculate interest-rate sensitivity for clients as they ponder purchase decisions.

Computerized formulas, "whose basis is calculus," says Nakazaki, help her factor the endless stream of numbers flowing across her desk.

On a typical day, Nakazaki might be given a bid on "10 million, GMAC, 8.5%, January 2003." Translation: Her customer wants her to buy General Motors Acceptance Corporation bonds with a face value of $10 million and an interest rate of 8.5%, maturing in January 2003.

After she calls her firm's trader to find out the yield on the bond in question, Nakazaki enters the price and other variables, such as the interest rate and date of maturity, and the computer prints out the answers. Nakazaki can then relay to her client the bond's current yield, accrued interest, and so on. In Nakazaki's rapid-fire work environment, such speed is essential. Nakazaki cautions that "computer users have to understand what's behind the formulas." The software "relies on the basics of calculus. If people don't understand the formula, it's useless for them to use the calculations."

With an MBA from New York University, Nakazaki typifies the younger generation of Japanese women who have chosen to succeed in the business world. Since earning her degree, she has sold bonds for a global securities firm in New York City.

Nakazaki's client list reads like a *who's who* of the leading Japanese banks, insurance companies, mutual funds, and corporations. As institutional buyers, her clients purchase large blocks of American corporate bonds and mortgage-backed securities such as Ginnie Maes.

SOLUTIONS TO SELF-CHECK EXERCISES 5.3

1. Using Formula (8) with $r = 0.12$ and $m = 12$, so that

$$i = \frac{0.12}{12} = 0.01, \qquad n = (12)(3) = 36, \qquad A = 20{,}000$$

we find the required present value to be

$$P = 20{,}000(1.01)^{-36}$$
$$= 13{,}978.50$$

or $13,978.50.

2. The accumulated amount of Glen's current investment is found by using Formula (6) with $P = 100{,}000$, $r = 0.116$, and $m = 360$. Thus,

$$i = \frac{0.116}{360} = 0.0003222 \qquad \text{and} \qquad n = 360$$

so the required accumulated amount is

$$A = 100{,}000(1.0003222)^{360}$$
$$= 112{,}296.59$$

or $112,296.59. Next, we compute the accumulated amount of Glen's reinvestment. Once again, using (6) with $P = 100{,}000$, $r = 0.092$, and $m = 360$ so that

$$i = \frac{0.092}{360} = 0.0002556 \qquad \text{and} \qquad n = 360$$

we find the required accumulated amount in this case to be

$$\overline{A} = 100{,}000(1.0002556)^{360}$$

or $109,636.95. Therefore, Glen can expect to experience a net decrease in yearly income of

$$112{,}296.59 - 109{,}636.95$$

or $2,659.64.

3. **a.** Using Formula (10) with $P = 10{,}000$, $r = 0.1$, and $t = 5$, we find that the required accumulated amount is given by

$$A = 10{,}000e^{(0.1)(5)}$$
$$= 16{,}487.21$$

or $16,487.21.
   b. Using Formula (11) with $A = 10{,}000$, $r = 0.1$, and $t = 5$, we see that the required present value is given by

$$P = 10{,}000e^{-(0.1)(5)}$$
$$= 6065.31$$

or $6065.31.

5.4 Differentiation of Exponential Functions

THE DERIVATIVE OF THE EXPONENTIAL FUNCTION

To study the effects of budget deficit-reduction plans at different income levels, it is important to know the income distribution of American families. Based on data from the House Budget Committee, the House Ways and Means Committee, and the U.S. Census Bureau, the graph of f shown in Figure 5.10 gives the number of American families y (in millions) as a function of their annual income x (in thousands of dollars) in 1990.

FIGURE 5.10
The graph of f shows the number of families versus their annual income.

Source: House Budget Committee, House Ways and Means Committee, and U.S. Census Bureau

Observe that the graph of f rises very quickly and then tapers off. From the graph of f, you can see that the bulk of American families earned less than $100,000 per year. In fact, 95% of U.S. families earned less than $102,358 per year in 1990. (We will refer to this model again in Using Technology at the end of this section.)

To analyze mathematical models involving exponential and logarithmic functions in greater detail, we need to develop rules for computing the derivative of these functions. We begin by looking at the rule for computing the derivative of the exponential function.

Rule 1: Derivative of the Exponential Function

$$\frac{d}{dx}e^x = e^x$$

■ EXPONENTIAL AND LOGARITHMIC FUNCTIONS

Thus, the derivative of the exponential function with base e is equal to the function itself. To demonstrate the validity of this rule, we compute

$$f'(x) = \lim_{h \to 0} \frac{f(x + h) - f(x)}{h}$$

$$= \lim_{h \to 0} \frac{e^{x+h} - e^x}{h}$$

$$= \lim_{h \to 0} \frac{e^x(e^h - 1)}{h} \qquad \text{(Writing } e^{x+h} = e^x e^h \text{ and factoring)}$$

$$= e^x \lim_{h \to 0} \frac{e^h - 1}{h} \qquad \text{(Why?)}$$

To evaluate

$$\lim_{h \to 0} \frac{e^h - 1}{h}$$

let's refer to Table 5.4, which is constructed with the aid of a calculator. From the table, we see that

$$\lim_{h \to 0} \frac{e^h - 1}{h} = 1$$

(Although a rigorous proof of this fact is possible, it is beyond the scope of this book. Also see Example 1, Using Technology, page 414.) Using this result, we conclude that

$$f'(x) = e^x \cdot 1 = e^x$$

as we set out to show.

Table 5.4

h	0.1	0.01	0.001	−0.1	−0.01	−0.001
$\frac{e^h - 1}{h}$	1.0517	1.0050	1.0005	0.9516	0.9950	0.9995

EXAMPLE 1   Compute the derivative of each of the following functions:

a. $f(x) = x^2 e^x$ **b.** $g(t) = (e^t + 2)^{3/2}$

SOLUTION ✔   **a.** The product rule gives

$$f'(x) = \frac{d}{dx}(x^2 e^x)$$

$$= x^2 \frac{d}{dx}(e^x) + e^x \frac{d}{dx}(x^2)$$

$$= x^2 e^x + e^x(2x)$$

$$= xe^x(x + 2)$$

b. Using the general power rule, we find

$$g'(t) = \frac{3}{2}(e^t + 2)^{1/2}\frac{d}{dt}(e^t + 2)$$

$$= \frac{3}{2}(e^t + 2)^{1/2}\,e^t = \frac{3}{2}e^t(e^t + 2)^{1/2}$$

■ ■ ■ ■

Exploring with Technology

Consider the exponential function $f(x) = b^x$ $(b > 0, b \neq 1)$.

1. Use the definition of the derivative of a function to show that

$$f'(x) = b^x \cdot \lim_{h \to 0}\frac{b^h - 1}{h}$$

2. Use the result of part 1 to show that

$$\frac{d}{dx}(2^x) = 2^x \cdot \lim_{h \to 0}\frac{2^h - 1}{h}$$

$$\frac{d}{dx}(3^x) = 3^x \cdot \lim_{h \to 0}\frac{3^h - 1}{h}$$

3. Use the technique in Using Technology, page 414, to show that (to two decimal places)

$$\lim_{h \to 0}\frac{2^h - 1}{h} = 0.69 \quad \text{and} \quad \lim_{h \to 0}\frac{3^h - 1}{h} = 1.10$$

4. Conclude from the results of parts 2 and 3 that

$$\frac{d}{dx}(2^x) \approx (0.69)2^x \quad \text{and} \quad \frac{d}{dx}(3^x) \approx (1.10)3^x$$

Thus,

$$\frac{d}{dx}(b^x) = k \cdot b^x$$

where k is an appropriate constant.

5. The results of part 4 suggest that, for convenience, we pick the base b, where $2 < b < 3$, so that $k = 1$. This value of b is $e \approx 2.718281828\ldots$. Thus,

$$\frac{d}{dx}(e^x) = e^x$$

This is why we prefer to work with the exponential function $f(x) = e^x$.

APPLYING THE CHAIN RULE TO EXPONENTIAL FUNCTIONS

To enlarge the class of exponential functions to be differentiated, we appeal to the chain rule to obtain the following rule for differentiating composite functions of the form $h(x) = e^{f(x)}$. An example of such a function is $h(x) = e^{x^2-2x}$. Here, $f(x) = x^2 - 2x$.

Rule 2: Chain Rule for Exponential Functions

If $f(x)$ is a differentiable function, then

$$\frac{d}{dx}(e^{f(x)}) = e^{f(x)}f'(x)$$

To see this, observe that if $h(x) = g[f(x)]$, where $g(x) = e^x$, then by virtue of the chain rule,

$$h'(x) = g'(f(x))f'(x) = e^{f(x)}f'(x)$$

since $g'(x) = e^x$.

As an aid to remembering the chain rule for exponential functions, observe that it has the following form:

$$\frac{d}{dx}(e^{f(x)}) = e^{f(x)} \cdot \text{derivative of exponent}$$
$$\underset{\llcorner \text{Same} \lrcorner}{}$$

EXAMPLE 2   Find the derivative of each of the following functions.

a. $f(x) = e^{2x}$ **b.** $y = e^{-3x}$ **c.** $g(t) = e^{2t^2+t}$

SOLUTION ✔

a. $f'(x) = e^{2x}\dfrac{d}{dx}(2x) = e^{2x} \cdot 2 = 2e^{2x}$

b. $\dfrac{dy}{dx} = e^{-3x}\dfrac{d}{dx}(-3x) = -3e^{-3x}$

c. $g'(t) = e^{2t^2+t} \cdot \dfrac{d}{dt}(2t^2 + t) = (4t + 1)e^{2t^2+t}$ ■ ■ ■ ■

EXAMPLE 3   Differentiate the function $y = xe^{-2x}$.

SOLUTION ✔   Using the product rule, followed by the chain rule, we find

$$\frac{dy}{dx} = x\frac{d}{dx}e^{-2x} + e^{-2x}\frac{d}{dx}(x)$$

$$= xe^{-2x}\frac{d}{dx}(-2x) + e^{-2x} \quad \text{(Using the chain rule on the first term)}$$

$$= -2xe^{-2x} + e^{-2x}$$

$$= e^{-2x}(1 - 2x)$$ ■ ■ ■ ■

EXAMPLE 4   Differentiate the function $g(t) = \dfrac{e^t}{e^t + e^{-t}}$.

SOLUTION ✔   Using the quotient rule, followed by the chain rule, we find

$$g'(t) = \frac{(e^t + e^{-t})\dfrac{d}{dt}(e^t) - e^t \dfrac{d}{dt}(e^t + e^{-t})}{(e^t + e^{-t})^2}$$

$$= \frac{(e^t + e^{-t})e^t - e^t(e^t - e^{-t})}{(e^t + e^{-t})^2}$$

$$= \frac{e^{2t} + 1 - e^{2t} + 1}{(e^t + e^{-t})^2} \quad (e^0 = 1)$$

$$= \frac{2}{(e^t + e^{-t})^2} \qquad ■ ■ ■ ■$$

EXAMPLE 5   In Section 5.6 we will discuss some practical applications of the exponential function

$$Q(t) = Q_0 e^{kt}$$

where Q_0 and k are positive constants and $t \in [0, \infty)$. A quantity $Q(t)$ growing according to this law experiences exponential growth. Show that for a quantity $Q(t)$ experiencing exponential growth, the rate of growth of the quantity $Q'(t)$ at any time t is directly proportional to the amount of the quantity present.

SOLUTION ✔   Using the chain rule for exponential functions, we compute the derivative Q' of the function Q. Thus,

$$Q'(t) = Q_0 e^{kt} \frac{d}{dt}(kt)$$

$$= Q_0 e^{kt}(k)$$

$$= k Q_0 e^{kt}$$

$$= kQ(t) \quad (Q(t) = Q_0 e^{kt})$$

which is the desired conclusion.   ■ ■ ■ ■

EXAMPLE 6   Find the points of inflection of the function $f(x) = e^{-x^2}$.

SOLUTION ✔   The first derivative of f is

$$f'(x) = -2xe^{-x^2}$$

Differentiating $f'(x)$ with respect to x yields

$$f''(x) = (-2x)(-2xe^{-x^2}) - 2e^{-x^2}$$

$$= 2e^{-x^2}(2x^2 - 1)$$

Setting $f''(x) = 0$ gives

$$2e^{-x^2}(2x^2 - 1) = 0$$

FIGURE 5.11
Sign diagram for f''

Since e^{-x^2} never equals zero for any real value of x, we see that $x = \pm 1/\sqrt{2}$ are the only candidates for inflection points of f. The sign diagram of f'', shown in Figure 5.11, tells us that both $x = -1/\sqrt{2}$ and $x = 1/\sqrt{2}$ give rise to inflection points of f.

Next,

$$f\left(-\frac{1}{\sqrt{2}}\right) = f\left(\frac{1}{\sqrt{2}}\right) = e^{-1/2}$$

and the inflection points of f are $(-1/\sqrt{2},\, e^{-1/2})$ and $(1/\sqrt{2},\, e^{-1/2})$. The graph of f appears in Figure 5.12.

FIGURE 5.12
The graph of $y = e^{-x^2}$ has two inflection points.

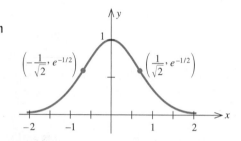

APPLICATION

Our final example involves finding the absolute maximum of an exponential function.

EXAMPLE 7

Refer to Example 6, Section 5.3. The present value of the market price of the Blakely Office Building is given by

$$P(t) = 300{,}000 e^{-0.09t + \sqrt{t}/2} \qquad (0 \le t \le 10)$$

Find the optimal present value of the building's market price.

SOLUTION ✔

To find the maximum value of P over $[0, 10]$, we compute

$$P'(t) = 300{,}000 e^{-0.09t + \sqrt{t}/2} \frac{d}{dt}\left(-0.09t + \frac{1}{2}t^{1/2}\right)$$

$$= 300{,}000 e^{-0.09t + \sqrt{t}/2}\left(-0.09 + \frac{1}{4}t^{-1/2}\right)$$

Setting $P'(t) = 0$ gives

$$-0.09 + \frac{1}{4t^{1/2}} = 0$$

since $e^{-0.09t+\sqrt{t}/2}$ is never zero for any value of t. Solving this equation, we find

$$\frac{1}{4t^{1/2}} = 0.09$$

$$t^{1/2} = \frac{1}{4(0.09)}$$

$$= \frac{1}{0.36}$$

$$t \approx 7.72$$

the sole critical point of the function P. Finally, evaluating $P(t)$ at the critical point as well as at the end points of $[0, 10]$, we have

t	$P(t)$
0	300,000
7.72	600,779
10	592,838

We conclude, accordingly, that the optimal present value of the property's market price is $600,779 and that this will occur 7.72 years from now.

■ ■ ■ ■

SELF-CHECK EXERCISES 5.4

1. Let $f(x) = xe^{-x}$.
   a. Find the first and second derivatives of f.
   b. Find the relative extrema of f.
   c. Find the inflection points of f.

2. An industrial asset is being depreciated at a rate so that its book value t yr from now will be

$$V(t) = 50,000e^{-0.4t}$$

dollars. How fast will the book value of the asset be changing 3 yr from now?

Solutions to Self-Check Exercises 5.4 can be found on page 416.

5.4 Exercises

In Exercises 1–28, find the derivative of the function.

1. $f(x) = e^{3x}$

2. $f(x) = 3e^x$

3. $g(t) = e^{-t}$

4. $f(x) = e^{-2x}$

5. $f(x) = e^x + x$

6. $f(x) = 2e^x - x^2$

7. $f(x) = x^3 e^x$

8. $f(u) = u^2 e^{-u}$

9. $f(x) = \frac{2e^x}{x}$

10. $f(x) = \frac{x}{e^x}$

11. $f(x) = 3(e^x + e^{-x})$

12. $f(x) = \frac{e^x + e^{-x}}{2}$

13. $f(w) = \dfrac{e^w + 1}{e^w}$

14. $f(x) = \dfrac{e^x}{e^x + 1}$

15. $f(x) = 2e^{3x-1}$

16. $f(t) = 4e^{3t+2}$

17. $h(x) = e^{-x^2}$

18. $f(x) = e^{x^2-1}$

19. $f(x) = 3e^{-1/x}$

20. $f(x) = e^{1/(2x)}$

21. $f(x) = (e^x + 1)^{25}$

22. $f(x) = (4 - e^{-3x})^3$

23. $f(x) = e^{\sqrt{x}}$

24. $f(t) = -e^{-\sqrt{2t}}$

25. $f(x) = (x - 1)e^{3x+2}$

26. $f(s) = (s^2 + 1)e^{-s^2}$

27. $f(x) = \dfrac{e^x - 1}{e^x + 1}$

28. $g(t) = \dfrac{e^{-t}}{1 + t^2}$

In Exercises 29–32, find the second derivative of the function.

29. $f(x) = e^{-4x} + 2e^{3x}$

30. $f(t) = 3e^{-2t} - 5e^{-t}$

31. $f(x) = 2xe^{3x}$

32. $f(t) = t^2 e^{-2t}$

33. Find an equation of the tangent line to the graph of $y = e^{2x-3}$ at the point $(\frac{3}{2}, 1)$.

34. Find an equation of the tangent line to the graph of $y = e^{-x^2}$ at the point $(1, 1/e)$.

35. Determine the intervals where the function $f(x) = e^{-x^2/2}$ is increasing and where it is decreasing.

36. Determine the intervals where the function $f(x) = x^2 e^{-x}$ is increasing and where it is decreasing.

37. Determine the intervals of concavity for the function $f(x) = \dfrac{e^x - e^{-x}}{2}$.

38. Determine the intervals of concavity for the function $f(x) = xe^x$.

39. Find the inflection point of the function $f(x) = xe^{-2x}$.

40. Find the inflection point(s) of the function $f(x) = 2e^{-x^2}$.

In Exercises 41–44, find the absolute extrema of the function.

41. $f(x) = e^{-x^2}$ on $[-1, 1]$

42. $h(x) = e^{x^2-4}$ on $[-2, 2]$

43. $g(x) = (2x - 1)e^{-x}$ on $[0, \infty)$

44. $f(x) = xe^{-x^2}$ on $[0, 2]$

In Exercises 45–48, use the curve-sketching guidelines of Chapter 4, page 327, to sketch the graph of the function.

45. $f(t) = e^t - t$

46. $h(x) = \dfrac{e^x + e^{-x}}{2}$

47. $f(x) = 2 - e^{-x}$

48. $f(x) = \dfrac{3}{1 + e^{-x}}$

A calculator is recommended for Exercises 49–59.

49. SALES PROMOTION The Lady Bug, a women's clothing chain store, found that t days after the end of a sales promotion the volume of sales was given by

$$S(t) = 20{,}000(1 + e^{-0.5t}) \qquad (0 \le t \le 5)$$

dollars. Find the rate of change of The Lady Bug's sales volume when $t = 1$, $t = 2$, $t = 3$, and $t = 4$.

50. ENERGY CONSUMPTION OF APPLIANCES The average energy consumption of the typical refrigerator/freezer manufactured by York Industries is approximately

$$C(t) = 1486e^{-0.073t} + 500 \qquad (0 \le t \le 20)$$

kilowatt-hours (kWh) per year, where t is measured in years, with $t = 0$ corresponding to 1972.
a. What was the average energy consumption of the York refrigerator/freezer at the beginning of 1972?
b. Prove that the average energy consumption of the York refrigerator/freezer is decreasing over the years in question.
c. All refrigerator/freezers manufactured as of January 1, 1990, must meet the 950-kWh/year maximum energy-consumption standard set by the National Appliance Conservation Act. Show that the York refrigerator/freezer satisfies this requirement.

51. POLIO IMMUNIZATION Polio, a once-feared killer, declined markedly in the United States in the 1950s after Jonas Salk developed the inactivated polio vaccine and mass immunization of children took place. The number of polio cases in the United States from the beginning of 1959 to the beginning of 1963 is approximated by the function

$$N(t) = 5.3e^{0.095t^2 - 0.85t} \qquad (0 \le t \le 4)$$

where $N(t)$ gives the number of polio cases (in thousands) and t is measured in years, with $t = 0$ corresponding to the beginning of 1959.
a. Show that the function N is decreasing over the time interval under consideration.
b. How fast was the number of polio cases decreasing

at the beginning of 1959? At the beginning of 1962? (*Comment:* Following the introduction of the oral vaccine developed by Dr. Albert B. Sabin in 1963, polio in the United States has, for all practical purposes, been eliminated.)

52. **BLOOD ALCOHOL LEVEL** The percentage of alcohol in a person's bloodstream t hr after drinking 8 fluid oz of whiskey is given by

$$A(t) = 0.23te^{-0.4t} \qquad (0 \le t \le 12)$$

a. What is the percentage of alcohol in a person's bloodstream after $\frac{1}{2}$ hr? After 8 hr?
b. How fast is the percentage of alcohol in a person's bloodstream changing after $\frac{1}{2}$ hr? After 8 hr?
Source: Encyclopedia Britannica

53. **PRICE OF PERFUME** The monthly demand for a certain brand of perfume is given by the demand equation

$$p = 100e^{-0.0002x} + 150$$

where p denotes the retail unit price (in dollars) and x denotes the quantity (in 1-oz bottles) demanded.
a. Find the rate of change of the price per bottle when $x = 1000$ and when $x = 2000$.
b. What is the price per bottle when $x = 1000$? When $x = 2000$?

54. **PRICE OF WINE** The monthly demand for a certain brand of table wine is given by the demand equation

$$p = 240 \left(1 - \frac{3}{3 + e^{-0.0005x}} \right)$$

where p denotes the wholesale price per case (in dollars) and x denotes the number of cases demanded.
a. Find the rate of change of the price per case when $x = 1000$.
b. What is the price per case when $x = 1000$?

55. **SPREAD OF AN EPIDEMIC** During a flu epidemic, the total number of students on a state university campus who had contracted influenza by the xth day was given by

$$N(x) = \frac{3000}{1 + 99e^{-x}} \qquad (x \ge 0)$$

a. How many students had influenza initially?
b. Derive an expression for the rate at which the disease was being spread and prove that the function N is increasing on the interval $(0, \infty)$.
c. Sketch the graph of N. What was the total number of students who contracted influenza during that particular epidemic?

56. **MAXIMUM OIL PRODUCTION** It has been estimated that the total production of oil from a certain oil well is given by

$$T(t) = -1000(t + 10)e^{-0.1t} + 10,000$$

thousand barrels t years after production has begun. Determine the year when the oil well will be producing at maximum capacity.

57. **OPTIMAL SELLING TIME** Refer to Exercise 26, page 402. The present value of a piece of waterfront property purchased by an investor is given by the function

$$P(t) = 80,000e^{\sqrt{t/2} - 0.09t} \qquad (0 \le t \le 8)$$

Determine the optimal time (based on present value) for the investor to sell the property. What is the property's optimal present value?

58. **OIL USED TO FUEL PRODUCTIVITY** A study on worldwide oil use was prepared for a major oil company. The study predicted that the amount of oil used to fuel productivity in a certain country is given by

$$f(t) = 1.5 + 1.8te^{-1.2t} \qquad (0 \le t \le 4)$$

where $f(t)$ denotes the number of barrels per \$1000 of economic output and t is measured in decades ($t = 0$ corresponds to 1965). Compute $f'(0)$, $f'(1)$, $f'(2)$, and $f'(3)$ and interpret your results.

59. **PERCENTAGE OF POPULATION RELOCATING** Based on data obtained from the Census Bureau, the manager of Plymouth Van Lines estimates that the percentage of the total population relocating in year t ($t = 0$ corresponds to the year 1960) may be approximated by the formula

$$P(t) = 20.6e^{-0.009t} \qquad (0 \le t \le 35)$$

Compute $P'(10)$, $P'(20)$, and $P'(30)$ and interpret your results.

60. **PRICE OF A COMMODITY** The price of a certain commodity in dollars per unit at time t (measured in weeks) is given by $p = 18 - 3e^{-2t} - 6e^{-t/3}$.
a. What is the price of the commodity at $t = 0$?
b. How fast is the price of the commodity changing at $t = 0$?
c. Find the equilibrium price of the commodity.
Hint: It is given by $\lim\limits_{t \to \infty} p$.

61. **PRICE OF A COMMODITY** The price of a certain commodity in dollars per unit at time t (measured in weeks) is given by $p = 8 + 4e^{-2t} + te^{-2t}$.
a. What is the price of the commodity at $t = 0$?
b. How fast is the price of the commodity changing at $t = 0$?
c. Find the equilibrium price of the commodity.
Hint: It's given by $\lim\limits_{t \to \infty} p$. Also, use the fact that $\lim\limits_{t \to \infty} te^{-2t} = 0$.

(continued on p. 416)

Using Technology

EXAMPLE **1** At the beginning of Section 5.4, we demonstrated via a table of values of $(e^h - 1)/h$ for selected values of h the plausibility of the result

$$\lim_{h \to 0} \frac{e^h - 1}{h} = 1$$

To obtain a visual confirmation of this result, we plot the graph of

$$f(x) = \frac{e^x - 1}{x}$$

in the viewing rectangle $[-1, 1] \times [0, 2]$ (Figure T1). From the graph of f, we see that $f(x)$ appears to approach 1 as x approaches 0.

FIGURE T1
The graph of f in the viewing rectangle $[-1, 1] \times [0, 2]$

The numerical derivative function of a graphing utility will yield the derivative of an exponential or logarithmic function for any value of x, just as it did for algebraic functions.*

---

* The rules for differentiating logarithmic functions will be covered in Section 5.5. However, the exercises given here can be done without using these rules.

Exercises

In Exercises 1–6, use the numerical derivative operation of a graphing utility to find the rate of change of $f(x)$ at the given value of x. Give your answer accurate to four decimal places.

1. $f(x) = x^3 e^{-1/x}$; $x = -1$

2. $f(x) = (\sqrt{x} + 1)^{3/2} e^{-x}$; $x = 0.5$

3. $f(x) = x^3 \sqrt{\ln x}$; $x = 2$

4. $f(x) = \dfrac{\sqrt{x} \ln x}{x + 1}$; $x = 3.2$

5. $f(x) = e^{-x} \ln(2x + 1)$; $x = 0.5$

6. $f(x) = \dfrac{e^{-\sqrt{x}}}{\ln(x^2 + 1)}$; $x = 1$

7. AN EXTINCTION SITUATION The number of saltwater crocodiles in a certain area of northern Australia is given by

$$P(t) = \frac{300 e^{-0.024t}}{5 e^{-0.024t} + 1}$$

a. How many crocodiles were in the population initially?

b. Show that $\lim_{t \to \infty} P(t) = 0$.

c. Use a graphing calculator to plot the graph of P in the viewing rectangle $[0, 200] \times [0, 70]$.
(*Comment:* This phenomenon is referred to as an *extinction situation*.)

8. INCOME OF AMERICAN FAMILIES Based on data compiled by the House Budget Committee, the House Ways and Means Committee, and the U.S. Census Bureau, it is estimated that the number of American families y (in millions) who earned x thousand dollars in 1990 is related by the equation

$$y = 0.1584xe^{-0.0000016x^3 + 0.00011x^2 - 0.04491x} \qquad (x > 0)$$

a. Use a graphing utility to plot the graph of the equation in the viewing rectangle $[0, 150] \times [0, 2]$.
b. How fast is y changing with respect to x when $x = 10$? When $x = 50$? Interpret your results.
Source: House Budget Committee, House Ways and Means Committee, and U.S. Census Bureau

9. **WORLD POPULATION GROWTH** According to a study conducted by the United Nations Population Division, the world population (in billions) is approximated by the function

$$f(t) = \frac{12}{1 + 3.74914e^{-1.42804t}} \qquad (0 \le t \le 4)$$

where t is measured in half-centuries, with $t = 0$ corresponding to the beginning of 1950.
a. Use a graphing utility to plot the graph of f in the viewing rectangle $[0, 5] \times [0, 14]$.
b. How fast was the world population expected to increase at the beginning of the year 2000?
Source: United Nations Population Division

10. **LOAN AMORTIZATION** The Sotos plan to secure a loan of $160,000 to purchase a house. They are considering a conventional 30-yr home mortgage at 9%/year on the unpaid balance. It can be shown that the Sotos will have an outstanding principal of

$$B(x) = \frac{160,000(1.0075^{360} - 1.0075^x)}{1.0075^{360} - 1}$$

dollars after making x monthly payments of $1287.40.
a. Use a graphing utility to plot the graph of $B(x)$, using the viewing rectangle $[0, 360] \times [0, 160,000]$.
b. Compute $B(0)$ and $B'(0)$ and interpret your results; compute $B(180)$ and $B'(180)$ and interpret your results.

11. **INCREASE IN JUVENILE OFFENDERS** The number of youths aged 15 to 19 will increase by 21% between 1994 and 2005, pushing up the crime rate. According to the National Council on Crime and Delinquency, the number of violent crime arrests of juveniles under age 18 in year t is given by

$$f(t) = -0.438t^2 + 9.002t + 107 \qquad (0 \le t \le 13)$$

where $f(t)$ is measured in thousands and t in years, with $t = 0$ corresponding to 1989. According to the same source, if trends like inner-city drug use and wider avail-

ability of guns continues, then the number of violent crime arrests of juveniles under age 18 in year t will be given by

$$g(t) = \begin{cases} -0.438t^2 + 9.002t + 107 & \text{if } 0 \le t < 4 \\ 99.456e^{0.07824t} & \text{if } 4 \le t \le 13 \end{cases}$$

where $g(t)$ is measured in thousands and $t = 0$ corresponds to 1989.
a. Compute $f(11)$ and $g(11)$ and interpret your results.
b. Compute $f'(11)$ and $g'(11)$ and interpret your results.
Source: National Council on Crime and Delinquency

12. **INCREASING CROP YIELDS** If left untreated on bean stems, aphids (small insects that suck plant juices) will multiply at an increasing rate during the summer months and reduce productivity and crop yield of cultivated crops. But if the aphids are treated in mid-June, the numbers decrease sharply to less than 100/bean stem, allowing for steep rises in crop yield. The function

$$F(t) = \begin{cases} 62e^{1.152t} & \text{if } 0 \le t < 1.5 \\ 349e^{-1.324(t-1.5)} & \text{if } 1.5 \le t \le 3 \end{cases}$$

gives the number of aphids in a typical bean stem at time t, where t is measured in months, with $t = 0$ corresponding to the beginning of May.
a. How many aphids are there on a typical bean stem at the beginning of June ($t = 1$)? At the beginning of July ($t = 2$)?
b. How fast is the population of aphids changing at the beginning of June? At the beginning of July?
Source: The Random House Encyclopedia

13. **PERCENTAGE OF FEMALES IN THE LABOR FORCE** Based on data from the U.S. Census Bureau, the chief economist of Manpower, Inc., constructed the following formula giving the percentage of the total female population in the civilian labor force, $P(t)$, at the beginning of the tth decade ($t = 0$ corresponds to the year 1900):

$$P(t) = \frac{74}{1 + 2.6e^{-0.166t + 0.04536t^2 - 0.0066t^3}} \qquad (0 \le t \le 11)$$

Assume this trend continued for the rest of the twentieth century.
a. What was the percentage of the total female population in the civilian labor force at the beginning of the year 2000?
b. What was the growth rate of the percentage of the total female population in the civilian labor force at the beginning of the year 2000?
Source: U.S. Census Bureau

62. ABSORPTION OF DRUGS A liquid carries a drug into an organ of volume V cm³ at the rate of a cm³/sec and leaves at the same rate. The concentration of the drug in the entering liquid is c g/cm³. Letting $x(t)$ denote the concentration of the drug in the organ at any time t, we have $x(t) = c(1 - e^{-at/V})$.

a. Show that x is an increasing function on $(0, \infty)$.

b. Sketch the graph of x.

63. ABSORPTION OF DRUGS Refer to Exercise 62. Suppose the maximum concentration of the drug in the organ must *not* exceed m g/cm³, where $m < c$. Show that the liquid must not be allowed to enter the organ for a time longer than

$$T = \left(\frac{V}{a}\right) \ln\left(\frac{c}{c - m}\right)$$

minutes.

In Exercises 64–67, determine whether the statement is true or false. If it is true, explain why it is true. If it is false, give an example to show why it is false.

64. If $f(x) = 3^x$, then $f'(x) = x \cdot 3^{x-1}$.

65. If $f(x) = e^{\pi}$, then $f'(x) = e^{\pi}$.

66. If $f(x) = \pi^x$, then $f'(x) = \pi^x$.

67. If $x^2 + e^y = 10$, then $y' = \dfrac{-2x}{e^y}$.

SOLUTIONS TO SELF-CHECK EXERCISES 5.4

1. a. Using the product rule, we obtain

$$f'(x) = x\frac{d}{dx}e^{-x} + e^{-x}\frac{d}{dx}x$$

$$= -xe^{-x} + e^{-x} = (1 - x)e^{-x}$$

Using the product rule once again, we obtain

$$f''(x) = (1 - x)\frac{d}{dx}e^{-x} + e^{-x}\frac{d}{dx}(1 - x)$$

$$= (1 - x)(-e^{-x}) + e^{-x}(-1)$$

$$= -e^{-x} + xe^{-x} - e^{-x} = (x - 2)e^{-x}$$

b. Setting $f'(x) = 0$ gives

$$(1 - x)e^{-x} = 0$$

Since $e^{-x} \neq 0$, we see that $1 - x = 0$, and this gives $x = 1$ as the only critical point of f. The sign diagram of f' shown in the accompanying figure tells us that the point $(1, e^{-1})$ is a relative maximum of f.

c. Setting $f''(x) = 0$ gives $x - 2 = 0$, so $x = 2$ is a candidate for an inflection point of f. The sign diagram of f'' (accompanying figure) shows that $(2, 2e^{-2})$ is an inflection point of f.

2. The rate of change of the book value of the asset t yr from now is

$$V'(t) = 50,000 \frac{d}{dt} e^{-0.4t}$$

$$= 50,000(-0.4)e^{-0.4t} = -20,000e^{-0.4t}$$

Therefore, 3 yr from now the book value of the asset will be changing at the rate of

$$V'(3) = -20,000e^{-0.4(3)} = -20,000e^{-1.2} \approx -6023.88$$

—that is, decreasing at the rate of approximately \$6024/year.

5.5 Differentiation of Logarithmic Functions

THE DERIVATIVE OF ln x

Let's now turn our attention to the differentiation of logarithmic functions.

Rule 3: Derivative of ln x

$$\frac{d}{dx} \ln |x| = \frac{1}{x} \qquad (x \neq 0)$$

To derive Rule 3, suppose $x > 0$ and write $f(x) = \ln x$ in the equivalent form

$$x = e^{f(x)}$$

Differentiating both sides of the equation with respect to x, we find, using the chain rule,

$$1 = e^{f(x)} \cdot f'(x)$$

from which we see that $\qquad f'(x) = \dfrac{1}{e^{f(x)}}$

or, since $e^{f(x)} = x$,

$$f'(x) = \frac{1}{x}$$

as we set out to show. You are asked to prove the rule for the case $x < 0$ in Exercise 61.

EXAMPLE 1   Compute the derivative of each of the following functions:

a. $f(x) = x \ln x$ **b.** $g(x) = \dfrac{\ln x}{x}$

SOLUTION ✔　**a.** Using the product rule, we obtain

$$f'(x) = \frac{d}{dx}(x \ln x) = x \frac{d}{dx}(\ln x) + (\ln x) \frac{d}{dx}(x)$$

$$= x\left(\frac{1}{x}\right) + \ln x = 1 + \ln x$$

b. Using the quotient rule, we obtain

$$g'(x) = \frac{x \dfrac{d}{dx}(\ln x) - (\ln x)\dfrac{d}{dx}(x)}{x^2} = \frac{x\left(\dfrac{1}{x}\right) - \ln x}{x^2} = \frac{1 - \ln x}{x^2}$$ ■ ■ ■ ■

 Group Discussion

You can derive the formula for the derivative of $f(x) = \ln x$ directly from the definition of the derivative, as follows.

1. Show that

$$f'(x) = \lim_{h \to 0} \frac{f(x+h) - f(x)}{h} = \lim_{h \to 0} \ln\left(1 + \frac{h}{x}\right)^{1/h}$$

2. Put $m = x/h$ and note that $m \to \infty$ as $h \to 0$. Furthermore, $f'(x)$ can be written in the form

$$f'(x) = \lim_{m \to \infty} \ln\left(1 + \frac{1}{m}\right)^{m/x}$$

3. Finally, use both the fact that the natural logarithmic function is continuous and the definition of the number e to show that

$$f'(x) = \frac{1}{x} \ln\left[\lim_{m \to \infty}\left(1 + \frac{1}{m}\right)^m\right] = \frac{1}{x}$$

THE CHAIN RULE FOR LOGARITHMIC FUNCTIONS

To enlarge the class of logarithmic functions to be differentiated, we appeal once more to the chain rule to obtain the following rule for differentiating composite functions of the form $h(x) = \ln f(x)$, where $f(x)$ is assumed to be a positive differentiable function.

Rule 4: Chain Rule for Logarithmic Functions

If $f(x)$ is a differentiable function, then

$$\frac{d}{dx}[\ln f(x)] = \frac{f'(x)}{f(x)} \qquad [f(x) > 0]$$

To see this, observe that $h(x) = g[f(x)]$, where $g(x) = \ln x$ $(x > 0)$. Since $g'(x) = 1/x$, we have, using the chain rule,

$$h'(x) = g'(f(x))f'(x)$$
$$= \frac{1}{f(x)}f'(x) = \frac{f'(x)}{f(x)}$$

Observe that in the special case $f(x) = x$, $h(x) = \ln x$, so the derivative of h is, by Rule 3, given by $h'(x) = 1/x$.

EXAMPLE 2   Find the derivative of the function $f(x) = \ln(x^2 + 1)$.

SOLUTION ✔   Using Rule 4, we see immediately that

$$f'(x) = \frac{\dfrac{d}{dx}(x^2 + 1)}{x^2 + 1} = \frac{2x}{x^2 + 1}$$
■■■■

When differentiating functions involving logarithms, the rules of logarithms may be used to advantage, as shown in Examples 3 and 4.

EXAMPLE 3   Differentiate the function $y = \ln[(x^2 + 1)(x^3 + 2)^6]$.

SOLUTION ✔   We first rewrite the given function using the properties of logarithms:

$$y = \ln[(x^2 + 1)(x^3 + 2)^6]$$
$$= \ln(x^2 + 1) + \ln(x^3 + 2)^6 \qquad (\ln mn = \ln m + \ln n)$$
$$= \ln(x^2 + 1) + 6\ln(x^3 + 2) \qquad (\ln m^n = n \ln m)$$

Differentiating and using Rule 4, we obtain

$$y' = \frac{\dfrac{d}{dx}(x^2 + 1)}{x^2 + 1} + \frac{6\dfrac{d}{dx}(x^3 + 2)}{x^3 + 2}$$
$$= \frac{2x}{x^2 + 1} + \frac{6(3x^2)}{x^3 + 2} = \frac{2x}{x^2 + 1} + \frac{18x^2}{x^3 + 2}$$
■■■■

Exploring with Technology

Use a graphing utility to plot the graphs of $f(x) = \ln x$; its first derivative function, $f'(x) = 1/x$; and its second derivative function, $f''(x) = -1/x^2$, using the same viewing rectangle $[0, 4] \times [-3, 3]$.

1. Describe the properties of the graph of f revealed by studying the graph of $f'(x)$. What can you say about the rate of increase of f for large values of x?
2. Describe the properties of the graph of f revealed by studying the graph of $f''(x)$. What can you say about the concavity of f for large values of x?

EXAMPLE **4**   Find the derivative of the function $g(t) = \ln(t^2 e^{-t^2})$.

SOLUTION ✔   Here again, to save a lot of work, we first simplify the given expression using the properties of logarithms. We have

$$
\begin{aligned}
g(t) &= \ln(t^2 e^{-t^2}) \\
&= \ln t^2 + \ln e^{-t^2} \quad (\ln mn = \ln m + \ln n) \\
&= 2 \ln t - t^2 \quad (\ln m^n = n \ln m \text{ and } \ln e = 1)
\end{aligned}
$$

Therefore,

$$
g'(t) = \frac{2}{t} - 2t = \frac{2(1 - t^2)}{t}
$$

■ ■ ■ ■

LOGARITHMIC DIFFERENTIATION

As we saw in the last two examples, the task of finding the derivative of a given function can be made easier by first applying the laws of logarithms to simplify the function. We now illustrate a process called **logarithmic differentiation,** which not only simplifies the calculation of the derivatives of certain functions but also enables us to compute the derivatives of functions we could not otherwise differentiate using the techniques developed thus far.

EXAMPLE **5**   Differentiate $y = x(x + 1)(x^2 + 1)$, using logarithmic differentiation.

SOLUTION ✔   First, we take the natural logarithm on both sides of the given equation, obtaining

$$
\ln y = \ln x(x + 1)(x^2 + 1)
$$

Next, we use the properties of logarithms to rewrite the right-hand side of this equation, obtaining

$$
\ln y = \ln x + \ln(x + 1) + \ln(x^2 + 1)
$$

If we differentiate both sides of this equation, we have

$$
\begin{aligned}
\frac{d}{dx} \ln y &= \frac{d}{dx} [\ln x + \ln(x + 1) + \ln(x^2 + 1)] \\
&= \frac{1}{x} + \frac{1}{x + 1} + \frac{2x}{x^2 + 1} \quad \text{(Using Rule 4)}
\end{aligned}
$$

To evaluate the expression on the left-hand side, note that y is a function of x. Therefore, writing $y = f(x)$ to remind us of this fact, we have

$$\frac{d}{dx} \ln y = \frac{d}{dx} \ln[f(x)] \qquad \text{[Writing } y = f(x)\text{]}$$

$$= \frac{f'(x)}{f(x)} \qquad \text{(Using Rule 4)}$$

$$= \frac{y'}{y} \qquad \text{[Returning to using } y \text{ instead of } f(x)\text{]}$$

Therefore, we have

$$\frac{y'}{y} = \frac{1}{x} + \frac{1}{x+1} + \frac{2x}{x^2+1}$$

Finally, solving for y', we have

$$y' = y\left(\frac{1}{x} + \frac{1}{x+1} + \frac{2x}{x^2+1}\right)$$

$$= x(x+1)(x^2+1)\left(\frac{1}{x} + \frac{1}{x+1} + \frac{2x}{x^2+1}\right) \qquad ■ ■ ■ ■$$

Before considering other examples, let's summarize the important steps involved in logarithmic differentiation.

Finding $\dfrac{dy}{dx}$ by Logarithmic Differentiation

1. Take the natural logarithm on both sides of the equation and use the properties of logarithms to write any "complicated expression" as a sum of simpler terms.
2. Differentiate both sides of the equation with respect to x.
3. Solve the resulting equation for $\dfrac{dy}{dx}$.

EXAMPLE 6

Differentiate $y = x^2(x - 1)(x^2 + 4)^3$.

 SOLUTION ✔

Taking the natural logarithm on both sides of the given equation and using the laws of logarithms, we obtain

$$\ln y = \ln x^2(x - 1)(x^2 + 4)^3$$

$$= \ln x^2 + \ln(x - 1) + \ln(x^2 + 4)^3$$

$$= 2 \ln x + \ln(x - 1) + 3 \ln(x^2 + 4)$$

Differentiating both sides of the equation with respect to x, we have

$$\frac{d}{dx} \ln y = \frac{y'}{y} = \frac{2}{x} + \frac{1}{x-1} + 3 \cdot \frac{2x}{x^2+4}$$

Finally, solving for y', we have

$$y' = y\left(\frac{2}{x} + \frac{1}{x-1} + \frac{6x}{x^2+4}\right)$$

$$= x^2(x-1)(x^2+4)^3\left(\frac{2}{x} + \frac{1}{x-1} + \frac{6x}{x^2+4}\right)$$

■ ■ ■ ■

EXAMPLE 7   Find the derivative of $f(x) = x^x(x > 0)$.

SOLUTION ✔   A word of caution! This function is neither a power function nor an exponential function. Taking the natural logarithm on both sides of the equation gives

$$\ln f(x) = \ln x^x = x \ln x$$

Differentiating both sides of the equation with respect to x, we obtain

$$\frac{f'(x)}{f(x)} = x\frac{d}{dx}\ln x + (\ln x)\frac{d}{dx}x$$

$$= x\left(\frac{1}{x}\right) + \ln x$$

$$= 1 + \ln x$$

Therefore,

$$f'(x) = f(x)(1 + \ln x) = x^x(1 + \ln x)$$

■ ■ ■ ■

Exploring with Technology

Refer to Example 7.

1. Use a graphing utility to plot the graph of $f(x) = x^x$, using the viewing rectangle $[0, 2] \times [0, 2]$. Then use ZOOM and TRACE to show that

$$\lim_{x\to0^+} f(x) = 1$$

2. Use the results of part 1 and Example 7 to show that $\lim\limits_{x\to0^+} f'(x) = -\infty$. Justify your answer.

SELF-CHECK EXERCISES 5.5

1. Find an equation of the tangent line to the graph of $f(x) = x \ln(2x + 3)$ at the point $(-1, 0)$.

2. Use logarithmic differentiation to compute y', given $y = (2x + 1)^3(3x + 4)^5$.

Solutions to Self-Check Exercises 5.5 can be found on page 424.

5.5 Exercises

In Exercises 1–32, find the derivative of the function.

1. $f(x) = 5 \ln x$

2. $f(x) = \ln 5x$

3. $f(x) = \ln(x + 1)$

4. $g(x) = \ln(2x + 1)$

5. $f(x) = \ln x^8$

6. $h(t) = 2 \ln t^5$

7. $f(x) = \ln \sqrt{x}$

8. $f(x) = \ln(\sqrt{x} + 1)$

9. $f(x) = \ln \dfrac{1}{x^2}$

10. $f(x) = \ln \dfrac{1}{2x^3}$

11. $f(x) = \ln(4x^2 - 6x + 3)$

12. $f(x) = \ln(3x^2 - 2x + 1)$

13. $f(x) = \ln \dfrac{2x}{x + 1}$

14. $f(x) = \ln \dfrac{x + 1}{x - 1}$

15. $f(x) = x^2 \ln x$

16. $f(x) = 3x^2 \ln 2x$

17. $f(x) = \dfrac{2 \ln x}{x}$

18. $f(x) = \dfrac{3 \ln x}{x^2}$

19. $f(u) = \ln(u - 2)^3$

20. $f(x) = \ln(x^3 - 3)^4$

21. $f(x) = \sqrt{\ln x}$

22. $f(x) = \sqrt{\ln x + x}$

23. $f(x) = (\ln x)^3$

24. $f(x) = 2(\ln x)^{3/2}$

25. $f(x) = \ln(x^3 + 1)$

26. $f(x) = \ln\sqrt{x^2 - 4}$

27. $f(x) = e^x \ln x$

28. $f(x) = e^x \ln\sqrt{x + 3}$

29. $f(t) = e^{2t} \ln(t + 1)$

30. $g(t) = t^2 \ln(e^{2t} + 1)$

31. $f(x) = \dfrac{\ln x}{x}$

32. $g(t) = \dfrac{t}{\ln t}$

In Exercises 33–36, find the second derivative of the function.

33. $f(x) = \ln 2x$

34. $f(x) = \ln(x + 5)$

35. $f(x) = \ln(x^2 + 2)$

36. $f(x) = (\ln x)^2$

In Exercises 37–46, use logarithmic differentiation to find the derivative of the function.

37. $y = (x + 1)^2(x + 2)^3$

38. $y = (3x + 2)^4(5x - 1)^2$

39. $y = (x - 1)^2(x + 1)^3(x + 3)^4$

40. $y = \sqrt{3x + 5}(2x - 3)^4$

41. $y = \dfrac{(2x^2 - 1)^5}{\sqrt{x + 1}}$

42. $y = \dfrac{\sqrt{4 + 3x^2}}{\sqrt[3]{x^2 + 1}}$

43. $y = 3^x$

44. $y = x^{x+2}$

45. $y = (x^2 + 1)^x$

46. $y = x^{\ln x}$

47. Find an equation of the tangent line to the graph of $y = x \ln x$ at the point $(1, 0)$.

48. Find an equation of the tangent line to the graph of $y = \ln x^2$ at the point $(2, \ln 4)$.

49. Determine the intervals where the function $f(x) = \ln x^2$ is increasing and where it is decreasing.

50. Determine the intervals where the function $f(x) = \dfrac{\ln x}{x}$ is increasing and where it is decreasing.

51. Determine the intervals of concavity for the function $f(x) = x^2 + \ln x^2$.

52. Determine the intervals of concavity for the function $f(x) = \dfrac{\ln x}{x}$.

53. Find the inflection points of the function $f(x) = \ln(x^2 + 1)$.

54. Find the inflection points of the function $f(x) = x^2 \ln x$.

55. Find the absolute extrema of the function $f(x) = x - \ln x$ on $[\frac{1}{2}, 3]$.

56. Find the absolute extrema of the function $g(x) = \dfrac{x}{\ln x}$ on $[2, \infty)$.

In Exercises 57 and 58, use the guidelines on page 327 to sketch the graph of the given function.

57. $f(x) = \ln(x - 1)$

58. $f(x) = 2x - \ln x$

In Exercises 59 and 60, determine whether the statement is true or false. If it is true, explain why it is true. If it is false, give an example to show why it is false.

59. If $f(x) = \ln 5$, then $f'(x) = 1/5$.

60. If $f(x) = \ln a^x$, then $f'(x) = \ln a$.

61. Prove that $\dfrac{d}{dx} \ln |x| = \dfrac{1}{x}$ $(x \neq 0)$ for the case $x < 0$.

62. Use the definition of the derivative to show that

$$\lim_{x \to 0} \frac{\ln(x + 1)}{x} = 1$$

SOLUTIONS TO SELF-CHECK EXERCISES 5.5

1. The slope of the tangent line to the graph of f at any point $(x, f(x))$ lying on the graph of f is given by $f'(x)$. Using the product rule, we find

$$f'(x) = \frac{d}{dx}[x \ln(2x + 3)]$$

$$= x \frac{d}{dx} \ln(2x + 3) + \ln(2x + 3) \cdot \frac{d}{dx}(x)$$

$$= x \left(\frac{2}{2x + 3}\right) + \ln(2x + 3) \cdot 1$$

$$= \frac{2x}{2x + 3} + \ln(2x + 3)$$

In particular, the slope of the tangent line to the graph of f at the point $(-1, 0)$ is

$$f'(-1) = \frac{-2}{-2 + 3} + \ln 1 = -2$$

Therefore, using the point-slope form of the equation of a line, we see that a required equation is

$$y - 0 = -2(x + 1)$$
$$y = -2x - 2$$

2. Taking the logarithm on both sides of the equation gives

$$\ln y = \ln(2x + 1)^3 (3x + 4)^5$$
$$= \ln(2x + 1)^3 + \ln(3x + 4)^5$$
$$= 3 \ln(2x + 1) + 5 \ln(3x + 4)$$

Differentiating both sides of the equation with respect to x, keeping in mind that y is a function of x, we obtain

$$\frac{d}{dx}(\ln y) = \frac{y'}{y} = 3 \cdot \frac{2}{2x + 1} + 5 \cdot \frac{3}{3x + 4}$$

$$= 3 \left[\frac{2}{2x + 1} + \frac{5}{3x + 4}\right]$$

$$= \left(\frac{6}{2x + 1} + \frac{15}{3x + 4}\right)$$

and

$$y' = (2x + 1)^3 (3x + 4)^5 \cdot \left(\frac{6}{2x + 1} + \frac{15}{3x + 4}\right)$$

5.6 Exponential Functions as Mathematical Models

FIGURE 5.13
Exponential growth

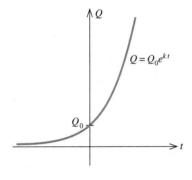

EXPONENTIAL GROWTH

Many problems arising from practical situations can be described mathematically in terms of exponential functions or functions closely related to the exponential function. In this section we look at some applications involving exponential functions from the fields of the life and social sciences.

In Section 5.1 we saw that the exponential function $f(x) = b^x$ is an increasing function when $b > 1$. In particular, the function $f(x) = e^x$ shares this property. From this result one may deduce that the function $Q(t) = Q_0 e^{kt}$, where Q_0 and k are positive constants, has the following properties:

1. $Q(0) = Q_0$
2. $Q(t)$ increases "rapidly" without bound as t increases without bound (Figure 5.13).

Property 1 follows from the computation

$$Q(0) = Q_0 e^0 = Q_0$$

Next, to study the rate of change of the function $Q(t)$, we differentiate it with respect to t, obtaining

$$Q'(t) = \frac{d}{dt}(Q_0 e^{kt})$$
$$= Q_0 \frac{d}{dt}(e^{kt})$$
$$= kQ_0 e^{kt}$$
$$= kQ(t) \qquad \textbf{(12)}$$

Since $Q(t) > 0$ (because Q_0 is assumed to be positive) and $k > 0$, we see that $Q'(t) > 0$ and so $Q(t)$ is an increasing function of t. Our computation has in fact shed more light on an important property of the function $Q(t)$. Equation (12) says that the rate of increase of the function $Q(t)$ is proportional to the amount $Q(t)$ of the quantity present at time t. The implication is that as $Q(t)$ increases, so does the *rate of increase* of $Q(t)$, resulting in a very rapid increase in $Q(t)$ as t increases without bound.

Thus, the exponential function

$$Q(t) = Q_0 e^{kt} \qquad (0 \le t < \infty) \qquad \textbf{(13)}$$

provides us with a mathematical model of a quantity $Q(t)$ that is initially present in the amount of $Q(0) = Q_0$ and whose rate of growth at any time t is directly proportional to the amount of the quantity present at time t. Such a quantity is said to exhibit **exponential growth,** and the constant k is called the **growth constant.** Interest earned on a fixed deposit when compounded continuously exhibits exponential growth. Other examples of exponential growth follow.

EXAMPLE 1

Under ideal laboratory conditions, the number of bacteria in a culture grows in accordance with the law $Q(t) = Q_0 e^{kt}$, where Q_0 denotes the number of bacteria initially present in the culture, k is some constant determined by the strain of bacteria under consideration, and t is the elapsed time measured in hours. Suppose 10,000 bacteria are present initially in the culture and 60,000 present 2 hours later.

a. How many bacteria will there be in the culture at the end of 4 hours?
b. What is the rate of growth of the population after 4 hours?

SOLUTION ✔   **a.** We are given that $Q(0) = Q_0 = 10{,}000$, so $Q(t) = 10{,}000 e^{kt}$. Next, the fact that 60,000 bacteria are present 2 hours later translates into $Q(2) = 60{,}000$. Thus,

$$60{,}000 = 10{,}000 e^{2k}$$

$$e^{2k} = 6$$

Taking the natural logarithm on both sides of the equation, we obtain

$$\ln e^{2k} = \ln 6$$

$$2k = \ln 6 \qquad \text{(Since } \ln e = 1\text{)}$$

$$k \approx 0.8959$$

Thus, the number of bacteria present at any time t is given by

$$Q(t) = 10{,}000 e^{0.8959t}$$

In particular, the number of bacteria present in the culture at the end of 4 hours is given by

$$Q(4) = 10{,}000 e^{0.8959(4)}$$

$$= 360{,}029$$

b. The rate of growth of the bacteria population at any time t is given by

$$Q'(t) = kQ(t)$$

Thus, using the result from part (a), we find that the rate at which the population is growing at the end of 4 hours is

$$Q'(4) = kQ(4)$$
$$\approx (0.8959)(360{,}029)$$
$$\approx 322{,}550$$

or approximately 322,550 bacteria per hour.   ■ ■ ■ ■

EXPONENTIAL DECAY

In contrast to exponential growth, a quantity exhibits **exponential decay** if it decreases at a rate that is directly proportional to its size. Such a quantity may be described by the exponential function

$$Q(t) = Q_0 e^{-kt} \qquad [t \in [0, \infty)] \qquad \textbf{(14)}$$

where the positive constant Q_0 measures the amount present initially ($t = 0$) and k is some suitable positive number, called the **decay constant.** The choice of this number is determined by the nature of the substance under consideration. The graph of this function is sketched in Figure 5.14.

To verify the properties ascribed to the function $Q(t)$, we simply compute

$$Q(0) = Q_0 e^0 = Q_0$$

$$Q'(t) = \frac{d}{dt}(Q_0 e^{-kt})$$

$$= Q_0 \frac{d}{dt}(e^{-kt})$$

$$= -kQ_0 e^{-kt} = -kQ(t)$$

FIGURE 5.14
Exponential decay

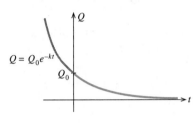

EXAMPLE **2**

Radioactive substances decay exponentially. For example, the amount of radium present at any time t obeys the law $Q(t) = Q_0 e^{-kt}$, where Q_0 is the initial amount present and k is a suitable positive constant. The **half-life of a radioactive substance** is the time required for a given amount to be reduced by one-half. Now, it is known that the half-life of radium is approximately 1600 years. Suppose initially there are 200 milligrams of pure radium. Find the amount left after t years. What is the amount left after 800 years?

SOLUTION ✔

The initial amount of radium present is 200 milligrams, so $Q(0) = Q_0 = 200$. Thus, $Q(t) = 200e^{-kt}$. Next, the datum concerning the half-life of radium implies that $Q(1600) = 100$, and this gives

$$100 = 200e^{-1600k}$$

$$e^{-1600k} = \frac{1}{2}$$

Taking the natural logarithm on both sides of this equation yields

$$-1600k \ln e = \ln \frac{1}{2}$$

$$-1600k = \ln \frac{1}{2} \qquad (\ln e = 1)$$

$$k = -\frac{1}{1600} \ln \left(\frac{1}{2}\right) = 0.0004332$$

Therefore, the amount of radium left after t years is

$$Q(t) = 200e^{-0.0004332t}$$

In particular, the amount of radium left after 800 years is

$$Q(800) = 200e^{-0.0004332(800)} \approx 141.42$$

or approximately 141 milligrams.   ■ ■ ■ ■

EXAMPLE 3   Carbon 14, a radioactive isotope of carbon, has a half-life of 5770 years. What is its decay constant?

SOLUTION ✔   We have $Q(t) = Q_0 e^{-kt}$. Since the half-life of the element is 5770 years, half of the substance is left at the end of that period. That is,

$$Q(5770) = Q_0 e^{-5770k} = \frac{1}{2} Q_0$$

$$e^{-5770k} = \frac{1}{2}$$

Taking the natural logarithm on both sides of this equation, we have

$$\ln e^{-5770k} = \ln \frac{1}{2}$$

$$-5770k = -0.693147$$

$$k \approx 0.00012 \qquad ■ ■ ■ ■$$

     Carbon-14 dating is a well-known method used by anthropologists to establish the age of animal and plant fossils. This method assumes that the proportion of carbon 14 (C-14) present in the atmosphere has remained constant over the past 50,000 years. Professor Willard Libby, recipient of the Nobel Prize in chemistry in 1960, proposed this theory.

     The amount of C-14 in the tissues of a living plant or animal is constant. However, when an organism dies, it stops absorbing new quantities of C-14, and the amount of C-14 in the remains diminishes because of the natural decay of the radioactive substance. Thus, the approximate age of a plant or animal fossil can be determined by measuring the amount of C-14 present in the remains.

EXAMPLE 4   A skull from an archeological site has one-tenth the amount of C-14 that it originally contained. Determine the approximate age of the skull.

SOLUTION ✔   Here,

$$Q(t) = Q_0 e^{-kt}$$
$$= Q_0 e^{-0.00012t}$$

where Q_0 is the amount of C-14 present originally and k, the decay constant, is equal to 0.00012 (see Example 3). Since $Q(t) = (1/10)Q_0$, we have

$$\frac{1}{10}Q_0 = Q_0 e^{-0.00012t}$$

$$\ln\frac{1}{10} = -0.00012t \qquad \text{(Taking the natural logarithm on both sides)}$$

$$t = \frac{\ln\dfrac{1}{10}}{-0.00012}$$

$$\approx 19{,}200$$

or approximately 19,200 years.   ■ ■ ■ ■

LEARNING CURVES

The next example shows how the exponential function may be applied to describe certain types of learning processes. Consider the function

$$Q(t) = C - Ae^{-kt}$$

where C, A, and k are positive constants. To sketch the graph of the function Q, observe that its y-intercept is given by $Q(0) = C - A$. Next, we compute

$$Q'(t) = kAe^{-kt}$$

Since both k and A are positive, we see that $Q'(t) > 0$ for all values of t. Thus, $Q(t)$ is an increasing function of t. Also,

$$\lim_{t\to\infty} Q(t) = \lim_{t\to\infty}(C - Ae^{-kt})$$

$$= \lim_{t\to\infty} C - \lim_{t\to\infty} Ae^{-kt}$$

$$= C$$

so $y = C$ is a horizontal asymptote of Q. Thus, $Q(t)$ increases and approaches the number C as t increases without bound. The graph of the function Q is shown in Figure 5.15, where that part of the graph corresponding to the negative values of t is drawn with a gray line since, in practice, one normally restricts the domain of the function to the interval $[0, \infty)$.

Observe that $Q(t)$ $(t > 0)$ increases rather rapidly initially but that the rate of increase slows down considerably after a while. To see this, we compute

$$\lim_{t\to\infty} Q'(t) = \lim_{t\to\infty} kAe^{-kt} = 0$$

This behavior of the graph of the function Q closely resembles the learning pattern experienced by workers engaged in highly repetitive work. For example, the productivity of an assembly-line worker increases very rapidly in the early stages of the training period. This productivity increase is a direct result

FIGURE 5.15

A learning curve

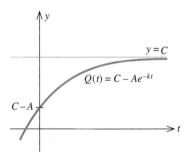

of the worker's training and accumulated experience. But the rate of increase of productivity slows as time goes by, and the worker's productivity level approaches some fixed level due to the limitations of the worker and the machine. Because of this characteristic, the graph of the function $Q(t) = C - Ae^{-kt}$ is often called a **learning curve.**

 EXAMPLE 5

The Camera Division of the Eastman Optical Company produces a 35-mm single-lens reflex camera. Eastman's training department determines that after completing the basic training program, a new, previously inexperienced employee will be able to assemble

$$Q(t) = 50 - 30e^{-0.5t}$$

model F cameras per day, t months after the employee starts work on the assembly line.

a. How many model F cameras can a new employee assemble per day after basic training?
b. How many model F cameras can an employee with 1 month of experience assemble per day? An employee with 2 months of experience? An employee with 6 months of experience?
c. How many model F cameras can the average experienced employee assemble per day?

SOLUTION ✔

a. The number of model F cameras a new employee can assemble is given by

$$Q(0) = 50 - 30 = 20$$

b. The number of model F cameras that an employee with 1 month of experience, 2 months of experience, and 6 months of experience can assemble per day is given by

$$Q(1) = 50 - 30e^{-0.5} \approx 31.80$$

$$Q(2) = 50 - 30e^{-1} \approx 38.96$$

$$Q(6) = 50 - 30e^{-3} \approx 48.51$$

or approximately 32, 39, and 49, respectively.
c. As t increases without bound, $Q(t)$ approaches 50. Hence, the average experienced employee can ultimately be expected to assemble 50 model F cameras per day. ■ ■ ■ ■

Other applications of the learning curve are found in models that describe the dissemination of information about a product or the velocity of an object dropped into a viscous medium.

LOGISTIC GROWTH FUNCTIONS

Our last example of an application of exponential functions to the description of natural phenomena involves the **logistic** (also called the **S-shaped,** or **sigmoi-**

FIGURE 5.16

A logistic curve

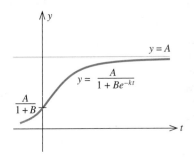

dal) curve, which is the graph of the function

$$Q(t) = \frac{A}{1 + Be^{-kt}}$$

where A, B, and k are positive constants. The function Q is called a **logistic growth function,** and the graph of the function Q is sketched in Figure 5.16.

Observe that $Q(t)$ increases rather rapidly for small values of t. In fact, for small values of t, the logistic curve resembles an exponential growth curve. However, the *rate of growth* of $Q(t)$ decreases quite rapidly as t increases and $Q(t)$ approaches the number A as t increases without bound.

Thus, the logistic curve exhibits both the property of rapid growth of the exponential growth curve as well as the "saturation" property of the learning curve. Because of these characteristics, the logistic curve serves as a suitable mathematical model for describing many natural phenomena. For example, if a small number of rabbits were introduced to a tiny island in the South Pacific, the rabbit population might be expected to grow very rapidly at first, but the growth rate would decrease quickly as overcrowding, scarcity of food, and other environmental factors affected it. The population would eventually stabilize at a level compatible with the life-support capacity of the environment. Models describing the spread of rumors and epidemics are other examples of the application of the logistic curve.

EXAMPLE 6   The number of soldiers at Fort MacArthur who contracted influenza after t days during a flu epidemic is approximated by the exponential model

$$Q(t) = \frac{5000}{1 + 1249e^{-kt}}$$

If 40 soldiers contracted the flu by day 7, find how many soldiers contracted the flu by day 15.

SOLUTION ✔   The given information implies that

$$Q(7) = 40 \quad \text{and} \quad Q(7) = \frac{5000}{1 + 1249e^{-7k}} = 40$$

Thus,

$$40(1 + 1249e^{-7k}) = 5000$$

$$1 + 1249e^{-7k} = \frac{5000}{40} = 125$$

$$e^{-7k} = \frac{124}{1249}$$

$$-7k = \ln\frac{124}{1249}$$

$$k = -\frac{\ln\frac{124}{1249}}{7} \approx 0.33$$

Therefore, the number of soldiers who contracted the flu after t days is given by

$$Q(t) = \frac{5000}{1 + 1249e^{-0.33t}}$$

In particular, the number of soldiers who contracted the flu by day 15 is given by

$$Q(15) = \frac{5000}{1 + 1249e^{-15(0.33)}}$$

$$\approx 508$$

or approximately 508 soldiers.   ■ ■ ■ ■

Exploring with Technology

Refer to Example 6.

1. Use a graphing utility to plot the graph of the function Q, using the viewing rectangle $[0, 40] \times [0, 5000]$.

2. Find how long it takes for the first 1000 soldiers to contract the flu.

Hint: Plot the graphs of $y_1 = Q(t)$ and $y_2 = 1000$ and find the point of intersection of the two graphs.

SELF-CHECK EXERCISE **5.6**

Suppose that the population (in millions) of a country at any time t grows in accordance with the rule

$$P = \left(P_0 + \frac{I}{k}\right)e^{kt} - \frac{I}{k}$$

where P denotes the population at any time t, k is a constant reflecting the natural growth rate of the population, I is a constant giving the (constant) rate of immigration into the country, and P_0 is the total population of the country at time $t = 0$. The population of the United States in the year 1980 ($t = 0$) was 226.5 million. If the natural growth rate is 0.8% annually ($k = 0.008$) and net immigration is allowed at the rate of half a million people per year ($I = 0.5$) until the end of the century, what is the population of the United States expected to be in the year 2005?

Solutions to Self-Check Exercise 5.6 can be found on page 435.

5.6 Exercises

A calculator is recommended for this exercise set.

1. **EXPONENTIAL GROWTH** Given that a quantity $Q(t)$ is described by the exponential growth function

$$Q(t) = 400e^{0.05t}$$

where t is measured in minutes, answer the following questions.
   a. What is the growth constant?
   b. What quantity is present initially?
   c. Using Table 1 in the appendix or a calculator, complete the following table of values:

t	0	10	20	100	1000
Q					

2. **EXPONENTIAL DECAY** Given that a quantity $Q(t)$ exhibiting exponential decay is described by the function

$$Q(t) = 2000e^{-0.06t}$$

where t is measured in years, answer the following questions.
   a. What is the decay constant?
   b. What quantity is present initially?
   c. Using Table 1 in the appendix or a calculator, complete the following table of values:

t	0	5	10	20	100
Q					

3. **GROWTH OF BACTERIA** The growth rate of the bacterium *Escherichia coli*, a common bacterium found in the human intestine, is proportional to its size. Under ideal laboratory conditions, when this bacterium is grown in a nutrient broth medium, the number of cells in a culture doubles approximately every 20 min.
   a. If the initial cell population is 100, determine the function $Q(t)$ that expresses the exponential growth of the number of cells of this bacterium as a function of time t (in minutes).
   b. How long will it take for a colony of 100 cells to increase to a population of 1 million?
   c. If the initial cell population were 1000, how would this alter our model?

4. **WORLD POPULATION** The world population at the beginning of 1990 was 5.3 billion. Assume that the population continues to grow at its present rate of approximately 2%/year and find the function $Q(t)$ that expresses the world population (in billions) as a function of time t (in years) where $t = 0$ corresponds to the beginning of 1990.
   a. Using this function, complete the following table of values and sketch the graph of the function Q.

Year	1990	1995	2000	2005
World Population				

Year	2010	2015	2020	2025
World Population				

   b. Find the estimated rate of growth in the year 2000.

5. **WORLD POPULATION** Refer to Exercise 4.
   a. If the world population continues to grow at its present rate of approximately 2%/year, find the length of time t_0 required for the world population to triple in size.
   b. Using the time t_0 found in part (a), what would be the world population if the growth rate were reduced to 1.8%?

6. **RESALE VALUE** A certain piece of machinery was purchased 3 yr ago by the Garland Mills Company for $500,000. Its present resale value is $320,000. Assuming that the machine's resale value decreases exponentially, what will it be 4 yr from now?

7. **ATMOSPHERIC PRESSURE** If the temperature is constant, then the atmospheric pressure P (in pounds per square inch) varies with the altitude above sea level h in accordance with the law

$$P = p_0 e^{-kh}$$

where p_0 is the atmospheric pressure at sea level and k is a constant. If the atmospheric pressure is 15 lb/in.^{2} at sea level and 12.5 lb/in.^{2} at 4000 ft, find the atmospheric pressure at an altitude of 12,000 ft. How fast is the atmospheric pressure changing with respect to altitude at an altitude of 12,000 ft?

8. **RADIOACTIVE DECAY** The radioactive element polonium decays according to the law

$$Q(t) = Q_0 \cdot 2^{-(t/140)}$$

where Q_0 is the initial amount and the time t is measured in days. If the amount of polonium left after 280 days is 20 mg, what was the initial amount present?

9. **RADIOACTIVE DECAY** Phosphorus 32 has a half-life of 14.2 days. If 100 g of this substance are present initially, find the amount present after t days. What amount will be left after 7.1 days? How fast is the phosphorus 32 decaying when $t = 7.1$?

10. **NUCLEAR FALLOUT** Strontium 90, a radioactive isotope of strontium, is present in the fallout resulting from nuclear explosions. It is especially hazardous to animal life, including humans, because, upon ingestion of contaminated food, it is absorbed into the bone structure. Its half-life is 27 yr. If the amount of strontium 90 in a certain area is found to be four times the "safe" level, find how much time must elapse before an "acceptable level" is reached.

11. **CARBON-14 DATING** Wood deposits recovered from an archeological site contain 20% of the carbon 14 they originally contained. How long ago did the tree from which the wood was obtained die?

12. **CARBON-14 DATING** Skeletal remains of the so-called "Pittsburgh Man," unearthed in Pennsylvania, had lost 82% of the carbon 14 they originally contained. Determine the approximate age of the bones.

13. **LEARNING CURVES** The American Court Reporting Institute finds that the average student taking Advanced Machine Shorthand, an intensive 20-wk course, progresses according to the function

$$Q(t) = 120(1 - e^{-0.05t}) + 60 \qquad (0 \le t \le 20)$$

where $Q(t)$ measures the number of words (per minute) of dictation that the student can take in machine shorthand after t wk in the course. Sketch the graph of the function Q and answer the following questions.
a. What is the beginning shorthand speed for the average student in this course?
b. What shorthand speed does the average student attain halfway through the course?
c. How many words per minute can the average student take after completing this course?

14. **EFFECT OF ADVERTISING ON SALES** The Metro Department Store found that t wk after the end of a sales promotion the volume of sales was given by a function of the form

$$S(t) = B + Ae^{-kt} \qquad (0 \le t \le 4)$$

where $B = 50,000$ and is equal to the average weekly volume of sales before the promotion. The sales volumes at the end of the first and third weeks were \$83,515 and \$65,055, respectively. Assume that the sales volume is decreasing exponentially.
a. Find the decay constant k.
b. Find the sales volume at the end of the fourth week.
c. How fast is the sales volume dropping at the end of the fourth week?

15. **DEMAND FOR COMPUTERS** The Universal Instruments Company found that the monthly demand for its new line of Galaxy Home Computers t mo after placing the line on the market was given by

$$D(t) = 2000 - 1500e^{-0.05t} \qquad (t > 0)$$

Graph this function and answer the following questions.
a. What is the demand after 1 mo? After 1 yr? After 2 yr? After 5 yr?
b. At what level is the demand expected to stabilize?
c. Find the rate of growth of the demand after the tenth month.

16. **NEWTON'S LAW OF COOLING** Newton's law of cooling states that the rate at which the temperature of an object changes is proportional to the difference in temperature between the object and that of the surrounding medium. Thus, the temperature $F(t)$ of an object that is greater than the temperature of its surrounding medium is given by

$$F(t) = T + Ae^{-kt}$$

where t is the time expressed in minutes, T is the temperature of the surrounding medium, and A and k are constants. Suppose a cup of instant coffee is prepared with boiling water (212°F) and left to cool on the counter in a room where the temperature is 72°F. If $k = 0.1865$, determine when the coffee will be cool enough to drink (say, 110°F).

17. **SPREAD OF AN EPIDEMIC** During a flu epidemic, the number of children in the Woodbridge Community School System who contracted influenza after t days was given by

$$Q(t) = \frac{1000}{1 + 199e^{-0.8t}}$$

a. How many children were stricken by the flu after the first day?
b. How many children had the flu after 10 days?
c. How many children eventually contracted the disease?

18. **GROWTH OF A FRUIT-FLY POPULATION** On the basis of data collected during an experiment, a biologist found that the growth of the fruit fly (*Drosophila*) with a limited food supply could be approximated by the exponential model

$$N(t) = \frac{400}{1 + 39e^{-0.16t}}$$

where t denotes the number of days since the beginning of the experiment.
a. What was the initial fruit-fly population in the experiment?

b. What was the maximum fruit-fly population that could be expected under this laboratory condition?
c. What was the population of the fruit-fly colony on the 20th day?
d. How fast was the population changing on the 20th day?

19. PERCENTAGE OF HOUSEHOLDS WITH VCRs According to estimates by Paul Kroger Associates, the percentage of households that own videocassette recorders (VCRs) is given by

$$P(t) = \frac{68}{1 + 21.67e^{-0.62t}} \qquad (0 \le t \le 12)$$

where t is measured in years, with $t = 0$ corresponding to the beginning of 1985. What percentage of households owned VCRs at the beginning of 1985? At the beginning of 1995?

20. POPULATION GROWTH IN THE TWENTY-FIRST CENTURY The U.S. population is approximated by the function

$$P(t) = \frac{616.5}{1 + 4.02e^{-0.5t}}$$

where $P(t)$ is measured in millions of people and t is measured in 30-yr intervals, with $t = 0$ corresponding to 1930. What is the expected population of the United States in 2020 ($t = 3$)?

21. SPREAD OF A RUMOR Three hundred students attended the dedication ceremony of a new building on a college campus. The president of the traditionally female college announced a new expansion program, which included plans to make the college coeducational. The number of students who learned of the new program t hr later is given by the function

$$f(t) = \frac{3000}{1 + Be^{-kt}}$$

If 600 students on campus had heard about the new program 2 hr after the ceremony, how many students had heard about the policy after 4 hr? How fast was the rumor spreading 4 hr after the ceremony?

22. CHEMICAL MIXTURES Two chemicals react to form another chemical. Suppose the amount of the chemical formed in time t (in hours) is given by

$$x(t) = \frac{15\left[1 - \left(\frac{2}{3}\right)^{3t}\right]}{1 - \frac{1}{4}\left(\frac{2}{3}\right)^{3t}}$$

where $x(t)$ is measured in pounds. How many pounds of the chemical are formed eventually?
Hint: You need to evaluate $\lim_{t\to\infty} x(t)$.

23. CONCENTRATION OF GLUCOSE IN THE BLOODSTREAM A glucose solution is administered intravenously into the bloodstream at a constant rate of r mg/hr. As the glucose is being administered, it is converted into other substances and removed from the bloodstream. Suppose the concentration of the glucose solution at time t is given by

$$C(t) = \frac{r}{k} - \left[\left(\frac{r}{k}\right) - C_0\right]e^{-kt}$$

where C_0 is the concentration at time $t = 0$ and k is a constant.
a. Assuming that $C_0 < r/k$, evaluate

$$\lim_{t\to\infty} C(t)$$

and interpret your result.
b. Sketch the graph of the function C.

24. GOMPERTZ GROWTH CURVE Consider the function

$$Q(t) = Ce^{-Ae^{-kt}}$$

where $Q(t)$ is the size of a quantity at time t and A, C, and k are positive constants. The graph of this function, called the **Gompertz growth curve,** is used by biologists to describe restricted population growth.
a. Show that the function Q is always increasing.
b. Find the time t at which the growth rate $Q'(t)$ is increasing most rapidly.
Hint: Find the inflection point of Q.
c. Show that $\lim_{t\to\infty} Q(t) = C$ and interpret your result.

SOLUTION TO SELF-CHECK EXERCISE 5.6

We are given that $P_0 = 226.5$, $k = 0.008$, and $I = 0.5$. So

$$P = \left(226.5 + \frac{0.5}{0.008}\right)e^{0.008t} - \frac{0.5}{0.008}$$

$$= 289e^{0.008t} - 62.5$$

Therefore, the population in the year 2005 will be given by

$$P(25) = 289e^{0.2} - 62.5$$
$$\approx 290.5$$

or approximately 290.5 million.

CHAPTER 5 Summary of Principal Formulas and Terms

Formulas

1. Exponential function with base b	$y = b^x$
2. The number e	$e = \lim\limits_{m \to \infty} \left(1 + \dfrac{1}{m}\right)^m = 2.71828$
3. Exponential function with base e	$y = e^x$
4. Logarithmic function with base b	$y = \log_b x$
5. Logarithmic function with base e	$y = \ln x$
6. Inverse properties of $\ln x$ and e	$\ln e^x = x$ and $e^{\ln x} = x$
7. Compound interest (accumulated amount)	$A = P(1 + i)^n$, where $i = r/m$ and $n = mt$
8. Effective rate of interest	$r_{\text{eff}} = \left(1 + \dfrac{r}{m}\right)^m - 1$
9. Compound interest (present value)	$P = A(1 + i)^{-n}$, where $i = r/m$ and $n = mt$
10. Continuous compound interest	$A = Pe^{rt}$
11. Derivative of the exponential function	$\dfrac{d}{dx}(e^x) = e^x$
12. Chain rule for exponential functions	$\dfrac{d}{dx}(e^u) = e^u \dfrac{du}{dx}$
13. Derivative of the logarithmic function	$\dfrac{d}{dx} \ln |x| = \dfrac{1}{x}$
14. Chain rule for logarithmic functions	$\dfrac{d}{dx}(\ln u) = \dfrac{1}{u}\dfrac{du}{dx}$

Terms

common logarithm

natural logarithm

logarithmic differentiation

exponential growth

growth constant

exponential decay

decay constant

half-life of a radioactive element

logistic growth function

CHAPTER 5 REVIEW EXERCISES

1. Sketch on the same set of coordinate axes the graphs of the exponential functions defined by the equations.

   a. $y = 2^{-x}$ **b.** $y = \left(\dfrac{1}{2}\right)^x$

In Exercises 2 and 3, express each in logarithmic form.

2. $\left(\dfrac{2}{3}\right)^{-3} = \dfrac{27}{8}$ 3. $16^{-3/4} = 0.125$

In Exercises 4 and 5, solve each equation for x.

4. $\log_4(2x + 1) = 2$

5. $\ln(x - 1) + \ln 4 = \ln(2x + 4) - \ln 2$

In Exercises 6–8, given that ln 2 = x, ln 3 = y, and ln 5 = z, express each of the given logarithmic values in terms of x, y, and z.

6. $\ln 30$ 7. $\ln 3.6$ 8. $\ln 75$

9. Sketch the graph of the function $y = \log_2(x + 3)$.

10. Sketch the graph of the function $y = \log_3(x + 1)$.

In Exercises 11–28, find the derivative of the function.

11. $f(x) = xe^{2x}$ 12. $f(t) = \sqrt{t}e^t + t$

13. $g(t) = \sqrt{t}e^{-2t}$ 14. $g(x) = e^x\sqrt{1 + x^2}$

15. $y = \dfrac{e^{2x}}{1 + e^{-2x}}$ 16. $f(x) = e^{2x^2 - 1}$

17. $f(x) = xe^{-x^2}$ 18. $g(x) = (1 + e^{2x})^{3/2}$

19. $f(x) = x^2 e^x + e^x$ 20. $g(t) = t\ln t$

21. $f(x) = \ln(e^{x^2} + 1)$ 22. $f(x) = \dfrac{x}{\ln x}$

23. $f(x) = \dfrac{\ln x}{x + 1}$ 24. $y = (x + 1)e^x$

25. $y = \ln(e^{4x} + 3)$ 26. $f(r) = \dfrac{re^r}{1 + r^2}$

27. $f(x) = \dfrac{\ln x}{1 + e^x}$ 28. $g(x) = \dfrac{e^{x^2}}{1 + \ln x}$

29. Find the second derivative of the function $y = \ln(3x + 1)$.

30. Find the second derivative of the function $y = x\ln x$.

31. Find $h'(0)$ if $h(x) = g(f(x))$, $g(x) = x + (1/x)$, and $f(x) = e^x$.

32. Find $h'(1)$ if $h(x) = g(f(x))$, $g(x) = \dfrac{x + 1}{x - 1}$, and $f(x) = \ln x$.

33. Use logarithmic differentiation to find the derivative of $f(x) = (2x^3 + 1)(x^2 + 2)^3$.

34. Use logarithmic differentiation to find the derivative of $f(x) = \dfrac{x(x^2 - 2)^2}{(x - 1)}$.

35. Find an equation of the tangent line to the graph of $y = e^{-2x}$ at the point $(1, e^{-2})$.

36. Find an equation of the tangent line to the graph of $y = xe^{-x}$ at the point $(1, e^{-1})$.

37. Sketch the graph of the function $f(x) = xe^{-2x}$.

38. Sketch the graph of the function $f(x) = x^2 - \ln x$.

39. Find the absolute extrema of the function $f(t) = te^{-t}$.

40. Find the absolute extrema of the function
$$g(t) = \dfrac{\ln t}{t}$$
on [1, 2].

41. A hotel was purchased by a conglomerate for $4.5 million and sold 5 yr later for $8.2 million. Find the annual rate of return (compounded continuously).

42. Find the present value of $119,346 due in 4 yr at an interest rate of 10%/year compounded continuously.

43. A culture of bacteria that initially contained 2000 bacteria has a count of 18,000 bacteria after 2 hr.
a. Determine the function $Q(t)$ that expresses the exponential growth of the number of cells of this bacterium as a function of time t (in minutes).
b. Find the number of bacteria present after 4 hr.

44. The radioactive element radium has a half-life of 1600 yr. What is its decay constant?

45. The VCA Television Company found that the monthly demand for its new line of video disc players t mo after placing the players on the market is given by
$$D(t) = 4000 - 3000e^{-0.06t} \qquad (t \geq 0)$$
Graph this function and answer the following questions.
a. What was the demand after 1 mo? After 1 yr? After 2 yr?
b. At what level is the demand expected to stabilize?

46. During a flu epidemic, the number of students at a certain university who contracted influenza after t days could be approximated by the exponential model
$$Q(t) = \dfrac{3000}{1 + 499e^{-kt}}$$
If 90 students contracted the flu by day 10, how many students contracted the flu by day 20?

INTEGRATION

6

Differential calculus is concerned with the problem of finding the rate of change of one quantity with respect to another. In this chapter we begin the study of the other branch of calculus, known as integral calculus. Here we are interested in precisely the opposite problem: If we know the rate of change of one quantity with respect to another, can we find the relationship between the two quantities? The principal tool used in the study of integral calculus is the *antiderivative* of a function, and we develop rules for antidifferentiation, or *integration*, as the process of finding the antiderivative is called. We also show that a link is established between differential and integral calculus—via the fundamental theorem of calculus.

How much will the solar cell panels cost? The head of Soloron Corporation's research and development department has projected that the cost of producing solar cell panels will drop at a certain rate in the next several years. In Example 7, page 460, you will see how this information can be used to predict the cost of solar cell panels in the coming years.

6.1 Antiderivatives and the Rules of Integration

ANTIDERIVATIVES

Let's return, once again, to the example involving the motion of the maglev (Figure 6.1).

FIGURE 6.1
A maglev moving along an elevated monorail track

In Chapter 2, we discussed the following problem:

If we know the position of the maglev at any time t, can we find its velocity at time t?

As it turns out, if the position of the maglev is described by the position function f, then its velocity at any time t is given by $f'(t)$. Here f'—the velocity function of the maglev—is just the derivative of f.

Now, in Chapters 6 and 7, we will consider precisely the opposite problem:

If we know the velocity of the maglev at any time t, can we find its position at time t?

Stated another way, if we know the velocity function f' of the maglev, can we find its position function f?

To solve this problem, we need the concept of an antiderivative of a function.

Antiderivative

> A function F is an **antiderivative** of f on an interval I if $F'(x) = f(x)$ for all x in I.

Thus, an antiderivative of a function f is a function F whose derivative is f. For example, $F(x) = x^2$ is an antiderivative of $f(x) = 2x$ because

$$F'(x) = \frac{d}{dx}(x^2) = 2x = f(x)$$

and $F(x) = x^3 + 2x + 1$ is an antiderivative of $f(x) = 3x^2 + 2$ because

$$F'(x) = \frac{d}{dx}(x^3 + 2x + 1) = 3x^2 + 2 = f(x)$$

EXAMPLE 1   Let $F(x) = \frac{1}{3}x^3 - 2x^2 + x - 1$. Show that F is an antiderivative of $f(x) = x^2 - 4x + 1$.

SOLUTION ✔   Differentiating the function F, we obtain

$$F'(x) = x^2 - 4x + 1 = f(x)$$

and the desired result follows.   ■ ■ ■ ■

EXAMPLE 2   Let $F(x) = x$, $G(x) = x + 2$, and $H(x) = x + C$, where C is a constant. Show that F, G, and H are all antiderivatives of the function f defined by $f(x) = 1$.

SOLUTION ✔   Since

$$F'(x) = \frac{d}{dx}(x) = 1 = f(x)$$

$$G'(x) = \frac{d}{dx}(x + 2) = 1 = f(x)$$

$$H'(x) = \frac{d}{dx}(x + C) = 1 = f(x)$$

we see that F, G, and H are indeed antiderivatives of f.   ■ ■ ■ ■

Example 2 shows that once an antiderivative G of a function f is known, then another antiderivative of f may be found by adding an arbitrary constant to the function G. The following theorem states that no function other than one obtained in this manner can be an antiderivative of f. (We omit the proof.)

THEOREM 1   Let G be an antiderivative of a function f. Then, every antiderivative F of f must be of the form $F(x) = G(x) + C$, where C is a constant.

Returning to Example 2, we see that there are infinitely many antiderivatives of the function $f(x) = 1$. We obtain each one by specifying the constant C in the function $F(x) = x + C$. Figure 6.2 shows the graphs of some of these

FIGURE 6.2

The graphs of some antiderivatives of $f(x) = 1$

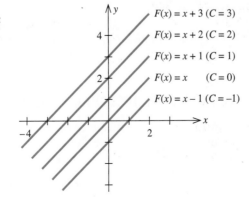

antiderivatives for selected values of C. These graphs constitute part of a family of infinitely many parallel straight lines, each having a slope equal to 1. This result is expected since there are infinitely many curves (straight lines) with a given slope equal to 1. The antiderivatives $F(x) = x + C$ (C, a constant) are precisely the functions representing this family of straight lines.

EXAMPLE 3   Prove that the function $G(x) = x^2$ is an antiderivative of the function $f(x) = 2x$. Write a general expression for the antiderivatives of f.

SOLUTION ✔   Since $G'(x) = 2x = f(x)$, we have shown that $G(x) = x^2$ is an antiderivative of $f(x) = 2x$. By Theorem 1, every antiderivative of the function $f(x) = 2x$ has the form $F(x) = x^2 + C$, where C is some constant. The graphs of a few of the antiderivatives of f are shown in Figure 6.3.

FIGURE 6.3
The graphs of some antiderivatives of $f(x) = 2x$

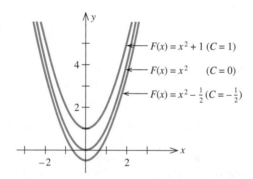

$F(x) = x^2 + 1$ ($C = 1$)
$F(x) = x^2$ ($C = 0$)
$F(x) = x^2 - \frac{1}{2}$ ($C = -\frac{1}{2}$)

Exploring with Technology

Let $f(x) = x^2 - 1$.

1. Show that $F(x) = \frac{1}{3}x^3 - x + C$, where C is an arbitrary constant, is an antiderivative of f.
2. Use a graphing utility to plot the graphs of the antiderivatives of f corresponding to $C = -2$, $C = -1$, $C = 0$, $C = 1$, and $C = 2$ on the same set of axes using the viewing rectangle $[-4, 4] \times [-4, 4]$.
3. If your graphing utility has the capability, draw the tangent line to each of the graphs in part 2 at the point whose x-coordinate is 2. What can you say about this family of tangent lines?
4. What is the slope of a tangent line in this family? Explain how you obtained your answer.

THE INDEFINITE INTEGRAL

The process of finding all antiderivatives of a function is called **antidifferentiation,** or **integration.** We use the symbol $\int$, called an **integral sign,** to indicate that the operation of integration is to be performed on some function f. Thus,

$$\int f(x)\, dx = F(x) + C$$

[read "the indefinite integral of $f(x)$ with respect to x equals $F(x)$ plus C"] tells us that the **indefinite integral** of f is the family of functions given by $F(x) + C$, where $F'(x) = f(x)$. The function f to be integrated is called the **integrand,** and the constant C is called a **constant of integration.** The expression dx following the integrand $f(x)$ reminds us that the operation is performed with respect to x. If the independent variable is t, we write $\int f(t)\, dt$ instead. In this sense both t and x are "dummy variables."

Using this notation, we can write the results of Examples 2 and 3 as

$$\int 1\, dx = x + C \qquad \text{and} \qquad \int 2x\, dx = x^2 + K$$

where C and K are arbitrary constants.

BASIC INTEGRATION RULES

Our next task is to develop some rules for finding the indefinite integral of a given function f. Because integration and differentiation are reverse operations, we discover many of the rules of integration by first making an "educated guess" at the antiderivative F of the function f to be integrated. Then this result is verified by demonstrating that $F' = f$.

Rule 1: The Indefinite Integral of a Constant

$$\int k\, dx = kx + C \qquad (k, \text{a constant})$$

To prove this result, observe that

$$F'(x) = \frac{d}{dx}(kx + C) = k$$

EXAMPLE 4

Find each of the following indefinite integrals:

a. $\displaystyle\int 2\, dx$ **b.** $\displaystyle\int \pi^2\, dx$

SOLUTION ✔

Each of the integrands has the form $f(x) = k$, where k is a constant. Applying Rule 1 in each case yields

a. $\displaystyle\int 2\, dx = 2x + C$ **b.** $\displaystyle\int \pi^2\, dx = \pi^2 x + C$ ■ ■ ■ ■

Next, from the rule of differentiation,

$$\frac{d}{dx}x^n = nx^{n-1}$$

we obtain the following rule of integration.

**Rule 2: The
Power Rule**

$$\int x^n \, dx = \frac{1}{n+1} x^{n+1} + C \qquad (n \ne -1)$$

An antiderivative of a power function is another power function obtained from the integrand by increasing its power by 1 and dividing the resulting expression by the new power.

To prove this result, observe that

$$
\begin{aligned}
F'(x) &= \frac{d}{dx}\left[\frac{1}{n+1} x^{n+1} + C\right] \\
&= \frac{n+1}{n+1} x^n \\
&= x^n \\
&= f(x)
\end{aligned}
$$

EXAMPLE 5   Find each of the following indefinite integrals:

a. $\int x^3 \, dx$ **b.** $\int x^{3/2} \, dx$ **c.** $\int \frac{1}{x^{3/2}} \, dx$

SOLUTION ✔   Each integrand is a power function with exponent $n \ne -1$. Applying Rule 2 in each case yields the following results:

a. $\int x^3 \, dx = \frac{1}{4} x^4 + C$

b. $\int x^{3/2} \, dx = \frac{1}{\frac{5}{2}} x^{5/2} + C = \frac{2}{5} x^{5/2} + C$

c. $\int \frac{1}{x^{3/2}} \, dx = \int x^{-3/2} \, dx = \frac{1}{-\frac{1}{2}} x^{-1/2} + C = -2x^{-1/2} + C$

These results may be verified by differentiating each of the antiderivatives and showing that the result is equal to the corresponding integrand.

■ ■ ■ ■

The next rule tells us that a constant factor may be moved through an integral sign.

**Rule 3: The Indefinite
Integral of a
Constant Multiple
of a Function**

$$\int cf(x) \, dx = c \int f(x) \, dx \qquad (c, \text{a constant})$$

The indefinite integral of a constant multiple of a function is equal to the constant multiple of the indefinite integral of the function.

This result follows from the corresponding rule of differentiation (see Rule 3, Section 3.1).

 Only a constant can be "moved out" of an integral sign. For example, it would be incorrect to write

$$\int x^2 \, dx = x^2 \int 1 \, dx$$

In fact, $\int x^2 \, dx = \frac{1}{3}x^3 + C$, whereas $x^2 \int 1 \, dx = x^2(x + C) = x^3 + Cx^2$.

EXAMPLE 6   Find each of the following indefinite integrals:

a. $\int 2t^3 \, dt$ **b.** $\int -3x^{-2} \, dx$

SOLUTION ✔   Each integrand has the form $cf(x)$, where c is a constant. Applying Rule 3, we obtain:

a. $\int 2t^3 \, dt = 2 \int t^3 \, dt = 2 \left[\frac{1}{4}t^4 + K \right] = \frac{1}{2}t^4 + 2K = \frac{1}{2}t^4 + C$

where $C = 2K$. From now on, we will write the constant of integration as C, since any nonzero multiple of an arbitrary constant is an arbitrary constant.

b. $\int -3x^{-2} \, dx = -3 \int x^{-2} \, dx = (-3)(-1)x^{-1} + C = \frac{3}{x} + C$ ■ ■ ■ ■

Rule 4: The Sum Rule

$$\int [f(x) + g(x)] \, dx = \int f(x) \, dx + \int g(x) \, dx$$

$$\int [f(x) - g(x)] \, dx = \int f(x) \, dx - \int g(x) \, dx$$

The indefinite integral of a sum (difference) of two integrable functions is equal to the sum (difference) of their indefinite integrals.

This result is easily extended to the case involving the sum and difference of any finite number of functions. As in Rule 3, the proof of Rule 4 follows from the corresponding rule of differentiation (see Rule 4, Section 3.1).

EXAMPLE 7   Find the indefinite integral

$$\int (3x^5 + 4x^{3/2} - 2x^{-1/2}) \, dx$$

SOLUTION ✔   Applying the extended version of Rule 4, we find that

$$\int (3x^5 + 4x^{3/2} - 2x^{-1/2})\, dx$$

$$= \int 3x^5\, dx + \int 4x^{3/2}\, dx - \int 2x^{-1/2}\, dx$$

$$= 3\int x^5\, dx + 4\int x^{3/2}\, dx - 2\int x^{-1/2}\, dx \qquad \text{(Rule 3)}$$

$$= (3)\left(\frac{1}{6}\right)x^6 + (4)\left(\frac{2}{5}\right)x^{5/2} - (2)(2)x^{1/2} + C \qquad \text{(Rule 2)}$$

$$= \frac{1}{2}x^6 + \frac{8}{5}x^{5/2} - 4x^{1/2} + C$$

■■■■

Observe that we have combined the three constants of integration, which arise from evaluating the three indefinite integrals, to obtain one constant C. After all, the sum of three arbitrary constants is also an arbitrary constant.

Rule 5: The Indefinite Integral of the Exponential Function

$$\int e^x\, dx = e^x + C$$

The indefinite integral of the exponential function with base e is equal to the function itself (except, of course, for the constant of integration).

EXAMPLE **8**   Find the indefinite integral

$$\int (2e^x - x^3)\, dx$$

SOLUTION ✔   We have

$$\int (2e^x - x^3)\, dx = \int 2e^x\, dx - \int x^3\, dx$$

$$= 2\int e^x\, dx - \int x^3\, dx$$

$$= 2e^x - \frac{1}{4}x^4 + C$$

■■■■

The last rule of integration in this section covers the integration of the function $f(x) = x^{-1}$. Remember that this function constituted the only exceptional case in the integration of the power function $f(x) = x^n$ (see Rule 2).

Rule 6: The Indefinite Integral of the Function $f(x) = x^{-1}$

$$\int x^{-1}\, dx = \int \frac{1}{x}\, dx = \ln |x| + C \qquad (x \neq 0)$$

To prove Rule 6, observe that

$$\frac{d}{dx}\ln|x| = \frac{1}{x} \qquad \text{(See Rule 3, Section 5.5.)}$$

EXAMPLE 9   Find the indefinite integral

$$\int\left(2x + \frac{3}{x} + \frac{4}{x^2}\right)dx$$

SOLUTION ✔

$$\int\left(2x + \frac{3}{x} + \frac{4}{x^2}\right)dx = \int 2x\,dx + \int\frac{3}{x}\,dx + \int\frac{4}{x^2}\,dx$$

$$= 2\int x\,dx + 3\int\frac{1}{x}\,dx + 4\int x^{-2}\,dx$$

$$= 2\left(\frac{1}{2}\right)x^2 + 3\ln|x| + 4(-1)x^{-1} + C$$

$$= x^2 + 3\ln|x| - \frac{4}{x} + C \qquad \blacksquare\blacksquare\blacksquare\blacksquare$$

DIFFERENTIAL EQUATIONS

Let's return to the problem posed at the beginning of the section: *Given the derivative of a function f′, can we find the function f?* As an example, suppose we are given the function

$$f'(x) = 2x - 1 \tag{1}$$

and we wish to find $f(x)$. From what we now know, we can find f by integrating Equation (1). Thus,

$$f(x) = \int f'(x)\,dx = \int(2x - 1)\,dx = x^2 - x + C \tag{2}$$

where C is an arbitrary constant. Thus, infinitely many functions have the derivative f', each differing from the other by a constant.

Equation (1) is called a differential equation. In general, a **differential equation** is an equation that involves the derivative or differential of an unknown function. [In the case of Equation (1), the unknown function is f.] A **solution** of a differential equation is any function that satisfies the differential equation. Thus, Equation (2) gives *all* the solutions of the differential Equation (1), and it is, accordingly, called the **general solution** of the differential equation $f'(x) = 2x - 1$.

The graphs of $f(x) = x^2 - x + C$ for selected values of C are shown in Figure 6.4. These graphs have one property in common: For any fixed value of x, the tangent lines to these graphs have the same slope. This follows

FIGURE 6.4
The graphs of some of the functions having the derivative $f'(x) = 2x - 1$. Observe that the slopes of the tangent lines to the graphs are the same for a fixed value of x.

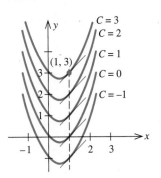

because any member of the family $f(x) = x^2 - x + C$ must have the same slope at x—namely, $2x - 1$!

Although there are infinitely many solutions to the differential equation $f'(x) = 2x - 1$, we can obtain a **particular solution** by specifying the value the function must assume at a certain value of x. For example, suppose we stipulate that the function f under consideration must satisfy the condition $f(1) = 3$ or, equivalently, the graph of f must pass through the point $(1, 3)$. Then, using the condition on the general solution $f(x) = x^2 - x + C$, we find that

$$f(1) = 1 - 1 + C = 3$$

and $C = 3$. Thus, the particular solution is $f(x) = x^2 - x + 3$ (see Figure 6.4).

The condition $f(1) = 3$ is an example of an initial condition. More generally, an **initial condition** is a condition imposed on the value of f at a point $x = a$.

INITIAL VALUE PROBLEMS

An **initial value problem** is one in which we are required to find a function satisfying (1) a differential equation and (2) one or more initial conditions. The following are examples of initial value problems.

EXAMPLE 10   Find the function f if it is known that

$$f'(x) = 3x^2 - 4x + 8 \qquad \text{and} \qquad f(1) = 9$$

SOLUTION ✔   We are required to solve the initial value problem

$$\left.\begin{array}{l} f'(x) = 3x^2 - 4x + 8 \\ f(1) = 9 \end{array}\right\}$$

Integrating the function f', we find

$$f(x) = \int f'(x)\, dx$$
$$= \int (3x^2 - 4x + 8)\, dx$$
$$= x^3 - 2x^2 + 8x + C$$

Using the condition $f(1) = 9$, we have

$$9 = f(1) = 1^3 - 2(1)^2 + 8(1) + C = 7 + C$$
$$C = 2$$

Therefore, the required function f is given by $f(x) = x^3 - 2x^2 + 8x + 2$.

■ ■ ■ ■

APPLICATIONS

EXAMPLE **11**

In a test run of a maglev along a straight elevated monorail track, data obtained from reading its speedometer indicate that the velocity of the maglev at time t can be described by the velocity function

$$v(t) = 8t \qquad (0 \le t \le 30)$$

Find the position function of the maglev. Assume that initially the maglev is located at the origin of a coordinate line.

SOLUTION ✔

Let $s(t)$ denote the position of the maglev at any time t $(0 \le t \le 30)$. Then, $s'(t) = v(t)$. So, we have the initial value problem

$$\left.\begin{array}{l} s'(t) = 8t \\ s(0) = 0 \end{array}\right\}$$

Integrating both sides of the differential equation $s'(t) = 8t$, we obtain

$$s(t) = \int s'(t)\, dt = \int 8t\, dt = 4t^2 + C$$

where C is an arbitrary constant. To evaluate C, we use the initial condition $s(0) = 0$ to write

$$s(0) = 4(0) + C = 0 \qquad \text{or} \qquad C = 0$$

Therefore, the required position function is $s(t) = 4t^2$ $(0 \le t \le 30)$.

■ ■ ■ ■

EXAMPLE **12**

The current circulation of the *Investor's Digest* is 3000 copies per week. The managing editor of the weekly projects a growth rate of

$$4 + 5t^{2/3}$$

copies per week, t weeks from now, for the next 3 years. Based on her projection, what will the circulation of the digest be 125 weeks from now?

SOLUTION ✔

Let $S(t)$ denote the circulation of the digest t weeks from now. Then $S'(t)$ is the rate of change in the circulation in the tth week and is given by

$$S'(t) = 4 + 5t^{2/3}$$

Furthermore, the current circulation of 3000 copies per week translates into the initial condition $S(0) = 3000$. Integrating the differential equation with respect to t gives

$$S(t) = \int S'(t)\, dt = \int (4 + 5t^{2/3})\, dt$$

$$= 4t + 5\left(\frac{t^{5/3}}{\frac{5}{3}}\right) + C = 4t + 3t^{5/3} + C$$

To determine the value of C, we use the condition $S(0) = 3000$ to write

$$S(0) = 4(0) + 3(0) + C = 3000$$

which gives $C = 3000$. Therefore, the circulation of the digest t weeks from now will be

$$S(t) = 4t + 3t^{5/3} + 3000$$

In particular, the circulation 125 weeks from now will be

$$S(125) = 4(125) + 3(125)^{5/3} + 3000 = 12{,}875$$

copies per week.   ■ ■ ■ ■

SELF-CHECK EXERCISES 6.1

1. Evaluate $\displaystyle\int \left(\frac{1}{\sqrt{x}} - \frac{2}{x} + 3e^x \right) dx$.

2. Find the rule for the function f given that (1) the slope of the tangent line to the graph of f at any point $P(x, f(x))$ is given by the expression $3x^2 - 6x + 3$ and (2) the graph of f passes through the point $(2, 9)$.

3. Suppose United Motors' share of the new cars sold in a certain country is changing at the rate of

$$f(t) = -0.01875t^2 + 0.15t - 1.2 \qquad (0 \le t \le 12)$$

percent at year t ($t = 0$ corresponds to the beginning of 1988). The company's market share at the beginning of 1988 was 48.4%. What was United Motors' market share at the beginning of 2000?

Solutions to Self-Check Exercises 6.1 can be found on page 454.

Exercises

In Exercises 1–4, verify directly that *F* is an antiderivative of *f*.

1. $F(x) = \dfrac{1}{3} x^3 + 2x^2 - x + 2; f(x) = x^2 + 4x - 1$

2. $F(x) = xe^x + \pi; f(x) = e^x(1 + x)$

3. $F(x) = \sqrt{2x^2 - 1}; f(x) = \dfrac{2x}{\sqrt{2x^2 - 1}}$

4. $F(x) = x \ln x - x; f(x) = \ln x$

In Exercises 5–8, (a) verify that *G* is an antiderivative of *f*, (b) find all antiderivatives of *f*, and (c) sketch the graphs of a few of the family of antiderivatives found in part (b).

5. $G(x) = 2x; f(x) = 2$ 6. $G(x) = 2x^2; f(x) = 4x$

7. $G(x) = \dfrac{1}{3} x^3; f(x) = x^2$ 8. $G(x) = e^x; f(x) = e^x$

In Exercises 9–50, find the indefinite integral.

9. $\displaystyle\int 6\, dx$

10. $\displaystyle\int \sqrt{2}\, dx$

11. $\displaystyle\int x^3\, dx$

12. $\displaystyle\int 2x^5\, dx$

13. $\displaystyle\int x^{-4}\, dx$

14. $\displaystyle\int 3t^{-7}\, dt$

15. $\displaystyle\int x^{2/3}\, dx$

16. $\displaystyle\int 2u^{3/4}\, du$

17. $\displaystyle\int x^{-5/4}\, dx$

18. $\displaystyle\int 3x^{-2/3}\, dx$

19. $\displaystyle\int \frac{2}{x^2}\, dx$

20. $\displaystyle\int \frac{1}{3x^5}\, dx$

21. $\int \pi \sqrt{t}\, dt$

22. $\int \dfrac{3}{\sqrt{t}}\, dt$

23. $\int (3 - 2x)\, dx$

24. $\int (1 + u + u^2)\, du$

25. $\int (x^2 + x + x^{-3})\, dx$

26. $\int (0.3t^2 + 0.02t + 2)\, dt$

27. $\int 4e^x\, dx$

28. $\int (1 + e^x)\, dx$

29. $\int (1 + x + e^x)\, dx$

30. $\int (2 + x + 2x^2 + e^x)\, dx$

31. $\int \left(4x^3 - \dfrac{2}{x^2} - 1\right) dx$

32. $\int \left(6x^3 + \dfrac{3}{x^2} - x\right) dx$

33. $\int (x^{5/2} + 2x^{3/2} - x)\, dx$

34. $\int (t^{3/2} + 2t^{1/2} - 4t^{-1/2})\, dt$

35. $\int \left(\sqrt{x} + \dfrac{3}{\sqrt{x}}\right) dx$

36. $\int \left(\sqrt[3]{x^2} - \dfrac{1}{x^2}\right) dx$

37. $\int \left(\dfrac{u^3 + 2u^2 - u}{3u}\right) du$

Hint: $\dfrac{u^3 + 2u^2 - u}{3u} = \dfrac{1}{3}u^2 + \dfrac{2}{3}u - \dfrac{1}{3}$

38. $\int \dfrac{x^4 - 1}{x^2}\, dx$

Hint: $\dfrac{x^4 - 1}{x^2} = x^2 - x^{-2}$

39. $\int (2t + 1)(t - 2)\, dt$

40. $\int u^{-2}(1 - u^2 + u^4)\, du$

41. $\int \dfrac{1}{x^2}(x^4 - 2x^2 + 1)\, dx$

42. $\int \sqrt{t}(t^2 + t - 1)\, dt$

43. $\int \dfrac{ds}{(s + 1)^{-2}}$

44. $\int \left(\sqrt{x} + \dfrac{3}{x} - 2e^x\right) dx$

45. $\int (e^t + t^e)\, dt$

46. $\int \left(\dfrac{1}{x^2} - \dfrac{1}{\sqrt[3]{x^2}} + \dfrac{1}{\sqrt{x}}\right) dx$

47. $\int \left(\dfrac{x^3 + x^2 - x + 1}{x^2}\right) dx$

Hint: Simplify the integrand first.

48. $\int \dfrac{t^3 + \sqrt[3]{t}}{t^2}\, dt$

Hint: Simplify the integrand first.

49. $\int \dfrac{(\sqrt{x} - 1)^2}{x^2}\, dx$

Hint: Simplify the integrand first.

50. $\int (x + 1)^2 \left(1 - \dfrac{1}{x}\right) dx$

Hint: Simplify the integrand first.

In Exercises 51–58, find f(x) by solving the initial value problem.

51. $f'(x) = 2x + 1;\ f(1) = 3$

52. $f'(x) = 3x^2 - 6x;\ f(2) = 4$

53. $f'(x) = 3x^2 + 4x - 1;\ f(2) = 9$

54. $f'(x) = \dfrac{1}{\sqrt{x}};\ f(4) = 2$

55. $f'(x) = 1 + \dfrac{1}{x^2};\ f(1) = 2$

56. $f'(x) = e^x - 2x;\ f(0) = 2$

57. $f'(x) = \dfrac{x + 1}{x};\ f(1) = 1$

58. $f'(x) = 1 + e^x + \dfrac{1}{x};\ f(1) = 3 + e$

In Exercises 59–62, find the function f given that the slope of the tangent line to the graph of f at any point (x, f(x)) is f'(x) and that the graph of f passes through the given point.

59. $f'(x) = \dfrac{1}{2} x^{-1/2};\ (2, \sqrt{2})$

60. $f'(t) = t^2 - 2t + 3;\ (1, 2)$

61. $f'(x) = e^x + x;\ (0, 3)$ **62.** $f'(x) = \dfrac{2}{x} + 1;\ (1, 2)$

63. VELOCITY OF A CAR The velocity of a car (in feet/second) t sec after starting from rest is given by the function

$$f(t) = 2\sqrt{t} \qquad (0 \le t \le 30)$$

Find the car's position at any time t.

64. VELOCITY OF A MAGLEV The velocity (in feet/second) of a maglev is

$$v(t) = 0.2t + 3 \qquad (0 \le t \le 120)$$

At $t = 0$, it is at the station. Find the function giving the position of the maglev at time t, assuming that the motion takes place along a straight stretch of track.

65. COST OF PRODUCING CLOCKS The Lorimar Watch Company manufactures travel clocks. The daily marginal cost function associated with producing these clocks is

$$C'(x) = 0.000009x^2 - 0.009x + 8$$

where $C'(x)$ is measured in dollars/unit and x denotes the number of units produced. Management has determined that the daily fixed cost incurred in producing these clocks is $120. Find the total cost incurred by Lorimar in producing the first 500 travel clocks/day.

66. **REVENUE FUNCTIONS** The management of the Lorimar Watch Company has determined that the daily marginal revenue function associated with producing and selling their travel clocks is given by

$$R'(x) = -0.009x + 12$$

where x denotes the number of units produced and sold and $R'(x)$ is measured in dollars/unit.
a. Determine the revenue function $R(x)$ associated with producing and selling these clocks.
b. What is the demand equation that relates the wholesale unit price with the quantity of travel clocks demanded?

67. **PROFIT FUNCTIONS** The Cannon Precision Instruments Corporation makes an automatic electronic flash with Thyrister circuitry. The estimated marginal profit associated with producing and selling these electronic flashes is

$$(-0.004x + 20)$$

dollars/unit/month when the production level is x units per month. Cannon's fixed cost for producing and selling these electronic flashes is $16,000/month. At what level of production does Cannon realize a maximum profit? What is the maximum monthly profit?

68. **COST OF PRODUCING GUITARS** The Carlota Music Company estimates that the marginal cost of manufacturing its Professional Series guitars is

$$C'(x) = 0.002x + 100$$

dollars/month when the level of production is x guitars/month. The fixed costs incurred by Carlota are $4000/month. Find the total monthly cost incurred by Carlota in manufacturing x guitars/month.

69. **QUALITY CONTROL** As part of a quality-control program, the chess sets manufactured by the Jones Brothers Company are subjected to a final inspection before packing. The rate of increase in the number of sets checked per hour by an inspector t hr into the 8 A.M. to 12 noon morning shift is approximately

$$N'(t) = -3t^2 + 12t + 45 \qquad (0 \le t \le 4)$$

a. Find an expression $N(t)$ that approximates the number of sets inspected at the end of t hours.
Hint: $N(0) = 0$.
b. How many sets does the average inspector check during a morning shift?

70. **BALLAST DROPPED FROM A BALLOON** A ballast is dropped from a stationary hot-air balloon that is hovering at an altitude of 400 ft. Its velocity after t sec is $-32t$ ft/sec.
a. Find the height $h(t)$ of the ballast from the ground at time t.
Hint: $h'(t) = -32t$ and $h(0) = 400$.
b. When will the ballast strike the ground?
c. Find the velocity of the ballast when it hits the ground.

Ballast

■ **A calculator is recommended for Exercises 71–76.**

71. **CABLE TV SUBSCRIBERS** A study conducted by Tele-Cable, Inc., estimates that the number of cable television subscribers will grow at the rate of

$$100 + 210t^{3/4}$$

new subscribers/month t mo from the start date of the service. If 5000 subscribers signed up for the service before the starting date, how many subscribers will there be 16 mo from that date?

72. **AIR POLLUTION** On an average summer day, the level of carbon monoxide (CO) in a city's air is 2 parts per million (ppm). An environmental protection agency's study predicts that, unless more stringent measures are taken to protect the city's atmosphere, the CO concentration present in the air will increase at the rate of

$$0.003t^2 + 0.06t + 0.1$$

ppm/year t yr from now. If no further pollution-control efforts are made, what will be the CO concentration on an average summer day 5 yr from now?

73. POPULATION GROWTH The development of Astro World ("The Amusement Park of the Future") on the outskirts of a city will increase the city's population at the rate of

$$4500\sqrt{t} + 1000$$

people/year t yr from the start of construction. The population before construction is 30,000. Determine the projected population 9 yr after construction of the park has begun.

74. OZONE POLLUTION The rate of change of the level of ozone, an invisible gas that is an irritant and impairs breathing, present in the atmosphere on a certain May day in the city of Riverside is given by

$$R(t) = 3.2922t^2 - 0.366t^3 \qquad (0 < t < 11)$$

(measured in pollutant standard index per hour). Here, t is measured in hours, with $t = 0$ corresponding to 7 A.M. Find the ozone level $A(t)$ at any time t assuming that at 7 A.M. it is zero.
Hint: $A'(t) = R(t)$ and $A(0) = 0$.

75. SURFACE AREA OF A HUMAN Empirical data suggest that the surface area of a 180-cm-tall human body changes at the rate of

$$S'(W) = 0.131773W^{-0.575}$$

square meters/kilogram, where W is the weight of the body in kilograms. If the surface area of a 180-cm-tall human body weighing 70 kg is 1.886277 m³, what is the surface area of a human body of the same height weighing 75 kg?

76. FLIGHT OF A ROCKET The velocity, in feet/second, of a rocket t sec into vertical flight is given by

$$v(t) = -3t^2 + 192t + 120$$

Find an expression $h(t)$ that gives the rocket's altitude, in feet, t sec after liftoff. What is the altitude of the rocket 30 sec after liftoff?
Hint: $h'(t) = v(t)$; $h(0) = 0$.

77. BLOOD FLOW IN AN ARTERY Nineteenth-century physician Jean Louis Marie Poiseuille discovered that the rate of change of the velocity of blood r cm from the central axis of an artery (in centimeters/second/centimeters) is given by

$$a(r) = -kr$$

where k is a constant. If the radius of an artery is R cm, find an expression for the velocity of blood as a function of r (see the accompanying figure).
Hint: $v'(r) = a(r)$ and $v(R) = 0$. (Why?)

Blood vessel

78. ACCELERATION OF A CAR A car traveling along a straight road at 66 ft/sec accelerated to a speed of 88 ft/sec over a distance of 440 ft. What was the acceleration of the car, assuming it was constant?

79. DECELERATION OF A CAR What constant deceleration would a car moving along a straight road have to be subjected to if it were brought to rest from a speed of 88 ft/sec in 9 sec? What would be the stopping distance?

80. A tank has a constant cross-sectional area of 50 ft² and an orifice of constant cross-sectional area of $\frac{1}{2}$ ft² located at the bottom of the tank (see the accompanying figure).

If the tank is filled with water to a height of h ft and allowed to drain, then the height of the water decreases at a rate that is described by the equation

$$\frac{dh}{dt} = -\frac{1}{25}\left(\sqrt{20} - \frac{t}{50}\right) \qquad (0 \le t \le 50\sqrt{20})$$

Find an expression for the height of the water at any time t if its height initially is 20 ft.

81. LAUNCHING A FIGHTER AIRCRAFT A fighter aircraft is launched from the deck of a Nimitz-class aircraft carrier with the help of a steam catapult. If the aircraft is to attain a takeoff speed of at least 240 ft/sec after traveling 800 ft along the flight deck, find the minimum acceleration it must be subjected to, assuming it is constant.

82. BANK DEPOSITS The Madison Finance Company opened two branches on September 1 ($t = 0$). Branch A is located in an established industrial park, and branch B is located in a fast-growing new development. The net rate at which money was deposited into branch A and branch B in the first 180 business days is given by the graphs of f and g, respectively (see the figure). Which branch has a larger amount on deposit at the end of 180 business days? Justify your answer.

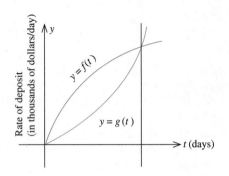

83. If F and G are antiderivatives of f on an interval I, then $F(x) = G(x) + C$ on I.

84. If F is an antiderivative of f on an interval I, then $\int f(x)\, dx = F(x)$.

85. If f and g are integrable, then $\int [2f(x) - 3g(x)]\, dx = 2 \int f(x)\, dx - 3 \int g(x)\, dx$.

86. If f and g are integrable, then $\int f(x)g(x)\, dx = [\int f(x)\, dx][\int g(x)\, dx]$.

SOLUTIONS TO SELF-CHECK EXERCISES 6.1

1.
$$\int \left(\frac{1}{\sqrt{x}} - \frac{2}{x} + 3e^x \right) dx = \int \left(x^{-1/2} - \frac{2}{x} + 3e^x \right) dx$$
$$= \int x^{-1/2}\, dx - 2 \int \frac{1}{x}\, dx + 3 \int e^x\, dx$$
$$= 2x^{1/2} - 2 \ln|x| + 3e^x + C$$
$$= 2\sqrt{x} - 2 \ln|x| + 3e^x + C$$

2. The slope of the tangent line to the graph of the function f at any point $P(x, f(x))$ is given by the derivative f' of f. Thus, the first condition implies that
$$f'(x) = 3x^2 - 6x + 3$$
which, upon integration, yields
$$f(x) = \int (3x^2 - 6x + 3)\, dx$$
$$= x^3 - 3x^2 + 3x + k$$
where k is the constant of integration.

To evaluate k, we use the initial condition (2), which implies that $f(2) = 9$, or
$$9 = f(2) = 2^3 - 3(2)^2 + 3(2) + k$$
or $k = 7$. Hence, the required rule of definition of the function f is
$$f(x) = x^3 - 3x^2 + 3x + 7$$

3. Let $M(t)$ denote United Motors' market share at year t. Then,
$$M(t) = \int f(t)\, dt$$
$$= \int (-0.01875t^2 + 0.15t - 1.2)\, dt$$
$$= -0.00625t^3 + 0.075t^2 - 1.2t + C$$

To determine the value of C, we use the initial condition $M(0) = 48.4$, obtaining $C = 48.4$. Therefore,

$$M(t) = -0.00625t^3 + 0.075t^2 - 1.2t + 48.4$$

In particular, United Motors' market share of new cars at the beginning of 2000 is given by

$$M(12) = -0.00625(12)^3 + 0.075(12)^2$$
$$-1.2(12) + 48.4 = 34$$

or 34%.

6.2 Integration by Substitution

In Section 6.1 we developed certain rules of integration that are closely related to the corresponding rules of differentiation in Chapters 3 and 5. In this section we introduce a method of integration called the **method of substitution,** which is related to the chain rule for differentiating functions. When used in conjunction with the rules of integration developed earlier, the method of substitution is a powerful tool for integrating a large class of functions.

HOW THE METHOD OF SUBSTITUTION WORKS

Consider the indefinite integral

$$\int 2(2x + 4)^5 \, dx \tag{3}$$

One way of evaluating this integral is to expand the expression $(2x + 4)^5$ and then integrate the resulting integrand term by term. As an alternative approach, let's see if we can simplify the integral by making a change of variable. Write

$$u = 2x + 4$$

with differential

$$du = 2 \, dx$$

If we substitute these quantities into Equation (3), we obtain

$$\int 2(2x + 4)^5 \, dx = \int (2x + 4)^5(2 \, dx) = \int u^5 \, du$$

Rewriting

$$\begin{cases} u = 2x + 4 \\ du = 2 \, dx \end{cases}$$

Now, the last integral involves a power function and is easily evaluated using Rule 2 of Section 6.1. Thus,

$$\int u^5 \, du = \frac{1}{6}u^6 + C$$

Therefore, using this result and replacing u by $u = 2x + 4$, we obtain

$$\int 2(2x + 4)^5 \, dx = \frac{1}{6}(2x + 4)^6 + C$$

We can verify that the foregoing result is indeed correct by computing

$$\frac{d}{dx}\left[\frac{1}{6}(2x + 4)^6\right] = \frac{1}{6} \cdot 6(2x + 4)^5(2) \quad \text{(Using the chain rule)}$$

$$= 2(2x + 4)^5$$

and observing that the last expression is just the integrand of (3).

THE METHOD OF INTEGRATION BY SUBSTITUTION

To see why the approach used in evaluating the integral in (3) is successful, write

$$f(x) = x^5 \quad \text{and} \quad g(x) = 2x + 4$$

Then, $g'(x) = 2 \, dx$. Furthermore, the integrand of (3) is just the composition of f and g. Thus,

$$(f \circ g)(x) = f(g(x))$$
$$= [g(x)]^5 = (2x + 4)^5$$

Therefore, (3) can be written as

$$\int f(g(x))g'(x) \, dx \qquad\qquad \textbf{(4)}$$

Next, let's show that an integral having the form (4) can always be written as

$$\int f(u) \, du \qquad\qquad \textbf{(5)}$$

Suppose F is an antiderivative of f. By the chain rule, we have

$$\frac{d}{dx}[F(g(x))] = F'(g(x))g'(x)$$

Therefore,

$$\int F'(g(x))g'(x) \, dx = F(g(x)) + C$$

Letting $F' = f$ and making the substitution $u = g(x)$, we have

$$\int f(g(x))g'(x) \, dx = F(u) + C = \int F'(u) \, du = \int f(u) \, du$$

as we wished to show. Thus, if the transformed integral is readily evaluated, as is the case with the integral (3), then the method of substitution will prove successful.

Before we look at more examples, let's summarize the steps involved in integration by substitution.

Integration by Substitution

Step 1	Let $u = g(x)$, where $g(x)$ is part of the integrand, usually the "inside function" of the composite function $f(g(x))$.
Step 2	Compute $du = g'(x)\, dx$.
Step 3	Use the substitution $u = g(x)$ and $du = g'(x)\, dx$ to convert the *entire* integral into one involving *only u*.
Step 4	Evaluate the resulting integral.
Step 5	Replace u by $g(x)$ to obtain the final solution as a function of x.

REMARK Sometimes we need to consider different choices of g for the substitution $u = g(x)$ in order to carry out step 3 and/or step 4.   ■■■

EXAMPLE 1   Find $\int 2x(x^2 + 3)^4\, dx$.

SOLUTION ✔

Step 1   Observe that the integrand involves the composite function $(x^2 + 3)^4$ with "inside function" $g(x) = x^2 + 3$. So, we choose $u = x^2 + 3$.

Step 2   Compute $du = 2x\, dx$.

Step 3   Making the substitution $u = x^2 + 3$ and $du = 2x\, dx$, we obtain

$$\int 2x(x^2 + 3)^4\, dx = \int (x^2 + 3)^4(2x\, dx) = \int u^4\, du$$

↑
Rewriting

an integral involving only the variable u.

Step 4   Evaluate

$$\int u^4\, du = \frac{1}{5} u^5 + C$$

Step 5   Replacing u by $x^2 + 3$, we obtain

$$\int 2x(x^2 + 3)^4\, dx = \frac{1}{5}(x^2 + 3)^5 + C$$ ■■■■

EXAMPLE 2   Find $\int 3\sqrt{3x + 1}\, dx$.

SOLUTION ✔

Step 1   The integrand involves the composite function $\sqrt{3x + 1}$ with "inside function" $g(x) = 3x + 1$. So, let $u = 3x + 1$.

Step 2   Compute $du = 3\, dx$.

Step 3   Making the substitution $u = 3x + 1$ and $du = 3\, dx$, we obtain

$$\int 3\sqrt{3x + 1}\, dx = \int \sqrt{3x + 1}(3\, dx) = \int \sqrt{u}\, du$$

an integral involving only the variable u.

Step 4   Evaluate

$$\int \sqrt{u}\,du = \int u^{1/2}\,du = \frac{2}{3}u^{3/2} + C$$

Step 5   Replacing u by $u = 3x + 1$, we obtain

$$\int 3\sqrt{3x + 1}\,dx = \frac{2}{3}(3x + 1)^{3/2} + C$$

■■■■

EXAMPLE 3   Find $\int x^2(x^3 + 1)^{3/2}\,dx$.

SOLUTION ✔

Step 1   The integrand contains the composite function $(x^3 + 1)^{3/2}$ with "inside function" $g(x) = x^3 + 1$. So, let $u = x^3 + 1$.

Step 2   Compute $du = 3x^2\,dx$.

Step 3   Making the substitution $u = x^3 + 1$ and $du = 3x^2\,dx$, or $x^2\,dx = \frac{1}{3}\,du$, we obtain

$$\int x^2(x^3 + 1)^{3/2}\,dx = \int (x^3 + 1)^{3/2}(x^2\,dx)$$
$$= \int u^{3/2}\left(\frac{1}{3}\,du\right) = \frac{1}{3}\int u^{3/2}\,du$$

an integral involving only the variable u.

Step 4   We evaluate

$$\frac{1}{3}\int u^{3/2}\,du = \frac{1}{3}\cdot\frac{2}{5}u^{5/2} + C = \frac{2}{15}u^{5/2} + C$$

Step 5   Replacing u by $x^3 + 1$, we obtain

$$\int x^2(x^3 + 1)^{3/2}\,dx = \frac{2}{15}(x^3 + 1)^{5/2} + C$$

■■■■

Group Discussion

Let $f(x) = x^2(x^3 + 1)^{3/2}$. Using the result of Example 3, we see that an antiderivative of f is $F(x) = \frac{2}{15}(x^3 + 1)^{5/2}$. However, in terms of u (where $u = x^3 + 1$), an antiderivative of f is $G(u) = \frac{2}{15}u^{5/2}$. Compute $F(2)$. Next, suppose we want to compute $F(2)$ using the function G instead. At what value of u should you evaluate $G(u)$ in order to obtain the desired result? Explain your answer.

In the remaining examples, we drop the practice of labeling the steps involved in evaluating each integral.

EXAMPLE 4   Find $\int e^{-3x}\, dx$.

SOLUTION ✔   Let $u = -3x$ so that $du = -3\, dx$, or $dx = -\frac{1}{3}\, du$. Then,

$$\int e^{-3x}\, dx = \int e^u \left(-\frac{1}{3}\, du \right) = -\frac{1}{3} \int e^u\, du$$

$$= -\frac{1}{3} e^u + C = -\frac{1}{3} e^{-3x} + C$$

■■■■

EXAMPLE 5   Find $\int \dfrac{x}{3x^2 + 1}\, dx$.

SOLUTION ✔   Let $u = 3x^2 + 1$. Then, $du = 6x\, dx$, or $x\, dx = \frac{1}{6}\, du$. Making the appropriate substitutions, we have

$$\int \frac{x}{3x^2 + 1}\, dx = \int \frac{\frac{1}{6}}{u}\, du$$

$$= \frac{1}{6} \int \frac{1}{u}\, du$$

$$= \frac{1}{6} \ln |u| + C$$

$$= \frac{1}{6} \ln(3x^2 + 1) + C \qquad \text{(Since } 3x^2 + 1 > 0)$$

■■■■

EXAMPLE 6   Find $\int \dfrac{(\ln x)^2}{2x}\, dx$.

SOLUTION ✔   Let $u = \ln x$. Then,

$$du = \frac{d}{dx}(\ln x)\, dx = \frac{1}{x}\, dx$$

$$\int \frac{(\ln x)^2}{2x}\, dx = \frac{1}{2} \int \frac{(\ln x)^2}{x}\, dx$$

$$= \frac{1}{2} \int u^2\, du$$

$$= \frac{1}{6} u^3 + C$$

$$= \frac{1}{6} (\ln x)^3 + C$$

■■■■

> **Group Discussion**
>
> Suppose $\int f(u)\,du = F(u) + C$.
>
> **1.** Show that $\int f(ax + b)\,dx = \dfrac{1}{a}F(ax + b) + C$.
>
> **2.** How can you use this result to facilitate the evaluation of integrals such as $\int (2x + 3)^5\,dx$ and $\int e^{3x-2}\,dx$? Explain your answer.

APPLICATIONS

Examples 7 and 8 show how the method of substitution can be used in practical situations.

EXAMPLE 7   In 1990 the head of the research and development department of the Soloron Corporation claimed that the cost of producing solar cell panels would drop at the rate of

$$\frac{58}{(3t + 2)^2} \qquad (0 \le t \le 10)$$

dollars per peak watt for the next t years, with $t = 0$ corresponding to the beginning of the year 1990. (A peak watt is the power produced at noon on a sunny day.) In 1990 the panels, which are used for photovoltaic power systems, cost $10 per peak watt. Find an expression giving the cost per peak watt of producing solar cell panels at the beginning of year t. What was the cost at the beginning of 2000?

SOLUTION ✔   Let $C(t)$ denote the cost per peak watt for producing solar cell panels at the beginning of year t. Then,

$$C'(t) = -\frac{58}{(3t + 2)^2}$$

Integrating, we find that

$$C(t) = \int \frac{-58}{(3t + 2)^2}\,dt$$

$$= -58 \int (3t + 2)^{-2}\,dt$$

Let $u = 3t + 2$ so that

$$du = 3\,dt \qquad \text{or} \qquad dt = \frac{1}{3}\,du$$

Then,

$$C(t) = -58\left(\frac{1}{3}\right) \int u^{-2}\,du$$

$$= -\frac{58}{3}(-1)u^{-1} + k$$

$$= \frac{58}{3(3t + 2)} + k$$

where k is an arbitrary constant. To determine the value of k, note that the cost per peak watt of producing solar cell panels at the beginning of 1990 ($t = 0$) was 10, or $C(0) = 10$. This gives

$$C(0) = \frac{58}{3(2)} + k = 10$$

or $k = \frac{1}{3}$. Therefore, the required expression is given by

$$C(t) = \frac{58}{3(3t + 2)} + \frac{1}{3}$$

$$= \frac{58 + (3t + 2)}{3(3t + 2)} = \frac{3t + 60}{3(3t + 2)}$$

$$= \frac{t + 20}{3t + 2}$$

The cost per peak watt for producing solar cell panels at the beginning of 2000 is given by

$$C(10) = \frac{10 + 20}{3(10) + 2} \approx 0.94$$

or approximately $.94 per peak watt.   ■ ■ ■ ■

Exploring with Technology

Refer to Example 7.

1. Use a graphing utility to plot the graph of

$$C(t) = \frac{t + 20}{3t + 2}$$

using the viewing rectangle $[0, 10] \times [0, 5]$. Then, use the numerical differentiation capability of the graphing utility to compute $C'(10)$.

2. Plot the graph of

$$C'(t) = -\frac{58}{(3t + 2)^2}$$

using the viewing rectangle $[0, 10] \times [-10, 0]$. Then, use the evaluation capability of the graphing utility to find $C'(10)$. Is this value of $C'(10)$ the same as that obtained in part 1? Explain your answer.

 EXAMPLE **8**   A study prepared by the marketing department of the Universal Instruments Company forecasts that, after its new line of Galaxy Home Computers is introduced into the market, sales will grow at the rate of

$$2000 - 1500e^{-0.05t} \qquad (0 \le t \le 60)$$

units per month. Find an expression that gives the total number of computers that will sell t months after they become available on the market. How many computers will Universal sell in the first year they are on the market?

SOLUTION ✔ Let $N(t)$ denote the total number of computers that may be expected to be sold t months after their introduction in the market. Then, the rate of growth of sales is given by $N'(t)$ units per month. Thus,

$$N'(t) = 2000 - 1500e^{-0.05t}$$

so that
$$N(t) = \int (2000 - 1500e^{-0.05t})\, dt$$
$$= \int 2000\, dt - 1500 \int e^{-0.05t}\, dt$$

Upon integrating the second integral by the method of substitution, we obtain

$$N(t) = 2000t + \frac{1500}{0.05} e^{-0.05t} + C \qquad \text{(Let } u = -0.05t,$$
$$\text{then } du = -0.05\, dt.)$$
$$= 2000t + 30{,}000e^{-0.05t} + C$$

To determine the value of C, note that the number of computers sold at the end of month 0 is nil, so $N(0) = 0$. This gives

$$N(0) = 30{,}000 + C = 0 \qquad \text{(Since } e^0 = 1)$$

or $C = -30{,}000$. Therefore, the required expression is given by

$$N(t) = 2000t + 30{,}000e^{-0.05t} - 30{,}000$$
$$= 2000t + 30{,}000(e^{-0.05t} - 1)$$

The number of computers that Universal can expect to sell in the first year is given by

$$N(12) = 2000(12) + 30{,}000(e^{-0.05(12)} - 1)$$
$$= 10{,}464 \text{ units} \qquad \blacksquare\blacksquare\blacksquare\blacksquare$$

SELF-CHECK EXERCISES 6.2

1. Evaluate $\int \sqrt{2x + 5}\, dx$.

2. Evaluate $\int \dfrac{x^2}{(2x^3 + 1)^{3/2}}\, dx$.

3. Evaluate $\int xe^{2x^2-1}\, dx$.

4. According to a joint study conducted by Oxnard's Environmental Management Department and a state government agency, the concentration of carbon monoxide (CO) in the air due to automobile exhaust is increasing at the rate given by

$$f(t) = \frac{8(0.1t + 1)}{300(0.2t^2 + 4t + 64)^{1/3}}$$

parts per million (ppm) per year t. Currently, the CO concentration due to automobile exhaust is 0.16 ppm. Find an expression giving the CO concentration t yr from now.

Solutions to Self-Check Exercises 6.2 can be found on page 465.

6.2 Exercises

In Exercises 1–50, find the indefinite integral.

1. $\int 4(4x+3)^4\,dx$

2. $\int 4x(2x^2+1)^7\,dx$

3. $\int (x^3-2x)^2(3x^2-2)\,dx$

4. $\int (3x^2-2x+1)(x^3-x^2+x)^4\,dx$

5. $\int \dfrac{4x}{(2x^2+3)^3}\,dx$

6. $\int \dfrac{3x^2+2}{(x^3+2x)^2}\,dx$

7. $\int 3t^2\sqrt{t^3+2}\,dt$

8. $\int 3t^2(t^3+2)^{3/2}\,dt$

9. $\int (x^2-1)^9 x\,dx$

10. $\int x^2(2x^3+3)^4\,dx$

11. $\int \dfrac{x^4}{1-x^5}\,dx$

12. $\int \dfrac{x^2}{\sqrt{x^3-1}}\,dx$

13. $\int \dfrac{2}{x-2}\,dx$

14. $\int \dfrac{x^2}{x^3-3}\,dx$

15. $\int \dfrac{0.3x-0.2}{0.3x^2-0.4x+2}\,dx$

16. $\int \dfrac{2x^2+1}{0.2x^3+0.3x}\,dx$

17. $\int \dfrac{x}{3x^2-1}\,dx$

18. $\int \dfrac{x^2-1}{x^3-3x+1}\,dx$

19. $\int e^{-2x}\,dx$

20. $\int e^{-0.02x}\,dx$

21. $\int e^{2-x}\,dx$

22. $\int e^{2t+3}\,dt$

23. $\int xe^{-x^2}\,dx$

24. $\int x^2 e^{x^3-1}\,dx$

25. $\int (e^x-e^{-x})\,dx$

26. $\int (e^{2x}+e^{-3x})\,dx$

27. $\int \dfrac{e^x}{1+e^x}\,dx$

28. $\int \dfrac{e^{2x}}{1+e^{2x}}\,dx$

29. $\int \dfrac{e^{\sqrt{x}}}{\sqrt{x}}\,dx$

30. $\int \dfrac{e^{-1/x}}{x^2}\,dx$

31. $\int \dfrac{e^{3x}+x^2}{(e^{3x}+x^3)^3}\,dx$

32. $\int \dfrac{e^x-e^{-x}}{(e^x+e^{-x})^{3/2}}\,dx$

33. $\int e^{2x}(e^{2x}+1)^3\,dx$

34. $\int e^{-x}(1+e^{-x})\,dx$

35. $\int \dfrac{\ln 5x}{x}\,dx$

36. $\int \dfrac{(\ln u)^3}{u}\,du$

37. $\int \dfrac{1}{x\ln x}\,dx$

38. $\int \dfrac{1}{x(\ln x)^2}\,dx$

39. $\int \dfrac{\sqrt{\ln x}}{x}\,dx$

40. $\int \dfrac{(\ln x)^{7/2}}{x}\,dx$

41. $\int \left(xe^{x^2}-\dfrac{x}{x^2+2}\right)dx$

42. $\int \left(xe^{-x^2}+\dfrac{e^x}{e^x+3}\right)dx$

43. $\int \dfrac{x+1}{\sqrt{x}-1}\,dx$

Hint: Let $u=\sqrt{x}-1$.

44. $\int \dfrac{e^{-u}-1}{e^{-u}+u}\,du$

Hint: Let $v=e^{-u}+u$.

45. $\int x(x-1)^5\,dx$

Hint: $u=x-1$ implies $x=u+1$.

46. $\int \dfrac{t}{t+1}\,dt$

Hint: $\dfrac{t}{t+1}=1-\dfrac{1}{t+1}$.

47. $\int \dfrac{1-\sqrt{x}}{1+\sqrt{x}}\,dx$

Hint: Let $u=1+\sqrt{x}$.

48. $\int \dfrac{1+\sqrt{x}}{1-\sqrt{x}}\,dx$

Hint: Let $u=1-\sqrt{x}$.

49. $\int v^2(1-v)^6\,dv$

Hint: Let $u=1-v$.

50. $\int x^3(x^2+1)^{3/2}\,dx$

Hint: Let $u=x^2+1$.

In Exercises 51–54, find the function f given that the slope of the tangent line to the graph of f at any point $(x, f(x))$ is $f'(x)$ and that the graph of f passes through the given point.

51. $f'(x)=5(2x-1)^4$; $(1,3)$

52. $f'(x)=\dfrac{3x^2}{2\sqrt{x^3-1}}$; $(1,1)$

53. $f'(x)=-2xe^{-x^2+1}$; $(1,0)$

54. $f'(x)=1-\dfrac{2x}{x^2+1}$; $(0,2)$

55. STUDENT ENROLLMENT The registrar of Kellogg University estimates that the total student enrollment in the Continuing Education division will grow at the rate of

$$N'(t)=2000(1+0.2t)^{-3/2}$$

students/year t yr from now. If the current student enrollment is 1000, find an expression giving the total student enrollment t yr from now. What will be the student enrollment 5 yr from now?

56. **TV VIEWERS: NEWSMAGAZINE SHOWS**   The number of viewers of a weekly TV newsmagazine show, introduced in the 1995 season, has been increasing at the rate of

$$3\left(2 + \frac{1}{2}t\right)^{-1/3} \qquad (1 \le t \le 6)$$

million viewers/year in its tth year on the air. The number of viewers of the program during its first year on the air is given by $9(5/2)^{2/3}$ million. Find how many viewers were expected in the 2000 season.

57. **DEMAND: LADIES' BOOTS**   The rate of change of the unit price p (in dollars) of Apex ladies' boots is given by

$$p'(x) = \frac{-250x}{(16 + x^2)^{3/2}}$$

where x is the quantity demanded daily in units of a hundred. Find the demand function for these boots if the quantity demanded daily is 300 pairs ($x = 3$) when the unit price is $50/pair.

■ **A calculator is recommended for the remaining exercises.**

58. **SUPPLY: LADIES' BOOTS**   The rate of change of the unit price p (in dollars) of Apex ladies' boots is given by

$$p'(x) = \frac{240x}{(5 - x)^2}$$

where x is the number of pairs that the supplier will make available in the market daily when the unit price is p/pair. Find the supply equation for these boots if the quantity the supplier is willing to make available is 200 pairs daily ($x = 2$) when the unit price is $50/pair.

59. **OIL SPILL**   In calm waters the oil spilling from the ruptured hull of a grounded tanker forms an oil slick that is circular in shape. If the radius r of the circle is increasing at the rate of

$$r'(t) = \frac{30}{\sqrt{2t + 4}}$$

feet/minute t min after the rupture occurs, find an expression for the radius at any time t. How large is the polluted area 16 min after the rupture occurred?
Hint: $r(0) = 0$.

60. **LIFE EXPECTANCY OF A FEMALE**   Suppose in a certain country the life expectancy at birth of a female is changing at

the rate of

$$g'(t) = \frac{5.45218}{(1 + 1.09t)^{0.9}}$$

years/year. Here, t is measured in years, and $t = 0$ corresponds to the beginning of 1900. Find an expression $g(t)$ giving the life expectancy at birth (in years) of a female in that country if the life expectancy at the beginning of 1900 is 50.02 yr. What is the life expectancy at birth of a female born in the year 2000 in that country?

61. **AVERAGE BIRTH HEIGHT OF BOYS**   Using data collected at Kaiser Hospital, pediatricians estimate that the average height of male children changes at the rate of

$$h'(t) = \frac{52.8706e^{-0.3277t}}{(1 + 2.449e^{-0.3277t})^2}$$

inches/year, where the child's height $h(t)$ is measured in inches and t, the child's age, is measured in years, with $t = 0$ corresponding to the age at birth. Find an expression $h(t)$ for the average height of a boy at age t if the height at birth of an average child is 19.4 in. What is the height of an average 8-yr-old boy?

62. **LEARNING CURVES**   The average student enrolled in the 20-wk Court Reporting I course at the American Institute of Court Reporting progresses according to the rule

$$N'(t) = 6e^{-0.05t} \qquad (0 \le t \le 20)$$

where $N'(t)$ measures the rate of change in the number of words per minute of dictation the student takes in machine shorthand after t wk in the course. Assuming that the average student enrolled in this course begins with a dictation speed of 60 words/minute, find an expression $N(t)$ that gives the dictation speed of the student after t wk in the course.

63. **SALES: LOUDSPEAKERS**   In the first year they appeared in the market, 2000 pairs of Acrosonic model F loudspeaker systems were sold. Since then, sales of these loudspeaker systems have been growing at the rate of

$$f'(t) = 2000(3 - 2e^{-t})$$

units/year, where t denotes the number of years these systems have been on the market. Determine the number of systems that were sold in the first 5 yr after their introduction.

64. **AMOUNT OF GLUCOSE IN THE BLOODSTREAM**   Suppose a patient is given a continuous intravenous infusion of glucose at a constant rate of r mg/min. Then, the rate at which the

amount of glucose in the bloodstream is changing at time t due to this infusion is given by

$$A'(t) = re^{-at}$$

mg/min, where a is a positive constant associated with the rate at which excess glucose is eliminated from the bloodstream and is dependent on the patient's metabolism rate. Derive an expression for the amount of glucose in the bloodstream at time t.

Hint: $A(0) = 0$.

65. **CONCENTRATION OF A DRUG IN AN ORGAN** A drug is carried into an organ of volume V cm³ by a liquid that enters the organ at the rate of a cm³/sec and leaves it at the rate of b cm³/sec. The concentration of the drug in the liquid entering the organ is c g/cm³. If the concentration of the drug in the organ at time t is increasing at the rate of

$$x'(t) = \frac{1}{V}(ac - bx_0)e^{-bt/V}$$

g/cm³/ sec, and the concentration of the drug in the organ initially is x_0 g/cm³, show that the concentration of the drug in the organ at time t is given by

$$x(t) = \frac{ac}{b} + \left(x_0 - \frac{ac}{b}\right)e^{-bt/V}$$

SOLUTIONS TO SELF-CHECK EXERCISES 6.2

1. Let $u = 2x + 5$. Then, $du = 2\,dx$, or $dx = \frac{1}{2}\,du$. Making the appropriate substitutions, we have

$$\int \sqrt{2x + 5}\,dx = \int \sqrt{u}\left(\frac{1}{2}\,du\right) = \frac{1}{2}\int u^{1/2}\,du$$

$$= \frac{1}{2}\left(\frac{2}{3}\right)u^{3/2} + C$$

$$= \frac{1}{3}(2x + 5)^{3/2} + C$$

2. Let $u = 2x^3 + 1$, so that $du = 6x^2\,dx$, or $x^2\,dx = \frac{1}{6}\,du$. Making the appropriate substitutions, we have

$$\int \frac{x^2}{(2x^3 + 1)^{3/2}}\,dx = \int \frac{\left(\frac{1}{6}\right)du}{u^{3/2}} = \frac{1}{6}\int u^{-3/2}\,du$$

$$= \left(\frac{1}{6}\right)(-2)u^{-1/2} + C$$

$$= -\frac{1}{3}(2x^3 + 1)^{-1/2} + C$$

$$= -\frac{1}{3\sqrt{2x^3 + 1}} + C$$

3. Let $u = 2x^2 - 1$, so that $du = 4x\,dx$, or $x\,dx = \frac{1}{4}\,du$. Then,

$$\int xe^{2x^2-1}\,dx = \frac{1}{4}\int e^u\,du$$

$$= \frac{1}{4}e^u + C$$

$$= \frac{1}{4}e^{2x^2-1} + C$$

4. Let $C(t)$ denote the CO concentration in the air due to automobile exhaust t yr from now. Then,

$$C'(t) = f(t) = \frac{8(0.1t + 1)}{300(0.2t^2 + 4t + 64)^{1/3}}$$

$$= \frac{8}{300}(0.1t + 1)(0.2t^2 + 4t + 64)^{-1/3}$$

Integrating, we find

$$C(t) = \int \frac{8}{300}(0.1t + 1)(0.2t^2 + 4t + 64)^{-1/3}\, dt$$

$$= \frac{8}{300}\int (0.1t + 1)(0.2t^2 + 4t + 64)^{-1/3}\, dt$$

Let $u = 0.2t^2 + 4t + 64$, so that $du = (0.4t + 4)\, dt = 4(0.1t + 1)\, dt$, or

$$(0.1t + 1)\, dt = \frac{1}{4}\, du$$

Then,

$$C(t) = \frac{8}{300}\left(\frac{1}{4}\right)\int u^{-1/3}\, du$$

$$= \frac{1}{150}\left(\frac{3}{2}u^{2/3}\right) + k$$

$$= 0.01(0.2t^2 + 4t + 64)^{2/3} + k$$

where k is an arbitrary constant. To determine the value of k, we use the condition $C(0) = 0.16$, obtaining

$$C(0) = 0.16 = 0.01(64)^{2/3} + k$$

$$0.16 = 0.16 + k$$

$$k = 0$$

Therefore,

$$C(t) = 0.01(0.2t^2 + 4t + 64)^{2/3}$$

6.3 Area and the Definite Integral

AN INTUITIVE LOOK

Suppose a certain state's annual rate of petroleum consumption over a 4-year period is constant and is given by the function

$$f(t) = 1.2 \qquad (0 \le t \le 4)$$

where t is measured in years and $f(t)$ in millions of barrels per year. Then, the state's total petroleum consumption over the period of time in question is

$$(1.2)(4 - 0) \qquad \text{(Rate of consumption · time elapsed)}$$

FIGURE 6.5
The total petroleum consumption is given by the area of the rectangular region.

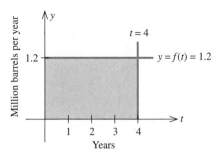

FIGURE 6.6
The daily petroleum consumption is given by the "area" of the shaded region.

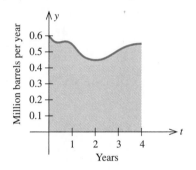

or 4.8 million barrels. If you examine the graph of f shown in Figure 6.5, you will see that this total is just the area of the rectangular region bounded above by the graph of f, below by the t-axis, and to the left and right by the vertical lines $t = 0$ (the y-axis) and $t = 4$, respectively.

Figure 6.6 shows the actual petroleum consumption of a certain New England state over a 4-year period from 1990 ($t = 0$) to 1994 ($t = 4$). Observe that the rate of consumption is not constant; that is, the function f is not a constant function. What is the state's total petroleum consumption over this 4-year period? It seems reasonable to conjecture that it is given by the "area" of the region bounded above by the graph of f, below by the t-axis, and to the left and right by the vertical lines $t = 0$ and $t = 4$, respectively.

This example raises two questions:

1. What is the "area" of the region shown in Figure 6.6?
2. How do we compute this area?

FIGURE 6.7
The area under the graph of f on $[a, b]$

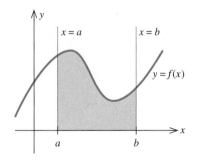

THE AREA PROBLEM

The preceding example touches on the second fundamental problem in calculus: Calculate the area of the region bounded by the graph of a nonnegative function f, the x-axis, and the vertical lines $x = a$ and $x = b$ (Figure 6.7). This area is called the **area under the graph of f** on the interval $[a, b]$, or from a to b.

DEFINING AREA—TWO EXAMPLES

Just as we used the slopes of secant lines (quantities that we could compute) to help us define the slope of the tangent line to a point on the graph of a function, we now adopt a parallel approach and use the areas of rectangles (quantities that we can compute) to help us define the area under the graph of a function. We begin by looking at a specific example.

EXAMPLE **1**

Let $f(x) = x^2$ and consider the region R under the graph of f on the interval $[0, 1]$ (Figure 6.8a). To obtain an approximation of the area of R, let's construct four nonoverlapping rectangles as follows: Divide the interval $[0, 1]$ into four subintervals

$$\left[0,\frac{1}{4}\right], \quad \left[\frac{1}{4},\frac{1}{2}\right], \quad \left[\frac{1}{2},\frac{3}{4}\right], \quad \left[\frac{3}{4},1\right]$$

of equal length $\frac{1}{4}$. Next, construct four rectangles with these subintervals as bases and with heights given by the values of the function at the midpoints

$$\frac{1}{8}, \quad \frac{3}{8}, \quad \frac{5}{8}, \quad \frac{7}{8}$$

of each subinterval. Then, each of these rectangles has width $\frac{1}{4}$ and height

$$f\left(\frac{1}{8}\right), \quad f\left(\frac{3}{8}\right), \quad f\left(\frac{5}{8}\right), \quad f\left(\frac{7}{8}\right)$$

respectively (Figure 6.8b).

FIGURE 6.8
The area of the region under the graph of f on [0, 1] in (a) is approximated by the sum of the areas of the four rectangles in (b).

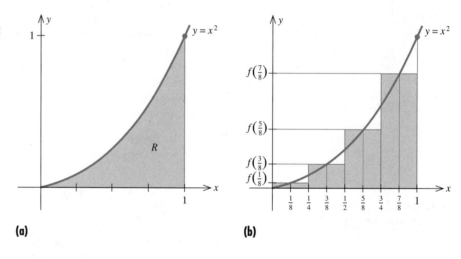

(a)   (b)

If we approximate the area A of S by the sum of the areas of the four rectangles, we obtain

$$A \approx \frac{1}{4}f\left(\frac{1}{8}\right) + \frac{1}{4}f\left(\frac{3}{8}\right) + \frac{1}{4}f\left(\frac{5}{8}\right) + \frac{1}{4}f\left(\frac{7}{8}\right)$$

$$= \frac{1}{4}\left[f\left(\frac{1}{8}\right) + f\left(\frac{3}{8}\right) + f\left(\frac{5}{8}\right) + f\left(\frac{7}{8}\right)\right]$$

$$= \frac{1}{4}\left[\left(\frac{1}{8}\right)^2 + \left(\frac{3}{8}\right)^2 + \left(\frac{5}{8}\right)^2 + \left(\frac{7}{8}\right)^2\right] \qquad \text{[Recall that } f(x) = x^2\text{.]}$$

$$= \frac{1}{4}\left(\frac{1}{64} + \frac{9}{64} + \frac{25}{64} + \frac{49}{64}\right) = \frac{21}{64}$$

or approximately 0.328125 square unit.   ■ ■ ■ ■

Following the procedure of Example 1, we can obtain approximations of the area of the region R using any number n of rectangles ($n = 4$ in Example 1). Figure 6.9a shows the approximation of the area A of R using 8 rectangles ($n = 8$), and Figure 6.9b shows the approximation of the area A of R using 16 rectangles.

FIGURE 6.9

As n increases, the number of rectangles increases, and the approximation improves.

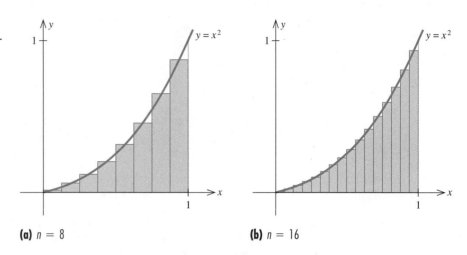

(a) $n = 8$ **(b)** $n = 16$

These figures suggest that the approximations seem to get better as n increases. This is borne out by the results given in Table 6.1, which were obtained using a computer.

Table 6.1

Number of Rectangles n	4	8	16	32	64	100	200
Approximation of A	0.328125	0.332031	0.333008	0.333252	0.333313	0.333325	0.333331

Our computations seem to suggest that the approximations approach the number $\frac{1}{3}$ as n gets larger and larger. This result suggests that we *define* the area of the region under the graph of $f(x) = x^2$ on the interval $[0, 1]$ to be $\frac{1}{3}$ square unit.

In Example 1 we chose the *midpoint* of each subinterval as the point at which to evaluate $f(x)$ to obtain the height of the approximating rectangle. Let's consider another example, this time choosing the *left end point* of each subinterval.

EXAMPLE 2   Let R be the region under the graph of $f(x) = 16 - x^2$ on the interval $[1, 3]$. Find an approximation of the area A of R using four subintervals of $[1, 3]$ of equal length and picking the left end point of each subinterval to evaluate $f(x)$ to obtain the height of the approximating rectangle.

FIGURE 6.10
The area of R in (a) is approximated by the sum of the areas of the four rectangles in (b).

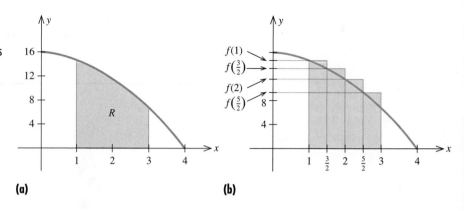

(a)   (b)

SOLUTION ✔   The graph of f is sketched in Figure 6.10a. Since the length of $[1, 3]$ is 2, we see that the length of each subinterval is $\frac{2}{4}$, or $\frac{1}{2}$. Therefore, the four subintervals are

$$\left[1, \frac{3}{2}\right], \quad \left[\frac{3}{2}, 2\right], \quad \left[2, \frac{5}{2}\right], \quad \left[\frac{5}{2}, 3\right]$$

The left end points of these subintervals are 1, $\frac{3}{2}$, 2, and $\frac{5}{2}$, respectively, so the heights of the approximating rectangles are $f(1)$, $f(\frac{3}{2})$, $f(2)$, and $f(\frac{5}{2})$, respectively (Figure 6.10b). Therefore, the required approximation is

$$A \approx \frac{1}{2} f(1) + \frac{1}{2} f\left(\frac{3}{2}\right) + \frac{1}{2} f(2) + \frac{1}{2} f\left(\frac{5}{2}\right)$$

$$= \frac{1}{2}\left[f(1) + f\left(\frac{3}{2}\right) + f(2) + f\left(\frac{5}{2}\right)\right]$$

$$= \frac{1}{2}\left\{[16 - (1)^2] + \left[16 - \left(\frac{3}{2}\right)^2\right]\right.$$

$$\left. + [16 - (2)^2] + \left[16 - \left(\frac{5}{2}\right)^2\right]\right\} \qquad \text{[Recall that } f(x) = 16 - x^2.\text{]}$$

$$= \frac{1}{2}\left(15 + \frac{55}{4} + 12 + \frac{39}{4}\right) = \frac{101}{4}$$

or approximately 25.25 square units.   ■ ■ ■ ■

Table 6.2 shows the approximations of the area A of the region R of Example 2 when n rectangles are used for the approximation and the heights of the approximating rectangles are found by evaluating $f(x)$ at the left end points.

Once again, we see that the approximations seem to approach a unique number as n gets larger and larger—this time the number is $23\frac{1}{3}$. This result

Table 6.2							
Number of Rectangles n	4	10	100	1,000	10,000	50,000	100,000
Approximation of A	25.2500	24.1200	23.4132	23.3413	23.3341	23.3335	23.3334

suggests that we *define* the area of the region under the graph of $f(x) = 16 - x^2$ on the interval $[1, 3]$ to be $23\frac{1}{3}$ square units.

DEFINING AREA—THE GENERAL CASE

Examples 1 and 2 point the way to defining the area A under the graph of an arbitrary but continuous and nonnegative function f on an interval $[a, b]$ (Figure 6.11a).

FIGURE 6.11
The area of the region under the graph of f on $[a, b]$ in (a) is approximated by the sum of the areas of the n rectangles shown in (b).

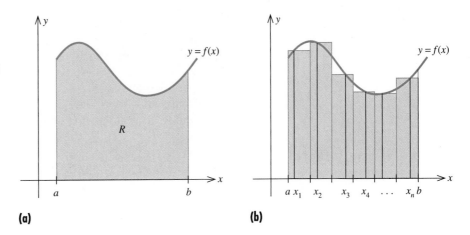

(a)   **(b)**

Divide the interval $[a, b]$ into n subintervals of equal length $\Delta x = (b - a)/n$. Next, pick n arbitrary points $x_1, x_2, \ldots, x_n$, called *representative points,* from the first, second, …, and nth subintervals, respectively (Figure 6.11b). Then, approximating the area A of the region R by the n rectangles of width Δx and heights $f(x_1), f(x_2), \ldots, f(x_n)$, so that the areas of the rectangles are $f(x_1)\Delta x, f(x_2)\Delta x, \ldots, f(x_n)\Delta x$, we have

$$A \approx f(x_1)\Delta x + f(x_2)\Delta x + \cdots + f(x_n)\Delta x$$

The sum on the right-hand side of this expression is called a **Riemann sum** in honor of the German mathematician Bernhard Riemann (1826–1866). Now, as the earlier examples seem to suggest, the Riemann sum will approach a unique number as n becomes arbitrarily large.* We define this number to be the area A of the region R.

The Area Under the Graph of a Function

Let f be a nonnegative continuous function on $[a, b]$. Then, the area of the region under the graph of f is

$$A = \lim_{n \to \infty} [f(x_1) + f(x_2) + \cdots + f(x_n)]\Delta x \qquad (6)$$

where $x_1, x_2, \ldots, x_n$ are arbitrary points in the n subintervals of $[a, b]$ of equal width $\Delta x = (b - a)/n$.

---

* Even though we chose the representative points to be the midpoints of the subintervals in Example 1 and the left end points in Example 2, it can be shown that each of the respective sums will always approach a unique number as n approaches infinity.

THE DEFINITE INTEGRAL

As we have just seen, the area under the graph of a continuous *nonnegative* function f on an interval $[a, b]$ is defined by the limit of the Riemann sum

$$\lim_{n \to \infty} [f(x_1)\Delta x + f(x_2)\Delta x + \cdots + f(x_n)\Delta x]$$

We now turn our attention to the study of limits of Riemann sums involving functions that are not necessarily nonnegative. Such limits arise in many applications of calculus.

For example, the calculation of the distance covered by a body traveling along a straight line involves evaluating a limit of this form. The computation of the total revenue realized by a company over a certain time period, the calculation of the total amount of electricity consumed in a typical home over a 24-hour period, the average concentration of a drug in a body over a certain interval of time, and the volume of a solid—all involve limits of this type.

We begin with the following definition.

The Definite Integral

Let f be defined on $[a, b]$. If

$$\lim_{n \to \infty} [f(x_1)\Delta x + f(x_2)\Delta x + \cdots + f(x_n)\Delta x]$$

exists for all choices of representative points $x_1, x_2, \ldots, x_n$ in the n subintervals of $[a, b]$ of equal width $\Delta x = (b - a)/n$, then this limit is called the **definite integral of f from a to b** and is denoted by $\int_a^b f(x)\, dx$. Thus,

$$\int_a^b f(x)\, dx = \lim_{n \to \infty} [f(x_1)\Delta x + f(x_2)\Delta x + \cdots + f(x_n)\Delta x] \qquad (7)$$

The number a is the **lower limit of integration**, and the number b is the **upper limit of integration**.

REMARKS

1. If f is nonnegative, then the limit in (7) is the same as the limit in (6); therefore, the definite integral gives the area under the graph of f on $[a, b]$.
2. The limit in (7) is denoted by the integral sign $\int$ because, as we will see later, the definite integral and the antiderivative of a function f are related.
3. It is important to realize that the definite integral $\int_a^b f(x)\, dx$ is a number, whereas the indefinite integral $\int f(x)\, dx$ represents a family of functions (the antiderivatives of f).
4. If the limit in (7) exists, we say that f is **integrable** on the interval $[a, b]$.

■ ■ ■

WHEN IS A FUNCTION INTEGRABLE?

The following theorem, which we state without proof, guarantees that a continuous function is integrable.

Integrability of a Function

Let f be continuous on $[a, b]$. Then, f is integrable on $[a, b]$; that is, the definite integral $\int_a^b f(x)\, dx$ exists.

GEOMETRIC INTERPRETATION OF THE DEFINITE INTEGRAL

If f is nonnegative and integrable on $[a, b]$, then we have the following geometric interpretation of the definite integral $\int_a^b f(x)\, dx$.

Geometric Interpretation of $\int_a^b f(x)\, dx$ for $f(x) \geq 0$ on $[a, b]$

If f is nonnegative and continuous on $[a, b]$, then

$$\int_a^b f(x)\, dx \qquad (8)$$

is equal to the area of the region under the graph of f on $[a, b]$ (Figure 6.12).

FIGURE 6.12
If $f(x) \geq 0$ on $[a, b]$, then $\int_a^b f(x)\, dx = $ area under the graph of f on $[a, b]$.

Group Discussion

Suppose f is nonpositive [that is, $f(x) \leq 0$] and continuous on $[a, b]$. Explain why the area of the region below the x-axis and above the graph of f is given by $-\int_a^b f(x)\, dx$.

Next, let's extend our geometric interpretation of the definite integral to include the case where f assumes both positive as well as negative values on $[a, b]$. Consider a typical Riemann sum of the function f,

$$f(x_1)\Delta x + f(x_2)\Delta x + \cdots + f(x_n)\Delta x$$

corresponding to a partition of $[a, b]$ into n subintervals of equal width $(b - a)/n$, where $x_1, x_2, \ldots, x_n$ are representative points in the subintervals. The sum consists of n terms in which a positive term corresponds to the area of a rectangle of height $f(x_k)$ (for some positive integer k) lying above the x-axis and a negative term corresponds to the area of a rectangle of height $-f(x_k)$ lying below the x-axis. (See Figure 6.13, which depicts a situation with $n = 6$).

As n gets larger and larger, the sums of the areas of the rectangles lying above the x-axis seem to give a better and better approximation of the area of the region lying above the x-axis (Figure 6.14). Similarly, the sums of the

FIGURE 6.13
The positive terms in the Riemann sum are associated with the areas of the rectangles that lie above the x-axis, and the negative terms are associated with the areas of those that lie below the x-axis.

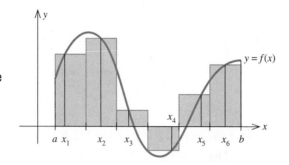

FIGURE 6.14
As n gets larger, the approximations get better. Here, $n = 12$ and we are approximating with twice as many rectangles as in Figure 6.13.

areas of those rectangles lying below the x-axis seem to give a better and better approximation of the area of the region lying below the x-axis.

These observations suggest the following geometric interpretation of the definite integral for an arbitrary continuous function on an interval $[a, b]$.

Geometric Interpretation of $\int_a^b f(x)\,dx$ on $[a, b]$

If f is continuous on $[a, b]$, then

$$\int_a^b f(x)\,dx$$

is equal to the area of the region above $[a, b]$ minus the area of the region below $[a, b]$ (Figure 6.15).

FIGURE 6.15
$\int_a^b f(x)\,dx =$
area of R_1 − area of R_2 + area of R_3

SELF-CHECK EXERCISE 6.3

Find an approximation of the area of the region R under the graph of $f(x) = 2x^2 + 1$ on the interval $[0, 3]$, using four subintervals of $[0, 3]$ of equal length and picking the midpoint of each subinterval as a representative point.

The solution to Self-Check Exercise 6.3 can be found on page 477.

6.3 Exercises

In Exercises 1 and 2, find an approximation of the area of the region *R* under the graph of *f* by computing the Riemann sum of *f* corresponding to the partition of the interval into the subintervals shown in the accompanying figures. In each case, use the midpoints of the subintervals as the representative points.

1.

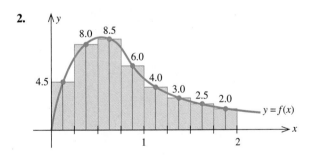

2.

3. Let $f(x) = 3x$.
   a. Sketch the region R under the graph of f on the interval $[0, 2]$ and find its exact area using geometry.
   b. Use a Riemann sum with four subintervals of equal length ($n = 4$) to approximate the area of R. Choose the representative points to be the left end points of the subintervals.

c. Repeat part (b) with eight subintervals of equal length ($n = 8$).
d. Compare the approximations obtained in parts (b) and (c) with the exact area found in part (a). Do the approximations improve with larger n?

4. Repeat Exercise 3, choosing the representative points to be the right end points of the subintervals.

5. Let $f(x) = 4 - 2x$.
   a. Sketch the region R under the graph of f on the interval $[0, 2]$ and find its exact area using geometry.
   b. Use a Riemann sum with five subintervals of equal length ($n = 5$) to approximate the area of R. Choose the representative points to be the left end points of the subintervals.
   c. Repeat part (b) with ten subintervals of equal length ($n = 10$).
   d. Compare the approximations obtained in parts (b) and (c) with the exact area found in part (a). Do the approximations improve with larger n?

6. Repeat Exercise 5, choosing the representative points to be the right end points of the subintervals.

7. Let $f(x) = x^2$ and compute the Riemann sum of f over the interval $[2, 4]$, using:
   a. Two subintervals of equal length ($n = 2$).
   b. Five subintervals of equal length ($n = 5$).
   c. Ten subintervals of equal length ($n = 10$).
   In each case, choose the representative points to be the midpoints of the subintervals.
   d. Can you guess at the area of the region under the graph of f on the interval $[2, 4]$?

8. Repeat Exercise 7, choosing the representative points to be the left end points of the subintervals.

9. Repeat Exercise 7, choosing the representative points to be the right end points of the subintervals.

10. Let $f(x) = x^3$ and compute the Riemann sum of f over the interval [0, 1], using:
a. Two subintervals of equal length ($n = 2$).
b. Five subintervals of equal length ($n = 5$).
c. Ten subintervals of equal length ($n = 10$).
In each case, choose the representative points to be the midpoints of the subintervals.
d. Can you guess at the area of the region under the graph of f on the interval [0, 1]?

11. Repeat Exercise 10, choosing the representative points to be the left end points of the subintervals.

12. Repeat Exercise 10, choosing the representative points to be the right end points of the subintervals.

In Exercises 13–16, find an approximation of the area of the region R under the graph of the function f on the interval [a, b]. In each case, use n subintervals and choose the representative points as indicated.

13. $f(x) = x^2 + 1$; [0, 2]; $n = 5$; midpoints

14. $f(x) = 4 - x^2$; [−1, 2]; $n = 6$; left end points

15. $f(x) = \dfrac{1}{x}$; [1, 3]; $n = 4$; right end points

16. $f(x) = e^x$; [0, 3]; $n = 5$; midpoints

17. REAL ESTATE Figure (a) shows a vacant lot with a 100-ft frontage in a development. To estimate its area, we introduce a coordinate system so that the x-axis coincides with the edge of the straight road forming the lower boundary of the property, as shown in Figure (b). Then, thinking of the upper boundary of the property as the graph of a continuous function f over the interval [0, 100], we see that the problem is mathematically equivalent to that of finding the area under the graph of f on [0, 100]. To estimate the area of the lot using a Riemann sum, we divide the interval [0, 100] into five equal subintervals of length 20 ft. Then, using surveyor's equipment, we measure the distance from the midpoint of each of these subintervals to the upper boundary of the property. These measurements give the values of $f(x)$ at $x = 10$, 30, 50, 70, and 90. What is the approximate area of the lot?

Road

(a)

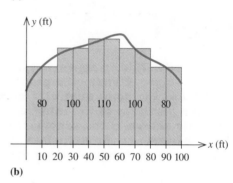

(b)

18. REAL ESTATE Use the technique of Exercise 17 to obtain an estimate of the area of the vacant lot shown in the accompanying figures.

Road

Road

(a)

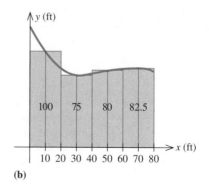

(b)

Solution to Self-Check Exercise 6.3

The length of each subinterval is $\frac{3}{4}$. Therefore, the four subintervals are

$$\left[0, \frac{3}{4}\right], \qquad \left[\frac{3}{4}, \frac{3}{2}\right], \qquad \left[\frac{3}{2}, \frac{9}{4}\right], \qquad \left[\frac{9}{4}, 3\right]$$

The representative points are $\frac{3}{8}$, $\frac{9}{8}$, $\frac{15}{8}$, and $\frac{21}{8}$, respectively. Therefore, the required approximation is

$$
\begin{aligned}
A &= \frac{3}{4} f\left(\frac{3}{8}\right) + \frac{3}{4} f\left(\frac{9}{8}\right) + \frac{3}{4} f\left(\frac{15}{8}\right) + \frac{3}{4} f\left(\frac{21}{8}\right) \\
&= \frac{3}{4}\left[f\left(\frac{3}{8}\right) + f\left(\frac{9}{8}\right) + f\left(\frac{15}{8}\right) + f\left(\frac{21}{8}\right)\right] \\
&= \frac{3}{4}\left\{ \left[2\left(\frac{3}{8}\right)^2 + 1\right] + \left[2\left(\frac{9}{8}\right)^2 + 1\right] + \left[2\left(\frac{15}{8}\right)^2 + 1\right] + \left[2\left(\frac{21}{8}\right)^2 + 1\right]\right\} \\
&= \frac{3}{4}\left(\frac{41}{32} + \frac{113}{32} + \frac{257}{32} + \frac{473}{32}\right) = \frac{663}{32}
\end{aligned}
$$

or approximately 20.72 square units.

6.4 The Fundamental Theorem of Calculus

The Fundamental Theorem of Calculus

In Section 6.3 we defined the definite integral of an arbitrary continuous function on an interval $[a, b]$ as a limit of Riemann sums. Calculating the value of a definite integral by actually taking the limit of such sums is tedious and in most cases impractical. It is important to realize that the numerical results we obtained in Examples 1 and 2 of Section 6.3 were *approximations* of the respective areas of the regions in question, even though these results enabled us to *conjecture* what the actual areas might be. Fortunately, there is a much better way of finding the exact value of a definite integral.

The following theorem shows how to evaluate the definite integral of a continuous function provided we can find an antiderivative of that function. Because of its importance in establishing the relationship between differentiation and integration, this theorem—discovered independently by Sir Isaac Newton (1642–1727) in England and Gottfried Wilhelm Leibniz (1646–1716) in Germany—is called the **fundamental theorem of calculus.**

THEOREM 2

The Fundamental Theorem of Calculus
Let f be continuous on $[a, b]$. Then,

$$\int_a^b f(x)\, dx = F(b) - F(a) \qquad (9)$$

where F is any antiderivative of f; that is, $F'(x) = f(x)$.

We will explain why this theorem is true at the end of this section.

When applying the fundamental theorem of calculus, it is convenient to use the notation

$$F(x)\, \bigg|_a^b = F(b) - F(a)$$

For example, using this notation, Equation (9) is written

$$\int_a^b f(x)\, dx = F(x)\, \bigg|_a^b = F(b) - F(a)$$

EXAMPLE 1 Let R be the region under the graph of $f(x) = x$ on the interval $[1, 3]$. Use the fundamental theorem of calculus to find the area A of R and verify your result by elementary means.

SOLUTION ✔ The region R is shown in Figure 6.16a. Since f is nonnegative on $[1, 3]$, the area of R is given by the definite integral of f from 1 to 3; that is,

$$A = \int_1^3 x\, dx$$

FIGURE 6.16
The area of R can be computed in two different ways.

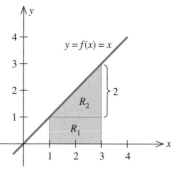

(a)   (b)

To evaluate the definite integral, observe that an antiderivative of $f(x) = x$ is $F(x) = \frac{1}{2}x^2 + C$, where C is an arbitrary constant. Therefore, by the fundamental theorem of calculus, we have

$$A = \int_1^3 x\, dx = \frac{1}{2}x^2 + C\, \bigg|_1^3$$

$$= \left(\frac{9}{2} + C\right) - \left(\frac{1}{2} + C\right) = 4 \text{ square units}$$

To verify this result by elementary means, observe that the area A is the area of the rectangle R_1 (width $\times$ height) plus the area of the triangle R_2 ($\frac{1}{2}$ base $\times$ height) (see Figure 6.16b); that is,

$$2(1) + \frac{1}{2}(2)(2) = 2 + 2 = 4$$

which agrees with the result obtained earlier.   ■ ■ ■ ■

Observe that in evaluating the definite integral in Example 1, the constant of integration "dropped out." This is true in general, for if $F(x) + C$ denotes an antiderivative of some function f, then

$$F(x) + C \,\Big|_a^b = [F(b) + C] - [F(a) + C]$$
$$= F(b) + C - F(a) - C$$
$$= F(b) - F(a)$$

With this fact in mind, we may, in all future computations involving the evaluations of a definite integral, drop the constant of integration from our calculations.

FINDING THE AREA UNDER A CURVE

Having seen how effective the fundamental theorem of calculus is in helping us find the area of simple regions, we now use it to find the area of more complicated regions.

EXAMPLE 2

In Section 6.3 we conjectured that the area of the region R under the graph of $f(x) = x^2$ on the interval $[0, 1]$ was $\frac{1}{3}$ square unit. Use the fundamental theorem of calculus to verify this conjecture.

SOLUTION ✔

The region R is reproduced in Figure 6.17. Observe that f is nonnegative on $[0, 1]$, so the area of R is given by $A = \int_0^1 x^2 \, dx$. Since an antide-

FIGURE 6.17
The area of R is $\int_0^1 x^2 \, dx = \frac{1}{3}$.

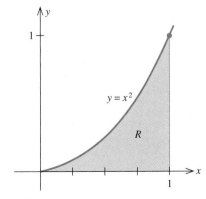

rivative of $f(x) = x^2$ is $F(x) = \frac{1}{3}x^3$, we see, using the fundamental theorem of calculus, that

$$A = \int_0^1 x^2\, dx = \frac{1}{3}x^3\Big|_0^1 = \frac{1}{3}(1) - \frac{1}{3}(0) = \frac{1}{3}\ \text{square unit}$$

as we wished to show.   ■ ■ ■ ■

EXAMPLE 3   Find the area of the region R under the graph of $y = x^2 + 1$ from $x = -1$ to $x = 2$.

SOLUTION ✔   The region R under consideration is shown in Figure 6.18. Using the fundamental theorem of calculus, we find that the required area is

$$\int_{-1}^2 (x^2 + 1)\, dx = \left(\frac{1}{3}x^3 + x\right)\Big|_{-1}^2$$

$$= \left[\frac{1}{3}(8) + 2\right] - \left[\frac{1}{3}(-1)^3 + (-1)\right] = 6$$

or 6 square units.

FIGURE 6.18
The area of R is $\int_{-1}^2 (x^2 + 1)\, dx.$

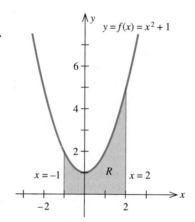

■ ■ ■ ■

EVALUATING DEFINITE INTEGRALS

In Examples 4 and 5 we use the rules of integration of Section 6.1 to help us evaluate the definite integrals.

EXAMPLE 4   Evaluate $\int_1^3 (3x^2 + e^x)\, dx.$

SOLUTION ✔

$$\int_1^3 (3x^2 + e^x)\, dx = x^3 + e^x\Big|_1^3$$

$$= (27 + e^3) - (1 + e) = 26 + e^3 - e \qquad ■ ■ ■ ■$$

EXAMPLE 5   Evaluate $\int_1^2 \left(\dfrac{1}{x} - \dfrac{1}{x^2} \right) dx$.

SOLUTION ✔

$$\int_1^2 \left(\frac{1}{x} - \frac{1}{x^2} \right) dx = \int_1^2 \left(\frac{1}{x} - x^{-2} \right) dx$$

$$= \ln |x| + \frac{1}{x} \bigg|_1^2$$

$$= \left(\ln 2 + \frac{1}{2} \right) - (\ln 1 + 1)$$

$$= \ln 2 - \frac{1}{2} \qquad \text{(Recall, } \ln 1 = 0.)$$

■ ■ ■ ■

 Group Discussion

Consider the definite integral $\int_{-1}^1 \dfrac{1}{x^2} dx$.

1. Show that a formal application of Equation (9) leads to

$$\int_{-1}^1 \frac{1}{x^2} dx = -\frac{1}{x} \bigg|_{-1}^1 = -1 - 1 = -2$$

2. Observe that $f(x) = 1/x^2$ is positive at each value of x in $[-1, 1]$ where it is defined. Therefore, one might expect that the definite integral with integrand f has a positive value, if it exists.

3. Explain this apparent contradiction in the result (1) and the observation (2).

APPLICATIONS

 EXAMPLE 6   The management of Staedtler Office Equipment has determined that the daily marginal cost function associated with producing battery-operated pencil sharpeners is given by

$$C'(x) = 0.000006x^2 - 0.006x + 4$$

where $C'(x)$ is measured in dollars per unit and x denotes the number of units produced. Management has also determined that the daily fixed cost incurred in producing these pencil sharpeners is $100. Find Staedtler's daily total cost for producing (a) the first 500 units and (b) the 201st through 400th units.

SOLUTION ✔   **a.** Since $C'(x)$ is the marginal cost function, its antiderivative $C(x)$ is the total cost function. The daily fixed cost incurred in producing the pencil sharpeners is $C(0)$ dollars. Since the daily fixed cost is given as $100, we have $C(0) = 100$. We are required to find $C(500)$. Let's compute $C(500) - C(0)$, the net change in the total cost function $C(x)$ over the interval $[0, 500]$. Using the fundamental theorem of calculus, we find

$$C(500) - C(0) = \int_0^{500} C'(x)\, dx$$

$$= \int_0^{500} (0.000006x^2 - 0.006x + 4)\, dx$$

$$= 0.000002x^3 - 0.003x^2 + 4x \Big|_0^{500}$$

$$= [0.000002(500)^3 - 0.003(500)^2 + 4(500)]$$
$$\quad - [0.000002(0)^3 - 0.003(0)^2 + 4(0)]$$

$$= 1500$$

Therefore, $C(500) = 1500 + C(0) = 1500 + 100 = 1600$, so the total cost incurred daily by Staedtler in producing 500 pencil sharpeners is $1600.

b. The daily total cost incurred by Staedtler in producing the 201st through 400th units of battery-operated pencil sharpeners is given by

$$C(400) - C(200) = \int_{200}^{400} C'(x)\, dx$$

$$= \int_{200}^{400} (0.000006x^2 - 0.006x + 4)\, dx$$

$$= 0.000002x^3 - 0.003x^2 + 4x \Big|_{200}^{400}$$

$$= 552$$

or $552.   ▪ ▪ ▪ ▪

Since $C'(x)$ is nonnegative for x in the interval $(0, \infty)$, we have the following geometric interpretation of the two definite integrals in Example 6: $\int_0^{500} C'(x)\, dx$ is the area of the region under the graph of the function C' from $x = 0$ to $x = 500$, shown in Figure 6.19a, and $\int_{200}^{400} C'(x)\, dx$ is the area of the region from $x = 200$ to $x = 400$, shown in Figure 6.19b.

FIGURE 6.19

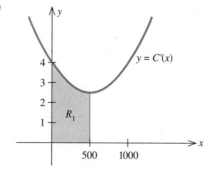

(a) Area of $R_1 = \int_0^{500} C'(x)\, dx$

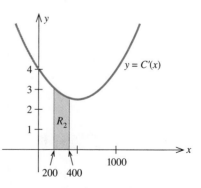

(b) Area of $R_2 = \int_{200}^{400} C'(x)\, dx$

EXAMPLE **7**

An efficiency study conducted for the Elektra Electronics Company showed that the rate at which Space Commander walkie-talkies are assembled by the average worker t hours after starting work at 8 A.M. is given by the function

$$f(t) = -3t^2 + 12t + 15 \qquad (0 \le t \le 4)$$

Determine how many walkie-talkies can be assembled by the average worker in the first hour of the morning shift.

SOLUTION ✔

Let $N(t)$ denote the number of walkie-talkies assembled by the average worker t hours after starting work in the morning shift. Then, we have

$$N'(t) = f(t) = -3t^2 + 12t + 15$$

Therefore, the number of units assembled by the average worker in the first hour of the morning shift is

$$N(1) - N(0) = \int_0^1 N'(t)\, dt = \int_0^1 (-3t^2 + 12t + 15)\, dt$$

$$= -t^3 + 6t^2 + 15t \, \Big|_0^1 = -1 + 6 + 15$$

$$= 20$$

or 20 units.

■ ■ ■ ■

Exploring with Technology

You can demonstrate graphically that $\int_0^x t\, dt = \frac{1}{2}x^2$ as follows:

1. Plot the graphs of $y1 = \int_0^x t\, dt$ and $y2 = \frac{1}{2}x^2$ on the same set of axes using the viewing rectangle $[-5, 5] \times [0, 10]$.
2. Compare the graphs of $y1$ and $y2$ and draw the desired conclusion.

 EXAMPLE **8**

A certain city's rate of electricity consumption is expected to grow exponentially with a growth constant of $k = 0.04$. If the present rate of consumption is 40 million kilowatt-hours (kWh) per year, what should the total production of electricity be over the next 3 years in order to meet the projected demand?

SOLUTION ✔

If $R(t)$ denotes the expected rate of consumption of electricity t years from now, then

$$R(t) = 40e^{0.04t}$$

million kWh per year. Next, if $C(t)$ denotes the expected total consumption of electricity over a period of t years, then

$$C'(t) = R(t)$$

Therefore, the total consumption of electricity expected over the next 3 years is given by

$$\int_0^3 C'(t)\, dt = \int_0^3 40 e^{0.04t}\, dt$$

$$= \frac{40}{0.04} e^{0.04t} \Big|_0^3$$

$$= 1000(e^{0.12} - 1)$$

$$= 127.5$$

or 127.5 million kWh, the amount that must be produced over the next 3 years in order to meet the demand.   ■ ■ ■ ■

> **Group Discussion**
>
> The definite integral $\int_{-3}^3 \sqrt{9 - x^2}\, dx$ cannot be evaluated using the fundamental theorem of calculus because the method of this section does not enable us to find an antiderivative of the integrand. But the integral can be evaluated by interpreting it as the area of a certain plane region. What is the region? And what is the value of the integral?

VALIDITY OF THE FUNDAMENTAL THEOREM OF CALCULUS

To demonstrate the plausibility of the fundamental theorem of calculus for the case where f is nonnegative on an interval $[a, b]$, let's define an "area function" A as follows. Let $A(t)$ denote the area of the region R under the graph of $y = f(x)$ from $x = a$ to $x = t$, where $a \le t \le b$ (Figure 6.20).

If h is a small positive number, then $A(t + h)$ is the area of the region under the graph of $y = f(x)$ from $x = a$ to $x = t + h$. Therefore, the difference

$$A(t + h) - A(t)$$

is the area under the graph of $y = f(x)$ from $x = t$ to $x = t + h$ (Figure 6.21).

FIGURE 6.20
$A(t)$ = area under the graph of f from $x = a$ to $x = t$

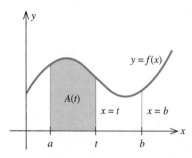

FIGURE 6.21
$A(t + h) - A(t)$ = area under the graph of f from $x = t$ to $x = t + h$

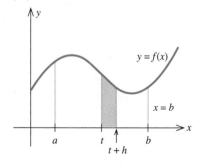

FIGURE 6.22
The area of the rectangle is $h \cdot f(t)$.

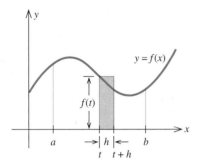

Now, the area of this last region can be approximated by the area of the rectangle of width h and height $f(t)$—that is, by the expression $h \cdot f(t)$ (Figure 6.22). Thus,

$$A(t + h) - A(t) \approx h \cdot f(t)$$

where the approximations improve as h is taken to be smaller and smaller.
Dividing both sides of the foregoing relationship by h, we obtain

$$\frac{A(t + h) - A(t)}{h} \approx f(t)$$

Taking the limit as h approaches zero, we find, by the definition of the derivative, that the left-hand side is

$$\lim_{h \to 0} \frac{A(t + h) - A(t)}{h} = A'(t)$$

The right-hand side, which is independent of h, remains constant throughout the limiting process. Because the approximation becomes exact as h approaches zero, we find that

$$A'(t) = f(t)$$

Since the foregoing equation holds for all values of t in the interval $[a, b]$, we have shown that the *area function* A is an antiderivative of the function $f(x)$. By Theorem 1 of Section 6.1, we conclude that $A(x)$ must have the form

$$A(x) = F(x) + C$$

where F is any antiderivative of f and C is an arbitrary constant. To determine the value of C, observe that $A(a) = 0$. This condition implies that

$$A(a) = F(a) + C = 0$$

FIGURE 6.23
The area of R is given by $A(b)$.

or $C = -F(a)$. Next, since the area of the region R is $A(b)$ (Figure 6.23), we see that the required area is

$$A(b) = F(b) + C$$
$$= F(b) - F(a)$$

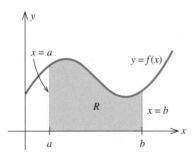

Since the area of the region R is

$$\int_a^b f(x)\, dx$$

we have

$$\int_a^b f(x)\, dx = F(b) - F(a)$$

as we set out to show.

SELF-CHECK EXERCISES 6.4

1. Evaluate $\int_0^2 (x + e^x)\, dx$.

2. The daily marginal profit function associated with producing and selling Texa-Pep hot sauce is

$$P'(x) = -0.000006x^2 + 6$$

where x denotes the number of cases (each case contains 24 bottles) produced and sold daily and $P'(x)$ is measured in dollars/unit. The fixed cost is $400.
a. What is the total profit realizable from producing and selling 1000 cases of Texa-Pep per day?
b. What is the additional profit realizable if the production and sale of Texa-Pep is increased from 1000 to 1200 cases/day?

Solutions to Self-Check Exercises 6.4 can be found on page 490.

6.4 Exercises

In Exercises 1–4, find the area of the region under the graph of the function *f* on the interval [*a*, *b*], using the fundamental theorem of calculus. Then verify your result using geometry.

1. $f(x) = 2;\ [1, 4]$

2. $f(x) = 4;\ [-1, 2]$

3. $f(x) = 2x;\ [1, 3]$

4. $f(x) = -\dfrac{1}{4}x + 1;\ [1, 4]$

In Exercises 5–16, find the area of the region under the graph of the function *f* on the interval [*a*, *b*].

5. $f(x) = 2x + 3;\ [-1, 2]$

6. $f(x) = 4x - 1;\ [2, 4]$

7. $f(x) = -x^2 + 4;\ [-1, 2]$

8. $f(x) = 4x - x^2;\ [0, 4]$

9. $f(x) = \dfrac{1}{x};\ [1, 2]$

10. $f(x) = \dfrac{1}{x^2};\ [2, 4]$

11. $f(x) = \sqrt{x};\ [1, 9]$

12. $f(x) = x^3;\ [1, 3]$

13. $f(x) = 1 - \sqrt[3]{x};\ [-8, -1]$

14. $f(x) = \dfrac{1}{\sqrt{x}};\ [1, 9]$

15. $f(x) = e^x;\ [0, 2]$

16. $f(x) = e^x - x;\ [1, 2]$

In Exercises 17–40, evaluate the definite integral.

17. $\int_2^4 3\, dx$

18. $\int_{-1}^2 -2\, dx$

19. $\int_1^3 (2x + 3)\, dx$

20. $\int_{-1}^0 (4 - x)\, dx$

21. $\int_{-1}^3 2x^2\, dx$

22. $\int_0^2 8x^3\, dx$

23. $\int_{-2}^2 (x^2 - 1)\, dx$

24. $\int_1^4 \sqrt{u}\, du$

25. $\int_1^8 4x^{1/3}\, dx$

26. $\int_1^4 2x^{-3/2}\, dx$

27. $\int_0^1 (x^3 - 2x^2 + 1)\, dx$

28. $\int_1^2 (t^5 - t^3 + 1)\, dt$

29. $\int_2^4 \dfrac{1}{x}\, dx$

30. $\int_1^3 \dfrac{2}{x}\, dx$

31. $\int_0^4 x(x^2 - 1)\, dx$

32. $\int_0^2 (x - 4)(x - 1)\, dx$

33. $\int_1^3 (t^2 - t)^2\, dt$

34. $\int_{-1}^1 (x^2 - 1)^2\, dx$

35. $\int_{-3}^{-1} \dfrac{1}{x^2}\, dx$

36. $\int_1^2 \dfrac{2}{x^3}\, dx$

37. $\int_1^4 \left(\sqrt{x} - \dfrac{1}{\sqrt{x}}\right) dx$

38. $\int_0^1 \sqrt{2x}(\sqrt{x} + \sqrt{2})\, dx$

39. $\int_1^4 \dfrac{3x^3 - 2x^2 + 4}{x^2}\, dx$

40. $\int_1^2 \left(1 + \dfrac{1}{u} + \dfrac{1}{u^2}\right) du$

A calculator is recommended for Exercises
41–47.

41. MARGINAL COST A division of Ditton Industries manufactures a deluxe toaster oven. Management has determined that the daily marginal cost function associated with producing these toaster ovens is given by

$$C'(x) = 0.0003x^2 - 0.12x + 20$$

where $C'(x)$ is measured in dollars/unit and x denotes the number of units produced. Management has also determined that the daily fixed cost incurred in the production is $800.
a. Find the total cost incurred by Ditton in producing the first 300 units of these toaster ovens per day.
b. What is the total cost incurred by Ditton in producing the 201st through 300th units/day?

42. MARGINAL REVENUE The management of Ditton Industries has determined that the daily marginal revenue function associated with selling x units of their deluxe toaster ovens is given by

$$R'(x) = -0.1x + 40$$

where $R'(x)$ is measured in dollars/unit.
a. Find the daily total revenue realized from the sale of 200 units of the toaster oven.
b. Find the additional revenue realized when the production (and sales) level is increased from 200 to 300 units.

43. MARGINAL PROFIT Refer to Exercise 41. The daily marginal profit function associated with the production and sales of the deluxe toaster ovens is known to be

$$P'(x) = -0.0003x^2 + 0.02x + 20$$

where x denotes the number of units manufactured and sold daily and $P'(x)$ is measured in dollars/unit.
a. Find the total profit realizable from the manufacture and sale of 200 units of the toaster ovens per day.

Hint: $P(200) - P(0) = \int_0^{200} P'(x)\, dx$, $P(0) = -800$.

b. What is the additional daily profit realizable if the production and sale of the toaster ovens are increased from 200 to 220 units/day?

44. EFFICIENCY STUDIES Tempco Electronics, a division of Tempco Toys, Inc., manufactures an electronic football game. An efficiency study showed that the rate at which the games are assembled by the average worker t hr

after starting work at 8 A.M. is

$$-\frac{3}{2}t^2 + 6t + 20 \qquad (0 \le t \le 4)$$

units/hour.
a. Find the total number of games the average worker can be expected to assemble in the 4-hr morning shift.
b. How many units can the average worker be expected to assemble in the first hour of the morning shift? In the second hour of the morning shift?

45. SPEEDBOAT RACING In a recent pretrial run for the world water speed record, the velocity of the *Sea Falcon II* t sec after firing the booster rocket was given by

$$v(t) = -t^2 + 20t + 440 \qquad (0 \le t \le 20)$$

ft/sec. Find the distance covered by the boat over the 20-sec period after the booster rocket was activated.
Hint: The distance is given by $\int_0^{20} v(t)\, dt$.

46. HAND-HELD COMPUTERS Annual sales (in millions of units) of hand-held computers are expected to grow in accordance with the function

$$f(t) = 0.18t^2 + 0.16t + 2.64 \qquad (0 \le t \le 6)$$

where t is measured in years, with $t = 0$ corresponding to 1997. How many hand-held computers will be sold over the 6-yr period between the beginning of 1997 and the end of 2002?
Source: Dataquest, Inc.

47. U.S. CENSUS According to the U.S. Census Bureau, the number of Americans aged 45 to 54 (which stood at 25 million at the beginning of 1990) grew at the rate of

$$R(t) = 0.00933t^3 + 0.019t^2 - 0.10833t + 1.3467$$

million people/year, t yr from the beginning of 1990. How many Americans aged 45 to 54 were added to the population between 1990 and the year 2000?
Source: U.S. Census Bureau

48. AIR PURIFICATION To test air purifiers, the engineers run a purifier in a smoke-filled 10×20-ft room. While conducting a test for a certain brand of air purifier, it was determined that the amount of smoke in the room was decreasing at the rate of

$$R(t) = 0.00032t^4 - 0.01872t^3 + 0.3948t^2$$
$$- 3.83t + 17.63 \qquad (0 \le t \le 20)$$

percent of the (original) amount of the smoke per minute, t min after the start of the test. How much smoke was left in the room 5 min after the start of the test? Ten minutes after the start of the test?
Source: Consumer Reports

(continued on p. 490)

EVALUATING DEFINITE INTEGRALS

Some graphing utilities have an operation for finding the definite integral of a function. If your graphing utility has this capability, use it to work through the example and exercises of this section.

EXAMPLE 1   Use the numerical integral operation of a graphing utility to evaluate

$$\int_{-1}^{2} \frac{2x+4}{(x^2+1)^{3/2}}\, dx$$

SOLUTION ✔   Using the numerical integral operation of a graphing utility, we find

$$\int_{-1}^{2} \frac{2x+4}{(x^2+1)^{3/2}}\, dx \approx 6.92592225992$$

■ ■ ■ ■

Exercises

In Exercises 1–4, use a graphing utility to find the area of the region under the graph of *f* on the interval [*a*, *b*]. Express your answer to four decimal places.

1. $f(x) = 0.002x^5 + 0.032x^4 - 0.2x^2 + 2;\ [-1.1, 2.2]$

2. $f(x) = x\sqrt{x^3 + 1};\ [1, 2]$

3. $f(x) = \sqrt{x}e^{-x};\ [0, 3]$

4. $f(x) = \dfrac{\ln x}{\sqrt{1 + x^2}};\ [1, 2]$

In Exercises 5–10, use a graphing utility to evaluate the definite integral.

5. $\displaystyle\int_{-1.2}^{2.3} (0.2x^4 - 0.32x^3 + 1.2x - 1)\ dx$

6. $\displaystyle\int_{1}^{3} x(x^4 - 1)^{3.2}\ dx$

7. $\displaystyle\int_{0}^{2} \frac{3x^3 + 2x^2 + 1}{2x^2 + 3}\ dx$ **8.** $\displaystyle\int_{1}^{2} \frac{\sqrt{x} + 1}{2x^2 + 1}\ dx$

9. $\displaystyle\int_{0}^{2} \frac{e^x}{\sqrt{x^2 + 1}}\ dx$ **10.** $\displaystyle\int_{1}^{3} e^{-x} \ln(x^2 + 1)\ dx$

11. Rework Exercise 47, Exercises 6.4.

12. Rework Exercise 48, Exercises 6.4.

13. THE GLOBAL EPIDEMIC The number of AIDS-related deaths/year in the United States is given by the function

$$f(t) = -53.254t^4 + 673.7t^3 - 2801.07t^2$$
$$+ 8833.379t + 20,000 \qquad (0 \le t \le 9)$$

where $t = 0$ corresponds to the beginning of 1988. Find the total number of AIDS-related deaths in the United States between the beginning of 1988 and the end of 1996.
Source: Centers for Disease Control

14. MARIJUANA ARRESTS The number of arrests for marijuana sales and possession in New York City grew at the rate of approximately

$$f(t) = 0.0125t^4 - 0.01389t^3 + 0.55417t^2$$
$$+ 0.53294t + 4.95238 \qquad (0 \le t \le 5)$$

thousand/year, where t is measured in years, with $t = 0$ corresponding to the beginning of 1992. Find the approximate number of marijuana arrests in the city from the beginning of 1992 to the end of 1997.
Source: State Division of Criminal Justice Services

15. POPULATION GROWTH The population of a certain city is projected to grow at the rate of $18\sqrt{t + 1}\ \ln \sqrt{t + 1}$ thousand people/year t yr from now. If the current population is 800,000, what will be the population 45 yr from now?

489

In Exercises 49–51, determine whether the statement is true or false. If it is true, explain why it is true. If it is false, give an example to show why it is false.

49. $\displaystyle\int_{-1}^{1} \frac{1}{x^3}\, dx = -\frac{1}{2x^2}\bigg|_{-1}^{1} = -\frac{1}{2} - \left(-\frac{1}{2}\right) = 0$

50. $\displaystyle\int_{0}^{2} (1 - x)\, dx$ gives the area of the region under the graph of $f(x) = 1 - x$ on the interval $[0, 2]$.

51. The total revenue realized in selling the first 5000 units of a product is given by

$$\int_{0}^{500} R'(x)\, dx = R(500) - R(0)$$

where $R(x)$ is the total revenue.

SOLUTIONS TO SELF-CHECK EXERCISES 6.4

1. $\displaystyle\int_{0}^{2} (x + e^x)\, dx = \frac{1}{2}x^2 + e^x\bigg|_{0}^{2}$

$$= \left[\frac{1}{2}(2)^2 + e^2\right] - \left[\frac{1}{2}(0) + e^0\right]$$

$$= 2 + e^2 - 1$$

$$= e^2 + 1$$

2. a. We want $P(1000)$. But

$$P(1000) - P(0) = \int_{0}^{1000} P'(x)\, dx = \int_{0}^{1000} (-0.000006x^2 + 6)\, dx$$

$$= -0.000002x^3 + 6x\bigg|_{0}^{1000}$$

$$= -0.000002(1000)^3 + 6(1000)$$

$$= 4000$$

So, $P(1000) = 4000 + P(0) = 4000 - 400$, or \$3600/day $[P(0) = -C(0)]$.
b. The additional profit realizable is given by

$$\int_{1000}^{1200} P'(x)\, dx = -0.000002x^3 + 6x\bigg|_{1000}^{1200}$$

$$= [-0.000002(1200)^3 + 6(1200)]$$

$$- [-0.000002(1000)^3 + 6(1000)]$$

$$= 3744 - 4000$$

$$= -256$$

That is, the company sustains a loss of \$256/day if production is increased to 1200 cases/day.

6.5 Evaluating Definite Integrals

This section continues our discussion of the applications of the fundamental theorem of calculus.

PROPERTIES OF THE DEFINITE INTEGRAL

Before going on, we list the following useful properties of the definite integral, some of which parallel the rules of integration of Section 6.1.

Properties of the Definite Integral

Let f and g be integrable functions; then,

1. $\int_a^a f(x)\, dx = 0$

2. $\int_a^b f(x)\, dx = -\int_b^a f(x)\, dx$

3. $\int_a^b cf(x)\, dx = c \int_a^b f(x)\, dx$ $\qquad$ (c, a constant)

4. $\int_a^b [f(x) \pm g(x)]\, dx = \int_a^b f(x)\, dx \pm \int_a^b g(x)\, dx$

5. $\int_a^b f(x)\, dx = \int_a^c f(x)\, dx + \int_c^b f(x)\, dx$ $\qquad$ ($a < c < b$)

Property 5 states that if c is a number lying between a and b so that the interval $[a, b]$ is divided into the intervals $[a, c]$ and $[c, b]$, then the integral of f over the interval $[a, b]$ may be expressed as the sum of the integral of f over the interval $[a, c]$ and the integral of f over the interval $[c, b]$.

Property 5 has the following geometric interpretation when f is nonnegative. By definition

$$\int_a^b f(x)\, dx$$

is the area of the region under the graph of $y = f(x)$ from $x = a$ to $x = b$ (Figure 6.24). Similarly, we interpret the definite integrals

$$\int_a^c f(x)\, dx \qquad \text{and} \qquad \int_c^b f(x)\, dx$$

as the areas of the regions under the graph of $y = f(x)$ from $x = a$ to $x = c$ and from $x = c$ to $x = b$, respectively. Since the two regions do not overlap, we see that

$$\int_a^b f(x)\, dx = \int_a^c f(x)\, dx + \int_c^b f(x)\, dx$$

FIGURE 6.24
$\int_a^b f(x)\, dx =$
$\int_a^c f(x)\, dx + \int_c^b f(x)\, dx$

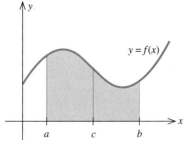

THE METHOD OF SUBSTITUTION FOR DEFINITE INTEGRALS

Our first example shows two approaches generally used when evaluating a definite integral using the method of substitution.

EXAMPLE 1   Evaluate $\int_0^4 x\sqrt{9+x^2}\,dx$.

SOLUTION ✔   **Method 1**   We first find the corresponding indefinite integral:

$$I = \int x\sqrt{9+x^2}\,dx$$

Make the substitution $u = 9 + x^2$ so that

$$du = \frac{d}{dx}(9+x^2)\,dx$$
$$= 2x\,dx$$
$$x\,dx = \frac{1}{2}\,du \qquad \text{(Dividing both sides by 2)}$$

Then,

$$I = \int \frac{1}{2}\sqrt{u}\,du = \frac{1}{2}\int u^{1/2}\,du$$
$$= \frac{1}{3}u^{3/2} + C = \frac{1}{3}(9+x^2)^{3/2} + C \qquad \begin{array}{l}\text{(Substituting}\\ 9+x^2 \text{ for } u)\end{array}$$

Using this result, we now evaluate the given definite integral:

$$\int_0^4 x\sqrt{9+x^2}\,dx = \frac{1}{3}(9+x^2)^{3/2}\Big|_0^4$$
$$= \frac{1}{3}[(9+16)^{3/2} - 9^{3/2}]$$
$$= \frac{1}{3}(125 - 27) = \frac{98}{3} = 32\frac{2}{3}$$

Method 2   *Changing the Limits of Integration.* As before, we make the substitution

$$u = 9 + x^2 \tag{10}$$

so that

$$du = 2x\,dx$$
$$x\,dx = \frac{1}{2}\,du$$

Next, observe that the given definite integral is evaluated *with respect to x* with the range of integration given by the interval [0, 4]. If we perform the integration *with respect to u* via the substitution (10), then we must adjust the range of integration to reflect the fact that the integration is being performed with respect to the new variable u. To determine the proper range of integration, note that when $x = 0$, Equation (10) implies that

$$u = 9 + 0^2 = 9$$

which gives the required lower limit of integration with respect to u. Similarly, when $x = 4$,

$$u = 9 + 16 = 25$$

is the required upper limit of integration with respect to u. Thus, the range of integration when the integration is performed with respect to u is given by the interval $[9, 25]$. Therefore, we have

$$\int_0^4 x\sqrt{9 + x^2}\,dx = \int_9^{25} \frac{1}{2}\sqrt{u}\,du = \frac{1}{2}\int_9^{25} u^{1/2}\,du$$

$$= \frac{1}{3}u^{3/2}\Big|_9^{25} = \frac{1}{3}(25^{3/2} - 9^{3/2})$$

$$= \frac{1}{3}(125 - 27) = \frac{98}{3} = 32\frac{2}{3}$$

which agrees with the result obtained using Method 1.   ■ ■ ■ ■

Exploring with Technology

Refer to Example 1. You can confirm the results obtained there by using a graphing utility as follows:

1. Use the numerical integration operation of the graphing utility to evaluate

$$\int_0^4 x\sqrt{9 + x^2}\,dx$$

2. Evaluate $\dfrac{1}{2}\displaystyle\int_9^{25} \sqrt{u}\,du$.

3. Conclude that $\displaystyle\int_0^4 x\sqrt{9 + x^2}\,dx = \frac{1}{2}\int_9^{25} \sqrt{u}\,du$.

EXAMPLE 2

Evaluate $\displaystyle\int_0^2 xe^{2x^2}\,dx$.

SOLUTION ✔

Let $u = 2x^2$ so that $du = 4x\,dx$, or $x\,dx = \frac{1}{4}\,du$. When $x = 0$, $u = 0$, and when $x = 2$, $u = 8$. This gives the lower and upper limits of integration with respect to u. Making the indicated substitutions, we find

$$\int_0^2 xe^{2x^2}\,dx = \int_0^8 \frac{1}{4}e^u\,du = \frac{1}{4}e^u\Big|_0^8 = \frac{1}{4}(e^8 - 1)$$

■ ■ ■ ■

EXAMPLE 3

Evaluate $\int_0^1 \dfrac{x^2}{x^3 + 1}\, dx$.

SOLUTION ✔

Let $u = x^3 + 1$ so that $du = 3x^2\, dx$, or $x^2\, dx = \frac{1}{3}\, du$. When $x = 0$, $u = 1$, and when $x = 1$, $u = 2$. This gives the lower and upper limits of integration with respect to u. Making the indicated substitutions, we find

$$\int_0^1 \frac{x^2}{x^3 + 1}\, dx = \frac{1}{3} \int_1^2 \frac{du}{u} = \frac{1}{3} \ln |u| \Big|_1^2$$

$$= \frac{1}{3}(\ln 2 - \ln 1) = \frac{1}{3} \ln 2$$
■■■■

FINDING THE AREA UNDER A CURVE

EXAMPLE 4

Find the area of the region R under the graph of $f(x) = e^{(1/2)x}$ from $x = -1$ to $x = 1$.

SOLUTION ✔

The region R is shown in Figure 6.25. Its area is given by

$$A = \int_{-1}^1 e^{(1/2)x}\, dx$$

FIGURE 6.25
Area of $R = \int_{-1}^1 e^{(1/2)x}dx$

To evaluate this integral, we make the substitution

$$u = \frac{1}{2}x$$

so that

$$du = \frac{1}{2}\, dx$$

$$dx = 2\, du$$

When $x = -1$, $u = -\frac{1}{2}$, and when $x = 1$, $u = \frac{1}{2}$. Making the indicated substitutions, we obtain

$$A = \int_{-1}^1 e^{(1/2)x}\, dx = 2 \int_{-1/2}^{1/2} e^u\, du$$

$$= 2e^u \Big|_{-1/2}^{1/2} = 2(e^{1/2} - e^{-1/2})$$

or approximately 2.08 square units.
■■■■

Group Discussion

Let f be a function defined piecewise by the rule

$$f(x) = \begin{cases} \sqrt{x} & \text{if } 0 \le x \le 1 \\ \dfrac{1}{x} & \text{if } 1 < x \le 2 \end{cases}$$

How would you use Property 5 of definite integrals to find the area of the region under the graph of f on $[0, 2]$? What is the area?

AVERAGE VALUE OF A FUNCTION

The *average value* of a function over an interval provides us with an application of the definite integral. Recall that the average value of a set of n numbers is the number

$$\frac{y_1 + y_2 + \cdots + y_n}{n}$$

Now, suppose f is a continuous function defined on $[a, b]$. Let's divide the interval $[a, b]$ into n subintervals of equal length $(b - a)/n$. Choose points $x_1, x_2, \ldots, x_n$ in the first, second, $\ldots$, and nth subintervals, respectively. Then, the average value of the numbers $f(x_1), f(x_2), \ldots, f(x_n)$, given by

$$\frac{f(x_1) + f(x_2) + \cdots + f(x_n)}{n}$$

is an approximation of the average of all the values of $f(x)$ on the interval $[a, b]$. This expression can be written in the form

$$\frac{(b - a)}{(b - a)}\left[f(x_1) \cdot \frac{1}{n} + f(x_2) \cdot \frac{1}{n} + \cdots + f(x_n) \cdot \frac{1}{n} \right]$$

$$= \frac{1}{b - a}\left[f(x_1) \cdot \frac{b - a}{n} + f(x_2) \cdot \frac{b - a}{n} + \cdots + f(x_n) \cdot \frac{b - a}{n} \right]$$

$$= \frac{1}{b - a}[f(x_1)\Delta x + f(x_2)\Delta x + \cdots + f(x_n)\Delta x] \tag{11}$$

As n gets larger and larger, the expression (11) approximates the average value of $f(x)$ over $[a, b]$ with increasing accuracy. But the sum inside the brackets in (11) is a Riemann sum of the function f over $[a, b]$. In view of this, we have

$$\lim_{n \to \infty}\left[\frac{f(x_1) + f(x_2) + \cdots + f(x_n)}{n} \right]$$

$$= \frac{1}{b - a} \lim_{n \to \infty} [f(x_1)\Delta x + f(x_2)\Delta x + \cdots + f(x_n)\Delta x]$$

$$= \frac{1}{b - a} \int_a^b f(x)\, dx$$

This discussion motivates the following definition.

The Average Value of a Function

Suppose f is integrable on $[a, b]$. Then the **average value of f** over $[a, b]$ is

$$\frac{1}{b - a} \int_a^b f(x)\, dx$$

EXAMPLE 5

Find the average value of the function $f(x) = \sqrt{x}$ over the interval $[0, 4]$.

SOLUTION ✔

The required average value is given by

$$\frac{1}{4-0}\int_0^4 \sqrt{x}\, dx = \frac{1}{4}\int_0^4 x^{1/2}\, dx$$

$$= \frac{1}{6}x^{3/2}\Big|_0^4 = \frac{1}{6}(4^{3/2})$$

$$= \frac{4}{3}$$ ■ ■ ■ ■

APPLICATIONS

EXAMPLE 6

The interest rates charged by Madison Finance on auto loans for used cars over a certain 6-month period in 2000 are approximated by the function

$$r(t) = -\frac{1}{12}t^3 + \frac{7}{8}t^2 - 3t + 12 \qquad (0 \le t \le 6)$$

where t is measured in months and $r(t)$ is the annual percentage rate. What is the average rate on auto loans extended by Madison over the 6-month period?

SOLUTION ✔

The average rate over the 6-month period in question is given by

$$\frac{1}{6-0}\int_0^6 \left(-\frac{1}{12}t^3 + \frac{7}{8}t^2 - 3t + 12\right) dt$$

$$= \frac{1}{6}\left(-\frac{1}{48}t^4 + \frac{7}{24}t^3 - \frac{3}{2}t^2 + 12t\right)\Big|_0^6$$

$$= \frac{1}{6}\left[-\frac{1}{48}(6^4) + \frac{7}{24}(6^3) - \frac{3}{2}(6^2) + 12(6)\right]$$

$$= 9$$

or 9% per year.   ■ ■ ■ ■

EXAMPLE 7

The amount of a certain drug in a patient's body t days after it has been administered is

$$C(t) = 5e^{-0.2t}$$

units. Determine the average amount of the drug present in the patient's body for the first 4 days after the drug has been administered.

SOLUTION ✔   The average amount of the drug present in the patient's body for the first 4 days after it has been administered is given by

$$\frac{1}{4-0}\int_0^4 5e^{-0.2t}\,dt = \frac{5}{4}\int_0^4 e^{-0.2t}\,dt$$

$$= \frac{5}{4}\left[\left(-\frac{1}{0.2}\right)e^{-0.2t}\Big|_0^4\right]$$

$$= \frac{5}{4}(-5e^{-0.8}+5)$$

$$\approx 3.44$$

or approximately 3.44 units.   ■ ■ ■ ■

FIGURE 6.26

The average value of *f* over [*a, b*] is *k*.

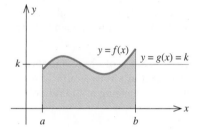

We now give a geometric interpretation of the average value of a function *f* over an interval [*a, b*]. Suppose *f*(*x*) is nonnegative so that the definite integral

$$\int_a^b f(x)\,dx$$

gives the area under the graph of *f* from *x* = *a* to *x* = *b* (Figure 6.26). Observe that, in general, the "height" *f*(*x*) varies from point to point. Can we replace *f*(*x*) by a constant function *g*(*x*) = *k* (which has constant height) such that the areas under each of the two functions *f* and *g* are the same? If so, since the area under the graph of *g* from *x* = *a* to *x* = *b* is *k*(*b* − *a*), we have

$$k(b-a) = \int_a^b f(x)\,dx$$

$$k = \frac{1}{b-a}\int_a^b f(x)\,dx$$

so that *k* is the average value of *f* over [*a, b*]. Thus, the average value of a function *f* over an interval [*a, b*] is the height of a rectangle with base of length (*b* − *a*) that has the same area as that of the region under the graph of *f* from *x* = *a* to *x* = *b*.

SELF-CHECK EXERCISES 6.5

1. Evaluate $\int_0^2 \sqrt{2x+5}\,dx$.
2. Find the average value of the function *f*(*x*) = 1 − *x*² over the interval [−1, 2].
3. The median price of a house in a southwestern state between January 1, 1995, and January 1, 2000, is approximated by the function

$$f(t) = t^3 - 7t^2 + 17t + 190 \qquad (0 \le t \le 5)$$

where *f*(*t*) is measured in thousands of dollars and *t* is expressed in years (*t* = 0 corresponds to the beginning of 1995). Determine the average median price of a house over that time interval.

Solutions to Self-Check Exercises 6.5 can be found on page 502.

6.5 Exercises

In Exercises 1–28, evaluate the given definite integral.

1. $\int_0^2 x(x^2 - 1)^3 \, dx$

2. $\int_0^1 x^2(2x^3 - 1)^4 \, dx$

3. $\int_0^1 x\sqrt{5x^2 + 4} \, dx$

4. $\int_1^3 x\sqrt{3x^2 - 2} \, dx$

5. $\int_0^2 x^2(x^3 + 1)^{3/2} \, dx$

6. $\int_1^5 (2x - 1)^{5/2} \, dx$

7. $\int_0^1 \frac{1}{\sqrt{2x + 1}} \, dx$

8. $\int_0^2 \frac{x}{\sqrt{x^2 + 5}} \, dx$

9. $\int_1^2 (2x - 1)^4 \, dx$

10. $\int_1^2 (2x + 4)(x^2 + 4x - 8)^3 \, dx$

11. $\int_{-1}^1 x^2(x^3 + 1)^4 \, dx$

12. $\int_{-1}^1 \left(x^3 + \frac{3}{4}\right)(x^4 + 3x)^{-2} \, dx$

13. $\int_1^5 x\sqrt{x - 1} \, dx$

14. $\int_1^4 x\sqrt{x + 1} \, dx$
Hint: Let $u = x + 1$.

15. $\int_0^2 xe^{x^2} \, dx$

16. $\int_0^1 e^{-x} \, dx$

17. $\int_0^1 (e^{2x} + x^2 + 1) \, dx$

18. $\int_0^2 (e^t - e^{-t}) \, dt$

19. $\int_{-1}^1 xe^{x^2 + 1} \, dx$

20. $\int_0^4 \frac{e^{\sqrt{x}}}{\sqrt{x}} \, dx$

21. $\int_3^6 \frac{2}{x - 2} \, dx$

22. $\int_0^1 \frac{x}{1 + 2x^2} \, dx$

23. $\int_1^2 \frac{x^2 + 2x}{x^3 + 3x^2 - 1} \, dx$

24. $\int_0^1 \frac{e^x}{1 + e^x} \, dx$

25. $\int_1^2 \left(4e^{2u} - \frac{1}{u}\right) du$

26. $\int_1^2 \left(1 + \frac{1}{x} + e^x\right) dx$

27. $\int_1^2 \left(2e^{-4x} - \frac{1}{x^2}\right) dx$

28. $\int_1^2 \frac{\ln x}{x} \, dx$

In Exercises 29–38, find the average value of the given function f over the indicated interval [a, b].

29. $f(x) = 2x + 3$; $[0, 2]$

30. $f(x) = 8 - x$; $[1, 4]$

31. $f(x) = 2x^2 - 3$; $[1, 3]$

32. $f(x) = 4 - x^2$; $[-2, 3]$

33. $f(x) = x^2 + 2x - 3$; $[-1, 2]$

34. $f(x) = x^3$; $[-1, 1]$

35. $f(x) = \sqrt{2x + 1}$; $[0, 4]$

36. $f(x) = e^{-x}$; $[0, 4]$

37. $f(x) = xe^{x^2}$; $[0, 2]$

38. $f(x) = \frac{1}{x + 1}$; $[0, 2]$

39. WORLD PRODUCTION OF COAL A study proposed in 1980 by researchers from the major producers and consumers of the world's coal concluded that coal could and must play an important role in fueling global economic growth over the next 20 yr. The world production of coal in 1980 was 3.5 billion metric tons. If output increased at the rate of $3.5e^{0.05t}$ billion metric tons /year in year t ($t = 0$ corresponding to 1980), determine how much coal was produced worldwide between 1980 and the end of the century.

40. NEWTON'S LAW OF COOLING A bottle of white wine at room temperature (68°F) is placed in a refrigerator at 4 P.M. Its temperature after t hr is changing at the rate of

$$-18e^{-0.6t}$$

degrees Fahrenheit/hour. By how many degrees will the temperature of the wine have dropped by 7 P.M.? What will the temperature of the wine be at 7 P.M.?

41. NET INVESTMENT FLOW The net investment flow (rate of capital formation) of the giant conglomerate LTF Incorporated is projected to be

$$t\sqrt{\frac{1}{2}t^2 + 1}$$

million dollars/year in year t. Find the accruement on the company's capital stock in the second year.

Hint: The amount is given by

$$\int_1^2 t\sqrt{\frac{1}{2}t^2 + 1} \, dt$$

42. OIL PRODUCTION Based on a preliminary report by a geological survey team, it is estimated that a newly discovered oil field can be expected to produce oil at the rate of

$$R(t) = \frac{600t^2}{t^3 + 32} + 5 \qquad (0 \le t \le 20)$$

thousand barrels/year, t years after production begins. Find the amount of oil that the field can be expected to yield during the first 5 yr of production, assuming that the projection holds true.

43. DEPRECIATION: DOUBLE DECLINING-BALANCE METHOD Suppose a tractor purchased at a price of $60,000 is to be depreciated by the *double declining-balance method* over a period of 10 yr. It can be shown that the rate at which the book value will be decreasing is given by

$$R(t) = 13388.61e^{-0.22314t} \qquad (0 \le t \le 10)$$

dollars/year at year t. Find the amount by which the book value of the tractor will depreciate over the first 5 yr of its life.

44. VELOCITY OF A CAR A car moves along a straight road in such a way that its velocity (in feet/second) at any time t (in seconds) is given by

$$v(t) = 3t\sqrt{16 - t^2} \qquad (0 \le t \le 4)$$

Find the distance traveled by the car in the 4 sec from $t = 0$ to $t = 4$.

45. AVERAGE TEMPERATURE The temperature (in °F) in Boston over a 12-hr period on a certain December day was given by

$$T = -0.05t^3 + 0.4t^2 + 3.8t + 5.6 \qquad (0 \le t \le 12)$$

where t is measured in hours, with $t = 0$ corresponding to 6 A.M. Determine the average temperature on that day over the 12-hr period from 6 A.M. to 6 P.M.

46. WHALE POPULATION A group of marine biologists estimates that if certain conservation measures are implemented, the population of an endangered species of whale will be

$$N(t) = 3t^3 + 2t^2 - 10t + 600 \qquad (0 \le t \le 10)$$

where $N(t)$ denotes the population at the end of year t. Find the average population of the whales over the next 10 yr.

47. CABLE TV SUBSCRIBERS The manager of the Tele-Star Cable Television Service estimates that the total number of subscribers to the service in a certain city t yr from now will be

$$N(t) = -\frac{40{,}000}{\sqrt{1 + 0.2t}} + 50{,}000$$

Find the average number of cable television subscribers over the next 5 yr if this prediction holds true.

48. AVERAGE YEARLY SALES The sales of the Universal Instruments Company in the first t yr of its operation are approximated by the function

$$S(t) = t\sqrt{0.2t^2 + 4}$$

where $S(t)$ is measured in millions of dollars. What were Universal's average yearly sales over its first 5 yr of operation?

49. CONCENTRATION OF A DRUG IN THE BLOODSTREAM The concentration of a certain drug in a patient's bloodstream t hr after injection is

$$C(t) = \frac{0.2t}{t^2 + 1}$$

mg/cm^3. Determine the average concentration of the drug in the patient's bloodstream over the first 4 hr after the drug is injected.

50. Refer to Exercise 44. Find the average velocity of the car over the time interval $[0, 4]$.

51. FLOW OF BLOOD IN AN ARTERY According to a law discovered by nineteenth-century physician Jean Louis Marie Poiseuille, the velocity of blood (in centimeters/second) r cm from the central axis of an artery is given by

$$v(r) = k(R^2 - r^2)$$

where k is a constant and R is the radius of the artery. Find the average velocity of blood along a radius of the artery (see accompanying figure).

Hint: Evaluate $\dfrac{1}{R}\displaystyle\int_0^R v(r)\, dr$.

Blood vessel

52. Prove Property 1 of the definite integral.
   Hint: Let F be an antiderivative of f, and use the definition of the definite integral.

53. Prove Property 2 of the definite integral.
   Hint: See Exercise 52.

54. Verify by direct computation that

$$\int_1^3 x^2\, dx = -\int_3^1 x^2\, dx$$

55. Prove Property 3 of the definite integral.
   Hint: See Exercise 52.

(continued on p. 502)

Evaluating Definite Integrals
for Piecewise-Defined Functions

We continue using of graphing utilities to find the definite integral of a function. But here we will make use of Property 5 of the properties of the definite integral (page 491).

EXAMPLE 1

Use the numerical integral operation of a graphing utility to evaluate

$$\int_{-1}^{2} f(x)\, dx$$

where

$$f(x) = \begin{cases} -x^2 & \text{if } x < 0 \\ \sqrt{x} & \text{if } x \ge 0 \end{cases}$$

SOLUTION ✔

Using Property 5 of the definite integral, we can write

$$\int_{-1}^{2} f(x)\, dx = \int_{-1}^{0} -x^2\, dx + \int_{0}^{2} x^{1/2}\, dx$$

Using a graphing utility, we find

$$\int_{-1}^{2} f(x)\, dx = \text{fnInt}(-x{\wedge}2, x, -1, 0) + \text{fnInt}(x{\wedge}0.5, x, 0, 2)$$

$$\approx -0.333333 + 1.885618$$

$$= 1.552285$$

■■■■

In Exercises 1–4, use Property 5 of the properties of the definite integral (page 491) to evaluate the definite integral accurate to six decimal places.

1. $\int_{-1}^{2} f(x)\, dx$, where

$$f(x) = \begin{cases} 2.3x^3 - 3.1x^2 + 2.7x + 3 & \text{if } x < 1 \\ -1.7x^2 + 2.3x + 4.3 & \text{if } x \geq 1 \end{cases}$$

2. $\int_{0}^{3} f(x)\, dx$, where $f(x) = \begin{cases} \dfrac{\sqrt{x}}{1 + x^2} & \text{if } 0 \leq x < 1 \\ 0.5e^{-0.1x^2} & \text{if } x \geq 1 \end{cases}$

3. $\int_{-2}^{2} f(x)\, dx$, where $f(x) = \begin{cases} x^4 - 2x^2 + 4 & \text{if } x < 0 \\ 2\ln(x + e^2) & \text{if } x \geq 0 \end{cases}$

4. $\int_{-2}^{6} f(x)\, dx$, where

$$f(x) = \begin{cases} 2x^3 - 3x^2 + x + 2 & \text{if } x < -1 \\ \sqrt{3x + 4} - 5 & \text{if } -1 \leq x \leq 4 \\ x^2 - 3x - 5 & \text{if } x > 4 \end{cases}$$

5. AIDS IN MASSACHUSETTS The rate of growth (and decline) of the number of AIDS cases diagnosed in Massachusetts from the beginning of 1989 ($t = 0$) through the end of 1997 ($t = 8$) is approximated by the function

$$f(t) = \begin{cases} 69.83333t^2 + 30.16667t + 1000 & \text{if } 0 \leq t < 3 \\ 1719 & \text{if } 3 \leq t < 4 \\ -28.79167t^3 + 491.37500t^2 & \text{if } 4 \leq t \leq 8 \\ \quad - 2985.083333t + 7640 \end{cases}$$

where $f(t)$ is measured in the number of cases/year. Estimate the total number of AIDS cases diagnosed in Massachusetts from the beginning of 1989 through the end of 1997.

Source: Massachusetts Department of Health

56. Verify by direct computation that

$$\int_1^9 2\sqrt{x}\,dx = 2\int_1^9 \sqrt{x}\,dx$$

57. Verify by direct computation that

$$\int_0^1 (1 + x - e^x)\,dx = \int_0^1 dx + \int_0^1 x\,dx - \int_0^1 e^x\,dx$$

What properties of the definite integral are demonstrated in this exercise?

58. Verify by direct computation that

$$\int_0^3 (1 + x^3)\,dx = \int_0^1 (1 + x^3)\,dx + \int_1^3 (1 + x^3)\,dx$$

What property of the definite integral is demonstrated here?

59. Verify by direct computation that

$$\int_0^3 (1 + x^3)\,dx$$
$$= \int_0^1 (1 + x^3)\,dx + \int_1^2 (1 + x^3)\,dx + \int_2^3 (1 + x^3)\,dx$$

hence showing that Property 5 may be extended.

60. Evaluate $\int_3^3 (1 + \sqrt{x})e^{-x}\,dx$.

61. Evaluate $\int_3^0 f(x)\,dx$, given that $\int_0^3 f(x)\,dx = 4$.

62. Evaluate $\int_0^3 4f(x)\,dx$, given that $\int_0^3 f(x)\,dx = -1$.

63. Given that $\int_{-1}^2 f(x)\,dx = -2$ and $\int_{-1}^2 g(x)\,dx = 3$, evaluate:

a. $\int_{-1}^2 [2f(x) + g(x)]\,dx$

b. $\int_{-1}^2 [g(x) - f(x)]\,dx$

c. $\int_{-1}^2 [2f(x) - 3g(x)]\,dx$

64. Given that $\int_{-1}^2 f(x)\,dx = 2$ and $\int_0^2 f(x)\,dx = 3$, evaluate:

a. $\int_{-1}^0 f(x)\,dx$

b. $\int_0^2 f(x)\,dx - \int_{-1}^0 f(x)\,dx$

In Exercises 65–69, determine whether the statement is true or false. If it is true, explain why it is true. If it is false, give an example to show why it is false.

65. $\int_2^2 \dfrac{e^x}{\sqrt{1+x}}\,dx = 0$

66. $\int_1^3 \dfrac{dx}{x-2} = -\int_3^1 \dfrac{dx}{x-2}$

67. $\int_0^1 x\sqrt{x+1}\,dx = \sqrt{x+1}\int_0^1 x\,dx = \dfrac{1}{2}x^2\sqrt{x+1}\Big|_0^1 = \dfrac{\sqrt{2}}{2}$

68. If f' is continuous on $[0, 2]$, then $\int_0^2 f'(x)\,dx = f(2) - f(0)$.

69. If f and g are continuous on $[a, b]$ and k is a constant, then

$$\int_a^b [kf(x) + g(x)]\,dx = k\int_a^b f(x)\,dx + \int_a^b g(x)\,dx$$

SOLUTIONS TO SELF-CHECK EXERCISES 6.5

1. Let $u = 2x + 5$. Then, $du = 2\,dx$, or $dx = \frac{1}{2}\,du$. Also, when $x = 0$, $u = 5$, and when $x = 2$, $u = 9$. Therefore,

$$\int_0^2 \sqrt{2x+5}\,dx = \int_0^2 (2x+5)^{1/2}\,dx$$
$$= \frac{1}{2}\int_5^9 u^{1/2}\,du$$
$$= \left(\frac{1}{2}\right)\left(\frac{2}{3}u^{3/2}\right)\Big|_5^9$$
$$= \frac{1}{3}[9^{3/2} - 5^{3/2}]$$
$$= \frac{1}{3}(27 - 5\sqrt{5})$$

2. The required average value is given by

$$\frac{1}{2 - (-1)} \int_{-1}^{2} (1 - x^2) \, dx = \frac{1}{3} \int_{-1}^{2} (1 - x^2) \, dx$$

$$= \frac{1}{3} \left(x - \frac{1}{3} x^3 \right) \Big|_{-1}^{2}$$

$$= \frac{1}{3} \left[\left(2 - \frac{8}{3} \right) - \left(-1 + \frac{1}{3} \right) \right] = 0$$

3. The average median price of a house over the stated time interval is given by

$$\frac{1}{5 - 0} \int_{0}^{5} (t^3 - 7t^2 + 17t + 190) \, dt = \frac{1}{5} \left(\frac{1}{4} t^4 - \frac{7}{3} t^3 + \frac{17}{2} t^2 + 190t \right) \Big|_{0}^{5}$$

$$= \frac{1}{5} \left[\frac{1}{4} (5)^4 - \frac{7}{3} (5)^3 + \frac{17}{2} (5)^2 + 190(5) \right]$$

$$= 205.417$$

or $205,417.

6.6 Area Between Two Curves

Suppose a certain country's petroleum consumption is expected to grow at the rate of $f(t)$ million barrels per year, t years from now, for the next 5 years. Then, the country's total petroleum consumption over the period of time in question is given by the area under the graph of f on the interval [0, 5] (Figure 6.27).

Next, suppose that because of the implementation of certain energy-conservation measures, the rate of growth of petroleum consumption is expected to be $g(t)$ million barrels per year instead. Then, the country's projected total petroleum consumption over the 5-year period is given by the area under the graph of g on the interval [0, 5] (Figure 6.28).

FIGURE 6.27
At a rate of consumption $f(t)$ million barrels per year, the total petroleum consumption is given by the area of the region under the graph of f.

FIGURE 6.28
At a rate of consumption of $g(t)$ million barrels per year, the total petroleum consumption is given by the area of the region under the graph of g.

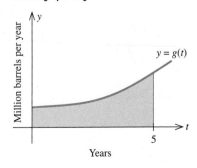

Therefore, the area of the shaded region S lying between the graphs of f and g on the interval $[0, 5]$ (Figure 6.29) gives the amount of petroleum that would be saved over the 5-year period because of the conservation measures.

FIGURE 6.29
The area of S gives the amount of petroleum that would be saved over the 5-year period.

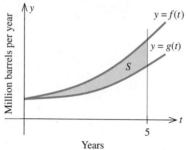

But the area of S is given by

Area under the graph of f on $[a, b]$ − Area under the graph of g on $[a, b]$

$$= \int_0^5 f(t)\, dt - \int_0^5 g(t)\, dt$$

$$= \int_0^5 [f(t) - g(t)]\, dt \qquad \text{(By Property 4, Section 6.5)}$$

This example shows that some practical problems can be solved by finding the area of a region between two curves, which in turn can be found by evaluating an appropriate definite integral.

FINDING THE AREA BETWEEN TWO CURVES

We now turn our attention to the general problem of finding the area of a plane region bounded both above and below by the graphs of functions. First, consider the situation in which the graph of one function lies above that of another. More specifically, let R be the region in the xy-plane (Figure 6.30) that is bounded above by the graph of a continuous function f, below by a continuous function g where $f(x) \geq g(x)$ on $[a, b]$ and to the left and right by the vertical lines $x = a$ and $x = b$, respectively. From the figure, we see that

Area of R = Area under $f(x)$ − Area under $g(x)$

$$= \int_a^b f(x)\, dx - \int_a^b g(x)\, dx$$

$$= \int_a^b [f(x) - g(x)]\, dx$$

upon using Property 4 of the definite integral.

FIGURE 6.30
Area of $R = \int_a^b [f(x) - g(x)]\, dx$

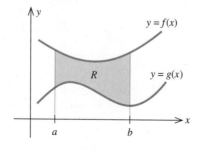

The Area Between Two Curves

Let f and g be continuous functions such that $f(x) \geq g(x)$ on the interval $[a, b]$. Then, the area of the region bounded above by $y = f(x)$ and below by $y = g(x)$ on $[a, b]$ is given by

$$\int_a^b [f(x) - g(x)]\, dx \qquad \textbf{(12)}$$

Even though we assumed that both f and g were nonnegative in the derivation of (12), it may be shown that this equation is valid if f and g are not nonnegative (see Exercise 52). Also, observe that if $g(x)$ is 0 for all x—that is, when the lower boundary of the region R is the x-axis—Equation (12) gives the area of the region under the curve $y = f(x)$ from $x = a$ to $x = b$, as we would expect.

EXAMPLE 1   Find the area of the region bounded by the x-axis, the graph of $y = -x^2 + 4x - 8$, and the lines $x = -1$ and $x = 4$.

SOLUTION ✔   The region R under consideration is shown in Figure 6.31. We can view R as the region bounded above by the graph of $f(x) = 0$ (the x-axis) and below by the graph of $g(x) = -x^2 + 4x - 8$ on $[-1, 4]$. Therefore, the area of R is given by

$$\int_a^b [f(x) - g(x)]\, dx = \int_{-1}^4 [0 - (-x^2 + 4x - 8)]\, dx$$

$$= \int_{-1}^4 (x^2 - 4x + 8)\, dx$$

$$= \frac{1}{3}x^3 - 2x^2 + 8x \Big|_{-1}^4$$

$$= \left[\frac{1}{3}(64) - 2(16) + 8(4)\right] - \left[\frac{1}{3}(-1) - 2(1) + 8(-1)\right]$$

$$= 31\frac{2}{3}$$

or $31\frac{2}{3}$ square units.   ■ ■ ■ ■

FIGURE 6.31
Area of $R = -\int_{-1}^4 g(x)\, dx$

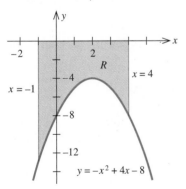

EXAMPLE 2   Find the area of the region R bounded by the graphs of

$$f(x) = 2x - 1 \qquad \text{and} \qquad g(x) = x^2 - 4$$

and the vertical lines $x = 1$ and $x = 2$.

SOLUTION ✔   We first sketch the graphs of the functions $f(x) = 2x - 1$ and $g(x) = x^2 - 4$ and the vertical lines $x = 1$ and $x = 2$, and then we identify the region R whose area is to be calculated (Figure 6.32).

FIGURE 6.32
Area of $R = \int_1^2 [f(x) - g(x)]\, dx$

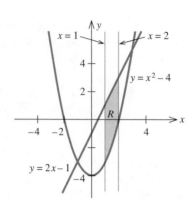

Since the graph of f always lies above that of g for x in the interval $[1, 2]$, we see by Equation (12) that the required area is given by

$$\int_1^2 [f(x) - g(x)]\, dx = \int_1^2 [(2x - 1) - (x^2 - 4)]\, dx$$

$$= \int_1^2 (-x^2 + 2x + 3)\, dx$$

$$= -\frac{1}{3}x^3 + x^2 + 3x \Big|_1^2$$

$$= \left(-\frac{8}{3} + 4 + 6\right) - \left(-\frac{1}{3} + 1 + 3\right) = \frac{11}{3}$$

or $\frac{11}{3}$ square units.   ■ ■ ■ ■

EXAMPLE 3   Find the area of the region R that is completely enclosed by the graphs of the functions

$$f(x) = 2x - 1 \qquad \text{and} \qquad g(x) = x^2 - 4$$

SOLUTION ✔   The region R is shown in Figure 6.33. First, we find the points of intersection of the two curves. To do this, we solve the system that consists of the two

FIGURE 6.33
Area of $R = \int_{-1}^3 [f(x) - g(x)]\, dx$

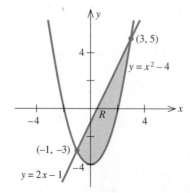

equations $y = 2x - 1$ and $y = x^2 - 4$. Equating the two values of y gives

$$x^2 - 4 = 2x - 1$$

$$x^2 - 2x - 3 = 0$$

$$(x + 1)(x - 3) = 0$$

so $x = -1$ or $x = 3$. That is, the two curves intersect when $x = -1$ and $x = 3$.

Observe that we could also view the region R as the region bounded above by the graph of the function $f(x) = 2x - 1$, below by the graph of the function $g(x) = x^2 - 4$, and to the left and right by the vertical lines $x = -1$ and $x = 3$, respectively.

Next, since the graph of the function f always lies above that of the function g on $[-1, 3]$, we can use (12) to compute the desired area:

$$\int_a^b [f(x) - g(x)]\, dx = \int_{-1}^3 [(2x - 1) - (x^2 - 4)]\, dx$$

$$= \int_{-1}^3 (-x^2 + 2x + 3)\, dx$$

$$= -\frac{1}{3}x^3 + x^2 + 3x \Big|_{-1}^3$$

$$= (-9 + 9 + 9) - \left(\frac{1}{3} + 1 - 3 \right) = \frac{32}{3}$$

$$= 10\tfrac{2}{3}$$

or $10\tfrac{2}{3}$ square units.   ■ ■ ■ ■

EXAMPLE 4

Find the area of the region R bounded by the graphs of the functions

$$f(x) = x^2 - 2x - 1 \quad \text{and} \quad g(x) = -e^x - 1$$

and the vertical lines $x = -1$ and $x = 1$.

SOLUTION ✔

The region R is shown in Figure 6.34. Since the graph of the function f always lies above that of the function g, the area of the region R is given by

$$\int_a^b [f(x) - g(x)]\, dx = \int_{-1}^1 [(x^2 - 2x - 1) - (-e^x - 1)]\, dx$$

$$= \int_{-1}^1 (x^2 - 2x + e^x)\, dx$$

$$= \frac{1}{3}x^3 - x^2 + e^x \Big|_{-1}^1$$

$$= \left(\frac{1}{3} - 1 + e \right) - \left(-\frac{1}{3} - 1 + e^{-1} \right)$$

$$= \frac{2}{3} + e - \frac{1}{e}, \text{ or } 3.02 \text{ square units} \qquad ■ ■ ■ ■$$

FIGURE 6.34
Area of $R = \int_{-1}^1 [f(x) - g(x)]\, dx$

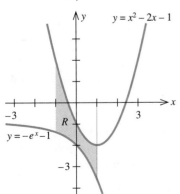

Equation (12), which gives the area of the region between the curves $y = f(x)$ and $y = g(x)$ for $a \le x \le b$, is valid when the graph of the function f lies above that of the function g over the interval $[a, b]$. Example 5 shows how to use (12) to find the area of a region when the latter condition does not hold.

EXAMPLE 5

Find the area of the region bounded by the graph of the function $f(x) = x^3$, the x-axis, and the lines $x = -1$ and $x = 1$.

SOLUTION ✔

The region R under consideration can be thought of as composed of the two subregions R_1 and R_2, as shown in Figure 6.35.

Recall that the x-axis is represented by the function $g(x) = 0$. Since $g(x) \geq f(x)$ on $[-1, 0]$, we see that the area of R_1 is given by

FIGURE 6.35
Area of R_1 = area of R_2

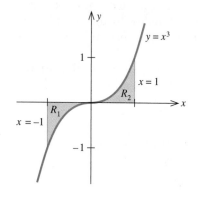

$$\int_a^b [g(x) - f(x)]\, dx = \int_{-1}^0 (0 - x^3)\, dx = -\int_{-1}^0 x^3\, dx$$

$$= -\frac{1}{4}x^4\Big|_{-1}^0 = 0 - \left(-\frac{1}{4}\right) = \frac{1}{4}$$

To find the area of R_2, we observe that $f(x) \geq g(x)$ on $[0, 1]$, so it is given by

$$\int_a^b [f(x) - g(x)]\, dx = \int_0^1 (x^3 - 0)\, dx = \int_0^1 x^3\, dx$$

$$= \frac{1}{4}x^4\Big|_0^1 = \left(\frac{1}{4}\right) - 0 = \frac{1}{4}$$

Therefore, the area of R is $\frac{1}{4} + \frac{1}{4}$, or $\frac{1}{2}$, square units.

By making use of symmetry, we could have obtained the same result by computing

$$-2\int_{-1}^0 x^3\, dx \qquad \text{or} \qquad 2\int_0^1 x^3\, dx$$

as you may verify.   ■ ■ ■ ■

Group Discussion

A function is *even* if it satisfies the condition $f(-x) = f(x)$, and it is *odd* if it satisfies the condition $f(-x) = -f(x)$. Show that the graph of an even function is symmetric with respect to the y-axis while the graph of an odd function is symmetric with respect to the origin. Explain why

$$\int_{-a}^a f(x)\, dx = 2\int_0^a f(x)\, dx \quad \text{if } f \text{ is even}$$

$$\int_{-a}^a f(x)\, dx = 0 \qquad\qquad \text{if } f \text{ is odd}$$

EXAMPLE 6

Find the area of the region completely enclosed by the graphs of the functions

$$f(x) = x^3 - 3x + 3 \qquad \text{and} \qquad g(x) = x + 3$$

SOLUTION ✔

First, sketch the graphs of $y = x^3 - 3x + 3$ and $y = x + 3$ and then identify the required region R. We can view the region R as being composed of the two subregions R_1 and R_2, as shown in Figure 6.36. By solving the equations $y = x + 3$ and $y = x^3 - 3x + 3$ simultaneously, we find the points of intersection

FIGURE 6.36

Area of R_1 + Area of R_2

$$= \int_{-2}^{0} [f(x) - g(x)]\, dx$$
$$+ \int_{0}^{2} [g(x) - f(x)]\, dx$$

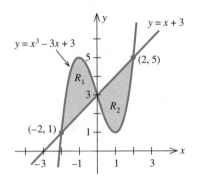

of the two curves. Equating the two values of y, we have

$$x^3 - 3x + 3 = x + 3$$
$$x^3 - 4x = 0$$
$$x(x^2 - 4) = 0$$
$$x(x + 2)(x - 2) = 0$$
$$x = 0, -2, 2$$

Hence, the points of intersection of the two curves are $(-2, 1)$, $(0, 3)$, and $(2, 5)$.

For $-2 \le x \le 0$, we see that the graph of the function f lies above that of the function g, so the area of the region R_1 is, by virtue of (12),

$$\int_{-2}^{0} [(x^3 - 3x + 3) - (x + 3)]\, dx = \int_{-2}^{0} (x^3 - 4x)\, dx$$
$$= \frac{1}{4}x^4 - 2x^2 \Big|_{-2}^{0}$$
$$= -(4 - 8)$$
$$= 4$$

or 4 square units. For $0 \le x \le 2$, the graph of the function g lies above that of the function f, and the area of R_2 is given by

$$\int_{0}^{2} [(x + 3) - (x^3 - 3x + 3)]\, dx = \int_{0}^{2} (-x^3 + 4x)\, dx$$
$$= -\frac{1}{4}x^4 + 2x^2 \Big|_{0}^{2}$$
$$= -4 + 8$$
$$= 4$$

or 4 square units. Therefore, the required area is the sum of the area of the two regions $R_1 + R_2$—that is, $4 + 4$, or 8 square units.   ■ ■ ■ ■

APPLICATION

EXAMPLE 7

In a 1994 study for a developing country's Economic Development Board, government economists and energy experts concluded that if the Energy Conservation Bill were implemented in 1995, the country's oil consumption for the next 5 years would be expected to grow in accordance with the model

$$R(t) = 20e^{0.05t}$$

where t is measured in years ($t = 0$ corresponding to the year 1995) and $R(t)$ in millions of barrels per year. Without the government-imposed conservation measures, however, the expected rate of growth of oil consumption would be given by

$$R_1(t) = 20e^{0.08t}$$

millions of barrels per year. Using these models, determine how much oil would have been saved from 1995 through 2000 if the bill had been implemented.

SOLUTION ✔ Under the Energy Conservation Bill, the total amount of oil that would have been consumed between 1995 and 2000 is given by

$$\int_0^5 R(t)\, dt = \int_0^5 20e^{0.05t}\, dt \qquad (13)$$

Without the bill, the total amount of oil that would have been consumed between 1995 and 2000 is given by

$$\int_0^5 R_1(t)\, dt = \int_0^5 20e^{0.08t}\, dt \qquad (14)$$

FIGURE 6.37
Area of $S = \int_0^5 [R_1(t) - R(t)]\, dt$

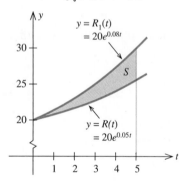

Equation (13) may be interpreted as the area of the region under the curve $y = R(t)$ from $t = 0$ to $t = 5$. Similarly, we interpret (14) as the area of the region under the curve $y = R_1(t)$ from $t = 0$ to $t = 5$. Furthermore, note that the graph of $y = R_1(t) = 20e^{0.08t}$ always lies on or above the graph of $\dot{y} = R(t) = 20e^{0.05t}$ ($t \geq 0$). Thus, the area of the shaded region S in Figure 6.37 shows the amount of oil that would have been saved from 1995 to 2000 if the Energy Conservation Bill had been implemented. But the area of the region S is given by

$$\int_0^5 [R_1(t) - R(t)]\, dt = \int_0^5 [20e^{0.08t} - 20^{0.05t}]\, dt$$

$$= 20 \int_0^5 (e^{0.08t} - e^{0.05t})\, dt$$

$$= 20 \left(\frac{e^{0.08t}}{0.08} - \frac{e^{0.05t}}{0.05} \right) \Bigg|_0^5$$

$$= 20 \left[\left(\frac{e^{0.4}}{0.08} - \frac{e^{0.25}}{0.05} \right) - \left(\frac{1}{0.08} - \frac{1}{0.05} \right) \right]$$

$$\approx 9.3$$

or approximately 9.3 square units. Thus, the amount of oil that would have been saved is 9.3 million barrels. ■ ■ ■ ■

Exploring with Technology

Refer to Example 7. Suppose we want to construct a mathematical model giving the amount of oil saved from 1995 through the year 1995 + x, where $x \geq 0$. For example, in Example 7, $x = 5$.

1. Show that this model is given by

$$F(x) = \int_0^x [R_1(t) - R(t)] \, dt$$
$$= 250e^{0.08x} - 400e^{0.05x} + 150$$

Hint: You may find it helpful to use some of the results of Example 7.

2. Use a graphing utility to plot the graph of F, using the viewing rectangle $[0, 10] \times [0, 50]$.
3. Find $F(5)$ and thus confirm the result of Example 7.
4. What is the main advantage of this model?

SELF-CHECK EXERCISES 6.6

1. Find the area of the region bounded by the graphs of $f(x) = x^2 + 2$ and $g(x) = 1 - x$ and the vertical lines $x = 0$ and $x = 1$.

2. Find the area of the region completely enclosed by the graphs of $f(x) = -x^2 + 6x + 5$ and $g(x) = x^2 + 5$.

3. The management of the Kane Corporation, which operates a chain of hotels, expects its profits to grow at the rate of $1 + t^{2/3}$ million dollars/year t yr from now. However, with renovations and improvements of existing hotels and proposed acquisitions of new hotels, Kane's profits are expected to grow at the rate of $t - 2\sqrt{t} + 4$ million dollars/year in the next decade. What additional profits are expected over the next 10 yr if the group implements the proposed plans?

Solutions to Self-Check Exercises 6.6 can be found on page 515.

6.6 Exercises

In Exercises 1–8, find the area of the shaded region.

1.

2.

3.

4.

5.

6.

7.

8.

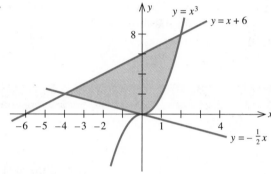

In Exercises 9–16, sketch the graph and find the area of the region bounded below by the graph of each of the given functions and above by the x-axis from $x = a$ to $x = b$.

9. $f(x) = -x^2$; $a = -1$, $b = 2$

10. $f(x) = x^2 - 4$; $a = -2$, $b = 2$

11. $f(x) = x^2 - 5x + 4$; $a = 1$, $b = 3$

12. $f(x) = x^3$; $a = -1$, $b = 0$

13. $f(x) = -1 - \sqrt{x}$; $a = 0$, $b = 9$

14. $f(x) = \dfrac{1}{2}x - \sqrt{x}$; $a = 0$, $b = 4$

15. $f(x) = -e^{(1/2)x}$; $a = -2$, $b = 4$

16. $f(x) = -xe^{-x^2}$; $a = 0$, $b = 1$

In Exercises 17–26, sketch the graphs of the functions f and g and find the area of the region enclosed by these graphs and the vertical lines $x = a$ and $x = b$.

17. $f(x) = x^2 + 3$, $g(x) = 1$; $a = 1$, $b = 3$

18. $f(x) = x + 2$, $g(x) = x^2 - 4$; $a = -1$, $b = 2$

19. $f(x) = -x^2 + 2x + 3$, $g(x) = -x + 3$; $a = 0$, $b = 2$

20. $f(x) = 9 - x^2$, $g(x) = 2x + 3$; $a = -1$, $b = 1$

21. $f(x) = x^2 + 1$, $g(x) = \frac{1}{3}x^3$; $a = -1$, $b = 2$

22. $f(x) = \sqrt{x}$, $g(x) = -\frac{1}{2}x - 1$; $a = 1$, $b = 4$

23. $f(x) = \frac{1}{x}$, $g(x) = 2x - 1$; $a = 1$, $b = 4$

24. $f(x) = x^2$, $g(x) = \frac{1}{x^2}$; $a = 1$, $b = 3$

25. $f(x) = e^x$, $g(x) = \frac{1}{x}$; $a = 1$, $b = 2$

26. $f(x) = x$, $g(x) = e^{2x}$; $a = 1$, $b = 3$

In Exercises 27–34, sketch the graph and find the area of the region bounded by the graph of the function f and the lines $y = 0$, $x = a$, and $x = b$.

27. $f(x) = x$; $a = -1$, $b = 2$

28. $f(x) = x^2 - 2x$; $a = -1$, $b = 1$

29. $f(x) = -x^2 + 4x - 3$; $a = -1$, $b = 2$

30. $f(x) = x^3 - x^2$; $a = -1$, $b = 1$

31. $f(x) = x^3 - 4x^2 + 3x$; $a = 0$, $b = 2$

32. $f(x) = 4x^{1/3} + x^{4/3}$; $a = -1$, $b = 8$

33. $f(x) = e^x - 1$; $a = -1$, $b = 3$

34. $f(x) = xe^{x^2}$; $a = 0$, $b = 2$

In Exercises 35–40, sketch the graph and find the area of the region completely enclosed by the graphs of the given functions f and g.

35. $f(x) = x + 2$ and $g(x) = x^2 - 4$

36. $f(x) = -x^2 + 4x$ and $g(x) = 2x - 3$

37. $f(x) = x^2$ and $g(x) = x^3$

38. $f(x) = x^3 - 6x^2 + 9x$ and $g(x) = x^2 - 3x$

39. $f(x) = \sqrt{x}$ and $g(x) = x^2$

40. $f(x) = 2x$ and $g(x) = x\sqrt{x + 1}$

41. EFFECT OF ADVERTISING ON REVENUE In the accompanying figure, the function f gives the rate of change of Odyssey Travel's revenue with respect to the amount x it spends on advertising with their current advertising agency. By engaging the services of a different advertising agency, it is expected that Odyssey's revenue will grow at the rate given by the function g. Give an interpretation of the area A of the region S and find an expression for A in terms of a definite integral involving f and g.

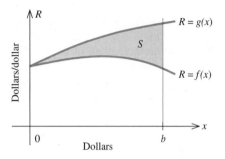

42. PULSE RATE DURING EXERCISE In the accompanying figure, the function f gives the rate of increase of an individual's pulse rate when he walked a prescribed course on a treadmill 6 mo ago. The function g gives the rate of increase of his pulse rate when he recently walked the same prescribed course. Give an interpretation of the area A of the region S and find an expression for A in terms of a definite integral involving f and g.

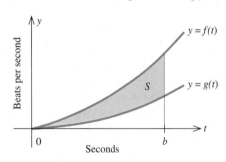

43. AIR PURIFICATION To study the effectiveness of air purifiers in removing smoke, engineers run each purifier in a smoke-filled 10×20-ft room. In the accompanying figure, the function f gives the rate of change of the smoke level/minute, t min after the start of the test, when a brand A purifier is used. The function g gives the rate of change of the smoke level/minute when a brand B purifier is used.
a. Give an interpretation of the area of the region S.
b. Find an expression for the area of S in terms of a definite integral involving f and g.

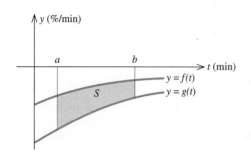

44. Two cars start out side by side and travel along a straight road. The velocity of car 1 is $f(t)$ ft/sec, the velocity of car 2 is $g(t)$ ft/sec over the interval $[0, T]$, and $0 < T_1 < T$. Furthermore, suppose the graphs of f and g are as depicted in the accompanying figure. Denote the area of region I by A_1 and the area of region II by A_2.

a. Write the number

$$\int_{T_1}^{T} [g(t) - f(t)]\, dt - \int_{0}^{T_1} [f(t) - g(t)]\, dt$$

in terms of A_1 and A_2.

b. What does the number obtained in part (a) represent?

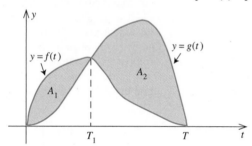

45. The rate of change of the revenue of company A over the (time) interval $[0, T]$ is $f(t)$ dollars/week, whereas the rate of change of the revenue of company B over the same period is $g(t)$ dollars/week. Suppose the graphs of f and g are depicted in the accompanying figure. Find an expression in terms of definite integrals involving f and g giving the additional revenue that company B will have over company A in the period $[0, T]$.

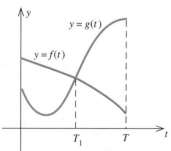

46. TURBO-CHARGED ENGINE VS. STANDARD ENGINE In tests conducted by *Auto Test Magazine* on two identical models of the Phoenix Elite—one equipped with a standard engine and the other with a turbo-charger—it was found that the acceleration of the former is given by

$$a = f(t) = 4 + 0.8t \qquad (0 \le t \le 12)$$

ft/sec/sec, t sec after starting from rest at full throttle, whereas the acceleration of the latter is given by

$$a = g(t) = 4 + 1.2t + 0.03t^2 \qquad (0 \le t \le 12)$$

ft/sec/sec. How much faster is the turbo-charged model moving than the model with the standard engine, at the end of a 10-sec test run at full throttle?

47. ALTERNATIVE ENERGY SOURCES Because of the increasingly important role played by coal as a viable alternative energy source, the production of coal has been growing at the rate of

$$3.5e^{0.05t}$$

billion metric tons/year t yr from 1980 (which corresponds to $t = 0$). Had it not been for the energy crisis, the rate of production of coal since 1980 might have been only

$$3.5e^{0.01t}$$

billion metric tons/year t yr from 1980. Determine how much additional coal was produced between 1980 and the end of the century as an alternate energy source.

48. EFFECT OF TV ADVERTISING ON CAR SALES Carl Williams, the new proprietor of Carl Williams Auto Sales, estimates that with extensive television advertising, car sales over the next several years could be increasing at the rate of

$$5e^{0.3t}$$

thousand cars/year t yr from now, instead of at the current rate of

$$(5 + 0.5t^{3/2})$$

thousand cars/year t yr from now. Find how many more cars Carl expects to sell over the next 5 yr by implementing his advertising plans.

49. POPULATION GROWTH In an endeavor to curb population growth in a Southeast Asian island state, the government has decided to launch an extensive propaganda campaign. Without curbs, the government expects the rate of population growth to have been

$$60e^{0.02t}$$

thousand people/year t yr from now, over the next 5 yr. However, successful implementation of the proposed campaign is expected to result in a population growth rate of

$$-t^2 + 60$$

thousand people/yr t yr from now, over the next 5 yr. Assuming that the campaign is mounted, how many fewer people will there be in that country 5 yr from now than there would have been if no curbs had been imposed?

In Exercises 50 and 51, determine whether the statement is true or false. If it is true, explain why it is true. If it is false, give an example to show why it is false.

50. If f and g are continuous on $[a, b]$ and either $f(x) \geq g(x)$ for all x in $[a, b]$ or $f(x) \leq g(x)$ for all x in $[a, b]$, then the area of the region bounded by the graphs of f and g and the vertical lines $x = a$ and $x = b$ is given by $\int_a^b |f(x) - g(x)| \, dx$.

51. The area of the region bounded by the graphs of $f(x) = 2 - x$ and $g(x) = 4 - x^2$ and the vertical lines $x = 0$ and $x = 2$ is given by $\int_0^2 [f(x) - g(x)] \, dx$.

52. Show that the area of a region R bounded above by the graph of a function f and below by the graph of a function g from $x = a$ to $x = b$ is given by

$$\int_a^b [f(x) - g(x)] \, dx$$

Hint: The validity of the formula was verified earlier for the case when both f and g were nonnegative. Now, let f and g be two functions such that $f(x) \geq g(x)$ for $a \leq x \leq b$. Then, there exists some nonnegative constant c such that the curves $y = f(x) + c$ and $y = g(x) + c$ are translated in such a way that the region R' has the same area as the region R (see the accompanying figures). Show that the area of R' is given by

$$\int_a^b \{[f(x) + c] - [g(x) + c]\} \, dx = \int_a^b [f(x) - g(x)] \, dx$$

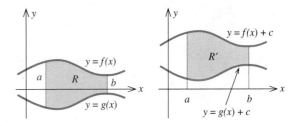

SOLUTIONS TO SELF-CHECK EXERCISES 6.6

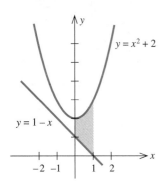

1. The region in question is shown in the accompanying figure. Since the graph of the function f lies above that of the function g for $0 \leq x \leq 1$, we see that the required area is given by

$$\int_0^1 [(x^2 + 2) - (1 - x)] \, dx = \int_0^1 (x^2 + x + 1) \, dx$$

$$= \frac{1}{3}x^3 + \frac{1}{2}x^2 + x \Big|_0^1$$

$$= \frac{1}{3} + \frac{1}{2} + 1$$

$$= \frac{11}{6}$$

or $\frac{11}{6}$ square units.

2. The region in question is shown in the figure on page 518. To find the points of intersection of the two curves, we solve the equation

$$-x^2 + 6x + 5 = x^2 + 5$$

$$2x^2 - 6x = 0$$

$$2x(x - 3) = 0$$

giving $x = 0$ or $x = 3$. Therefore, the points of intersection are $(0, 5)$ and $(3, 14)$.

(continued on p. 518)

Using Technology

Finding the Area Between Two Curves

The numerical integral operation can also be used to find the area between two curves. We do this by using the numerical integral operation to evaluate an appropriate definite integral or the sum (difference) of appropriate definite integrals. In the following example, the intersection operation is also used to advantage to help us find the limits of integration.

EXAMPLE 1 Use a graphing utility to find the area of the region R that is completely enclosed by the graphs of the functions

$$f(x) = 2x^3 - 8x^2 + 4x - 3 \quad \text{and} \quad g(x) = 3x^2 + 10x - 11$$

SOLUTION ✔ The graphs of f and g in the viewing rectangle $[-3, 4] \times [-20, 5]$ are shown in Figure T1.

FIGURE T1
The region R is completely enclosed by the graphs of f and g.

Using the intersection operation of a graphing utility, we find the x-coordinates of the points of intersection of the two graphs to be approximately -1.04 and 0.65, respectively. Since the graph of f lies above that of g on the interval $[-1.04, 0.65]$, we see that the area of R is given by

$$A = \int_{-1.04}^{0.65} [(2x^3 - 8x^2 + 4x - 3) - (3x^2 + 10x - 11)] \, dx$$

$$= \int_{-1.04}^{0.65} (2x^3 - 11x^2 - 6x + 8) \, dx$$

Using the numerical integral function of a graphing utility, we find $A \approx 9.87$, and so the area of R is approximately 9.87 square units. ■ ■ ■ ■

In Exercises 1–6, use a graphing utility to (a) plot the graphs of the functions f and g and (b) find the area of the region enclosed by these graphs and the vertical lines $x = a$ and $x = b$. Express your answers accurate to four decimal places.

1. $f(x) = x^3(x - 5)^4$, $g(x) = 0$; $a = 1$, $b = 3$

2. $f(x) = x - \sqrt{1 - x^2}$, $g(x) = 0$; $a = -\dfrac{1}{2}$, $b = \dfrac{1}{2}$

3. $f(x) = x^{1/3}(x + 1)^{1/2}$, $g(x) = x^{-1}$; $a = 1.2$, $b = 2$

4. $f(x) = 2$, $g(x) = \ln(1 + x^2)$; $a = -1$, $b = 1$

5. $f(x) = \sqrt{x}$, $g(x) = \dfrac{x^2 - 3}{x^2 + 1}$; $a = 0$, $b = 3$

6. $f(x) = \dfrac{4}{x^2 + 1}$, $g(x) = x^4$; $a = -1$, $b = 1$

In Exercises 7–12, use a graphing utility to (a) plot the graphs of the functions f and g and (b) find the area of the region totally enclosed by the graphs of these functions.

7. $f(x) = 2x^3 - 8x^2 + 4x - 3$ and $g(x) = -3x^2 + 10x - 10$

8. $f(x) = x^4 - 2x^2 + 2$ and $g(x) = 4 - 2x^2$

9. $f(x) = 2x^3 - 3x^2 + x + 5$ and $g(x) = e^{2x} - 3$

10. $f(x) = \dfrac{1}{2}x^2 - 3$ and $g(x) = \ln x$

11. $f(x) = xe^{-x}$ and $g(x) = x - 2\sqrt{x}$

12. $f(x) = e^{-x^2}$ and $g(x) = x^4$

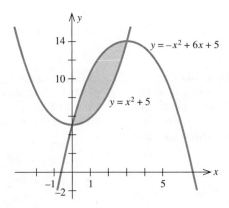

Since the graph of f always lies above that of g for $0 \leq x \leq 3$, we see that the required area is given by

$$\int_0^3 [(-x^2 + 6x + 5) - (x^2 + 5)]\, dx = \int_0^3 (-2x^2 + 6x)\, dx$$

$$= -\frac{2}{3}x^3 + 3x^2 \Big|_0^3$$

$$= -18 + 27$$

$$= 9$$

or 9 square units.

3. The additional profits realizable over the next 10 yr are given by

$$\int_0^{10} [(t - 2\sqrt{t} + 4) - (1 + t^{2/3})]\, dt$$

$$= \int_0^{10} (t - 2t^{1/2} + 3 - t^{2/3})\, dt$$

$$= \frac{1}{2}t^2 - \frac{4}{3}t^{3/2} + 3t - \frac{3}{5}t^{5/3} \Big|_0^{10}$$

$$= \frac{1}{2}(10)^2 - \frac{4}{3}(10)^{3/2} + 3(10) - \frac{3}{5}(10)^{5/3}$$

$$\approx 9.99$$

or approximately $10 million.

6.7 Applications of the Definite Integral to Business and Economics

In this section we consider several applications of the definite integral in the fields of business and economics.

CONSUMERS' AND PRODUCERS' SURPLUS

FIGURE 6.38
$D(x)$ is a demand function.

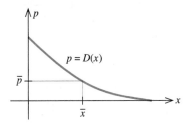

We begin by deriving a formula for computing the consumers' surplus. Suppose $p = D(x)$ is the demand function that relates the unit price p of a commodity to the quantity x demanded of it. Furthermore, suppose a fixed unit market price has been established for the commodity and corresponding to this unit price the quantity demanded is $\bar{x}$ units (Figure 6.38). Then, those consumers who would be willing to pay a unit price higher than $\bar{p}$ for the commodity would in effect experience a savings. This difference between what the consumers *would* be willing to pay for $\bar{x}$ units of the commodity and what they *actually* pay for them is called the **consumers' surplus.**

To derive a formula for computing the consumers' surplus, divide the interval $[0, \bar{x}]$ into n subintervals, each of length $\Delta x = \bar{x}/n$, and denote the right end points of these subintervals by $x_1, x_2, \ldots, x_n = \bar{x}$ (Figure 6.39).

We observe in Figure 6.39 that there are consumers who would pay a unit price of at least $D(x_1)$ dollars for the first Δx units of the commodity instead of the market price of $\bar{p}$ dollars per unit. The savings to these consumers is approximated by

$$D(x_1)\Delta x - \bar{p}\Delta x = [D(x_1) - \bar{p}]\Delta x$$

FIGURE 6.39
Approximating consumers' surplus by the sum of the rectangles $r_1, r_2, \ldots, r_n$

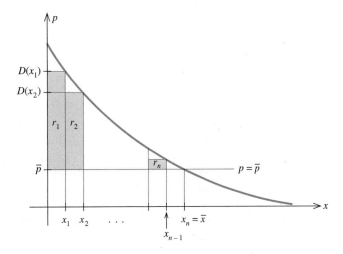

which is the area of the rectangle r_1. Pursuing the same line of reasoning, we find that the savings to the consumers who would be willing to pay a unit price of at least $D(x_2)$ dollars for the next Δx units (from x_1 through x_2) of the commodity, instead of the market price of $\bar{p}$ dollars per unit, is approximated by

$$D(x_2)\Delta x - \bar{p}\Delta x = [D(x_2) - \bar{p}]\Delta x$$

Continuing, we approximate the total savings to the consumers in purchasing $\bar{x}$ units of the commodity by the sum

$$[D(x_1) - \bar{p}]\Delta x + [D(x_2) - \bar{p}]\Delta x + \cdots + [D(x_n) - \bar{p}]\Delta x$$
$$= [D(x_1) + D(x_2) + \cdots + D(x_n)]\Delta x - \underbrace{[\bar{p}\Delta x + \bar{p}\Delta x + \cdots + \bar{p}\Delta x]}_{n\ \text{terms}}$$

$$= [D(x_1) + D(x_2) + \cdots + D(x_n)]\Delta x - n\bar{p}\Delta x$$

$$= [D(x_1) + D(x_2) + \cdots + D(x_n)]\Delta x - \bar{p}\bar{x}$$

Now, the first term in the last expression is the Riemann sum of the demand function $p = D(x)$ over the interval $[0, \bar{x}]$ with representative points $x_1, x_2, \ldots, x_n$. Letting n approach infinity, we obtain the following formula for the consumers' surplus CS.

Consumers' Surplus

The **consumers' surplus** is given by

$$CS = \int_0^{\bar{x}} D(x)\,dx - \bar{p}\bar{x} \tag{15}$$

where D is the demand function, $\bar{p}$ is the unit market price, and $\bar{x}$ is the quantity sold.

The consumers' surplus is given by the area of the region bounded above by the demand curve $p = D(x)$ and below by the straight line $p = \bar{p}$ from $x = 0$ to $x = \bar{x}$ (Figure 6.40). We can also see this if we rewrite Equation (15) in the form

$$\int_0^{\bar{x}} [D(x) - \bar{p}]\,dx$$

and interpret the result geometrically.

Analogously, we can derive a formula for computing the producers' surplus. Suppose $p = S(x)$ is the supply equation that relates the unit price p of a certain commodity to the quantity x that the supplier will make available in the market at that price.

Again, suppose a fixed market price $\bar{p}$ has been established for the commodity and, corresponding to this unit price, a quantity of $\bar{x}$ units will be made

FIGURE 6.40
Consumers' surplus

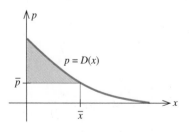

available in the market by the supplier (Figure 6.41). Then, the suppliers who would be willing to make the commodity available at a lower price stand to gain from the fact that the market price is set as such. The difference between what the suppliers actually receive and what they would be willing to receive is called the **producers' surplus.** Proceeding in a manner similar to the derivation of the equation for computing the consumers' surplus, we find that the producers' surplus PS is defined as follows:

Producers' Surplus

The **producers' surplus** is given by

$$PS = \bar{p}\bar{x} - \int_0^{\bar{x}} S(x)\, dx \qquad (16)$$

where $S(x)$ is the supply function, $\bar{p}$ is the unit market price, and $\bar{x}$ is the quantity supplied.

Geometrically, the producers' surplus is given by the area of the region bounded above by the straight line $p = \bar{p}$ and below by the supply curve $p = S(x)$ from $x = 0$ to $x = \bar{x}$ (Figure 6.42).

FIGURE 6.41
S(x) is a supply function.

FIGURE 6.42
Producers' surplus

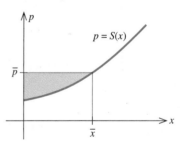

We can also show that the last statement is true by converting Equation (16) to the form

$$\int_0^{\bar{x}} [\bar{p} - S(x)]\, dx$$

and interpreting the definite integral geometrically.

EXAMPLE 1   The demand function for a certain make of 10-speed bicycle is given by

$$p = D(x) = -0.001x^2 + 250$$

where p is the unit price in dollars and x is the quantity demanded in units of a thousand. The supply function for these bicycles is given by

$$p = S(x) = 0.0006x^2 + 0.02x + 100$$

where p stands for the unit price in dollars and x stands for the number of bicycles that the supplier will put on the market, in units of a thousand. Determine the consumers' surplus and the producers' surplus if the market price of a bicycle is set at the equilibrium price.

SOLUTION ✔

FIGURE 6.43

Consumers' surplus and producers' surplus when market price = equilibrium price

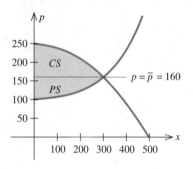

Recall that the equilibrium price is the unit price of the commodity when market equilibrium occurs. We determine the equilibrium price by solving for the point of intersection of the demand curve and the supply curve (Figure 6.43). To solve the system of equations

$$p = -0.001x^2 \qquad\quad + 250$$
$$p = 0.0006x^2 + 0.02x + 100$$

we simply substitute the first equation into the second, obtaining

$$0.0006x^2 + 0.02x + 100 = -0.001x^2 + 250$$
$$0.0016x^2 + 0.02x - 150 = 0$$
$$16x^2 + 200x - 1{,}500{,}000 = 0$$
$$2x^2 + 25x - 187{,}500 = 0$$

Factoring this last equation, we obtain

$$(2x + 625)(x - 300) = 0$$

Thus, $x = -625/2$ or $x = 300$. The first number lies outside the interval of interest, so we are left with the solution $x = 300$, with a corresponding value of

$$p = -0.001(300)^2 + 250 = 160$$

Thus, the equilibrium point is (300, 160); that is, the equilibrium quantity is 300,000, and the equilibrium price is $160. Setting the market price at $160 per unit and using Formula (15) with $\bar{p} = 160$ and $\bar{x} = 300$, we find that the consumers' surplus is given by

$$CS = \int_0^{300} (-0.001x^2 + 250)\, dx - (160)(300)$$

$$= \left(-\frac{1}{3000}x^3 + 250x \right) \Big|_0^{300} - 48{,}000$$

$$= -\frac{300^3}{3000} + (250)(300) - 48{,}000$$

$$= 18{,}000$$

or $18,000,000.

(Recall that x is measured in units of a thousand.) Next, using (16), we find that the producers' surplus is given by

$$PS = (160)(300) - \int_0^{300} (0.0006x^2 + 0.02x + 100)\, dx$$

$$= 48,000 - (0.0002x^3 + 0.01x^2 + 100x)\Big|_0^{300}$$

$$= 48,000 - [(0.0002)(300)^3 + (0.01)(300)^2 + 100(300)]$$

$$= 11,700$$

or \$11,700,000.   ■ ■ ■ ■

THE FUTURE AND PRESENT VALUE OF AN INCOME STREAM

To introduce the notion of the future and the present value of an income stream, suppose a firm generates a stream of income over a period of time—the revenue generated by a large chain of retail stores over a 5-year period, for example. As the income is realized, it is reinvested and earns interest at a fixed rate. The **accumulated future income stream** over the 5-year period is the amount of money the firm ends up with at the end of that period.

The definite integral can be used to determine this accumulated, or total, future income stream over a period of time. The total future value of an income stream gives us a way to measure the value of such a stream. To find the **total future value of an income stream,** suppose

$R(t)$ = Rate of income generation at any time t (Dollars per year)

r = Interest rate compounded continuously

T = Term   (In years)

FIGURE 6.44
The time interval $[0, T]$ is partitioned into n subintervals.

Let's divide the time interval $[0, T]$ into n subintervals of equal length $\Delta t = T/n$ and denote the right end points of these intervals by $t_1, t_2, \ldots, t_n = T$, as shown in Figure 6.44.

If R is a continuous function on $[0, T]$, then $R(t)$ will not differ by much from $R(t_1)$ in the subinterval $[0, t_1]$ provided that the subinterval is small (which is true if n is large). Therefore, the income generated over the time interval $[0, t_1]$ is approximately

$$R(t_1)\Delta t \qquad \text{(Constant rate of income · Length of time)}$$

dollars. The future value of this amount, T years from now, calculated as if it were earned at time t_1, is

$$[R(t_1)\Delta t]e^{r(T-t_1)} \qquad \text{[Equation (10), Section 5.3]}$$

dollars. Similarly, the income generated over the time interval $[t_1, t_2]$ is approximately $P(t_2)\Delta t$ dollars and has a future value, T years from now, of approximately

$$[R(t_2)\Delta t]e^{r(T-t_2)}$$

dollars. Therefore, the sum of the future values of the income stream generated over the time interval $[0, T]$ is approximately

$$R(t_1)e^{r(T-t_1)}\Delta t + R(t_2)e^{r(T-t_2)}\Delta t + \cdots + R(t_n)e^{r(T-t_n)}\Delta t$$
$$= e^{rT}[R(t_1)e^{-rt_1}\Delta t + R(t_2)e^{-rt_2}\Delta t + \cdots + R(t_n)e^{-rt_n}\Delta t]$$

dollars. But this sum is just the Riemann sum of the function $e^{rT}R(t)e^{-rt}$ over the interval $[0, T]$ with representative points $t_1, t_2, \ldots, t_n$. Letting n approach infinity, we obtain the following result.

Accumulated or Total Future Value of an Income Stream

The **accumulated, or total, future value** after T years of an income stream of $R(t)$ dollars per year, earning interest at the rate of r per year compounded continuously, is given by

$$A = e^{rT}\int_0^T R(t)e^{-rt}\,dt \tag{17}$$

EXAMPLE 2   Crystal Car Wash recently bought an automatic car-washing machine that is expected to generate $40,000 in revenue per year, t years from now, for the next 5 years. If the income is reinvested in a business earning interest at the rate of 12% per year compounded continuously, find the total accumulated value of this income stream at the end of 5 years.

SOLUTION ✔   We are required to find the total future value of the given income stream after 5 years. Using Equation (17) with $R(t) = 40,000$, $r = 0.12$, and $T = 5$, we see that the required value is given by

$$e^{0.12(5)}\int_0^5 40{,}000e^{-0.12t}\,dt$$

$$= e^{0.6}\left[-\frac{40{,}000}{0.12}e^{-0.12t}\right]\Bigg|_0^5 \qquad \text{(Integrate using the substitution } u = -0.12t.\text{)}$$

$$= -\frac{40{,}000e^{0.6}}{0.12}(e^{-0.6}-1) \approx 274{,}039.60$$

or approximately $274,040.   ■■■■

Another way of measuring the value of an income stream is by considering its present value. The **present value of an income stream** of $R(t)$ dollars per year over a term of T years, earning interest at the rate of r per year compounded continuously, is the principal P that will yield the same accumulated value as the income stream itself when P is invested today for a period of T years at the same rate of interest. In other words,

$$Pe^{rT} = e^{rT}\int_0^T R(t)e^{-rt}\,dt$$

Dividing both sides of the equation by e^{rT} gives the following result.

Present Value of an Income Stream

The **present value of an income stream** of $R(t)$ dollars per year, earning interest at the rate of r per year compounded continuously, is given by

$$PV = \int_0^T R(t)e^{-rt}\, dt \qquad\qquad (18)$$

 EXAMPLE 3

The owner of a local cinema is considering two alternative plans for renovating and improving the theater. Plan A calls for an immediate cash outlay of $250,000, whereas plan B requires an immediate cash outlay of $180,000. It has been estimated that adopting plan A would result in a net income stream generated at the rate of

$$f(t) = 630{,}000$$

dollars per year, whereas adopting plan B would result in a net income stream generated at the rate of

$$g(t) = 580{,}000$$

dollars per year for the next 3 years. If the prevailing interest rate for the next 5 years is 10% per year, which plan will generate a higher net income by the end of 3 years?

SOLUTION ✔ Since the initial outlay is $250,000, we find—using Equation (18) with $R(t) = 630{,}000$, $r = 0.1$, and $T = 3$—that the present value of the net income under plan A is given by

$$\int_0^3 630{,}000 e^{-0.1t}\, dt - 250{,}000$$

$$= \frac{630{,}000}{-0.1} e^{-0.1t}\Big|_0^3 - 250{,}000 \qquad \text{(Integrate using the substitution } u = -0.1t.)$$

$$= -6{,}300{,}000 e^{-0.3} + 6{,}300{,}000 - 250{,}000$$

$$\approx 1{,}382{,}845$$

or approximately $1,382,845.

To find the present value of the net income under plan B, we use (18) with $R(t) = 580{,}000$, $r = 0.1$, and $T = 3$, obtaining

$$\int_0^3 580{,}000 e^{-0.1t}\, dt - 180{,}000$$

dollars. Proceeding as in the previous computation, we see that the required value is $1,323,254 (Exercise 8).

Comparing the present value of each plan, we conclude that plan A would generate a higher net income by the end of 3 years. ■ ■ ■ ■

REMARK The function R in Example 3 is a constant function. If R is not a constant function, then we may need more sophisticated techniques of integration to evaluate the integral in (18). Exercises 7.1 and 7.2 in Chapter 7 contain problems of this type. ■■■

THE AMOUNT AND PRESENT VALUE OF AN ANNUITY

An annuity is a sequence of payments made at regular time intervals. The time period in which these payments are made is called the *term* of the annuity. Although the payments need not be equal in size, they are equal in many important applications, and we will assume that they are equal in our discussion. Examples of annuities are regular deposits to a savings account, monthly home mortgage payments, and monthly insurance payments.

The **amount of an annuity** is the sum of the payments plus the interest earned. A formula for computing the amount of an annuity A can be derived with the help of (17). Let

$$P = \text{Size of each payment in the annuity}$$
$$r = \text{Interest rate compounded continuously}$$
$$T = \text{Term of the annuity (in years)}$$
$$m = \text{Number of payments per year}$$

The payments into the annuity constitute a constant income stream of $R(t) = mP$ dollars per year. With this value of $R(t)$, (17) yields

$$A = e^{rT} \int_0^T R(t) e^{-rt} \, dt = e^{rT} \int_0^T mP e^{-rt} \, dt$$

$$= mPe^{rT} \left[-\frac{e^{-rt}}{r} \right]\Bigg|_0^T$$

$$= mPe^{rT} \left[-\frac{e^{-rT}}{r} + \frac{1}{r} \right] = \frac{mP}{r} (e^{rT} - 1)$$

This leads us to the following formula.

Amount of an Annuity

> The **amount of an annuity** is
>
> $$A = \frac{mP}{r} (e^{rT} - 1) \tag{19}$$
>
> where P, r, T, and m are as defined earlier.

 EXAMPLE 4

On January 1, 1990, Marcus Chapman deposited $2000 into an Individual Retirement Account (IRA) paying interest at the rate of 10% per year compounded continuously. Assuming that he deposits $2000 annually into the account, how much will he have in his IRA at the beginning of the year 2006?

SOLUTION ✔   We use (19), with $P = 2000$, $r = 0.1$, $T = 16$, and $m = 1$, obtaining

$$A = \frac{2000}{0.1}(e^{1.6} - 1)$$

$$\approx 79{,}060.65$$

Thus, Marcus will have approximately $79,061 in his account at the beginning of the year 2006.   ■ ■ ■ ■

Exploring with Technology

Refer to Example 4. Suppose Marcus wishes to know how much he will have in his IRA at any time in the future, not just at the beginning of 2006, as you were asked to compute in the example.

1. Using Formula (17) and the relevant data from Example 4, show that the required amount at any time x (x measured in years, $x > 0$) is given by

$$A = f(x) = 20{,}000(e^{0.1x} - 1)$$

2. Use a graphing utility to plot the graph of f, using the viewing rectangle $[0, 30] \times [2000, 400{,}000]$.
3. Using **ZOOM** and **TRACE**, or using the function evaluation capability of your graphing utility, use the result of part 2 to verify the result obtained in Example 4. Comment on the advantage of the mathematical model found in part 1.

Using (18), we can derive the following formula for the present value of an annuity.

Present Value of an Annuity

The **present value of an annuity** is given by

$$PV = \frac{mP}{r}(1 - e^{-rT}) \qquad (20)$$

where P, r, T, and m are as defined earlier.

 EXAMPLE **5**   Tomas Perez, the proprietor of a hardware store, wants to establish a fund from which he will withdraw $1000 per month for the next 10 years. If the fund earns interest at the rate of 9% per year compounded continuously, how much money does he need to establish the fund?

SOLUTION ✔   We want to find the present value of an annuity with $P = 1000$, $r = 0.09$, $T = 10$, and $m = 12$. Using Equation (20), we find

$$PV = \frac{12{,}000}{0.09}(1 - e^{-(0.09)(10)})$$

$$\approx 79{,}124.04$$

Thus, Tomas needs approximately $79,124 to establish the fund.   ■ ■ ■ ■

LORENTZ CURVES AND INCOME DISTRIBUTIONS

One method used by economists to study the distribution of income in a society is based on the **Lorentz curve,** named after American statistician M. D. Lorentz. To describe the Lorentz curve, let $f(x)$ denote the proportion of the total income received by the poorest $100x\%$ of the population for $0 \le x \le 1$. Using this terminology, $f(0.3) = 0.1$ simply states that the lowest 30% of the income recipients receive 10% of the total income.

The function f has the following properties:

FIGURE 6.45
A Lorentz curve

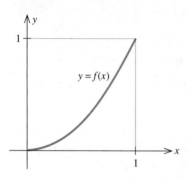

1. The domain of f is $[0, 1]$.
2. The range of f is $[0, 1]$.
3. $f(0) = 0$ and $f(1) = 1$.
4. $f(x) \le x$ for every x in $[0, 1]$.
5. f is increasing on $[0, 1]$.

The first two properties follow from the fact that both x and $f(x)$ are fractions of a whole. Property 3 is a statement that 0% of the income recipients receive 0% of the total income and 100% of the income recipients receive 100% of the total income. Property 4 follows from the fact that the lowest $100x\%$ of the income recipients cannot receive more than $100x\%$ of the total income. A typical Lorentz curve is shown in Figure 6.45.

EXAMPLE 6

A developing country's income distribution is described by the function

$$f(x) = \frac{19}{20}x^2 + \frac{1}{20}x$$

a. Sketch the Lorentz curve for the given function.
b. Compute $f(0.2)$ and $f(0.8)$ and interpret your results.

SOLUTION ✔

FIGURE 6.46

The Lorentz curve $f(x) = \frac{19}{20}x^2 + \frac{1}{20}x$.

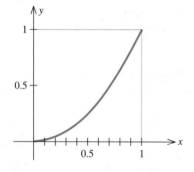

a. The Lorentz curve is shown in Figure 6.46.

b.
$$f(0.2) = \frac{19}{20}(0.2)^2 + \frac{1}{20}(0.2) = 0.048$$

Thus, the lowest 20% of the people receive 4.8% of the total income.

$$f(0.8) = \frac{19}{20}(0.8)^2 + \frac{1}{20}(0.8) = 0.648$$

Thus, the lowest 80% of the people receive 64.8% of the total income.

■ ■ ■ ■

Next, let's consider the Lorentz curve described by the function $y = f(x) = x$. Since exactly $100x\%$ of the total income is received by the lowest $100x\%$ of income recipients, the line $y = x$ is called the **line of complete equality.** For example, 10% of the total income is received by the lowest 10% of income recipients, 20% of the total income is received by the lowest 20%

FIGURE 6.47
The closer the Lorentz curve is to the line, the more equitable the income distribution.

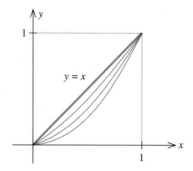

of income recipients, and so on. Now, it is evident that the closer a Lorentz curve is to this line, the more equitable the income distribution is among the income recipients. But the proximity of a Lorentz curve to the line of complete equality is reflected by the area between the Lorentz curve and the line $y = x$ (Figure 6.47). The closer the curve is to the line, the smaller the enclosed area.

This observation suggests that we may define a number, called the coefficient of inequality of a Lorentz curve, as the ratio of the area between the line of complete equality and the Lorentz curve to the area under the line of complete equality. Since the area under the line of complete equality is $\frac{1}{2}$, we see that the coefficient of inequality is given by the following formula.

Coefficient of Inequality of a Lorentz Curve

> The **coefficient of inequality**, or **Gini index**, of a Lorentz curve is
>
> $$L = 2 \int_0^1 [x - f(x)]\, dx \qquad (21)$$

The coefficient of inequality is a number between 0 and 1. For example, a coefficient of zero implies that the income distribution is perfectly uniform.

EXAMPLE 7   In a study conducted by a certain country's Economic Development Board with regard to the income distribution of certain segments of the country's workforce, it was found that the Lorentz curves for the distribution of income of medical doctors and of movie actors are described by the functions

$$f(x) = \frac{14}{15}x^2 + \frac{1}{15}x \qquad \text{and} \qquad g(x) = \frac{5}{8}x^4 + \frac{3}{8}x$$

respectively. Compute the coefficient of inequality for each Lorentz curve. Which profession has a more equitable income distribution?

SOLUTION ✔ The required coefficients of inequality are, respectively,

$$L_1 = 2 \int_0^1 \left[x - \left(\frac{14}{15}x^2 + \frac{1}{15}x \right) \right] dx = 2 \int_0^1 \left(\frac{14}{15}x - \frac{14}{15}x^2 \right) dx$$

$$= \frac{28}{15} \int_0^1 (x - x^2) \, dx = \frac{28}{15} \left(\frac{1}{2}x^2 - \frac{1}{3}x^3 \right) \Big|_0^1$$

$$= \frac{14}{45} \approx 0.311$$

$$L_2 = 2 \int_0^1 \left[x - \left(\frac{5}{8}x^4 + \frac{3}{8}x \right) \right] dx = 2 \int_0^1 \left(\frac{5}{8}x - \frac{5}{8}x^4 \right) dx$$

$$= \frac{5}{4} \int_0^1 (x - x^4) \, dx = \frac{5}{4} \left(\frac{1}{2}x^2 - \frac{1}{5}x^5 \right) \Big|_0^1$$

$$= \frac{15}{40} \approx 0.375$$

We conclude that in this country the incomes of medical doctors are more evenly distributed than the incomes of movie actors. ■ ■ ■ ■

SELF-CHECK EXERCISE **6.7**

The demand function for a certain make of exercise bicycle that is sold exclusively through cable television is

$$p = d(x) = \sqrt{9 - 0.02x}$$

where p is the unit price in hundreds of dollars and x is the quantity demanded/week. The corresponding supply function is given by

$$p = s(x) = \sqrt{1 + 0.02x}$$

where p has the same meaning as before and x is the number of exercise bicycles the supplier will make available at price p. Determine the consumers' surplus and the producers' surplus if the unit price is set at the equilibrium price.

The solution to Self-Check Exercise 6.7 can be found on page 534.

6.7 Exercises

📇 **A calculator is recommended for this exercise set.**

1. **CONSUMERS' SURPLUS** The demand function for a certain make of replacement cartridges for a water purifier is given by

$$p = -0.01x^2 - 0.1x + 6$$

where p is the unit price in dollars and x is the quantity demanded each week, measured in units of a thousand. Determine the consumers' surplus if the market price is set at $4/cartridge.

2. **CONSUMERS' SURPLUS** The demand function for a certain brand of compact disc is given by

$$p = -0.01x^2 - 0.2x + 8$$

where p is the wholesale unit price in dollars and x is the quantity demanded each week, measured in units of a thousand. Determine the consumers' surplus if the wholesale market price is set at $5/disc.

3. **CONSUMERS' SURPLUS** It is known that the quantity demanded of a certain make of portable hair dryer is x hundred units/week and the corresponding wholesale unit price is

$$p = \sqrt{225 - 5x}$$

dollars. Determine the consumers' surplus if the wholesale market price is set at $10/unit.

4. **PRODUCERS' SURPLUS** The supplier of the portable hair dryers in Exercise 3 will make x hundred units of hair dryers available in the market when the wholesale unit price is

$$p = \sqrt{36 + 1.8x}$$

dollars. Determine the producers' surplus if the wholesale market price is set at $9/unit.

5. **PRODUCERS' SURPLUS** The supply function for the compact discs of Exercise 2 is given by

$$p = 0.01x^2 + 0.1x + 3$$

where p is the unit wholesale price in dollars and x stands for the quantity that will be made available in the market by the supplier, measured in units of a thousand. Determine the producers' surplus if the wholesale market price is set at the equilibrium price.

6. **CONSUMERS' AND PRODUCERS' SURPLUS** The management of the Titan Tire Company has determined that the quantity demanded x of their Super Titan tires/week is related to the unit price p by the relation

$$p = 144 - x^2$$

where p is measured in dollars and x is measured in units of a thousand. Titan will make x units of the tires available in the market if the unit price is

$$p = 48 + \frac{1}{2}x^2$$

dollars. Determine the consumers' surplus and the producers' surplus when the market unit price is set at the equilibrium price.

7. **CONSUMERS' AND PRODUCERS' SURPLUS** The quantity demanded x (in units of a hundred) of the Mikado miniature cameras/week is related to the unit price p (in dollars) by

$$p = -0.2x^2 + 80$$

and the quantity x (in units of a hundred) that the supplier is willing to make available in the market is related to the unit price p (in dollars) by

$$p = 0.1x^2 + x + 40$$

If the market price is set at the equilibrium price, find the consumers' surplus and the producers' surplus.

8. Refer to Example 3, page 525. Verify that

$$\int_0^3 580{,}000e^{-0.1t}\,dt - 180{,}000 \approx 1{,}323{,}254$$

9. **PRESENT VALUE OF AN INVESTMENT** Suppose an investment is expected to generate income at the rate of

$$R(t) = 200{,}000$$

dollars/year for the next 5 yr. Find the present value of this investment if the prevailing interest rate is 8%/year compounded continuously.

10. **FRANCHISES** Camille purchased a 15-yr franchise for a computer outlet store that is expected to generate income at the rate of

$$R(t) = 400{,}000$$

dollars/year. If the prevailing interest rate is 10%/year compounded continuously, find the present value of the franchise.

11. THE AMOUNT OF AN ANNUITY   Find the amount of an annuity if \$250/month is paid into it for a period of 20 yr earning interest at the rate of 8%/year compounded continuously.

12. THE AMOUNT OF AN ANNUITY   Find the amount of an annuity if \$400/month is paid into it for a period of 20 yr earning interest at the rate of 8%/year compounded continuously.

13. THE AMOUNT OF AN ANNUITY   Aiso deposits \$150/month in a savings account paying 8%/year compounded continuously. Estimate the amount that will be in his account after 15 yr.

14. CUSTODIAL ACCOUNTS   The Armstrongs wish to establish a custodial account to finance their children's education. If they deposit \$200 monthly for 10 yr in a savings account paying 9%/year compounded continuously, how much will their savings account be worth at the end of this period?

15. IRA ACCOUNTS   Refer to Example 4, page 526. Suppose Marcus makes his IRA payment on April 1, 1990, and annually thereafter. If interest is paid at the same initial rate, approximately how much will Marcus have in his account at the beginning of the year 2006?

16. PRESENT VALUE OF AN ANNUITY   Estimate the present value of an annuity if payments are \$800 monthly for 12 yr and the account earns interest at the rate of 10%/year compounded continuously.

17. PRESENT VALUE OF AN ANNUITY   Estimate the present value of an annuity if payments are \$1200 monthly for 15 yr and the account earns interest at the rate of 10%/year compounded continuously.

18. LOTTERY PAYMENTS   A state lottery commission pays the winner of the "Million Dollar" lottery 20 annual installments of \$50,000 each. If the prevailing interest rate is 8%/year compounded continuously, find the present value of the winning ticket.

19. REVERSE ANNUITY MORTGAGES   Sinclair wishes to supplement his retirement income by \$300/month for the next 10 yr. He plans to obtain a reverse annuity mortgage (RAM) on his home to meet this need. Estimate the amount of the mortgage he will require if the prevailing interest rate is 12%/year compounded continuously.

20. REVERSE ANNUITY MORTGAGE   Refer to Exercise 19. Leah wishes to supplement her retirement income by \$400/month for the next 15 yr by obtaining a RAM. Estimate the amount of the mortgage she will require if the prevailing interest rate is 9%/year compounded continuously.

21. LORENTZ CURVES   A certain country's income distribution is described by the function

$$f(x) = \frac{15}{16}x^2 + \frac{1}{16}x$$

a. Sketch the Lorentz curve for this function.
b. Compute $f(0.4)$ and $f(0.9)$ and interpret your results.

22. LORENTZ CURVES   In a study conducted by a certain country's Economic Development Board, it was found that the Lorentz curve for the distribution of income of college teachers was described by the function

$$f(x) = \frac{13}{14}x^2 + \frac{1}{14}x$$

and that of lawyers by the function

$$g(x) = \frac{9}{11}x^4 + \frac{2}{11}x$$

a. Compute the coefficient of inequality for each Lorentz curve.
b. Which profession has a more equitable income distribution?

23. LORENTZ CURVES   A certain country's income distribution is described by the function

$$f(x) = \frac{14}{15}x^2 + \frac{1}{15}x$$

a. Sketch the Lorentz curve for this function.
b. Compute $f(0.3)$ and $f(0.7)$.

24. LORENTZ CURVES   In a study conducted by a certain country's Economic Development Board, it was found that the Lorentz curve for the distribution of income of stockbrokers was described by the function

$$f(x) = \frac{11}{12}x^2 + \frac{1}{12}x$$

and that of high school teachers by the function

$$g(x) = \frac{5}{6}x^2 + \frac{1}{6}x$$

a. Compute the coefficient of inequality for each Lorentz curve.
b. Which profession has a more equitable income distribution?

Consumers' Surplus and Producers' Surplus

1. Resolve Example 1, Section 6.7, using a graphing utility.

Hint: Use the intersection operation to find the equilibrium quantity and the equilibrium price. Use the numerical integral operation to evaluate the definite integral.

2. Resolve Exercise 7, Section 6.7, using a graphing utility.

Hint: See Exercise 1.

3. The demand function for a certain brand of travel alarm clocks is given by

$$p = -0.01x^2 - 0.3x + 10$$

where p is the wholesale unit price in dollars and x is the quantity demanded each month, measured in units of a thousand. The supply function for this brand of clocks is given by

$$p = -0.01x^2 + 0.2x + 4$$

where p has the same meaning as before and x is the quantity, in thousands, the supplier will make available in the marketplace per month. Determine the consumers' surplus and the producers' surplus when the market unit price is set at the equilibrium price.

4. The quantity demanded of a certain make of compact disc organizer is x thousand units per week, and the corresponding wholesale unit price is

$$p = \sqrt{400 - 8x}$$

dollars. The supplier of the organizers will make x thousand units available in the market when the unit wholesale price is

$$p = 0.02x^2 + 0.04x + 5$$

dollars. Determine the consumers' surplus and the producers' surplus when the market unit price is set at the equilibrium price.

SOLUTION TO SELF-CHECK EXERCISE 6.7

We find the equilibrium price and equilibrium quantity by solving the system of equations

$$p = \sqrt{9 - 0.02x}$$
$$p = \sqrt{1 + 0.02x}$$

simultaneously. Substituting the first equation into the second, we have

$$\sqrt{9 - 0.02x} = \sqrt{1 + 0.02x}$$

Squaring both sides of the equation then leads to

$$9 - 0.02x = 1 + 0.02x$$
$$x = 200$$

Therefore,

$$p = \sqrt{9 - 0.02(200)}$$
$$= \sqrt{5} \approx 2.24$$

The equilibrium price is $224, and the equilibrium quantity is 200. The consumers' surplus is given by

$$CS = \int_0^{200} \sqrt{9 - 0.02x}\, dx - (2.24)(200)$$

$$= \int_0^{200} (9 - 0.02x)^{1/2}\, dx - 448$$

$$= -\frac{1}{0.02}\left(\frac{2}{3}\right)(9 - 0.02x)^{3/2}\, \Big|_0^{200} - 448 \qquad \text{(Integrating by substitution)}$$

$$= -\frac{1}{0.03}(5^{3/2} - 9^{3/2}) - 448$$

$$\approx 79.32$$

or approximately $7932.

Next, the producers' surplus is given by

$$PS = (2.24)(200) - \int_0^{200} \sqrt{1 + 0.02x}\, dx$$

$$= 448 - \int_0^{200} (1 + 0.02x)^{1/2}\, dx$$

$$= 448 - \frac{1}{0.02}\left(\frac{2}{3}\right)(1 + 0.02x)^{3/2}\, \Big|_0^{200}$$

$$= 448 - \frac{1}{0.03}(5^{3/2} - 1)$$

$$\approx 108.66$$

or approximately $10,866.

CHAPTER 6 Summary of Principal Formulas and Terms

Formulas

1. Indefinite integral of a constant

$$\int k\,du = ku + C$$

2. Power rule

$$\int u^n\,du = \frac{u^{n+1}}{n+1} + C$$

3. Constant multiple rule

$$\int kf(u)\,du = k\int f(u)\,du$$
$$(k, \text{ a constant})$$

4. Sum rule

$$\int [f(u) \pm g(u)]\,du$$
$$= \int f(u)\,du \pm \int g(u)\,du$$

5. Indefinite integral of the exponential function

$$\int e^u\,du = e^u + C$$

6. Indefinite integral of $f(u) = \dfrac{1}{u}$

$$\int \frac{du}{u} = \ln|u| + C$$

7. Method of substitution

$$\int f'(g(x))g'(x)\,dx = \int f'(u)\,du$$

8. Fundamental theorem of calculus

$$\int_a^b f(x)\,dx$$
$$= F(b) - F(a), F'(x) = f(x)$$

9. Area between two curves

$$\int_a^b [f(x) - g(x)]\,dx, f(x) \ge g(x)$$

10. Definite integral as the limit of a sum

$$\int_a^b f(x)\,dx = \lim_{n \to \infty} S_n,$$
$$\text{where } S_n \text{ is a Riemann sum}$$

11. Average value of f over $[a, b]$

$$\frac{1}{b-a}\int_a^b f(x)\,dx$$

12. Consumers' surplus

$$CS = \int_0^{\bar{x}} D(x)\,dx - \overline{p}\overline{x}$$

13. Producers' surplus

$$PS = \overline{p}\overline{x} - \int_0^{\bar{x}} S(x)\,dx$$

14. Accumulated (future) value of an income stream

$$A = e^{rT}\int_0^T R(t)e^{-rt}\,dt$$

15. Present value of an income stream

$$PV = \int_0^T R(t)e^{-rt}\,dt$$

16. Present value of an annuity	$PV = \dfrac{mP}{r}(1 - e^{-rT})$
17. Amount of an annuity	$A = \dfrac{mP}{r}(e^{rT} - 1)$
18. Coefficient of inequality of a Lorentz curve	$L = 2\displaystyle\int_0^1 [x - f(x)]\,dx$

Terms

antiderivative	initial value problem
antidifferentiation	Riemann sum
integration	definite integral
indefinite integral	lower limit of integration
integrand	upper limit of integration
constant of integration	Lorentz curve
differential equation	line of complete equality

Chapter 6 Review Exercises

In Exercises 1–20, find each indefinite integral.

1. $\displaystyle\int (x^3 + 2x^2 - x)\,dx$

2. $\displaystyle\int \left(\frac{1}{3}x^3 - 2x^2 + 8\right) dx$

3. $\displaystyle\int \left(x^4 - 2x^3 + \frac{1}{x^2}\right) dx$

4. $\displaystyle\int (x^{1/3} - \sqrt{x} + 4)\,dx$

5. $\displaystyle\int x(2x^2 + x^{1/2})\,dx$

6. $\displaystyle\int (x^2 + 1)(\sqrt{x} - 1)\,dx$

7. $\displaystyle\int \left(x^2 - x + \frac{2}{x} + 5\right) dx$

8. $\displaystyle\int \sqrt{2x + 1}\,dx$

9. $\displaystyle\int (3x - 1)(3x^2 - 2x + 1)^{1/3}\,dx$

10. $\displaystyle\int x^2(x^3 + 2)^{10}\,dx$

11. $\displaystyle\int \frac{x - 1}{x^2 - 2x + 5}\,dx$

12. $\displaystyle\int 2e^{-2x}\,dx$

13. $\displaystyle\int \left(x + \frac{1}{2}\right) e^{x^2 + x + 1}\,dx$

14. $\displaystyle\int \frac{e^{-x} - 1}{(e^{-x} + x)^2}\,dx$

15. $\displaystyle\int \frac{(\ln x)^5}{x}\,dx$

16. $\displaystyle\int \frac{\ln x^2}{x}\,dx$

17. $\displaystyle\int x^3(x^2 + 1)^{10}\,dx$

18. $\displaystyle\int x\sqrt{x + 1}\,dx$

19. $\displaystyle\int \frac{x}{\sqrt{x - 2}}\,dx$

20. $\displaystyle\int \frac{3x}{\sqrt{x + 1}}\,dx$

In Exercises 21–32, evaluate each definite integral.

21. $\displaystyle\int_0^1 (2x^3 - 3x^2 + 1)\,dx$

22. $\displaystyle\int_0^2 (4x^3 - 9x^2 + 2x - 1)\,dx$

23. $\displaystyle\int_1^4 (\sqrt{x} + x^{-3/2})\,dx$

24. $\displaystyle\int_0^1 20x(2x^2 + 1)^4\,dx$

25. $\displaystyle\int_{-1}^0 12(x^2 - 2x)(x^3 - 3x^2 + 1)^3\,dx$

26. $\int_4^7 x\sqrt{x-3}\,dx$

27. $\int_0^2 \frac{x}{x^2+1}\,dx$

28. $\int_0^1 \frac{dx}{(5-2x)^2}$

29. $\int_0^2 \frac{4x}{\sqrt{1+2x^2}}\,dx$

30. $\int_0^2 xe^{(-1/2)x^2}\,dx$

31. $\int_{-1}^0 \frac{e^{-x}}{(1+e^{-x})^2}\,dx$

32. $\int_1^e \frac{\ln x}{x}\,dx$

In Exercises 33–36, find the function f given that the slope of the tangent line to the graph at any point (x, f(x)) is f′(x) and that the graph of f passes through the given point.

33. $f'(x) = 3x^2 - 4x + 1;\ (1, 1)$

34. $f'(x) = \dfrac{x}{\sqrt{x^2+1}};\ (0, 1)$

35. $f'(x) = 1 - e^{-x};\ (0, 2)$

36. $f'(x) = \dfrac{\ln x}{x};\ (1, -2)$

37. Let $f(x) = -2x^2 + 1$ and compute the Riemann sum of f over the interval $[1, 2]$ by partitioning the interval into five subintervals of the same length ($n = 5$), where the points p_i ($1 \le i \le 5$) are taken to be the *right* end points of the respective subintervals.

38. The management of the National Electric Corporation has determined that the daily marginal cost function associated with producing their automatic drip coffeemakers is given by

$$C'(x) = 0.00003x^2 - 0.03x + 20$$

where $C'(x)$ is measured in dollars/unit and x denotes the number of units produced. Management has also determined that the daily fixed cost incurred in producing these coffeemakers is $500. What is the total cost incurred by National in producing the first 400 coffeemakers/day?

39. Refer to Exercise 38. Management has also determined that the daily marginal revenue function associated with producing and selling their coffeemakers is given by

$$R'(x) = -0.03x + 60$$

where x denotes the number of units produced and sold and $R'(x)$ is measured in dollars/unit.

a. Determine the revenue function $R(x)$ associated with producing and selling these coffeemakers.
b. What is the demand equation relating the wholesale unit price to the quantity of coffeemakers demanded?

40. The Franklin National Life Insurance Company purchased a new computer for $200,000. If the rate at which the computer's resale value changes is given by the function

$$V'(t) = 3800(t - 10)$$

where t is the length of time since the purchase date and $V'(t)$ is measured in dollars/year, find an expression $V(t)$ that gives the resale value of the computer after t yr. How much would the computer cost after 6 yr?

41. The marketing department of the Vista Vision Corporation forecasts that sales of their new line of projection television systems will grow at the rate of

$$3000 - 2000e^{-0.04t} \qquad (0 \le t \le 24)$$

units/month once they are introduced into the market. Find an expression giving the total number of units of the projection television systems that Vista may expect to sell t mo from the time they are put on the market. How many units of the television systems can Vista expect to sell during the first year?

42. Due to the increasing cost of fuel, the manager of the City Transit Authority estimates that the number of commuters using the city subway system will increase at the rate of

$$3000(1 + 0.4t)^{-1/2} \qquad (0 \le t \le 36)$$

per month t mo from now. If 100,000 commuters are currently using the system, find an expression giving the total number of commuters who will be using the subway t mo from now. How many commuters will be using the subway 6 mo from now?

43. The management of a division of Ditton Industries has determined that the daily marginal cost function associated with producing their hot-air corn poppers is given by

$$C'(x) = 0.00003x^2 - 0.03x + 10$$

where $C'(x)$ is measured in dollars/unit and x denotes the number of units manufactured. Management has also determined that the daily fixed cost incurred in producing these corn poppers is $600. Find the total cost incurred by Ditton in producing the first 500 corn poppers.

44. In 1980 the world produced 3.5 billion metric tons of coal. If output increased at the rate of

$$3.5e^{0.04t}$$

billion metric tons/year in year t ($t = 0$ corresponds to 1980), determine how much coal was produced worldwide between 1980 and the end of 1985.

45. Find the area of the region under the curve $y = 3x^2 + 2x + 1$ from $x = -1$ to $x = 2$.

46. Find the area of the region under the curve $y = e^{2x}$ from $x = 0$ to $x = 2$.

47. Find the area of the region bounded by the graph of the function $y = 1/x^2$, the x-axis, and the lines $x = 1$ and $x = 3$.

48. Find the area of the region bounded by the curve $y = -x^2 - x + 2$ and the x-axis.

49. Find the area of the region bounded by the graphs of the functions $f(x) = e^x$ and $g(x) = x$ and the vertical lines $x = 0$ and $x = 2$.

50. Find the area of the region that is completely enclosed by the graphs of $f(x) = x^4$ and $g(x) = x$.

51. Find the area of the region between the curve $y = x(x - 1)(x - 2)$ and the x-axis.

52. Based on current production techniques, the rate of oil production from a certain oil well t yr from now is estimated to be

$$R_1(t) = 100e^{0.05t}$$

thousand barrels/year. Based on a new production technique, however, it is estimated that the rate of oil production from that oil well t yr from now will be

$$R_2(t) = 100e^{0.08t}$$

thousand barrels/year. Determine how much additional oil will be produced over the next 10 yr if the new technique is adopted.

53. Find the average value of the function

$$f(x) = \frac{x}{\sqrt{x^2 + 16}}$$

over the interval [0, 3].

54. The demand function for a brand of blank digital camcorder tapes is given by

$$p = -0.01x^2 - 0.2x + 23$$

where p is the wholesale unit price in dollars and x is the quantity demanded each week, measured in units of a thousand. Determine the consumers' surplus if the wholesale unit price is $8/tape.

55. The quantity demanded x (in units of a hundred) of the Sportsman 5 × 7 tents, per week, is related to the unit price p (in dollars) by the relation

$$p = -0.1x^2 - x + 40$$

The quantity x (in units of a hundred) that the supplier is willing to make available in the market is related to the unit price by the relation

$$p = 0.1x^2 + 2x + 20$$

If the market price is set at the equilibrium price, find the consumers' surplus and the producers' surplus.

56. Chi-Tai plans to deposit $4000/year in his Keogh Retirement Account. If interest is compounded continuously at the rate of 8%/year, how much will he have in his retirement account after 20 yr?

57. Glenda sold her house under an installment contract whereby the buyer gave her a down payment of $9000 and agreed to make monthly payments of $925/month for 30 yr. If the prevailing interest rate is 12%/year compounded continuously, find the present value of the purchase price of the house.

58. Alicia purchased a 10-yr franchise for a health spa that is expected to generate income at the rate of

$$P(t) = 80,000$$

dollars/year. If the prevailing interest rate is 10%/year compounded continuously, find the present value of the franchise.

59. A certain country's income distribution is described by the function

$$f(x) = \frac{17}{18}x^2 + \frac{1}{18}x$$

a. Sketch the Lorentz curve for this function.
b. Compute $f(0.3)$ and $f(0.6)$ and interpret your results.
c. Compute the coefficient of inequality for this Lorentz curve.

60. The population of a certain Sun Belt city, currently 80,000, is expected to grow exponentially in the next 5 yr with a growth constant of 0.05. If the prediction comes true, what will be the average population of the city over the next 5 yr?

7

ADDITIONAL TOPICS IN INTEGRATION

Besides the basic rules of integration developed in Chapter 6, there are more sophisticated techniques for finding the antiderivatives of functions. We begin this chapter by looking at the method of integration by parts. We then look at a technique of integration that involves using tables of integrals that have been compiled for this purpose. We also look at numerical methods of integration, which enable us to obtain approximate solutions to definite integrals, especially those whose exact value cannot be found otherwise. More specifically, we study the trapezoidal rule and Simpson's rule. Numerical integration methods are especially useful when the integrand is known only at discrete points. Finally, we learn how to evaluate integrals in which the intervals of integration are unbounded. Such integrals, called *improper integrals*, play an important role in the study of probability, the last topic of this chapter.

What is the area of the oil spill caused by a grounded tanker? In Example 5, page 566, you will see how to determine the area of the oil spill.

7.1 Integration by Parts

THE METHOD OF INTEGRATION BY PARTS

Integration by parts is another technique of integration that, like the method of substitution discussed in Chapter 6, is based on a corresponding rule of differentiation. In this case, the rule of differentiation is the product rule, which asserts that if f and g are differentiable functions, then

$$\frac{d}{dx}[f(x)g(x)] = f(x)g'(x) + g(x)f'(x) \tag{1}$$

If we integrate both sides of Equation (1) with respect to x, we obtain

$$\int \frac{d}{dx} f(x)g(x)\, dx = \int f(x)g'(x)\, dx + \int g(x)f'(x)\, dx$$

$$f(x)g(x) = \int f(x)g'(x)\, dx + \int g(x)f'(x)\, dx$$

This last equation, which may be written in the form

$$\int f(x)g'(x)\, dx = f(x)g(x) - \int g(x)f'(x)\, dx \tag{2}$$

is called the formula for **integration by parts.** This formula is useful since it enables us to express one indefinite integral in terms of another that may be easier to evaluate. Formula (2) may be simplified by letting

$$u = f(x) \qquad dv = g'(x)\, dx$$
$$du = f'(x)\, dx \qquad v = g(x)$$

giving the following version of the formula for integration by parts.

Integration by Parts Formula

$$\int u\, dv = uv - \int v\, du \tag{3}$$

EXAMPLE 1   Evaluate $\int xe^x\, dx$.

SOLUTION ✔   No method of integration developed thus far enables us to evaluate the given indefinite integral in its present form. Therefore, we attempt to write it in terms of an indefinite integral that will be easier to evaluate. Let's use the integration by parts Formula (3) by letting

$$u = x \qquad \text{and} \qquad dv = e^x\, dx$$

so that

$$du = dx \qquad \text{and} \qquad v = e^x$$

Therefore,

$$\int xe^x \, dx = \int u \, dv$$

$$= uv - \int v \, du$$

$$= xe^x - \int e^x \, dx$$

$$= xe^x - e^x + C$$

$$= (x - 1)e^x + C \qquad\qquad ■■■■$$

The success of the method of integration by parts depends on the proper choice of u and dv. For example, if we had chosen

$$u = e^x \qquad \text{and} \qquad dv = x \, dx$$

in the last example, then

$$du = e^x \, dx \qquad \text{and} \qquad v = \frac{1}{2}x^2$$

Thus, (3) would have yielded

$$\int xe^x \, dx = \int u \, dv$$

$$= uv - \int v \, du$$

$$= \frac{1}{2}x^2e^x - \int \frac{1}{2}x^2e^x \, dx$$

Since the indefinite integral on the right-hand side of this equation is not readily evaluated (it is in fact more complicated than the original integral!), choosing u and dv as shown has not helped us evaluate the given indefinite integral.

In general, we can use the following guidelines.

Guidelines for Choosing _u_ and _dv_

Choose u and dv so that

1. du is simpler than u.

2. dv is easy to integrate.

EXAMPLE **2**   Evaluate $\int x \ln x \, dx$.

SOLUTION ✔   Letting

$$u = \ln x \qquad \text{and} \qquad dv = x \, dx$$

we have

$$du = \frac{1}{x} \, dx \qquad \text{and} \qquad v = \frac{1}{2}x^2$$

Therefore,

$$\int x \ln x \, dx = \int u \, dv = uv - \int v \, du$$

$$= \frac{1}{2} x^2 \ln x - \int \frac{1}{2} x^2 \cdot \left(\frac{1}{x} \right) dx$$

$$= \frac{1}{2} x^2 \ln x - \frac{1}{2} \int x \, dx$$

$$= \frac{1}{2} x^2 \ln x - \frac{1}{4} x^2 + C$$

$$= \frac{1}{4} x^2 (2 \ln x - 1) + C$$

■■■■

EXAMPLE 3   Evaluate $\int \frac{xe^x}{(x+1)^2} \, dx$.

SOLUTION ✔   Let

$$u = xe^x \qquad \text{and} \qquad dv = \frac{1}{(x+1)^2} \, dx$$

Then,

$$du = (xe^x + e^x) \, dx = e^x(x+1) \, dx \qquad \text{and} \qquad v = -\frac{1}{x+1}$$

Therefore,

$$\int \frac{xe^x}{(x+1)^2} \, dx = \int u \, dv = uv - \int v \, du$$

$$= xe^x \left(\frac{-1}{x+1} \right) - \int \left(-\frac{1}{x+1} \right) e^x(x+1) \, dx$$

$$= -\frac{xe^x}{x+1} + \int e^x \, dx$$

$$= -\frac{xe^x}{x+1} + e^x + C$$

$$= \frac{e^x}{x+1} + C$$

■■■■

The next example shows that repeated applications of the technique of integration by parts is sometimes required to evaluate an integral.

EXAMPLE 4   Evaluate $\int x^2 e^x \, dx$.

SOLUTION ✔   Let

$$u = x^2 \qquad \text{and} \qquad dv = e^x \, dx$$

so that

$$du = 2x \, dx \qquad \text{and} \qquad v = e^x$$

Therefore,

$$\int x^2 e^x \, dx = \int u \, dv = uv - \int v \, du$$

$$= x^2 e^x - \int e^x (2x) \, dx = x^2 e^x - 2 \int x e^x \, dx$$

To complete the solution of the problem, we need to evaluate the integral

$$\int x e^x \, dx$$

But this integral may be found using integration by parts. In fact, you will recognize that this integral is precisely that of Example 1. Using the results obtained there, we now find

$$\int x^2 e^x \, dx = x^2 e^x - 2[(x-1)e^x] + C = e^x (x^2 - 2x + 2) + C \quad \blacksquare\blacksquare\blacksquare\blacksquare$$

Group Discussion

1. Use the method of integration by parts to derive the formula

$$\int x^n e^{ax} \, dx = \frac{1}{a} x^n e^{ax} - \frac{n}{a} \int x^{n-1} e^{ax} \, dx$$

where n is a positive integer and a is a real number.
2. Use the formula of part 1 to evaluate

$$\int x^3 e^x \, dx.$$

Hint: You may find the results of Example 4 helpful.

APPLICATION

EXAMPLE 5

The estimated rate at which oil will be produced from a certain oil well t years after production has begun is given by

$$R(t) = 100te^{-0.1t}$$

thousand barrels per year. Find an expression that describes the total production of oil at the end of year t.

SOLUTION ✔

Let $T(t)$ denote the total production of oil from the well at the end of year t ($t \geq 0$). Then, the rate of oil production will be given by $T'(t)$ thousand barrels per year. Thus,

$$T'(t) = R(t) = 100te^{-0.1t}$$

so

$$T(t) = \int 100te^{-0.1t} \, dt$$

$$= 100 \int te^{-0.1t} \, dt$$

We use the technique of integration by parts to evaluate this integral. Let

$$u = t \qquad \text{and} \qquad dv = e^{-0.1t}\, dt$$

so that

$$du = dt \qquad \text{and} \qquad v = -\frac{1}{0.1}e^{-0.1t} = -10e^{-0.1t}$$

Therefore,

$$T(t) = 100\left[-10te^{-0.1t} + 10\int e^{-0.1t}\, dt\right]$$
$$= 100[-10te^{-0.1t} - 100e^{-0.1t}] + C$$
$$= -1000e^{-0.1t}(t + 10) + C$$

To determine the value of C, note that the total quantity of oil produced at the end of year 0 is nil, so $T(0) = 0$. This gives

$$T(0) = -1000(10) + C = 0$$
$$C = 10{,}000$$

Thus, the required production function is given by

$$T(t) = -1000e^{-0.1t}(t + 10) + 10{,}000 \qquad\qquad ■■■■$$

Exploring with Technology

Refer to Example 5.

1. Use a graphing utility to plot the graph of

$$T(t) = -1000e^{-0.1t}(t + 10) + 10{,}000$$

using the viewing rectangle $[0, 100] \times [0, 12{,}000]$. Use **TRACE** to see what happens when t is very large.

2. Verify the result of part 1 by evaluating $\lim_{t \to \infty} T(t)$. Interpret your results.

SELF-CHECK EXERCISES 7.1

1. Evaluate $\int x^2 \ln x\, dx$.

2. Since the inauguration of Ryan's Express at the beginning of 1998, the number of passengers (in millions) flying on this commuter airline has been growing at the rate of

$$R(t) = 0.1 + 0.2te^{-0.4t}$$

passengers/year ($t = 0$ corresponds to the beginning of 1998). Assuming that this trend continues through 2002, determine how many passengers will have flown on Ryan's Express by that time.

Solutions to Self-Check Exercises 7.1 can be found on page 548.

7.1 Exercises

In Exercises 1–26, evaluate each indefinite integral.

1. $\int xe^{2x}\,dx$

2. $\int xe^{-x}\,dx$

3. $\int xe^{x/4}\,dx$

4. $\int 6xe^{3x}\,dx$

5. $\int (e^x - x)^2\,dx$

6. $\int (e^{-x} + x)^2\,dx$

7. $\int (x + 1)e^x\,dx$

8. $\int (x - 3)e^{3x}\,dx$

9. $\int x(x + 1)^{-3/2}\,dx$

10. $\int x(x + 4)^{-2}\,dx$

11. $\int x\sqrt{x - 5}\,dx$

12. $\int \dfrac{x}{\sqrt{2x + 3}}\,dx$

13. $\int x\ln 2x\,dx$

14. $\int x^2\ln 2x\,dx$

15. $\int x^3\ln x\,dx$

16. $\int \sqrt{x}\ln x\,dx$

17. $\int \sqrt{x}\ln\sqrt{x}\,dx$

18. $\int \dfrac{\ln x}{\sqrt{x}}\,dx$

19. $\int \dfrac{\ln x}{x^2}\,dx$

20. $\int \dfrac{\ln x}{x^3}\,dx$

21. $\int \ln x\,dx$

Hint: Let $u = \ln x$ and $dv = dx$.

22. $\int \ln(x + 1)\,dx$

23. $\int x^2e^{-x}\,dx$

Hint: Integrate by parts twice.

24. $\int e^{-\sqrt{x}}\,dx$

Hint: First, make the substitution $u = \sqrt{x}$; then, integrate by parts.

25. $\int x(\ln x)^2\,dx$

Hint: Integrate by parts twice.

26. $\int x\ln(x + 1)\,dx$

Hint: First, make the substitution $u = x + 1$; then, integrate by parts.

In Exercises 27–32, evaluate each definite integral by using the method of integration by parts.

27. $\int_0^{\ln 2} xe^x\,dx$

28. $\int_0^2 xe^{-x}\,dx$

29. $\int_1^4 \ln x\,dx$

30. $\int_1^2 x\ln x\,dx$

31. $\int_0^2 xe^{2x}\,dx$

32. $\int_0^1 x^2e^{-x}\,dx$

33. Find the function f given that the slope of the tangent line to the graph of f at any point $(x, f(x))$ is xe^{-2x} and that the graph passes through the point $(0, 3)$.

34. Find the function f given that the slope of the tangent line to the graph of f at any point $(x, f(x))$ is $x\sqrt{x + 1}$ and that the graph passes through the point $(3, 6)$.

35. Find the area of the region under the graph of $f(x) = \ln x$ from $x = 1$ to $x = 5$.

36. Find the area of the region under the graph of $f(x) = xe^{-x}$ from $x = 0$ to $x = 3$.

37. **VELOCITY OF A DRAGSTER** The velocity of a dragster t sec after leaving the starting line is

$$100te^{-0.2t}$$

ft/sec. What is the distance covered by the dragster in the first 10 sec of its run?

38. **PRODUCTION OF STEAM COAL** In keeping with the projected increase in worldwide demand for steam coal, the boiler-firing fuel used for generating electricity, the management of Consolidated Mining has decided to step up its mining operations. Plans call for increasing the yearly production of steam coal by

$$2te^{-0.05t}$$

million metric tons/year for the next 20 yr. The current yearly production is 20 million metric tons. Find a function that describes Consolidated's total production of steam coal at the end of t yr. How much coal will Consolidated have produced over the next 20 yr if this plan is carried out?

39. **CONCENTRATION OF A DRUG IN THE BLOODSTREAM** The concentration (in milligrams per milliliter) of a certain drug in a patient's bloodstream t hr after it has been administered is given by $C(t) = 3te^{-t/3}$ mg/mL. Find the average concentration of the drug in the patient's bloodstream over the first 12 hr after administration.

40. ALCOHOL-RELATED TRAFFIC ACCIDENTS As a result of increasingly stiff laws aimed at reducing the number of alcohol-related traffic accidents in a certain state, preliminary data indicate that the number of such accidents has been changing at the rate of

$$R(t) = -10 - te^{0.1t}$$

accidents/month t mo after the laws took effect. There were 982 alcohol-related accidents for the year before the enactment of the laws. Determine how many alcohol-related accidents were expected during the first year the laws were in effect.

41. COMPACT DISC SALES Sales of the latest recording by Brittania, a British rock group, are currently $2te^{-0.1t}$ units/week (each unit representing 10,000 discs), where t denotes the number of weeks since the recording's release. Find an expression that gives the total number of discs sold as a function of t.

42. RATE OF RETURN ON AN INVESTMENT Suppose an investment is expected to generate income at the rate of

$$P(t) = 30,000 + 800t$$

dollars/year for the next 5 yr. Find the present value of this investment if the prevailing interest rate is 8%/year compounded continuously.
Hint: Use Formula (18), Section 6.7 (page 525).

43. PRESENT VALUE OF A FRANCHISE Tracy purchased a 15-yr franchise for a computer outlet store that is expected to generate income at the rate of

$$P(t) = 50,000 + 3000t$$

dollars/year. If the prevailing interest rate is 10%/year compounded continuously, find the present value of the franchise.
Hint: Use Formula (18), Section 6.7 (page 525).

44. GROWTH OF HMOs The membership of the Cambridge Community Health Plan, Inc. (a health maintenance organization), is projected to grow at the rate of $9\sqrt{t+1} \ln \sqrt{t+1}$ thousand people/year, t yr from now. If the HMO's current membership is 50,000, what will be the membership 5 yr from now?

In Exercises 45 and 46, determine whether the statement is true or false. If it is true, explain why it is true. If it is false, give an example to show why it is false.

45. $\int u \, dv + \int v \, du = uv$

46. $\int e^x g'(x) \, dx = e^x g(x) - \int e^x g(x) \, dx$

SOLUTIONS TO SELF-CHECK EXERCISES 7.1

1. Let $u = \ln x$ and $dv = x^2 \, dx$ so that $du = \dfrac{1}{x} \, dx$ and $v = \dfrac{1}{3} x^3$. Therefore,

$$\int x^2 \ln x \, dx = \int u \, dv = uv - \int v \, du$$

$$= \frac{1}{3} x^3 \ln x - \int \frac{1}{3} x^2 \, dx$$

$$= \frac{1}{3} x^3 \ln x - \frac{1}{9} x^3 + C$$

$$= \frac{1}{9} x^3 (3 \ln x - 1) + C$$

2. Let $N(t)$ denote the total number of passengers who will have flown on Ryan's Express by the end of year t. Then, $N'(t) = R(t)$, so that

$$N(t) = \int R(t) \, dt$$

$$= \int (0.1 + 0.2te^{-0.4t}) \, dt$$

$$= \int 0.1 \, dt + 0.2 \int te^{-0.4t} \, dt$$

We now use the technique of integration by parts on the second integral. Letting $u = t$ and $dv = e^{-0.4t}\, dt$, we have

$$du = dt \quad \text{and} \quad v = -\frac{1}{0.4}e^{-0.4t} = -2.5e^{-0.4t}$$

Therefore,

$$N(t) = 0.1t + 0.2\left[-2.5te^{-0.4t} + 2.5\int e^{-0.4t}\, dt\right]$$

$$= 0.1t - 0.5te^{-0.4t} - \frac{0.5}{0.4}e^{-0.4t} + C$$

$$= 0.1t - 0.5(t + 2.5)e^{-0.4t} + C$$

To determine the value of C, note that $N(0) = 0$, which gives

$$N(0) = -0.5(2.5) + C = 0$$
$$C = 1.25$$

Therefore,

$$N(t) = 0.1t - 0.5(t + 2.5)e^{-0.4t} + 1.25$$

The number of passengers who will have flown on Ryan's Express by the end of 2002 is given by

$$N(5) = 0.1(5) - 0.5(5 + 2.5)e^{-0.4(5)} + 1.25$$
$$= 1.242493$$

—that is, 1,242,493 passengers.

7.2 Integration Using Tables of Integrals

A TABLE OF INTEGRALS

We have studied several techniques for finding an antiderivative of a function. However, useful as they are, these techniques are not always applicable. There are of course numerous other methods for finding an antiderivative of a function. Extensive lists of integration formulas have been compiled based on these methods.

A small sample of the integration formulas that can be found in many mathematical handbooks is given in the following table of integrals. The formulas are grouped according to the basic form of the integrand. Note that it may be necessary to modify the integrand of the integral to be evaluated in order to use one of these formulas.

Table of Integrals

Forms Involving $a + bu$

1. $\displaystyle\int \frac{u\,du}{a+bu} = \frac{1}{b^2}[a+bu-a\ln|a+bu|] + C$

2. $\displaystyle\int \frac{u^2\,du}{a+bu} = \frac{1}{2b^3}[(a+bu)^2 - 4a(a+bu) + 2a^2\ln|a+bu|] + C$

3. $\displaystyle\int \frac{u\,du}{(a+bu)^2} = \frac{1}{b^2}\left[\frac{a}{a+bu} + \ln|a+bu|\right] + C$

4. $\displaystyle\int u\sqrt{a+bu}\,du = \frac{2}{15b^2}(3bu-2a)(a+bu)^{3/2} + C$

5. $\displaystyle\int \frac{u\,du}{\sqrt{a+bu}} = \frac{2}{3b^2}(bu-2a)\sqrt{a+bu} + C$

6. $\displaystyle\int \frac{du}{u\sqrt{a+bu}} = \frac{1}{\sqrt{a}}\ln\left|\frac{\sqrt{a+bu}-\sqrt{a}}{\sqrt{a+bu}+\sqrt{a}}\right| + C$ (if $a > 0$)

Forms Involving $\sqrt{a^2 + u^2}$

7. $\displaystyle\int \sqrt{a^2+u^2}\,du = \frac{u}{2}\sqrt{a^2+u^2} + \frac{a^2}{2}\ln|u+\sqrt{a^2+u^2}| + C$

8. $\displaystyle\int u^2\sqrt{a^2+u^2}\,du = \frac{u}{8}(a^2+2u^2)\sqrt{a^2+u^2} - \frac{a^4}{8}\ln|u+\sqrt{a^2+u^2}| + C$

9. $\displaystyle\int \frac{du}{\sqrt{a^2+u^2}} = \ln|u+\sqrt{a^2+u^2}| + C$

10. $\displaystyle\int \frac{du}{u\sqrt{a^2+u^2}} = -\frac{1}{a}\ln\left|\frac{\sqrt{a^2+u^2}+a}{u}\right| + C$

11. $\displaystyle\int \frac{du}{u^2\sqrt{a^2+u^2}} = -\frac{\sqrt{a^2+u^2}}{a^2u} + C$

12. $\displaystyle\int \frac{du}{(a^2+u^2)^{3/2}} = \frac{u}{a^2\sqrt{a^2+u^2}} + C$

Forms Involving $\sqrt{u^2 - a^2}$

13. $\displaystyle\int \sqrt{u^2-a^2}\,du = \frac{u}{2}\sqrt{u^2-a^2} - \frac{a^2}{2}\ln|u+\sqrt{u^2-a^2}| + C$

14. $\displaystyle\int u^2\sqrt{u^2-a^2}\,du = \frac{u}{8}(2u^2-a^2)\sqrt{u^2-a^2} - \frac{a^4}{8}\ln|u+\sqrt{u^2-a^2}| + C$

15. $\displaystyle\int \frac{\sqrt{u^2-a^2}}{u^2}\,du = -\frac{\sqrt{u^2-a^2}}{u} + \ln|u+\sqrt{u^2-a^2}| + C$

16. $\displaystyle\int \frac{du}{\sqrt{u^2-a^2}} = \ln|u+\sqrt{u^2-a^2}| + C$

17. $\int \dfrac{du}{u^2 \sqrt{u^2 - a^2}} = \dfrac{\sqrt{u^2 - a^2}}{a^2 u} + C$

18. $\int \dfrac{du}{(u^2 - a^2)^{3/2}} = - \dfrac{u}{a^2 \sqrt{u^2 - a^2}} + C$

Forms Involving $\sqrt{a^2 - u^2}$

---

19. $\int \dfrac{\sqrt{a^2 - u^2}}{u}\, du = \sqrt{a^2 - u^2} - a \ln \left| \dfrac{a + \sqrt{a^2 - u^2}}{u} \right| + C$

20. $\int \dfrac{du}{u \sqrt{a^2 - u^2}} = -\dfrac{1}{a} \ln \left| \dfrac{a + \sqrt{a^2 - u^2}}{u} \right| + C$

21. $\int \dfrac{du}{u^2 \sqrt{a^2 - u^2}} = - \dfrac{\sqrt{a^2 - u^2}}{a^2 u} + C$

22. $\int \dfrac{du}{(a^2 - u^2)^{3/2}} = \dfrac{u}{a^2 \sqrt{a^2 - u^2}} + C$

Forms Involving e^{au} and $\ln u$

---

23. $\int u e^{au}\, du = \dfrac{1}{a^2}(au - 1)e^{au} + C$

24. $\int u^n e^{au}\, du = \dfrac{1}{a} u^n e^{au} - \dfrac{n}{a} \int u^{n-1} e^{au}\, du$

25. $\int \dfrac{du}{1 + be^{au}} = u - \dfrac{1}{a} \ln(1 + be^{au}) + C$

26. $\int \ln u\, du = u \ln u - u + C$

27. $\int u^n \ln u\, du = \dfrac{u^{n+1}}{(n + 1)^2}[(n + 1)\ln u - 1] + C \qquad (n \neq -1)$

28. $\int \dfrac{du}{u \ln u} = \ln|\ln u| + C$

29. $\int (\ln u)^n\, du = u(\ln u)^n - n \int (\ln u)^{n-1}\, du$

USING A TABLE OF INTEGRALS

We now consider several examples that illustrate how the table of integrals can be used to evaluate an integral.

EXAMPLE 1

Use the table of integrals to evaluate $\int \dfrac{2x\, dx}{\sqrt{3 + x}}$.

SOLUTION ✓

We first write

$$\int \dfrac{2x\, dx}{\sqrt{3 + x}} = 2 \int \dfrac{x\, dx}{\sqrt{3 + x}}$$

Since $\sqrt{3 + x}$ is of the form $\sqrt{a + bu}$, with $a = 3$, $b = 1$, and $u = x$, we use Formula 5,

$$\int \frac{u \, du}{\sqrt{a + bu}} = \frac{2}{3b^2}(bu - 2a)\sqrt{a + bu} + C$$

obtaining

$$2\int \frac{x}{\sqrt{3 + x}} \, dx = 2\left[\frac{2}{3(1)}(x - 6)\sqrt{3 + x}\right] + C$$

$$= \frac{4}{3}(x - 6)\sqrt{3 + x} + C$$

■■■■

Group Discussion

All formulas given in the table of integrals can be verified by direct computation. Describe a method you would use and apply it to verify a formula of your choice.

EXAMPLE 2   Use the table of integrals to evaluate $\int x^2 \sqrt{3 + x^2} \, dx$.

SOLUTION ✔   Observe that if we write 3 as $(\sqrt{3})^2$, then $3 + x^2$ has the form $\sqrt{a^2 + u^2}$, with $a = \sqrt{3}$ and $u = x$. Using Formula 8,

$$\int u^2 \sqrt{a^2 + u^2} \, du = \frac{u}{8}(a^2 + 2u^2)\sqrt{a^2 + u^2} - \frac{a^4}{8} \ln\left|u + \sqrt{a^2 + u^2}\right| + C$$

we obtain

$$\int x^2 \sqrt{3 + x^2} \, dx = \frac{x}{8}(3 + 2x^2)\sqrt{3 + x^2} - \frac{9}{8} \ln\left|x + \sqrt{3 + x^2}\right| + C$$

■■■■

EXAMPLE 3   Use the table of integrals to evaluate

$$\int_3^4 \frac{dx}{x^2 \sqrt{50 - 2x^2}}$$

SOLUTION ✔   We first evaluate the indefinite integral

$$I = \int \frac{dx}{x^2 \sqrt{50 - 2x^2}}$$

Observe that $\sqrt{50 - 2x^2} = \sqrt{2(25 - x^2)} = \sqrt{2}\sqrt{25 - x^2}$, so we can write I as

$$I = \frac{1}{\sqrt{2}}\int \frac{dx}{x^2 \sqrt{25 - x^2}} = \frac{\sqrt{2}}{2}\int \frac{dx}{x^2 \sqrt{25 - x^2}} \qquad \left(\frac{1}{\sqrt{2}} = \frac{\sqrt{2}}{2}\right)$$

Next, using Formula 21,

$$\int \frac{du}{u^2 \sqrt{a^2 - u^2}} = -\frac{\sqrt{a^2 - u^2}}{a^2 u} + C$$

with $a = 5$ and $u = x$, we find

$$I = \frac{\sqrt{2}}{2}\left[-\frac{\sqrt{25 - x^2}}{25x}\right]$$

$$= -\left(\frac{\sqrt{2}}{50}\right)\frac{\sqrt{25 - x^2}}{x}$$

Finally, using this result, we obtain

$$\int_3^4 \frac{dx}{x^2\sqrt{50 - 2x^2}} = -\frac{\sqrt{2}}{50}\frac{\sqrt{25 - x^2}}{x}\bigg|_3^4$$

$$= -\frac{\sqrt{2}\sqrt{25 - 16}}{50}{4} - \left(-\frac{\sqrt{2}\sqrt{25 - 9}}{50}{3}\right)$$

$$= -\frac{3\sqrt{2}}{200} + \frac{2\sqrt{2}}{75} = \frac{7\sqrt{2}}{600}$$ ■ ■ ■ ■

Group Discussion

The formulas given in the table of integrals were derived using various techniques, including the method of substitution and the method of integration by parts studied earlier. For example, Formula 1,

$$\int \frac{u\,du}{a + bu} = \frac{1}{b^2}[a + bu - a\ln|a + bu|] + C$$

can be derived using the method of substitution. Show how this is done.

As illustrated in the next example, we may need to apply a formula more than once in order to evaluate an integral.

EXAMPLE 4 Use the table of integrals to evaluate $\int x^2 e^{(-1/2)x}\,dx$.

SOLUTION ✔ Scanning the table of integrals for a formula involving e^{ax} in the integrand, we are led to Formula 24,

$$\int u^n e^{au}\,du = \frac{1}{a}u^n e^{au} - \frac{n}{a}\int u^{n-1}e^{au}\,du$$

With $n = 2$, $a = -\frac{1}{2}$, and $u = x$, we have

$$\int x^2 e^{(-1/2)x}\,dx = \left(\frac{1}{-\frac{1}{2}}\right)x^2 e^{(-1/2)x} - \frac{2}{(-\frac{1}{2})}\int xe^{(-1/2)x}\,dx$$

$$= -2x^2 e^{(-1/2)x} + 4\int xe^{(-1/2)x}\,dx$$

If we use Formula 24 once again, with $n = 1$, $a = -\frac{1}{2}$, and $u = x$, to evaluate the integral on the right, we obtain

$$\int x^2 e^{(-1/2)x}\, dx = -2x^2 e^{(-1/2)x} + 4\left[\left(\frac{1}{-\frac{1}{2}}\right)xe^{(-1/2)x} - \frac{1}{(-\frac{1}{2})}\int e^{(-1/2)x}\, dx\right]$$

$$= -2x^2 e^{(-1/2)x} + 4\left[-2xe^{(-1/2)x} + 2\cdot\frac{1}{(-\frac{1}{2})}e^{(-1/2)x}\right] + C$$

$$= -2e^{(-1/2)x}(x^2 + 4x + 8) + C \qquad \blacksquare\blacksquare\blacksquare\blacksquare$$

EXAMPLE 5   A study prepared for the National Association of Realtors estimated that the mortgage rate over the next t months will be

$$r(t) = \frac{8t + 100}{t + 10} \qquad (0 \le t \le 24)$$

percent per year. If the prediction holds true, what will be the average mortgage rate over the next 12 months?

SOLUTION ✔   The average mortgage rate over the next 12 months will be given by

$$A = \frac{1}{12 - 0}\int_0^{12}\frac{8t + 100}{t + 10}\, dt = \frac{1}{12}\left[\int_0^{12}\frac{8t}{t + 10}\, dt + \int_0^{12}\frac{100}{t + 10}\, dt\right]$$

$$= \frac{8}{12}\int_0^{12}\frac{t}{t + 10}\, dt + \frac{100}{12}\int_0^{12}\frac{1}{t + 10}\, dt$$

Using Formula 1,

$$\int\frac{u\, du}{a + bu} = \frac{1}{b^2}[a + bu - a\ln|a + bu|] + C \qquad (a = 10, b = 1, u = t)$$

to evaluate the first integral, we have

$$A = \left(\frac{2}{3}\right)[10 + t - 10\ln(10 + t)]\Big|_0^{12} + \left(\frac{25}{3}\right)\ln(10 + t)\Big|_0^{12}$$

$$= \left(\frac{2}{3}\right)[(22 - 10\ln 22) - (10 - 10\ln 10)] + \left(\frac{25}{3}\right)[\ln 22 - \ln 10]$$

$$\approx 9.31$$

or approximately 9.31% per year. $\qquad \blacksquare\blacksquare\blacksquare\blacksquare$

1. Use the table of integrals to evaluate

$$\int_0^2 \frac{dx}{(5-x^2)^{3/2}}$$

2. During a flu epidemic, the number of children in the Easton Middle School who contracted influenza t days after the outbreak began was given by

$$N(t) = \frac{200}{1+9e^{-0.8t}}$$

Determine the average number of children who contracted the flu in the first 10 days of the epidemic.

Solutions to Self-Check Exercises 7.2 can be found on page 557.

7.2 Exercises

In Exercises 1–32, use the table of integrals in this section to evaluate each of the given integrals.

1. $\int \frac{2x}{2+3x}\,dx$

2. $\int \frac{x}{(1+2x)^2}\,dx$

3. $\int \frac{3x^2}{2+4x}\,dx$

4. $\int \frac{x^2}{3+x}\,dx$

5. $\int x^2\sqrt{9+4x^2}\,dx$

6. $\int x^2\sqrt{4+x^2}\,dx$

7. $\int \frac{dx}{x\sqrt{1+4x}}$

8. $\int_0^2 \frac{x+1}{\sqrt{2+3x}}\,dx$

9. $\int_0^2 \frac{dx}{\sqrt{9+4x^2}}$

10. $\int \frac{dx}{x\sqrt{4+8x^2}}$

11. $\int \frac{dx}{(9-x^2)^{3/2}}$

12. $\int \frac{dx}{(2-x^2)^{3/2}}$

13. $\int x^2\sqrt{x^2-4}\,dx$

14. $\int_3^5 \frac{dx}{x^2\sqrt{x^2-9}}$

15. $\int \frac{\sqrt{4-x^2}}{x}\,dx$

16. $\int_0^1 \frac{dx}{(4-x^2)^{3/2}}$

17. $\int xe^{2x}\,dx$

18. $\int \frac{dx}{1+e^{-x}}$

19. $\int \frac{dx}{(x+1)\ln(1+x)}$
Hint: First use the substitution $u = x + 1$.

20. $\int \frac{x}{(x^2+1)\ln(x^2+1)}\,dx$

Hint: First use the substitution $u = x^2 + 1$.

21. $\int \frac{e^{2x}}{(1+3e^x)^2}\,dx$

22. $\int \frac{e^{2x}}{\sqrt{1+3e^x}}\,dx$

23. $\int \frac{3e^x}{1+e^{(1/2)x}}\,dx$

24. $\int \frac{dx}{1-2e^{-x}}$

25. $\int \frac{\ln x}{x(2+3\ln x)}\,dx$

26. $\int_1^e (\ln x)^2\,dx$

27. $\int_0^1 x^2 e^x\,dx$

28. $\int x^3 e^{2x}\,dx$

29. $\int x^2 \ln x\,dx$

30. $\int x^3 \ln x\,dx$

31. $\int (\ln x)^3\,dx$

32. $\int (\ln x)^4\,dx$

33. CONSUMERS' SURPLUS Refer to Section 6.7. The demand function for Apex ladies' boots is

$$p = \frac{250}{\sqrt{16+x^2}}$$

where p is the wholesale unit price in dollars and x is the quantity demanded daily, in units of a hundred. Find the consumers' surplus if the wholesale price is set at $50/pair.

34. PRODUCERS' SURPLUS Refer to Section 6.7. The supplier of Apex ladies' boots will make x hundred pairs of the boots available in the market daily when the wholesale unit price is

$$p = \frac{30x}{5 - x}$$

dollars. Find the producers' surplus if the wholesale price is set at $50/pair.

35. AMUSEMENT PARK ATTENDANCE The management of Astro World ("The Amusement Park of the Future") estimates that the number of visitors (in thousands) entering the amusement park t hr after opening time at 9 A.M. is given by

$$R(t) = \frac{60}{(2 + t^2)^{3/2}}$$

per hour. Determine the number of visitors admitted by noon.

36. VOTER REGISTRATION The number of voters in a certain district of a city is expected to grow at the rate of

$$R(t) = \frac{3000}{\sqrt{4 + t^2}}$$

people/year t yr from now. If the number of voters at present is 20,000, how many voters will be in the district 5 yr from now?

37. GROWTH OF FRUIT FLIES Based on data collected during an experiment, a biologist found that the number of fruit flies (*Drosophila*) with a limited food supply could be approximated by the exponential model

$$N(t) = \frac{1000}{1 + 24e^{-0.02t}}$$

where t denotes the number of days since the beginning of the experiment. Find the average number of fruit flies in the colony in the first 10 days of the experiment and in the first 20 days.

38. VCR OWNERSHIP According to estimates by Paul Kroger Associates, the percentage of households that own videocassette recorders (VCRs) is given by

$$P(t) = \frac{68}{1 + 21.67e^{-0.62t}} \qquad (0 \le t \le 12)$$

where t is measured in years, with $t = 0$ corresponding to the beginning of 1981. Find the average percentage of households owning VCRs from the beginning of 1981 to the beginning of 1993.
Source: Paul Kroger Associates

39. RECYCLING PROGRAMS The commissioner of the City of Newton Department of Public Works estimates that the number of people in the city who have been recycling their magazines in year t following the introduction of the recycling program at the beginning of 1990 is

$$N(t) = \frac{100,000}{2 + 3e^{-0.2t}}$$

Find the average number of people who will have recycled their magazines during the first 5 yr since the program was introduced.

40. FRANCHISES Elaine purchased a 10-yr franchise for a fast-food restaurant that is expected to generate income at the rate of $R(t) = 250,000 + 2000t^2$ dollars/year, t yr from now. If the prevailing interest rate is 10%/year compounded continuously, find the present value of the franchise.

Hint: Use Formula (18), Section 6.7.

41. ACCUMULATED VALUE OF AN INCOME STREAM The revenue of Virtual Reality, a video-game arcade, is generated at the rate of $R(t) = 20,000t$ dollars. If the revenue is invested t yr from now in a business earning interest at the rate of 15%/year compounded continuously, find the accumulated value of this stream of income at the end of 5 yr.

Hint: Use Formula (18), Section 6.7.

42. LORENTZ CURVES In a study conducted by a certain country's Economic Development Board regarding the income distribution of certain segments of the country's workforce, it was found that the Lorentz curve for the distribution of income of college professors is described by the function

$$g(x) = \frac{1}{3}x\sqrt{1 + 8x}$$

Compute the coefficient of inequality of the Lorentz curve.

Hint: Use Formula (21), Section 6.7.

Solutions to Self-Check Exercises 7.2

1. Using Formula 22, page 551, with $a^2 = 5$ and $u = x$, we see that

$$\int_0^2 \frac{dx}{(5 - x^2)^{3/2}} = \frac{x}{5\sqrt{5 - x^2}}\bigg|_0^2$$

$$= \frac{2}{5\sqrt{5 - 4}}$$

$$= \frac{2}{5}$$

2. The average number of children who contracted the flu in the first 10 days of the epidemic is given by

$$A = \frac{1}{10}\int_0^{10} \frac{200}{1 + 9e^{-0.8t}}\, dt = 20\int_0^{10} \frac{dt}{1 + 9e^{-0.8t}}$$

$$= 20\left[t + \frac{1}{0.8}\ln(1 + 9e^{-0.8t})\right]\bigg|_0^{10} \qquad \text{(Formula 25, } a = -0.8,$$
$$\qquad\qquad b = 9, u = t)$$

$$= 20\left[10 + \frac{1}{0.8}\ln(1 + 9e^{-8})\right] - 20\left(\frac{1}{0.8}\right)\ln 10$$

$$\approx 200.07537 - 57.56463$$

$$\approx 143$$

or 143 students.

7.3 Numerical Integration

APPROXIMATING DEFINITE INTEGRALS

One method of measuring cardiac output is to inject 5 to 10 milligrams (mg) of a dye into a vein leading to the heart. After making its way through the lungs, the dye returns to the heart and is pumped into the aorta, where its concentration is measured at equal time intervals. The graph of the function c in Figure 7.1 shows the concentration of dye in a person's aorta, measured at 2-second intervals after 5 mg of dye have been injected. The person's cardiac output, measured in liters per minute (L/min), is computed using the formula

$$R = \frac{60D}{\displaystyle\int_0^{28} c(t)\, dt} \qquad\qquad\qquad \textbf{(4)}$$

where D is the quantity of dye injected (see Exercise 44).

Now, to use Formula (4), we need to evaluate the definite integral

$$\int_0^{28} c(t)\, dt$$

FIGURE 7.1

The function c gives the concentration of a dye measured at the aorta. The graph is constructed by drawing a smooth curve through a set of discrete points.

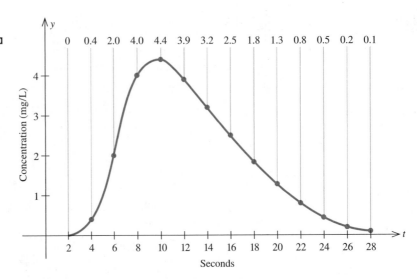

But we do not have the algebraic rule defining the integrand c for all values of t in $[0, 28]$. In fact, we are given its values only at a set of discrete points in that interval. In situations such as this, the fundamental theorem of calculus proves useless because we cannot find an antiderivative of c. (We will complete the solution to this problem in Example 4.)

Other situations also arise in which an integrable function has an antiderivative that cannot be found in terms of elementary functions (functions that can be expressed as a finite combination of algebraic, exponential, logarithmic, and trigonometric functions). Examples of such functions are

$$f(x) = e^{x^2}, \qquad g(x) = x^{-1/2}e^x, \qquad h(x) = \frac{1}{\ln x}$$

Riemann sums provide us with a good approximation of a definite integral, provided the number of subintervals in the partitions is large enough. But there are better techniques and formulas, called *quadrature formulas,* that give a more efficient way of computing approximate values of definite integrals. In this section we look at two rather simple but effective ways of approximating definite integrals.

THE TRAPEZOIDAL RULE

We assume that $f(x) \geq 0$ on $[a, b]$ in order to simplify the derivation of the trapezoidal rule, but the result is valid without this restriction. We begin by subdividing the interval $[a, b]$ into n subintervals of equal length Δx, by means of the $(n + 1)$ points $x_0 = a, x_1, x_2, \ldots, x_n = b$, where n is a positive integer (Figure 7.2).

Then, the length of each subinterval is given by

$$\Delta x = \frac{b - a}{n}$$

FIGURE 7.2
The area under the curve is equal to the sum of the n subregions $R_1, R_2, \ldots, R_n$.

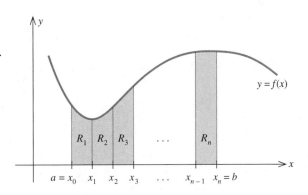

FIGURE 7.3
The area of R_1 is approximated by the area of the trapezoid.

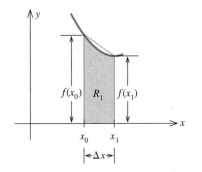

Furthermore, as we saw earlier, we may view the definite integral

$$\int_a^b f(x)\,dx$$

as the area of the region R under the curve $y = f(x)$ between $x = a$ and $x = b$. This area is given by the sum of the areas of the n nonoverlapping subregions $R_1, R_2, \ldots, R_n$, such that R_1 represents the region under the curve $y = f(x)$ from $x = x_0$ to $x = x_1$, and so on.

The basis for the trapezoidal rule lies in the approximation of each of the regions $R_1, R_2, \ldots, R_n$ by a suitable trapezoid. This often leads to a much better approximation than one obtained by means of rectangles (a Riemann sum).

Let's consider the subregion R_1, shown magnified for the sake of clarity in Figure 7.3. Observe that the area of the region R_1 may be approximated by the trapezoid of width Δx whose parallel sides are of lengths $f(x_0)$ and $f(x_1)$. The area of the trapezoid is given by

$$\left[\frac{f(x_0) + f(x_1)}{2} \right] \Delta x \qquad \text{(Average of the lengths of the parallel sides times the width)}$$

square units. Similarly, the area of the region R_2 may be approximated by the trapezoid of width Δx and sides of lengths $f(x_1)$ and $f(x_2)$. The area of the trapezoid is given by

$$\left[\frac{f(x_1) + f(x_2)}{2} \right] \Delta x$$

Similarly, we see that the area of the last (nth) approximating trapezoid is given by

$$\left[\frac{f(x_{n-1}) + f(x_n)}{2} \right] \Delta x$$

Then, the area of the region R is approximated by the sum of the areas of the n trapezoids—that is,

$$\left[\frac{f(x_0) + f(x_1)}{2}\right]\Delta x + \left[\frac{f(x_1) + f(x_2)}{2}\right]\Delta x + \cdots + \left[\frac{f(x_{n-1}) + f(x_n)}{2}\right]\Delta x$$

$$= \frac{\Delta x}{2}[f(x_0) + f(x_1) + f(x_1) + f(x_2) + \cdots + f(x_{n-1}) + f(x_n)]$$

$$= \frac{\Delta x}{2}[f(x_0) + 2f(x_1) + 2f(x_2) + \cdots + 2f(x_{n-1}) + f(x_n)]$$

Since the area of the region R is given by the value of the definite integral we wished to approximate, we are led to the following approximation formula, which is called the **trapezoidal rule.**

Trapezoidal Rule

$$\int_a^b f(x)\,dx \approx \frac{\Delta x}{2}[f(x_0) + 2f(x_1) + 2f(x_2)$$

$$+ \cdots + 2f(x_{n-1}) + f(x_n)]$$

(5)

where $\Delta x = \dfrac{b - a}{n}$.

The approximation generally improves with larger values of n.

EXAMPLE 1   Approximate the value of

$$\int_1^2 \frac{1}{x}\,dx$$

using the trapezoidal rule with $n = 10$. Compare this result with the exact value of the integral.

SOLUTION ✔   Here, $a = 1$, $b = 2$, and $n = 10$, so

$$\Delta x = \frac{b - a}{n} = \frac{1}{10} = 0.1$$

and

$$x_0 = 1, \qquad x_1 = 1.1, \qquad x_2 = 1.2, \qquad x_3 = 1.3, \ldots, x_9 = 1.9, \qquad x_{10} = 2$$

The trapezoidal rule yields

$$\int_1^2 \frac{1}{x}\,dx \approx \frac{0.1}{2}\left[1 + 2\left(\frac{1}{1.1}\right) + 2\left(\frac{1}{1.2}\right) + 2\left(\frac{1}{1.3}\right) + \cdots + 2\left(\frac{1}{1.9}\right) + \frac{1}{2}\right]$$

$$\approx 0.693771$$

In this case we can easily compute the actual value of the definite integral under consideration. In fact,

$$\int_1^2 \frac{1}{x}\, dx = \ln x \Big|_1^2 = \ln 2 - \ln 1 = \ln 2$$
$$\approx 0.693147$$

Thus, the trapezoidal rule with $n = 10$ yields a result with an error of 0.000624 to six decimal places.   ■ ■ ■ ■

EXAMPLE 2 The demand function for a certain brand of perfume is given by

$$p = D(x) = \sqrt{10{,}000 - 0.01x^2}$$

where p is the unit price in dollars and x is the quantity demanded each week, measured in ounces. Find the consumers' surplus if the market price is set at $60 per ounce.

SOLUTION ✔ When $p = 60$, we have

$$\sqrt{10{,}000 - 0.01x^2} = 60$$
$$10{,}000 - 0.01x^2 = 3{,}600$$
$$x^2 = 640{,}000$$

or $x = 800$ since x must be nonnegative. Next, using the consumers' surplus formula (page 520) with $\bar{p} = 60$ and $\bar{x} = 800$, we see that the consumers' surplus is given by

$$CS = \int_0^{800} \sqrt{10{,}000 - 0.01x^2}\, dx - (60)(800)$$

It is not easy to evaluate this definite integral by finding an antiderivative of the integrand. Instead, let's use the trapezoidal rule with $n = 10$.

With $a = 0$ and $b = 800$, we find that

$$\Delta x = \frac{b - a}{n} = \frac{800}{10} = 80$$

and

$$x_0 = 0, \quad x_1 = 80, \quad x_2 = 160, \quad x_3 = 240, \ldots, x_9 = 720, \quad x_{10} = 800$$

so

$$\int_0^{800} \sqrt{10{,}000 - 0.01x^2}\, dx$$

$$\approx \frac{80}{2}[100 + 2\sqrt{10{,}000 - (0.01)(80)^2}$$
$$+ 2\sqrt{10{,}000 - (0.01)(160)^2} + \cdots + 2\sqrt{10{,}000 - (0.01)(720)^2}$$
$$+ \sqrt{10{,}000 - (0.01)(800)^2}]$$

$$= 40[100 + 199.3590 + 197.4234 + 194.1546 + 189.4835$$
$$+ 183.3030 + 175.4537 + 165.6985$$
$$+ 153.6750 + 138.7948 + 60]$$
$$\approx 70{,}293.82$$

Therefore, the consumers' surplus is approximately $70{,}294 - 48{,}000$, or $22,294.   ■ ■ ■ ■

Group Discussion

Explain how you would approximate the value of $\int_0^2 f(x)\,dx$ using the trapezoidal rule with $n = 10$, where

$$f(x) = \begin{cases} \sqrt{1 + x^2} & \text{if } 0 \le x \le 1 \\[2mm] \dfrac{2}{\sqrt{1 + x^2}} & \text{if } 1 < x \le 2 \end{cases}$$

and find the value.

SIMPSON'S RULE

Before stating Simpson's rule, let's review the two rules we have used in approximating a definite integral. Let f be a continuous nonnegative function defined on the interval $[a, b]$. Suppose the interval $[a, b]$ is partitioned by means of the $n + 1$ equally spaced points $x_0 = a, x_1, x_2, \ldots, x_n = b$, where n is a positive integer, so that the length of each subinterval is $\Delta x = (b - a)/n$ (Figure 7.4).

FIGURE 7.4
The area under the curve is equal to the sum of the n subregions $R_1, R_2, \ldots, R_n$.

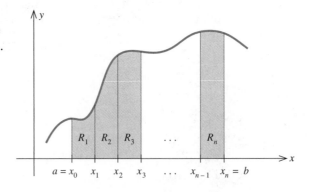

Let's concentrate on the portion of the graph of $y = f(x)$ defined on the interval $[x_0, x_2]$. In using a Riemann sum to approximate the definite integral, we are in effect approximating the function $f(x)$ on $[x_0, x_1]$ by the *constant* function $y = f(p_1)$, where p_1 is chosen to be a point in $[x_0, x_1]$; the function

$f(x)$ on $[x_1, x_2]$ by the constant function $y = f(p_2)$, where p_2 lies in $[x_1, x_2]$; and so on. Using a Riemann sum, we see that the area of the region under the curve $y = f(x)$ between $x = a$ and $x = b$ is approximated by the area under the approximating "step" function (Figure 7.5a).

FIGURE 7.5

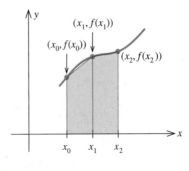

(a) The area under the curve is approximated by the area of the rectangles.

(b) The area under the curve is approximated by the area of the trapezoids.

When we use the trapezoidal rule, we are in effect approximating the function $f(x)$ on the interval $[x_0, x_1]$ by a *linear* function through the two points $(x_0, f(x_0))$ and $(x_1, f(x_1))$; the function $f(x)$ on $[x_1, x_2]$ by a *linear* function through the two points $(x_1, f(x_1))$ and $(x_2, f(x_2))$; and so on. Thus, the trapezoidal rule simply approximates the actual area of the region under the curve $y = f(x)$ from $x = a$ to $x = b$ by the area under the approximating polygonal curve (Figure 7.5b).

A natural extension of the preceding idea is to approximate portions of the graph of $y = f(x)$ by means of portions of the graphs of second-degree polynomials (parts of parabolas). It can be shown that given any three noncollinear points there is a unique parabola that passes through the given points. Choose the points $(x_0, f(x_0))$, $(x_1, f(x_1))$, and $(x_2, f(x_2))$ corresponding to the first three points of the partition. Then, we can approximate the function $f(x)$ on $[x_0, x_2]$ by means of a quadratic function whose graph contains these three points (Figure 7.6).

FIGURE 7.6

Simpson's rule approximates the area under the curve by the area under the parabola.

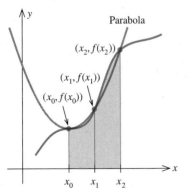

Although we will not do so here, it can be shown that the area under the parabola between $x = x_0$ and $x = x_2$ is given by

$$\frac{\Delta x}{3} [f(x_0) + 4f(x_1) + f(x_2)]$$

square units. Repeating this argument on the interval $[x_2, x_4]$, we see that the area under the curve between $x = x_2$ and $x = x_4$ is approximated by the area under the parabola between x_2 and x_4—that is, by

$$\frac{\Delta x}{3} [f(x_2) + 4f(x_3) + f(x_4)]$$

square units. Proceeding, we conclude that if n is even (Why?), then the area under the curve $y = f(x)$ from $x = a$ to $x = b$ may be approximated by the sum of the areas under the $n/2$ approximating parabolas—that is,

$$\frac{\Delta x}{3}[f(x_0) + 4f(x_1) + f(x_2)] + \frac{\Delta x}{3}[f(x_2) + 4f(x_3) + f(x_4)] + \cdots$$

$$+ \frac{\Delta x}{3}[f(x_{n-2}) + 4f(x_{n-1}) + f(x_n)]$$

$$= \frac{\Delta x}{3}[f(x_0) + 4f(x_1) + f(x_2) + f(x_2) + 4f(x_3) + f(x_4) + \cdots$$

$$+ f(x_{n-2}) + 4f(x_{n-1}) + f(x_n)]$$

$$= \frac{\Delta x}{3}[f(x_0) + 4f(x_1) + 2f(x_2) + 4f(x_3)$$

$$+ 2f(x_4) + \cdots + 4f(x_{n-1}) + f(x_n)]$$

The preceding is the derivation of the approximation formula known as **Simpson's rule.**

Simpson's Rule

$$\int_a^b f(x)\, dx \approx \frac{\Delta x}{3}[f(x_0) + 4f(x_1) + 2f(x_2) + 4f(x_3) + 2f(x_4)$$

$$+ \cdots + 4f(x_{n-1}) + f(x_n)] \tag{6}$$

where $\Delta x = \dfrac{b-a}{n}$ and n is even.

In using this rule, remember that n must be even.

 EXAMPLE **3**   Find an approximation of

$$\int_1^2 \frac{1}{x}\, dx$$

using Simpson's rule with $n = 10$. Compare this result with that of Example 1 and also with the exact value of the integral.

SOLUTION ✔   We have $a = 1$, $b = 2$, $f(x) = \dfrac{1}{x}$, and $n = 10$, so

$$\Delta x = \frac{b-a}{n} = \frac{1}{10} = 0.1$$

and

$$x_0 = 1, \quad x_1 = 1.1, \quad x_2 = 1.2, \quad x_3 = 1.3, \ldots, x_9 = 1.9, \quad x_{10} = 2$$

Simpson's rule yields

$$\int_1^2 \frac{1}{x}\,dx \approx \frac{0.1}{3}[f(1) + 4f(1.1) + 2f(1.2) + \cdots + 4f(1.9) + f(2)]$$

$$= \frac{0.1}{3}\left[1 + 4\left(\frac{1}{1.1}\right) + 2\left(\frac{1}{1.2}\right) + 4\left(\frac{1}{1.3}\right) + 2\left(\frac{1}{1.4}\right) + 4\left(\frac{1}{1.5}\right)\right.$$

$$\left. + 2\left(\frac{1}{1.6}\right) + 4\left(\frac{1}{1.7}\right) + 2\left(\frac{1}{1.8}\right) + 4\left(\frac{1}{1.9}\right) + \frac{1}{2}\right]$$

$$\approx 0.693150$$

The trapezoidal rule with $n = 10$ yielded an approximation of 0.693771, which is 0.000624 off the value of ln 2 ≈ 0.693147 to six decimal places. Simpson's rule yields an approximation with an error of 0.000003, a definite improvement over the trapezoidal rule. ■ ■ ■ ■

EXAMPLE 4   Solve the problem posed at the beginning of this section. Recall that we wished to find a person's cardiac output by using the formula

$$R = \frac{60D}{\displaystyle\int_0^{28} c(t)\,dt}$$

where D (the quantity of dye injected) is equal to 5 milligrams and the function c has the graph shown in Figure 7.7. Use Simpson's rule with $n = 14$ to estimate the value of the integral.

FIGURE 7.7
The function c gives the concentration of a dye measured at the aorta. The graph is constructed by drawing a smooth curve through a set of discrete points.

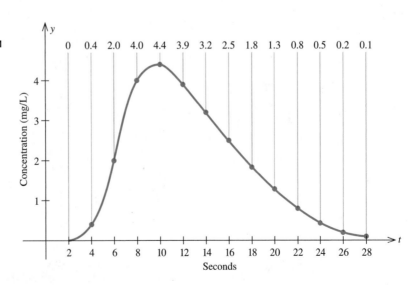

SOLUTION ✔   Using Simpson's rule with $n = 14$ and $\Delta t = 2$ so that

$$t_0 = 0, \quad t_1 = 2, \quad t_2 = 4, \quad t_3 = 6, \ldots, t_{14} = 28$$

we obtain

$$\int_0^{28} c(t)\, dt \approx \frac{2}{3}[c(0) + 4c(2) + 2c(4) + 4c(6) + \cdots$$
$$+ 4c(26) + c(28)]$$
$$= \frac{2}{3}[0 + 4(0) + 2(0.4) + 4(2.0) + 2(4.0)$$
$$+ 4(4.4) + 2(3.9) + 4(3.2) + 2(2.5) + 4(1.8)$$
$$+ 2(1.3) + 4(0.8) + 2(0.5) + 4(0.2) + 0.1]$$
$$\approx 49.9$$

Therefore, the person's cardiac output is

$$R \approx \frac{60(5)}{49.9} \approx 6.0$$

or 6.0 L/min.   ■ ■ ■ ■

EXAMPLE 5   An oil spill off the coastline was caused by a ruptured tank in a grounded oil tanker. Using aerial photographs, the Coast Guard was able to obtain the dimensions of the oil spill (Figure 7.8). Using Simpson's rule with $n = 10$, estimate the area of the oil spill.

FIGURE 7.8
Simpson's rule can be used to calculate the area of the oil spill.

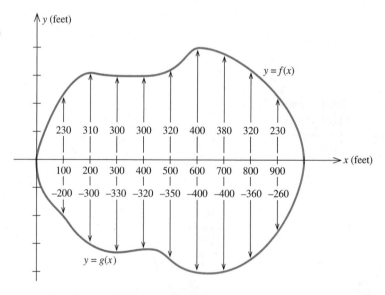

SOLUTION ✔   We may think of the area affected by the oil spill as the area of the plane region bounded above by the graph of the function $f(x)$ and below by the graph of the function $g(x)$ between $x = 0$ and $x = 1000$ (Figure 7.8). Then, the required area is given by

$$A = \int_0^{1000} [f(x) - g(x)]\, dx$$

Using Simpson's rule with $n = 10$ and $\Delta x = 100$ so that

$$x_0 = 0, \qquad x_1 = 100, \qquad x_2 = 200, \ldots, x_{10} = 1000$$

we have

$$
\begin{aligned}
A &= \int_0^{1000} [f(x) - g(x)]\, dx \\
&\approx \frac{\Delta x}{3} \{[f(x_0) - g(x_0)] + 4[f(x_1) - g(x_1)] + 2[f(x_2) - g(x_2)] \\
&\quad + \cdots + 4[f(x_9) - g(x_9)] + [f(x_{10}) - g(x_{10})]\} \\
&= \frac{100}{3} \{[0 - 0] + 4[230 - (-200)] + 2[310 - (-300)] \\
&\quad + 4[300 - (-330)] + 2[300 - (-320)] + 4[320 - (-350)] \\
&\quad + 2[400 - (-400)] + 4[380 - (-400)] + 2[320 - (-360)] \\
&\quad + 4[230 - (-260)] + [0 - 0]\} \\
&= \frac{100}{3} [0 + 4(430) + 2(610) + 4(630) + 2(620) + 4(670) \\
&\quad + 2(800) + 4(780) + 2(680) + 4(490) + 0] \\
&= \frac{100}{3} (17{,}420) \\
&\approx 580{,}667
\end{aligned}
$$

or approximately 580,667 square feet.   ■ ■ ■ ■

Group Discussion

Explain how you would approximate the value of $\int_0^2 f(x)\, dx$ using Simpson's rule with $n = 10$, where

$$
f(x) = \begin{cases} \sqrt{1 + x^2} & \text{if } 0 \le x \le 1 \\[2mm] \dfrac{2}{\sqrt{1 + x^2}} & \text{if } 1 < x \le 2 \end{cases}
$$

and find the value.

ERROR ANALYSIS

The following results give the bounds on the errors incurred when the trapezoidal rule and Simpson's rule are used to approximate a definite integral (proof omitted).

Errors in the Trapezoidal and Simpson Approximations

Suppose the definite integral

$$\int_a^b f(x)\, dx$$

is approximated with n subintervals.

1. The *maximum* error incurred in using the trapezoidal rule is

$$\frac{M(b-a)^3}{12n^2} \tag{7}$$

where M is a number such that $|f''(x)| \leq M$ for all x in $[a, b]$.

2. The *maximum* error incurred in using Simpson's rule is

$$\frac{M(b-a)^5}{180n^4} \tag{8}$$

where M is a number such that $|f^{(4)}(x)| \leq M$ for all x in $[a, b]$.

REMARK In many instances, the actual error is less than the upper error bounds given. ■ ■ ■

EXAMPLE 6 Find bounds on the errors incurred when

$$\int_1^2 \frac{1}{x}\, dx$$

is approximated using (a) the trapezoidal rule and (b) Simpson's rule with $n = 10$. Compare these with the actual errors found in Examples 1 and 3.

SOLUTION ✔ **a.** Here, $a = 1$, $b = 2$, and $f(x) = 1/x$. Next, to find a value for M, we compute

$$f'(x) = -\frac{1}{x^2} \qquad \text{and} \qquad f''(x) = \frac{2}{x^3}$$

Since $f''(x)$ is positive and decreasing on $(1, 2)$ (Why?), it attains its maximum value of 2 at $x = 1$, the left end point of the interval. Therefore, if we take $M = 2$, then $|f''(x)| \leq 2$. Using (7), we see that the maximum error incurred is

$$\frac{2(2-1)^3}{12(10)^2} = \frac{2}{1200} = 0.0016667$$

The actual error found in Example 1, 0.000624, is much less than the upper bound just found.

b. We compute

$$f'''(x) = \frac{-6}{x^4} \qquad \text{and} \qquad f^{(4)}(x) = \frac{24}{x^5}$$

Since $f^{(4)}(x)$ is positive and decreasing on $(1, 2)$ (just look at $f^{(5)}$ to verify this fact), it attains its maximum at the left end point of $[1, 2]$. Now,

$$f^{(4)}(1) = 24$$

and so we may take $M = 24$. Using (8), we obtain the maximum error of

$$\frac{24(2 - 1)^5}{180(10)^4} = 0.0000133$$

The actual error is 0.000003 (see Example 3).   ■ ■ ■ ■

> **Group Discussion**
>   Refer to the Group Discussions on pages 562 and 567. Explain how you would find the maximum error incurred in using (1) the trapezoidal rule and (2) Simpson's rule with $n = 10$ to approximate $\int_0^2 f(x)\, dx$.

SELF-CHECK EXERCISES 7.3

1. Use the trapezoidal rule and Simpson's rule with $n = 8$ to approximate the value of the definite integral

$$\int_0^2 \frac{1}{\sqrt{1 + x^2}}\, dx$$

2. The graph in the accompanying figure shows the consumption of petroleum in the United States in quadrillion BTU, from 1976 to 1990. Using Simpson's rule with $n = 14$, estimate the average consumption during the 14-yr period.

Source: The World Almanac

Solutions to Self-Check Exercises 7.3 can be found on page 573.

7.3 Exercises

A calculator is recommended for this exercise set. In Exercises 1–14, use the trapezoidal rule and Simpson's rule to approximate the value of each definite integral. Compare your result with the exact value of the integral.

1. $\int_0^2 x^2 \, dx; n = 6$ **2.** $\int_1^3 (x^2 - 1) \, dx; n = 4$

3. $\int_0^1 x^3 \, dx; n = 4$ **4.** $\int_1^2 x^3 \, dx; n = 6$

5. $\int_1^2 \frac{1}{x} \, dx; n = 4$ **6.** $\int_1^2 \frac{1}{x} \, dx; n = 8$

7. $\int_1^2 \frac{1}{x^2} \, dx; n = 4$ **8.** $\int_0^1 \frac{1}{1+x} \, dx; n = 4$

9. $\int_0^4 \sqrt{x} \, dx; n = 8$ **10.** $\int_0^2 x\sqrt{2x^2 + 1} \, dx; n = 6$

11. $\int_0^1 e^{-x} \, dx; n = 6$ **12.** $\int_0^1 xe^{-x^2} \, dx; n = 6$

13. $\int_1^2 \ln x \, dx; n = 4$ **14.** $\int_0^1 x \ln(x^2 + 1) \, dx; n = 8$

In Exercises 15–22, use the trapezoidal rule and Simpson's rule to approximate the value of each definite integral.

15. $\int_0^1 \sqrt{1 + x^3} \, dx; n = 4$ **16.** $\int_0^2 x\sqrt{1 + x^3} \, dx; n = 4$

17. $\int_0^2 \frac{1}{\sqrt{x^3 + 1}} \, dx; n = 4$ **18.** $\int_0^1 \sqrt{1 - x^2} \, dx; n = 4$

19. $\int_0^2 e^{-x^2} \, dx; n = 4$ **20.** $\int_0^1 e^{x^2} \, dx; n = 6$

21. $\int_1^2 x^{-1/2} e^x \, dx; n = 4$ **22.** $\int_2^4 \frac{dx}{\ln x}; n = 6$

In Exercises 23–28, find a bound on the error in approximating the given definite integral using (a) the trapezoidal rule and (b) Simpson's rule with *n* intervals.

23. $\int_{-1}^2 x^5 \, dx; n = 10$ **24.** $\int_0^1 e^{-x} \, dx; n = 8$

25. $\int_1^3 \frac{1}{x} \, dx; n = 10$ **26.** $\int_1^3 \frac{1}{x^2} \, dx; n = 8$

27. $\int_0^2 \frac{1}{\sqrt{1+x}} \, dx; n = 8$ **28.** $\int_1^3 \ln x \, dx; n = 10$

29. TRIAL RUN OF AN ATTACK SUBMARINE In a submerged trial run of an attack submarine, a reading of the sub's velocity was made every quarter hour, as shown in the accompanying table. Use the trapezoidal rule to estimate the distance traveled by the submarine during the 2-hr period.

Time, t (hr)	0	$\frac{1}{4}$	$\frac{1}{2}$	$\frac{3}{4}$
Velocity, $V(t)$ (mph)	19.5	24.3	34.2	40.5

Time, t (hr)	1	$\frac{5}{4}$	$\frac{3}{2}$	$\frac{7}{4}$	2
Velocity, $V(t)$ (mph)	38.4	26.2	18	16	8

30. REAL ESTATE Cooper Realty is considering development of a time-sharing condominium resort complex along the oceanfront property illustrated in the accompanying graph. To obtain an estimate of the area of this property, measurements of the distances from the edge of a straight road, which defines one boundary of the property, to the corresponding points on the shoreline are made at 100-ft intervals. Using Simpson's rule with $n = 10$, estimate the area of the oceanfront property.

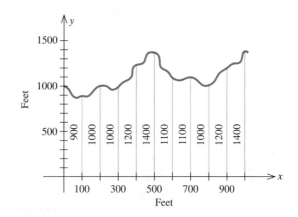

31. FUEL CONSUMPTION OF DOMESTIC CARS Thanks to smaller and more fuel-efficient models, American carmakers have doubled their average fuel economy over a 13-yr period, from 1974 to 1987. The graph depicted in the figure on p. 571 gives the average fuel consumption in miles per gallon (mpg) of domestic-built cars over the period under consideration ($t = 0$ corresponds to the beginning of 1974). Use the trapezoidal rule to estimate the average fuel consumption of the domestic car built during this period.

Hint: Approximate the integral $\frac{1}{13}\int_0^{13} f(t)\, dt$.

32. **AVERAGE TEMPERATURE**   The graph depicted in the accompanying figure shows the daily mean temperatures recorded during one September in Cameron Highlands. Using (a) the trapezoidal rule and (b) Simpson's rule with $n = 10$, estimate the average temperature during that month.

33. **CONSUMERS' SURPLUS**   Refer to Section 6.7. The demand equation for the Sicard wristwatch is given by

$$p = \frac{50}{0.01x^2 + 1} \qquad (0 \le x \le 20)$$

where x (measured in units of a thousand) is the quantity demanded per week and p is the unit price in dollars. Use (a) the trapezoidal rule and (b) Simpson's rule (take $n = 8$) to estimate the consumers' surplus if the market price is \$25/watch.

34. **PRODUCERS' SURPLUS**   Refer to Section 6.7. The supply function for the audio compact disc manufactured by the Herald Record Company is given by

$$p = S(x) = \sqrt{0.01x^2 + 0.11x + 38}$$

where p is the unit wholesale price in dollars and x stands for the quantity that will be made available in the market by the supplier, measured in units of a thousand. Use (a) the trapezoidal rule and (b) Simpson's rule (take $n = 8$) to estimate the producers' surplus if the wholesale price is \$8/disc.

35. **AIR POLLUTION**   The amount of nitrogen dioxide, a brown gas that impairs breathing, present in the atmosphere on a certain May day in the city of Long Beach has been approximated by

$$A(t) = \frac{136}{1 + 0.25(t - 4.5)^2} + 28 \qquad (0 \le t \le 11)$$

where $A(t)$ is measured in pollutant standard index (PSI), t is measured in hours, and $t = 0$ corresponds to 7 A.M. Use the trapezoidal rule with $n = 10$ to estimate the average PSI between 7 A.M. and noon.

Hint: $\frac{1}{5}\int_0^5 A(t)\, dt$.

36. **GROWTH OF SERVICE INDUSTRIES**   It has been estimated that service industries, which currently make up 30% of the nonfarm work force in a certain country, will continue to grow at the rate of

$$R(t) = 5e^{1/(t+1)}$$

percent/decade t decades from now. Estimate the percentage of the nonfarm workforce in the service industries one decade from now.

Hint: (a) Show that the desired answer is given by $30 + \int_0^1 5e^{1/(t+1)}\, dt$ and (b) use Simpson's rule with $n = 10$ to approximate the definite integral.

37. **LENGTH OF INFANTS AT BIRTH**   Medical records of infants delivered at the Kaiser Memorial Hospital show that the percentage of infants whose length at birth is between 19 and 21 in. is given by

$$P = 100 \int_{19}^{21} \frac{1}{2.6\sqrt{2\pi}} e^{-1/2[(x-20)/2.6]^2}\, dx$$

Use Simpson's rule with $n = 10$ to estimate P.

JAMES H. CHESEBRO, M.D.

TITLE: Professor of Medicine
INSTITUTION: Harvard Medical School, Massachusetts General Hospital

For over 20 years, James Chesebro has worked as an investigative cardiologist, diagnosing and treating patients suffering from heart and blood vessel diseases. He specializes in researching ways to prevent blood clots from narrowing the heart's arteries. Even when patients have coronary-bypass operations to correct problems, stresses Chesebro, "the piece of vein used to detour around blocked arteries is prone to plug up with clots."

Throughout his career, Chesebro has investigated the body's clotting mechanisms, as well as substances that may prevent, or even dissolve, clots. A study conducted in 1982 showed that bypass patients taking dipyridamole and aspirin were nearly three times less likely to develop clots in vein grafts than patients given placebos.

Recently, Chesebro has been working with a substance called hirudin derived from leech saliva. Hirudin has proved quite beneficial in blocking arterial blood-clot formations.

Determining the correct medication and dosage involves intense quantitative research. Colleagues from other branches of science such as nuclear medicine, molecular biology, and biochemistry play key roles in the research process. Chesebro and his colleagues have to understand how the heart's physiology and pharmacological variables such as distribution rates, concentration levels, elimination rates, and biological effects of particular substances dictate the choice and dosage of medication.

In considering whether a particular medication will bind effectively with specific body cells, researchers must determine bond tightness, speed, and length of time to reach optimal concentration. Researchers rely on complicated equations such as "integrating the area under a curve to find the amount of medication in the body at a given time," notes Chesebro. Without calculus, these equations can't be solved.

Whether physicians rely on medication or surgical procedures such as balloon angioplasty to open clogged passages, other factors play a major role in the outcome. Cholesterol and blood sugar levels, patient age, blood pressure, the geometry of the blockage, and the velocity and turbulence of blood flow all warrant consideration. According to Chesebro "linear modeling can be used to predict the contribution of patient variables and local blood vessel variables in the outcome of opening blocked arteries."

The bottom line? Calculus has been a key contributor to Chesebro's success in preventing disabling arterial blood clots.

In 1992 Chesebro was appointed professor of medicine at Harvard Medical School, after having conducted cardiovascular research at the Mayo Clinic for a number of years. In addition, he was made Associate Director for Research in the Cardiac Unit at Massachusetts General Hospital, one of Harvard's teaching affiliates.

38. TREAD LIVES OF TIRES   Under normal driving conditions the percentage of Super Titan radial tires expected to have a useful tread life of between 30,000 and 40,000 mi is given by

$$P = 100 \int_{30,000}^{40,000} \frac{1}{2000\sqrt{2\pi}} e^{-1/2[(x-40,000)/2000]^2} \, dx$$

Use Simpson's rule with $n = 10$ to estimate P.

39. MEASURING CARDIAC OUTPUT   Eight milligrams of a dye are injected into a vein leading to an individual's heart. The concentration of the dye in the aorta measured at 2-sec intervals is shown in the accompanying table. Use Simpson's rule and the formula of Example 4 to estimate the person's cardiac output.

t	0	2	4	6	8	10	12
$C(t)$	0	0	2.8	6.1	9.7	7.6	4.8

t	14	16	18	20	22	24
$C(t)$	3.7	1.9	0.8	0.3	0.1	0

In Exercises 40–43, determine whether the statement is true or false. If it is true, explain why it is true. If it is false, give an example to show why it is false.

40. In using the trapezoidal rule, the number of subintervals n must be even.

41. In using Simpson's rule, the number of subintervals n may be chosen to be odd or even.

42. Simpson's rule is more accurate than the trapezoidal rule.

43. If f is a polynomial function of degree less than or equal to 3, then the approximation $\int_a^b f(x) \, dx$ using Simpson's rule is exact.

44. Derive the formula

$$R = \frac{60D}{\int_0^T c(t) \, dt}$$

for calculating the cardiac output of a person in L/min. Here, $c(t)$ is the concentration of dye in the aorta (in mg/L) at time t (in seconds) for t in $[0, T]$, and D is the amount of dye (in mg) injected into a vein leading to the heart.

Hint: Partition the interval $[0, T]$ into n subintervals of equal length Δt. The amount of dye that flows past the measuring point in the aorta during the time interval $[0, \Delta t]$ is approximately $c(t_i)(R\Delta t)/60$ (concentration times volume). Therefore, the total amount of dye measured at the aorta is

$$\frac{[c(t_1)R\Delta t + c(t_2)R\Delta t + \cdots + c(t_n)R\Delta t]}{60} = D$$

Take the limit of the Riemann sum to obtain

$$R = \frac{60D}{\int_0^T c(t) \, dt}$$

SOLUTIONS TO SELF-CHECK EXERCISES 7.3

1. We have $x = 0$, $b = 2$, and $n = 8$, so

$$\Delta x = \frac{b-a}{n} = \frac{2}{8} = 0.25$$

and $x_0 = 0$, $x_1 = 0.25$, $x_2 = 0.50$, $x_3 = 0.75, \ldots, x_7 = 1.75$, and $x_8 = 2$. The trapezoidal rule gives

$$\int_0^2 \frac{1}{\sqrt{1+x^2}} \, dx \approx \frac{0.25}{2} \left[1 + \frac{2}{\sqrt{1+(0.25)^2}} + \frac{2}{\sqrt{1+(0.5)^2}} + \cdots \right.$$

$$\left. + \frac{2}{\sqrt{1+(1.75)^2}} + \frac{1}{\sqrt{5}} \right]$$

$$\approx 0.125(1 + 1.9403 + 1.7889 + 1.6000 + 1.4142 + 1.2494$$

$$+ 1.1094 + 0.9923 + 0.4472)$$

$$\approx 1.4427$$

Using Simpson's rule with $n = 8$ gives

$$\int_0^2 \frac{1}{\sqrt{1+x^2}}\, dx \approx \frac{0.25}{3}\left[1 + \frac{4}{\sqrt{1+(0.25)^2}} + \frac{2}{\sqrt{1+(0.5)^2}} + \frac{4}{\sqrt{1+(0.75)^2}}\right.$$

$$\left. + \cdots + \frac{4}{\sqrt{1+(1.75)^2}} + \frac{1}{\sqrt{5}}\right]$$

$$\approx \frac{0.25}{3}(1 + 3.8806 + 1.7889 + 3.2000 + 1.4142 + 2.4988 + 1.1094$$

$$+ 1.9846 + 0.4472)$$

$$\approx 1.4436$$

2. The average consumption of petroleum during the 14-yr period is given by

$$\frac{1}{14}\int_0^{14} f(x)\, dx$$

where f is the function describing the given graph. Using Simpson's rule with $a = 0$, $b = 14$, and $n = 14$ so that $\Delta x = 1$ and

$$x_0 = 0, \qquad x_1 = 1, \qquad x_2 = 2, \ldots, x_{14} = 14$$

we have

$$\frac{1}{14}\int_0^{14} f(x)\, dx$$

$$\approx \left(\frac{1}{14}\right)\left(\frac{1}{3}\right)[f(x_0) + 4f(x_1) + 2f(x_2) + 4f(x_3) + \cdots + 4f(x_{13}) + f(x_{14})]$$

$$= \frac{1}{42}[35 + 4(36.6) + 2(37.6) + 4(36.6) + 2(34.2)$$

$$+ 4(32.0) + 2(30.2) + 4(30.2) + 2(31.0) + 4(31.0)$$

$$+ 2(32.2) + 4(32.9) + 2(34.2) + 4(34.2) + 33.6]$$

$$\approx 33.4$$

or approximately 33.4 quadrillion BTU/year.

7.4 Improper Integrals

IMPROPER INTEGRALS

All the definite integrals we have encountered have had finite intervals of integration. In many applications, however, we are concerned with integrals that have unbounded intervals of integration. Such integrals are called **improper integrals.**

FIGURE 7.9
The area of the unbounded region R can be approximated by a definite integral.

FIGURE 7.10

Area of shaded region $= \int_1^b \frac{1}{x^2}\, dx$

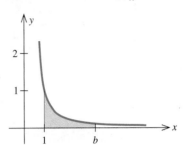

To lead us to the definition of an improper integral of a function f over an infinite interval, consider the problem of finding the area of the region R under the curve $y = f(x) = 1/x^2$ and to the right of the vertical line $x = 1$, as shown in Figure 7.9. Because the interval over which the integration must be performed is unbounded, the method of integration presented previously cannot be applied directly in solving this problem. However, we can approximate the region R by the definite integral

$$\int_1^b \frac{1}{x^2}\, dx \qquad\qquad (9)$$

which gives the area of the region under the curve $y = f(x) = 1/x^2$ from $x = 1$ to $x = b$ (Figure 7.10). You can see that the approximation of the region R by the definite integral (9) improves as the upper limit of integration, b, becomes larger and larger. Figure 7.11 illustrates the situation for $b = 2, 3,$ and 4, respectively.

FIGURE 7.11
As b increases, the approximation of R by the definite integral improves.

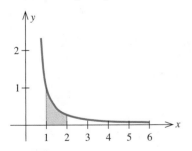

(a) Area of region under the graph of f on [1, 2]

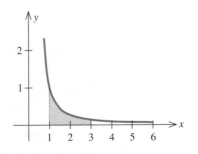

(b) Area of region under the graph of f on [1, 3]

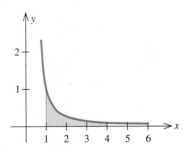

(c) Area of region under the graph of f on [1, 4]

This observation suggests that if we define a function $I(b)$ by

$$I(b) = \int_1^b \frac{1}{x^2}\,dx \tag{10}$$

then we can find the area of the required region R by evaluating the limit of $I(b)$ as b tends to infinity; that is, the area of R is given by

$$\lim_{b\to\infty} I(b) = \lim_{b\to\infty} \int_1^b \frac{1}{x^2}\,dx \tag{11}$$

EXAMPLE 1

a. Evaluate the definite integral $I(b)$ in Equation (10).
b. Compute $I(b)$ for $b = 10, 100, 1000, 10{,}000$.
c. Evaluate the limit in Equation (11).
d. Interpret the results of parts (b) and (c).

SOLUTION ✔

a.
$$I(b) = \int_1^b \frac{1}{x^2}\,dx = -\frac{1}{x}\Big|_1^b = -\frac{1}{b} + 1$$

b. From the result of part (a),

$$I(b) = 1 - \frac{1}{b}$$

Therefore,

$$I(10) = 1 - \frac{1}{10} = 0.9$$

$$I(100) = 1 - \frac{1}{100} = 0.99$$

$$I(1000) = 1 - \frac{1}{1000} = 0.999$$

$$I(10{,}000) = 1 - \frac{1}{10{,}000} = 0.9999$$

c. Once again, using the result of part (a), we find

$$\lim_{b\to\infty} I(b) = \lim_{b\to\infty} \int_1^b \frac{1}{x^2}\,dx$$
$$= \lim_{b\to\infty}\left(1 - \frac{1}{b}\right)$$
$$= 1$$

d. The result of part (c) tells us that the area of the region R is 1 square unit. The results of the computations performed in part (b) reinforce our expectation that $I(b)$ should approach 1, the area of the region R, as b approaches infinity. ■ ■ ■ ■

The preceding discussion and the results of Example 1 suggest that we define the improper integral of a continuous function f over the unbounded interval $[a, \infty)$ as follows.

Improper Integral of f over $[a, \infty)$

Let f be a continuous function on the unbounded interval $[a, \infty)$. Then, the improper integral of f over $[a, \infty)$ is defined by

$$\int_a^\infty f(x)\, dx = \lim_{b \to \infty} \int_a^b f(x)\, dx \qquad (12)$$

if the limit exists.

If the limit exists, the improper integral is said to be **convergent**. An improper integral for which the limit in Equation (12) fails to exist is said to be **divergent**.

EXAMPLE 2   Evaluate $\int_2^\infty \dfrac{1}{x}\, dx$ if it converges.

SOLUTION ✔

$$\int_2^\infty \frac{1}{x}\, dx = \lim_{b \to \infty} \int_2^b \frac{1}{x}\, dx$$

$$= \lim_{b \to \infty} \ln x \Big|_2^b$$

$$= \lim_{b \to \infty} (\ln b - \ln 2)$$

Since $\ln b \to \infty$ as $b \to \infty$, the limit does not exist, and we conclude that the given improper integral is divergent.   ■ ■ ■ ■

Group Discussion
1. Suppose f is continuous and nonnegative on $[0, \infty)$. Furthermore, suppose $\lim_{x \to \infty} f(x) = L$, where L is a positive number. What can you say about the convergence of the improper integral $\int_0^\infty f(x)\, dx$? Explain your answer and illustrate with an example.
2. Suppose f is continuous and nonnegative on $[0, \infty)$ and satisfies the condition $\lim_{x \to \infty} f(x) = 0$. What can you say about $\int_0^\infty f(x)\, dx$? Explain and illustrate your answer with examples.

EXAMPLE 3   Find the area of the region R under the curve $y = e^{-x/2}$ for $x \geq 0$.

SOLUTION ✔   The region R is shown in Figure 7.12. Taking $b > 0$, we compute the area of the region under the curve $y = e^{-x/2}$ from $x = 0$ to $x = b$—namely,

FIGURE 7.12

Area of $R = \int_0^\infty e^{-x/2}\, dx$

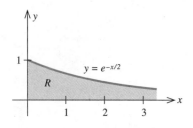

$$I(b) = \int_0^b e^{-x/2}\, dx = -2e^{-x/2}\Big|_0^b = -2e^{-b/2} + 2$$

Then, the area of the region R is given by

$$\lim_{b \to \infty} I(b) = \lim_{b \to \infty} (2 - 2e^{-b/2}) = 2 - 2 \lim_{b \to \infty} \frac{1}{e^{b/2}}$$
$$= 2$$

or 2 square units.   ■ ■ ■ ■

Exploring with Technology

You can see how fast the improper integral in Example 3 converges, as follows:

1. Use a graphing utility to plot the graph of $I(b) = 2 - 2e^{-b/2}$, using the viewing window $[0, 50] \times [0, 3]$.
2. Use TRACE to follow the values of y for increasing values of x starting at the origin.

The improper integral defined in Equation (12) has an interval of integration that is unbounded on the right. Improper integrals with intervals of integration that are unbounded on the left also arise in practice and are defined in a similar manner.

Improper Integral of f over $(-\infty, b]$

Let f be a continuous function on the unbounded interval $(-\infty, b]$. Then, the improper integral of f over $(-\infty, b]$ is defined by

$$\int_{-\infty}^b f(x)\, dx = \lim_{a \to -\infty} \int_a^b f(x)\, dx \tag{13}$$

if the limit exists.

In this case, the improper integral is said to be convergent. Otherwise, the improper integral is said to be divergent.

EXAMPLE 4   Find the area of the region R bounded above by the x-axis, below by the curve $y = -e^{2x}$, and on the right by the vertical line $x = 1$.

SOLUTION ✔ The region R is shown in Figure 7.13. Taking $a < 1$, compute the area of the region bounded above by the x-axis ($y = 0$), and below by the curve $y = -e^{2x}$ from $x = a$ to $x = 1$—namely,

$$I(a) = \int_a^1 [0 - (-e^{2x})]\, dx = \int_a^1 e^{2x}\, dx$$

$$= \frac{1}{2} e^{2x} \bigg|_a^1 = \frac{1}{2} e^2 - \frac{1}{2} e^{2a}$$

FIGURE 7.13

Area of $R = -\displaystyle\int_{-\infty}^1 -e^{2x}\, dx$

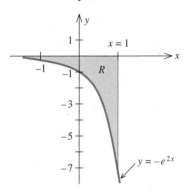

Then, the area of the required region is given by

$$\lim_{a \to -\infty} I(a) = \lim_{a \to -\infty} \left(\frac{1}{2} e^2 - \frac{1}{2} e^{2a} \right)$$

$$= \frac{1}{2} e^2 - \frac{1}{2} \lim_{a \to -\infty} e^{2a}$$

$$= \frac{1}{2} e^2$$

■■■■

Another improper integral found in practical applications involves the integration of a function f over the unbounded interval $(-\infty, \infty)$.

Improper Integral of f over $(-\infty, \infty)$

Let f be a continuous function over the unbounded interval $(-\infty, \infty)$. Let c be any real number and suppose both the improper integrals

$$\int_{-\infty}^c f(x)\, dx \qquad \text{and} \qquad \int_c^\infty f(x)\, dx$$

are convergent. Then, the improper integral of f over $(-\infty, \infty)$ is defined by

$$\int_{-\infty}^\infty f(x)\, dx = \int_{-\infty}^c f(x)\, dx + \int_c^\infty f(x)\, dx \qquad \textbf{(14)}$$

In this case we say that the improper integral on the left in Equation (14) is convergent. If either one of the two improper integrals on the right in (14) is divergent, then the improper integral on the left is not defined.

REMARK Usually, we choose $c = 0$. ■■■

EXAMPLE 5 Evaluate the improper integral

$$\int_{-\infty}^\infty x e^{-x^2}\, dx$$

and give a geometric interpretation of the results.

SOLUTION ✔   Take the point c in Equation (14) to be $c = 0$. Let's first evaluate

$$\int_{-\infty}^{0} xe^{-x^2}\, dx = \lim_{a\to-\infty} \int_{a}^{0} xe^{-x^2}\, dx$$

$$= \lim_{a\to-\infty} -\frac{1}{2} e^{-x^2}\Big|_{a}^{0}$$

$$= \lim_{a\to-\infty} \left[-\frac{1}{2} + \frac{1}{2} e^{-a^2} \right] = -\frac{1}{2}$$

Next, we evaluate

$$\int_{0}^{\infty} xe^{-x^2}\, dx = \lim_{b\to\infty} \int_{0}^{b} xe^{-x^2}\, dx$$

$$= \lim_{b\to\infty} -\frac{1}{2} e^{-x^2}\Big|_{0}^{b}$$

$$= \lim_{b\to\infty} \left[-\frac{1}{2} e^{-b^2} + \frac{1}{2} \right] = \frac{1}{2}$$

Therefore,

$$\int_{-\infty}^{\infty} xe^{-x^2}\, dx = \int_{-\infty}^{0} xe^{-x^2}\, dx + \int_{0}^{\infty} xe^{-x^2}\, dx$$

$$= -\frac{1}{2} + \frac{1}{2}$$

$$= 0$$

FIGURE 7.14

$$\int_{-\infty}^{\infty} xe^{-x^2}\, dx$$

$$= \int_{-\infty}^{0} xe^{-x^2}\, dx + \int_{0}^{\infty} xe^{-x^2}\, dx$$

The graph of $y = xe^{-x^2}$ is sketched in Figure 7.14. A glance at the figure tells us that the improper integral

$$\int_{-\infty}^{0} xe^{-x^2}\, dx$$

gives the negative of the area of the region R_1, bounded above by the x-axis, below by the curve $y = xe^{-x^2}$, and on the right by the y-axis ($x = 0$).

However, the improper integral

$$\int_{0}^{\infty} xe^{-x^2}\, dx$$

gives the area of the region R_2 under the curve $y = xe^{-x^2}$ for $x \geq 0$. Since the graph of f is symmetric with respect to the origin, the area of R_1 is equal to the area of R_2. In other words,

$$\int_{-\infty}^{0} xe^{-x^2}\, dx = -\int_{0}^{\infty} xe^{-x^2}\, dx$$

Therefore,

$$\int_{-\infty}^{\infty} xe^{-x^2}\, dx = \int_{-\infty}^{0} xe^{-x^2}\, dx + \int_{0}^{\infty} xe^{-x^2}\, dx$$

$$= -\int_{0}^{\infty} xe^{-x^2}\, dx + \int_{0}^{\infty} xe^{-x^2}\, dx$$

$$= 0$$

as was shown earlier.   ■ ■ ■ ■

* only ok
since each exists
individually

PERPETUITIES

Recall from Section 6.7 that the present value of an annuity is given by

$$PV \approx mP \int_0^T e^{-rt}\, dt = \frac{mP}{r}(1 - e^{-rT}) \tag{15}$$

Now, if the payments of an annuity are allowed to continue indefinitely, we have what is called a **perpetuity.** The present value of a perpetuity may be approximated by the improper integral

$$PV \approx mP \int_0^\infty e^{-rt}\, dt$$

obtained from Formula (15) by allowing the term of the annuity, T, to approach infinity. Thus,

$$mP \int_0^\infty e^{-rt}\, dt = \lim_{b\to\infty} mP \int_0^b e^{-rt}\, dt$$

$$= mP \lim_{b\to\infty} \int_0^b e^{-rt}\, dt$$

$$= mP \lim_{b\to\infty} \left[-\frac{1}{r} e^{-rt} \Big|_0^b \right]$$

$$= mP \lim_{b\to\infty} \left(-\frac{1}{r} e^{-rb} + \frac{1}{r} \right) = \frac{mP}{r}$$

The Present Value of a Perpetuity

The **present value PV of a perpetuity** is given by

$$PV = \frac{mP}{r} \tag{16}$$

where m is the number of payments per year, P is the size of each payment, and r is the interest rate (compounded continuously).

EXAMPLE 6   The Robinson family wishes to create a scholarship fund at a college. If a scholarship in the amount of $5000 is awarded annually beginning 1 year from now, find the amount of the endowment they are required to make now. Assume that this fund will earn interest at a rate of 8% per year compounded continuously.

SOLUTION ✔   The amount of the endowment, A, is given by the present value of a perpetuity, with $m = 1$, $P = 5000$, and $r = 0.08$. Using Formula (16), we find

$$A = \frac{(1)(5000)}{0.08}$$

$$= 62,500$$

or $62,500.

The improper integral also plays an important role in the study of probability theory, as we will see in Section 7.5.

1. Evaluate $\int_{-\infty}^{\infty} \dfrac{x^3}{(1+x^4)^{3/2}}\,dx$.

2. Suppose an income stream is expected to continue indefinitely. Then, the present value of such a stream can be calculated from the formula for the present value of an income stream by letting T approach infinity. Thus, the required present value is given by

$$PV = \int_0^{\infty} P(t)e^{-rt}\,dt$$

Suppose Marcia has an oil well in her backyard that generates a stream of income given by

$$P(t) = 20e^{-0.02t}$$

where $P(t)$ is expressed in thousands of dollars per year and t is the time in years from the present. Assuming that the prevailing interest rate in the foreseeable future is 10%/year compounded continuously, what is the present value of the income stream?

Solutions to Self-Check Exercises 7.4 can be found on page 584.

7.4 Exercises

In Exercises 1–10, find the area of the region under the given curve $y = f(x)$ over the indicated interval.

1. $f(x) = \dfrac{2}{x^2}; x \geq 3$ 2. $f(x) = \dfrac{2}{x^3}; x \geq 2$

3. $f(x) = \dfrac{1}{(x-2)^2}; x \geq 3$

4. $f(x) = \dfrac{2}{(x+1)^3}; x \geq 0$

5. $f(x) = \dfrac{1}{x^{3/2}}; x \geq 1$ 6. $f(x) = \dfrac{3}{x^{5/2}}; x \geq 4$

7. $f(x) = \dfrac{1}{(x+1)^{5/2}}; x \geq 0$

8. $f(x) = \dfrac{1}{(1-x)^{3/2}}; x \leq 0$

9. $f(x) = e^{2x}; x \leq 2$ 10. $f(x) = xe^{-x^2}; x \geq 0$

11. Find the area of the region bounded by the x-axis and the graph of the function

$$f(x) = \dfrac{x}{(1+x^2)^2}$$

12. Find the area of the region bounded by the x-axis and the graph of the function

$$f(x) = \dfrac{e^x}{(1+e^x)^2}$$

13. Consider the improper integral

$$\int_0^{\infty} \sqrt{x}\,dx$$

a. Evaluate $I(b) = \int_0^b \sqrt{x}\,dx$.

b. Show that

$$\lim_{b \to \infty} I(b) = \infty$$

thus proving that the given improper integral is divergent.

14. Consider the improper integral

$$\int_1^\infty x^{-2/3}\, dx$$

a. Evaluate $I(b) = \int_1^b x^{-2/3}\, dx$.

b. Show that

$$\lim_{b\to\infty} I(b) = \infty$$

thus proving that the given improper integral is divergent.

In Exercises 15–40, evaluate each improper integral whenever it is convergent.

15. $\int_1^\infty \dfrac{3}{x^4}\, dx$ **16.** $\int_1^\infty \dfrac{1}{x^3}\, dx$

17. $\int_4^\infty \dfrac{2}{x^{3/2}}\, dx$ **18.** $\int_1^\infty \dfrac{1}{\sqrt{x}}\, dx$

19. $\int_1^\infty \dfrac{4}{x}\, dx$ **20.** $\int_2^\infty \dfrac{3}{x}\, dx$

21. $\int_{-\infty}^0 \dfrac{1}{(x-2)^3}\, dx$ **22.** $\int_2^\infty \dfrac{1}{(x+1)^2}\, dx$

23. $\int_1^\infty \dfrac{1}{(2x-1)^{3/2}}\, dx$ **24.** $\int_{-\infty}^0 \dfrac{1}{(4-x)^{3/2}}\, dx$

25. $\int_0^\infty e^{-x}\, dx$ **26.** $\int_0^\infty e^{-x/2}\, dx$

27. $\int_{-\infty}^0 e^{2x}\, dx$ **28.** $\int_{-\infty}^0 e^{3x}\, dx$

29. $\int_1^\infty \dfrac{e^{\sqrt{x}}}{\sqrt{x}}\, dx$ **30.** $\int_1^\infty \dfrac{e^{-\sqrt{x}}}{\sqrt{x}}\, dx$

31. $\int_{-\infty}^0 x e^x\, dx$ **32.** $\int_0^\infty x e^{-2x}\, dx$

33. $\int_{-\infty}^\infty x\, dx$ **34.** $\int_{-\infty}^\infty x^3\, dx$

35. $\int_{-\infty}^\infty x^3(1+x^4)^{-2}\, dx$ **36.** $\int_{-\infty}^\infty x(x^2+4)^{-3/2}\, dx$

37. $\int_{-\infty}^\infty x e^{1-x^2}\, dx$ **38.** $\int_{-\infty}^\infty \left(x-\dfrac{1}{2}\right) e^{-x^2+x-1}\, dx$

39. $\int_{-\infty}^\infty \dfrac{e^{-x}}{1+e^{-x}}\, dx$ **40.** $\int_{-\infty}^\infty \dfrac{x e^{-x^2}}{1+e^{-x^2}}\, dx$

41. $\int_e^\infty \dfrac{1}{x\ln^3 x}\, dx$ **42.** $\int_{e^2}^\infty \dfrac{1}{x\ln x}\, dx$

43. THE AMOUNT OF AN ENDOWMENT A university alumni group wishes to provide an annual scholarship in the amount of \$1500 beginning next year. If the scholarship fund will earn interest at the rate of 8%/year compounded continuously, find the amount of the endowment the alumni are required to make now.

44. THE AMOUNT OF AN ENDOWMENT Mel Thompson wishes to establish a fund to provide a university medical center with an annual research grant of \$50,000 beginning next year. If the fund will earn interest at the rate of 9%/year compounded continuously, find the amount of the endowment he is required to make now.

45. PERPETUAL NET INCOME STREAMS The present value of a perpetual stream of income that flows continually at the rate of $P(t)$ dollars/year is given by the formula

$$PV \approx \int_0^\infty P(t)e^{-rt}\, dt$$

where r is the rate at which interest is compounded continuously. Using this formula, find the present value of a perpetual net income stream that is generated at the rate of

$$P(t) = 10{,}000 + 4000t$$

dollars/year.

Hint: $\displaystyle\lim_{b\to\infty} \dfrac{b}{e^{rb}} = 0.$

46. ESTABLISHING A TRUST FUND Becky Wilkinson wants to establish a trust fund that will provide her children and heirs with a perpetual annuity in the amount of

$$P(t) = (20 + t)$$

thousand dollars/year beginning next year. If the trust fund will earn interest at the rate of 10%/year compounded continuously, find the amount that she must place in the trust fund now.

Hint: Use the formula given in Exercise 45.

In Exercises 47–49, determine whether the statement is true or false. If it is true, explain why it is true. If it is false, give an example to show why it is false.

47. If $\int_a^\infty f(x)\, dx$ exists, then $\int_b^\infty f(x)\, dx$ exists for every real number $b > a$.

48. If $\int_a^\infty f(x)\, dx$ exists, then $\int_{-\infty}^{-a} f(x)\, dx$ exists, and $\int_a^\infty f(x)\, dx = -\int_{-\infty}^{-a} f(x)\, dx$ exists.

49. If $\int_{-\infty}^\infty f(x)\, dx$ exists, then $\int_0^\infty f(x)\, dx$ exists, and $\int_{-\infty}^\infty f(x)\, dx = 2\int_0^\infty f(x)\, dx.$

50. CAPITAL VALUE The capital value (present sale value) CV of property that can be rented on a perpetual basis for R dollars annually is approximated by the formula

$$CV \approx \int_0^\infty R e^{-it} \, dt$$

where i is the prevailing continuous interest rate.
a. Show that $CV = R/i$.
b. Find the capital value of property that can be rented at $10,000 annually when the prevailing continuous interest rate is 12%/per year.

51. Show that an integral of the form $\int_a^\infty e^{-px} \, dx$ is convergent if $p > 0$ and divergent if $p < 0$.

52. Show that an integral of the form $\int_{-\infty}^b e^{px} \, dx$ is convergent if $p > 0$ and divergent if $p < 0$.

SOLUTIONS TO SELF-CHECK EXERCISES 7.4

1. Write

$$\int_{-\infty}^\infty \frac{x^3}{(1+x^4)^{3/2}} \, dx = \int_{-\infty}^0 \frac{x^3}{(1+x^4)^{3/2}} \, dx + \int_0^\infty \frac{x^3}{(1+x^4)^{3/2}} \, dx$$

Now,

$$\int_{-\infty}^0 \frac{x^3}{(1+x^4)^{3/2}} \, dx = \lim_{a \to -\infty} \int_a^0 x^3 (1+x^4)^{-3/2} \, dx$$

$$= \lim_{a \to -\infty} \frac{1}{4}(-2)(1+x^4)^{-1/2} \Big|_a^0 \quad \begin{array}{l}\text{(Integrating}\\\text{by substitution)}\end{array}$$

$$= -\frac{1}{2} \lim_{a \to -\infty} \left[1 - \frac{1}{(1+a^4)^{1/2}} \right]$$

$$= -\frac{1}{2}$$

Similarly, you can show that

$$\int_0^\infty \frac{x^3}{(1+x^4)^{3/2}} \, dx = \frac{1}{2}$$

Therefore,

$$\int_{-\infty}^\infty \frac{x^3}{(1+x^4)^{3/2}} \, dx = -\frac{1}{2} + \frac{1}{2}$$

$$= 0$$

2. The required present value is given by

$$PV = \int_0^\infty 20 e^{-0.02t} e^{-0.10t} \, dt$$

$$= 20 \int_0^\infty e^{-0.12t} \, dt$$

$$= 20 \lim_{b \to \infty} \int_0^b e^{-0.12t} \, dt$$

$$= -\frac{20}{0.12} \lim_{b \to \infty} e^{-0.12t} \Big|_0^b$$

$$= -\frac{500}{3} \lim_{b \to \infty} (e^{-0.12b} - 1)$$

$$= \frac{500}{3}$$

or approximately $166,667.

7.5 Applications of Probability to Calculus

The systematic study of probability began in the seventeenth century, when certain aristocrats wanted to discover superior strategies to use in the gaming rooms of Europe. Some of the best mathematicians of the period were engaged in this pursuit. Since then, probability has evolved into an important branch of mathematics with widespread applications in virtually every sphere of human endeavor in which an element of uncertainty is present. In this section we take a brief look at the role of the integral in the study of probability theory.

PROBABILITY DENSITY FUNCTIONS

We begin by mentioning some elementary terms important in the study of probability. For our purpose, an **experiment** is an activity with observable results called **outcomes,** or **sample points.** The totality of all outcomes makes up the **sample space** of the experiment. A subset of the sample space is called an **event** of the experiment.

Now, given an event associated with an experiment, our primary objective is to determine the likelihood that this event will occur. This likelihood, or **probability of an event,** is a number between 0 and 1 and may be viewed as the proportionate number of times that the event will occur if the experiment associated with the event is repeated indefinitely under independent and similar conditions. As an example, let's consider the simple experiment of tossing an unbiased coin and observing whether it lands "heads" (H) or "tails" (T). Since the coin is *unbiased*, we see that the probability of each outcome is $\frac{1}{2}$, abbreviated

$$P(\mathrm{H}) = \frac{1}{2} \quad \text{and} \quad P(\mathrm{T}) = \frac{1}{2}$$

In many situations it is desirable to assign numerical values to the outcomes of an experiment. For example, suppose an experiment consists of casting a die and observing the face that lands uppermost. If we let X denote the outcome of the experiment, then X assumes one of the values 1, 2, 3, 4, 5, or 6. Because the values assumed by X depend on the outcomes of a chance experiment, the outcome X is referred to as a **random variable.** In this case, the random variable X is also said to be **finite discrete** since it can assume only a finite number of integer values.

A random variable x that can assume any value in an interval is called a **continuous random variable.** Examples of continuous random variables are the life span of a light bulb, the length of a telephone call, the length of an infant at birth, the daily amount of rainfall in Boston, and the life span of a certain plant species. For the remainder of this section, we will be interested primarily in continuous, random variables.

Consider an experiment in which the associated random variable x has the interval $[a, b]$ as its sample space. Then, an event of the experiment is any subset of $[a, b]$. For example, if x denotes the life span of a light bulb, then the sample space associated with the experiment is $[0, \infty)$, and the event that a light bulb selected at random has a life span between 500 and 600 hours, inclusive, is described by the interval $[500, 600]$ or, equivalently, by the inequality $500 \leq x \leq 600$. The probability that the light bulb will have a life span of between 500 and 600 hours is denoted by $P(500 \leq x \leq 600)$.

In general, we will be interested in computing $P(a \leq x \leq b)$, the probability that a random variable x assumes a value in the interval $a \leq x \leq b$. This computation is based on the notion of a probability density function, which we now introduce.

Probability Density Function

A **probability density function** of a random variable x in an interval I, where I may be bounded or unbounded, is a nonnegative function f having the following properties:

1. The total area of the region under the graph of f is equal to 1 (Figure 7.15a).
2. The probability that an observed value of the random variable x lies in the interval $[a, b]$ is given by

$$P(a \leq x \leq b) = \int_a^b f(x)\, dx \qquad \text{(Figure 7.15b)}$$

FIGURE 7.15

(a) Area of $R = 1$

(b) $P(a < x < b)$ is the probability that an outcome of an experiment will lie between a and b.

A few comments are in order. First, a probability density function of a random variable x may be constructed using methods that range from theoretical considerations of the problem on the one extreme to an interpretation of data associated with the experiment on the other. Second, Property 1 states that the probability that a continuous random variable takes on a value lying in its range is 1, a certainty, which is expected. Third, Property 2 states that the probability that the random variable x assumes a value in an interval $a \leq x \leq b$ is given by the area of the region between the graph of f and the x-axis from $x = a$ to $x = b$. Because the area under one point of the graph of f is equal to zero, we see immediately that $P(a \leq x \leq b) = P(a < x \leq b) = P(a \leq x < b) = P(a < x < b)$.

EXAMPLE 1   Show that each of the following functions satisfies the nonnegativity condition and Property 1 of probability density functions.

$$\textbf{a. } f(x) = \frac{2}{27}x(x-1) \qquad (1 \le x \le 4)$$

$$\textbf{b. } f(x) = \frac{1}{3}e^{(-1/3)x} \qquad (0 \le x < \infty)$$

SOLUTION ✔   **a.** Since the factors x and $(x-1)$ are both nonnegative, we see that $f(x) \ge 0$ on $[1, 4]$. Next, we compute

$$\int_1^4 \frac{2}{27}x(x-1)\, dx = \frac{2}{27}\int_1^4 (x^2 - x)\, dx$$

$$= \frac{2}{27}\left(\frac{1}{3}x^3 - \frac{1}{2}x^2\right)\Bigg|_1^4$$

$$= \frac{2}{27}\left[\left(\frac{64}{3} - 8\right) - \left(\frac{1}{3} - \frac{1}{2}\right)\right]$$

$$= \frac{2}{27}\left(\frac{27}{2}\right)$$

$$= 1$$

showing that Property 1 of probability density functions holds as well.
b. First, $f(x) = \frac{1}{3}e^{(-1/3)x} \ge 0$ for all values of x in $[0, \infty)$. Next,

$$\int_0^\infty \frac{1}{3}e^{(-1/3)x}\, dx = \lim_{b\to\infty}\int_0^b \frac{1}{3}e^{(-1/3)x}\, dx$$

$$= \lim_{b\to\infty} -e^{(-1/3)x}\Bigg|_0^b$$

$$= \lim_{b\to\infty}(-e^{(-1/3)b} + 1)$$

$$= 1$$

so the area under the graph of $f(x) = \frac{1}{3}e^{(-1/3)x}$ is equal to 1, as we set out to show.   ■ ■ ■ ■

EXAMPLE 2   **a.** Determine the value of the constant k so that the function $f(x) = kx^2$ is a probability density function on the interval $[0, 5]$.
b. If x is a continuous random variable with the probability density function given in part (a), compute the probability that x will assume a value between $x = 1$ and $x = 2$.

SOLUTION ✔   **a.** We compute

$$\int_0^5 kx^2\, dx = k \int_0^5 x^2\, dx$$

$$= \frac{k}{3} x^3 \bigg|_0^5$$

$$= \frac{125}{3} k$$

FIGURE 7.16

$P(1 \le x \le 2)$ for the probability density function $y = \dfrac{3}{125}\, x^2$

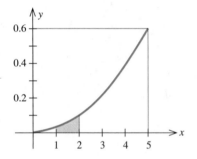

Since this value must be equal to 1, we find that $k = 3/125$.

b. The required probability is given by

$$P(1 \le x \le 2) = \int_1^2 f(x)\, dx = \int_1^2 \frac{3}{125} x^2\, dx$$

$$= \frac{1}{125} x^3 \bigg|_1^2 = \frac{1}{125}(8 - 1)$$

$$= \frac{7}{125}$$

The graph of the probability density function f and the area corresponding to the probability $P(1 \le x \le 2)$ are shown in Figure 7.16.   ■ ■ ■ ■

EXAMPLE 3   The TKK Products Corporation manufactures a 200-watt electric light bulb. Laboratory tests show that the life spans of these light bulbs have a distribution described by the probability density function

$$f(x) = .001e^{-.001x} \qquad (0 \le x < \infty)$$

Determine the probability that a light bulb will have a life span of (a) 500 hours or less, (b) more than 500 hours, and (c) more than 1000 hours but less than 1500 hours.

SOLUTION ✔   Let x denote the life span of a light bulb.

a. The probability that a light bulb will have a life span of 500 hours or less is given by

$$P(0 \le x \le 500) = \int_0^{500} .001e^{-.001x}\, dx$$

$$= -e^{-.001x} \bigg|_0^{500} = -e^{-.5} + 1$$

$$\approx .3935$$

b. The probability that a light bulb will have a life span of more than 500 hours is given by

$$P(x > 500) = \int_{500}^{\infty} .001e^{-.001x}\, dx$$

$$= \lim_{b \to \infty} \int_{500}^{b} .001e^{-.001x}\, dx$$

$$= \lim_{b \to \infty} -e^{-.001x}\Big|_{500}^{b}$$

$$= \lim_{b \to \infty} (-e^{-.001b} + e^{-.5})$$

$$= e^{-.5} \approx .6065$$

This result may also be obtained by observing that

$$P(x > 500) = 1 - P(x \le 500)$$

$$= 1 - .3935 \qquad \text{[Using the result from part (a)]}$$

$$\approx .6065$$

c. The probability that a light bulb will have a life span of more than 1000 hours but less than 1500 hours is given by

$$P(1000 < x < 1500) = \int_{1000}^{1500} .001e^{-.001x}\, dx$$

$$= -e^{-.001x}\Big|_{1000}^{1500}$$

$$= -e^{-1.5} + e^{-1}$$

$$\approx -.2231 + .3679$$

$$= .1448 \qquad\qquad ■ ■ ■ ■$$

The probability density function of Example 3 has the form

$$f(x) = ke^{-kx}$$

where $x \ge 0$ and k is a positive constant. Its graph is shown in Figure 7.17. Such a probability function is called an **exponential density function,** and a random variable associated with such a probability density function is said

FIGURE 7.17
The area under the graph of the exponential density function is equal to 1.

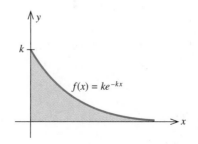

$$f(x) = ke^{-kx}$$

to be **exponentially distributed.** Exponential random variables are used to represent the life span of electronic components, the duration of telephone calls, the waiting time in a doctor's office, and the time between successive flight arrivals and departures in an airport, to mention but a few applications.

Another probability density function, and the one most widely used, is the **normal density function,** defined by

$$f(x) = \frac{1}{\sigma\sqrt{2\pi}}\, e^{-(1/2)[(x-\mu)/\sigma]^2}$$

where μ and σ are constants. The graph of the normal distribution is bell-shaped (Figure 7.18). Many phenomena, such as the heights of people in a given population, the weights of newborn infants, the IQs of college students, the actual weights of 16-ounce packages of cereals, and so on, have probability distributions that are normal.

FIGURE 7.18
The area under the bell-shaped normal distribution curve

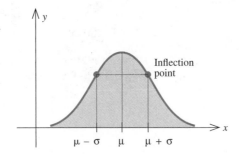

Areas under the standard normal curve (the normal curve with $\mu = 0$ and $\sigma = 1$) have been extensively computed and tabulated. Most problems involving the normal distribution can be solved with the aid of these tables.

EXPECTED VALUE

The average value, or expected value, of a discrete variable X that takes on values $x_1, x_2, \ldots, x_n$ with associated probabilities $p_1, p_2, \ldots, p_n$ is defined by

$$E(X) = x_1 p_1 + x_2 p_2 + \cdots + x_n p_n$$

If each of the values $x_1, x_2, \ldots, x_n$ occurs with equal frequency, then $p_1 = p_2 = \cdots = p_n = 1/n$ and

$$E(X) = x_1\left(\frac{1}{n}\right) + x_2\left(\frac{1}{n}\right) + \cdots + x_n\left(\frac{1}{n}\right)$$
$$= \frac{x_1 + x_2 + \cdots + x_n}{n}$$

giving the familiar formula for computing the average value of the n numbers $x_1, x_2, \ldots, x_n$.

Now, suppose x is a continuous random variable and f is the probability density function associated with it. For simplicity, let's first assume that $a \leq x \leq b$. Divide the interval $[a, b]$ into n subintervals of equal length $\Delta x = (b - a)/n$ by means of the $(n + 1)$ points $x_0 = a, x_1, x_2, \ldots, x_n = b$ (Figure 7.19). To find an approximation of the average value, or expected value, of

FIGURE 7.19

Approximating the expected value of a random variable x on $[a, b]$ by a Riemann sum

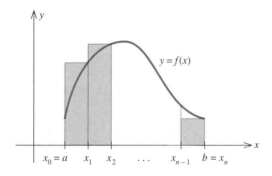

x on the interval $[a, b]$, let's treat x as if it were a discrete random variable that takes on the values $x_1, x_2, \ldots, x_n$ with probabilities $p_1, p_2, \ldots, p_n$. Then,

$$E(x) \approx x_1 p_1 + x_2 p_2 + \cdots + x_n p_n$$

But p_1 is the probability that x is in the interval $[x_0, x_1]$, and this is just the area under the graph of f from $x = x_0$ to $x = x_1$, which may be approximated by $f(x_1)\Delta x$. The probabilities $p_2, \ldots, p_n$ may be approximated in a similar manner. Thus,

$$E(x) \approx x_1 f(x_1)\Delta x + x_2 f(x_2)\Delta x + \cdots + x_n f(x_n)\Delta x$$

which is seen to be the Riemann sum of the function $g(x) = xf(x)$ over the interval $[a, b]$. Letting n approach infinity, we obtain the following formula:

Expected Value of a Continuous Random Variable

> Suppose the function f defined on the interval $[a, b]$ is the probability density function associated with a continuous random variable x. Then, the **expected value** of x is
>
> $$E(x) = \int_a^b xf(x)\, dx \qquad (17)$$

If either $a = -\infty$ or $b = \infty$, then the integral in (17) becomes an improper integral.

The expected value of a random variable plays an important role in many practical applications. For example, if x represents the life span of a certain brand of electronic components, then the expected value of x gives the average life span of these components. If x measures the waiting time in a doctor's office, then $E(x)$ gives the average waiting time, and so on.

EXAMPLE **4**

Show that if a continuous random variable x is exponentially distributed with the probability density function

$$f(x) = ke^{-kx} \qquad (0 \leq x < \infty)$$

then the expected value $E(x)$ is equal to $1/k$. Using this result, determine the average life span of a 200-watt electric light bulb manufactured by the TKK Products Corporation of Example 3.

SOLUTION ✔

We compute

$$E(x) = \int_0^\infty xf(x)\, dx$$

$$= \int_0^\infty kxe^{-kx}\, dx$$

$$= k \lim_{b \to \infty} \int_0^b xe^{-kx}\, dx$$

Integrating by parts with

$$u = x \qquad \text{and} \qquad dv = e^{-kx}\, dx$$

so that

$$du = dx \qquad \text{and} \qquad v = -\frac{1}{k}e^{-kx}$$

we have

$$E(x) = k \lim_{b \to \infty} \left[-\frac{1}{k}xe^{-kx}\Big|_0^b + \frac{1}{k}\int_0^b e^{-kx}\, dx \right]$$

$$= k \lim_{b \to \infty} \left[-\left(\frac{1}{k}\right)be^{-kb} - \frac{1}{k^2}e^{-kx}\Big|_0^b \right]$$

$$= k \lim_{b \to \infty} \left[-\left(\frac{1}{k}\right)be^{-kb} - \frac{1}{k^2}e^{-kb} + \frac{1}{k^2} \right]$$

$$= -\lim_{b \to \infty} \frac{b}{e^{kb}} - \frac{1}{k}\lim_{b \to \infty} \frac{1}{e^{kb}} + \frac{1}{k}\lim_{b \to \infty} 1$$

Now, by taking a sequence of values of b that approaches infinity—for example, $b = 10, 100, 1000, 10{,}000, \ldots$—we see that, for a fixed k,

$$\lim_{b \to \infty} \frac{b}{e^{kb}} = 0$$

Therefore,

$$E(x) = \frac{1}{k}$$

as we set out to show. Next, since $k = .001$ in Example 3, we see that the average life span of the TKK light bulb is $1/(.001) = 1000$ hours.   ■ ■ ■ ■

Before considering another example, let's summarize the important result obtained in Example 4.

**The Expected Value
of an Exponential
Density Function**

If a continuous random variable x is exponentially distributed with probability density function

$$f(x) = ke^{-kx} \ (0 \le x < \infty)$$

then the **expected (average) value** of x is given by

$$E = \frac{1}{k}$$

EXAMPLE 5

On a typical Monday morning, the time between successive arrivals of planes at Jackson International Airport is an exponentially distributed random variable x with expected value of 10 (minutes).

a. Find the probability density function associated with x.
b. What is the probability that between 6 and 8 minutes will elapse between successive arrivals of planes?
c. What is the probability that the time between successive arrivals of planes will be more than 15 minutes?

SOLUTION ✔

a. Since x is exponentially distributed, the associated probability density function has the form $f(x) = ke^{-kx}$. Next, since the expected value of x is 10, we see that

$$E(x) = \frac{1}{k}$$

$$k = \frac{1}{10}$$

$$= .1$$

so the required probability density function is

$$f(x) = .1e^{-.1x}$$

b. The probability that between 6 and 8 minutes will elapse between successive arrivals is given by

$$P(6 \le x \le 8) = \int_6^8 .1e^{-.1x}\,dx = -e^{-.1x}\Big|_6^8$$

$$= -e^{-.8} + e^{-.6}$$

$$\approx .10$$

c. The probability that the time between successive arrivals will be more than 15 minutes is given by

$$P(x > 15) = \int_{15}^{\infty} .1e^{-.1x}\, dx$$

$$= \lim_{b \to \infty} \int_{15}^{b} .1e^{-.1x}\, dx$$

$$= \lim_{b \to \infty} \left[-e^{-.1x} \Big|_{15}^{b} \right]$$

$$= \lim_{b \to \infty} (-e^{-.1b} + e^{-1.5}) = e^{-1.5}$$

$$\approx .22 \qquad\qquad ■■■■$$

SELF-CHECK EXERCISES 7.5

1. Determine the value of the constant k such that the function $f(x) = k(4x - x^2)$ is a probability density function on the interval $[0, 4]$.

2. Suppose x is a continuous random variable with the probability density function of Self-Check Exercise 1. Find the probability that x will assume a value between $x = 1$ and $x = 3$.

Solutions to Self-Check Exercises 7.5 can be found on page 596.

7.5 Exercises

In Exercises 1–8, show that the given function is a probability density function on the specified interval.

1. $f(x) = \dfrac{2}{32}x;\ (2 \le x \le 6)$

2. $f(x) = \dfrac{2}{9}(3x - x^2);\ (0 \le x \le 3)$

3. $f(x) = \dfrac{3}{8}x^2;\ (0 \le x \le 2)$

4. $f(x) = \dfrac{3}{32}(x - 1)(5 - x);\ (1 \le x \le 5)$

5. $f(x) = 20(x^3 - x^4);\ (0 \le x \le 1)$

6. $f(x) = \dfrac{8}{7x^2};\ (1 \le x \le 8)$

7. $f(x) = \dfrac{3}{14}\sqrt{x};\ (1 \le x \le 4)$

8. $f(x) = \dfrac{12 - x}{72};\ (0 \le x \le 12)$

9. a. Determine the value of the constant k so that the function $f(x) = k(4 - x)$ is a probability density function on the interval $[0, 4]$.
  b. If x is a continuous random variable with the probability density function given in part (a), compute the probability that x will assume a value between $x = 1$ and $x = 3$.

10. a. Determine the value of the constant k so that the function $f(x) = k/x^2$ is a probability density function on the interval $[1, 10]$.
  b. If x is a continuous random variable with the probability density function of part (a), compute the probability that x will assume a value between $x = 2$ and $x = 6$.

11. a. Determine the value of the constant k so that the function $f(x) = kxe^{-2x^2}$ is a probability density function on the interval $[0, \infty)$.
b. If x is a continuous random variable with the probability density function of part (a), find the probability that x will assume a value greater than 1.

12. LIFE SPAN OF A PLANT SPECIES The life span of a certain plant species (in days) is described by the probability density function

$$f(x) = \frac{1}{100} e^{-x/100} \qquad (0 \le x < \infty)$$

a. Find the probability that a plant of this species will live for 100 days or less.
b. Find the probability that a plant of this species will live more than 120 days.
c. Find the probability that a plant of this species will live more than 60 days but less than 140 days.

13. AVERAGE WAITING TIME FOR PATIENTS The average waiting time for patients arriving at the Newtown Health Clinic between 1 P.M. and 4 P.M. on a weekday is an exponentially distributed random variable x with expected value of 15 min.
a. Find the probability density function associated with x.
b. What is the probability that a patient arriving at the clinic between 1 P.M. and 4 P.M. will have to wait between 10 and 12 min?
c. What is the probability that a patient arriving at the clinic between 1 P.M. and 4 P.M. will have to wait more than 15 min?

14. FREQUENCY OF ROAD REPAIRS The number of streets in the downtown section of a certain city that need repairs in a given year is a random variable with a distribution described by the probability density function

$$f(x) = 12x^2(1 - x) \qquad (0 \le x \le 1)$$

Find the probability that at most half of the streets will need repairs in any given year.

15. EXPECTED NUMBER OF CHOCOLATE CHIPS The number of chocolate chips in each cookie of a certain brand has a distribution described by the probability density function

$$f(x) = \frac{1}{36}(6x - x^2) \qquad (0 \le x \le 6)$$

Find the expected number of chips in each cookie.

16. RELIABILITY OF ROBOTS The National Welding Company uses industrial robots in some of its assembly line operations. Management has determined that, on average, a robot breaks down after 1000 hr of use and that the lengths of time between breakdowns are exponentially distributed.
a. What is the probability that a robot selected at random will break down between 600 and 800 hr of use?
b. What is the probability that a robot will break down after 1200 hr of use?

17. EXPRESSWAY TOLLBOOTHS Suppose the time intervals between arrivals of successive cars at an expressway tollbooth during rush hour are exponentially distributed and that the average time interval between arrivals is 8 sec. Find the probability that the average time interval between arrivals of successive cars is more than 8 sec.

18. TIME INTERVALS BETWEEN PHONE CALLS A study conducted by Uni-Mart, a mail-order department store, reveals that the time intervals between incoming telephone calls on its toll-free 800 line between 10 A.M. and 2 P.M. are exponentially distributed and that the average time interval is 30 sec. What is the probability that the time interval between successive calls is more than 2 min?

19. RELIABILITY OF MICROPROCESSORS The microprocessors manufactured by the United Motor Works Company, which are used in automobiles to regulate fuel consumption, are guaranteed against defects for 20,000 mi of use. Tests conducted in the laboratory under simulated driving conditions reveal that the distances driven before the microprocessors break down are exponentially distributed and that the average distance driven before the microprocessors fail is 100,000 mi. What is the probability that a microprocessor selected at random will fail during the warranty period?

In Exercises 20–22, determine whether the statement is true or false. If it is true, explain why it is true. If it is false, give an example to show why it is false.

20. If $\int_a^b f(x)\,dx = 1$, then f is a probability density function on $[a, b]$.

21. If f is a probability function on an interval $[a, b]$, then f is a probability function on $[c, d]$ for any real numbers c and d satisfying $a < c < d < b$.

22. If f is a probability density function of a continuous random variable x in the interval $[a, b]$, then the expected value of x is given by $\int_a^b x^2 f(x)\,dx$.

SOLUTIONS TO SELF-CHECK EXERCISES 7.5

1. We compute

$$\int_0^4 k(4x - x^2)\,dx = k\left(2x^2 - \frac{1}{3}x^3\right)\bigg|_0^4$$

$$= k\left(32 - \frac{64}{3}\right)$$

$$= \frac{32}{3}k$$

Since this value must be equal to 1, we see that $k = 3/32$.

2. The required probability is given by

$$P(1 \le x \le 3) = \int_1^3 f(x)\,dx$$

$$= \int_1^3 \frac{3}{32}(4x - x^2)\,dx$$

$$= \frac{3}{32}\left(2x^2 - \frac{1}{3}x^3\right)\bigg|_1^3$$

$$= \frac{3}{32}\left[(18 - 9) - \left(2 - \frac{1}{3}\right)\right] = \frac{11}{16}$$

CHAPTER 7 Summary of Principal Formulas and Terms

Formulas

1. Integration by parts $\int u\,dv = uv - \int v\,du$

2. Improper integral of f over $[a, \infty)$ $\int_a^\infty f(x)\,dx = \lim\limits_{b \to \infty} \int_a^b f(x)\,dx$

3. Improper integral of f over $(-\infty, b]$ $\int_{-\infty}^b f(x)\,dx = \lim\limits_{a \to -\infty} \int_a^b f(x)\,dx$

4. Improper integral of f over $(-\infty, \infty)$ $\int_{-\infty}^\infty f(x)\,dx = \int_{-\infty}^c f(x)\,dx$

$$+ \int_c^\infty f(x)\,dx$$

5. Present value of a perpetuity $PV = \dfrac{mP}{r}$

6. Trapezoidal rule

$$\int_a^b f(x)\,dx \approx \frac{\Delta x}{2}[f(x_0) + 2f(x_1)$$
$$+ 2f(x_2) + \cdots$$
$$+ 2f(x_{n-1}) + f(x_n)]$$

where $\Delta x = \dfrac{b-a}{n}$

7. Simpson's rule

$$\int_a^b f(x)\,dx \approx \frac{\Delta x}{3}[f(x_0) + 4f(x_1)$$
$$+ 2f(x_2) + 4f(x_3)$$
$$+ 2f(x_4) + \cdots$$
$$+ 4f(x_{n-1}) + f(x_n)]$$

where $\Delta x = \dfrac{b-a}{n}$

8. Maximum error for trapezoidal rule

$\dfrac{M(b-a)^3}{12n^2}$, where $|f''(x)| \le M$
$(a \le x \le b)$

9. Maximum error for Simpson's rule

$\dfrac{M(b-a)^5}{180n^4}$, where $|f^{(4)}(x)| \le M$
$(a \le x \le b)$

10. Probability an outcome of an experiment lies between a and b

$P(a \le x \le b) = \int_a^b f(x)\,dx$

11. Exponential density function

$f(x) = ke^{-kx}$

12. Expected value

$E(x) = x_1p_1 + x_2p_2 + \cdots + x_np_n$

13. Expected value of a continuous random variable

$E(x) = \int_a^b xf(x)\,dx$

Terms

improper integral
convergent integral
divergent integral
perpetuity
experiment
outcome
sample space
event

probability of an event
random variable
finite discrete random variable
continuous random variable
probability density function
exponential density function
normal density function
expected value

CHAPTER 7 REVIEW EXERCISES

In Exercises 1-6, evaluate the integral.

1. $\int 2xe^{-x}\,dx$ **2.** $\int xe^{4x}\,dx$

3. $\int \ln 5x\,dx$ **4.** $\int_1^4 \ln 2x\,dx$

5. $\int_0^1 xe^{-2x}\,dx$ **6.** $\int_0^2 xe^{2x}\,dx$

7. Find the function f given that the slope of the tangent line to the graph of f at any point $(x, f(x))$ is

$$f'(x) = \frac{\ln x}{\sqrt{x}}$$

and that the graph of f passes through the point $(1, -2)$.

8. Find the function f given that the slope of the tangent line to the graph of f at any point $(x, f(x))$ is

$$f'(x) = \frac{e^x}{1 + e^x}$$

and that the graph of f passes through the point $(0, 0)$.

In Exercises 9-14, use the table of integrals in Section 7.2 to evaluate the integral.

9. $\int \frac{x^2\,dx}{(3 + 2x)^2}$ **10.** $\int \frac{2x}{\sqrt{2x + 3}}\,dx$

11. $\int x^2 e^{4x}\,dx$ **12.** $\int \frac{dx}{(x^2 - 25)^{3/2}}$

13. $\int \frac{dx}{x^2\sqrt{x^2 - 4}}$ **14.** $\int 8x^3 \ln 2x\,dx$

In Exercises 15-20, evaluate each improper integral whenever it is convergent.

15. $\int_0^\infty e^{-2x}\,dx$ **16.** $\int_{-\infty}^0 e^{3x}\,dx$

17. $\int_3^\infty \frac{2}{x}\,dx$ **18.** $\int_2^\infty \frac{1}{(x + 2)^{3/2}}\,dx$

19. $\int_2^\infty \frac{dx}{(1 + 2x)^2}$ **20.** $\int_1^\infty 3e^{1-x}\,dx$

In Exercises 21-24, use the trapezoidal rule and Simpson's rule to approximate the value of the definite integral.

21. $\int_1^3 \frac{dx}{1 + \sqrt{x}}; n = 4$ **22.** $\int_0^1 e^{x^2}\,dx; n = 4$

23. $\int_{-1}^1 \sqrt{1 + x^4}\,dx; n = 4$ **24.** $\int_1^3 \frac{e^x}{x}\,dx; n = 4$

25. Show that the function $f(x) = (3/128)(16 - x^2)$ is a probability density function on the interval $[0, 4]$.

26. Show that the function $f(x) = (1/9)x\sqrt{9 - x^2}$ is a probability density function on the interval $[0, 3]$.

27. a. Determine the value of the constant k such that the function $f(x) = kx\sqrt{4 - x^2}$ is a probability density function on the interval $[0, 2]$.
b. If x is a continuous random variable with the probability density function given in part (a), compute the probability that x will assume a value between $x = 1$ and $x = 2$.

28. a. Determine the value of the constant k such that the function $f(x) = k/\sqrt{x}$ is a probability density function on the interval $[1, 4]$.
b. If x is a continuous random variable with the probability density function given in part (a), compute the probability that x will assume a value between $x = 2$ and $x = 3$.

29. a. Determine the value of the constant k such that the function $f(x) = kx^2(3 - x)$ is a probability density function on the interval $[0, 3]$.
b. If x is a continuous random variable with the probability density function given in part (a), compute the probability that x will assume a value between $x = 1$ and $x = 2$.

30. Records at the Centerville Hospital indicate that the length of time in days that a maternity patient stays in the hospital has a probability density function given by

$$P(t) = \frac{1}{4}e^{(-1/4)t}$$

a. What is the probability that a woman entering the maternity wing will be there more than 6 days?
b. What is the probability that a woman entering the maternity wing will be there less than 2 days?
c. What is the average length of time that a woman entering the maternity wing stays in the hospital?

31. The supply equation for the GTC Slim-Phone is given by

$$p = 2\sqrt{25 + x^2}$$

where p is the unit price in dollars and x is the quantity demanded per month in units of 10,000. Find the producers' surplus if the market price is $26. Use the table of integrals in Section 7.2 to evaluate the definite integral.

32. The sales of the Starr Communication Company's newest computer game, Laser Beams, are currently

$$te^{-0.05t}$$

units/month (each unit representing 1000 games), where t denotes the number of months since the release of the game. Find an expression that gives the total number of games sold as a function of t. How many games will be sold by the end of the first year?

33. The demand equation for a computer software program is given by

$$p = 2\sqrt{325 - x^2}$$

where p is the unit price in dollars and x is the quantity demanded per month in units of a thousand. Find the consumers' surplus if the market price is $30. Evaluate the definite integral using Simpson's rule with $n = 10$.

34. Using aerial photographs, the Coast Guard was able to determine the dimensions of an oil spill along an embankment on a coastline, as shown in the accompanying figure. Using (a) the trapezoidal rule and (b) Simpson's rule with $n = 10$, estimate the area of the oil spill.

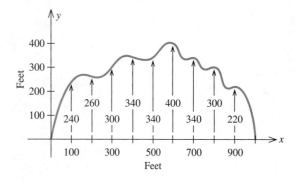

35. Lindsey wishes to establish a memorial fund at the Newtown Hospital in the amount of $10,000/year beginning next year. If the fund earns interest at a rate of 9%/year compounded continuously, find the amount of endowment that he is required to make now.

CALCULUS OF SEVERAL VARIABLES

8

Up to now we have dealt with functions involving one variable. In many real-life situations, however, we encounter quantities that depend on two or more quantities. For example, the Consumer Price Index (CPI) compiled every month by the Bureau of Labor Statistics depends on the price of more than 95,000 consumer items from gas to groceries. To study such relationships, we need the notion of a function of several variables, the first topic in this chapter. Next, generalizing the concept of the derivative of a function of one variable, we study the *partial derivatives* of a function of two or more variables. Partial derivatives enable us to study the rate of change of a function with respect to one variable while holding all other variables constant. We then learn how to find the extremum values of a function of several variables. As an application of optimization theory, we learn how to find an equation of the straight line that "best" fits a set of data points scattered about a straight line. Finally, we generalize the notion of the integral to the case involving a function of two variables.

What should the dimensions of the swimming pool be? The operators of the *Viking Princess*, a luxury cruise ship, are thinking about adding another swimming pool to the *Princess*. The chief engineer has suggested that an area in the form of an ellipse, located in the rear of the promenade deck, would be suitable for this purpose. Subject to this constraint, what are the dimensions of the largest pool that can be built? See Example 5, page 661, to see how to solve this problem.

8.1 Functions of Several Variables

Up to now, our study of calculus has been restricted to functions of one variable. In many practical situations, however, the formulation of a problem results in a mathematical model that involves a function of two or more variables. For example, suppose the Ace Novelty Company determines that the profits are $6, $5, and $4 for three types of souvenirs it produces. Let x, y, and z denote the number of type-A, type-B, and type-C souvenirs to be made; then the company's profit is given by

$$P = 6x + 5y + 4z$$

and P is a function of the three variables x, y, and z.

FUNCTIONS OF TWO VARIABLES

Although this chapter deals with real-valued functions of several variables, most of our definitions and results are stated in terms of a function of two variables. One reason for adopting this approach, as you will soon see, is that there is a geometric interpretation for this special case, which serves as an important visual aid. We can then draw upon the experience gained from studying the two-variable case to help us understand the concepts and results connected with the more general case, which, by and large, is just a simple extension of the lower-dimensional case.

A Function of Two Variables

A **real-valued function of two variables,** f, consists of

1. A set A of ordered pairs of real numbers (x, y) called the **domain** of the function.
2. A rule that associates with each ordered pair in the domain of f one and only one real number, denoted by $z = f(x, y)$.

The variables x and y are called **independent variables,** and the variable z, which is dependent on the values of x and y, is referred to as a **dependent variable.**

As in the case of a real-valued function of one real variable, the number $z = f(x, y)$ is called the **value of f** at the point (x, y). And, unless specified, the domain of the function f will be taken to be the largest possible set for which the rule defining f is meaningful.

EXAMPLE 1   Let f be the function defined by

$$f(x, y) = x + xy + y^2 + 2$$

Compute $f(0, 0)$, $f(1, 2)$, and $f(2, 1)$.

SOLUTION ✔   We have

$$f(0, 0) = 0 + (0)(0) + 0^2 + 2 = 2$$
$$f(1, 2) = 1 + (1)(2) + 2^2 + 2 = 9$$
$$f(2, 1) = 2 + (2)(1) + 1^2 + 2 = 7$$ ■ ■ ■ ■

The domain of a function of two variables $f(x, y)$ is a set of ordered pairs of real numbers and may therefore be viewed as a subset of the xy-plane.

EXAMPLE **2**   Find the domain of each of the following functions.

a. $f(x, y) = x^2 + y^2$ **b.** $g(x, y) = \dfrac{2}{x - y}$

c. $h(x, y) = \sqrt{1 - x^2 - y^2}$

◦ SOLUTION ✔   **a.** $f(x, y)$ is defined for all real values of x and y, so the domain of the function f is the set of all points (x, y) in the xy-plane.

b. $g(x, y)$ is defined for all $x \neq y$, so the domain of the function g is the set of all points in the xy-plane except those lying on the line $y = x$ (Figure 8.1a).

c. We require that $1 - x^2 - y^2 \geq 0$ or $x^2 + y^2 \leq 1$, which is just the set of all points (x, y) lying on and inside the circle of radius 1 with center at the origin (Figure 8.1b).

FIGURE **8.1**

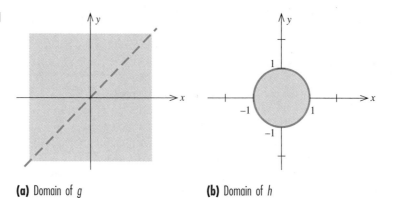

(a) Domain of g **(b)** Domain of h ■ ■ ■ ■

APPLICATIONS

EXAMPLE **3**   The Acrosonic Company manufactures a bookshelf loudspeaker system that may be bought fully assembled or in a kit. The demand equations that relate the unit prices, p and q, to the quantities demanded weekly, x and y, of the assembled and kit versions of the loudspeaker systems are given by

$$p = 300 - \frac{1}{4}x - \frac{1}{8}y \quad \text{and} \quad q = 240 - \frac{1}{8}x - \frac{3}{8}y$$

a. What is the weekly total revenue function $R(x, y)$?

b. What is the domain of the function R?

SOLUTION ✔   **a.** The weekly revenue realizable from the sale of x units of the assembled speaker systems at p dollars per unit is given by xp dollars. Similarly, the weekly revenue realizable from the sale of y units of the kits at q dollars per unit is given by yq dollars. Therefore, the weekly total revenue function R is given by

$$R(x, y) = xp + yq$$

$$= x\left(300 - \frac{1}{4}x - \frac{1}{8}y\right) + y\left(240 - \frac{1}{8}x - \frac{3}{8}y\right)$$

$$= -\frac{1}{4}x^2 - \frac{3}{8}y^2 - \frac{1}{4}xy + 300x + 240y$$

FIGURE 8.2

The domain of $R(x, y)$

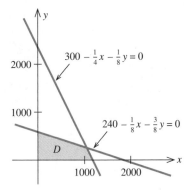

b. To find the domain of the function R, let's observe that the quantities x, y, p, and q must be nonnegative. This observation leads to the following system of linear inequalities:

$$300 - \frac{1}{4}x - \frac{1}{8}y \geq 0$$

$$240 - \frac{1}{8}x - \frac{3}{8}y \geq 0$$

$$x \geq 0$$

$$y \geq 0$$

The domain of the function R is sketched in Figure 8.2.   ■ ■ ■ ■

Group Discussion

Suppose the total profit of a two-product company is given by $P(x, y)$, where x denotes the number of units of the first product produced and sold and y denotes the number of units of the second product produced and sold. Fix $x = a$, where a is a positive number so that (a, y) is in the domain of P. Describe and give an economic interpretation of the function $f(y) = P(a, y)$. Next, fix $y = b$, where b is a positive number so that (x, b) is in the domain of P. Describe and give an economic interpretation of the function $g(x) = P(x, b)$.

EXAMPLE 4   The monthly payment that amortizes a loan of A dollars in t years when the interest rate is r per year is given by

$$P = f(A, r, t) = \frac{Ar}{12\left[1 - \left(1 + \dfrac{r}{12}\right)^{-12t}\right]}$$

Find the monthly payment for a home mortgage of \$90,000 to be amortized over 30 years when the interest rate is 10% per year.

SOLUTION ✔ Letting $A = 90{,}000$, $r = 0.1$, and $t = 30$, we find the required monthly payment to be

$$P = f(90{,}000, 0.1, 30) = \frac{90{,}000(0.1)}{12\left[1 - \left(1 + \dfrac{0.1}{12}\right)^{-360}\right]}$$

$$\approx 789.81$$

or approximately \$789.81.   ■ ■ ■ ■

FIGURE 8.3
The three-dimensional Cartesian coordinate system

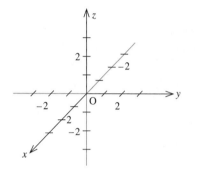

GRAPHS OF FUNCTIONS OF TWO VARIABLES

To graph a function of two variables, we need a three-dimensional coordinate system. This is readily constructed by adding a third axis to the plane Cartesian coordinate system in such a way that the three resulting axes are mutually perpendicular and intersect at O. Observe that, by construction, the zeros of the three number scales coincide at the origin of the **three-dimensional Cartesian coordinate system** (Figure 8.3).

FIGURE 8.4

 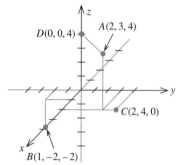

(a) A point in three-dimensional space

(b) Some sample points in three-dimensional space

FIGURE 8.5
The graph of a function in three-dimensional space

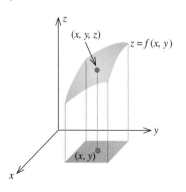

A point in three-dimensional space can now be represented uniquely in this coordinate system by an **ordered triple** of numbers (x, y, z), and, conversely, every ordered triple of real numbers (x, y, z) represents a point in three-dimensional space (Figure 8.4a). For example, the points $A(2, 3, 4)$, $B(1, -2, -2)$, $C(2, 4, 0)$, and $D(0, 0, 4)$ are shown in Figure 8.4b.

Now, if $f(x, y)$ is a function of two variables x and y, the domain of f is a subset of the xy-plane. Let $z = f(x, y)$ so that there is one and only one point $(x, y, z) \equiv (x, y, f(x, y))$ associated with each point (x, y) in the domain of f. The totality of all such points makes up the **graph** of the function f and is, except for certain degenerate cases, a surface in three-dimensional space (Figure 8.5).

In interpreting the graph of a function $f(x, y)$, one often thinks of the value $z = f(x, y)$ of the function at the point (x, y) as the "height" of the point (x, y, z) on the graph of f. If $f(x, y) > 0$, then the point (x, y, z) is

$f(x, y)$ units above the xy-plane; if $f(x, y) < 0$, then the point (x, y, z) is $|f(x, y)|$ units below the xy-plane.

In general, it is quite difficult to draw the graph of a function of two variables. But techniques have been developed that enable us to generate such graphs with minimum effort, using a computer. Figure 8.6 shows the computer-generated graphs of two functions.

FIGURE 8.6
Two computer-generated graphs of functions of two variables

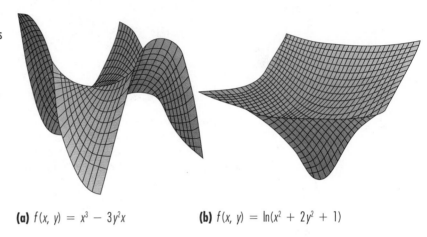

(a) $f(x, y) = x^3 - 3y^2x$ **(b)** $f(x, y) = \ln(x^2 + 2y^2 + 1)$

LEVEL CURVES

As mentioned earlier, the graph of a function of two variables is often difficult to sketch, and we will not develop a systematic procedure for sketching it. Instead, we describe a method that is used in constructing topographic maps. This method is relatively easy to apply and conveys sufficient information to enable one to obtain a feel for the graph of the function.

Suppose that $f(x, y)$ is a function of two variables x and y, with a graph as shown in Figure 8.7. If c is some value of the function f, then the equation $f(x, y) = c$ describes a curve lying on the plane $z = c$ called the **trace** of the

FIGURE 8.7
The graph of the function $z = f(x, y)$ and its intersection with the plane $z = c$

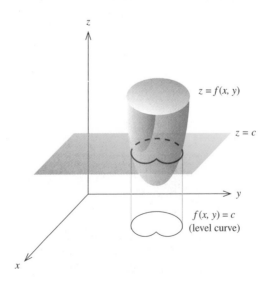

graph of f in the plane $z = c$. If this trace is projected onto the xy-plane, the resulting curve in the xy-plane is called a **level curve.** By drawing the level curves corresponding to several admissible values of c, we obtain a **contour map.** Observe that, by construction, every point on a particular level curve corresponds to a point on the surface $z = f(x, y)$ that is a certain fixed distance from the xy-plane. Thus, by elevating or depressing the level curves that make up the contour map in one's mind, it is possible to obtain a feel for the general shape of the surface represented by the function f. Figure 8.8a shows a part of a mountain range with one peak; Figure 8.8b is the associated contour map.

FIGURE 8.8

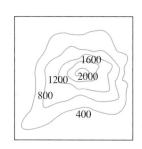

(a) A peak on a mountain range

(b) A contour map for the mountain peak

EXAMPLE 5 Sketch a contour map for the function $f(x, y) = x^2 + y^2$.

SOLUTION ✔ The level curves are the graphs of the equation $x^2 + y^2 = c$ for nonnegative numbers c. Taking $c = 0, 1, 4, 9$, and 16, for example, we obtain

$$c = \ 0: x^2 + y^2 = 0$$
$$c = \ 1: x^2 + y^2 = 1$$
$$c = \ 4: x^2 + y^2 = 4 = 2^2$$
$$c = \ 9: x^2 + y^2 = 9 = 3^2$$
$$c = 16: x^2 + y^2 = 16 = 4^2$$

FIGURE 8.9

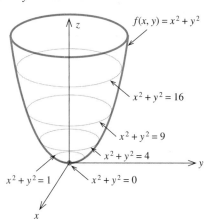

(a) Level curves of $f(x, y) = x^2 + y^2$

(b) The graph of $f(x, y) = x^2 + y^2$

The five level curves are concentric circles with center at the origin and radius given by $r = 0, 1, 2, 3,$ and 4, respectively (Figure 8.9a). A sketch of the graph of $f(x, y) = x^2 + y^2$ is included for your reference in Figure 8.9b.   ■ ■ ■ ■

EXAMPLE 6   Sketch the level curves for the function $f(x, y) = 2x^2 - y$ corresponding to $z = -2, -1, 0, 1,$ and 2.

SOLUTION ✔   The level curves are the graphs of the equation $2x^2 - y = k$ or $y = 2x^2 - k$ for $k = -2, -1, 0, 1,$ and 2. The required level curves are shown in Figure 8.10.   ■ ■ ■ ■

FIGURE 8.10
Level curves for $f(x, y) = 2x^2 - y$

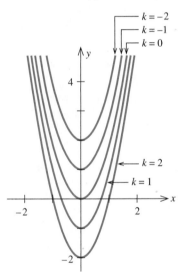

- $k = -2$
- $k = -1$
- $k = 0$
- $k = 2$
- $k = 1$

Level curves of functions of two variables are found in many practical applications. For example, if $f(x, y)$ denotes the temperature at a location within the continental United States with longitude x and latitude y at a certain time of day, then the temperature at the point (x, y) is given by the "height" of the surface, represented by $z = f(x, y)$. In this situation the level curve $f(x, y) = k$ is a curve superimposed on a map of the United States, connecting points having the same temperature at a given time (Figure 8.11). These level curves are called **isotherms.**

Similarly, if $f(x, y)$ gives the barometric pressure at the location (x, y), then the level curves of the function f are called **isobars,** lines connecting points having the same barometric pressure at a given time.

As a final example, suppose $P(x, y, z)$ is a function of three variables x, y, and z giving the profit realized when x, y, and z units of three products A, B, and C, respectively, are produced and sold. Then, the equation $P(x, y, z) = k$, where k is a constant, represents a surface in three-dimensional space called a **level surface** of P. In this situation, the level surface represented by $P(x, y, z) = k$ represents the product mix that results in a profit of exactly k dollars. Such a level surface is called an **isoprofit surface.**

FIGURE 8.11
Isotherms: curves connecting points that have the same temperature

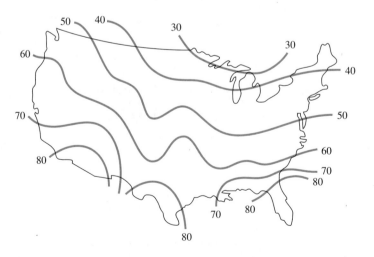

SELF-CHECK EXERCISES 8.1

1. Let $f(x, y) = x^2 - 3xy + \sqrt{x + y}$. Compute $f(1, 3)$ and $f(-1, 1)$. Is the point $(-1, 0)$ in the domain of f?

2. Find the domain of $f(x, y) = \dfrac{1}{x} + \dfrac{1}{x - y} - e^{x+y}$.

 3. The Odyssey Travel Agency has a monthly advertising budget of $20,000. Odyssey's management estimates that if they spend x dollars on newspaper advertising and y dollars on television advertising, then the monthly revenue will be

$$f(x, y) = 30x^{1/4}y^{3/4}$$

dollars. What will be the monthly revenue if Odyssey spends $5000/month on newspaper ads and $15,000/month on television ads? If Odyssey spends $4000/month on newspaper ads and $16,000/month on television ads?

Solutions to Self-Check Exercises 8.1 can be found on page 612.

8.1 Exercises

1. Let $f(x, y) = 2x + 3y - 4$. Compute $f(0, 0)$, $f(1, 0)$, $f(0, 1)$, $f(1, 2)$, and $f(2, -1)$.

2. Let $g(x, y) = 2x^2 - y^2$. Compute $g(1, 2)$, $g(2, 1)$, $g(1, 1)$, $g(-1, 1)$, and $g(2, -1)$.

3. Let $f(x, y) = x^2 + 2xy - x + 3$. Compute $f(1, 2)$, $f(2, 1)$, $f(-1, 2)$, and $f(2, -1)$.

4. Let $h(x, y) = (x + y)/(x - y)$. Compute $h(0, 1)$, $h(-1, 1)$, $h(2, 1)$, and $h(\pi, -\pi)$.

5. Let $g(s, t) = 3s\sqrt{t} + t\sqrt{s} + 2$. Compute $g(1, 2)$, $g(2, 1)$, $g(0, 4)$, and $g(4, 9)$.

6. Let $f(x, y) = xye^{x^2+y^2}$. Compute $f(0, 0)$, $f(0, 1)$, $f(1, 1)$, and $f(-1, -1)$.

7. Let $h(s, t) = s \ln t - t \ln s$. Compute $h(1, e)$, $h(e, 1)$, and $h(e, e)$.

8. Let $f(u, v) = (u^2 + v^2)e^{uv^2}$. Compute $f(0, 1)$, $f(-1, -1)$, $f(a, b)$, and $f(b, a)$.

9. Let $g(r, s, t) = re^{s/t}$. Compute $g(1, 1, 1)$, $g(1, 0, 1)$, and $g(-1, -1, -1)$.

10. Let $g(u, v, w) = (ue^{vw} + ve^{uw} + we^{uv})/(u^2 + v^2 + w^2)$. Compute $g(1, 2, 3)$ and $g(3, 2, 1)$.

In Exercises 11–18, find the domain of the function.

11. $f(x, y) = 2x + 3y$

12. $g(x, y, z) = x^2 + y^2 + z^2$

13. $h(u, v) = \dfrac{uv}{u - v}$ **14.** $f(s, t) = \sqrt{s^2 + t^2}$

15. $g(r, s) = \sqrt{rs}$ **16.** $f(x, y) = e^{-xy}$

17. $h(x, y) = \ln(x + y - 5)$

18. $h(u, v) = \sqrt{4 - u^2 - v^2}$

In Exercises 19–24, sketch the level curves of the function corresponding to the given values of z.

19. $f(x, y) = 2x + 3y$; $z = -2, -1, 0, 1, 2$

20. $f(x, y) = -x^2 + y$; $z = -2, -1, 0, 1, 2$

21. $f(x, y) = 2x^2 + y$; $z = -2, -1, 0, 1, 2$

22. $f(x, y) = xy$; $z = -4, -2, 2, 4$

23. $f(x, y) = \sqrt{16 - x^2 - y^2}$; $z = 0, 1, 2, 3, 4$

24. $f(x, y) = e^x - y$; $z = -2, -1, 0, 1, 2$

25. The volume of a cylindrical tank of radius r and height h is given by

$$V = f(r, h) = \pi r^2 h$$

Find the volume of a cylindrical tank of radius 1.5 ft and height 4 ft.

26. IQs The IQ (intelligence quotient) of a person whose mental age is m yr and whose chronological age is c yr is defined as

$$f(m, c) = \frac{100m}{c}$$

What is the IQ of a 9-yr-old child who has a mental age of 13.5 yr?

27. POISEUILLE'S LAW Poiseuille's law states that the resistance R, measured in dynes, of blood flowing in a blood vessel of length l and radius r (both in centimeters) is given by

$$R = f(l, r) = \frac{kl}{r^4}$$

where k is the viscosity of blood (in dyne-sec/cm^2). What is the resistance, in terms of k, of blood flowing through an arteriole 4 cm long and of radius 0.1 cm?

28. REVENUE FUNCTIONS The Country Workshop manufactures both finished and unfinished furniture for the home. The estimated quantities demanded each week of its rolltop desks in the finished and unfinished versions are x and y units when the corresponding unit prices are

$$p = 200 - \frac{1}{5}x - \frac{1}{10}y$$

$$q = 160 - \frac{1}{10}x - \frac{1}{4}y$$

dollars, respectively.
a. What is the weekly total revenue function $R(x, y)$?
b. Find the domain of the function R.

29. For the total revenue function $R(x, y)$ of Exercise 28, compute $R(100, 60)$ and $R(60, 100)$. Interpret your results.

30. REVENUE FUNCTIONS The Weston Publishing Company publishes a deluxe edition and a standard edition of its English language dictionary. Weston's management estimates that the number of deluxe editions demanded is x copies/day and the number of standard editions demanded is y copies/day when the unit prices are

$$p = 20 - 0.005x - 0.001y$$

$$q = 15 - 0.001x - 0.003y$$

dollars, respectively.
a. Find the daily total revenue function $R(x, y)$.
b. Find the domain of the function R.

31. For the total revenue function $R(x, y)$ of Exercise 30, compute $R(300, 200)$ and $R(200, 300)$. Interpret your results.

32. VOLUME OF A GAS The volume of a certain mass of gas is related to its pressure and temperature by the formula

$$V = \frac{30.9T}{P}$$

where the volume V is measured in liters, the temperature T is measured in degrees Kelvin (obtained by adding 273° to the Celsius temperature), and the pressure P is measured in millimeters of mercury pressure.
a. Find the domain of the function V.
b. Calculate the volume of the gas at standard temperature and pressure—that is, when $T = 273$ K and $P = 760$ mm of mercury.

33. SURFACE AREA OF A HUMAN BODY An empirical formula by E. F. Dubois relates the surface area S of a human body (in square meters) to its weight W (in kilograms) and its height H (in centimeters). The formula, given by

$$S = 0.007184W^{0.425}H^{0.725}$$

is used by physiologists in metabolism studies.
a. Find the domain of the function S.
b. What is the surface area of a human body that weighs 70 kg and has a height of 178 cm?

34. ARSON FOR PROFIT A study of arson for profit was conducted by a team of paid civilian experts and police detectives appointed by the mayor of a large city. It was found that the number of suspicious fires in that city in 1992 was very closely related to the concentration of tenants in the city's public housing and to the level of reinvestment in the area in conventional mortgages by the ten largest banks. In fact, the number of fires was

closely approximated by the formula

$$N(x, y) = \frac{100(1000 + 0.03x^2y)^{1/2}}{(5 + 0.2y)^2} \qquad (0 \le x \le 150; \\ 5 \le y \le 35)$$

where x denotes the number of persons/census tract and y denotes the level of reinvestment in the area in cents/dollar deposited. Using this formula, estimate the total number of suspicious fires in the districts of the city where the concentration of public housing tenants was 100/census tract and the level of reinvestment was 20 cents/dollar deposited.

35. CONTINUOUSLY COMPOUNDED INTEREST   If a principal of P dollars is deposited in an account earning interest at the rate of r/year compounded continuously, then the accumulated amount at the end of t yr is given by

$$A = f(P, r, t) = Pe^{rt}$$

dollars. Find the accumulated amount at the end of 3 yr if a sum of $10,000 is deposited in an account earning interest at the rate of 10%/year.

36. HOME MORTGAGES   The monthly payment that amortizes a loan of A dollars in t years when the interest rate is r per year is given by

$$P = f(A, r, t) = \frac{Ar}{12\left[1 - \left(1 + \dfrac{r}{12}\right)^{-12t}\right]}$$

a. What is the monthly payment for a home mortgage of $100,000 that will be amortized over 30 yr with an interest rate of 8%/year? With an interest rate of 10%/year?
b. Find the monthly payment for a home mortgage of $100,000 that will be amortized over 20 yr with an interest rate of 8%/year.

37. HOME MORTGAGES   Suppose a home buyer secures a bank loan of A dollars to purchase a house. If the interest rate charged is r/year and the loan is to be amortized in t yr, then the principal repayment at the end of i mo is given by

$$B = f(A, r, t, i)$$
$$= A\left[\frac{\left(1 + \dfrac{r}{12}\right)^i - 1}{\left(1 + \dfrac{r}{12}\right)^{12t} - 1}\right] \qquad (0 \le i \le 12t)$$

Suppose the Blakelys borrow a sum of $80,000 from a bank to help finance the purchase of a house and the bank charges interest at a rate of 9%/year. If the Blakelys agree to repay the loan in equal installments over 30 yr, how much will they owe the bank after the 60th payment (5 yr)? After the 240th payment (20 yr)?

38. FORCE GENERATED BY A CENTRIFUGE   A centrifuge is a machine designed for the specific purpose of subjecting materials to a sustained centrifugal force. The actual amount of centrifugal force, F, expressed in dynes (1 gram of force = 980 dynes) is given by

$$F = f(M, S, R) = \frac{\pi^2 S^2 MR}{900}$$

where S is in revolutions per minute (rpm), M is in grams, and R is in centimeters. Show that an object revolving at the rate of 600 rpm in a circle with radius of 10 cm generates a centrifugal force that is approximately 40 times gravity.

39. WILSON LOT-SIZE FORMULA   The Wilson lot-size formula in economics states that the optimal quantity Q of goods for a store to order is given by

$$Q = f(C, N, h) = \sqrt{\frac{2CN}{h}}$$

where C is the cost of placing an order, N is the number of items the store sells per week, and h is the weekly holding cost for each item. Find the most economical quantity of 10-speed bicycles to order if it costs the store $20 to place an order, $5 to hold a bicycle for a week, and the store expects to sell 40 bicycles a week.

In Exercises 40–43, determine whether the statement is true or false. If it is true, explain why it is true. If it is false, give an example to show why it is false.

40. If h is a function of x and y, then there are functions f and g of one variable such that

$$h(x, y) = f(x) + g(y).$$

41. If f is a function of x and y and a is a real number, then

$$f(ax, ay) = af(x, y).$$

42. The domain of $f(x, y) = 1/(x^2 - y^2)$ is $\{(x, y) \mid y \ne x\}$.

43. Every point on the level curve $f(x, y) = c$ is c units above the xy-plane if $c > 0$ and $|c|$ units below the xy-plane if $c < 0$.

SOLUTIONS TO SELF-CHECK EXERCISES 8.1

1. $f(1, 3) = 1^2 - 3(1)(3) + \sqrt{1 + 3} = -6$
$f(-1, 1) = (-1)^2 - 3(-1)(1) + \sqrt{-1 + 1} = 4$

The point $(-1, 0)$ is not in the domain of f because the term $\sqrt{x + y}$ is not defined when $x = -1$ and $y = 0$. In fact, the domain of f consists of all real values of x and y that satisfy the inequality $x + y \geq 0$, the shaded half-plane shown in the accompanying figure.

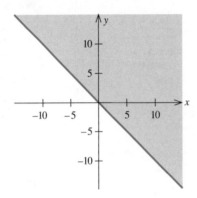

2. Since division by zero is not permitted, we see that $x \neq 0$ and $x - y \neq 0$. Therefore, the domain of f is the set of all points in the xy-plane not containing the y-axis ($x = 0$) and the straight line $x = y$.

3. If Odyssey spends \$5000/month on newspaper ads ($x = 5000$) and \$15,000/month on television ads ($y = 15,000$), then its monthly revenue will be given by

$$f(5000, 15,000) = 30(5000)^{1/4}(15,000)^{3/4}$$
$$\approx 341,926.06$$

or approximately \$341,926. If the agency spends \$4000/month on newspaper ads and \$16,000/month on television ads, then its monthly revenue will be given by

$$f(4000, 16,000) = 30(4000)^{1/4}(16,000)^{3/4}$$
$$\approx 339,411.25$$

or approximately \$339,411.

8.2 Partial Derivatives

PARTIAL DERIVATIVES

For a function $f(x)$ of one variable x, there is no ambiguity when we speak about the rate of change of $f(x)$ with respect to x since x must be constrained to move along the x-axis. The situation becomes more complicated, however,

FIGURE 8.12
We can approach a point in the plane from infinitely many directions.

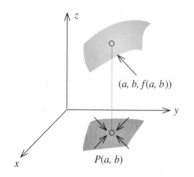

FIGURE 8.13
The curve C is formed by the intersection of the plane $y = b$ with the surface $z = f(x, y)$.

when we study the rate of change of a function of two or more variables. For example, the domain D of a function of two variables $f(x, y)$ is a subset of the plane (Figure 8.12), so if $P(a, b)$ is any point in the domain of f, there are infinitely many directions from which one can approach the point P. We may therefore ask for the rate of change of f at P along any of these directions.

However, we will not deal with this general problem. Instead, we will restrict ourselves to studying the rate of change of the function $f(x, y)$ at a point $P(a, b)$ in each of two *preferred directions*—namely, the direction parallel to the x-axis and the direction parallel to the y-axis. Let $y = b$, where b is a constant, so that $f(x, b)$ is a function of the one variable x. Since the equation $z = f(x, y)$ is the equation of a surface, the equation $z = f(x, b)$ is the equation of the curve C on the surface formed by the intersection of the surface and the plane $y = b$ (Figure 8.13).

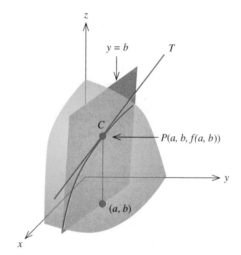

Because $f(x, b)$ is a function of one variable x, we may compute the derivative of f with respect to x at $x = a$. This derivative, obtained by keeping the variable y fixed and differentiating the resulting function $f(x, y)$ with respect to x, is called the **first partial derivative of f with respect to x** at (a, b), written

$$\frac{\partial z}{\partial x}(a, b) \quad \text{or} \quad \frac{\partial f}{\partial x}(a, b) \quad \text{or} \quad f_x(a, b)$$

Thus,

$$\frac{\partial z}{\partial x}(a, b) = \frac{\partial f}{\partial x}(a, b) = f_x(a, b) = \lim_{h \to 0} \frac{f(a + h, b) - f(a, b)}{h}$$

provided that the limit exists. The first partial derivative of f with respect to x at (a, b) measures both the slope of the tangent line T to the curve C and the rate of change of the function f in the x-direction when $x = a$ and $y = b$. We also write

$$\frac{\partial f}{\partial x}\bigg|_{(a,b)} \equiv f_x(a, b)$$

Similarly, we define the **first partial derivative of f with respect to y** at (a, b), written

$$\frac{\partial z}{\partial y}(a, b) \quad \text{or} \quad \frac{\partial f}{\partial y}(a, b) \quad \text{or} \quad f_y(a, b)$$

as the derivative obtained by keeping the variable x fixed and differentiating the resulting function $f(x, y)$ with respect to y. That is,

$$\frac{\partial z}{\partial y}(a, b) = \frac{\partial f}{\partial y}(a, b) = f_y(a, b)$$

$$= \lim_{k \to 0} \frac{f(a, b + k) - f(a, b)}{k}$$

if the limit exists. The first partial derivative of f with respect to y at (a, b) measures both the slope of the tangent line T to the curve C, obtained by holding x constant (Figure 8.14), and the rate of change of the function f in the y-direction when $x = a$ and $y = b$. We write

$$\frac{\partial f}{\partial y}\bigg|_{(a,b)} \equiv f_y(a, b)$$

Before looking at some examples, let's summarize these definitions.

FIGURE 8.14
The first partial derivative of f with respect to y at (a, b) measures the slope of the tangent line T to the curve C with x held constant.

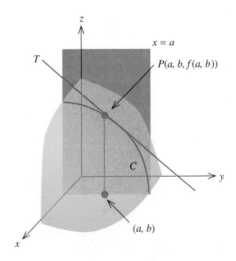

First Partial Derivatives of $f(x, y)$

Suppose $f(x, y)$ is a function of the two variables x and y. Then, the **first partial derivative of** f with respect to x at the point (x, y) is

$$\frac{\partial f}{\partial x} = \lim_{h \to 0} \frac{f(x + h, y) - f(x, y)}{h}$$

provided the limit exists. The first partial derivative of f with respect to y at the point (x, y) is

$$\frac{\partial f}{\partial y} = \lim_{k \to 0} \frac{f(x, y + k) - f(x, y)}{k}$$

provided the limit exists.

EXAMPLE 1   Find the partial derivatives $\partial f / \partial x$ and $\partial f / \partial y$ of the function

$$f(x, y) = x^2 - xy^2 + y^3$$

What is the rate of change of the function f in the x-direction at the point $(1, 2)$? What is the rate of change of the function f in the y-direction at the point $(1, 2)$?

SOLUTION ✔   To compute $\partial f / \partial x$, think of the variable y as a constant and differentiate the resulting function of x with respect to x. Let's write

$$f(x, y) = x^2 - xy^2 + y^3$$

where the variable y to be treated as a constant is shown in color. Then,

$$\frac{\partial f}{\partial x} = 2x - y^2$$

To compute $\partial f / \partial y$, think of the variable x as being fixed—that is, as a constant—and differentiate the resulting function of y with respect to y. In this case,

$$f(x, y) = x^2 - xy^2 + y^3$$

so that

$$\frac{\partial f}{\partial y} = -2xy + 3y^2$$

The rate of change of the function f in the x-direction at the point $(1, 2)$ is given by

$$f_x(1, 2) = \left. \frac{\partial f}{\partial x} \right|_{(1, 2)} = 2(1) - 2^2 = -2$$

That is, f decreases 2 units for each unit increase in the x-direction, y being kept constant ($y = 2$). The rate of change of the function f in the y-direction at the point $(1, 2)$ is given by

$$f_y(1, 2) = \left.\frac{\partial f}{\partial y}\right|_{(1, 2)} = -2(1)(2) + 3(2)^2 = 8$$

That is, f increases 8 units for each unit increase in the y-direction, x being kept constant $(x = 1)$.   ■ ■ ■ ■

Group Discussion

Refer to the Group Discussion on page 604. Suppose the management of the company has decided that the projected sales of the first product is a units. Describe how you might help management decide how many units of the second product the company should produce and sell in order to maximize the company's total profit. Justify your method to management. Suppose, however, management feels that b units of the second product can be manufactured and sold. How would you help management decide how many units of the first product to manufacture in order to maximize the company's total profit?

EXAMPLE 2   Compute the first partial derivatives of each of the following functions.

a. $f(x, y) = \dfrac{xy}{x^2 + y^2}$ **b.** $g(s, t) = (s^2 - st + t^2)^5$

c. $h(u, v) = e^{u^2 - v^2}$ **d.** $f(x, y) = \ln(x^2 + 2y^2)$

SOLUTION ✔   **a.** To compute $\partial f/\partial x$, think of the variable y as a constant. Thus,

$$f(x, y) = \frac{xy}{x^2 + y^2}$$

so that, upon using the quotient rule, we have

$$\frac{\partial f}{\partial x} = \frac{(x^2 + y^2)y - xy(2x)}{(x^2 + y^2)^2}$$

$$= \frac{y(y^2 - x^2)}{(x^2 + y^2)^2}$$

upon simplification and factorization. To compute $\partial f/\partial y$, think of the variable x as a constant. Thus,

$$f(x, y) = \frac{xy}{x^2 + y^2}$$

so that, upon using the quotient rule once again, we obtain

$$\frac{\partial f}{\partial y} = \frac{(x^2 + y^2)x - xy(2y)}{(x^2 + y^2)^2}$$

$$= \frac{x(x^2 - y^2)}{(x^2 + y^2)^2}$$

b. To compute $\partial g/\partial s$, we treat the variable t as if it were a constant. Thus,

$$g(s, t) = (s^2 - st + t^2)^5$$

Using the general power rule, we find

$$\frac{\partial g}{\partial s} = 5(s^2 - st + t^2)^4 \cdot (2s - t)$$
$$= 5(2s - t)(s^2 - st + t^2)^4$$

To compute $\partial g/\partial t$, we treat the variable s as if it were a constant. Thus,

$$g(s, t) = (s^2 - st + t^2)^5$$
$$\frac{\partial g}{\partial t} = 5(s^2 - st + t^2)^4(-s + 2t)$$
$$= 5(2t - s)(s^2 - st + t^2)^4$$

c. To compute $\partial h/\partial u$, think of the variable v as a constant. Thus,

$$h(u, v) = e^{u^2 - v^2}$$

Using the chain rule for exponential functions, we have

$$\frac{\partial h}{\partial u} = e^{u^2 - v^2} \cdot 2u$$
$$= 2ue^{u^2 - v^2}$$

Next, we treat the variable u as if it were a constant,

$$h(u, v) = e^{u^2 - v^2}$$

and we obtain

$$\frac{\partial h}{\partial v} = e^{u^2 - v^2} \cdot (-2v)$$
$$= -2ve^{u^2 - v^2}$$

d. To compute $\partial f/\partial x$, think of the variable y as a constant. Thus,

$$f(x, y) = \ln(x^2 + 2y^2)$$

so that the chain rule for logarithmic functions gives

$$\frac{\partial f}{\partial x} = \frac{2x}{x^2 + 2y^2}$$

Next, treating the variable x as if it were a constant, we find

$$f(x, y) = \ln(x^2 + 2y^2)$$
$$\frac{\partial f}{\partial y} = \frac{4y}{x^2 + 2y^2}$$

To compute the partial derivative of a function of several variables with respect to one variable—say, x—we think of the other variables as if they were constants and differentiate the resulting function with respect to x.

Group Discussion

1. Let (a, b) be a point in the domain of $f(x, y)$. Put $g(x) = f(x, b)$ and suppose g is differentiable at $x = a$. Explain why you can find $f_x(a, b)$ by computing $g'(a)$. How would you go about calculating $f_y(a, b)$ using a similar technique? Give a geometric interpretation of these processes.

2. Let $f(x, y) = x^2y^3 - 3x^2y + 2$. Use the method of Problem 1 to find $f_x(1, 2)$ and $f_y(1, 2)$.

EXAMPLE 3

Compute the first partial derivatives of the function

$$w = f(x, y, z) = xyz - xe^{yz} + x \ln y$$

SOLUTION ✔

Here we have a function of three variables, x, y, and z, and we are required to compute

$$\frac{\partial f}{\partial x}, \quad \frac{\partial f}{\partial y}, \quad \frac{\partial f}{\partial z}$$

To compute f_x, we think of the other two variables, y and z, as fixed, and we differentiate the resulting function of x with respect to x, thereby obtaining

$$f_x = yz - e^{yz} + \ln y$$

To compute f_y, we think of the other two variables, x and z, as constants, and we differentiate the resulting function of y with respect to y. We then obtain

$$f_y = xz - xze^{yz} + \frac{x}{y}$$

Finally, to compute f_z, we treat the variables x and y as constants and differentiate the function f with respect to z, obtaining

$$f_z = xy - xye^{yz}$$ ■■■■

Exploring with Technology

Refer to the Group Discussion on this page. Let

$$f(x, y) = \frac{e^{\sqrt{xy}}}{(1 + xy^2)^{3/2}}$$

1. Compute $g(x) = f(x, 1)$ and use a graphing utility to plot the graph of g in the viewing rectangle $[0, 2] \times [0, 2]$.
2. Use the differentiation operation of your graphing utility to find $g'(1)$ and hence $f_x(1, 1)$.
3. Compute $h(y) = f(1, y)$ and use a graphing utility to plot the graph of g in the viewing rectangle $[0, 2] \times [0, 2]$.
4. Use the differentiation operation of your graphing utility to find $h'(1)$ and hence $f_y(1, 1)$.

THE COBB–DOUGLAS PRODUCTION FUNCTION

For an economic interpretation of the first partial derivatives of a function of two variables, let's turn our attention to the function

$$f(x, y) = ax^b y^{1-b} \tag{1}$$

where a and b are positive constants with $0 < b < 1$. This function is called the **Cobb–Douglas production function.** Here, x stands for the amount of money expended for labor, y stands for the cost of capital equipment (buildings, machinery, and other tools of production), and the function f measures the output of the finished product (in suitable units) and is called, accordingly, the production function.

The partial derivative f_x is called the **marginal productivity of labor.** It measures the rate of change of production with respect to the amount of money expended for labor, with the level of capital expenditure held constant. Similarly, the partial derivative f_y, called the **marginal productivity of capital,** measures the rate of change of production with respect to the amount expended on capital, with the level of labor expenditure held fixed.

EXAMPLE 4 A certain country's production in the early years following World War II is described by the function

$$f(x, y) = 30x^{2/3}y^{1/3}$$

units, when x units of labor and y units of capital were used.

a. Compute f_x and f_y.
b. What is the marginal productivity of labor and the marginal productivity of capital when the amounts expended on labor and capital are 125 units and 27 units, respectively?
c. Should the government have encouraged capital investment rather than increasing expenditure on labor to increase the country's productivity?

SOLUTION ✔ **a.** $f_x = 30 \cdot \dfrac{2}{3} x^{-1/3} y^{1/3} = 20 \left(\dfrac{y}{x}\right)^{1/3}$

$f_y = 30x^{2/3} \cdot \dfrac{1}{3} y^{-2/3} = 10 \left(\dfrac{x}{y}\right)^{2/3}$

b. The required marginal productivity of labor is given by

$$f_x(125, 27) = 20 \left(\frac{27}{125}\right)^{1/3} = 20 \left(\frac{3}{5}\right)$$

or 12 units per unit increase in labor expenditure (capital expenditure is held constant at 27 units). The required marginal productivity of capital is given by

$$f_y(125, 27) = 10 \left(\frac{125}{27}\right)^{2/3} = 10 \left(\frac{25}{9}\right)$$

or $27\frac{7}{9}$ units per unit increase in capital expenditure (labor outlay is held constant at 125 units).

c. From the results of part (b), we see that a unit increase in capital expenditure resulted in a much faster increase in productivity than a unit increase in labor expenditure would have. Therefore, the government should have encouraged increased spending on capital rather than on labor during the early years of reconstruction.   ■ ■ ■ ■

SUBSTITUTE AND COMPLEMENTARY COMMODITIES

For another application of the first partial derivatives of a function of two variables in the field of economics, let's consider the relative demands of two commodities. We say that the two commodities are **substitute** (competitive) **commodities** if a decrease in the demand for one results in an increase in the demand for the other. Examples of competitive commodities are coffee and tea. Conversely, two commodities are referred to as **complementary commodities** if a decrease in the demand for one results in a decrease in the demand for the other as well. Examples of complementary commodities are automobiles and tires.

We now derive a criterion for determining whether two commodities A and B are substitute or complementary. Suppose the demand equations that relate the quantities demanded, x and y, to the unit prices, p and q, of the two commodities are given by

$$x = f(p, q) \qquad \text{and} \qquad y = g(p, q)$$

Let's consider the partial derivative $\partial f/\partial p$. Since f is the demand function for commodity A, we see that, for fixed q, f is typically a decreasing function of p—that is, $\partial f/\partial p < 0$. Now, if the two commodities were substitute commodities, then the quantity demanded of commodity B would increase with respect to p—that is, $\partial g/\partial p > 0$. A similar argument with p fixed shows that if A and B are substitute commodities, then $\partial f/\partial q > 0$. Thus, the two commodities A and B are substitute commodities if

$$\frac{\partial f}{\partial q} > 0 \qquad \text{and} \qquad \frac{\partial g}{\partial p} > 0$$

Similarly, A and B are complementary commodities if

$$\frac{\partial f}{\partial q} < 0 \qquad \text{and} \qquad \frac{\partial g}{\partial p} < 0$$

Substitute and Complementary Commodities

Two commodities A and B are substitute commodities if

$$\frac{\partial f}{\partial q} > 0 \quad \text{and} \quad \frac{\partial g}{\partial p} > 0 \tag{2}$$

Two commodities A and B are complementary commodities if

$$\frac{\partial f}{\partial q} < 0 \quad \text{and} \quad \frac{\partial g}{\partial p} < 0 \tag{3}$$

EXAMPLE 5 Suppose that the daily demand for butter is given by

$$x = f(p, q) = \frac{3q}{1 + p^2}$$

and the daily demand for margarine is given by

$$y = g(p, q) = \frac{2p}{1 + \sqrt{q}} \qquad (p > 0, q > 0)$$

where p and q denote the prices per pound (in dollars) of butter and margarine, respectively, and x and y are measured in millions of pounds. Determine whether these two commodities are substitute, complementary, or neither.

SOLUTION ✔ We compute

$$\frac{\partial f}{\partial q} = \frac{3}{1 + p^2} \quad \text{and} \quad \frac{\partial g}{\partial p} = \frac{2}{1 + \sqrt{q}}$$

Since

$$\frac{\partial f}{\partial q} > 0 \quad \text{and} \quad \frac{\partial g}{\partial p} > 0$$

for all values of $p > 0$ and $q > 0$, we conclude that butter and margarine are substitute commodities.   ■ ■ ■ ■

SECOND-ORDER PARTIAL DERIVATIVES

The first partial derivatives $f_x(x, y)$ and $f_y(x, y)$ of a function $f(x, y)$ of the two variables x and y are also functions of x and y. As such, we may differentiate each of the functions f_x and f_y to obtain the **second-order partial derivatives of** f (Figure 8.15). Thus, differentiating the function f_x with respect to x leads to the second partial derivative

$$f_{xx} \equiv \frac{\partial^2 f}{\partial x^2} = \frac{\partial}{\partial x}(f_x)$$

FIGURE 8.15
A schematic showing the four second-order partial derivatives of f

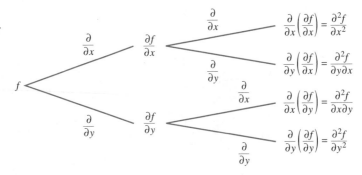

However, differentiation of f_x with respect to y leads to the second partial derivative

$$f_{xy} \equiv \frac{\partial^2 f}{\partial y \partial x} = \frac{\partial}{\partial y}(f_x)$$

Similarly, differentiation of the function f_y with respect to x and with respect to y leads to

$$f_{yx} \equiv \frac{\partial^2 f}{\partial x \partial y} = \frac{\partial}{\partial x}(f_y)$$

$$f_{yy} \equiv \frac{\partial^2 f}{\partial y^2} = \frac{\partial}{\partial y}(f_y)$$

respectively. Note that, in general, it is not true that $f_{xy} = f_{yx}$. However, in most practical applications f_{xy} and f_{yx} are equal, but $f_{xy}(a, b)$ and $f_{yx}(a, b)$ are equal if they are both continuous at (a, b).

EXAMPLE 6 Find the second-order partial derivatives of the function

$$f(x, y) = x^3 - 3x^2y + 3xy^2 + y^2$$

SOLUTION ✔ The first partial derivatives of f are

$$f_x = \frac{\partial}{\partial x}(x^3 - 3x^2y + 3xy^2 + y^2)$$

$$= 3x^2 - 6xy + 3y^2$$

$$f_y = \frac{\partial}{\partial y}(x^3 - 3x^2y + 3xy^2 + y^2)$$

$$= -3x^2 + 6xy + 2y$$

Therefore,

$$f_{xx} = \frac{\partial}{\partial x}(f_x) = \frac{\partial}{\partial x}(3x^2 - 6xy + 3y^2)$$

$$= 6x - 6y = 6(x - y)$$

$$f_{xy} = \frac{\partial}{\partial y}(f_x) = \frac{\partial}{\partial y}(3x^2 - 6xy + 3y^2)$$
$$= -6x + 6y = 6(y - x)$$
$$f_{yx} = \frac{\partial}{\partial x}(f_y) = \frac{\partial}{\partial x}(-3x^2 + 6xy + 2y)$$
$$= -6x + 6y = 6(y - x)$$
$$f_{yy} = \frac{\partial}{\partial y}(f_y) = \frac{\partial}{\partial y}(-3x^2 + 6xy + 2y)$$
$$= 6x + 2$$

EXAMPLE 7 Find the second-order partial derivatives of the function

$$f(x, y) = e^{xy^2}$$

SOLUTION ✔ We have

$$f_x = \frac{\partial}{\partial x}(e^{xy^2})$$
$$= y^2 e^{xy^2}$$
$$f_y = \frac{\partial}{\partial y}(e^{xy^2})$$
$$= 2xy e^{xy^2}$$

so the required second-order partial derivatives of f are

$$f_{xx} = \frac{\partial}{\partial x}(f_x) = \frac{\partial}{\partial x}(y^2 e^{xy^2})$$
$$= y^4 e^{xy^2}$$
$$f_{xy} = \frac{\partial}{\partial y}(f_x) = \frac{\partial}{\partial y}(y^2 e^{xy^2})$$
$$= 2y e^{xy^2} + 2xy^3 e^{xy^2}$$
$$= 2y e^{xy^2}(1 + xy^2)$$
$$f_{yx} = \frac{\partial}{\partial x}(f_y) = \frac{\partial}{\partial x}(2xy e^{xy^2})$$
$$= 2y e^{xy^2} + 2xy^3 e^{xy^2}$$
$$= 2y e^{xy^2}(1 + xy^2)$$
$$f_{yy} = \frac{\partial}{\partial y}(f_y) = \frac{\partial}{\partial y}(2xy e^{xy^2})$$
$$= 2x e^{xy^2} + (2xy)(2xy)e^{xy^2}$$
$$= 2x e^{xy^2}(1 + 2xy^2)$$

SELF-CHECK EXERCISES 8.2

1. Compute the first partial derivatives of $f(x, y) = x^3 - 2xy^2 + y^2 - 8$.
2. Find the first partial derivatives of $f(x, y) = x \ln y + ye^x - x^2$ at $(0, 1)$ and interpret your results.
3. Find the second-order partial derivatives of the function of Self-Check Exercise 1.
4. A certain country's production is described by the function

$$f(x, y) = 60x^{1/3}y^{2/3}$$

when x units of labor and y units of capital are used.
a. What is the marginal productivity of labor and the marginal productivity of capital when the amounts expended on labor and capital are 125 units and 8 units, respectively?
b. Should the government encourage capital investment rather than increased expenditure on labor at this time in order to increase the country's productivity?

Solutions to Self-Check Exercises 8.2 can be found on page 629.

8.2 Exercises

In Exercises 1–22, find the first partial derivatives of the function.

1. $f(x, y) = 2x + 3y + 5$

2. $f(x, y) = 2xy$

3. $g(x, y) = 2x^2 + 4y + 1$

4. $f(x, y) = 1 + x^2 + y^2$

5. $f(x, y) = \dfrac{2y}{x^2}$

6. $f(x, y) = \dfrac{x}{1 + y}$

7. $g(u, v) = \dfrac{u - v}{u + v}$

8. $f(x, y) = \dfrac{x^2 - y^2}{x^2 + y^2}$

9. $f(s, t) = (s^2 - st + t^2)^3$

10. $g(s, t) = s^2t + st^{-3}$

11. $f(x, y) = (x^2 + y^2)^{2/3}$

12. $f(x, y) = x\sqrt{1 + y^2}$

13. $f(x, y) = e^{xy+1}$

14. $f(x, y) = (e^x + e^y)^5$

15. $f(x, y) = x \ln y + y \ln x$

16. $f(x, y) = x^2 e^{y^2}$

17. $g(u, v) = e^u \ln v$

18. $f(x, y) = \dfrac{e^{xy}}{x + y}$

19. $f(x, y, z) = xyz + xy^2 + yz^2 + zx^2$

20. $g(u, v, w) = \dfrac{2uvw}{u^2 + v^2 + w^2}$

21. $h(r, s, t) = e^{rst}$

22. $f(x, y, z) = xe^{y/z}$

In Exercises 23–32, evaluate the first partial derivatives of the function at the given point.

23. $f(x, y) = x^2y + xy^2$; $(1, 2)$

24. $f(x, y) = x^2 + xy + y^2 + 2x - y$; $(-1, 2)$

25. $f(x, y) = x\sqrt{y} + y^2$; (2, 1)

26. $g(x, y) = \sqrt{x^2 + y^2}$; (3, 4)

27. $f(x, y) = \dfrac{x}{y}$; (1, 2)

28. $f(x, y) = \dfrac{x + y}{x - y}$; (1, −2)

29. $f(x, y) = e^{xy}$; (1, 1)

30. $f(x, y) = e^x \ln y$; (0, e)

31. $f(x, y, z) = x^2yz^3$; (1, 0, 2)

32. $f(x, y, z) = x^2y^2 + z^2$; (1, 1, 2)

In Exercises 33–40, find the second-order partial derivatives of the given function. In each case, show that the mixed partial derivatives f_{xy} and f_{yx} are equal.

33. $f(x, y) = x^2y + xy^3$

34. $f(x, y) = x^3 + x^2y + x + 4$

35. $f(x, y) = x^2 - 2xy + 2y^2 + x - 2y$

36. $f(x, y) = x^3 + x^2y^2 + y^3 + x + y$

37. $f(x, y) = \sqrt{x^2 + y^2}$

38. $f(x, y) = x\sqrt{y} + y\sqrt{x}$

39. $f(x, y) = e^{-x/y}$

40. $f(x, y) = \ln(1 + x^2y^2)$

41. PRODUCTIVITY OF A COUNTRY The productivity of a South American country is given by the function

$$f(x, y) = 20x^{3/4}y^{1/4}$$

when x units of labor and y units of capital are used.
a. What is the marginal productivity of labor and the marginal productivity of capital when the amounts expended on labor and capital are 256 units and 16 units, respectively?
b. Should the government encourage capital investment rather than increased expenditure on labor at this time in order to increase the country's productivity?

42. PRODUCTIVITY OF A COUNTRY The productivity of a country in Western Europe is given by the function

$$f(x, y) = 40x^{4/5}y^{1/5}$$

when x units of labor and y units of capital are used.
a. What is the marginal productivity of labor and the marginal productivity of capital when the amounts ex-

pended on labor and capital are 32 units and 243 units, respectively?
b. Should the government encourage capital investment rather than increased expenditure on labor at this time in order to increase the country's productivity?

43. LAND PRICES The rectangular region R shown in the accompanying figure represents a city's financial district. The price of land within the district is approximated by the function

$$p(x, y) = 200 - 10\left(x - \frac{1}{2}\right)^2 - 15(y - 1)^2$$

where $p(x, y)$ is the price of land at the point (x, y) in dollars per square foot and x and y are measured in miles. Compute

$$\frac{\partial p}{\partial x}(0, 1) \quad \text{and} \quad \frac{\partial p}{\partial y}(0, 1)$$

and interpret your results.

44. COMPLEMENTARY AND SUBSTITUTE COMMODITIES In a survey conducted by *Home Entertainment* magazine, it was determined that the demand equation for videocassette recorders (VCRs) is given by

$$x = f(p, q) = 10,000 - 10p + 0.2q^2$$

and the demand equation for videodisc players is given by

$$y = g(p, q) = 5000 + 0.8p^2 - 20q$$

where p and q denote the unit prices (in dollars) for the VCRs and disc players, respectively, and x and y denote

Finding Partial Derivatives at a Given Point

Suppose $f(x, y)$ is a function of two variables, and we wish to compute

$$f_x(a, b) = \frac{\partial f}{\partial x}\bigg|_{(a,b)}$$

Recall that in computing $\partial f/\partial x$, we think of y as being fixed. But in this situation, we are evaluating $\partial f/\partial x$ at (a, b). Therefore, we set y equal to b. Doing this leads to the function g of one variable, x, defined by

$$g(x) = f(x, b)$$

It follows from the definition of the partial derivative that

$$f_x(a, b) = g'(a)$$

Thus, the value of the partial derivative $\partial f/\partial x$ at a given point (a, b) can be found by evaluating the derivative of a function of one variable. In particular, the latter can be found by using the numerical derivative operation of a graphing utility. We find $f_y(a, b)$ in a similar manner.

EXAMPLE 1 Let $f(x, y) = (1 + xy^2)^{3/2}e^{x^2 y}$. Find (a) $f_x(1, 2)$ and (b) $f_y(1, 2)$.

SOLUTION ✔

a. Define $g(x) = f(x, 2) = (1 + 4x)^{3/2}e^{2x^2}$. Using the numerical derivative operation to find $g'(1)$, we obtain

$$f_x(1, 2) = g'(1) \approx 429.5832249$$

b. Define $h(y) = f(1, y) = (1 + y^2)^{3/2}e^y$. Using the numerical derivative operation to find $h'(2)$, we obtain

$$f_y(1, 2) = h'(2) \approx 181.74674899 \quad \blacksquare\blacksquare\blacksquare\blacksquare$$

In Exercises 1–6, use a graphing utility to compute

$$\frac{\partial f}{\partial x} \quad \text{and} \quad \frac{\partial f}{\partial y}$$

at the given point.

1. $f(x, y) = \sqrt{x}(2 + xy^2)^{1/3}$; $(1, 2)$

2. $f(x, y) = \sqrt{xy}(1 + 2xy)^{2/3}$; $(1, 4)$

3. $f(x, y) = \dfrac{x + y^2}{1 + x^2y}$; $(1, 2)$

4. $f(x, y) = \dfrac{xy^2}{(\sqrt{x} + \sqrt{y})^2}$; $(4, 1)$

5. $f(x, y) = e^{-xy^2}(x + y)^{1/3}$; $(1, 1)$

6. $f(x, y) = \dfrac{\ln(\sqrt{x} + y^2)}{x^2 + y^2}$; $(4, 1)$

the number of VCRs and disc players demanded per week. Determine whether these two products are substitute, complementary, or neither.

45. COMPLEMENTARY AND SUBSTITUTE COMMODITIES   In a survey it was determined that the demand equation for VCRs is given by

$$x = f(p, q) = 10,000 - 10p - e^{0.5q}$$

The demand equation for blank tapes for the VCRs is given by

$$y = g(p, q) = 50,000 - 4000q - 10p$$

where p and q denote the unit prices, respectively, and x and y denote the number of VCRs and the number of blank VCR tapes demanded per week. Determine whether these two products are substitute, complementary, or neither.

46. COMPLEMENTARY AND SUBSTITUTE COMMODITIES   Refer to Exercise 28, Exercises 8.1. Show that the finished and unfinished home furniture manufactured by the Country Workshop are substitute commodities.

Hint: Solve the system of equations for x and y in terms of p and q.

47. REVENUE FUNCTIONS   The total weekly revenue (in dollars) of the Country Workshop associated with manufacturing and selling their rolltop desks is given by the function

$$R(x, y) = -0.2x^2 - 0.25y^2 - 0.2xy + 200x + 160y$$

where x denotes the number of finished units and y denotes the number of unfinished units manufactured and sold per week. Compute $\partial R/\partial x$ and $\partial R/\partial y$ when $x = 300$ and $y = 250$. Interpret your results.

48. PROFIT FUNCTIONS   The monthly profit (in dollars) of the Bond and Barker Department Store depends on the level of inventory x (in thousands of dollars) and the floor space y (in thousands of square feet) available for display of the merchandise, as given by the equation

$$P(x, y) = -0.02x^2 - 15y^2 + xy$$
$$+ 39x + 25y - 20,000$$

Compute $\partial P/\partial x$ and $\partial P/\partial y$ when $x = 4000$ and $y = 150$. Interpret your results. Repeat with $x = 5000$ and $y = 150$.

49. VOLUME OF A GAS   The volume V (in liters) of a certain mass of gas is related to its pressure P (in millimeters of mercury) and its temperature T (in degrees Kelvin) by the law

$$V = \frac{30.9T}{P}$$

Compute $\partial V/\partial T$ and $\partial V/\partial P$ when $T = 300$ and $P = 800$. Interpret your results.

50. SURFACE AREA OF A HUMAN BODY   The formula

$$S = 0.007184W^{0.425}H^{0.725}$$

gives the surface area S of a human body (in square meters) in terms of its weight W (in kilograms) and its height H (in centimeters). Compute $\partial S/\partial W$ and $\partial S/\partial H$ when $W = 70$ kg and $H = 180$ cm. Interpret your results.

51. According to the *ideal gas law*, the volume V (in liters) of an ideal gas is related to its pressure P (in pascals) and temperature T (in degrees Kelvin) by the formula

$$V = \frac{kT}{P}$$

where k is a constant. Show that

$$\frac{\partial V}{\partial T} \cdot \frac{\partial T}{\partial P} \cdot \frac{\partial P}{\partial V} = -1$$

In Exercises 52–55, determine whether the statement is true or false. If it is true, explain why it is true. If it is false, give an example to show why it is false.

52. If $f_x(x, y)$ is defined at (a, b), then $f_y(x, y)$ must also be defined at (a, b).

53. If $f_x(a, b) < 0$, then f is decreasing with respect to x near (a, b).

54. If $f_{xy}(x, y)$ and $f_{yx}(x, y)$ are both continuous for all values of x and y, then $f_{xy} = f_{yx}$ for all values of x and y.

55. If both f_{xy} and f_{yx} are defined at (a, b), then f_{xx} and f_{yy} must be defined at (a, b).

SOLUTIONS TO SELF-CHECK EXERCISES 8.2

1. $f_x = \dfrac{\partial f}{\partial x} = 3x^2 - 2y^2$

   $f_y = \dfrac{\partial f}{\partial y} = -2x(2y) + 2y$

   $\quad = 2y(1 - 2x)$

2. $f_x = \ln y + ye^x - 2x$; $f_y = \dfrac{x}{y} + e^x$

   In particular,

   $$f_x(0, 1) = \ln 1 + 1e^0 - 2(0) = 1$$
   $$f_y(0, 1) = \frac{0}{1} + e^0 = 1$$

   The results tell us that at the point $(0, 1)$, $f(x, y)$ increases 1 unit for each unit increase in the x-direction, y being kept constant; $f(x, y)$ also increases 1 unit for each unit increase in the y-direction, x being kept constant.

3. From the results of Self-Check Exercise 1,

   $$f_x = 3x^2 - 2y^2$$

   Therefore,

   $$f_{xx} = \frac{\partial}{\partial x}(3x^2 - 2y^2) = 6x$$

   $$f_{xy} = \frac{\partial}{\partial y}(3x^2 - 2y^2) = -4y$$

   Also, from the results of Self-Check Exercise 1,

   $$f_y = 2y(1 - 2x)$$

   Thus,

   $$f_{yx} = \frac{\partial}{\partial x}[2y(1 - 2x)] = -4y$$

   $$f_{yy} = \frac{\partial}{\partial y}[2y(1 - 2x)] = 2(1 - 2x)$$

4. **a.** The marginal productivity of labor when the amounts expended on labor and capital are x and y units, respectively, is given by

   $$f_x(x, y) = 60\left(\frac{1}{3}x^{-2/3}\right)y^{2/3} = 20\left(\frac{y}{x}\right)^{2/3}$$

   In particular, the required marginal productivity of labor is given by

   $$f_x(125, 8) = 20\left(\frac{8}{125}\right)^{2/3} = 20\left(\frac{4}{25}\right)$$

or 3.2 units/unit increase in labor expenditure, capital expenditure being held constant at 8 units. Next, we compute

$$f_y(x, y) = 60x^{1/3}\left(\frac{2}{3}y^{-1/3}\right) = 40\left(\frac{x}{y}\right)^{1/3}$$

and deduce that the required marginal productivity of capital is given by

$$f_y(125, 8) = 40\left(\frac{125}{8}\right)^{1/3} = 40\left(\frac{5}{2}\right)$$

or 100 units/unit increase in capital expenditure, labor expenditure being held constant at 125 units.

b. The results of part (a) tell us that the government should encourage increased spending on capital rather than on labor.

8.3 Maxima and Minima of Functions of Several Variables

MAXIMA AND MINIMA

In Chapter 4 we saw that the solution of a problem often reduces to finding the extreme values of a function of one variable. In practice, however, situations also arise in which a problem is solved by finding the absolute maximum or absolute minimum value of a function of two or more variables.

For example, suppose the Scandi Company manufactures computer desks in both assembled and unassembled versions. Its profit P is therefore a function of the number of assembled units, x, and the number of unassembled units, y, manufactured and sold per week; that is, $P = f(x, y)$. A question of paramount importance to the manufacturer is, How many assembled and unassembled desks should the company manufacture per week in order to maximize its weekly profit? Mathematically, the problem is solved by finding the values of x and y that will make $f(x, y)$ a maximum.

In this section we will focus our attention on finding the extrema of a function of two variables. As in the case of a function of one variable, we distinguish between the relative (or local) extrema and the absolute extrema of a function of two variables.

Relative Extrema of a Function of Two Variables

Let f be a function defined on a region R containing the point (a, b). Then, f has a **relative maximum** at (a, b) if $f(x, y) \le f(a, b)$ for all points (x, y) that are sufficiently close to (a, b). The number $f(a, b)$ is called a **relative maximum value**. Similarly, f has a **relative minimum** at (a, b), with **relative minimum value** $f(a, b)$ if $f(x, y) \ge f(a, b)$ for all points (x, y) that are sufficiently close to (a, b).

Loosely speaking, f has a relative maximum at (a, b) if the point $(a, b, f(a, b))$ is the highest point on the graph of f when compared with all nearby points. A similar interpretation holds for a relative minimum.

If the inequalities in this last definition hold for *all* points (x, y) in the domain of f, then f has an **absolute maximum** or **(absolute minimum)** at (a, b) with **absolute maximum value** (or **absolute minimum value**) $f(a, b)$. Figure 8.16 shows the graph of a function with relative maxima at (a, b) and (e, f) and a relative minimum at (c, d). The absolute maximum of f occurs at (e, f) and the absolute minimum of f occurs at (g, h).

FIGURE 8.16

Just as in the case of a function of one variable, a relative extremum (relative maximum or relative minimum) may or may not be an absolute extremum. However, to simplify matters, we will assume that whenever an absolute extremum exists, it will occur at a point where f has a relative extremum.

Just as the first and second derivatives play an important role in determining the relative extrema of a function of one variable, the first and second partial derivatives are powerful tools for locating and classifying the relative extrema of functions of several variables.

Suppose now that a differentiable function $f(x, y)$ of two variables has a relative maximum (relative minimum) at a point (a, b) in the domain of f. From Figure 8.17 it is clear that at the point (a, b) the slope of the "tangent lines" to the surface in any direction must be zero. In particular, this implies that both

$$\frac{\partial f}{\partial x}(a, b) \qquad \text{and} \qquad \frac{\partial f}{\partial y}(a, b)$$

must be zero.

The point (a, b) is called a **critical point** of the function f. In case we are tempted to conclude that a critical point of a function f must automatically be a relative extremum of f, let's consider the graph of the function f in Figure 8.18. Here, we have both

$$\frac{\partial f}{\partial x}(a, b) = 0 \qquad \text{and} \qquad \frac{\partial f}{\partial y}(a, b) = 0$$

The point $(a, b, f(a, b))$ is neither a relative maximum nor a relative minimum of the function f since there are points nearby that are higher and others that

FIGURE 8.17

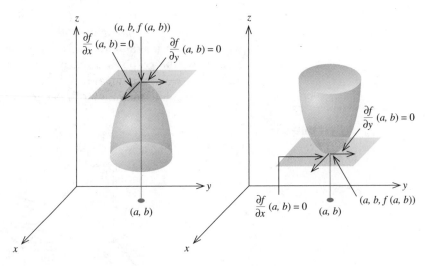

(a) f has a relative maximum at (a, b).

(b) f has a relative minimum at (a, b).

are lower than it. Such a point is called a **saddle point.** Thus, as in the case of a function of one variable, we may conclude that a critical point of a function of two (or more) variables is only a candidate for a relative extremum of f.

FIGURE 8.18
The point $(a, b, f(a, b))$ is called a saddle point.

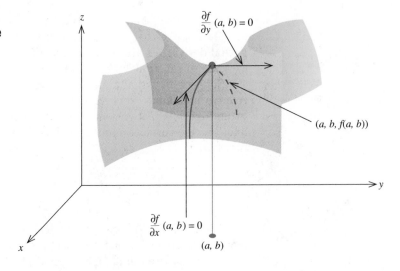

To determine the nature of a critical point of a function $f(x, y)$ of two variables, we use the second partial derivatives of f. The resulting test, which helps us classify these points, is called the **second derivative test** and is incorporated in the following procedure for finding and classifying the relative extrema of f.

Determining Relative Extrema

1. Find the critical points of $f(x, y)$ by solving the system of simultaneous equations

$$f_x = 0$$
$$f_y = 0$$

2. The second derivative test: Let

$$D(x, y) = f_{xx}f_{yy} - f_{xy}^2$$

Then,

a. $D(a, b) > 0$ and $f_{xx}(a, b) < 0$ implies that $f(x, y)$ has a **relative maximum** at the point (a, b).
b. $D(a, b) > 0$ and $f_{xx}(a, b) > 0$ implies that $f(x, y)$ has a **relative minimum** at the point (a, b).
c. $D(a, b) < 0$ implies that $f(x, y)$ has neither a relative maximum nor a relative minimum at the point (a, b).
d. $D(a, b) = 0$ implies that the test is inconclusive, so some other technique must be used to solve the problem.

Find the relative extrema of the function

$$f(x, y) = x^2 + y^2$$

SOLUTION ✔   We have

$$f_x = 2x$$
$$f_y = 2y$$

To find the critical point(s) of f, we set $f_x = 0$ and $f_y = 0$ and solve the resulting system of simultaneous equations

$$2x = 0$$
$$2y = 0$$

FIGURE 8.19
The graph of $f(x, y) = x^2 + y^2$

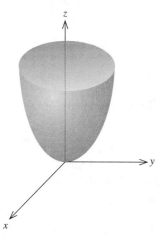

obtaining $x = 0$, $y = 0$, or $(0, 0)$, as the sole critical point of f. Next, we apply the second derivative test to determine the nature of the critical point $(0, 0)$. We compute

$$f_{xx} = 2, \quad f_{xy} = 0, \quad f_{yy} = 2$$

and

$$D(x, y) = f_{xx}f_{yy} - f_{xy}^2 = (2)(2) - 0 = 4$$

In particular, $D(0, 0) = 4$. Since $D(0, 0) > 0$ and $f_{xx}(0, 0) = 2 > 0$, we conclude that $f(x, y)$ has a relative minimum at the point $(0, 0)$. The relative minimum value, 0, also happens to be the absolute minimum of f. The graph of the function f, shown in Figure 8.19, confirms these results.   ■ ■ ■ ■

EXAMPLE 2

Find the relative extrema of the function

$$f(x, y) = 3x^2 - 4xy + 4y^2 - 4x + 8y + 4$$

SOLUTION ✔

We have

$$f_x = 6x - 4y - 4$$

$$f_y = -4x + 8y + 8$$

To find the critical points of f, we set $f_x = 0$ and $f_y = 0$ and solve the resulting system of simultaneous equations

$$6x - 4y = 4$$

$$-4x + 8y = -8$$

Multiplying the first equation by 2 and the second equation by 3, we obtain the equivalent system

$$12x - 8y = 8$$

$$-12x + 24y = -24$$

Adding the two equations gives $16y = -16$, or $y = -1$. We substitute this value for y into either equation in the system to get $x = 0$. Thus, the only critical point of f is the point $(0, -1)$. Next, we apply the second derivative test to determine whether the point $(0, -1)$ gives rise to a relative extremum of f. We compute

$$f_{xx} = 6, \quad f_{xy} = -4, \quad f_{yy} = 8$$

and

$$D(x, y) = f_{xx}f_{yy} - f_{xy}^2 = (6)(8) - (-4)^2 = 32$$

Since $D(0, -1) = 32 > 0$ and $f_{xx}(0, -1) = 6 > 0$, we conclude that $f(x, y)$ has a relative minimum at the point $(0, -1)$. The value of $f(x, y)$ at the point $(0, -1)$ is given by

$$f(0, -1) = 3(0)^2 - 4(0)(-1) + 4(-1)^2 - 4(0) + 8(-1) + 4 = 0$$

■■■■

Group Discussion

Suppose $f(x, y)$ has a relative extremum (relative maximum or relative minimum) at a point (a, b). Let $g(x) = f(x, b)$ and $h(y) = f(a, y)$. Assuming that f and g are differentiable, explain why $g'(a) = 0$ and $h'(b) = 0$. Explain why these results are equivalent to the conditions $f_x(a, b) = 0$ and $f_y(a, b) = 0$.

EXAMPLE 3

Find the relative extrema of the function

$$f(x, y) = 4y^3 + x^2 - 12y^2 - 36y + 2$$

SOLUTION ✔

To find the critical points of f, we set $f_x = 0$ and $f_y = 0$ simultaneously, obtaining

$$f_x = 2x = 0$$

$$f_y = 12y^2 - 24y - 36 = 0$$

The first equation implies that $x = 0$. The second equation implies that

$$y^2 - 2y - 3 = 0$$

$$(y + 1)(y - 3) = 0$$

—that is, $y = -1$ or 3. Therefore, there are two critical points of the function f—namely, $(0, -1)$ and $(0, 3)$.

Next, we apply the second derivative test to determine the nature of each of the two critical points. We compute

$$f_{xx} = 2, \quad f_{xy} = 0, \quad f_{yy} = 24y - 24 = 24(y - 1)$$

Therefore,

$$D(x, y) = f_{xx}f_{yy} - f_{xy}^2 = 48(y - 1)$$

For the point $(0, -1)$,

$$D(0, -1) = 48(-1 - 1) = -96 < 0$$

Since $D(0, -1) < 0$, we conclude that the point $(0, -1)$ gives a saddle point of f. For the point $(0, 3)$,

$$D(0, 3) = 48(3 - 1) = 96 > 0$$

Since $D(0, 3) > 0$ and $f_{xx}(0, 3) > 0$, we conclude that the function f has a relative minimum at the point $(0, 3)$. Furthermore, since

$$f(0, 3) = 4(3)^3 + (0)^2 - 12(3)^2 - 36(3) + 2$$
$$= -106$$

we see that the relative minimum value of f is -106.   ■ ■ ■ ■

APPLICATIONS

As in the case of a practical optimization problem involving a function of one variable, the solution to an optimization problem involving a function of several variables calls for finding the *absolute* extremum of the function. Determining the absolute extremum of a function of several variables is more difficult than merely finding the relative extrema of the function. However, in many situations, the absolute extremum of a function actually coincides with the largest relative extremum of the function that occurs in the interior of its domain. We assume that the problems considered here belong to this category. Furthermore, the existence of the absolute extremum (solution) of a practical problem is often deduced from the geometric or physical nature of the problem.

EXAMPLE 4   The total weekly revenue (in dollars) that the Acrosonic Company realizes in producing and selling its bookshelf loudspeaker systems is given by

$$R(x, y) = -\frac{1}{4}x^2 - \frac{3}{8}y^2 - \frac{1}{4}xy + 300x + 240y$$

where x denotes the number of fully assembled units and y denotes the number of kits produced and sold per week. The total weekly cost attributable to the production of these loudspeakers is

$$C(x, y) = 180x + 140y + 5000$$

dollars, where x and y have the same meaning as before. Determine how many assembled units and how many kits Acrosonic should produce per week to maximize its profit.

SOLUTION ✔

The contribution to Acrosonic's weekly profit stemming from the production and sale of the bookshelf loudspeaker systems is given by

$$P(x, y) = R(x, y) - C(x, y)$$

$$= \left(-\frac{1}{4}x^2 - \frac{3}{8}y^2 - \frac{1}{4}xy + 300x + 240y \right) - (180x + 140y + 5000)$$

$$= -\frac{1}{4}x^2 - \frac{3}{8}y^2 - \frac{1}{4}xy + 120x + 100y - 5000$$

To find the relative maximum of the profit function $P(x, y)$, we first locate the critical point(s) of P. Setting $P_x(x, y)$ and $P_y(x, y)$ equal to zero, we obtain

$$P_x = -\frac{1}{2}x - \frac{1}{4}y + 120 = 0$$

$$P_y = -\frac{3}{4}y - \frac{1}{4}x + 100 = 0$$

Solving the first of these equations for y yields

$$y = -2x + 480$$

which, upon substitution into the second equation, yields

$$-\frac{3}{4}(-2x + 480) - \frac{1}{4}x + 100 = 0$$

$$6x - 1440 - x + 400 = 0$$

$$x = 208$$

We substitute this value of x into the equation $y = -2x + 480$ to get

$$y = 64$$

Therefore, the function P has the sole critical point $(208, 64)$. To show that the point $(208, 64)$ is a solution to our problem, we use the second derivative test. We compute

$$P_{xx} = -\frac{1}{2}, \qquad P_{xy} = -\frac{1}{4}, \qquad P_{yy} = -\frac{3}{4}$$

So,

$$D(x, y) = \left(-\frac{1}{2} \right)\left(-\frac{3}{4} \right) - \left(-\frac{1}{4} \right)^2 = \frac{3}{8} - \frac{1}{16} = \frac{5}{16}$$

In particular, $D(208, 64) = 5/16 > 0$.

Since $D(208, 64) > 0$ and $P_{xx}(208, 64) < 0$, the point $(208, 64)$ yields a relative maximum of P. This relative maximum is also the absolute maximum of P. We conclude that Acrosonic can maximize its weekly profit by manufacturing 208 assembled units and 64 kits of their bookshelf loudspeaker systems. The maximum weekly profit realizable from the production and sale of these loudspeaker systems is given by

$$P(208, 64) = -\frac{1}{4}(208)^2 - \frac{3}{8}(64)^2 - \frac{1}{4}(208)(64)$$
$$+ 120(208) + 100(64) - 5000$$
$$= 10,680$$

or $10,680.   ■ ■ ■ ■

EXAMPLE 5   A television relay station will serve towns A, B, and C, whose relative locations are shown in Figure 8.20. Determine a site for the location of the station if the sum of the squares of the distances from each town to the site is minimized.

SOLUTION ✔   Suppose the required site is located at the point $P(x, y)$. With the aid of the distance formula, we find that the square of the distance from town A to the site is

$$(x - 30)^2 + (y - 20)^2$$

FIGURE 8.20
Locating a site for a television relay station

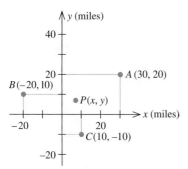

The respective distances from towns B and C to the site are found in a similar manner, so the sum of the squares of the distances from each town to the site is given by

$$f(x, y) = (x - 30)^2 + (y - 20)^2 + (x + 20)^2$$
$$+ (y - 10)^2 + (x - 10)^2 + (y + 10)^2$$

To find the relative minimum of $f(x, y)$, we first find the critical point(s) of f. Using the chain rule to find $f_x(x, y)$ and $f_y(x, y)$ and setting each equal to zero, we obtain

$$f_x = 2(x - 30) + 2(x + 20) + 2(x - 10) = 6x - 40 = 0$$
$$f_y = 2(y - 20) + 2(y - 10) + 2(y + 10) = 6y - 40 = 0$$

from which we deduce that $(20/3, 20/3)$ is the sole critical point of f. Since

$$f_{xx} = 6, \qquad f_{xy} = 0, \qquad f_{yy} = 6$$

we have

$$D(x, y) = f_{xx}f_{yy} - f_{xy}^2 = (6)(6) - 0 = 36$$

Since $D(20/3, 20/3) > 0$ and $f_{xx}(20/3, 20/3) > 0$, we conclude that the point $(20/3, 20/3)$ yields a relative minimum of f. Thus, the required site has coordinates $x = 20/3$ and $y = 20/3$.   ■ ■ ■ ■

1. Let $f(x, y) = 2x^2 + 3y^2 - 4xy + 4x - 2y + 3$.
 a. Find the critical point of f.
 b. Use the second derivative test to classify the nature of the critical point.
 c. Find the relative extremum of f, if it exists.

2. The Robertson Controls Company manufactures two basic models of setback thermostats: a standard mechanical thermostat and a deluxe electronic thermostat. Robertson's monthly revenue (in hundreds of dollars) is

$$R(x, y) = -\frac{1}{8}x^2 - \frac{1}{2}y^2 - \frac{1}{4}xy + 20x + 60y$$

where x (in units of a hundred) denotes the number of mechanical thermostats manufactured and y (in units of a hundred) denotes the number of electronic thermostats manufactured per month. The total monthly cost incurred in producing these thermostats is

$$C(x, y) = 7x + 20y + 280$$

hundred dollars. Find how many thermostats of each model Robertson should manufacture per month in order to maximize its profits. What is the maximum profit?

Solutions to Self-Check Exercises 8.3 can be found on page 640.

8.3 Exercises

In Exercises 1–20, find the critical point(s) of the function. Then use the second derivative test to classify the nature of each point, if possible. Finally, determine the relative extrema of the function.

1. $f(x, y) = 1 - 2x^2 - 3y^2$

2. $f(x, y) = x^2 - xy + y^2 + 1$

3. $f(x, y) = x^2 - y^2 - 2x + 4y + 1$

4. $f(x, y) = 2x^2 + y^2 - 4x + 6y + 3$

5. $f(x, y) = x^2 + 2xy + 2y^2 - 4x + 8y - 1$

6. $f(x, y) = x^2 - 4xy + 2y^2 + 4x + 8y - 1$

7. $f(x, y) = 2x^3 + y^2 - 9x^2 - 4y + 12x - 2$

8. $f(x, y) = 2x^3 + y^2 - 6x^2 - 4y + 12x - 2$

9. $f(x, y) = x^3 + y^2 - 2xy + 7x - 8y + 4$

10. $f(x, y) = 2y^3 - 3y^2 - 12y + 2x^2 - 6x + 2$

11. $f(x, y) = x^3 - 3xy + y^3 - 2$

12. $f(x, y) = x^3 - 2xy + y^2 + 5$

13. $f(x, y) = xy + \frac{4}{x} + \frac{2}{y}$

14. $f(x, y) = \frac{x}{y^2} + xy$

15. $f(x, y) = x^2 - e^{y^2}$

16. $f(x, y) = e^{x^2 - y^2}$

17. $f(x, y) = e^{x^2 + y^2}$

18. $f(x, y) = e^{xy}$

19. $f(x, y) = \ln(1 + x^2 + y^2)$

20. $f(x, y) = xy + \ln x + 2y^2$

21. MAXIMIZING PROFIT The total weekly revenue (in dollars) of the Country Workshop realized in manufacturing and selling its rolltop desks is given by

$$R(x, y) = -0.2x^2 - 0.25y^2 - 0.2xy + 200x + 160y$$

where x denotes the number of finished units and y denotes the number of unfinished units manufactured and sold per week. The total weekly cost attributable to the manufacture of these desks is given by

$$C(x, y) = 100x + 70y + 4000$$

dollars. Determine how many finished units and how many unfinished units the company should manufacture per week in order to maximize its profit. What is the maximum profit realizable?

22. **MAXIMIZING PROFIT** The total daily revenue (in dollars) that the Weston Publishing Company realizes in publishing and selling its English-language dictionaries is given by

$$R(x, y) = -0.005x^2 - 0.003y^2 - 0.002xy$$
$$+ 20x + 15y$$

where x denotes the number of deluxe copies and y denotes the number of standard copies published and sold daily. The total daily cost of publishing these dictionaries is given by

$$C(x, y) = 6x + 3y + 200$$

dollars. Determine how many deluxe copies and how many standard copies Weston should publish per day to maximize its profits. What is the maximum profit realizable?

23. **MAXIMUM PRICE** The rectangular region R shown in the accompanying figure represents the financial district of a city. The price of land within the district is approximated by the function

$$p(x, y) = 200 - 10\left(x - \frac{1}{2}\right)^2 - 15(y - 1)^2$$

where $p(x, y)$ is the price of land at the point (x, y) in dollars per square foot and x and y are measured in miles. At what point within the financial district is the price of land highest?

24. **MAXIMIZING PROFIT** C&G Imports, Inc., imports two brands of white wine, one from Germany and the other from Italy. The German wine costs $4/bottle, and the Italian wine can be obtained for $3/bottle. It has been estimated that if the German wine retails at p dollars/bottle and the Italian wine is sold for q dollars/bottle, then

$$2000 - 150p + 100q$$

bottles of the German wine and

$$1000 + 80p - 120q$$

bottles of the Italian wine will be sold per week. Determine the unit price for each brand that will allow C&G to realize the largest possible weekly profit.

25. **DETERMINING THE OPTIMAL SITE** An auxiliary electric power station will serve three communities, A, B, and C, whose relative locations are shown in the accompanying figure. Determine where the power station should be located if the sum of the squares of the distances from each community to the site is minimized.

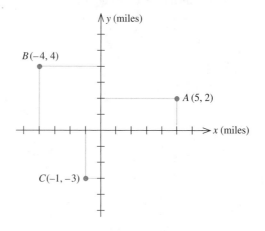

26. **PACKAGING** An open rectangular box having a volume of 108 in.³ is to be constructed from a tin sheet. Find the dimensions of such a box if the amount of material used in its construction is to be minimal.
Hint: Let the dimensions of the box be x'' by y'' by z''. Then, $xyz = 108$ and the amount of material used is given by $S = xy + 2yz + 2xz$. Show that

$$S = f(x, y) = xy + \frac{216}{x} + \frac{216}{y}$$

Minimize $f(x, y)$.

27. PACKAGING Postal regulations specify that the combined length and girth of a parcel sent by parcel post may not exceed 108 in. Find the dimensions of the rectangular package that would have the greatest possible volume under these regulations.

Hint: Let the dimensions of the box be x'' by y'' by z'' (see the figure below). Then, $2x + 2z + y = 108$, and the volume $V = xyz$. Show that

$$V = f(x, z) = 108xz - 2x^2z - 2xz^2$$

Maximize $f(x, z)$.

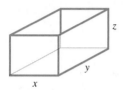

28. MINIMIZING HEATING AND COOLING COSTS A building in the shape of a rectangular box is to have a volume of 12,000 ft³ (see the figure). It is estimated that the annual heating and cooling costs will be $2/square foot for the top, $4/square foot for the front and back, and $3/square foot for the sides. Find the dimensions of the building that will result in a minimal annual heating and cooling cost. What is the minimal annual heating and cooling cost?

29. PACKAGING An open box having a volume of 48 in.³ is to be constructed. If the box is to include a partition that is parallel to a side of the box, as shown in the figure, and the amount of material used is to be minimal, what should be the dimensions of the box?

In Exercises 30 and 31, determine whether the statement is true or false. If it is true, explain why it is true. If it is false, give an example to show why it is false.

30. If $f_x(a, b) = 0$ and $f_y(a, b) = 0$, then f must have a relative extremum at (a, b).

31. If (a, b) is a critical point of f and both the conditions $f_{xx}(a, b) < 0$ and $f_{yy}(a, b) < 0$ hold, then f has a relative maximum at (a, b).

SOLUTIONS TO SELF-CHECK EXERCISES 8.3

1. a. To find the critical point(s) of f, we solve the system of equations

$$f_x = 4x - 4y + 4 = 0$$
$$f_y = -4x + 6y - 2 = 0$$

obtaining $x = -2$ and $y = -1$. Thus, the only critical point of f is the point $(-2, -1)$.

b. We have $f_{xx} = 4$, $f_{xy} = -4$, and $f_{yy} = 6$, so

$$D(x, y) = f_{xx}f_{yy} - f_{xy}^2$$
$$= (4)(6) - (-4)^2 = 8$$

Since $D(-2, -1) > 0$ and $f_{xx}(-2, -1) > 0$, we conclude that f has a relative minimum at the point $(-2, -1)$.

c. The relative minimum value of $f(x, y)$ at the point $(-2, -1)$ is

$$f(-2, -1) = 2(-2)^2 + 3(-1)^2 - 4(-2)(-1) + 4(-2) - 2(-1) + 3$$
$$= 0$$

2. Robertson's monthly profit is

$$P(x, y) = R(x, y) - C(x, y)$$
$$= \left(-\frac{1}{8}x^2 - \frac{1}{2}y^2 - \frac{1}{4}xy + 20x + 60y \right) - (7x + 20y + 280)$$
$$= -\frac{1}{8}x^2 - \frac{1}{2}y^2 - \frac{1}{4}xy + 13x + 40y - 280$$

The critical point of P is found by solving the system

$$P_x = -\frac{1}{4}x - \frac{1}{4}y + 13 = 0$$

$$P_y = -\frac{1}{4}x - \ y + 40 = 0$$

giving $x = 16$ and $y = 36$. Thus, $(16, 36)$ is the critical point of P. Next,

$$P_{xx} = -\frac{1}{4}, \qquad P_{xy} = -\frac{1}{4}, \qquad P_{yy} = -1$$

and

$$D(x, y) = f_{xx}f_{yy} - f_{xy}^2$$
$$= \left(-\frac{1}{4} \right)(-1) - \left(-\frac{1}{4} \right)^2 = \frac{3}{16}$$

Since $D(16, 36) > 0$ and $P_{xx}(16, 36) < 0$, the point $(16, 36)$ yields a relative maximum of P. We conclude that the monthly profit is maximized by manufacturing 1600 mechanical and 3600 electronic setback thermostats per month. The maximum monthly profit realizable is

$$P(16, 36) = -\frac{1}{8}(16)^2 - \frac{1}{2}(36)^2 - \frac{1}{4}(16)(36) + 13(16) + 40(36) - 280$$
$$= 544$$

or $54,400.

8.4 The Method of Least Squares

THE METHOD OF LEAST SQUARES

In Section 1.4, Example 10, we saw how a linear equation can be used to approximate the sales trend for a local sporting goods store. As we saw there, one use of a **trend line** is to predict a store's future sales. Recall that we obtained the line by requiring that it pass through two data points, the rationale being that such a line seems to *fit* the data reasonably well.

FIGURE 8.21

A scatter diagram

In this section we describe a general method, known as the **method of least squares,** for determining a straight line that, in some sense, *best* fits a set of data points when the points are scattered about a straight line. To illustrate the principle behind the method of least squares, suppose, for simplicity, that we are given five data points,

$$P_1(x_1, y_1), \qquad P_2(x_2, y_2), \qquad P_3(x_3, y_3), \qquad P_4(x_4, y_4), \qquad P_5(x_5, y_5)$$

that describe the relationship between the two variables x and y. By plotting these data points, we obtain a graph called a **scatter diagram** (Figure 8.21).

If we try to fit a straight line to these data points, the line will miss the first, second, third, fourth, and fifth data points by the amounts $d_1, d_2, d_3, d_4,$ and d_5, respectively (Figure 8.22).

The **principle of least squares** states that the straight line L that fits the data points best is the one chosen by requiring that the sum of the squares of $d_1, d_2, \ldots, d_5$—that is,

$$d_1^2 + d_2^2 + d_3^2 + d_4^2 + d_5^2$$

FIGURE 8.22

The approximating line misses each point by the amounts $d_1, d_2, \ldots, d_5$.

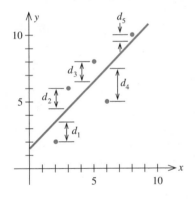

be made as small as possible. If we think of the amount d_1 as the error made when the value y_1 is approximated by the corresponding value of y lying on the straight line L, and d_2 as the error made when the value y_2 is approximated by the corresponding value of y, and so on, then it can be seen that the least-squares criterion calls for minimizing the sum of the squares of the errors. The line L obtained in this manner is called the **least-squares line,** or **regression line.**

To find a method for computing the regression line L, suppose L has representation $y = f(x) = mx + b$, where m and b are to be determined. Observe that

$$
\begin{aligned}
d_1^2 &+ d_2^2 + d_3^2 + d_4^2 + d_5^2 \\
&= [f(x_1) - y_1]^2 + [f(x_2) - y_2]^2 + [f(x_3) - y_3]^2 \\
&\quad + [f(x_4) - y_4]^2 + [f(x_5) - y_5]^2 \\
&= (mx_1 + b - y_1)^2 + (mx_2 + b - y_2)^2 + (mx_3 + b - y_3)^2 \\
&\quad + (mx_4 + b - y_4)^2 + (mx_5 + b - y_5)^2
\end{aligned}
$$

and may be viewed as a function of the two variables m and b. Thus, the least-squares criterion is equivalent to minimizing the function

$$
\begin{aligned}
f(m, b) &= (mx_1 + b - y_1)^2 + (mx_2 + b - y_2)^2 + (mx_3 + b - y_3)^2 \\
&\quad + (mx_4 + b - y_4)^2 + (mx_5 + b - y_5)^2
\end{aligned}
$$

with respect to m and b. Using the chain rule, we compute

$$
\begin{aligned}
\frac{\partial f}{\partial m} &= 2(mx_1 + b - y_1)x_1 + 2(mx_2 + b - y_2)x_2 + 2(mx_3 + b - y_3)x_3 \\
&\quad + 2(mx_4 + b - y_4)x_4 + 2(mx_5 + b - y_5)x_5 \\
&= 2[mx_1^2 + bx_1 - x_1y_1 + mx_2^2 + bx_2 - x_2y_2 + mx_3^2 + bx_3 - x_3y_3 \\
&\quad + mx_4^2 + bx_4 - x_4y_4 + mx_5^2 + bx_5 - x_5y_5] \\
&= 2[(x_1^2 + x_2^2 + x_3^2 + x_4^2 + x_5^2)m + (x_1 + x_2 + x_3 + x_4 + x_5)b \\
&\quad - (x_1y_1 + x_2y_2 + x_3y_3 + x_4y_4 + x_5y_5)]
\end{aligned}
$$

and

$$\frac{\partial f}{\partial b} = 2(mx_1 + b - y_1) + 2(mx_2 + b - y_2) + 2(mx_3 + b - y_3)$$
$$+ 2(mx_4 + b - y_4) + 2(mx_5 + b - y_5)$$
$$= 2[(x_1 + x_2 + x_3 + x_4 + x_5)m + 5b - (y_1 + y_2 + y_3 + y_4 + y_5)]$$

Setting

$$\frac{\partial f}{\partial m} = 0 \quad \text{and} \quad \frac{\partial f}{\partial b} = 0$$

gives

$$(x_1^2 + x_2^2 + x_3^2 + x_4^2 + x_5^2)m + (x_1 + x_2 + x_3 + x_4 + x_5)b$$
$$= x_1 y_1 + x_2 y_2 + x_3 y_3 + x_4 y_4 + x_5 y_5$$

and

$$(x_1 + x_2 + x_3 + x_4 + x_5)m + 5b = y_1 + y_2 + y_3 + y_4 + y_5$$

Solving these two simultaneous equations for m and b then leads to an equation $y = mx + b$ of a straight line.

Before looking at an example, we state a more general result whose derivation is identical to the special case involving the five data points just discussed.

The Method of Least Squares

Suppose we are given n data points:

$$P_1(x_1, y_1), \quad P_2(x_2, y_2), \quad P_3(x_3, y_3), \dots, P_n(x_n, y_n)$$

Then, the least-squares (regression) line for the data is given by the linear equation

$$y = f(x) = mx + b$$

where the constants m and b satisfy the equations

$$(x_1^2 + x_2^2 + x_3^2 + \cdots + x_n^2)m + (x_1 + x_2 + x_3 + \cdots + x_n)b$$
$$= x_1 y_1 + x_2 y_2 + x_3 y_3 + \cdots + x_n y_n \tag{4}$$

and

$$(x_1 + x_2 + x_3 + \cdots + x_n)m + nb$$
$$= y_1 + y_2 + y_3 + \cdots + y_n \tag{5}$$

simultaneously. Equations (4) and (5) are called **normal equations**.

EXAMPLE 1 Find an equation of the least-squares line for the data

$$P_1(1, 1), \qquad P_2(2, 3), \qquad P_3(3, 4), \qquad P_4(4, 3), \qquad P_5(5, 6)$$

SOLUTION ✔ Here, we have $n = 5$ and

$$x_1 = 1, \qquad x_2 = 2, \qquad x_3 = 3, \qquad x_4 = 4, \qquad x_5 = 5$$
$$y_1 = 1, \qquad y_2 = 3, \qquad y_3 = 4, \qquad y_4 = 3, \qquad y_5 = 6$$

so Equation (4) becomes

$$(1 + 4 + 9 + 16 + 25)m + (1 + 2 + 3 + 4 + 5)b = 1 + 6 + 12 + 12 + 30$$

or

$$55m + 15b = 61 \tag{6}$$

and (5) becomes

$$(1 + 2 + 3 + 4 + 5)m + 5b = 1 + 3 + 4 + 3 + 6$$

or

$$15m + 5b = 17 \tag{7}$$

Solving Equation (7) for b gives

$$b = -3m + \frac{17}{5} \tag{8}$$

which, upon substitution into (6), gives

$$15\left(-3m + \frac{17}{5}\right) + 55m = 61$$
$$-45m + 51 + 55m = 61$$
$$10m = 10$$
$$m = 1$$

Substituting this value of m into (8) gives

$$b = -3 + \frac{17}{5} = \frac{2}{5} = 0.4$$

Therefore, the required least-squares line is

$$y = x + 0.4$$

The scatter diagram and the regression line are shown in Figure 8.23.   ■ ■ ■ ■

FIGURE 8.23
The scatter diagram and the least-squares line $y = x + 0.4$

APPLICATIONS

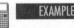 **EXAMPLE 2** The proprietor of the Leisure Travel Service compiled the following data relating the firm's annual profit to its annual advertising expenditure (both measured in thousands of dollars).

Annual Advertising Expenditure, x	12	14	17	21	26	30
Annual Profit, y	60	70	90	100	100	120

a. Determine an equation of the least-squares line for these data.
b. Draw a scatter diagram and the least-squares line for these data.
c. Use the result obtained in part (a) to predict Leisure Travel's annual profit if the annual advertising budget is $20,000.

SOLUTION ✔

a. The calculations required for obtaining the normal equations may be summarized as follows:

	x	y	x^2	xy
	12	60	144	720
	14	70	196	980
	17	90	289	1,530
	21	100	441	2,100
	26	100	676	2,600
	30	120	900	3,600
Sum	120	540	2,646	11,530

The normal equations are

$$6b + 120m = 540 \tag{9}$$

$$120b + 2646m = 11,530 \tag{10}$$

Solving Equation (9) for b gives

$$b = -20m + 90 \tag{11}$$

which, upon substitution into Equation (10), gives

$$120(-20m + 90) + 2646m = 11,530$$
$$-2400m + 10,800 + 2646m = 11,530$$
$$246m = 730$$
$$m \approx 2.97$$

Substituting this value of m into Equation (11) gives

$$b = -20(2.97) + 90 = 30.6$$

Therefore, the required least-squares line is given by

$$y = f(x) = 2.97x + 30.6$$

b. The scatter diagram and the least-squares line are shown in Figure 8.24.
c. Leisure Travel's predicted annual profit corresponding to an annual budget of $20,000 is given by

$$f(20) = 2.97(20) + 30.6$$
$$= 90$$

or $90,000.

FIGURE 8.24
The scatter diagram and the least-squares line $y = 2.97x + 30.6$

■ ■ ■ ■

EXAMPLE 3   A market research study conducted for the Century Communications Company provided the following data based on the projected monthly sales x (in thousands) of Century's videocassette version of a box-office hit adventure movie with a proposed wholesale unit price of p dollars.

p	38	36	34.5	30	28.5
x	2.2	5.4	7.0	11.5	14.6

a. Find the demand equation if the demand curve is the least-squares line for these data.
b. Suppose the total monthly cost function associated with producing and distributing the videocassette movies is given by

$$C(x) = 4x + 25$$

where x denotes the number of units (in thousands) produced and sold and $C(x)$ is in thousands of dollars. Determine the unit wholesale price that will maximize Century's monthly profit.

SOLUTION ✔   **a.** The calculations required for obtaining the normal equations may be summarized as follows:

	x	p	x^2	xp
	2.2	38	4.84	83.6
	5.4	36	29.16	194.4
	7.0	34.5	49	241.5
	11.5	30	132.25	345
	14.6	28.5	213.16	416.1
Sum	40.7	167	428.41	1280.6

The normal equations are

$$5b + \quad 40.7m = 167$$
$$40.7b + 428.41m = 1280.6$$

Solving this system of linear equations simultaneously, we find that

$$m \approx -0.81 \quad \text{and} \quad b \approx 39.99$$

Therefore, the required least-squares line is given by

$$p = f(x) = -0.81x + 39.99$$

which is the required demand equation, provided $0 \leq x \leq 49.37$.
b. The total revenue function in this case is given by

$$R(x) = xp = -0.81x^2 + 39.99x$$

and since the total cost function is

$$C(x) = 4x + 25$$

we see that the profit function is

$$P(x) = -0.81x^2 + 39.99x - (4x + 25)$$
$$= -0.81x^2 + 35.99x - 25$$

To find the absolute minimum of $P(x)$ over the closed interval $[0, 49.37]$, we compute

$$P'(x) = -1.62x + 35.99$$

Since $P'(x) = 0$, we find $x \approx 22.22$ as the only critical point of P. Finally, from the table

x	0	22.22	49.37
$P(x)$	−25	374.78	−222.47

we see that the optimal wholesale price is $22.22 per videocassette.

■ ■ ■ ■

SELF-CHECK EXERCISES 8.4

1. Find an equation of the least-squares line for the data

$$P_1(0, 3), \quad P_2(2, 6.5), \quad P_3(4, 10), \quad P_4(6, 16), \quad P_5(7, 16.5)$$

2. The following data obtained from the U.S. Department of Commerce give the percentage of people over age 65 who have high school diplomas.

Year, x	0	6	11	16	22	26
Percentage with Diplomas, y	19	25	30	35	44	48

Here, $x = 0$ corresponds to the beginning of the year 1959.
a. Find an equation of the least-squares line for the given data.
b. Assuming that this trend continues, what percentage of people over 65 years of age will have high school diplomas at the beginning of the year 2003 ($x = 44$)?
Source: U.S. Department of Commerce

Solutions to Self-Check Exercises 8.4 can be found on page 653.

8.4 Exercises

A calculator is recommended for this exercise set. In Exercises 1–6, (a) find the equation of the least-squares line for the given data and (b) draw a scatter diagram for the given data and graph the least-squares line.

1.

x	1	2	3	4
y	4	6	8	11

2.

x	1	3	5	7	9
y	9	8	6	3	2

3.

x	1	2	3	4	4	6
y	4.5	5	3	2	3.5	1

4.

x	1	1	2	3	4	4	5
y	2	3	3	3.5	3.5	4	5

5. $P_1(1, 3)$, $P_2(2, 5)$, $P_3(3, 5)$, $P_4(4, 7)$, $P_5(5, 8)$

6. $P_1(1, 8)$, $P_2(2, 6)$, $P_3(5, 6)$, $P_4(7, 4)$, $P_5(10, 1)$

7. COLLEGE ADMISSIONS The following data were compiled by the admissions office at Faber College during the past 5 yr. The data relate the number of college brochures and follow-up letters (x) sent to a preselected list of high school juniors who had taken the PSAT and the number of completed applications (y) received from these students (both measured in units of 1000).

x	4	4.5	5	5.5	6
y	0.5	0.6	0.8	0.9	1.2

a. Determine the equation of the least-squares line for these data.
b. Draw a scatter diagram and the least-squares line for these data.

c. Use the result obtained in part (a) to predict the number of completed applications that might be expected if 6400 brochures and follow-up letters are sent out during the next year.

8. NET SALES The management of Kaldor, Inc., a manufacturer of electric motors, submitted the following data in the annual report to its stockholders. The table shows the net sales (in millions of dollars) during the 5 yr that have elapsed since the new management team took over. (The first year the firm operated under the new management corresponds to the time period $x = 1$, and the four subsequent years correspond to $x = 2, 3, 4, 5$.)

Year, x	1	2	3	4	5
Net Sales, y	426	437	460	473	477

a. Determine the equation of the least-squares line for these data.
b. Draw a scatter diagram and the least-squares line for these data.
c. Use the result obtained in part (a) to predict the net sales for the upcoming year.

9. SAT VERBAL SCORES The following data were compiled by the superintendent of schools in a large metropolitan area. The table shows the average SAT verbal scores of high school seniors during the 5 yr since the district implemented the "back-to-basics" program.

Year, x	1	2	3	4	5
Average Score, y	436	438	428	430	426

a. Determine the equation of the least-squares line for these data.
b. Draw a scatter diagram and the least-squares line for these data.
c. Use the result obtained in part (a) to predict the average SAT verbal score of high school seniors 2 yr from now ($x = 7$).

10. AUTO OPERATING COSTS The following figures were compiled by Clarke, Kingsley, and Company, a consulting firm that specializes in auto operating costs, relating the annual mileage (in thousands of miles) that an average new compact car is driven to the cost per mile (in cents) of operating the car.

Annual Mileage, x	5	10	15	20	25	30
Cost per Mile, y	50.3	34.8	30.1	27.4	25.6	23.5

a. Determine an equation of the least-squares line for these data.
b. Draw a scatter diagram and the least-squares line for these data.
c. Use the result obtained in part (a) to estimate the cost per mile of operating a new company car if it is driven 8000 mi during the first year of ownership.
Source: Clarke, Kingsley, and Company

11. **SIZE OF AVERAGE FARM** The size of the average farm in the United States has been growing steadily over the years. The following data, obtained from the U.S. Department of Agriculture, give the size of the average farm y (in acres) from 1940 through 1991. (Here, x = 0 corresponds to the beginning of the year 1940).

Year, x	0	10	20	30	40	51
Size of Farm, y	168	213	297	374	427	467

a. Find the equation of the least-squares line for these data.
b. Use the result of part (a) to estimate the size of the average farm in the year 2003.
Source: The World Almanac

12. **WELFARE COSTS** According to the Massachusetts Department of Welfare, the spending (in billions of dollars) by Medicaid, the national health-care plan for the poor, over the 5-yr period from 1988 to 1992 is summarized in the following table. (Here, x = 0 represents the beginning of the year 1988.)

Year, x	0	1	2	3	4
Expenditure, y	1.550	1.662	1.786	1.888	2.009

a. Find an equation of the least-squares line for these data.
b. Use the result of part (a) to estimate Medicaid spending for the year 2002, assuming the trend continued.
Source: Massachusetts Department of Welfare

13. **MASS TRANSIT SUBSIDIES** The following table gives the projected state subsidies (in millions of dollars) to the Massachusetts Bay Transit Authority (MBTA) over a 5-yr period.

Year, x	1	2	3	4	5
Subsidy, y	20	24	26	28	32

a. Find an equation of the least-squares line for these data.
b. Use the result of part (a) to estimate the state subsidy to the MBTA for the eighth year (x = 8).
Source: Massachusetts Bay Transit Authority

14. **SOCIAL SECURITY WAGE BASE** The Social Security (FICA) wage base (in thousands of dollars) from 1994 to 1999 is given in the following table.

Year	1994	1995	1996	1997	1998	1999
Wage Base, y	60.6	61.2	62.7	65.4	68.4	72.6

a. Find an equation of the least-squares line for these data. (Let x = 1 represent the year 1994.)
b. Use your result of part (a) to estimate the FICA wage base in the year 2003.
Source: The World Almanac

15. **PRODUCTION OF ALL-ALUMINUM CANS** Steel has been playing a decreasing role in the manufacture of beverage cans in the United States. According to the Can Manufacture Institute, the use of bimetallic cans has been dwindling, whereas the use of all-aluminum cans has been growing steadily. The accompanying table gives the production of all-aluminum cans (in billions) over the period from 1975 through 1989.

Year	1975	1977	1979	1981
Number of Cans	16.7	26	33.3	48.3

Year	1983	1985	1987	1989
Number of Cans	57	65.8	74.2	83.3

a. Find an equation of the least-squares line for these data. (Let x = 1 represent 1975.)
b. Use the result of part (a) to estimate the number of cans produced in 1993, assuming the trend continued.
Source: Can Manufacturing Institute

FINDING AN EQUATION OF A LEAST-SQUARES LINE

A graphing utility is especially useful in calculating an equation of the least-squares line for a set of data. We simply enter the given data in the form of lists into the calculator and then use the linear regression function to obtain the coefficients of the required equation.

EXAMPLE 1 Find an equation of the least-squares line for the following data:

x	1.1	2.3	3.2	4.6	5.8	6.7	8
y	-5.8	-5.1	-4.8	-4.4	-3.7	-3.2	-2.5

SOLUTION ✔ First, we enter the data as

$x_1 = 1.1,$ $\quad y_1 = -5.8,$ $\quad x_2 = 2.3,$ $\quad y_2 = -5.1,$ $\quad x_3 = 3.2,$ $\quad y_3 = -4.8,$ $\quad x_4 = 4.6$

$y_4 = -4.4,$ $\quad x_5 = 5.8,$ $\quad y_5 = -3.7,$ $\quad x_6 = 6.7,$ $\quad y_6 = -3.2,$ $\quad x_7 = 8,$ $\quad y_7 = -2.5$

Then, using the linear regression function from the statistics menu, we find

$a = -6.29996900666,$ $\quad b = 0.460560979389,$ $\quad$ corr. $= .994488871079,$ $\quad n = 7$

Therefore, an equation of the least-squares line ($y = a + bx$) is

$$y = -6.3333 + 0.46x$$

The correlation coefficient of .99449 attests to the excellent fit of the regression line. ∎∎∎∎

EXAMPLE 2 According to Pacific Gas and Electric, the nation's largest utility company, the demand for electricity from 1990 through the year 2000 is as follows:

t	0	2	4	6	8	10
y	333	917	1500	2117	2667	3292

Here, $t = 0$ corresponds to 1990, and y gives the amount of electricity demanded in year t, measured in megawatts. Find an equation of the least-squares line for these data.

Source: Pacific Gas and Electric

SOLUTION ✔ First, we enter the data as

$$x_1 = 0, \quad y_1 = 333, \quad x_2 = 2, \quad y_2 = 917, \quad x_3 = 4, \quad y_3 = 1500,$$

$$x_4 = 6, \quad y_4 = 2117, \quad x_5 = 8, \quad y_5 = 2667, \quad x_6 = 10, \quad y_6 = 3292$$

Then, using the linear regression function from the statistics menu, we find

$$a = 328.476190476 \quad \text{and} \quad b = 295.171428571$$

Therefore, an equation of the least-squares line is

$$y = 328 + 295t$$

■ ■ ■ ■

Exercises

In Exercises 1–4, find an equation of the least-squares line for the data.

1.

x	2.1	3.4	4.7	5.6	6.8	7.2
y	8.8	12.1	14.8	16.9	19.8	21.1

2.

x	1.1	2.4	3.2	4.7	5.6	7.2
y	−0.5	1.2	2.4	4.4	5.7	8.1

3.

x	−2.1	−1.1	0.1	1.4	2.5	4.2	5.1
y	6.2	4.7	3.5	1.9	0.4	−1.4	−2.5

4.

x	−1.12	0.1	1.24	2.76	4.21	6.82
y	7.61	4.9	2.74	−0.47	−3.51	−8.94

5. WASTE GENERATION According to data from the Council on Environmental Quality, the amount of waste (in millions of tons per year) generated in the United States from 1960 to 1990 was:

Year	1960	1965	1970	1975
Amount, y	81	100	120	124

Year	1980	1985	1990
Amount, y	140	152	164

a. Find an equation of the least-squares line for these data. (Let x be in units of 5 and let $x = 1$ represent 1960.)
b. Use the result of part (a) to estimate the amount of waste generated in the year 2000, assuming the trend continues.
Source: Council on Environmental Quality

6. MEDIAN PRICE OF HOMES According to data from the Association of Realtors, the median price (in thousands of dollars) of existing homes in a certain metropolitan area from 1982 to 1992 was:

Year	1982	1983	1984	1985	1986	1987
Price, y	66.4	69.8	72.8	76.0	79.6	83.1

Year	1988	1989	1990	1991	1992
Price, y	86.3	89.5	92.3	96.0	99.5

a. Find an equation of the least-squares line for these data. (Let $x = 1$ represent 1982.)
b. Use the result of part (a) to estimate the median price of a house in the year 1997, assuming the trend continued.
Source: Association of Realtors

16. ON-LINE BANKING According to industry sources, on-line banking is expected to take off in the near future. The projected number of households (in millions) using this service is given in the following table. (Here, $x = 0$ corresponds to the beginning of 1997.)

Year, x	0	1	2	3	4	5
Number of Households, y	4.5	7.5	10.0	13.0	15.6	18.0

a. Find an equation of the least-squares line for these data.
b. Use the result of part (a) to estimate the number of households using on-line banking at the beginning of 2003, assuming the projection is accurate.
Source: Jupiter Communications, Forrester Research Inc.

17. NET-CONNECTED COMPUTERS IN EUROPE The projected number of computers (in millions) connected to the Internet in Europe from 1998 through 2002 is summarized in the accompanying table. (Here, $x = 0$ corresponds to the beginning of 1998.)

Year, x	0	1	2	3	4
Net-Connected Computers, y	21.7	32.1	45.0	58.3	69.6

a. Find an equation of the least-squares line for these data.
b. Use the result of part (a) to estimate the projected number of computers connected to the Internet in Europe at the beginning of 2003, assuming the trend continues.
Source: Dataquest, Inc.

18. PORTABLE-PHONE SERVICES The projected number of wireless subscribers y (in millions) from the year 2000 through 2006 is summarized in the accompanying table. (Here, $x = 0$ corresponds to the beginning of the year 2000.)

Year, x	0	1	2	3	4	5	6
Number of Subscribers, y	90.4	100.0	110.4	120.4	130.8	140.4	150.0

a. Find an equation of the least-squares line for these data.

b. Use the result of part (a) to estimate the projected number of wireless subscribers at the beginning of 2006. How does this result compare with the given data for that year?
Source: BancAmerica Robertson Stephens

19. HEALTH-CARE SPENDING The following data, compiled by the Organization for Economic Cooperation and Development (OECD) in 1990, gives the per capita Gross Domestic Product (GDP) (in thousands of dollars) and the corresponding per capita spending on health care (in dollars) for selected countries.

Country	Turkey	Spain	Netherlands
GDP	4.25	10	14
Health-Care Spending	178	667	1194

Country	Sweden	Switzerland	Canada
GDP	15.5	17.8	19.5
Health-Care Spending	1500	1388	1640

a. Letting x denote a country's GDP (in thousands of dollars per capita) and y denote the per capita health-care spending (in dollars), find an equation of the least-squares line for these data giving the typical relationship between GDP and health-care spending for the selected countries.
b. The per capita GDP of the United States is $20,000. If the health-care spending of the United States were in line with that of these sample OECD countries, what would it be? (*Note:* The actual per capita health-care spending of the United States in 1990 was $2444.)
Source: Organization for Economic Cooperation and Development

In Exercises 20 and 21, determine whether the statement is true or false. If it is true, explain why it is true. If it is false, give an example to show why it is false.

20. The least-squares line must pass through at least one of the data points.

21. The sum of the squares of the errors incurred in approximating n data points using the least-squares linear function is zero if and only if the n data points lie along a straight line.

SOLUTIONS TO SELF-CHECK EXERCISES 8.4

1. We first construct the table:

x	y	x^2	xy
0	3	0	0
2	6.5	4	13
4	10	16	40
6	16	36	96
7	16.5	49	115.5
Sum 19	52	105	264.5

The normal equations are

$$5b + 19m = 52$$
$$19b + 105m = 264.5$$

Solving the first equation for b gives

$$b = -3.8m + 10.4$$

which, upon substitution into the second equation, gives

$$19(-3.8m + 10.4) + 105m = 264.5$$
$$-72.2m + 197.6 + 105m = 264.5$$
$$32.8m = 66.9$$
$$m \approx 2.04$$

Substituting this value of m into the expression for b found earlier gives

$$b = -3.8(2.04) + 10.4 \approx 2.65$$

Therefore, the required least-squares line has the equation given by

$$y = 2.04x + 2.65$$

2. a. The calculations required for obtaining the normal equations may be summarized as follows:

x	y	x^2	xy
0	19	0	0
6	25	36	150
11	30	121	330
16	35	256	560
22	44	484	968
26	48	676	1248
Sum 81	201	1573	3256

The normal equations are

$$6b + \quad 81m = \quad 201$$
$$81b + 1573m = 3256$$

Solving this system of linear equations simultaneously, we find

$$m \approx 1.13 \quad \text{and} \quad b \approx 18.23$$

Therefore, the required least-squares line has the equation given by

$$y = f(x) = 1.13x + 18.23$$

b. The percentage of people over the age of 65 who will have high school diplomas at the beginning of the year 2003 is given by

$$f(44) = 1.13(44) + 18.23$$
$$= 67.95$$

or approximately 68%.

8.5 Constrained Maxima and Minima and the Method of Lagrange Multipliers

CONSTRAINED RELATIVE EXTREMA

In Section 8.3 we studied the problem of determining the relative extremum of a function $f(x, y)$ without placing any restrictions on the independent variables x and y—except, of course, that the point (x, y) lies in the domain of f. Such a relative extremum of a function f is referred to as an **unconstrained relative extremum** of f. However, in many practical optimization problems, we must maximize or minimize a function in which the independent variables are subjected to certain further constraints.

In this section we discuss a powerful method for determining the relative extrema of a function $f(x, y)$ whose independent variables x and y are required to satisfy one or more constraints of the form $g(x, y) = 0$. Such a relative extremum of a function f is called a **constrained relative extremum of f**. We can see the difference between an unconstrained extremum of a function $f(x, y)$ of two variables and a constrained extremum of f, where the independent variables x and y are subjected to a constraint of the form $g(x, y) = 0$, by considering the geometry of the two cases. Figure 8.25a depicts the graph of a function $f(x, y)$ that has an unconstrained relative minimum at the point $(0, 0)$. However, when the independent variables x and y are subjected to an equality constraint of the form $g(x, y) = 0$, the points (x, y, z) that satisfy both $z = f(x, y)$ and the constraint equation $g(x, y) = 0$ lie on a curve C. Therefore, the constrained relative minimum of f must also lie on C (Figure 8.25b).

Our first example involves an equality constraint $g(x, y) = 0$ in which we solve for the variable y explicitly in terms of x. In this case we may apply the

FIGURE 8.25

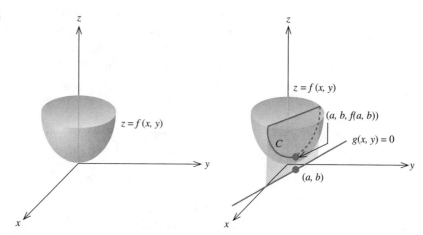

(a) $f(x, y)$ has an unconstrained relative extremum at $(0, 0)$.

(b) $f(x, y)$ has a constrained relative extremum at $(a, b, f(a, b))$.

technique used in Chapter 4 to find the relative extrema of a function of one variable.

EXAMPLE 1   Find the relative minimum of the function

$$f(x, y) = 2x^2 + y^2$$

subject to the constraint $g(x, y) = x + y - 1 = 0$.

SOLUTION ✔   Solving the constraint equation for y explicitly in terms of x, we obtain $y = -x + 1$. Substituting this value of y into the function $f(x, y) = 2x^2 + y^2$ results in a function of x,

$$h(x) = 2x^2 + (-x + 1)^2 = 3x^2 - 2x + 1$$

The function h describes the curve C lying on the graph of f on which the constrained relative minimum of f occurs. To find this point, use the technique developed in Chapter 4 to determine the relative extrema of a function of one variable:

$$h'(x) = 6x - 2 = 2(3x - 1)$$

Setting $h'(x) = 0$ gives $x = \frac{1}{3}$ as the sole critical point of the function h. Next, we find

$$h''(x) = 6$$

and, in particular,

$$h''\left(\frac{1}{3}\right) = 6 > 0$$

Therefore, by the second derivative test, the point $x = \frac{1}{3}$ gives rise to a relative minimum of h. Substitute this value of x into the constraint equation $x + y - 1 = 0$ to get $y = \frac{2}{3}$. Thus, the point $(\frac{1}{3}, \frac{2}{3})$ gives rise to the required constrained relative minimum of f. Since

$$f\left(\frac{1}{3}, \frac{2}{3}\right) = 2\left(\frac{1}{3}\right)^2 + \left(\frac{2}{3}\right)^2 = \frac{2}{3}$$

the required constrained relative minimum value of f is $\frac{2}{3}$ at the point $(\frac{1}{3}, \frac{2}{3})$. It may be shown that $\frac{2}{3}$ is in fact a constrained absolute minimum value of f (Figure 8.26).

FIGURE 8.26
f has a constrained absolute minimum
of $\frac{2}{3}$ at $(\frac{1}{3}, \frac{2}{3})$.

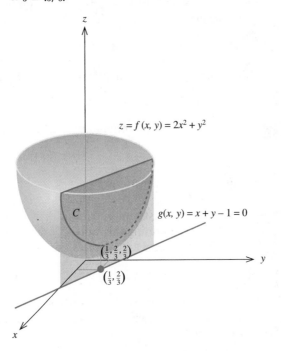

THE METHOD OF LAGRANGE MULTIPLIERS

The major drawback of the technique used in Example 1 is that it relies on our ability to solve the constraint equation $g(x, y) = 0$ for y explicitly in terms of x. This is not always an easy task. Moreover, even when we can solve the constraint equation $g(x, y) = 0$ for y explicitly in terms of x, the resulting function of one variable that is to be optimized may turn out to be unnecessarily complicated. Fortunately, an easier method exists. This method, called the **method of Lagrange multipliers** (Joseph Lagrange, 1736–1813), is as follows:

The Method of Lagrange Multipliers

To find the relative extremum of the function $f(x, y)$ subject to the constraint $g(x, y) = 0$ (assuming that these extreme values exist),

1. Form an auxiliary function

$$F(x, y, \lambda) = f(x, y) + \lambda g(x, y)$$

called the Lagrangian function (the variable λ is called the Lagrange multiplier).

2. Solve the system that consists of the equations

$$F_x = 0, \qquad F_y = 0, \qquad F_\lambda = 0$$

for all values of x, y, and λ.

3. Evaluate f at each of the points (x, y) found in step 2. The largest (smallest) of these values is the maximum (minimum) value of f.

Let's re-solve Example 1 using the method of Lagrange multipliers.

EXAMPLE 2   Using the method of Lagrange multipliers, find the relative minimum of the function

$$f(x, y) = 2x^2 + y^2$$

subject to the constraint $x + y = 1$.

SOLUTION ✔   Write the constraint equation $x + y = 1$ in the form $g(x, y) = x + y - 1 = 0$. Then, form the Lagrangian function

$$F(x, y, \lambda) = f(x, y) + \lambda g(x, y)$$
$$= 2x^2 + y^2 + \lambda(x + y - 1)$$

To find the critical point(s) of the function F, solve the system composed of the equations

$$F_x = 4x + \lambda = 0$$
$$F_y = 2y + \lambda = 0$$
$$F_\lambda = x + y - 1 = 0$$

Solving the first and second equations in this system for x and y in terms of λ, we obtain

$$x = -\frac{1}{4}\lambda \qquad \text{and} \qquad y = -\frac{1}{2}\lambda$$

which, upon substitution into the third equation, yields

$$-\frac{1}{4}\lambda - \frac{1}{2}\lambda - 1 = 0 \qquad \text{or} \qquad \lambda = -\frac{4}{3}$$

Therefore, $x = \frac{1}{3}$ and $y = \frac{2}{3}$, and $(\frac{1}{3}, \frac{2}{3})$ affords a constrained minimum of the function f, in agreement with the result obtained earlier.   ■ ■ ■ ■

The method of Lagrange multipliers may be used to solve a problem involving a function of three or more variables, as illustrated in the next example.

EXAMPLE 3   Use the method of Lagrange multipliers to find the minimum of the function

$$f(x, y, z) = 2xy + 6yz + 8xz$$

subject to the constraint

$$xyz = 12{,}000$$

(*Note:* The existence of the minimum is suggested by the geometry of the problem.)

SOLUTION ✔   Write the constraint equation $xyz = 12{,}000$ in the form $g(x, y, z) = xyz - 12{,}000$. Then, the Lagrangian function is

$$\begin{aligned} F(x, y, z, \lambda) &= f(x, y, z) + \lambda g(x, y, z) \\ &= 2xy + 6yz + 8xz + \lambda(xyz - 12{,}000) \end{aligned}$$

To find the critical point(s) of the function F, we solve the system composed of the equations

$$\begin{aligned} F_x &= 2y + 8z + \lambda yz = 0 \\ F_y &= 2x + 6z + \lambda xz = 0 \\ F_z &= 6y + 8x + \lambda xy = 0 \\ F_\lambda &= xyz - 12{,}000 = 0 \end{aligned}$$

Solving the first three equations of the system for λ in terms of x, y, and z, we have

$$\lambda = -\frac{2y + 8z}{yz}$$

$$\lambda = -\frac{2x + 6z}{xz}$$

$$\lambda = -\frac{6y + 8x}{xy}$$

Equating the first two expressions for λ leads to

$$\frac{2y + 8z}{yz} = \frac{2x + 6z}{xz}$$

$$2xy + 8xz = 2xy + 6yz$$

$$x = \frac{3}{4}y$$

Next, equating the second and third expressions for λ in the same system yields

$$\frac{2x + 6z}{xz} = \frac{6y + 8x}{xy}$$

$$2xy + 6yz = 6yz + 8xz$$

$$z = \frac{1}{4}y$$

Finally, substituting these values of x and z into the equation $xyz - 12{,}000 = 0$, the fourth equation of the first system of equations, we have

$$\left(\frac{3}{4}y\right)(y)\left(\frac{1}{4}y\right) - 12{,}000 = 0$$

$$y^3 = \frac{(12{,}000)(4)(4)}{3} = 64{,}000$$

or

$$y = 40$$

The corresponding values of x and z are given by $x = \frac{3}{4}(40) = 30$ and $z = \frac{1}{4}(40) = 10$. Therefore, we see that the point $(30, 40, 10)$ gives the constrained minimum of f. The minimum value is

$$f(30, 40, 10) = 2(30)(40) + 6(40)(10) + 8(30)(10) = 7200$$

■ ■ ■ ■

APPLICATIONS

EXAMPLE 4

Refer to Example 3, Section 8.1. The total weekly profit (in dollars) that the Acrosonic Company realized in producing and selling its bookshelf loudspeaker systems is given by the profit function

$$P(x, y) = -\frac{1}{4}x^2 - \frac{3}{8}y^2 - \frac{1}{4}xy + 120x + 100y - 5000$$

where x denotes the number of fully assembled units and y denotes the number of kits produced and sold per week. Acrosonic's management decides that production of these loudspeaker systems should be restricted to a total of exactly 230 units per week. Under this condition, how many fully assembled units and how many kits should be produced per week to maximize Acrosonic's weekly profit?

SOLUTION ✔

The problem is equivalent to the problem of maximizing the function

$$P(x, y) = -\frac{1}{4}x^2 - \frac{3}{8}y^2 - \frac{1}{4}xy + 120x + 100y - 5000$$

subject to the constraint

$$g(x, y) = x + y - 230 = 0$$

The Lagrangian function is

$$F(x, y, \lambda) = P(x, y) + \lambda g(x, y)$$

$$= -\frac{1}{4}x^2 - \frac{3}{8}y^2 - \frac{1}{4}xy + 120x + 100y$$

$$-5000 + \lambda(x + y - 230)$$

To find the critical point(s) of F, solve the following system of equations:

$$F_x = -\frac{1}{2}x - \frac{1}{4}y + 120 + \lambda = 0$$

$$F_y = -\frac{3}{4}y - \frac{1}{4}x + 100 + \lambda = 0$$

$$F_\lambda = x + y - 230 = 0$$

Solving the first equation of this system for λ, we obtain

$$\lambda = \frac{1}{2}x + \frac{1}{4}y - 120$$

which, upon substitution into the second equation, yields

$$-\frac{3}{4}y - \frac{1}{4}x + 100 + \frac{1}{2}x + \frac{1}{4}y - 120 = 0$$

$$-\frac{1}{2}y + \frac{1}{4}x - 20 = 0$$

Solving the last equation for y gives

$$y = \frac{1}{2}x - 40$$

When we substitute this value of y into the third equation of the system, we have

$$x + \frac{1}{2}x - 40 - 230 = 0$$

$$x = 180$$

The corresponding value of y is $(\frac{1}{2})(180) - 40$, or 50. Thus, the required constrained relative maximum of P occurs at the point (180, 50). Again, we can show that the point (180, 50) in fact yields a constrained absolute maximum for P. Thus, Acrosonic's profit is maximized by producing 180 assembled and 50 kit versions of their bookshelf loudspeaker systems. The maximum weekly profit realizable is given by

$$P(180, 50) = -\frac{1}{4}(180)^2 - \frac{3}{8}(50)^2 - \frac{1}{4}(180)(50)$$

$$+ 120(180) + 100(50) - 5000$$

$$= 10{,}312.5$$

or $10,312.50.

| EXAMPLE **5** |

The operators of the *Viking Princess,* a luxury cruise liner, are contemplating the addition of another swimming pool to the ship. The chief engineer has suggested that an area in the form of an ellipse located in the rear of the promenade deck would be suitable for this purpose. This location would provide a poolside area with sufficient space for passenger movement and placement of deck chairs (Figure 8.27). It has been determined that the shape of the ellipse may be described by the equation $x^2 + 4y^2 = 3600$, where x and y are measured in feet. *Viking*'s operators would like to know the dimensions of the rectangular pool with the largest possible area that would meet these requirements.

FIGURE 8.27
A rectangular-shaped pool will be built in the elliptical-shaped poolside area.

FIGURE 8.28
We want to find the largest rectangle that can be inscribed in the ellipse described by $x^2 + 4y^2 = 3600$.

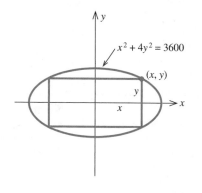

| SOLUTION ✔ |

To solve this problem, we need to find the rectangle inscribed in the ellipse with equation $x^2 + 4y^2 = 3600$ and having the largest area. Letting the sides of the rectangle be $2x$ and $2y$ feet, we see that the area of the rectangle is $A = 4xy$ (Figure 8.28). Furthermore, the point (x, y) must be constrained to move along the ellipse so that it satisfies the equation $x^2 + 4y^2 = 3600$. Thus, the problem is equivalent to the problem of maximizing the function

$$f(x, y) = 4xy$$

subject to the constraint $g(x, y) = x^2 + 4y^2 - 3600 = 0$. The Lagrangian function is

$$F(x, y, \lambda) = f(x, y) + \lambda g(x, y)$$
$$= 4xy + \lambda(x^2 + 4y^2 - 3600)$$

To find the critical point(s) of F, we solve the following system of equations:

$$F_x = 4y + 2\lambda x = 0$$
$$F_y = 4x + 8\lambda y = 0$$
$$F_\lambda = x^2 + 4y^2 - 3600 = 0$$

Solving the first equation of this system for λ, we obtain

$$\lambda = -\frac{2y}{x}$$

which, upon substitution into the second equation, yields

$$4x + 8\left(-\frac{2y}{x}\right)(y) = 0 \quad \text{or} \quad x^2 - 4y^2 = 0$$

—that is, $x = \pm 2y$. Substituting these values of x into the third equation of the system, we have

$$4y^2 + 4y^2 - 3600 = 0$$

or, upon solving $y = \pm\sqrt{450} = \pm 15\sqrt{2}$. The corresponding values of x are $\pm 30\sqrt{2}$. Because both x and y must be nonnegative, we have $x = 30\sqrt{2}$ and $y = 15\sqrt{2}$. Thus, the dimensions of the pool with maximum area are $30\sqrt{2}$ feet by $60\sqrt{2}$ feet, or approximately 42 feet $\times$ 85 feet.   ■ ■ ■ ■

EXAMPLE 6 Suppose x units of labor and y units of capital are required to produce

$$f(x, y) = 100x^{3/4}y^{1/4}$$

units of a certain product (recall that this is a Cobb–Douglas production function). If each unit of labor costs \$200 and each unit of capital costs \$300 and a total of \$60,000 is available for production, determine how many units of labor and how many units of capital should be used in order to maximize production.

SOLUTION ✔ The total cost of x units of labor at \$200 per unit and y units of capital at \$300 per unit is equal to $200x + 300y$ dollars. But \$60,000 is budgeted for production, so $200x + 300y = 60,000$, which we rewrite as

$$g(x, y) = 200x + 300y - 60,000 = 0$$

To maximize $f(x, y) = 100x^{3/4}y^{1/4}$ subject to the constraint $g(x, y) = 0$, we form the Lagrangian function

$$\begin{aligned}F(x, y, \lambda) &= f(x, y) + \lambda g(x, y) \\ &= 100x^{3/4}y^{1/4} + \lambda(200x + 300y - 60,000)\end{aligned}$$

To find the critical point(s) of F, we solve the following system of equations:

$$\begin{aligned}F_x &= 75x^{-1/4}y^{1/4} + 200\lambda = 0 \\ F_y &= 25x^{3/4}y^{-3/4} + 300\lambda = 0 \\ F_\lambda &= 200x + 300y - 60,000 = 0\end{aligned}$$

Solving the first equation for λ, we have

$$\lambda = -\frac{75x^{-1/4}y^{1/4}}{200} = -\frac{3}{8}\left(\frac{y}{x}\right)^{1/4}$$

which, when substituted into the second equation, yields

$$25\left(\frac{x}{y}\right)^{3/4} + 300\left(-\frac{3}{8}\right)\left(\frac{y}{x}\right)^{1/4} = 0$$

Multiplying the last equation by $(x/y)^{1/4}$ then gives

$$25\left(\frac{x}{y}\right) - \frac{900}{8} = 0$$

$$x = \left(\frac{900}{8}\right)\left(\frac{1}{25}\right)y = \frac{9}{2}y$$

Substituting this value of x into the third equation of the first system of equations, we have

$$200\left(\frac{9}{2}y\right) + 300y - 60{,}000 = 0$$

from which we deduce that $y = 50$. Hence, $x = 225$. Thus, maximum production is achieved when 225 units of labor and 50 units of capital are used. ■ ■ ■ ■

When used in the context of Example 6, the negative of the Lagrange multiplier λ is called the **marginal productivity of money.** That is, if one additional dollar is available for production, then approximately $-\lambda$ units of a product can be produced. Here,

$$\lambda = -\frac{3}{8}\left(\frac{y}{x}\right)^{1/4} = -\frac{3}{8}\left(\frac{50}{225}\right)^{1/4} \approx -0.257$$

so, in this case, the marginal productivity of money is 0.257. For example, if $65,000 is available for production instead of the originally budgeted figure of $60,000, then the maximum production may be boosted from the original

$$f(225, 50) = 100(225)^{3/4}(50)^{1/4}$$

or 15,448 units, to

$$15{,}448 + 5000(0.257)$$

or 16,733 units.

SELF-CHECK EXERCISES 8.5

1. Use the method of Lagrange multipliers to find the relative maximum of the function

$$f(x, y) = -2x^2 - y^2$$

subject to the constraint $3x + 4y = 12$.

8.5 Exercises

In Exercises 1–16, use the method of Lagrange multipliers to optimize the given function subject to the given constraint.

1. Minimize the function $f(x, y) = x^2 + 3y^2$ subject to the constraint $x + y - 1 = 0$.

2. Minimize the function $f(x, y) = x^2 + y^2 - xy$ subject to the constraint $x + 2y - 14 = 0$.

3. Maximize the function $f(x, y) = 2x + 3y - x^2 - y^2$ subject to the constraint $x + 2y = 9$.

4. Maximize the function $f(x, y) = 16 - x^2 - y^2$ subject to the constraint $x + y - 6 = 0$.

5. Minimize the function $f(x, y) = x^2 + 4y^2$ subject to the constraint $xy = 1$.

6. Minimize the function $f(x, y) = xy$ subject to the constraint $x^2 + 4y^2 = 4$.

7. Maximize the function $f(x, y) = x + 5y - 2xy - x^2 - 2y^2$ subject to the constraint $2x + y = 4$.

8. Maximize the function $f(x, y) = xy$ subject to the constraint $2x + 3y - 6 = 0$.

9. Maximize the function $f(x, y) = xy^2$ subject to the constraint $9x^2 + y^2 = 9$.

10. Minimize the function $f(x, y) = \sqrt{y^2 - x^2}$ subject to the constraint $x + 2y - 5 = 0$.

find the maximum and minimum values of the function $f(x, y) = xy$ subject to the constraint $x^2 + y^2 = 16$.

12. Find the maximum and minimum values of the function $f(x, y) = e^{xy}$ subject to the constraint $x^2 + y^2 = 8$.

13. Find the maximum and minimum values of the function $f(x, y) = xy^2$ subject to the constraint $x^2 + y^2 = 1$.

14. Maximize the function $f(x, y, z) = xyz$ subject to the constraint $2x + 2y + z = 84$.

15. Minimize the function $f(x, y, z) = x^2 + y^2 + z^2$ subject to the constraint $3x + 2y + z = 6$.

16. Find the maximum value of the function $f(x, y, z) = x + 2y - 3z$ subject to the constraint $z = 4x^2 + y^2$.

17. MAXIMIZING PROFIT The total weekly profit (in dollars) realized by the Country Workshop in manufacturing and selling its rolltop desks is given by the profit function

$$P(x, y) = -0.2x^2 - 0.25y^2 - 0.2xy + 100x + 90y - 4000$$

where x stands for the number of finished units and y denotes the number of unfinished units manufactured and sold per week. The company's management decides to restrict the manufacture of these desks to a total of exactly 200 units/week. How many finished and how many unfinished units should be manufactured per week to maximize the company's weekly profit?

18. MAXIMIZING PROFIT The total daily profit (in dollars) realized by the Weston Publishing Company in publishing and selling its dictionaries is given by the profit function

$$P(x, y) = -0.005x^2 - 0.003y^2 - 0.002xy$$
$$+ 14x + 12y - 200$$

where x stands for the number of deluxe editions and y denotes the number of standard editions sold daily. Weston's management decides that publication of these dictionaries should be restricted to a total of exactly 400 copies/day. How many deluxe copies and how many standard copies should be published per day to maximize Weston's daily profit?

19. **MINIMIZING CONSTRUCTION COSTS** The management of UNICO Department Store decides to enclose an 800-ft^2 area outside their building to display potted plants. The enclosed area will be a rectangle, one side of which is provided by the external walls of the store. Two sides of the enclosure will be made of pine board, and the fourth side will be made of galvanized steel fencing material. If the pine board fencing costs \$6/running foot and the steel fencing costs \$3/running foot, determine the dimensions of the enclosure that will cost the least to erect.

20. **MINIMIZING CONTAINER COSTS** The Betty Moore Company requires that its corned beef hash containers have a capacity of 64 in.^{3}, be right circular cylinders, and be made of a tin alloy. Find the radius and height of the least expensive container that can be made if the metal for the side and bottom costs 4 cents/square inch and the metal for the pull-off lid costs 2 cents/square inch.

Hint: Let the radius and height of the container be r and h inches, respectively. Then, the volume of the container is $\pi r^2 h = 64$, and the cost is given by $C(r, h) = 8\pi rh + 6\pi r^2$.

21. **MINIMIZING CONSTRUCTION COSTS** An open rectangular box is to be constructed from material that costs \$3/square foot for the bottom and \$1/square foot for its sides. Find the dimensions of the box of greatest volume that can be constructed for \$36.

22. **MINIMIZING CONSTRUCTION COSTS** A closed rectangular box having a volume of 4 ft^3 is to be constructed. If the material for the sides costs \$1/square foot and the material for the top and bottom costs \$1.50/square foot, find the dimensions of the box that can be constructed with minimum cost.

23. **MAXIMIZING SALES** The Ross–Simons Company has a monthly advertising budget of \$60,000. Their marketing department estimates that if they spend x dollars on newspaper advertising and y dollars on television advertising, then the monthly sales will be given by

$$z = f(x, y) = 90x^{1/4}y^{3/4}$$

dollars. Determine how much money Ross–Simons should spend on newspaper ads and on television ads per month to maximize its monthly sales.

24. **MAXIMIZING PRODUCTION** John Mills, the proprietor of the Mills Engine Company, a manufacturer of model airplane engines, finds that it takes x units of labor and y units of capital to produce

$$f(x, y) = 100x^{3/4}y^{1/4}$$

units of the product. If a unit of labor costs \$100 and a unit of capital costs \$200 and \$200,000 is budgeted for production, determine how many units should be expended on labor and how many units should be expended on capital in order to maximize production.

25. Use the method of Lagrange multipliers to solve Exercise 28, Exercises 8.3.

In Exercises 26 and 27, determine whether the statement is true or false. If it is true, explain why it is true. If it is false, give an example to show why it is false.

26. If (a, b) gives rise to a (constrained) relative extremum of f subject to the constraint $g(x, y) = 0$, then (a, b) also gives rise to the unconstrained relative extremum of f.

27. If (a, b) gives rise to a (constrained) relative extremum of f subject to the constraint $g(x, y) = 0$, then $f_x(a, b) = 0$ and $f_y(a, b) = 0$, simultaneously.

SOLUTIONS TO SELF-CHECK EXERCISES 8.5

1. Write the constraint equation in the form $g(x, y) = 3x + 4y - 12 = 0$. Then, the Lagrangian function is

$$F(x, y, \lambda) = -2x^2 - y^2 + \lambda(3x + 4y - 12)$$

To find the critical point(s) of F, we solve the system

$$F_x = -4x + 3\lambda = 0$$
$$F_y = -2y + 4\lambda = 0$$
$$F_\lambda = 3x + 4y - 12 = 0$$

Solving the first two equations for x and y in terms of λ, we find $x = \frac{3}{4}\lambda$ and $y = 2\lambda$. Substituting these values of x and y into the third equation of the system yields

$$3\left(\frac{3}{4}\lambda\right) + 4(2\lambda) - 12 = 0$$

or $\lambda = \frac{48}{41}$. Therefore, $x = \left(\frac{3}{4}\right)\left(\frac{48}{41}\right) = \frac{36}{41}$ and $y = 2\left(\frac{48}{41}\right) = \frac{96}{41}$, and we see that the point $\left(\frac{36}{41}, \frac{96}{41}\right)$ gives the constrained maximum of f. The maximum value is

$$f\left(\frac{36}{41}, \frac{96}{41}\right) = -2\left(\frac{36}{41}\right)^2 - \left(\frac{96}{41}\right)^2$$
$$= -\frac{11{,}808}{1681} = -\frac{288}{41}$$

2. We want to maximize

$$P(x, y) = -\frac{1}{8}x^2 - \frac{1}{2}y^2 - \frac{1}{4}xy + 13x + 40y - 280$$

subject to the constraint

$$g(x, y) = x + y - 40 = 0$$

The Lagrangian function is

$$F(x, y, \lambda) = P(x, y) + \lambda g(x, y)$$
$$= -\frac{1}{8}x^2 - \frac{1}{2}y^2 - \frac{1}{4}xy + 13x$$
$$+ 40y - 280 + \lambda(x + y - 40)$$

To find the critical points of F, solve the following system of equations:

$$F_x = -\frac{1}{4}x - \frac{1}{4}y + 13 + \lambda = 0$$

$$F_y = -\frac{1}{4}x - y + 40 + \lambda = 0$$

$$F_\lambda = x + y - 40 = 0$$

Subtracting the first equation from the second gives

$$-\frac{3}{4}y + 27 = 0 \quad \text{or} \quad y = 36$$

Substituting this value of y into the third equation yields $x = 4$. Therefore, to maximize its monthly profits, Robertson should manufacture 400 standard and 3600 deluxe thermostats. The maximum monthly profit is given by

$$P(4, 36) = -\frac{1}{8}(4)^2 - \frac{1}{2}(36)^2 - \frac{1}{4}(4)(36)$$
$$+ 13(4) + 40(36) - 280$$
$$= 526$$

or \$52,600.

8.6 Double Integrals

A GEOMETRIC INTERPRETATION OF THE DOUBLE INTEGRAL

To introduce the notion of the integral of a function of two variables, let's first recall the definition of the definite integral of a continuous function of one variable $y = f(x)$ over the interval $[a, b]$. We first divide the interval $[a, b]$ into n subintervals, each of equal length, by the points $x_0 = a < x_1 < x_2 < \cdots < x_n = b$ and define the **Riemann sum** by

$$S_n = f(p_1)h + f(p_2)h + \cdots + f(p_n)h$$

where $h = (b - a)/n$ and p_i is an arbitrary point in the interval $[x_{i-1}, x_i]$. The definite integral of f over $[a, b]$ is defined as the limit of the Riemann sum S_n as n tends to infinity, whenever it exists. Furthermore, recall that when f is a nonnegative continuous function on $[a, b]$, then the ith term of the Riemann sum, $f(p_i)h$, is an approximation (by the area of a rectangle) of the area under that part of the graph of $y = f(x)$ between $x = x_{i-1}$ and $x = x_i$, so that the Riemann sum S_n provides us with an approximation of the area under the curve $y = f(x)$ from $x = a$ to $x = b$. The integral

$$\int_a^b f(x)\, dx = \lim_{n \to \infty} S_n$$

gives the *actual* area under the curve from $x = a$ to $x = b$.

Now suppose $f(x, y)$ is a continuous function of two variables defined over a region R. For simplicity, we assume for the moment that R is a rectangular region in the plane (Figure 8.29). Let's construct a Riemann sum for this function over the rectangle R by following a procedure that parallels the case for a function of one variable over an interval I. We begin by observing that the analogue of a *partition* in the two-dimensional case is a rectangular **grid** composed of mn rectangles, each of length h and width k, as a result of partitioning the side of the rectangle R of length $(b - a)$ into m segments

FIGURE 8.29
$f(x, y)$ is a function defined over a rectangular region R.

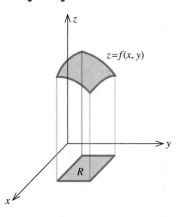

FIGURE 8.30
Grid with $m = 5$ and $n = 4$

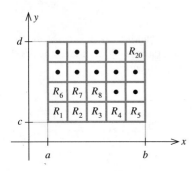

and the side of length $(d - c)$ into n segments. By construction

$$h = \frac{b - a}{m} \quad \text{and} \quad k = \frac{d - c}{n}$$

A sample grid with $m = 5$ and $n = 4$ is shown in Figure 8.30.

Let's label the rectangles $R_1, R_2, R_3, \ldots, R_{mn}$. If (x_i, y_i) is *any* point in R_i $(1 \le i \le mn)$, then the **Riemann sum of $f(x, y)$ over the region R** is defined as

$$S(m, n) = f(x_1, y_1)hk + f(x_2, y_2)hk + \cdots + f(x_{mn}, y_{mn})hk$$

If the limit of $S(m, n)$ exists as both m and n tend to infinity, we call this limit the value of the **double integral of $f(x, y)$ over the region R** and denote it by

$$\iint_R f(x, y) \, dA$$

Group Discussion
Using a geometric interpretation, evaluate
$$\iint_R \sqrt{4 - x^2 - y^2} \, dA$$
where $R = \{(x, y) \mid x^2 + y^2 \le 4\}$.

If $f(x, y)$ is a nonnegative function, then the number $f(x_i, y_i)hk$ is the volume of a prism with base R_i and height $f(x_i, y_i)$ that provides us with an approximation of the volume of the solid under the surface $z = f(x, y)$ and bounded below by the rectangular region R_i (Figure 8.31).

Therefore, the Riemann sum $S(m, n)$ gives us an approximation of the volume of the solid bounded above by the surface $z = f(x, y)$ and below by the plane region R. As both m and n tend to infinity, the Riemann sum $S(m, n)$ approaches the *actual* volume under the solid.

FIGURE 8.31

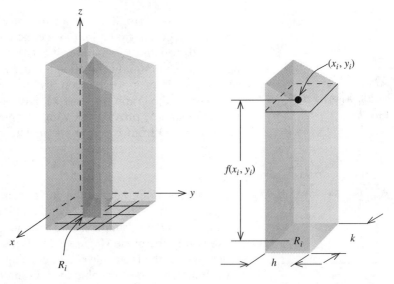

(a) Approximating the volume of a solid with rectangular prisms

(b) A prism with base R_i and height $f(x_i, y_i)$

EVALUATING A DOUBLE INTEGRAL OVER A RECTANGULAR REGION

Let's turn our attention to the evaluation of the double integral

$$\iint_R f(x, y)\, dA$$

where R is the rectangular region shown in Figure 8.29. As in the case of the definite integral of a function of one variable, it turns out that the double integral can be evaluated without our having to first find an appropriate Riemann sum and then take the limit of that sum. Instead, as we will now see, the technique calls for evaluating two single integrals—the so-called **iterated integrals**—in succession, using a process that might be called "anti-partial differentiation." The technique is described in the following result, which we state without proof.

Let R be the rectangle defined by the inequalities $a \leq x \leq b$ and $c \leq y \leq d$ (see Figure 8.30). Then,

$$\iint_R f(x, y)\, dA = \int_c^d \left[\int_a^b f(x, y)\, dx \right] dy \qquad \textbf{(12)}$$

where the iterated integrals on the right-hand side are evaluated as follows. We first compute the integral

$$\int_a^b f(x, y)\, dx$$

by treating y as if it were a constant and integrating the resulting function of x with respect to x (the "dx" reminds us that we are integrating with respect to x). In this manner we obtain a value for the integral that may contain the variable y. Thus,

$$\int_a^b f(x, y)\, dx = g(y)$$

for some function g. Substituting this value into Equation (12) gives

$$\int_c^d g(y)\, dy$$

which may be integrated in the usual manner.

EXAMPLE 1

Evaluate $\iint_R f(x, y)\, dA$, where $f(x, y) = x + 2y$ and R is the rectangle defined by $1 \le x \le 4$ and $1 \le y \le 2$.

SOLUTION ✔

Using Equation (12), we find

$$\iint_R f(x, y)\, dA = \int_1^2 \left[\int_1^4 (x + 2y)\, dx \right] dy$$

To compute

$$\int_1^4 (x + 2y)\, dx$$

we treat y as if it were a constant (remember that the "dx" reminds us that we are integrating with respect to x). We obtain

$$\int_1^4 (x + 2y)\, dx = \frac{1}{2}x^2 + 2xy \Big|_{x=1}^{x=4}$$

$$= \left[\frac{1}{2}(16) + 2(4)y \right] - \left[\frac{1}{2}(1) + 2(1)y \right]$$

$$= \frac{15}{2} + 6y$$

Thus,

$$\iint_R f(x, y)\, dA = \int_1^2 \left(\frac{15}{2} + 6y \right) dy = \left(\frac{15}{2}y + 3y^2 \right) \Big|_1^2$$

$$= (15 + 12) - \left(\frac{15}{2} + 3 \right) = 16\frac{1}{2}$$ ■■■■

EVALUATING A DOUBLE INTEGRAL OVER A PLANE REGION

Up to now we have assumed that the region over which a double integral is to be evaluated is rectangular. In fact, however, it is possible to compute the double integral of functions over rather arbitrary regions. The next theorem, which we state without proof, expands the number of types of regions over which we may integrate.

THEOREM 1

a. Suppose $g_1(x)$ and $g_2(x)$ are continuous functions on $[a, b]$ and the region R is defined by $R = \{(x, y) \mid g_1(x) \leq y \leq g_2(x); a \leq x \leq b\}$. Then,

$$\iint_R f(x, y)\, dA = \int_a^b \left[\int_{g_1(x)}^{g_2(x)} f(x, y)\, dy \right] dx \qquad \textbf{(13)}$$

(Figure 8.32a).

b. Suppose $h_1(y)$ and $h_2(y)$ are continuous functions on $[c, d]$ and the region R is defined by $R = \{(x, y) \mid h_1(y) \leq x \leq h_2(y); c \leq y \leq d\}$. Then,

$$\iint_R f(x, y)\, dA = \int_c^d \left[\int_{h_1(y)}^{h_2(y)} f(x, y)\, dx \right] dy \qquad \textbf{(14)}$$

(Figure 8.32b).

FIGURE 8.32

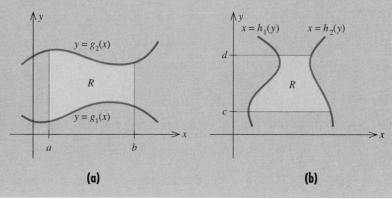

(a)   (b)

REMARK **1.** Observe that in (13) the lower and upper limits of integration with respect to y are given by $y = g_1(x)$ and $y = g_2(x)$. This is to be expected since, for a fixed value of x lying between $x = a$ and $x = b$, y runs between the lower curve defined by $y = g_1(x)$ and the upper curve defined by $y = g_2(x)$ (see Figure 8.32a). Observe, too, that in the special case when $g_1(x) = c$ and $g_2(x) = d$, the region R is rectangular, and (13) reduces to (12).
2. For a fixed value of y, x runs between $x = h_1(y)$ and $x = h_2(y)$, giving the indicated limits of integration with respect to x in (14) (see Figure 8.32b).
3. Note that the two curves in Figure 8.32b are not graphs of functions of x (use the vertical-line test), but they are graphs of functions of y. It is this observation that justifies the approach leading to (14). ■■■

We now look at several examples.

EXAMPLE **2**

Evaluate $\iint_R f(x, y)\, dA$ given that $f(x, y) = x^2 + y^2$ and R is the region bounded by the graphs of $g_1(x) = x$ and $g_2(x) = 2x$ for $0 \le x \le 2$.

FIGURE 8.33
R is the region bounded by $g_1(x) = x$ and $g_2(x) = 2x$ for $0 \le x \le 2$.

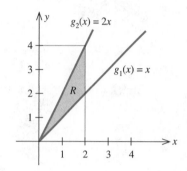

SOLUTION ✔

The region under consideration is shown in Figure 8.33. Using Equation (13), we find

$$\iint_R f(x, y)\, dA = \int_0^2 \left[\int_x^{2x} (x^2 + y^2)\, dy \right] dx$$

$$= \int_0^2 \left[\left(x^2 y + \frac{1}{3}y^3 \right) \Big|_x^{2x} \right] dx$$

$$= \int_0^2 \left[\left(2x^3 + \frac{8}{3}x^3 \right) - \left(x^3 + \frac{1}{3}x^3 \right) \right] dx$$

$$= \int_0^2 \frac{10}{3} x^3\, dx = \frac{5}{6}x^4 \Big|_0^2 = 13\frac{1}{3}$$

■ ■ ■ ■

EXAMPLE **3**

Evaluate $\iint_R f(x, y)\, dA$, where $f(x, y) = xe^y$ and R is the plane region bounded by the graphs of $y = x^2$ and $y = x$.

FIGURE 8.34
R is the region bounded by $y = x^2$ and $y = x$.

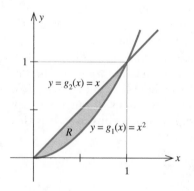

SOLUTION ✔

The region in question is shown in Figure 8.34. The point of intersection of the two curves is found by solving the equation $x^2 = x$, giving $x = 0$ and $x = 1$. Using Equation (13), we find

$$\iint_R f(x, y)\, dA = \int_0^1 \left[\int_{x^2}^x xe^y\, dy \right] dx = \int_0^1 \left[xe^y \Big|_{x^2}^x \right] dx$$

$$= \int_0^1 (xe^x - xe^{x^2})\, dx = \int_0^1 xe^x\, dx - \int_0^1 xe^{x^2}\, dx$$

and integrating the first integral on the right-hand side by parts,

$$= \left[(x - 1)e^x - \frac{1}{2}e^{x^2} \right] \Big|_0^1$$

$$= -\frac{1}{2}e - \left(-1 - \frac{1}{2} \right) = \frac{1}{2}(3 - e)$$

■ ■ ■ ■

Group Discussion
Refer to Example 3.

1. You can also view the region R as an example of the region shown in Figure 8.32b. Doing so, find the functions h_1 and h_2 and the numbers c and d.

2. Find an expression for $\iint_R f(x, y)\, dA$ in terms of iterated integrals using Formula (14).

3. Evaluate the iterated integrals of part 2 and hence verify the result of Example 3.

Hint: Integrate by parts twice.

4. Does viewing the region R in two different ways make a difference?

FINDING THE VOLUME OF A SOLID BY DOUBLE INTEGRALS

As we saw earlier, the double integral

$$\iint_R f(x, y)\, dA$$

gives the volume of the solid bounded by the graph of $f(x, y)$ over the region R.

The Volume of a Solid Under a Surface

> Let R be a region in the xy-plane and let f be continuous and nonnegative on R. Then, the volume of the solid bounded above by the surface $z = f(x, y)$ and below by R is given by
>
> $$V = \iint_R f(x, y)\, dA$$

EXAMPLE 4   Find the volume of the solid bounded above by the plane $z = f(x, y) = y$ and below by the plane region R defined by $y = \sqrt{1 - x^2}$, $0 \le x \le 1$.

SOLUTION ✔   The region R is sketched in Figure 8.35. Observe that $f(x, y) = y \ge 0$ for $y \in R$. Therefore, the required volume is given by

$$\iint_R y\, dA = \int_0^1 \left[\int_0^{\sqrt{1-x^2}} y\, dy \right] dx = \int_0^1 \left[\frac{1}{2} y^2 \Big|_0^{\sqrt{1-x^2}} \right] dx$$

$$= \int_0^1 \frac{1}{2}(1 - x^2)\, dx = \frac{1}{2}\left(x - \frac{1}{3}x^3 \right)\Big|_0^1 = \frac{1}{3}$$

or $\frac{1}{3}$ cubic unit. The solid is shown in Figure 8.36. Note that it is not necessary to make a sketch of the solid in order to compute its volume.

FIGURE 8.35
The plane region R defined by
$y = \sqrt{1 - x^2}$, $0 \le x \le 1$

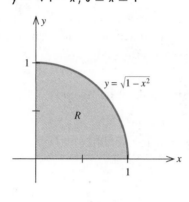

FIGURE 8.36
The solid bounded above by the plane
$z = y$ and below by the plane region
defined by $y = \sqrt{1 - x^2}$, $0 \le x \le 1$

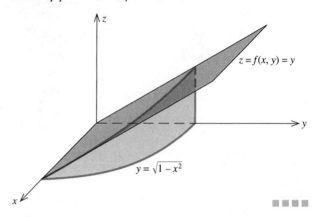

POPULATION OF A CITY

FIGURE 8.37

The rectangular region R representing a certain district of a city is enclosed by a rectangular grid.

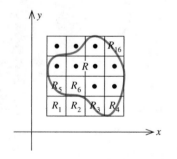

Suppose the plane region R represents a certain district of a city and $f(x, y)$ gives the population density (the number of people per square mile) at any point (x, y) in R. Enclose the set R by a rectangle and construct a grid for it in the usual manner. In any rectangular region of the grid that has no point in common with R, set $f(x_i, y_i)hk = 0$ (Figure 8.37). Then, corresponding to any grid covering the set R, the general term of the Riemann sum $f(x_i, y_i)hk$ (population density times area) gives the number of people living in that part of the city corresponding to the rectangular region R_i. Therefore, the Riemann sum gives an approximation of the number of people living in the district represented by R and, in the limit, the double integral

$$\iint_R f(x, y) \, dA$$

gives the actual number of people living in the district under consideration.

EXAMPLE 5

The population density of a certain city is described by the function

$$f(x, y) = 10{,}000e^{-0.2|x|-0.1|y|}$$

where the origin $(0, 0)$ gives the location of the city hall. What is the population inside the rectangular area described by

$$R = \{(x, y) \mid -10 \le x \le 10; \ -5 \le y \le 5\}$$

if x and y are in miles? (See Figure 8.38.)

FIGURE 8.38
The rectangular region R represents a certain district of a city.

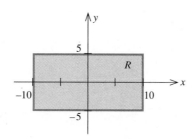

By symmetry, it suffices to compute the population in the first quadrant. (Why?) Then, upon observing that in this quadrant

$$f(x, y) = 10{,}000e^{-0.2x-0.1y} = 10{,}000e^{-0.2x}e^{-0.1y}$$

we see that the population in R is given by

$$\iint_R f(x, y)\, dA = 4\int_0^{10}\left[\int_0^5 10{,}000e^{-0.2x}\, e^{-0.1y}\, dy\right]dx$$

$$= 4\int_0^{10}\left[-100{,}000e^{-0.2x}e^{-0.1y}\Big|_0^5\right]dx$$

$$= 400{,}000(1 - e^{-0.5})\int_0^{10} e^{-0.2x}\, dx$$

$$= 2{,}000{,}000(1 - e^{-0.5})(1 - e^{-2})$$

or approximately 680,438 people.   ■ ■ ■ ■

 Group Discussion

1. Consider the improper double integral $\iint_D f(x, y)\, dA$ of the continuous function f of two variables defined over the plane region

$$D = \{(x, y) \mid 0 \le x < \infty; 0 \le y < \infty\}$$

Using the definition of improper integrals of functions of one variable (Section 7.4), explain why it makes sense to define

$$\iint_D f(x, y)\, dA = \lim_{N\to\infty}\int_0^N\left[\lim_{M\to\infty}\int_0^M f(x, y)\, dx\right]dy$$

$$= \lim_{M\to\infty}\int_0^M\left[\lim_{N\to\infty}\int_0^N f(x, y)\, dy\right]dx$$

provided the limits exist.

(*continued on p. 676*)

> **Group Discussion** (*continued*)
> **2.** Refer to Example 5. Assuming that the population density of the city is described by
>
> $$f(x, y) = 10{,}000e^{-0.2|x|-0.1|y|}$$
>
> for $-\infty < x < \infty$ and $-\infty < y < \infty$, show that the population outside the rectangular region
>
> $$R = \{(x, y)|\ -10 < x < 10;\ -5 < y \le 5\}$$
>
> of Example 5 is given by
>
> $$4 \iint_D f(x, y)\, dx\, dy - 680{,}438$$
>
> (recall that 680,438 is the approximate population inside R).
> **3.** Use the results of parts 1 and 2 to determine the population of the city outside the rectangular area R.

AVERAGE VALUE OF A FUNCTION

In Section 6.5 we showed that the average value of a continuous function $f(x)$ over an interval $[a, b]$ is given by

$$\frac{1}{b - a} \int_a^b f(x)\, dx$$

That is, the average value of a function over $[a, b]$ is the integral of f over $[a, b]$ divided by the length of the interval. An analogous result holds for a function of two variables $f(x, y)$ over a plane region R. To see this, we enclose R by a rectangle and construct a rectangular grid. Let (x_i, y_i) be any point in the rectangle R_i of area hk. Now, the average value of the mn numbers $f(x_1, y_1), f(x_2, y_2), \ldots, f(x_{mn}, y_{mn})$ is given by

$$\frac{f(x_1, y_1) + f(x_2, y_2) + \cdots + f(x_{mn}, y_{mn})}{mn}$$

which can also be written as

$$\frac{hk}{hk} \left[\frac{f(x_1, y_1) + f(x_2, y_2) + \cdots + f(x_{mn}, y_{mn})}{mn} \right]$$

$$= \frac{1}{(mn)hk} [f(x_1, y_1) + f(x_2, y_2) + \cdots + f(x_{mn}, y_{mn})]hk$$

Now the area of R is approximated by the sum of the mn rectangles (*omitting those having no points in common with R*), each of area hk. Note that this is the denominator of the previous expression. Therefore, taking the limit as m and n both tend to infinity, we obtain the following formula for the *average value of $f(x, y)$ over R.*

Average Value of $f(x, y)$ over the Region R

If f is integrable over the plane region R, then its average value over R is given by

$$\frac{\displaystyle\iint_R f(x, y)\, dA}{\text{Area of } R} \quad \text{or} \quad \frac{\displaystyle\iint_R f(x, y)\, dA}{\displaystyle\iint_R dA} \tag{15}$$

REMARK If we let $f(x, y) = 1$ for all (x, y) in R, then

$$\iint_R f(x, y)\, dA = \iint_R dA = \text{Area of } R$$

■ ■ ■

EXAMPLE 6

Find the average value of the function $f(x, y) = xy$ over the plane region defined by $y = e^x, 0 \leq x \leq 1$.

SOLUTION ✔

The region R is shown in Figure 8.39. The area of the region R is given by

$$\int_0^1 \left[\int_0^{e^x} dy \right] dx = \int_0^1 \left[y \Big|_0^{e^x} \right] dx$$

$$= \int_0^1 e^x\, dx$$

$$= e^x \Big|_0^1$$

$$= e - 1$$

FIGURE **8.39**
The plane region R defined by $y = e^x$, $0 \leq x \leq 1$

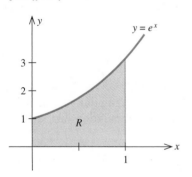

square units. We would obtain the same result had we viewed the area of this region as the area of the region under the curve $y = e^x$ from $x = 0$ to $x = 1$. Next, we compute

$$\iint_R f(x, y)\, dA = \int_0^1 \left[\int_0^{e^x} xy\, dy \right] dx$$

$$= \int_0^1 \left[\frac{1}{2} xy^2 \Big|_0^{e^x} \right] dx$$

$$= \int_0^1 \frac{1}{2} xe^{2x}\, dx$$

$$= \frac{1}{4} xe^{2x} - \frac{1}{8} e^{2x} \Big|_0^1 \quad \text{(Integrating by parts)}$$

$$= \left(\frac{1}{4} e^2 - \frac{1}{8} e^2 \right) + \frac{1}{8}$$

$$= \frac{1}{8}(e^2 + 1)$$

square units. Therefore, the required average value is given by

$$\frac{\iint_R f(x, y)\, dA}{\iint_R dA} = \frac{\frac{1}{8}(e^2 + 1)}{e - 1} = \frac{e^2 + 1}{8(e - 1)}$$

■ ■ ■ ■

EXAMPLE 7

(Refer to Example 5.) The population density of a certain city (number of people per square mile) is described by the function

$$f(x, y) = 10{,}000e^{-0.2|x|-0.1|y|}$$

where the origin gives the location of the city hall. What is the average population density inside the rectangular area described by

$$R = \{(x, y) \mid -10 \le x \le 10;\ -5 \le y \le 5\}$$

where x and y are measured in miles?

SOLUTION ✔

From the results of Example 5, we know that

$$\iint_R f(x, y)\, dA \approx 680{,}438$$

From Figure 8.38, we see that the area of the plane rectangular region R is $(20)(10)$, or 200, square miles. Therefore, the average population inside R is

$$\frac{\iint_R f(x, y)\, dA}{\iint_R dA} = \frac{680{,}438}{200} = 3402.19$$

or approximately 3402 people per square mile.

■ ■ ■ ■

SELF-CHECK EXERCISES 8.6

1. Evalute $\iint_R (x + y)\, dA$, where R is the region bounded by the graphs of $g_1(x) = x$ and $g_2(x) = x^{1/3}$.

2. The population density of a coastal town located on an island is described by the function

$$f(x, y) = \frac{5000xe^y}{1 + 2x^2} \qquad (0 \le x \le 4; -2 \le y \le 0)$$

where x and y are measured in miles (see the accompanying figure).

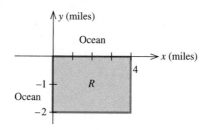

What is the population inside the rectangular area defined by $R = \{(x, y) \mid 0 \le x \le 4; -2 \le y \le 0\}$? What is the average population density in the area?

Solutions to Self-Check Exercises 8.6 can be found on page 683.

8.6 Exercises

In Exercises 1–25, evaluate the double integral $\iint\limits_{R} f(x, y) \, dA$ for the given function $f(x, y)$ and the region R.

1. $f(x, y) = y + 2x$; R is the rectangle defined by $1 \le x \le 2$ and $0 \le y \le 1$.

2. $f(x, y) = x + 2y$; R is the rectangle defined by $-1 \le x \le 2$ and $0 \le y \le 2$.

3. $f(x, y) = xy^2$; R is the rectangle defined by $-1 \le x \le 1$ and $0 \le y \le 1$.

4. $f(x, y) = 12xy^2 + 8y^3$; R is the rectangle defined by $0 \le x \le 1$ and $0 \le y \le 2$.

5. $f(x, y) = \dfrac{x}{y}$; R is the rectangle defined by $-1 \le x \le 2$ and $1 \le y \le e^3$.

6. $f(x, y) = \dfrac{xy}{1 + y^2}$; R is the rectangle defined by $-2 \le x \le 2$ and $0 \le y \le 1$.

7. $f(x, y) = 4xe^{2x^2+y}$; R is the rectangle defined by $0 \le x \le 1$ and $-2 \le y \le 0$.

8. $f(x, y) = \dfrac{y}{x^2}e^{y/x}$; R is the rectangle defined by $1 \le x \le 2$ and $0 \le y \le 1$.

9. $f(x, y) = \ln y$; R is the rectangle defined by $0 \le x \le 1$ and $1 \le y \le e$.

10. $f(x, y) = \dfrac{\ln y}{x}$; R is the rectangle defined by $1 \le x \le e^2$ and $1 \le y \le e$.

11. $f(x, y) = x + 2y$; R is bounded by $x = 0$, $x = 1$, $y = 0$, and $y = x$.

12. $f(x, y) = xy$; R is bounded by $x = 0$, $x = 1$, $y = 0$, and $y = x$.

13. $f(x, y) = 2x + 4y$; R is bounded by $x = 1$, $x = 3$, $y = 0$, and $y = x + 1$.

14. $f(x, y) = 2 - y$; R is bounded by $x = -1$, $x = 1 - y$, $y = 0$, and $y = 2$.

15. $f(x, y) = x + y$; R is bounded by $x = 0$, $x = \sqrt{y}$, $y = 0$, and $y = 4$.

16. $f(x, y) = x^2y^2$; R is bounded by $x = 0$, $x = 1$, $y = x^2$, and $y = x^3$.

17. $f(x, y) = y$; R is bounded by $x = 0$, $x = \sqrt{4 - y^2}$, $y = 0$, and $y = 2$.

18. $f(x, y) = \dfrac{y}{x^3 + 2}$; R is bounded by $x = 0$, $x = 1$, $y = 0$, and $y = x$.

19. $f(x, y) = 2xe^y$; R is bounded by $x = 0$, $x = 1$, $y = 0$, and $y = x$.

wrong order

20. $f(x, y) = 2x$; R is bounded by $x = e^{2y}$, $x = y$, $y = 0$, and $y = 1$.

$\int e^{2y}$
$\int \, y$

never discussed in ch.

21. $f(x, y) = ye^x$; R is bounded by $y = \sqrt{x}$ and $y = x$.

22. $f(x, y) = xe^{-y^2}$; R is bounded by $x = 0$, $y = x^2$, and $y = 4$. *must change*

23. $f(x, y) = e^{y^2}$; R is bounded by $x = 0$, $x = 1$, $y = 2x$, and $y = 2$. *must change*

24. $f(x, y) = y$; R is bounded by $x = 1$, $x = e$, $y = 0$, and $y = \ln x$.

25. $f(x, y) = ye^{x^3}$; R is bounded by $x = y/2$, $x = 1$, $y = 0$, and $y = 2$. *switch*

In Exercises 26–33, use a double integral to find the volume of the solid shown in the figure.

26.

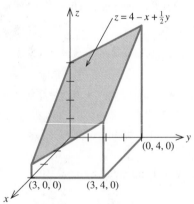

$z = 4 - x + \tfrac{1}{2}y$

$(0, 4, 0)$

$(3, 0, 0)$ $(3, 4, 0)$

27.

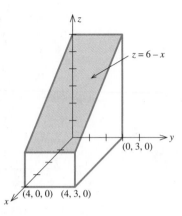

$z = 6 - x$

$(0, 3, 0)$

$(4, 0, 0)$ $(4, 3, 0)$

28.

$z = 5$

$y = -x + 4$

$y = x$

29.

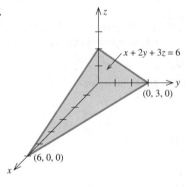

$x + 2y + 3z = 6$

$(0, 3, 0)$

$(6, 0, 0)$

30.

$z = 4$

$y = 4 - x^2$

31.

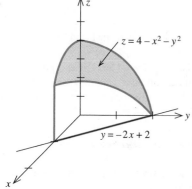

$z = 4 - x^2 - y^2$

$y = -2x + 2$

32.

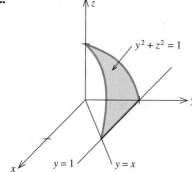

$y^2 + z^2 = 1$

$y = 1$ $y = x$

33.

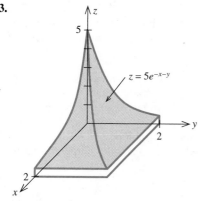

5

$z = 5e^{-x-y}$

2

2

In Exercises 34–41, find the volume of the solid bounded above by the surface $z = f(x, y)$ and below by the plane region R.

34. $f(x, y) = 4 - 2x - y$; $R = \{(x, y) \mid 0 \le x \le 1; 0 \le y \le 2\}$

35. $f(x, y) = 2x + y$; R is the triangle bounded by $y = 2x$, $y = 0$, and $x = 2$.

36. $f(x, y) = x^2 + y^2$; R is the rectangle with vertices $(0, 0)$, $(1, 0)$, $(1, 2)$, and $(0, 2)$.

37. $f(x, y) = e^{x+2y}$; R is the triangle with vertices $(0, 0)$, $(1, 0)$, and $(0, 1)$.

38. $f(x, y) = 2xe^y$; R is the triangle bounded by $y = x$, $y = 2$, and $x = 0$.

39. $f(x, y) = \dfrac{2y}{1 + x^2}$; R is the region bounded by $y = \sqrt{x}$, $y = 0$, and $x = 4$.

40. $f(x, y) = 2x^2y$; R is the region bounded by the graphs of $y = x$ and $y = x^2$.

41. $f(x, y) = x$; R is the region in the first quadrant bounded by the semicircle $y = \sqrt{16 - x^2}$, the x-axis, and the y-axis.

In Exercises 42–47, find the average value of the given function $f(x, y)$ over the plane region R.

42. $f(x, y) = 6x^2y^3$; $R = \{(x, y) \mid 0 \le x \le 2; 0 \le y \le 3\}$

43. $f(x, y) = x + 2y$; R is the triangle with vertices $(0, 0)$, $(1, 0)$, and $(1, 1)$.

44. $f(x, y) = xy$; R is the triangle bounded by $y = x$, $y = 2 - x$, and $y = 0$.

45. $f(x, y) = e^{-x^2}$; R is the triangle with vertices $(0, 0)$, $(1, 0)$, and $(1, 1)$.

46. $f(x, y) = xe^y$; R is the triangle with vertices $(0, 0)$, $(1, 0)$, and $(1, 1)$.

47. $f(x, y) = \ln x$; R is the region bounded by the graphs of $y = 2x$ and $y = 0$ from $x = 1$ to $x = 3$.
Hint: Use integration by parts.

48. POPULATION DENSITY The population density of a coastal town is described by the function

$$f(x, y) = \frac{10,000e^y}{1 + 0.5|x|} \qquad (-10 \le x \le 10; -4 \le y \le 0)$$

where x and y are measured in miles (see the accompanying figure). Find the population inside the rectangular area described by

$$R = \{(x, y) \mid -5 \le x \le 5; -2 \le y \le 0\}$$

49. AVERAGE POPULATION DENSITY Refer to Exercise 48. Find the average population density inside the rectangular area R.

50. AVERAGE PROFIT The Country Workshop's total weekly profit (in dollars) realized in manufacturing and selling its rolltop desks is given by the profit function

$$P(x, y) = -0.2x^2 - 0.25y^2 - 0.2xy$$
$$+ 100x + 90y - 4000$$

where x stands for the number of finished units and y stands for the number of unfinished units manufactured and sold per week. Find the average weekly profit if the number of finished units manufactured and sold varies between 180 and 200 and the number of unfinished units varies between 100 and 120/week.

51. AVERAGE PRICE OF LAND The rectangular region R shown in the accompanying figure represents a city's financial district. The price of land in the district is approximated

by the function

$$p(x, y) = 200 - 10\left(x - \frac{1}{2}\right)^2 - 15(y - 1)^2$$

where $p(x, y)$ is the price of land at the point (x, y) in dollars per square foot and x and y are measured in miles. What is the average price of land per square foot in the district?

In Exercises 52–55, determine whether the statement is true or false. If it is true, explain why it is true. If it is false, give an example to show why it is false.

52. If $h(x, y) = f(x)g(y)$, where f is continuous on $[a, b]$ and g is continuous on $[c, d]$, then $\iint\limits_{R} h(x, y)\, dA = [\int_a^b f(x)\, dx][\int_c^d g(y)\, dy]$, where $R = \{(x, y) \,|\, a \le x \le b; c \le y \le d\}$.

53. If $\iint\limits_{R_1} f(x, y)\, dA$ exists, where $R_1 = \{(x, y) \,|\, a \le x \le b; c \le y \le d\}$, then $\iint\limits_{R_2} f(x, y)\, dA$ exists, where $R_2 = \{(x, y) \,|\, c \le x \le d; a \le y \le b\}$.

54. Let R be a region in the xy-plane and let f and g be continuous functions on R that satisfy the condition $f(x, y) \le g(x, y)$ for all (x, y) in R. Then, $\iint\limits_{R} g(x, y) - f(x, y)]\, dA$ gives the volume of the solid bounded above by the surface $z = g(x, y)$ and below by the surface $z = f(x, y)$.

55. Suppose f is nonnegative and integrable over the plane region R. Then, the average value of f over R can be thought of as the (constant) height of the cylinder with base R and volume that is exactly equal to the volume of the solid under the graph of $z = f(x, y)$. (*Note:* The cylinder referred to here has sides perpendicular to R.)

SOLUTIONS TO SELF-CHECK EXERCISES 8.6

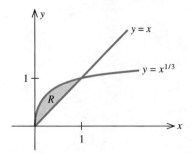

1. The region R is shown in the accompanying figure. The points of intersection of the two curves are found by solving the equation $x = x^{1/3}$, giving $x = 0$ and $x = 1$. Using Equation (13), we find

$$\iint_R (x + y)\, dA = \int_0^1 \left[\int_x^{x^{1/3}} (x + y)\, dy \right] dx$$

$$= \int_0^1 \left[xy + \frac{1}{2} y^2 \Big|_x^{x^{1/3}} \right] dx$$

$$= \int_0^1 \left[\left(x^{4/3} + \frac{1}{2} x^{2/3} \right) - \left(x^2 + \frac{1}{2} x^2 \right) \right] dx$$

$$= \int_0^1 \left(x^{4/3} + \frac{1}{2} x^{2/3} - \frac{3}{2} x^2 \right) dx$$

$$= \frac{3}{7} x^{7/3} + \frac{3}{10} x^{5/3} - \frac{1}{2} x^3 \Big|_0^1$$

$$= \frac{3}{7} + \frac{3}{10} - \frac{1}{2} = \frac{8}{35}$$

2. The population in R is given by

$$\iint_R f(x, y)\, dA = \int_0^4 \left[\int_{-2}^0 \frac{5000 x e^y}{1 + 2x^2}\, dy \right] dx$$

$$= \int_0^4 \left[\frac{5000 x e^y}{1 + 2x^2} \Big|_{-2}^0 \right] dx$$

$$= 5000(1 - e^{-2}) \int_0^4 \frac{x}{1 + 2x^2}\, dx$$

$$= 5000(1 - e^{-2}) \left[\frac{1}{4} \ln(1 + 2x^2) \Big|_0^4 \right]$$

$$= 5000(1 - e^{-2}) \left(\frac{1}{4} \right) \ln 33$$

or approximately 3779 people. The average population density inside R is

$$\frac{\displaystyle \iint_R f(x, y)\, dA}{\displaystyle \iint_R dA} = \frac{3779}{(2)(4)}$$

or approximately 472 people/square mile.

CHAPTER 8 Summary of Principal Terms

Terms

function of two variables	absolute maximum value
domain	absolute minimum
three-dimensional cartesian coordinate system	absolute minimum value
	critical point
level curve	saddle point
first partial derivative	second derivative test
Cobb–Douglas production function	method of least squares
marginal productivity of labor	least-squares line (regression line)
marginal productivity of capital	normal equations
substitute commodities	constrained relative extremum
complementary commodities	method of Lagrange multipliers
second-order partial derivative	Riemann sum
relative maximum	double integral
relative minimum	volume of a solid under a surface
absolute maximum	

CHAPTER 8 REVIEW EXERCISES

1. Let $f(x, y) = \dfrac{xy}{x^2 + y^2}$. Compute $f(0, 1)$, $f(1, 0)$, and $f(1, 1)$. Does $f(0, 0)$ exist?

2. Let $f(x, y) = \dfrac{xe^y}{1 + \ln xy}$. Compute $f(1, 1)$, $f(1, 2)$, and $f(2, 1)$. Does $f(1, 0)$ exist?

3. Let $h(x, y, z) = xye^z + \dfrac{x}{y}$. Compute $h(1, 1, 0)$, $h(-1, 1, 1)$, and $h(1, -1, 1)$.

4. Find the domain of the function $f(u, v) = \dfrac{\sqrt{u}}{u - v}$.

5. Find the domain of the function $f(x, y) = \dfrac{x - y}{x + y}$.

6. Find the domain of the function $f(x, y) = x\sqrt{y} + y\sqrt{1 - x}$.

7. Find the domain of the function $f(x, y, z) = \dfrac{xy\sqrt{z}}{(1 - x)(1 - y)(1 - z)}$.

In Exercises 8–11, sketch the level curves of the function corresponding to the given values of z.

8. $z = f(x, y) = 2x + 3y$; $z = -2, -1, 0, 1, 2$

9. $z = f(x, y) = y - x^2$; $z = -2, -1, 0, 1, 2$

10. $z = f(x, y) = \sqrt{x^2 + y^2}$; $z = 0, 1, 2, 3, 4$

11. $z = f(x, y) = e^{xy}$; $z = 1, 2, 3$

In Exercises 12–21, compute the first partial derivatives of the function.

12. $f(x, y) = x^2y^3 + 3xy^2 + \dfrac{x}{y}$

13. $f(x, y) = x\sqrt{y} + y\sqrt{x}$

14. $f(u, v) = \sqrt{uv^2 - 2u}$

15. $f(x, y) = \dfrac{x - y}{y + 2x}$ 16. $g(x, y) = \dfrac{xy}{x^2 + y^2}$

17. $h(x, y) = (2xy + 3y^2)^5$ 18. $f(x, y) = (xe^y + 1)^{1/2}$

19. $f(x, y) = (x^2 + y^2)e^{x^2+y^2}$

20. $f(x, y) = \ln(1 + 2x^2 + 4y^4)$

21. $f(x, y) = \ln\left(1 + \dfrac{x^2}{y^2}\right)$

In Exercises 22–27, compute the second-order partial derivatives of the function.

22. $f(x, y) = x^3 - 2x^2y + y^2 + x - 2y$

23. $f(x, y) = x^4 + 2x^2y^2 - y^4$

24. $f(x, y) = (2x^2 + 3y^2)^3$ 25. $g(x, y) = \dfrac{x}{x + y^2}$

26. $g(x, y) = e^{x^2+y^2}$ 27. $h(s, t) = \ln\left(\dfrac{s}{t}\right)$

28. Let $f(x, y, z) = x^3y^2z + xy^2z + 3xy - 4z$. Compute $f_x(1, 1, 0), f_y(1, 1, 0)$, and $f_z(1, 1, 0)$ and interpret your results.

In Exercises 29–34, find the critical point(s) of the functions. Then use the second derivative test to classify the nature of each of these points, if possible. Finally, determine the relative extrema of each function.

29. $f(x, y) = 2x^2 + y^2 - 8x - 6y + 4$

30. $f(x, y) = x^2 + 3xy + y^2 - 10x - 20y + 12$

31. $f(x, y) = x^3 - 3xy + y^2$

32. $f(x, y) = x^3 + y^2 - 4xy + 17x - 10y + 8$

33. $f(x, y) = e^{2x^2+y^2}$

34. $f(x, y) = \ln(x^2 + y^2 - 2x - 2y + 4)$

In Exercises 35–38, use the method of Lagrange multipliers to optimize the function subject to the given constraints.

35. Maximize the function $f(x, y) = -3x^2 - y^2 + 2xy$ subject to the constraint $2x + y = 4$.

36. Minimize the function $f(x, y) = 2x^2 + 3y^2 - 6xy + 4x - 9y + 10$ subject to the constraint $x + y = 1$.

37. Find the maximum and minimum values of the function $f(x, y) = 2x - 3y + 1$ subject to the constraint $2x^2 + 3y^2 - 125 = 0$.

38. Find the maximum and minimum values of the function $f(x, y) = e^{x-y}$ subject to the constraint $x^2 + y^2 = 1$.

In Exercises 39–42, evaluate the double integrals.

39. $f(x, y) = 3x - 2y$; R is the rectangle defined by $2 \leq x \leq 4$ and $-1 \leq y \leq 2$.

40. $f(x, y) = e^{-x-2y}$; R is the rectangle defined by $0 \leq x \leq 2$ and $0 \leq y \leq 1$.

41. $f(x, y) = 2x^2y$; R is bounded by $x = 0$, $x = 1$, $y = x^2$, and $y = x^3$.

42. $f(x, y) = \dfrac{y}{x}$; R is bounded by $x = 1$, $x = 2$, $y = 1$, and $y = x$.

In Exercises 43 and 44, find the volume of the solid bounded above by the surface $z = f(x, y)$ and below by the plane region R.

43. $f(x, y) = 4x^2 + y^2$; $R = \{0 \leq x \leq 2; 0 \leq y \leq 1\}$

44. $f(x, y) = x + y$; R is the region bounded by $y = x^2$, $y = 4x$, and $y = 4$.

45. Find the average value of the function

$$f(x, y) = xy + 1$$

over the plane region R bounded by $y = x^2$ and $y = 2x$.

46. A division of Ditton Industries makes a 16-speed and a 10-speed electric blender. The company's management estimates that x units of the 16-speed model and y units of the 10-speed model are demanded daily when the unit prices are

$$p = 80 - 0.02x - 0.1y$$
$$q = 60 - 0.1x - 0.05y$$

dollars, respectively.

a. Find the daily total revenue function $R(x, y)$.
b. Find the domain of the function R.
c. Compute $R(100, 300)$ and interpret your result.

47. In a survey conducted by *Home Entertainment* magazine, it was determined that the demand equation for compact disc (CD) players is given by

$$x = f(p, q) = 900 - 9p - e^{0.4q}$$

whereas the demand equation for audio CDs is given by

$$y = g(p, q) = 20,000 - 3000q - 4p$$

where p and q denote the unit prices (in dollars) for the CD players and audio CDs, respectively, and x and y denote the number of CD players and audio CDs demanded per week. Determine whether these two products are substitute, complementary, or neither.

48. The following data were compiled by the Bureau of Television Advertising in a large metropolitan area, giving the average daily TV-viewing time per household in that area over the years 1992 to 2000.

Year	1992	1994
Daily Viewing Time, y	6 hr 9 min	6 hr 30 min

Year	1996	1998	2000
Daily Viewing Time, y	6 hr 36 min	7 hr	7 hr 16 min

a. Find the least-squares line for these data. (Let $x = 1$ represent the year 1992.)
b. Estimate the average daily TV-viewing time per household in the year 2002.

49. The Odyssey Travel Agency's monthly revenue depends on the amount of money x (in thousands of dollars) spent on advertising per month and the number of agents y in its employ in accordance with the rule

$$R(x, y) = -x^2 - 0.5y^2 + xy + 8x + 3y + 20$$

Determine the amount of money the agency should spend per month and the number of agents it should employ in order to maximize its monthly revenue.

50. The owner of the Rancho Grande wants to enclose a rectangular piece of grazing land along the straight portion of a river and then subdivide it using a fence running parallel to the sides. No fencing is required along the river. If the material for the sides costs \$3/running yard and the material for the divider costs \$2/running yard, what will be the dimensions of a 303,750-yd pasture if the cost of fencing material is kept to a minimum?

51. The production of Q units of a commodity is related to the amount of labor x and the amount of capital y (in suitable units) expended by the equation

$$Q = f(x, y) = x^{3/4} y^{1/4}$$

If an expenditure of 100 units is available for production, how should it be apportioned between labor and capital so that Q is maximized?

Hint: Use the method of Lagrange multipliers to maximize the function Q subject to the constraint $x + y = 100$.

TABLES

Table 1	Exponential Functions				
x	e^x	e^{-x}	x	e^x	e^{-x}
0.00	1.0000	1.0000	1.5	4.4817	0.2231
0.01	1.0101	0.9901	1.6	4.9530	0.2019
0.02	1.0202	0.9802	1.7	5.4739	0.1827
0.03	1.0305	0.9705	1.8	6.0496	0.1653
0.04	1.0408	0.9608	1.9	6.6859	0.1496
0.05	1.0513	0.9512	2.0	7.3891	0.1353
0.06	1.0618	0.9418	2.1	8.1662	0.1225
0.07	1.0725	0.9324	2.2	9.0250	0.1108
0.08	1.0833	0.9231	2.3	9.9742	0.1003
0.09	1.0942	0.9139	2.4	11.023	0.0907
0.10	1.1052	0.9048	2.5	12.182	0.0821
0.11	1.1163	0.8958	2.6	13.464	0.0743
0.12	1.1275	0.8869	2.7	14.880	0.0672
0.13	1.1388	0.8781	2.8	16.445	0.0608
0.14	1.1503	0.8694	2.9	18.174	0.0550
0.15	1.1618	0.8607	3.0	20.086	0.0498
0.16	1.1735	0.8521	3.1	22.198	0.0450
0.17	1.1853	0.8437	3.2	24.533	0.0408
0.18	1.1972	0.8353	3.3	27.113	0.0369
0.19	1.2092	0.8270	3.4	29.964	0.0334
0.20	1.2214	0.8187	3.5	33.115	0.0302
0.21	1.2337	0.8106	3.6	36.598	0.0273
0.22	1.2461	0.8025	3.7	40.447	0.0247
0.23	1.2586	0.7945	3.8	44.701	0.0224
0.24	1.2712	0.7866	3.9	49.402	0.0202
0.25	1.2840	0.7788	4.0	54.598	0.0183
0.30	1.3499	0.7408	4.1	60.340	0.0166
0.35	1.4191	0.7047	4.2	66.686	0.0150
0.40	1.4918	0.6703	4.3	73.700	0.0136
0.45	1.5683	0.6376	4.4	81.451	0.0123
0.50	1.6487	0.6065	4.5	90.017	0.0111
0.55	1.7333	0.5769	4.6	99.484	0.0101
0.60	1.8221	0.5488	4.7	109.95	0.0091
0.65	1.9155	0.5220	4.8	121.51	0.0082
0.70	2.0138	0.4966	4.9	134.29	0.0074
0.75	2.1170	0.4724	5.0	148.41	0.0067
0.80	2.2255	0.4493	5.5	244.69	0.0041
0.85	2.3396	0.4274	6.0	403.43	0.0025
0.90	2.4596	0.4066	6.5	665.14	0.0015
0.95	2.5857	0.3867	7.0	1096.6	0.0009
1.0	2.7183	0.3679	7.5	1808.0	0.0006
1.1	3.0042	0.3329	8.0	2981.0	0.0003
1.2	3.3201	0.3012	8.5	4914.8	0.0002
1.3	3.6693	0.2725	9.0	8103.1	0.0001
1.4	4.0552	0.2466	10.0	22026	0.00005

From *Modern College Algebra and Trigonometry,* Fourth Edition, by Edwin F. Beckenbach, Irving Drooyan, and Michael D. Grady. © 1981 by Wadsworth, Inc. Reprinted by permission of Wadsworth Publishing Company, Belmont, California 94002.

Table 2 Common Logarithms

x	0	1	2	3	4	5	6	7	8	9
1.0	.0000	.0043	.0086	.0128	.0170	.0212	.0253	.0294	.0334	.0374
1.1	.0414	.0453	.0492	.0531	.0569	.0607	.0645	.0682	.0719	.0755
1.2	.0792	.0828	.0864	.0899	.0934	.0969	.1004	.1038	.1072	.1106
1.3	.1139	.1173	.1206	.1239	.1271	.1303	.1335	.1367	.1399	.1430
1.4	.1461	.1492	.1523	.1553	.1584	.1614	.1644	.1673	.1703	.1732
1.5	.1761	.1790	.1818	.1847	.1875	.1903	.1931	.1959	.1987	.2014
1.6	.2041	.2068	.2095	.2122	.2148	.2175	.2201	.2227	.2253	.2279
1.7	.2304	.2330	.2355	.2380	.2405	.2430	.2455	.2480	.2504	.2529
1.8	.2553	.2577	.2601	.2625	.2648	.2672	.2695	.2718	.2742	.2765
1.9	.2788	.2810	.2833	.2856	.2878	.2900	.2923	.2945	.2967	.2989
2.0	.3010	.3032	.3054	.3075	.3096	.3118	.3139	.3160	.3181	.3201
2.1	.3222	.3243	.3263	.3284	.3304	.3324	.3345	.3365	.3385	.3404
2.2	.3424	.3444	.3464	.3483	.3502	.3522	.3541	.3560	.3579	.3598
2.3	.3617	.3636	.3655	.3674	.3692	.3711	.3729	.3747	.3766	.3784
2.4	.3802	.3820	.3838	.3856	.3874	.3892	.3909	.3927	.3945	.3962
2.5	.3979	.3997	.4014	.4031	.4048	.4065	.4082	.4099	.4116	.4133
2.6	.4150	.4166	.4183	.4200	.4216	.4232	.4249	.4265	.4281	.4298
2.7	.4314	.4330	.4346	.4362	.4378	.4393	.4409	.4425	.4440	.4456
2.8	.4472	.4487	.4502	.4518	.4533	.4548	.4564	.4579	.4594	.4609
2.9	.4624	.4639	.4654	.4669	.4683	.4698	.4713	.4728	.4742	.4757
3.0	.4771	.4786	.4800	.4814	.4829	.4843	.4857	.4871	.4886	.4900
3.1	.4914	.4928	.4942	.4955	.4969	.4983	.4997	.5011	.5024	.5038
3.2	.5051	.5065	.5079	.5092	.5105	.5119	.5132	.5145	.5159	.5172
3.3	.5184	.5198	.5211	.5224	.5237	.5250	.5263	.5276	.5289	.5302
3.4	.5315	.5328	.5340	.5353	.5366	.5378	.5391	.5403	.5416	.5428
3.5	.5441	.5453	.5465	.5478	.5490	.5502	.5514	.5527	.5539	.5551
3.6	.5563	.5575	.5587	.5599	.5611	.5623	.5635	.5647	.5658	.5670
3.7	.5682	.5694	.5705	.5717	.5729	.5740	.5752	.5763	.5775	.5786
3.8	.5798	.5809	.5821	.5832	.5843	.5855	.5866	.5877	.5888	.5899
3.9	.5911	.5922	.5933	.5944	.5955	.5966	.5977	.5988	.5999	.6010
4.0	.6021	.6031	.6042	.6043	.6064	.6075	.6085	.6096	.6107	.6117
4.1	.6128	.6138	.6149	.6160	.6170	.6180	.6191	.6201	.6212	.6222
4.2	.6232	.6243	.6253	.6263	.6274	.6284	.6294	.6304	.6314	.6325
4.3	.6335	.6345	.6355	.6365	.6375	.6385	.6395	.6405	.6415	.6425
4.4	.6435	.6444	.6454	.6464	.6474	.6484	.6493	.6503	.6513	.6522
4.5	.6532	.6542	.6551	.6561	.6571	.6580	.6590	.6599	.6609	.6618
4.6	.6628	.6637	.6646	.6656	.6665	.6675	.6684	.6693	.6702	.6712
4.7	.6721	.6730	.6739	.6749	.6758	.6767	.6776	.6785	.6794	.6803
4.8	.6812	.6821	.6830	.6839	.6848	.6857	.6866	.6875	.6884	.6893
4.9	.6902	.6911	.6920	.6928	.6937	.6946	.6955	.6964	.6972	.6981
5.0	.6990	.6998	.7007	.7016	.7024	.7033	.7042	.7050	.7059	.7067
5.1	.7076	.7084	.7093	.7101	.7110	.7118	.7126	.7135	.7143	.7152
5.2	.7160	.7168	.7177	.7185	.7193	.7202	.7210	.7218	.7226	.7235
5.3	.7243	.7251	.7259	.7267	.7275	.7284	.7292	.7300	.7308	.7316
5.4	.7324	.7332	.7340	.7348	.7356	.7364	.7372	.7380	.7388	.7396
x	0	1	2	3	4	5	6	7	8	9

(*continued*)

From *Modern College Algebra and Trigonometry*, Fourth Edition, by Edwin F. Beckenbach, Irving Drooyan, and Michael D. Grady. © 1981 by Wadsworth, Inc. Reprinted by permission of Wadsworth Publishing Company, Belmont, California 94002.

Table 2 Common Logarithms (*continued*)

x	0	1	2	3	4	5	6	7	8	9
5.5	.7404	.7412	.7419	.7427	.7435	.7443	.7451	.7459	.7466	.7474
5.6	.7482	.7490	.7497	.7505	.7513	.7520	.7528	.7536	.7543	.7551
5.7	.7559	.7566	.7574	.7582	.7589	.7597	.7604	.7612	.7619	.7627
5.8	.7634	.7642	.7649	.7657	.7664	.7672	.7679	.7686	.7694	.7701
5.9	.7709	.7716	.7723	.7731	.7738	.7745	.7752	.7760	.7767	.7774
6.0	.7782	.7789	.7796	.7803	.7810	.7818	.7825	.7832	.7839	.7846
6.1	.7853	.7860	.7868	.7875	.7882	.7889	.7896	.7903	.7910	.7917
6.2	.7924	.7931	.7938	.7945	.7952	.7959	.7966	.7973	.7980	.7987
6.3	.7993	.8000	.8007	.8014	.8021	.8028	.8035	.8041	.8048	.8055
6.4	.8062	.8069	.8075	.8082	.8089	.8096	.8102	.8109	.8116	.8122
6.5	.8129	.8136	.8142	.8149	.8156	.8162	.8169	.8176	.8182	.8189
6.6	.8195	.8202	.8209	.8215	.8222	.8228	.8235	.8241	.8248	.8254
6.7	.8261	.8267	.8274	.8280	.8287	.8293	.8299	.8306	.8312	.8319
6.8	.8325	.8331	.8338	.8344	.8351	.8357	.8363	.8370	.8376	.8382
6.9	.8388	.8395	.8401	.8407	.8414	.8420	.8426	.8432	.8439	.8445
7.0	.8451	.8457	.8463	.8470	.8476	.8482	.8488	.8494	.8500	.8506
7.1	.8513	.8519	.8525	.8531	.8537	.8543	.8549	.8555	.8561	.8567
7.2	.8573	.8579	.8585	.8591	.8597	.8603	.8609	.8615	.8621	.8627
7.3	.8633	.8639	.8645	.8651	.8657	.8663	.8669	.8675	.8681	.8686
7.4	.8692	.8698	.8704	.8710	.8716	.8722	.8727	.8733	.8739	.8745
7.5	.8751	.8756	.8762	.8768	.8774	.8779	.8785	.8791	.8797	.8802
7.6	.8808	.8814	.8820	.8825	.8831	.8837	.8842	.8848	.8854	.8859
7.7	.8865	.8871	.8876	.8882	.8887	.8893	.8899	.8904	.8910	.8915
7.8	.8921	.8927	.8932	.8938	.8943	.8949	.8954	.8960	.8965	.8971
7.9	.8976	.8982	.8987	.8993	.8998	.9004	.9009	.9015	.9020	.9025
8.0	.9031	.9036	.9042	.9047	.9053	.9058	.9063	.9069	.9074	.9079
8.1	.9085	.9090	.9096	.9101	.9106	.9112	.9117	.9122	.9128	.9133
8.2	.9138	.9143	.9149	.9154	.9159	.9165	.9170	.9175	.9180	.9186
8.3	.9191	.9196	.9201	.9206	.9212	.9217	.9222	.9227	.9232	.9238
8.4	.9243	.9248	.9253	.9258	.9263	.9269	.9274	.9279	.9284	.9289
8.5	.9294	.9299	.9304	.9309	.9315	.9320	.9325	.9330	.9335	.9340
8.6	.9345	.9350	.9355	.9360	.9365	.9370	.9375	.9380	.9385	.9390
8.7	.9395	.9400	.9405	.9410	.9415	.9420	.9425	.9430	.9435	.9440
8.8	.9445	.9450	.9455	.9460	.9465	.9469	.9474	.9479	.9484	.9489
8.9	.9494	.9499	.9504	.9509	.9513	.9518	.9523	.9528	.9533	.9538
9.0	.9542	.9547	.9552	.9557	.9562	.9566	.9571	.9576	.9581	.9586
9.1	.9590	.9595	.9600	.9605	.9609	.9614	.9619	.9624	.9628	.9633
9.2	.9638	.9643	.9647	.9652	.9657	.9661	.9666	.9671	.9675	.9680
9.3	.9685	.9689	.9694	.9699	.9703	.9708	.9713	.9717	.9722	.9727
9.4	.9731	.9736	.9741	.9745	.9750	.9754	.9759	.9763	.9768	.9773
9.5	.9777	.9782	.9786	.9791	.9795	.9800	.9805	.9809	.9814	.9818
9.6	.9823	.9827	.9832	.9836	.9841	.9845	.9850	.9854	.9859	.9863
9.7	.9868	.9872	.9877	.9881	.9886	.9890	.9894	.9899	.9903	.9908
9.8	.9912	.9917	.9921	.9926	.9930	.9934	.9939	.9943	.9948	.9952
9.9	.9956	.9961	.9965	.9969	.9974	.9978	.9983	.9987	.9991	.9996
x	0	1	2	3	4	5	6	7	8	9

Table 3 Natural Logarithms

n	$\log_e n$	n	$\log_e n$	n	$\log_e n$
		4.5	1.5041	9.0	2.1972
0.1	7.6974 ⎤ *	4.6	1.5261	9.1	2.2083
0.2	8.3906	4.7	1.5476	9.2	2.2192
0.3	8.7960	4.8	1.5686	9.3	2.2300
0.4	9.0837	4.9	1.5892	9.4	2.2407
0.5	9.3069	5.0	1.6094	9.5	2.2513
0.6	9.4892	5.1	1.6292	9.6	2.2618
0.7	9.6433	5.2	1.6487	9.7	2.2721
0.8	9.7769	5.3	1.6677	9.8	2.2824
0.9	9.8946 ⎦	5.4	1.6864	9.9	2.2925
1.0	0.0000	5.5	1.7047	10	2.3026
1.1	0.0953	5.6	1.7228	11	2.3979
1.2	0.1823	5.7	1.7405	12	2.4849
1.3	0.2624	5.8	1.7579	13	2.5649
1.4	0.3365	5.9	1.7750	14	2.6391
1.5	0.4055	6.0	1.7918	15	2.7081
1.6	0.4700	6.1	1.8083	16	2.7726
1.7	0.5306	6.2	1.8245	17	2.8332
1.8	0.5878	6.3	1.8405	18	2.8904
1.9	0.6419	6.4	1.8563	19	2.9444
2.0	0.6931	6.5	1.8718	20	2.9957
2.1	0.7419	6.6	1.8871	25	3.2189
2.2	0.7885	6.7	1.9021	30	3.4012
2.3	0.8329	6.8	1.9169	35	3.5553
2.4	0.8755	6.9	1.9315	40	3.6889
2.5	0.9163	7.0	1.9459	45	3.8067
2.6	0.9555	7.1	1.9601	50	3.9120
2.7	0.9933	7.2	1.9741	55	4.0073
2.8	1.0296	7.3	1.9879	60	4.0943
2.9	1.0647	7.4	2.0015	65	4.1744
3.0	1.0986	7.5	2.0149	70	4.2485
3.1	1.1314	7.6	2.0281	75	4.3175
3.2	1.1632	7.7	2.0412	80	4.3820
3.3	1.1939	7.8	2.0541	85	4.4427
3.4	1.2238	7.9	2.0669	90	4.4998
3.5	1.2528	8.0	2.0794	100	4.6052
3.6	1.2809	8.1	2.0919	110	4.7005
3.7	1.3083	8.2	2.1041	120	4.7875
3.8	1.3350	8.3	2.1163	130	4.8676
3.9	1.3610	8.4	2.1282	140	4.9416
4.0	1.3863	8.5	2.1401	150	5.0106
4.1	1.4110	8.6	2.1518	160	5.0752
4.2	1.4351	8.7	2.1633	170	5.1358
4.3	1.4586	8.8	2.1748	180	5.1930
4.4	1.4816	8.9	2.1861	190	5.2470

* Subtract 10 for $n < 1$. Thus, $\log_e 0.1 = 7.6974 - 10 = -2.3026$.

From *Modern College Algebra and Trigonometry*, Fourth Edition, by Edwin F. Beckenbach, Irving Drooyan, and Michael D. Grady. © 1981 by Wadsworth, Inc. Reprinted by permission of Wadsworth Publishing Company, Belmont, California 94002.

Answers to Odd-Numbered Exercises

Chapter 1

Exercises 1.1, page 12

1. False   3. False

5.
$$\begin{array}{ccc} 0 & 3 & 6 \end{array}$$

7.
$$\begin{array}{cc} -1 \quad 0 & 4 \end{array}$$

9.
$$0$$

11. $(-\infty, 2)$ 13. $(-\infty, -5]$ 15. $(-4, 6)$

17. $(-\infty, -3) \cup (3, \infty)$ 19. $(-2, 3)$ 21. 4

23. 2   25. $5\sqrt{3}$ 27. $\pi + 1$ 29. 2

31. False   33. False   35. True   37. False

39. True   41. False   43. 9   45. 1   47. 4

49. 7   51. $\frac{1}{5}$ 53. 2   55. 2   57. 1

59. True   61. False   63. False   65. False

67. False   69. $\dfrac{1}{(xy)^2}$ 71. $\dfrac{1}{x^{5/6}}$ 73. $\dfrac{1}{(s + t)^3}$

75. $x^{13/3}$ 77. $\dfrac{1}{x^3}$ 79. x 81. $\dfrac{9}{x^2 y^4}$ 83. $\dfrac{y^8}{x^{10}}$

85. $2x^{11/6}$ 87. $-2xy^2$ 89. $2x^{4/3}y^{1/2}$ 91. 2.828

93. 5.196   95. 31.62   97. 316.2   99. $\dfrac{3\sqrt{x}}{2x}$

101. $\dfrac{2\sqrt{3y}}{3}$ 103. $\dfrac{\sqrt[3]{x^2}}{x}$ 105. $\dfrac{2x}{3\sqrt{x}}$ 107. $\dfrac{2y}{\sqrt{2xy}}$

109. $\dfrac{xz}{y\sqrt[3]{xz^2}}$ 111. [362, 488.7]   113. \$12,300

115. \$22,000   117. $|x - 0.5| < 0.01$

119. Between 1000 and 4000 units

121. False   123. True

Exercises 1.2, page 26

1. $9x^2 + 3x + 1$ 3. $4y^2 + y + 8$ 5. $-x - 1$

7. $\frac{2}{3} + e - e^{-1}$ 9. $6\sqrt{2} + 8 + \frac{1}{2}\sqrt{x} - \frac{11}{4}\sqrt{y}$

11. $x^2 + 6x - 16$ 13. $a^2 + 10a + 25$

15. $x^2 + 4xy + 4y^2$ 17. $4x^2 - y^2$ 19. $-2x$

21. $2t(2\sqrt{t} + 1)$ 23. $2x^3(2x^2 - 6x - 3)$

25. $7a^2(a^2 + 7ab - 6b^2)$ 27. $e^{-x}(1 - x)$

29. $\frac{1}{2}x^{-5/2}(4 - 3x)$ 31. $(2a + b)(3c - 2d)$

33. $(2a + b)(2a - b)$ 35. $-2(3x + 5)(2x - 1)$

37. $3(x - 4)(x + 2)$ 39. $2(3x - 5)(2x + 3)$

41. $(3x - 4y)(3x + 4y)$ 43. $(x^2 + 5)(x^4 - 5x^2 + 25)$

45. $x^3 - xy^2$ 47. $4(x - 1)(3x - 1)(2x + 2)^3$

49. $4(x - 1)(3x - 1)(2x + 2)^3$

51. $2x(x^2 + 2)^2(5x^4 + 20x^2 + 17)$

53. -4 and 3   55. -1 and $\frac{1}{2}$ 57. 2 and 2

59. -2 and $\frac{3}{4}$ 61. $\frac{1}{2} + \frac{1}{4}\sqrt{10}$ and $\frac{1}{2} - \frac{1}{4}\sqrt{10}$

63. $-1 + \frac{1}{2}\sqrt{10}$ and $-1 - \frac{1}{2}\sqrt{10}$ 65. $\dfrac{x - 1}{x - 2}$

67. $\dfrac{3(2t + 1)}{2t - 1}$ 69. $-\dfrac{7}{(4x - 1)^2}$ 71. -8

73. $\dfrac{3x - 1}{2}$ 75. $\dfrac{t + 20}{3t + 2}$ 77. $-\dfrac{x(2x - 13)}{(2x - 1)(2x + 5)}$

79. $-\dfrac{x + 27}{(x - 3)^2(x + 3)}$ 81. $\dfrac{x + 1}{x - 1}$ 83. $\dfrac{4x^2 + 7}{\sqrt{2x^2 + 7}}$

85. $\dfrac{x - 1}{x^2\sqrt{x + 1}}$ 87. $\dfrac{x - 1}{(2x + 1)^{3/2}}$ 89. $\dfrac{\sqrt{3} + 1}{2}$

91. $\dfrac{\sqrt{x} + \sqrt{y}}{x - y}$ 93. $\dfrac{(\sqrt{a} + \sqrt{b})^2}{a - b}$ 95. $\dfrac{x}{3\sqrt{x}}$

97. $-\dfrac{2}{3(1 + \sqrt{3})}$ 99. $-\dfrac{x + 1}{\sqrt{x + 2}(1 - \sqrt{x + 2})}$

101. True   103. False

Exercises 1.3, page 34

1. (3, 3); Quadrant I   3. (2, −2); Quadrant IV

5. (−4, −6); Quadrant III   7. A 9. E, F, and G

11. F 13.–19. See accompanying figure.

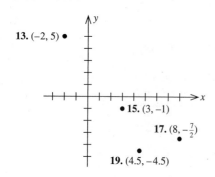

13. (−2, 5) •

• 15. (3, −1)

17. $(8, -\frac{7}{2})$

19. (4.5, −4.5)

21. 5   23. $\sqrt{61}$ 25. (−8, −6) and (8, −6)

29. $(x - 2)^2 + (y + 3)^2 = 25$

31. $x^2 + y^2 = 25$

33. $(x - 2)^2 + (y + 3)^2 = 34$

35. 3400 miles   37. Route 1   39. Model C

41. $10\sqrt{13}t$; 72.1 mi   43. True   45. True

Exercises 1.4, page 48

1. e   3. a   5. f   7. $\frac{1}{2}$ 9. Not defined

11. 5   13. $\frac{5}{6}$ 15. $\dfrac{d - b}{c - a}$ $(a \neq c)$

17. **a.** 4   **b.** −8   19. Parallel   21. Perpendicular

23. −5   25. $y = -3$ 27. $y = 2x - 10$

29. $y = 2$ 31. $y = 3x - 2$ 33. $y = x + 1$

35. $y = 3x + 4$ 37. $y = 5$

39. $y = \frac{1}{2}x$; $m = \frac{1}{2}$; $b = 0$

41. $y = \frac{2}{3}x - 3$; $m = \frac{2}{3}$; $b = -3$

43. $y = -\frac{1}{2}x + \frac{7}{2}$; $m = -\frac{1}{2}$; $b = \frac{7}{2}$

45. $y = \frac{1}{2}x + 3$ 47. $y = -6$ 49. $y = b$

51. $y = \frac{2}{3}x - \frac{2}{3}$ 53. $k = 8$

55.

57.

59.

63. $y = -2x - 4$ 65. $y = \frac{1}{8}x - \frac{1}{2}$ 67. Yes

69. **a.** $y = 0.55x$ **b.** 2000   71. 82.4% of men's wages

73. **a.** and **b.**

(100, 250)

Dollars

250
240
230
220
210
200

20 40 60 80 100

Units

c. $y = \frac{1}{2}x + 200$ **d.** \$227

75. a. and **b.**

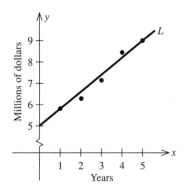

c. $y = 0.8x + 5$ **d.** \$12.2 million

77. True   **79.** True   **81.** Yes

Chapter 1 Review Exercises, page 56

1. $[-2, \infty)$ **3.** $(-\infty, -4) \cup (5, \infty)$ **5.** 4

7. $\pi - 6$ **9.** $\frac{27}{8}$ **11.** $\frac{1}{144}$ **13.** $\frac{1}{4}$

15. $4(x^2 + y)^2$ **17.** $\frac{2x}{3z}$ **19.** $6xy^7$

21. $2vw(v^2 + w^2 + u^2)$ **23.** $6t(t + 1)(2t - 3)$

25. -2 and $\frac{1}{3}$ **27.** $\frac{\sqrt{2}}{2}$ and $-\frac{\sqrt{2}}{2}$

29. $-2 + \frac{\sqrt{2}}{2}$ and $-2 - \frac{\sqrt{2}}{2}$ **31.** $\frac{15x^2 + 24x + 2}{4(x + 2)(3x^2 + 2)}$

33. $\frac{2(x + 2)}{\sqrt{x + 1}}$ **35.** $\frac{x - \sqrt{x}}{2x}$ **37.** 2   **39.** $y = 4$

41. $y = -\frac{4}{5}x + \frac{12}{5}$ **43.** $y = \frac{3}{4}x + \frac{11}{2}$

45. $y = -\frac{3}{5}x + \frac{12}{5}$

47.

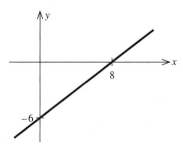

49. \$100

CHAPTER 2

Exercises 2.1, page 70

1. $21, -9, 5a + 6, -5a + 6, 5a + 21$

3. $-3, 6, 3a^2 - 6a - 3, 3a^2 + 6a - 3, 3x^2 - 6$

5. $\frac{8}{15}, 0, \frac{2a}{a^2 - 1}, \frac{2(2 + a)}{a^2 + 4a + 3}, \frac{2(t + 1)}{t(t + 2)}$

7. $8, \frac{2a^2}{\sqrt{a - 1}}, \frac{2(x + 1)^2}{\sqrt{x}}, \frac{2(x - 1)^2}{\sqrt{x - 2}}$

9. $5, 1, 1$ **11.** $\frac{5}{2}, 3, 3, 9$

13. a. -2 **b.** (i) $x = 2$; (ii) $x = 1$ **c.** $[0, 6]$ **d.** $[-2, 6]$

15. Yes   **17.** Yes

19. $(-\infty, \infty)$ **21.** $(-\infty, 0) \cup (0, \infty)$ **23.** $(-\infty, \infty)$

25. $(-\infty, 5]$ **27.** $(-\infty, -1) \cup (-1, 1) \cup (1, \infty)$

29. $[-3, \infty)$ **31.** $(-\infty, -2) \cup (-2, 1]$

33. a. $(-\infty, \infty)$
b. $6, 0, -4, -6, -\frac{25}{4}, -6, -4, 0$
c.

35.

$(-\infty, \infty); \; [1, \infty)$

37.
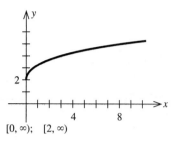
$[0, \infty);$ $[2, \infty)$

39.

$(-\infty, 1];$ $[0, \infty)$

41.
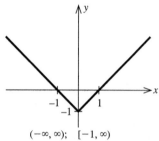
$(-\infty, \infty);$ $[-1, \infty)$

43.
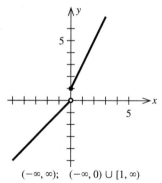
$(-\infty, \infty);$ $(-\infty, 0) \cup [1, \infty)$

45.
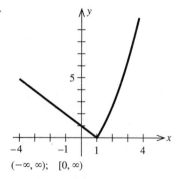
$(-\infty, \infty);$ $[0, \infty)$

47. Yes   **49.** No   **51.** Yes   **53.** Yes

55. 10π in.   **57.** 8   **59. a.** From 1985 to 1990
b. From 1990 on
c. 1990; $3.5 billion

61. a.
$$f(t) = \begin{cases} 0.0185t + 0.58 & \text{if } 0 \leq t \leq 20 \\ 0.015t + 0.65 & \text{if } 20 < t \leq 30 \end{cases}$$
b. 0.0185/yr from 1960 through 1980; 0.015/yr from 1980 through 1990
c. 1983

63. a. $0.06x$ **b.** $12.00; $0.34   **65.** 160 mg

67. a. $30 + 0.15x$; $25 + 0.20x$
b.

c. Acme

69. $700,000; $620,000; $540,000

71. a. $0 \leq r \leq 0.2$
b. 40; 30; 0; as the distance r increases, the velocity of the blood decreases.

73. 20; 26

75. 0.77. When the proportion of popular votes won by the Democratic presidential candidate is 0.60, the proportion of seats in the House of Representatives won by Democratic candidates is 0.77.

77. a. $(0, 12]$
b.

79. True   **81.** False

Using Technology Exercises 2.1, page 79

1.

3.

5.

7.

9. a.

b.

11. a.

b.

13. a.

b.

15. a.

b.

17. a.

b.

19. a.

b.

21.

23.

25.

27.

29.

31. 18   **33.** 2   **35.** 18.5505   **37.** 17.3850

39. 4.1616   **41.** 1.7214

43. a.

b. 2.1762%; 1.9095%

45. a.

b. 44.7; 52.7; 129.2

Exercises 2.2, page 87

1. $f(x) + g(x) = x^3 + x^2 + 3$

3. $f(x)g(x) = x^5 - 2x^3 + 5x^2 - 10$

5. $\dfrac{f(x)}{g(x)} = \dfrac{x^3 + 5}{x^2 - 2}$

7. $\dfrac{f(x)g(x)}{h(x)} = \dfrac{x^5 - 2x^3 + 5x^2 - 10}{2x + 4}$

9. $f(x) + g(x) = x - 1 + \sqrt{x + 1}$

11. $f(x)g(x) = (x - 1)\sqrt{x + 1}$

13. $\dfrac{g(x)}{h(x)} = \dfrac{\sqrt{x + 1}}{2x^3 - 1}$

15. $\dfrac{f(x)g(x)}{h(x)} = \dfrac{(x - 1)\sqrt{x + 1}}{2x^3 - 1}$

17. $\dfrac{f(x) - h(x)}{g(x)} = \dfrac{x - 2x^3}{\sqrt{x + 1}}$

19. $f(x) + g(x) = x^2 + \sqrt{x} + 3$;
$f(x) - g(x) = x^2 - \sqrt{x} + 7$;
$f(x)g(x) = (x^2 + 5)(\sqrt{x} - 2)$; $\dfrac{f(x)}{g(x)} = \dfrac{x^2 + 5}{\sqrt{x} - 2}$

21. $f(x) + g(x) = \dfrac{(x - 1)\sqrt{x + 3} + 1}{x - 1}$;
$f(x) - g(x) = \dfrac{(x - 1)\sqrt{x + 3} - 1}{x - 1}$;
$f(x)g(x) = \dfrac{\sqrt{x + 3}}{x - 1}$; $\dfrac{f(x)}{g(x)} = (x - 1)\sqrt{x + 3}$

23. $f(x) + g(x) = \dfrac{2(x^2 - 2)}{(x - 1)(x - 2)}$;
$f(x) - g(x) = \dfrac{-2x}{(x - 1)(x - 2)}$;
$f(x)g(x) = \dfrac{(x + 1)(x + 2)}{(x - 1)(x - 2)}$; $\dfrac{f(x)}{g(x)} = \dfrac{(x + 1)(x - 2)}{(x - 1)(x + 2)}$

25. $f(g(x)) = x^4 + x^2 + 1$; $g(f(x)) = (x^2 + x + 1)^2$

27. $f(g(x)) = \sqrt{x^2 - 1} + 1$; $g(f(x)) = x + 2\sqrt{x}$

29. $f(g(x)) = \dfrac{x}{x^2 + 1}$; $g(f(x)) = \dfrac{x^2 + 1}{x}$

31. 49   **33.** $\dfrac{\sqrt{5}}{5}$

35. $f(x) = 2x^3 + x^2 + 1$ and $g(x) = x^5$

37. $f(x) = x^2 - 1$ and $g(x) = \sqrt{x}$

39. $f(x) = x^2 - 1$ and $g(x) = \dfrac{1}{x}$

41. $f(x) = 3x^2 + 2$ and $g(x) = \dfrac{1}{x^{3/2}}$

43. $3h$ **45.** $-h(2a + h)$ **47.** $2a + h$

49. $C(x) = 0.6x + 12{,}100$

51. a. $P(x) = -0.000003x^3 - 0.07x^2 + 300x - 100{,}000$
  b. \$182,375

53. a. $N(t) = \dfrac{7}{1 + 0.02\left[\dfrac{10t + 150}{t + 10}\right]^2}$
  b. 1.27 million units; 1.74 million units; 1.85 million units

55. $N(x(t)) = 9.94\left[\dfrac{(t + 10)^2}{(t + 10)^2 + 2(t + 15)^2}\right]$; 2.24 million jobs; 2.48 million jobs

57. False   **59.** False

Exercises 2.3, page 99

1. Yes; $y = -\frac{2}{3}x + 2$ **3.** Yes, $y = \frac{1}{2}x + 2$

5. Yes; $y = \frac{1}{2}x + \frac{9}{4}$ **7.** No

9. Polynomial function; degree 6

11. Polynomial function; degree 6

13. Some other function   **15.** $m = -1$; $b = 2$

17. a. $C(x) = 8x + 40{,}000$
  b. $R(x) = 12x$
  c. $P(x) = 4x - 40{,}000$
  d. A loss of \$8000; a profit of \$8000

19. \$28,800   **21.** 104 mg   **23.** \$400,000

25. a.
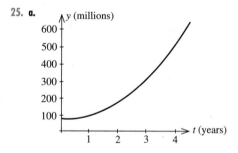

  b. 184.84 million

27. a. $R(x) = \dfrac{100x}{40 + x}$ **b.** 60%

29. a. $T = \frac{1}{4}N + 40$ **b.** $N = 4T - 160$; 248 times/min

31. \$72,000   **33. a.** 714,300   **b.** 8,327,800

35. $\dfrac{110}{\frac{1}{2}t + 1} - 26\left(\frac{1}{4}t^2 - 1\right)^2 - 52$; \$32, \$6.71, \$3; the gap was closing.

37. a.

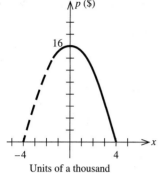

Units of a thousand

b. 3000 units

39. a.

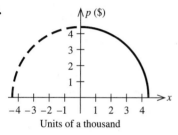

Units of a thousand

b. 3000

41. a.

Units of a thousand

b. $76

43. a.

Units of a thousand

b. $15

45. L_2; for each dollar increase in the price of a clock radio, more model A clock radios than model B clock radios will be made available in the marketplace.

47. $p = \sqrt{-x^2 + 100}$; 6614 units

49. $p = \frac{1}{10}\sqrt{x} + 10$; $30   **51.** 2500; $67.50

Units of a thousand

53. 11,000; $3   **55.** 500 tents; $32.50

57. a.

b. 5.6 mph; 71.6 mL/lb/min
c. The oxygen consumption of the walker is greater than that of the runner.

59. $f(x) = 2x + \dfrac{500}{x}; \, x > 0$

61. $f(x) = 0.5x^2 + \dfrac{8}{x}$

63. $f(x) = (22 + x)(36 - 2x)$ bushels/acre

65. True   **67.** False

Using Technology Exercises 2.3, page 106

1. $(-3.0414, 0.1503); (3.0414, 7.4497)$

3. $(-2.3371, 2.4117); (6.0514, -2.5015)$

5. $(-1.0219, -6.3461); (1.2414, -1.5931); (5.7805, 7.9391)$

7. a.

b. 438 wall clocks; $40.92

9. a. $y = 0.1375t^2 + 0.675t + 3.1$

b.

c. 3.1; 3.9; 5; 6.4; 8; 9.9

11. a. $y = -0.02028t^3 + 0.31393t^2 + 0.40873t + 0.66024$

b.

c. 0.66; 1.36; 2.57; 4.16; 6.02; 8.02; 10.03

13. a. $y = 0.05833t^3 - 0.325t^2 + 1.8881t + 5.07143$

b.

c. 6.7; 8.0; 9.4; 11.2; 13.7

15. a. $y = 0.0125t^4 - 0.01389t^3 + 0.55417t^2 + 0.53294t + 4.95238$ $(0 \le t \le 5)$

b.

c. 5.0; 6.0; 8.3; 12.2; 18.3; 27.5

Exercises 2.4, page 127

1. $\lim\limits_{x \to -2} f(x) = 3$ **3.** $\lim\limits_{x \to 3} f(x) = 3$ **5.** $\lim\limits_{x \to -2} f(x) = 3$

7. The limit does not exist.

9.

x	1.9	1.99	1.999
$f(x)$	4.61	4.9601	4.9960

x	2.001	2.01	2.1
$f(x)$	5.004	5.0401	5.41

$\lim\limits_{x \to 2} (x^2 + 1) = 5$

11.

x	-0.1	-0.01	-0.001
$f(x)$	-1	-1	-1

x	0.001	0.01	0.1
$f(x)$	1	1	1

The limit does not exist.

13.

x	0.9	0.99	0.999
$f(x)$	100	10,000	1,000,000

x	1.001	1.01	1.1
$f(x)$	1,000,000	10,000	100

The limit does not exist.

15.

x	0.9	0.99	0.999	1.001	1.01	1.1
$f(x)$	2.9	2.99	2.999	3.001	3.01	3.1

$\lim\limits_{x \to 1} \dfrac{x^2 + x - 2}{x - 1} = 3$

17.

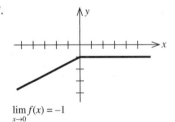

$\lim\limits_{x \to 0} f(x) = -1$

19.

$\lim\limits_{x \to 1} f(x) = 1$

21.

$$\lim_{x\to 0} f(x) = 0$$

23. 3

25. 3   **27.** −1   **29.** 2   **31.** −4   **33.** $\frac{5}{4}$

35. 2   **37.** $\sqrt{171} = 3\sqrt{19}$ **39.** $\frac{3}{2}$ **41.** −1

43. −6   **45.** 2   **47.** $\frac{1}{6}$ **49.** 2   **51.** −1

53. −10   **55.** The limit does not exist.   **57.** $\frac{5}{3}$

59. $\frac{1}{2}$ **61.** $\frac{1}{3}$ **63.** $\lim_{x\to\infty} f(x) = \infty$; $\lim_{x\to-\infty} f(x) = \infty$

65. 0; 0   **67.** $\lim_{x\to\infty} f(x) = -\infty$; $\lim_{x\to-\infty} f(x) = -\infty$

69.

x	1	10	100	1000
$f(x)$	0.5	0.009901	0.0001	0.000001

x	−1	−10	−100	−1000
$f(x)$	0.5	0.009901	0.0001	0.000001

$\lim_{x\to\infty} f(x) = 0$ and $\lim_{x\to-\infty} f(x) = 0$

71.

x	1	5	10	100
$f(x)$	12	360	2910	2.99×10^6

x	1000	−1	−5
$f(x)$	2.999×10^9	6	−390

x	−10	−100	−1000
$f(x)$	−3090	-3.01×10^6	-3.0×10^9

$\lim_{x\to\infty} f(x) = \infty$ and $\lim_{x\to-\infty} f(x) = -\infty$

73. 3   **75.** 3   **77.** $\lim_{x\to-\infty} f(x) = -\infty$ **79.** 0

81. a. $0.5 million; $0.75 million; $1,166,667; $2 million; $4.5 million; $9.5 million
b. The limit does not exist; as the percentage of pollutant to be removed approaches 100, the cost becomes astronomical.

83. $2.20; the average cost of producing x video discs will approach $2.20/disc in the long run.

85. a. $24 million; $60 million; $83.1 million
b. $120 million

87. a. 76.1 cents/mile; 30.5 cents/mile; 23 cents/mile; 20.6 cents/mile; 19.5 cents/mile
b.

c. It approaches 17.8 cents/mile

89. False   **91.** True   **93.** True

95. a moles/liter/second   **97.** No

Using Technology Exercises 2.4, page 132

1. 5   **3.** 3   **5.** $\frac{2}{3}$ **7.** $\frac{1}{2}$ **9.** e^2

13. a.

b. 25,000

Exercises 2.5, page 145

1. 3; 2; the limit does not exist.

3. The limit does not exist; 2; the limit does not exist.

5. 0; 2; the limit does not exist.

7. −2; 2; the limit does not exist.

9. True   **11.** True   **13.** False   **15.** True

17. False   **19.** True   **21.** 6   **23.** $-\frac{1}{4}$

25. The limit does not exist.   **27.** −1   **29.** 0

31. −4   **33.** The limit does not exist.   **35.** 4

37. 0   **39.** 0; 0   **41.** 2; 3

43. $x = 0$; conditions 2 and 3

45. Continuous everywhere   **47.** $x = 0$; condition 3

49. $x = 0$; condition 3   **51.** $(-\infty, \infty)$ **53.** $(-\infty, \infty)$

55. $(-\infty, \frac{1}{2}) \cup (\frac{1}{2}, \infty)$ **57.** $(-\infty, -2) \cup (-2, 1) \cup (1, \infty)$

59. $(-\infty, \infty)$ **61.** $(-\infty, \infty)$ **63.** $(-\infty, \infty)$

65. $(-\infty, \infty)$ **67.** -1 and 1 **69.** 1 and 2

71. f is discontinuous at $x = 1, 2, \ldots, 11$.

73. Michael makes progress toward solving the problem until $x = x_1$. Between $x = x_1$ and $x = x_2$, he makes no further progress. But at $x = x_2$ he suddenly achieves a breakthrough, and at $x = x_3$ he proceeds to complete the problem.

75. Conditions 2 and 3 are not satisfied at each of these points.

77.

79.

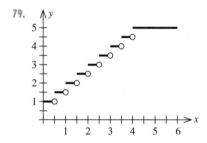

f is discontinuous at $x = \frac{1}{2}, 1, 1\frac{1}{2}, \ldots, 4$.

81. a. ∞; if the speed of the fish is very close to that of the current, the energy expended by the fish is enormous.
b. ∞; if the speed of the fish increases greatly, so does the amount of energy required to swim L feet.

83. $k = -4$ **85. a.** No   **b.** No

87. a. f is a polynomial of degree 3.
b. $f(-1) = -4$ and $f(1) = 4$

89. f is continuous on $[14, 16]$, $f(14) \approx -6.06$, and $f(16) \approx 1.60$.

91. $x = 2$ **93.** ≈ 1.34 **95.** False   **97.** False

99. True   **101. c.** No

Using Technology Exercises 2.5, page 152

1. $x = 0, 1$ **3.** $x = 2$ **5.** $x = 0, \frac{1}{2}$

7. $x = -\frac{1}{2}, 2$ **9.** $x = -2, 1$

11.

13.

15.

Exercises 2.6, page 169

1. 1.5 lb/month; 0.5833 lb/month; 1.25 lb/month

3. 3.075%/hr; -21.15%/hr

5. a. Car A
b. They are traveling at the same speed.
c. Car B
d. Both cars covered the same distance.

7. a. P_2 **b.** P_1 **c.** Bactericide B; bactericide A

9. 0   **11.** 2   **13.** $6x$ **15.** $-2x + 3$

17. $2; y = 2x + 7$ **19.** $6; y = 6x - 3$

21. $\frac{1}{9}; y = \frac{1}{9}x - \frac{2}{3}$

23. a. $4x$
b. $y = 4x - 1$
c.

25. a. $2x - 2$
b. $(1, 0)$
c.

d. 0

27. a. 6; 5.5; 5.1
b. 5
c. The computations in part (a) show that as h approaches zero, the average velocity approaches the instantaneous velocity.

29. a. 130 ft/sec; 128.2 ft/sec; 128.02 ft/sec
b. 128 ft/sec
c. The computations in part (a) show that as the time intervals over which the average velocity are computed become smaller and smaller, the average velocity approaches the instantaneous velocity of the car at $t = 20$.

31. a. 5 sec   **b.** 80 ft/sec   **c.** 160 ft/sec

33. a. $-\frac{1}{6}$ liter/atmosphere   **b.** $-\frac{1}{4}$ liter/atmosphere

35. a. $-\frac{2}{3}x + 7$ **b.** \$333/quarter; $-\$13,000$/quarter

37. \$6 billion/yr; \$10 billion/yr

39. Average rate of change of the seal population over $[a, a + h]$; instantaneous rate of change of the seal population at $x = a$

41. Average rate of change of the country's industrial production over $[a, a + h]$; instantaneous rate of change of the country's industrial production at $x = a$

43. Average rate of change of atmospheric pressure over $[a, a + h]$; instantaneous rate of change of atmospheric pressure at $x = a$

45. a. Yes   **b.** No   **c.** No

47. a. Yes   **b.** Yes   **c.** No

49. a. No   **b.** No   **c.** No

51. 5.06060; 5.06006; 5.060006; 5.0600006; 5.06000006; \$5.06/case

53. True

55.

57.

Yes; no; the graph of f has a kink at $x = 0$.

Using Technology Exercises 2.6, page 178

1. a. $y = 4x - 3$
b.

3. a. $y = -7x - 8$
b.

5. a. $y = 9x - 11$
b.

7. a. $y = 2$
b.

9. a. $y = \frac{1}{4}x + 1$
b.

11. a. 4
b. $y = 4x - 1$
c.

13. a. 20
b. $y = 20x - 35$
c.

15. a. 0.75
b. $y = 0.75x - 1$
c.

17. a. −0.25
   b. $y = -0.25x + 0.75$
   c.

19. a. 4.02
   b. $y = 4.02x - 3.57$
   c.

Chapter 2 Review Exercises, page 179

1. a. $(-\infty, 9]$
   b. $(-\infty, -1) \cup (-1, \frac{3}{2}) \cup (\frac{3}{2}, \infty)$

3. a.

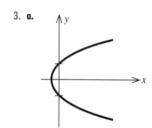

   b. No
   c. Yes

5. a. $\dfrac{2x + 3}{x}$ **b.** $\dfrac{1}{x(2x + 3)}$ **c.** $\dfrac{1}{2x + 3}$ **d.** $\dfrac{2}{x} + 3$

7. 2   **9.** 0   **11.** The limit does not exist.

13. $\frac{9}{2}$ **15.** $\frac{1}{2}$ **17.** 1   **19.** The limit does not exist.

21.

   4; 2; the limit does not exist.

23. $x = -\frac{1}{2}, 1$ **25.** $x = 0$ **27.** 3

29. $\frac{3}{2}$; $y = \frac{3}{2}x + 5$ **31. a.** Yes   **b.** No

33. a. $S(t) = t + 2.4$ **b.** $5.4 million

35. $(6, \frac{21}{2})$ **37.** 6000; $22   **39.** $45,000

41.

As the length of the list increases, the time taken to learn the list increases by a very large amount.

43. $C(x) = \begin{cases} 5 & \text{if } 1 \le x \le 100 \\ 9 & \text{if } 100 < x \le 200 \\ 12.50 & \text{if } 200 < x \le 300 \\ 15.00 & \text{if } 300 < x \le 400 \\ 7 + 0.02x & \text{if } x > 400 \end{cases}$

The function is discontinuous at $x = 100, 200,$ and 300.

CHAPTER 3

Exercises 3.1, page 191

1. 0   **3.** $5x^4$ **5.** $2.1x^{1.1}$ **7.** $6x$ **9.** $2\pi r$

11. $\dfrac{3}{x^{2/3}}$ **13.** $\dfrac{3}{2\sqrt{x}}$ **15.** $-84x^{-13}$ **17.** $10x - 3$

19. $-3x^2 + 4x$ **21.** $0.06x - 0.4$ **23.** $2x - 4 - \dfrac{3}{x^2}$

25. $16x^3 - 7.5x^{3/2}$ **27.** $-\dfrac{3}{x^2} - \dfrac{8}{x^3}$ **29.** $-\dfrac{16}{t^5} + \dfrac{9}{t^4} - \dfrac{2}{t^2}$

31. $2 - \dfrac{5}{2\sqrt{x}}$ **33.** $-\dfrac{4}{x^3} + \dfrac{1}{x^{4/3}}$

35. a. 20   **b.** −4   **c.** 20   **37.** 3   **39.** 11

41. $m = 5$; $y = 5x - 4$ **43.** $m = -2$; $y = -2x + 2$

45. a. (0, 0)   **b.**

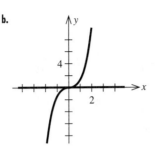

47. a. (−2, −7), (2, 9)
   b. $y = 12x + 17$ and $y = 12x − 15$
   c.

49. a. (0, 0); $(1, −\frac{13}{12})$
   b. (0, 0); $(2, −\frac{8}{3})$; $(−1, −\frac{5}{12})$
   c. (0, 0); $(4, \frac{80}{3})$; $(−3, \frac{81}{4})$

51. a. $\dfrac{16\pi}{9}$ cm³/cm   **b.** $\dfrac{25\pi}{4}$ cm³/cm

53. −89.01; −4.36; if you make 0.25 stop/mile, your average speed will decrease at the rate of approximately 89.01 mph/stop/mile. If you make 2 stops/mile, your average speed will decrease at the rate of approximately 4.36 mph/stop/mile.

55. a. 15 pts/yr; 12.6 pts/yr; 0 pts/yr   **b.** 10 pts/yr

57. 155/mo; 200/mo

59. 32 turtles/yr; 428 turtles/yr; 3260 turtles

61. a. $120 − 30t$ **b.** 120 ft/sec   **c.** 240 ft

63. a. 2.495 million   **b.** 405,000/yr

65. a. $(0.0001)(\frac{5}{4}) x^{1/4}$ **b.** $0.00125/radio

67. a. $20\left(1 − \dfrac{1}{\sqrt{t}}\right)$
   b. 50 mph; 30 mph; 33.43 mph
   c. −8.28; 0; 5.86; at 6:30 A.M., the average velocity is decreasing at the rate of 8.28 mph/hr; at 7 A.M., it is unchanged; and at 8 A.M., it is increasing at the rate of 5.86 mph.

69. a. $f'(t) = 0.225t^2 + 0.05t + 2.45$
   b. \$4.625 billion/yr
   c. \$12 billion/yr

71. False

Using Technology Exercises 3.1, page 193

1. 1   **3.** 0.4226   **5.** 0.1613

7. a.

   b. 3.4295 ppm; 105.4332 ppm

9. a.

   b. Decreasing at the rate of 9 days/yr; increasing at the rate of 13 days/yr

11. a.

   b. Decreasing at the rate of 0.188887%/yr; increasing at the rate of 0.0777812%/yr

Exercises 3.2, page 205

1. $6x^2 + 2$ **3.** $4t − 1$ **5.** $9x^2 + 2x − 6$

7. $4x^3 + 3x^2 − 1$ **9.** $5w^4 − 4w^3 + 9w^2 − 6w + 2$

11. $\dfrac{25x^2 − 10x^{3/2} + 1}{x^{1/2}}$ **13.** $\dfrac{3x^4 − 10x^3 + 4}{x^2}$

15. $\dfrac{−1}{(x − 2)^2}$ **17.** $\dfrac{3}{(2x + 1)^2}$ **19.** $−\dfrac{2x}{(x^2 + 1)^2}$

21. $\dfrac{s^2 + 2s + 4}{(s + 1)^2}$ **23.** $\dfrac{1 − 3x^2}{2\sqrt{x}(x^2 + 1)^2}$ **25.** $\dfrac{x^2 − 2x − 2}{(x^2 + x + 1)^2}$

27. $\dfrac{2x^3 − 5x^2 − 4x − 3}{(x − 2)^2}$

29. $\dfrac{−2(x^5 + 12x^4 − 8x^3 + 16x + 64)}{(x^4 − 16)^2}$ **31.** 8   **33.** −9

35. $2(3x^2 − x + 3)$; 10   **37.** $\dfrac{−3x^4 + 2x^2 − 1}{(x^4 − 2x^2 −1)^2}$; $−\dfrac{1}{2}$

39. 60; $y = 60x - 102$ **41.** $-\frac{1}{2}$; $y = -\frac{1}{2}x + \frac{3}{2}$

43. $y = 7x - 5$ **45.** $(\frac{1}{3}, \frac{50}{27})$; $(1, 2)$

47. $(\frac{4}{3}, -\frac{770}{27})$; $(2, -30)$ **49.** $y = -\frac{1}{2}x + 1$; $y = 2x - \frac{3}{2}$

51. 0.125; 0.5; 2; 50

53. -5000/min; -1600/min; 7000; 4000

55. a. $\dfrac{180}{(t + 6)^2}$ **b.** 3.7; 2.2; 1.8; 1.1

c.

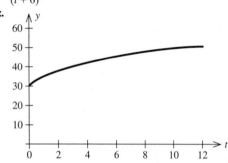

d. 50 words/min

57. Dropping at the rate of 0.0375 ppm/yr; dropping at the rate of 0.006 ppm/yr

59. False   **61.** False

Using Technology Exercises 3.2, page 209

1. 0.8750 **3.** 0.0774 **5.** -0.5000

7. $87{,}322$/yr

Exercises 3.3, page 220

1. $8(2x - 1)^3$ **3.** $10x(x^2 + 2)^4$

5. $6x^2(1 - x)(2 - x)^2$ **7.** $\dfrac{-4}{(2x + 1)^3}$ **9.** $3x\sqrt{x^2 - 4}$

11. $\dfrac{3}{2\sqrt{3x - 2}}$ **13.** $\dfrac{-2x}{3(1 - x^2)^{2/3}}$ **15.** $-\dfrac{6}{(2x + 3)^4}$

17. $\dfrac{-1}{(2t - 3)^{3/2}}$ **19.** $-\dfrac{3(16x^3 + 1)}{2(4x^4 + x)^{5/2}}$

21. $\dfrac{-4(3x + 1)}{(3x^2 + 2x + 1)^3}$ **23.** $6x(2x^2 - x + 1)$

25. $3(t^{-1} - t^{-2})^2(-t^{-2} + 2t^{-3})$ **27.** $\dfrac{1}{2\sqrt{x - 1}} + \dfrac{1}{2\sqrt{x + 1}}$

29. $-12x(4x - 1)(3 - 4x)^3$

31. $6(x - 1)(2x - 1)(2x + 1)^3$ **33.** $-\dfrac{15(x + 3)^2}{(x - 2)^4}$

35. $\dfrac{3\sqrt{t}}{2(2t + 1)^{5/2}}$ **37.** $\dfrac{-1}{2\sqrt{u + 1}\,(3u + 2)^{3/2}}$

39. $-\dfrac{2x(3x^2 + 1)}{(x^2 - 1)^5}$ **41.** $-\dfrac{2x(3x^2 + 1)^2(3x^2 + 13)}{(x^2 - 1)^5}$

43. $-\dfrac{3x^2 + 2x + 1}{\sqrt{2x + 1}\,(x^2 - 1)^2}$ **45.** $-\dfrac{t^2 + 2t - 1}{2\sqrt{t + 1}\,(t^2 + 1)^{3/2}}$

47. $\frac{4}{3}u^{1/3}$; $6x$; $8x(3x^2 - 1)^{1/3}$

49. $-\dfrac{2}{3u^{5/3}}$; $6x^2 - 1$; $-\dfrac{2(6x^2 - 1)}{3(2x^3 - x + 1)^{5/3}}$

51. $\frac{1}{2}u^{-1/2} - \frac{1}{2}u^{-3/2}$; $3x^2 - 1$; $\dfrac{(3x^2 - 1)(x^3 - x - 1)}{2(x^3 - x)^{3/2}}$

53. -12 **55.** 6 **57.** Yes

59. $y = -33x + 57$ **61.** $y = \frac{43}{5}x - \frac{54}{5}$

63. 0.333 million/wk; 0.305 million/wk; 16 million; 22.7 million

65. a. $0.027(0.2t^2 + 4t + 64)^{-1/3}(0.1t + 1)$
b. 0.0091 ppm

67. a. $0.21t^2(t - 3)(t - 7)^3$
b. 90.72; 0; -90.72; at 8 A.M., the level of nitrogen dioxide is increasing; at 10 A.M., the level stops increasing; and at 11 A.M., the level is decreasing.

69. $\dfrac{3450t}{(t + 25)^2\sqrt{\frac{1}{2}t^2 + 2t + 25}}$; 2.9 beats/min²; 0.7 beats/min²; 0.2 beats/min²; 179 beats/min

71. 160π ft²/sec   **73.** -27 mph/decade; 19 mph

75. $\dfrac{1.42(140t^2 + 3500t + 21{,}000)}{(3t^2 + 80t + 550)^2}$; $87{,}322$ jobs/yr

77. Decreasing at the rate of $\$5.86$/passenger/yr

79. True   **81.** False

Using Technology Exercises 3.3, page 219

1. 0.5774 **3.** 0.9390 **5.** -4.9498

7. $10{,}146{,}200$/decade; $7{,}810{,}520$/decade

Exercises 3.4, page 236

1. a. $C(x)$ is always increasing because as the number of units x produced increases, the greater the amount of money that must be spent on production.
b. 4000

3. a. $\$114$; $\$120.16$; $\$138.12$ **b.** $\$114$; $\$120$; $\$138$

5. a. $\dfrac{5000}{x} + 2$ **b.** $-\dfrac{5000}{x^2}$

7. $0.0002x^2 - 0.06x + 120 + \dfrac{5000}{x}$; $0.0004x - 0.06 - \dfrac{5000}{x^2}$

9. a. $-0.04x^2 + 800x$ **b.** $-0.08x + 800$
c. 400; when the level of production is 5000 units, the production of the next speaker system will bring in additional revenue of $400.

11. a. $-0.04x^2 + 600x - 300,000$ **b.** $-0.08x + 600$ **c.** 200; -40
d. The profit increases as production increases, peaking at 7500 units; beyond this level, profit falls.

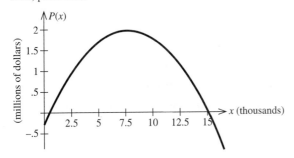

13. a. $-0.006x^2 + 180x$; $-0.000002x^3 + 0.014x^2 + 60x - 60,000$
b. $0.000006x^2 - 0.04x + 120$; $-0.012x + 180$; $-0.000006x^2 + 0.028x + 60$
c. 64; 156; 92   **d.**

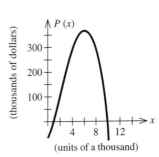

15. a. $0.000004x - 0.02 - \dfrac{60,000}{x^2}$
b. -0.0024; 0.0194; the average cost is decreasing when 5000 TV sets are produced and increasing when 10,000 units are produced.

17. 0.712 billion/billion dollars   **21.** $-\$0.209$ billion/billion dollars

23. $\frac{5}{3}$; elastic   **25.** 1; unitary   **27.** 0.104; inelastic

29. a. Inelastic; elastic   **b.** When $p = 8.66$
c. Increase   **d.** Increase

31. a. Inelastic   **b.** Increase

33. $\dfrac{25}{50 - p}$; for $p > 25$, demand is elastic; for $p = 25$, demand is unitary; and for $p < 25$, demand is inelastic.

35. False

Exercises 3.5, page 245

1. $8x - 2$; 8   **3.** $6x^2 - 6x$; $6(2x - 1)$

5. $4t^3 - 6t^2 + 12t - 3$; $12(t^2 - t + 1)$

7. $10x(x^2 + 2)^4$; $10(x^2 + 2)^3(9x^2 + 2)$

9. $6t(2t^2 - 1)(6t^2 - 1)$; $6(60t^4 - 24t^2 + 1)$

11. $14x(2x^2 + 2)^{5/2}$; $28(2x^2 + 2)^{3/2}(6x^2 + 1)$

13. $(x^2 + 1)(5x^2 + 1)$; $4x(5x^2 + 3)$

15. $\dfrac{1}{(2x + 1)^2}$; $-\dfrac{4}{(2x + 1)^3}$

17. $\dfrac{2}{(s + 1)^2}$; $-\dfrac{4}{(s + 1)^3}$

19. $-\dfrac{3}{2(4-3u)^{1/2}};\ -\dfrac{9}{4(4-3u)^{3/2}}$

21. $72x-24$ **23.** $-\dfrac{6}{x^4}$

25. $\frac{81}{8}(3s-2)^{-5/2}$ **27.** $192(2x-3)$

29. 128 ft/sec; 32 ft/sec^2

31. a. and b.

t	0	1	2	3	4	5	6	7
$N'(t)$	0	2.7	4.8	6.3	7.2	7.5	7.2	6.3
$N''(t)$				0.6	0	−0.6	−1.2	

33. a. $\frac{1}{4}t(t^2-12t+32)$ **b.** 0 ft/sec; 0 ft/sec; 0 ft/sec
c. $\frac{3}{4}t^2-6t+8$ **d.** 8 ft/sec^2; −4 ft/sec^2; 8 ft/sec^2
e. 0 ft; 16 ft; 0 ft

35. −3.09; 0.35; 10 min after the test began, the smoke was being removed at the rate of 3%/min and the rate of the rate of change of the amount of smoke remaining was .35%/min^2.

37. True   **39.** True

41. $f(x)=x^{n+1/2}$

Using Technology Exercises 3.5, page 247

1. −18   **3.** 15.2762   **5.** −0.6255   **7.** 0.1973

9. −68.46214; at the beginning of 1988, the rate of the rate of the rate at which banks were failing was 68 banks/yr/yr/yr.

Exercises 3.6, page 257

1. a. $-\frac{1}{2}$ **b.** $-\frac{1}{2}$ **3. a.** $-\dfrac{1}{x^2}$ **b.** $-\dfrac{y}{x}$

5. a. $2x-1+\dfrac{4}{x^2}$ **b.** $3x-2-\dfrac{y}{x}$

7. a. $\dfrac{1-x^2}{(1+x^2)^2}$ **b.** $-2y^2+\dfrac{y}{x}$

9. $-\dfrac{x}{y}$ **11.** $\dfrac{x}{2y}$ **13.** $1-\dfrac{y}{x}$ **15.** $-\dfrac{y}{x}$

17. $-\dfrac{\sqrt{y}}{\sqrt{x}}$ **19.** $2\sqrt{x+y}-1$ **21.** $-\dfrac{y^3}{x^3}$

23. $\dfrac{2\sqrt{xy}-y}{x-2\sqrt{xy}}$ **25.** $\dfrac{6x-3y-1}{3x+1}$ **27.** $\dfrac{2(2x-y^{3/2})}{3x\sqrt{y}-4y}$

29. $-\dfrac{2x^2+2xy+y^2}{x^2+2xy+2y^2}$ **31.** $y=2$ **33.** $y=-\frac{3}{2}x+\frac{5}{2}$

35. $\dfrac{2y}{x^2}$ **37.** $\dfrac{2y(y-x)}{(2y-x)^3}$

39. a. $\dfrac{dV}{dt}=\pi r\left(r\dfrac{dh}{dt}+2h\dfrac{dr}{dt}\right)$ **b.** 3.6π cu in./sec

41. Dropping at the rate of 111 tires/wk

43. Increasing at the rate of 44 ten packs/wk

45. Dropping at the rate of 3.7 cents/carton/wk

47. 0.37; inelastic   **49.** 160π ft^2/sec   **51.** 17 ft/sec

53. 7.69 ft/sec

57. 196.8 ft/sec   **59.** 9 ft/sec   **61.** 19.2 ft/sec   **63.** True

Exercises 3.7, page 268

1. $4x\,dx$ **3.** $(3x^2-1)\,dx$ **5.** $\dfrac{dx}{2\sqrt{x+1}}$

7. $\dfrac{6x+1}{2\sqrt{x}}dx$ **9.** $\dfrac{x^2-2}{x^2}dx$

11. $\dfrac{-x^2+2x+1}{(x^2+1)^2}dx$ **13.** $\dfrac{6x-1}{2\sqrt{3x^2-x}}dx$

15. a. $2x\,dx$ **b.** 0.04   **c.** 0.0404

17. a. $-\dfrac{dx}{x^2}$ **b.** −0.05   **c.** −0.05263

19. 3.167   **21.** 7.0358   **23.** 1.983   **25.** 0.298

27. 2.50146   **29.** ±8.64 cm^3 **31.** 18.85 ft^3

33. It will drop by 40%.   **35.** 274 sec   **37.** 111,595

39. Decrease of $1.33   **41.** ±64,800

43. Decrease of 11 crimes/yr   **45.** True

Using Technology Exercises 3.7, page 271

1. 7.5787   **3.** 0.031220185778   **5.** −0.0198761598

7. $26.60/month; $35.47/month; $44.34/month

9. 625

Chapter 3 Review Exercises, page 274

1. $15x^4-8x^3+6x-2$ **3.** $\dfrac{6}{x^4}-\dfrac{3}{x^2}$

5. $-\dfrac{1}{t^{3/2}}-\dfrac{6}{t^{5/2}}$ **7.** $1-\dfrac{2}{t^2}-\dfrac{6}{t^3}$

9. $2x+\dfrac{3}{x^{5/2}}$ **11.** $\dfrac{2t}{(2t^2+1)^2}$

13. $\dfrac{1}{\sqrt{x}(\sqrt{x}+1)^2}$ 15. $\dfrac{2x(x^4-2x^2-1)}{(x^2-1)^2}$

17. $72x^2(3x^3-2)^7$ 19. $\dfrac{2t}{\sqrt{2t^2+1}}$

21. $\dfrac{-4(3t-1)}{(3t^2-2t+5)^3}$ 23. $\dfrac{2(x^2+1)(x^2-1)}{x^3}$

25. $4t^2(5t+3)(t^2+t)^3$, or $4t^5(5t+3)(t+1)^3$

27. $\dfrac{1}{2\sqrt{x}}(x^2-1)^2(13x^2-1)$

29. $-\dfrac{12x+25}{2\sqrt{3x+2}\,(4x-3)^2}$

31. $2(12x^2-9x+2)$ 33. $\dfrac{2t(t^2-12)}{(t^2+4)^3}$

35. $\dfrac{2}{(2x^2+1)^{3/2}}$ 37. $\dfrac{2x}{y}$

39. $-\dfrac{2x}{y^2-1}$ 41. $\dfrac{x-2y}{2x+y}$

43. **a.** $(2,-25)$ and $(-1,14)$
 b. $y=-4x-17$; $y=-4x+10$

45. $y=-\dfrac{\sqrt{3}}{3}x+\dfrac{4}{3}\sqrt{3}$

47. $-\dfrac{48}{(2x-1)^4}$; $(-\infty,\frac{1}{2})\cup(\frac{1}{2},\infty)$

49. 200/wk

51. **a.** $-0.02x^2+600x$
 b. $-0.04x+600$
 c. 200; the sale of the 10,001st phone will bring in a revenue of \$200.

CHAPTER 4

Exercises 4.1, page 292

1. Decreasing on $(-\infty,0)$ and increasing on $(0,\infty)$

3. Increasing on $(-\infty,-1)\cup(1,\infty)$ and decreasing on $(-1,1)$

5. Decreasing on $(-\infty,0)\cup(2,\infty)$ and increasing on $(0,2)$

7. Decreasing on $(-\infty,-1)\cup(1,\infty)$ and increasing on $(-1,1)$

9. Increasing on $(20.2, 20.6)\cup(21.7, 21.8)$, constant on $(19.6, 20.2)\cup(20.6, 21.1)$, and decreasing on $(21.1, 21.7)\cup(21.8, 22.7)$

11. Increasing on $(-\infty,\infty)$

13. Decreasing on $(-\infty,\frac{3}{2})$ and increasing on $(\frac{3}{2},\infty)$

15. Decreasing on $(-\infty,-\sqrt{3}/3)\cup(\sqrt{3}/3,\infty)$ and increasing on $(-\sqrt{3}/3,\sqrt{3}/3)$

17. Increasing on $(-\infty,-2)\cup(0,\infty)$ and decreasing on $(-2,0)$

19. Increasing on $(-\infty,3)\cup(3,\infty)$

21. Decreasing on $(-\infty,0)\cup(0,3)$ and increasing on $(3,\infty)$

23. Decreasing on $(-\infty,2)\cup(2,\infty)$

25. Decreasing on $(-\infty,1)\cup(1,\infty)$

27. Increasing on $(-\infty,0)\cup(0,\infty)$

29. Increasing on $(-1,\infty)$

31. Increasing on $(-4,0)$; decreasing on $(0,4)$

33. Increasing on $(-\infty,0)\cup(0,\infty)$

35. Increasing on $(-\infty,1)$; decreasing on $(1,\infty)$

37. Relative maximum: $f(0)=1$; relative minima: $f(-1)=0$ and $f(1)=0$

39. Relative maximum: $f(-1)=2$; relative minimum: $f(1)=-2$

41. Relative maximum: $f(1)=3$; relative minimum: $f(2)=2$

43. Relative minimum: $f(0)=2$

45. a   47. d

49. Relative minimum: $f(2)=-4$

51. Relative minimum: $f(2)=2$

53. Relative minimum: $f(0)=2$

55. Relative maximum: $g(0)=4$; relative minimum: $g(2)=0$

57. Relative minimum: $F(3)=-5$; relative maximum: $F(-1)=\frac{17}{3}$

59. Relative minimum: $g(3)=-19$

61. Relative minimum: $f(\frac{1}{2})=\frac{63}{16}$

63. None

65. Relative maximum: $f(-3)=-4$; relative minimum: $f(3)=8$

67. Relative maximum: $f(1)=\frac{1}{2}$; relative minimum: $f(-1)=-\frac{1}{2}$

69. Relative maximum: $f(0)=0$

71. Relative minimum: $f(1)=0$

73. Increasing on $(0, 4000)$ and decreasing on $(4000,\infty)$

75. Decreasing on $(0, 5)$ and increasing on $(5, 10)$; after declining from 1984 through 1989, the index began to increase after 1989.

79. Increasing on $(0, 4)$ and decreasing on $(4, 5)$; the cash in the Central Provident Fund will be increasing from 1995 to 2035 and decreasing from 2035 to 2045.

81. Increasing on $(0, 1)$ and decreasing on $(1, 4)$

83. Increasing on $(0, 4.5)$ and decreasing on $(4.5, 11)$; the pollution is increasing from 7 A.M. to 11:30 A.M. and decreasing from 11:30 A.M. to 6 P.M.

87. False   **89.** False   **91.** False

93. a. $f'(x) = \begin{cases} -3 & \text{if } x < 0 \\ 2 & \text{if } x > 0 \end{cases}$ **b.** No

Using Technology Exercises 4.1, page 298

1. a. f is decreasing on $(-\infty, -0.2934)$ and increasing on $(-0.2934, \infty)$
   b. Relative minimum: $f(-0.2934) = -2.5435$

3. a. f is increasing on $(-\infty, -1.6144) \cup (0.2390, \infty)$ and decreasing on $(-1.6144, 0.2390)$
   b. Relative maximum: $f(-1.6144) = 26.7991$; relative minimum: $f(0.2390) = 1.6733$

5. a. f is decreasing on $(-\infty, -1) \cup (0.33, \infty)$ and increasing on $(-1, 0.33)$
   b. Relative maximum: $f(0.33) = 1.11$; relative minimum: $f(-1) = -0.63$

7. a. f is decreasing on $(-1, -0.71)$ and increasing on $(-0.71, 1)$
   b. Relative minimum: $f(-0.71) = -1.41$

9. a.

   b. f is decreasing on $(0, 0.2398) \cup (6.8758, 12)$ and increasing on $(0.2398, 6.8758)$
   c. $(6.8758, 200.14)$

11. a.

   b. f is decreasing on $(0, 0.8343) \cup (7.6726, 12)$ and increasing on $(0.8343, 7.6726)$

13. Increasing from 7 A.M. to 11:30 A.M. and decreasing from 11:30 A.M. to 6 P.M.; highest at 11:30 A.M. when it reaches 164 PSI.

Exercises 4.2, page 313

1. Concave downward on $(-\infty, 0)$ and concave upward on $(0, \infty)$; inflection point: $(0, 0)$

3. Concave downward on $(-\infty, 0) \cup (0, \infty)$

5. Concave upward on $(-\infty, 0) \cup (1, \infty)$ and concave downward on $(0, 1)$; inflection points: $(0, 0)$ and $(1, -1)$

7. Concave downward on $(-\infty, -2) \cup (-2, 2) \cup (2, \infty)$

9. a   **11.** b

13. a. $D_1'(t) > 0$, $D_2'(t) > 0$, $D_1''(t) > 0$, and $D_2''(t) < 0$ on $(0, 12)$
   b. With or without the proposed promotional campaign, the deposits will increase; with the promotion, the deposits will increase at an increasing rate; without the promotion, the deposits will increase at a decreasing rate.

15. At the time t_0, corresponding to its t-coordinate, the restoration process is working at its peak.

23. Concave upward on $(-\infty, \infty)$

25. Concave downward on $(-\infty, 0)$; concave upward on $(0, \infty)$

27. Concave upward on $(-\infty, 0) \cup (3, \infty)$; concave downward on $(0, 3)$

29. Concave downward on $(-\infty, 0) \cup (0, \infty)$

31. Concave downward on $(-\infty, 4)$

33. Concave downward on $(-\infty, 2)$; concave upward on $(2, \infty)$

35. Concave upward on $(-\infty, -\sqrt{6}/3) \cup (\sqrt{6}/3, \infty)$; concave downward on $(-\sqrt{6}/3, \sqrt{6}/3)$

37. Concave downward on $(-\infty, 1)$; concave upward on $(1, \infty)$

39. Concave upward on $(-\infty, 0) \cup (0, \infty)$

41. Concave upward on $(-\infty, 2)$; concave downward on $(2, \infty)$

43. Concave upward on $(-\infty, -1) \cup (1, \infty)$; concave downward on $(-1, 1)$

45. $(0, -2)$ **47.** $(1, -15)$ **49.** $(0, 1)$ and $(\frac{2}{3}, \frac{11}{27})$

51. $(0, 0)$ **53.** $(1, 2)$ **55.** $(-\sqrt{3}/3, 3/2)$ and $(\sqrt{3}/3, 3/2)$

57. Relative maximum: $f(1) = 5$

59. None

61. Relative maximum: $f(-1) = -\frac{22}{3}$; relative minimum: $f(5) = -\frac{130}{3}$

63. Relative maximum: $f(-3) = -6$; relative minimum: $f(3) = 6$

65. None

67. Relative minimum: $f(-2) = 12$

69. Relative maximum: $g(1) = \frac{1}{2}$; relative minimum: $g(-1) = -\frac{1}{2}$

71. Relative maximum: $f(0) = 0$; relative minimum: $f(\frac{4}{3}) = \frac{256}{27}$

73. Relative minimum: $g(4) = -\frac{1}{108}$

75. a. N is increasing on $(0, 12)$.
b. $N''(t) < 0$ on $(0, 6)$ and $N''(t) > 0$ on $(6, 12)$
c. The rate of growth of the number of help-wanted advertisements was decreasing over the first 6 mo of the year and increasing over the last 6 mo.

77.

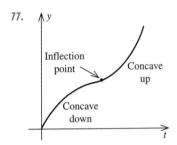

79. $(100, 4600)$; the sales increase rapidly until $100,000 is spent on advertising; after that, any additional expenditure results in increased sales but at a slower rate of increase.

81. 10 A.M.   **83.** $(18, 20,310)$; 1452 ft/sec   **85.** True

87. True   **89.** Yes

Using Technology Exercises 4.2, page 319

1. a. f is concave upward on $(-\infty, 0) \cup (1.1667, \infty)$ and concave downward on $(0, 1.1667)$.
b. $(1.1667, 1.1153)$; $(0, 2)$

3. a. f is concave downward on $(-\infty, 0)$ and concave upward on $(0, \infty)$.
b. $(0, 2)$

5. a. f is concave downward on $(-\infty, 0)$ and concave upward on $(0, \infty)$.
b. $(0, 0)$

7. a. f is concave downward on $(-\infty, -2.4495) \cup (0, 2.4495)$ and concave upward on $(-2.4495, 0) \cup (2.4495, \infty)$.
b. $(2.4495, 0.3402)$; $(-2.4495, -0.3402)$

9. a.

b. $(5.5318, 35.9483)$
c. $t = 5.5318$

11. a.

b. $(3.9024, 77.0919)$

Exercises 4.3, page 331

1. Horizontal asymptote: $y = 0$

3. Horizontal asymptote: $y = 0$; vertical asymptote: $x = 0$

5. Horizontal asymptote: $y = 0$; vertical asymptotes: $x = -1$ and $x = 1$

7. Horizontal asymptote: $y = 3$; vertical asymptote: $x = 0$

9. Horizontal asymptotes: $y = 1$ and $y = -1$

11. Horizontal asymptote: $y = 0$; vertical asymptote: $x = 0$

13. Horizontal asymptote: $y = 0$; vertical asymptote: $x = 0$

15. Horizontal asymptote: $y = 1$; vertical asymptote: $x = -1$

17. None

19. Horizontal asymptote: $y = 1$: vertical asymptotes: $t = -3$ and $t = 3$

21. Horizontal asymptote: $y = 0$; vertical asymptotes: $x = -2$ and $x = 3$

23. Horizontal asymptote: $y = 2$; vertical asymptote: $t = 2$

25. Horizontal asymptote: $y = 1$; vertical asymptotes: $x = -2$ and $x = 2$

27. None

29. f is the derivative function of the function g.

31.

33.

35.

37.

39.

41.

43.

45.

47.

49.

51.

53.

55.

57.

59.

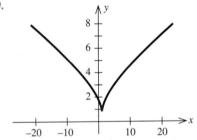

61. a. $y = 2.2$ **b.** \$2.20/disc

63. a. a
   b. The initial speed of the reaction approaches a moles/liter/sec as the amount of substrate becomes arbitrarily large.

65.

67.

69.

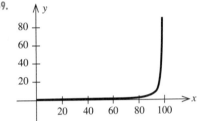

Using Technology Exercises 4.3, page 337

1.

3.

5. -0.9733; 2.3165, 4.6569 **7.** -1.1301; 2.9267

9. 1.5142

Exercises 4.4, page 350

1. None

3. Absolute minimum value: 0

5. Absolute maximum value: 3; absolute minimum value: -2

7. Absolute maximum value: 3; absolute minimum value: $-\frac{27}{16}$

9. Absolute minimum value: $-\frac{41}{8}$

11. No absolute extrema

13. Absolute maximum value: 1

15. Absolute maximum value: 5; absolute minimum value: -4

17. Absolute maximum value: 10; absolute minimum value: 1

19. Absolute maximum value: 19; absolute minimum value: −1

21. Absolute maximum value: 16; absolute minimum value: −1

23. Absolute maximum value: 3; absolute minimum value: $\frac{5}{3}$

25. Absolute maximum value: $\frac{37}{3}$; absolute minimum value: 5

27. Absolute maximum value ≈ 1.04; absolute minimum value: −1.5

29. No absolute extrema

31. Absolute maximum value: 1; absolute minimum value: 0

33. Absolute maximum value: 0; absolute minimum value: −3

35. Absolute maximum value: $\sqrt{2}/4 \approx 0.35$; absolute minimum value: $-\frac{1}{3}$

37. Absolute maximum value: $\sqrt{2}/2$; absolute minimum value: $-\sqrt{2}/2$

39. 144 ft   **41.** 3000

43. $f(6) = 3.60$, $f(0.5) = 1.13$; the number of nonfarm, full-time, self-employed women over the time interval from 1963 to 1993 reached its highest level, 3.6 million, in 1993.

45. 5000   **47.** 3333

49. a. $0.0025x + 80 + \dfrac{10,000}{x}$ **b.** 2000
   c. 2000   **d.** Same

51. 533

53. a. 2 days after the organic waste was dumped into the pond
   b. 3.5 days after the organic waste was dumped into the pond

63. False   **65.** False

69. c.

Using Technology Exercises 4.4, page 355

1. Absolute maximum value: 145.8985; absolute minimum value: −4.3834

3. Absolute maximum value: 16; absolute minimum value: −0.1257

5. Absolute maximum value: 2.8889; absolute minimum value: 0

7. a.

b. 200.1410 banks/yr

9. a.

b. Absolute maximum value: 108.8756; absolute minimum value: 49.7773

Exercises 4.5, page 365

1. 750 yd × 1500 yd; 1,125,000 yd²

3. $10\sqrt{2}$ ft × $40\sqrt{2}$ ft

5. $\frac{16}{3}$ in. × $\frac{16}{3}$ in. × $\frac{4}{3}$ in.

7. 5.04 in. × 5.04 in. × 5.04 in.

9. 18 in. × 18 in. × 36 in.; 11,664 in.³

11. $r = \dfrac{36}{\pi}$ in.; $l = 36$ in.; $\dfrac{46,656}{\pi}$ in.³

13. $\frac{2}{3}\sqrt[3]{9}$ ft × $\sqrt[3]{9}$ ft × $\frac{2}{3}\sqrt[3]{9}$ ft   **15.** 250; $62,500; $250

17. $w \approx 13.86$ in.; $h \approx 19.60$ in.

19. $x = 750\sqrt{2}$ ft   **21.** 56.57 mph   **23.** 45; 44,445

Chapter 4 Review Exercises, page 369

1. a. f is increasing on $(-\infty, 1) \cup (1, \infty)$.
   b. No relative extrema
   c. Concave down on $(-\infty, 1)$; concave up on $(1, \infty)$
   d. $(1, -\frac{17}{3})$

3. a. f is increasing on $(-1, 0) \cup (1, \infty)$; decreasing on $(-\infty, -1) \cup (0, 1)$
   b. Relative maximum: 0; relative minimum: −1
   c. Concave up on $(-\infty, -\sqrt{3}/3) \cup (\sqrt{3}/3, \infty)$; concave down on $(-\sqrt{3}/3, \sqrt{3}/3)$
   d. $(-\sqrt{3}/3, -5/9)$; $(\sqrt{3}/3, -5/9)$

5. a. f is increasing on $(-\infty, 0) \cup (2, \infty)$ and decreasing on $(0, 1) \cup (1, 2)$.
   b. Relative maximum: 0; relative minimum: 4
   c. Concave up on $(1, \infty)$; concave down on $(-\infty, 1)$
   d. None

7. a. f is decreasing on $(-\infty, 1) \cup (1, \infty)$.
  b. No relative extrema
  c. Concave down on $(-\infty, 1)$; concave up on $(1, \infty)$
  d. $(1, 0)$

9. a. f is increasing on $(-\infty, -1) \cup (-1, \infty)$.
  b. No relative extrema
  c. Concave down on $(-1, \infty)$; concave up on $(-\infty, -1)$
  d. No inflection point

11.

13.

15.

17.

19. Vertical asymptote: $x = -\frac{3}{2}$; horizontal asymptote: $y = 0$

21. Vertical asymptotes: $x = -2$ and $x = 4$; horizontal asymptote: $y = 0$

23. Absolute minimum value: $-\frac{25}{8}$

25. Absolute maximum value: 5; absolute minimum value: 0

27. Absolute maximum value: -16; absolute minimum value: -32

29. Absolute maximum value: $\frac{8}{3}$; absolute minimum value: 0

31. Absolute maximum value: $\frac{1}{2}$; absolute minimum value: $-\frac{1}{2}$

33. $4000   **35.** 168   **37.** 10 A.M.

39. 20,000 cases

CHAPTER 5

Exercises 5.1, page 379

1. a. 16   **b.** 27   **3. a.** 3   **b.** $\sqrt{5}$

5. a. -3 **b.** 8   **7. a.** 25   **b.** $4^{1.8}$

9. a. $4x^3$ **b.** $5xy^2\sqrt{x}$ **11. a.** $\dfrac{2}{a^2}$ **b.** $\frac{1}{3}b^2$

13. a. $8x^9y^6$ **b.** $16x^4y^4z^6$

15. a. $\dfrac{64x^6}{y^4}$ **b.** $(x-y)(x+y)$

17. 2   **19.** 3   **21.** 3   **23.** $\frac{5}{4}$ **25.** 1 or 2

27.

29.

31.

33.

35.

37. False   **39.** True

Using Technology Exercises 5.1, page 381

1.

3.

5.

7.

9.

Exercises 5.2, page 389

1. $\log_2 64 = 6$ **3.** $\log_3 \frac{1}{9} = -2$ **5.** $\log_{1/3} \frac{1}{3} = 1$

7. $\log_{32} 8 = \frac{3}{5}$ **9.** $\log_{10} 0.001 = -3$ **11.** 1.0792

13. 1.2042   **15.** 1.6813   **17.** $\log x + 4 \log(x + 1)$

19. $\frac{1}{2} \log(x + 1) - \log(x^2 + 1)$ **21.** $\ln x - x^2$

23. $-\frac{3}{2} \ln x - \frac{1}{2} \ln(1 + x^2)$ **25.** $x \ln x$

27.

29.

31.

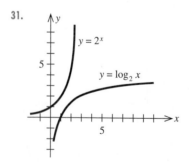

33. 5.1986   **35.** −0.0912   **37.** −8.0472

39. −4.9041   **41.** $-2 \ln\left(\dfrac{A}{B}\right)$ **43.** 105.7 mm

45. a. $10^3 I_0$
    b. 100,000 times greater
    c. 10,000,000 times greater

47. $34\frac{1}{2}$ hr earlier, at 1:30 P.M.

49. False   **51.** True

Exercises 5.3, page 401

1. \$4974.47   **3.** \$223,403.11

5. a. 10.25%/yr   **b.** 9.31%/yr

7. a. \$29,227.61   **b.** \$29,137.83   **9.** \$6885.64

11. \$112,926.52   **13.** \$3.795 million   **15.** \$23,329.48

17. 9.58 yr   **19.** 12.75%   **21.** \$40,000

23. a. \$33,885.14   **b.** \$33,565.38

25. a. \$16,262.79   **b.** \$12,047.77   **c.** \$6611.96

29. Bank B   **31.** 9.531%

Exercises 5.4, page 411

1. $3e^{3x}$ **3.** $-e^{-t}$ **5.** $e^x + 1$ **7.** $x^2 e^x(x + 3)$

9. $\dfrac{2e^x(x - 1)}{x^2}$ **11.** $3(e^x - e^{-x})$ **13.** $-\dfrac{1}{e^w}$

15. $6e^{3x-1}$ **17.** $-2xe^{-x^2}$ **19.** $\dfrac{3e^{-1/x}}{x^2}$

21. $25e^x(e^x + 1)^{24}$ **23.** $\dfrac{e^{\sqrt{x}}}{2\sqrt{x}}$ **25.** $e^{3x+2}(3x - 2)$

27. $\dfrac{2e^x}{(e^x + 1)^2}$ **29.** $2(8e^{-4x} + 9e^{3x})$ **31.** $6e^{3x}(3x + 2)$

33. $y = 2x - 2$

35. f is increasing on $(-\infty, 0)$ and decreasing on $(0, \infty)$.

37. Concave downward on $(-\infty, 0)$; concave upward on $(0, \infty)$

39. $(1, e^{-2})$

41. Absolute maximum value: 1; absolute minimum value: e^{-1}

43. Absolute minimum value: −1; absolute maximum value: $2e^{-3/2}$

45.   **47.**

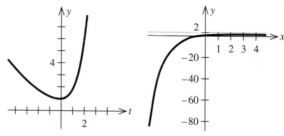

49. −$6065/day; −$3679/day; −$2231/day; −$1353/day

51. b. 4505/yr; 273 cases/yr

53. a. A decrease of 1.63 cents/bottle; a decrease of 1.34 cents/bottle
b. $231.87; $217.03

55. a. 30   **b.** $N'(x) = \dfrac{297,000e^{-x}}{(1 + 99e^{-x})^2}$

c. 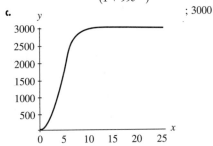   ; 3000

57. 7.72 yr; $160,207.69

59. −0.1694, −0.1549, −0.1415; the percentage of the total population relocating was decreasing at the rate of 0.17%/yr in 1970, 0.15%/yr in 1980, and 0.14%/yr in 1990.

61. a. $12/unit
b. Decreasing at the rate of $7/wk
c. $8/unit

65. False   **67.** True

Using Technology Exercises 5.4, page 414

1. 5.4366   **3.** 12.3929   **5.** 0.1861

7. a. 50   **c.**

9. a.

b. 4.2720 billion/half century

11. a. 153.024; 235.181
b. −0.634; 18.401

13. a. 69.63%   **b.** 5.094%/decade

Exercises 5.5, page 423

1. $\dfrac{5}{x}$ **3.** $\dfrac{1}{x+1}$ **5.** $\dfrac{8}{x}$ **7.** $\dfrac{1}{2x}$ **9.** $\dfrac{-2}{x}$

11. $\dfrac{2(4x-3)}{4x^2-6x+3}$ **13.** $\dfrac{1}{x(x+1)}$ **15.** $x(1 + 2\ln x)$

17. $\dfrac{2(1-\ln x)}{x^2}$ **19.** $\dfrac{3}{u-2}$ **21.** $\dfrac{1}{2x\sqrt{\ln x}}$

23. $\dfrac{3(\ln x)^2}{x}$ **25.** $\dfrac{3x^2}{x^3+1}$ **27.** $\dfrac{(x\ln x + 1)e^x}{x}$

29. $\dfrac{e^{2t}[2(t+1)\ln(t+1)+1]}{t+1}$ **31.** $\dfrac{1-\ln x}{x^2}$ **33.** $-\dfrac{1}{x^2}$

35. $\dfrac{2(2-x^2)}{(x^2+2)^2}$ **37.** $(x+1)(5x+7)(x+2)^2$

39. $(x-1)(x+1)^2(x+3)^3(9x^2+14x-7)$

41. $\dfrac{(2x^2-1)^4(38x^2+40x+1)}{2(x+1)^{3/2}}$ **43.** $3^x \ln 3$

45. $(x^2+1)^{x-1}[2x^2 + (x^2+1)\ln(x^2+1)]$

47. $y = x - 1$

49. f is decreasing on $(-\infty, 0)$ and increasing on $(0, \infty)$.

51. Concave up: $(-\infty, -1) \cup (1, \infty)$; concave down: $(-1, 0) \cup (0, 1)$

53. $(-1, \ln 2)$ and $(1, \ln 2)$

55. Absolute minimum value: 1; absolute maximum value: $3 - \ln 3$

57.

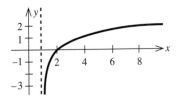

59. False

Exercises 5.6, page 433

1. a. 0.05
b. 400
c.

t	0	10	20	100	1000
Q	400	660	1087	59,365	2.07×10^{24}

3. a. $Q(t) = 100e^{0.035t}$
b. 266 min
c. $Q(t) = 1000e^{0.035t}$

5. a. 55.5 yr   **b.** 14.25 billion   **7.** 8.7 lb/in.²; 0.0004 lb/in.²/ft

9. $Q(t) = 100e^{-0.049t}$; 70.6 g; 3.451 g/day   **11.** 13,412 yr ago

13.

a. 60 words/min
b. 107 words/min
c. 136 words/min

15.

a. 573 computers; 1177 computers; 1548 computers; 1925 computers
b. 2000 computers
c. 46 computers/mo

17. a. 11   **b.** 937   **c.** 1000   **19.** 3%; 65%

21. 1080; 280

23. a. r/k
b.

Chapter 5 Review Exercises, page 437

1. a.–b.

$y = 2^{-x}$
$= \left(\frac{1}{2}\right)^x$

3. $\log_{16} 0.125 = -\frac{3}{4}$ **5.** 2   **7.** $x + 2y - z$

9.

$y = \log_2 (x + 3)$

11. $(2x + 1)e^{2x}$ **13.** $\dfrac{(1 - 4t)}{2\sqrt{t}e^{2t}}$ **15.** $\dfrac{2(e^{2x} + 2)}{(1 + e^{-2x})^2}$

17. $(1 - 2x^2)e^{-x^2}$ **19.** $(x + 1)^2 e^x$ **21.** $\dfrac{2xe^{x^2}}{e^{x^2} + 1}$

23. $\dfrac{x + 1 - x \ln x}{x(x + 1)^2}$ **25.** $\dfrac{4e^{4x}}{e^{4x} + 3}$ **27.** $\dfrac{1 + e^x(1 - x \ln x)}{x(1 + e^x)^2}$

29. $-\dfrac{9}{(3x + 1)^2}$ **31.** 0

33. $6x(x^2 + 2)^2(3x^3 + 2x + 1)$ **35.** $y = -(2x - 3)e^{-2}$

37.

$y = xe^{-2x}$

39. Absolute maximum value: $1/e$ **41.** 12%/yr

43. a. $Q(t) = 2000e^{0.01831t}$ **b.** 161,992

45.

a. 1175; 2540; 3289   **b.** 4000

CHAPTER 6

Exercises 6.1, page 450

5. b. $y = 2x + C$

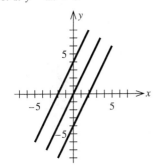

7. b. $y = \frac{1}{3}x^3 + C$

9. $6x + C$ **11.** $\frac{1}{4}x^4 + C$ **13.** $-\dfrac{1}{3x^3} + C$

15. $\frac{3}{5}x^{5/3} + C$ **17.** $-\dfrac{4}{x^{1/4}} + C$ **19.** $-\dfrac{2}{x} + C$

21. $\frac{2}{3}\pi t^{3/2} + C$ **23.** $3x - x^2 + C$

25. $\dfrac{1}{3}x^3 + \dfrac{1}{2}x^2 - \dfrac{1}{2x^2} + C$ **27.** $4e^x + C$

29. $x + \frac{1}{2}x^2 + e^x + C$ **31.** $x^4 + \dfrac{2}{x} - x + C$

33. $\frac{2}{7}x^{7/2} + \frac{4}{5}x^{5/2} - \frac{1}{2}x^2 + C$ **35.** $\frac{2}{3}x^{3/2} + 6\sqrt{x} + C$

37. $\frac{1}{9}u^3 + \frac{1}{3}u^2 - \frac{1}{3}u + C$ **39.** $\frac{2}{3}t^3 - \frac{3}{2}t^2 - 2t + C$

41. $\dfrac{1}{3}x^3 - 2x - \dfrac{1}{x} + C$ **43.** $\frac{1}{3}s^3 + s^2 + s + C$

45. $e^t + \dfrac{t^{e+1}}{e+1} + C$ **47.** $\dfrac{1}{2}x^2 + x - \ln|x| - \dfrac{1}{x} + C$

49. $\ln|x| + \dfrac{4}{\sqrt{x}} - \dfrac{1}{x} + C$ **51.** $x^2 + x + 1$

53. $x^3 + 2x^2 - x - 5$ **55.** $x - \dfrac{1}{x} + 2$ **57.** $x + \ln|x|$

59. $\sqrt{x}$ **61.** $e^x + \frac{1}{2}x^2 + 2$ **63.** $f(t) = \frac{4}{3}t^{3/2}$

65. \$3370   **67.** 5000 units; \$34,000

69. a. $-t^3 + 6t^2 + 45t$ **b.** 212

71. 21,960   **73.** 120,000

75. 1.9424 m²   **77.** $\frac{1}{2}k(R^2 - r^2)$

79. $9\frac{7}{9}$ ft/sec²; 396 ft   **81.** 36 ft/sec²   **83.** True   **85.** True

Exercises 6.2, page 463

1. $\frac{1}{5}(4x + 3)^5 + C$ **3.** $\frac{1}{3}(x^3 - 2x)^3 + C$

5. $-\dfrac{1}{2(2x^2 + 3)^2} + C$ **7.** $\frac{2}{3}(t^3 + 2)^{3/2} + C$

9. $\frac{1}{20}(x^2 - 1)^{10} + C$ **11.** $-\frac{1}{5}\ln|1 - x^5| + C$

13. $\ln(x - 2)^2 + C$ **15.** $\frac{1}{2}\ln(0.3x^2 - 0.4x + 2) + C$

17. $\frac{1}{6}\ln|3x^2 - 1| + C$ **19.** $-\frac{1}{2}e^{-2x} + C$

21. $-e^{2-x} + C$ **23.** $-\frac{1}{2}e^{-x^2} + C$ **25.** $e^x + e^{-x} + C$

27. $\ln(1 + e^x) + C$ **29.** $2e^{\sqrt{x}} + C$

31. $-\dfrac{1}{6(e^{3x} + x^3)^2} + C$ **33.** $\frac{1}{8}(e^{2x} + 1)^4 + C$

35. $\frac{1}{2}(\ln 5x)^2 + C$ **37.** $\ln|\ln x| + C$

39. $\frac{2}{3}(\ln x)^{3/2} + C$ **41.** $\frac{1}{2}e^{x^2} - \frac{1}{2}\ln(x^2 + 2) + C$

43. $\frac{2}{3}(\sqrt{x} - 1)^3 + 3(\sqrt{x} - 1)^2 + 8(\sqrt{x} - 1)$
$\quad + 4\ln|\sqrt{x} - 1| + C$

45. $\dfrac{(6x + 1)(x - 1)^6}{42} + C$

47. $5 + 4\sqrt{x} - x - 4\ln(1 + \sqrt{x}) + C$

49. $-\frac{1}{252}(1 - v)^7(28v^2 + 7v + 1) + C$

51. $\frac{1}{2}[(2x - 1)^5 + 5]$ **53.** $e^{-x^2+1} - 1$

55. $21,000 - \dfrac{20,000}{\sqrt{1 + 0.2t}}$; 6858   **57.** $p(x) = \dfrac{250}{\sqrt{16 + x^2}}$

59. $30(\sqrt{2t + 4} - 2)$; 14,400 π ft²

61. $\dfrac{71.86887}{1 + 2.449e^{-0.3277t}} - 1.43760$; 59.6 in.

63. 24,555 pairs

Exercises 6.3, page 475

1. 4.27 sq units

3. a. 6 sq units

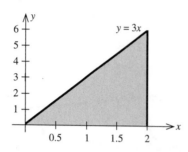

b. 4.5 sq units   **c.** 5.25 sq units   **d.** Yes

5. a. 4 sq units

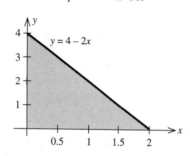

b. 4.8 sq units   **c.** 4.4 sq units   **d.** Yes

7. a. 18.5 sq units   **b.** 18.64 sq units
c. 18.66 sq units   **d.** $\approx$18.7 sq units

9. a. 25 sq units   **b.** 21.12 sq units
c. 19.88 sq units   **d.** $\approx$19.9 sq units

11. a. 0.0625 sq unit   **b.** 0.16 sq unit
c. 0.2025 sq unit   **d.** $\approx$0.2 sq unit

13. 4.64 sq units   **15.** 0.95 sq unit   **17.** 9400 sq ft

Exercises 6.4, page 486

1. 6 sq units   **3.** 8 sq units   **5.** 12 sq units

7. 9 sq units   **9.** ln 2 sq units   **11.** $17\frac{1}{3}$ sq units

13. $18\frac{1}{4}$ sq units   **15.** $(e^2 - 1)$ sq units   **17.** 6

19. 14   **21.** $18\frac{2}{3}$ **23.** $\frac{4}{3}$ **25.** 45   **27.** $\frac{7}{12}$

29. ln 2   **31.** 56   **33.** $\frac{256}{15}$ **35.** $\frac{2}{3}$ **37.** $2\frac{2}{3}$

39. $19\frac{1}{2}$ **41. a.** $4100   **b.** $900

43. a. $2800   **b.** $219.20   **45.** $10{,}133\frac{1}{3}$ ft

47. 37.7 million   **49.** False   **51.** False

Using Technology Exercises 6.4, page 489

1. 6.1787   **3.** 0.7873   **5.** −0.5888   **7.** 2.7044

9. 3.9973   **11.** 37.7 million   **13.** 333,209   **15.** 903,213

Exercises 6.5, page 498

1. 10   **3.** $\frac{19}{15}$ **5.** $32\frac{4}{15}$ **7.** $\sqrt{3} - 1$ **9.** $24\frac{1}{5}$

11. $\frac{32}{15}$ **13.** $18\frac{2}{15}$ **15.** $\frac{1}{2}(e^4 - 1)$ **17.** $\frac{1}{2}e^2 + \frac{5}{6}$

19. 0   **21.** 2 ln 4   **23.** $\frac{1}{3}(\ln 19 - \ln 3)$

25. $2e^4 - 2e^2 - \ln 2$ **27.** $\frac{1}{2}(e^{-4} - e^{-8} - 1)$ **29.** 5

31. $\frac{17}{3}$ **33.** −1   **35.** $\frac{13}{6}$ **37.** $\frac{1}{4}(e^4 - 1)$

39. 120.3 billion metric tons   **41.** $\approx$$2.24 million

43. $40,339.50   **45.** 26°F   **47.** 16,863

49. 0.071 mg/cm³   **51.** $\dfrac{2k}{3} R^2$ cm/sec

61. −4   **63. a.** −1   **b.** 5   **c.** −13

65. True   **67.** False   **69.** True

Using Technology Exercises 6.5, page 501

1. 7.71667   **3.** 17.56487   **5.** 10,140

Exercises 6.6, page 511

1. 108 sq units   **3.** $\frac{2}{3}$ sq unit   **5.** $2\frac{2}{3}$ sq units

7. $1\frac{1}{2}$ sq units   **9.** 3   **11.** $3\frac{1}{3}$ **13.** 27

15. $2(e^2 - e^{-1})$ **17.** $12\frac{2}{3}$ **19.** $3\frac{1}{3}$ **21.** $4\frac{3}{4}$

23. 12 − ln 4   **25.** $e^2 - e - \ln 2$ **27.** $2\frac{1}{2}$ **29.** $7\frac{1}{3}$

31. $\frac{3}{2}$ **33.** $e^3 - 4 + \dfrac{1}{e}$ **35.** $20\frac{5}{6}$ **37.** $\frac{1}{12}$ **39.** $\frac{1}{3}$

41. S is the additional revenue that Odyssey Travel could realize by switching to the new agency;

$$S = \int_0^b [g(x) - f(x)]\, dx$$

43. S is the additional amount of smoke that brand B will remove over brand A in the time interval $[a, b]$;

$$S = \int_a^b [f(t) - g(t)]\, dt$$

45. $\displaystyle\int_{T_1}^T [g(t) - f(t)]\, dt - \int_0^{T_1} [f(t) - g(t)]\, dt$

47. 42.79 million metric tons   **49.** 57,179 people

51. False

Using Technology Exercises 6.6, page 517

1. a.

b. 1074.2857

3. a.

b. 0.9961

5. a.

b. 5.4603

7. a.

b. 25.8549

9. a.

b. 10.5144

11. a.

b. 3.5799

Exercises 6.7, page 531

1. $11,667   **3.** $6667   **5.** $11,667

7. Consumers' surplus: $13,333;
producers' surplus: $11,667

9. $824,200   **11.** $148,239   **13.** $52,203

15. $76,615   **17.** $111,869   **19.** $20,964

21. a.

b. 0.175; 0.816

23. a.

b. 0.104; 0.504

Using Technology Exercises 6.7, page 533

1. Consumers' surplus: $18,000,000;
producers' surplus: $11,700,000

3. Consumers' surplus: $33,120;
producers' surplus: $2,880

Chapter 6 Review Exercises, page 536

1. $\frac{1}{4}x^4 + \frac{2}{3}x^3 - \frac{1}{2}x^2 + C$ **3.** $\frac{1}{5}x^5 - \frac{1}{2}x^4 - \frac{1}{x} + C$

5. $\frac{1}{2}x^4 + \frac{2}{5}x^{5/2} + C$ **7.** $\frac{1}{3}x^3 - \frac{1}{2}x^2 + 2\ln|x| + 5x + C$

9. $\frac{3}{8}(3x^2 - 2x + 1)^{4/3} + C$

11. $\frac{1}{2}\ln(x^2 - 2x + 5) + C$ **13.** $\frac{1}{2}e^{x^2+x+1} + C$

15. $\frac{1}{6}(\ln x)^6 + C$ **17.** $\frac{(11x^2 - 1)(x^2 + 1)^{11}}{264} + C$

19. $\frac{2}{3}(x + 4)\sqrt{x - 2} + C$ **21.** $\frac{1}{2}$ **23.** $5\frac{2}{3}$

25. -80 **27.** $\frac{1}{2}\ln 5$ **29.** 4   **31.** $\frac{e - 1}{2(1 + e)}$

33. $f(x) = x^3 - 2x^2 + x + 1$ **35.** $f(x) = x + e^{-x} + 1$

37. -4.28

39. a. $-0.015x^2 + 60x$ **b.** $p = -0.015x + 60$

41. $3000t - 50,000(1 - e^{-0.04t})$; 16,939

43. $3100   **45.** 15 sq units   **47.** $\frac{2}{3}$ sq unit

49. $e^2 - 3$ sq units   **51.** $\frac{1}{2}$ sq unit   **53.** $\frac{1}{3}$ sq unit

55. $2083; $3333   **57.** $98,973

59. a.

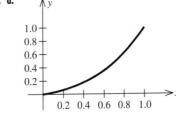

b. 0.1017; 0.3733   **c.** 0.315

CHAPTER 7

Exercises 7.1, page 547

1. $\frac{1}{4}e^{2x}(2x - 1) + C$ **3.** $4(x - 4)e^{x/4} + C$

5. $\frac{1}{2}e^{2x} - 2(x - 1)e^x + \frac{1}{3}x^3 + C$ **7.** $xe^x + C$

9. $\frac{2(x + 2)}{\sqrt{x + 1}} + C$ **11.** $\frac{2}{3}x(x - 5)^{3/2} - \frac{4}{15}(x - 5)^{5/2} + C$

13. $\frac{x^2}{4}(2 \ln 2x - 1) + C$ **15.** $\frac{x^4}{16}(4 \ln x - 1) + C$

17. $\frac{2}{9}x^{3/2}(3 \ln \sqrt{x} - 1) + C$ **19.** $-\frac{1}{x}(\ln x + 1) + C$

21. $x(\ln x - 1) + C$ **23.** $-(x^2 + 2x + 2)e^{-x} + C$

25. $\frac{1}{4}x^2[2(\ln x)^2 - 2 \ln x + 1] + C$ **27.** $2 \ln 2 - 1$

29. $4 \ln 4 - 3$ **31.** $\frac{1}{4}(3e^4 + 1)$

33. $-\frac{1}{2}xe^{-2x} - \frac{1}{4}e^{-2x} + \frac{13}{4}$ **35.** $5 \ln 5 - 4$

37. 1485 ft **39.** 2.04 mg/mL

41. $-20e^{-0.1t}(t + 10) + 200$ **43.** \$521,087 **45.** True

Exercises 7.2, page 555

1. $\frac{2}{9}[2 + 3x - 2 \ln|2 + 3x|] + C$

3. $\frac{3}{32}[(1 + 2x)^2 - 4(1 + 2x) + 2 \ln|1 + 2x|] + C$

5. $2\left[\frac{x}{8}\left(\frac{9}{4} + 2x^2\right)\sqrt{\frac{9}{4} + x^2} - \frac{81}{128} \ln\left(x + \sqrt{\frac{9}{4} + x^2}\right)\right] + C$

7. $\ln\left(\frac{\sqrt{1 + 4x} - 1}{\sqrt{1 + 4x} + 1}\right) + C$ **9.** $\frac{1}{2} \ln 3$

11. $\frac{x}{9\sqrt{9 - x^2}} + C$

13. $\frac{x}{8}(2x^2 - 4)\sqrt{x^2 - 4} - 2 \ln|x + \sqrt{x^2 - 4}| + C$

15. $\sqrt{4 - x^2} - 2 \ln\left|\frac{2 + \sqrt{4 - x^2}}{x}\right| + C$

17. $\frac{1}{4}(2x - 1)e^{2x} + C$ **19.** $\ln|\ln(1 + x)| + C$

21. $\frac{1}{9}\left[\frac{1}{1 + 3e^x} + \ln(1 + 3e^x)\right] + C$

23. $6[e^{(1/2)x} - \ln(1 + e^{(1/2)x})] + C$

25. $\frac{1}{9}(2 + 3 \ln x - 2 \ln|2 + 3 \ln x|) + C$

27. $e - 2$ **29.** $\frac{x^3}{9}(3 \ln x - 1) + C$

31. $x[(\ln x)^3 - 3(\ln x)^2 + 6 \ln x - 6] + C$

33. $\approx$\$2329 **35.** 27,136 **37.** 44; 49 **39.** 26,157

41. \$418,444

Exercises 7.3, page 570

1. 2.7037; 2.6667; $2\frac{2}{3}$ **3.** 0.2656; 0.2500; $\frac{1}{4}$

5. 0.6970; 0.6933; $\approx$0.6931 **7.** 0.5090; 0.5004; $\frac{1}{2}$

9. 5.2650; 5.3046; $\frac{16}{3}$ **11.** 0.6336; 0.6321; $\approx$0.6321

13. 0.3837; 0.3863; $\approx$0.3863 **15.** 1.1170; 1.1114

17. 1.3973; 1.4052 **19.** 0.8806; 0.8818

21. 3.7757; 3.7625 **23. a.** 3.6 **b.** 0.0324

25. a. 0.013 **b.** 0.00043

27. a. 0.0078125 **b.** 0.0002848

29. 52.84 mi **31.** 21.65 mpg

33. a. \$142,374 **b.** \$142,698

35. 103.9 PSI **37.** $\approx$30% **39.** $\approx$6.42 L/min

41. False **43.** True

Exercises 7.4, page 582

1. $\frac{2}{3}$ sq unit **3.** 1 sq unit **5.** 2 sq units

7. $\frac{2}{3}$ sq unit **9.** $\frac{1}{2}e^4$ sq units **11.** 1 sq unit

13. a. $\frac{2}{3}b^{3/2}$ **15.** 1 **17.** 2 **19.** Divergent

21. $-\frac{1}{8}$ **23.** 1 **25.** 1 **27.** $\frac{1}{2}$ **29.** Divergent

31. -1 **33.** Divergent **35.** 0 **37.** 0

39. Divergent **41.** Convergent **43.** \$18,750

45. $\frac{10,000r + 4000}{r^2}$ dollars **47.** True **49.** False

Exercises 7.5, page 594

9. a. $k = \frac{1}{8}$ **b.** $\frac{1}{2}$ **11. a.** $k = 4$ **b.** .1353

13. a. $f(x) = \frac{1}{15}e^{-x/15}$ **b.** .06 **c.** .37

15. 3 **17.** 0.37 **19.** 0.18 **21.** False

Chapter 7 Review Exercises, page 598

1. $-2(1 + x)e^{-x} + C$ **3.** $x(\ln 5x - 1) + C$

5. $\frac{1}{4}(1 - 3e^{-2})$ **7.** $2\sqrt{x}(\ln x - 2) + 2$

9. $\frac{1}{8}\left[3 + 2x - \frac{9}{3 + 2x} - 6 \ln|3 + 2x|\right] + C$

11. $\frac{1}{32}e^{4x}(8x^2 - 4x + 1) + C$

13. $\dfrac{1}{4}\dfrac{\sqrt{x^2-4}}{x}+C$ 15. $\frac{1}{2}$ 17. Divergent   19. $\frac{1}{10}$

21. 0.8421; 0.8404   23. 2.2379; 2.1791

27. **a.** $k=\frac{3}{8}$ **b.** 0.6495

29. **a.** $k=\frac{4}{27}$ **b.** 0.4815

31. \$1,157,641   33. \$41,100   35. \$111,111

CHAPTER 8

Exercises 8.1, page 609

1. $f(0,0)=-4; f(1,0)=-2; f(0,1)=-1; f(1,2)=4;$
$f(2,-1)=-3$

3. $f(1,2)=7; f(2,1)=9; f(-1,2)=1; f(2,-1)=1$

5. $g(1,2)=4+3\sqrt{2}; g(2,1)=8+\sqrt{2}; g(0,4)=2;$
$g(4,9)=56$

7. $h(1,e)=1; h(e,1)=-1; h(e,e)=0$

9. $g(1,1,1)=e; g(1,0,1)=1; g(-1,-1,-1)=-e$

11. All real values of x and y

13. All real values of u and v except those satisfying the equation $u=v$

15. All real values of r and s satisfying $rs\geq 0$

17. All real values of x and y satisfying $x+y>5$

19. '   21.

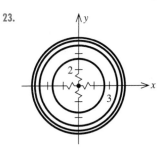

23.

25. 9π ft^3 27. 40,000k dynes

29. \$25,500; \$23,580   31. \$8310; \$7910

33. **a.** The set of all nonnegative values of W and H
b. 1.871 m^2

35. \$13,498.59   37. \$76,704.11; \$50,814.62   39. 18

41. False   43. True

Exercises 8.2, page 624

1. 2; 3   3. $4x$; 4   5. $-\dfrac{4y}{x^3}; \dfrac{2}{x^2}$

7. $\dfrac{2v}{(u+v)^2}; -\dfrac{2u}{(u+v)^2}$

9. $3(2s-t)(s^2-st+t^2)^2; 3(2t-s)(s^2-st+t^2)^2$

11. $\dfrac{4x}{3(x^2+y^2)^{1/3}}; \dfrac{4y}{3(x^2+y^2)^{1/3}}$ 13. $ye^{xy+1}; xe^{xy+1}$

15. $\ln y+\dfrac{y}{x}; \dfrac{x}{y}+\ln x$ 17. $e^u\ln v; \dfrac{e^u}{v}$

19. $yz+y^2+2xz; xz+2xy+z^2; xy+2yz+x^2$

21. $ste^{rst}; rte^{rst}; rse^{rst}$ 23. $f_x(1,2)=8; f_y(1,2)=5$

25. $f_x(2,1)=1; f_y(2,1)=3$

27. $f_x(1,2)=\frac{1}{2}; f_y(1,2)=-\frac{1}{4}$

29. $f_x(1,1)=e; f_y(1,1)=e$

31. $f_x(1,0,2)=0; f_y(1,0,2)=8; f_z(1,0,2)=0$

33. $f_{xx}=2y; f_{xy}=2x+3y^2=f_{yx}; f_{yy}=6xy$

35. $f_{xx}=2; f_{xy}=f_{yx}=-2; f_{yy}=4$

37. $f_{xx}=\dfrac{y^2}{(x^2+y^2)^{3/2}}; f_{xy}=f_{yx}=-\dfrac{xy}{(x^2+y^2)^{3/2}};$
$f_{yy}=\dfrac{x^2}{(x^2+y^2)^{3/2}}$

39. $f_{xx}=\dfrac{1}{y^2}e^{-x/y}; f_{xy}=\dfrac{y-x}{y^3}e^{-x/y}=f_{yx};$
$f_{yy}=\dfrac{x}{y^3}\left(\dfrac{x}{y}-2\right)e^{-x/y}$

41. **a.** $f_x=7.5; f_y=40$ **b.** Yes

43. $P_x=10$—at $(0,1)$, the price of land is changing at the rate of \$10/ft^2/mile change to the right; $P_y=0$—at $(0,1)$, the price of land is constant/mile change upward.

45. Complementary commodities

47. $30/unit change in finished desks; −$25/unit change in unfinished desks. The weekly revenue increases by $30/unit for each additional finished desk produced (beyond 300) when the level of production of unfinished desks remains fixed at 250; the revenue decreases by $25/unit when each additional unfinished desk (beyond 250) is produced and the level of production of finished desks remains fixed at 300.

49. 0.039 L/degree; −0.015 L/mm of mercury. The volume increases by 0.039 L when the temperature increases by 1 degree (beyond 300 K) and the pressure is fixed at 800 mm of mercury. The volume decreases by 0.015 L when the pressure increases by 1 mm of mercury (beyond 800 mm) and the temperature is fixed at 300 K.

53. True   **55.** False

Using Technology Exercises 8.2, page 627

1. 1.3124; 0.4038   **3.** −1.8889; 0.7778

5. −0.3863; −0.8497

Exercises 8.3, page 638

1. (0, 0); relative maximum value: $f(0, 0) = 1$

3. (1, 2); saddle point: $f(1, 2) = 4$

5. (8, −6); relative minimum value: $f(8, −6) = −41$

7. (1, 2) and (2, 2); saddle point: $f(1, 2) = −1$; relative minimum value: $f(2, 2) = −2$

9. $\left(-\frac{1}{3}, \frac{11}{3}\right)$ and (1, 5); saddle point: $f\left(-\frac{1}{3}, \frac{11}{3}\right) = -\frac{319}{27}$; relative minimum value: $f(1, 5) = −13$

11. (0, 0) and (1, 1); saddle point: $f(0, 0) = −2$; relative minimum value: $f(1, 1) = −3$

13. (2, 1); relative minimum value: $f(2, 1) = 6$

15. (0, 0); saddle point: $f(0, 0) = −1$

17. (0, 0); relative minimum value: $f(0, 0) = 1$

19. (0, 0); relative minimum value: $f(0, 0) = 0$

21. 200 finished units and 100 unfinished units; $10,500

23. Price of land ($200/ft²) is highest at $\left(\frac{1}{2}, 1\right)$

25. (0, 1) gives desired location.

27. 18 in. × 36 in. × 18 in.

29. 6 in. × 4 in. × 2 in.   **31.** False

Exercises 8.4, page 648

1. a. $y = 2.3x + 1.5$
b.

3. a. $y = −0.77x + 5.74$
b.

5. a. $y = 1.2x + 2$
b.

7. a. $y = 0.34x − 0.9$
b.

c. 1276 applications

9. a. $y = −2.8x + 440$ **c.** 420
b.

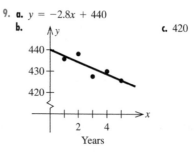

11. a. $y = 6.23x + 167.6$ **b.** 560 acres

13. a. $y = 2.8x + 17.6$ **b.** $40,000,000

15. a. $y = 4.842x + 11.842$ **b.** 103.8 billion cans

17. a. $y = 12.2x + 20.9$ **b.** 81.9 million

19. a. $y = 98.176x - 231.7$ **b.** $1732   **21.** True

Using Technology Exercises 8.4, page 651

1. $y = 2.3596x + 3.8639$ **3.** $y = -1.1948x + 3.5525$

5. a. $y = 13.321x + 72.571$ **b.** 192 million tons

Exercises 8.5, page 664

1. Min. of $\frac{3}{4}$ at $(\frac{3}{4}, \frac{1}{4})$ **3.** Max. of $-\frac{7}{4}$ at $(2, \frac{7}{2})$

5. Min. of 4 at $(\sqrt{2}, \sqrt{2}/2)$ and $(-\sqrt{2}, -\sqrt{2}/2)$

7. Max. of $-\frac{3}{4}$ at $(\frac{3}{2}, 1)$

9. Max. of $2\sqrt{3}$ at $(\sqrt{3}/3, -\sqrt{6})$ and $(\sqrt{3}/3, \sqrt{6})$

11. Max. of 8 at $(2\sqrt{2}, 2\sqrt{2})$ and $(-2\sqrt{2}, -2\sqrt{2})$;
min. of -8 at $(2\sqrt{2}, -2\sqrt{2})$ and $(-2\sqrt{2}, 2\sqrt{2})$

13. Max.: $\frac{2\sqrt{3}}{9}$; min.: $-\frac{2\sqrt{3}}{9}$

15. Min. of $\frac{18}{7}$ at $(\frac{9}{7}, \frac{6}{7}, \frac{3}{7})$

17. 140 finished and 60 unfinished units

19. $10\sqrt{2}$ ft $\times 40\sqrt{2}$ ft   **21.** 2 ft $\times$ 2 ft $\times$ 3 ft

23. $15,000 on newspaper and $45,000 on TV

25. 30 ft $\times$ 40 ft $\times$ 10 ft; $7200   **27.** False

Exercises 8.6, page 679

1. $\frac{7}{2}$ **3.** 0   **5.** $4\frac{1}{2}$

7. $(e^2 - 1)(1 - e^{-2})$ **9.** 1

11. $\frac{2}{3}$ **13.** $\frac{188}{3}$ **15.** $\frac{84}{5}$

17. $2\frac{2}{3}$ **19.** 1   **21.** $\frac{1}{2}(3 - e)$

23. $\frac{1}{4}(e^4 - 1)$ **25.** $\frac{2}{3}(e - 1)$ **27.** 48

29. 6   **31.** $\frac{19}{6}$

33. $5(1 - 2e^{-2} + e^{-4})$ **35.** 16

37. $\frac{1}{2}(e - 1)^2$ **39.** $\frac{1}{2} \ln 17$

41. $\frac{64}{3}$ **43.** $\frac{4}{3}$

45. $1 - \dfrac{1}{e}$ **47.** $\frac{1}{8}(9 \ln 3 - 4)$

49. ≈ 2166 people/mi^2 **51.** $194/ft^2 **53.** False   **55.** True

Chapter 8 Review Exercises, page 684

1. 0, 0, $\frac{1}{2}$; no   **3.** 2, $-(e + 1)$, $-(e + 1)$

5. The set of all ordered pairs (x, y) such that $y \neq -x$

7. The set of all (x, y, z) such that $z \geq 0$ and $x \neq 1$,
$y \neq 1$, and $z \neq 1$

9.   **11.**

13. $f_x = \sqrt{y} + \dfrac{y}{2\sqrt{x}}; f_y = \dfrac{x}{2\sqrt{y}} + \sqrt{x}$

15. $f_x = \dfrac{3y}{(y + 2x)^2}; f_y = -\dfrac{3x}{(y + 2x)^2}$

17. $10y(2xy + 3y^2)^4$; $10(x + 3y)(2xy + 3y^2)^4$

19. $2x(1 + x^2 + y^2)e^{x^2+y^2}$; $2y(1 + x^2 + y^2)e^{x^2+y^2}$

21. $\dfrac{2x}{x^2 + y^2}; -\dfrac{2x^2}{y(x^2 + y^2)}$

23. $f_{xx} = 12x^2 + 4y^2$; $f_{xy} = 8xy = f_{yx}$; $f_{yy} = 4x^2 - 12y^2$

25. $g_{xx} = \dfrac{-2y^2}{(x + y^2)^3}$; $g_{yy} = \dfrac{2x(3y^2 - x)}{(x + y^2)^3}$;
$g_{xy} = \dfrac{2y(x - y^2)}{(x + y^2)^3} = g_{yx}$

27. $h_{ss} = -\dfrac{1}{s^2}$; $h_{st} = h_{ts} = 0$; $h_{tt} = \dfrac{1}{t^2}$

29. $(2, 3)$; relative minimum value: $f(2, 3) = -13$

31. $(0, 0)$ and $(\frac{3}{2}, \frac{9}{4})$; saddle point at $f(0, 0) = 0$; relative minimum value: $f(\frac{3}{2}, \frac{9}{4}) = -\frac{27}{16}$

33. $(0, 0)$; relative minimum value: $f(0, 0) = 1$

35. $f(\frac{12}{11}, \frac{20}{11}) = -\frac{32}{11}$

37. Relative maximum value: $f(5, -5) = 26$; relative minimum value: $f(-5, 5) = -24$

39. 48   **41.** $\frac{2}{63}$ **43.** $11\frac{1}{3}$ cu units

45. 3   **47.** Complementary

49. $11,000; 14 agents

51. 75 units on labor; 25 units on capital